Grid Homology for Knots and Links

Mathematical
Surveys
and
Monographs

Volume 208

Grid Homology for Knots and Links

Peter S. Ozsváth
András I. Stipsicz
Zoltán Szabó

American Mathematical Society
Providence, Rhode Island

2010 *Mathematics Subject Classification.* Primary 57M25, 57M27, 57R58, 53D10.

For additional information and updates on this book, visit
www.ams.org/bookpages/surv-208

Library of Congress Cataloging-in-Publication Data

Ozsváth, Peter Steven, 1967-
Grid homology for knots and links / Peter S. Ozsváth, András I. Stipsicz, Zoltán Szabó.
pages cm. – (Mathematical surveys and monographs ; volume 208)
Includes bibliographical references and index.
ISBN 978-1-4704-3442-7 (alk. paper)
1. Knot theory. 2. Link theory. 3. Homology theory. I. Stipsicz, András. II. Szabó, Zoltán, 1965- III. Title.
QA612.2.O97 2015
514′.2242–dc23 2015025903

∞ The paper used in this book is acid-free and falls within the guidelines
established to ensure permanence and durability.
Visit the AMS home page at `http://www.ams.org/`

10 9 8 7 6 5 4 3 2 20 19 18 17 16 15

To the memory of our fathers,
Ozsváth István, Stipsicz István, and Dr. Szabó István

Contents

Chapter 1. Introduction 1
1.1. Grid homology and the Alexander polynomial 1
1.2. Applications of grid homology 3
1.3. Knot Floer homology 5
1.4. Comparison with Khovanov homology 7
1.5. On notational conventions 7
1.6. Necessary background 9
1.7. The organization of this book 9
1.8. Acknowledgements 11

Chapter 2. Knots and links in S^3 13
2.1. Knots and links 13
2.2. Seifert surfaces 20
2.3. Signature and the unknotting number 21
2.4. The Alexander polynomial 25
2.5. Further constructions of knots and links 30
2.6. The slice genus 32
2.7. The Goeritz matrix and the signature 37

Chapter 3. Grid diagrams 43
3.1. Planar grid diagrams 43
3.2. Toroidal grid diagrams 49
3.3. Grids and the Alexander polynomial 51
3.4. Grid diagrams and Seifert surfaces 56
3.5. Grid diagrams and the fundamental group 63

Chapter 4. Grid homology 65
4.1. Grid states 65
4.2. Rectangles connecting grid states 66
4.3. The bigrading on grid states 68
4.4. The simplest version of grid homology 72
4.5. Background on chain complexes 73
4.6. The grid chain complex GC^- 75
4.7. The Alexander grading as a winding number 82
4.8. Computations 86
4.9. Further remarks 90

Chapter 5. The invariance of grid homology 91
5.1. Commutation invariance 91
5.2. Stabilization invariance 99

5.3. Completion of the invariance proof for grid homology 107
5.4. The destabilization maps, revisited 108
5.5. Other variants of the grid complex 110
5.6. On the holomorphic theory 110
5.7. Further remarks on stabilization maps 110

Chapter 6. The unknotting number and τ 113
6.1. The definition of τ and its unknotting estimate 113
6.2. Construction of the crossing change maps 115
6.3. The Milnor conjecture for torus knots 120
6.4. Canonical grid cycles and estimates on τ 122

Chapter 7. Basic properties of grid homology 127
7.1. Symmetries of the simply blocked grid homology 127
7.2. Genus bounds 129
7.3. General properties of unblocked grid homology 130
7.4. Symmetries of the unblocked theory 132

Chapter 8. The slice genus and τ 135
8.1. Slice genus bounds from τ and their consequences 135
8.2. A version of grid homology for links 136
8.3. Grid homology and saddle moves 139
8.4. Adding unknots to a link 143
8.5. Assembling the pieces: τ bounds the slice genus 146
8.6. The existence of an exotic structure on $\mathbb{R}^4$ 147
8.7. Slice bounds vs. unknotting bounds 149

Chapter 9. The oriented skein exact sequence 151
9.1. The skein exact sequence 151
9.2. The skein relation on the chain level 153
9.3. Proofs of the skein exact sequences 160
9.4. First computations using the skein sequence 162
9.5. Knots with identical grid homologies 163
9.6. The skein exact sequence and the crossing change map 164
9.7. Further remarks 166

Chapter 10. Grid homologies of alternating knots 167
10.1. Properties of the determinant of a link 167
10.2. The unoriented skein exact sequence 176
10.3. Grid homology groups for alternating knots 183
10.4. Further remarks 185

Chapter 11. Grid homology for links 187
11.1. The definition of grid homology for links 188
11.2. The Alexander multi-grading on grid homology 192
11.3. First examples 194
11.4. Symmetries of grid homology for links 196
11.5. The multi-variable Alexander polynomial 199
11.6. The Euler characteristic of multi-graded grid homology 203
11.7. Seifert genus bounds from grid homology for links 204

11.8. Further examples 205
11.9. Link polytopes and the Thurston norm 210

Chapter 12. Invariants of Legendrian and transverse knots 215
12.1. Legendrian knots in $\mathbb{R}^3$ 216
12.2. Grid diagrams for Legendrian knots 220
12.3. Legendrian grid invariants 223
12.4. Applications of the Legendrian invariants 228
12.5. Transverse knots in $\mathbb{R}^3$ 231
12.6. Applications of the transverse invariant 236
12.7. Invariants of Legendrian and transverse links 240
12.8. Transverse knots, grid diagrams, and braids 244
12.9. Further remarks 245

Chapter 13. The filtered grid complex 247
13.1. Some algebraic background 247
13.2. Defining the invariant 252
13.3. Topological invariance of the filtered quasi-isomorphism type 254
13.4. Filtered homotopy equivalences 268

Chapter 14. More on the filtered chain complex 273
14.1. Information in the filtered grid complex 273
14.2. Examples of filtered grid complexes 278
14.3. Refining the Legendrian and transverse invariants: definitions 281
14.4. Applications of the refined Legendrian and transverse invariants 285
14.5. Filtrations in the case of links 287
14.6. Remarks on three-manifold invariants 289

Chapter 15. Grid homology over the integers 291
15.1. Signs assignments and grid homology over $\mathbb{Z}$ 291
15.2. Existence and uniqueness of sign assignments 295
15.3. The invariance of grid homology over $\mathbb{Z}$ 302
15.4. Invariance in the filtered theory 308
15.5. Other grid homology constructions over $\mathbb{Z}$ 319
15.6. On the τ-invariant 321
15.7. Relations in the spin group 321
15.8. Further remarks 323

Chapter 16. The holomorphic theory 325
16.1. Heegaard diagrams 325
16.2. From Heegaard diagrams to holomorphic curves 327
16.3. Multiple basepoints 333
16.4. Equivalence of knot Floer homology with grid homology 335
16.5. Further remarks 338

Chapter 17. Open problems 339
17.1. Open problems in grid homology 339
17.2. Open problems in knot Floer homology 341

Appendix A. Homological algebra 347
A.1. Chain complexes and their homology 347

A.2. Exact sequences 350
A.3. Mapping cones 352
A.4. On the structure of homology 357
A.5. Dual complexes 359
A.6. On filtered complexes 362
A.7. Small models for filtered grid complexes 363
A.8. Filtered quasi-isomorphism versus filtered homotopy type 365

Appendix B. Basic theorems in knot theory 367
B.1. The Reidemeister Theorem 367
B.2. Reidemeister moves in contact knot theory 373
B.3. The Reidemeister-Singer Theorem 382
B.4. Cromwell's Theorem 387
B.5. Normal forms of cobordisms between knots 394

Bibliography 399

Index 407

CHAPTER 1

Introduction

Heegaard Floer homology is an invariant of closed, oriented three-manifolds, defined by Ozsváth and Szabó by adapting methods from symplectic geometry to Heegaard diagrams [**174**]. Knot Floer homology is a variation of this construction, discovered in 2003 by Ozsváth and Szabó [**172**] and independently by Jacob Rasmussen [**191**], giving an invariant for knots and links in three-manifolds. It is the homology of a chain complex whose generators can be easily read off from a Heegaard diagram, and whose differential counts pseudo-holomorphic disks. Thanks to the contributions of many researchers, the resulting theory has evolved into an active and successful tool for studying knots, links, and three-manifolds.

In 2006, Sucharit Sarkar found a method that rendered some of the pseudo-holomorphic disk counts in the Heegaard Floer differential combinatorial; this was further pursued in a joint paper with Jiajun Wang [**206**]. Inspired by these ideas, Ciprian Manolescu, Ozsváth, and Sarkar [**135**] found a class of Heegaard diagrams for explicitly computing knot Floer homology for knots and links in S^3. These Heegaard diagrams are naturally associated to *grid diagrams*, which are simple combinatorial presentations of knots in S^3, dating back to the 19th century [**17**]. In fact, the above reformulation of knot Floer homology can be used as a *definition* of invariants of knots in S^3: their invariance can also be verified using elementary methods, following the paper of Manolescu, Ozsváth, Szabó, and Dylan Thurston [**136**]. To emphasize this simplicity of the definition, we will call the resulting theory the *grid homology* for knots and links, to distinguish it from its holomorphic antecedent. Of course, grid homology is isomorphic to knot Floer homology; but owing to its elegance and simplicity, grid homology deserves a purely self-contained treatment. This is the goal of the present book.

Before describing the contents of this book in detail, we explain the key features of grid homology.

1.1. Grid homology and the Alexander polynomial

Grid homology is defined in terms of a *grid presentation* of a knot. Such a presentation is given by an $n \times n$ grid of squares, n of which are marked with an O and n of which are marked with an X. These are distributed by the rule that each row and each column contains exactly one square marked with an O and exactly one square marked with an X. By connecting the two markings in every row and in every column (orienting the intervals from X to O in the columns and from O to X in the rows) and following the convention that vertical segments pass over horizontal ones, a grid $\mathbb{G}$ gives rise to a projection of a knot or link together with an orientation on it. In this case we say that $\mathbb{G}$ is a grid presentation of the knot or link. In fact, it is not hard to see that any link type can be presented in this way, cf. Figure 1.1.

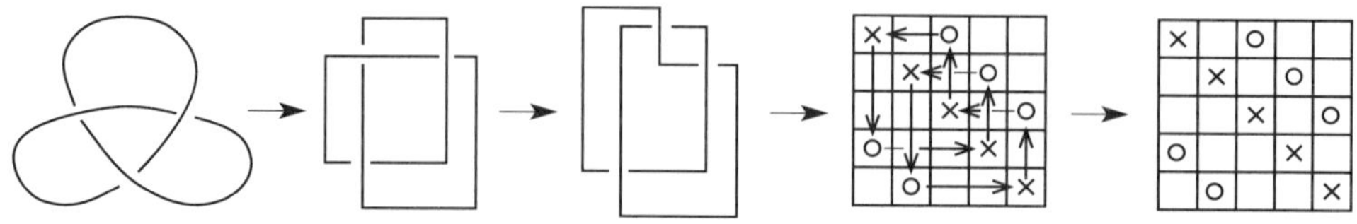

FIGURE 1.1. **Grid diagram of a knot from its projection.** We approximate the diagram by one that involves only horizontal and vertical segments, adjust crossings where horizontal passes over vertical, and mark the turns by O's and X's in an alternating fashion, resulting in a grid diagram.

The simplest version of grid homology for a knot K in S^3, $\widehat{GH}(K)$, is a finite-dimensional, bigraded vector space over the field $\mathbb{F} = \mathbb{Z}/2\mathbb{Z}$ of two elements. This means that $\widehat{GH}(K)$ is a vector space, equipped with a splitting as a direct sum, indexed by pairs of integers:

$$\widehat{GH}(K) = \bigoplus_{d,s\in\mathbb{Z}} \widehat{GH}_d(K,s).$$

In the above decomposition the index d is called the "Maslov grading" and the index s is called the "Alexander grading". Grid homology $\widehat{GH}(K)$ is defined as the homology of a bigraded chain complex $(\widehat{GC}(\mathbb{G}), \widehat{\partial})$ associated to a grid presentation $\mathbb{G}$ of the knot K. The underlying vector space $\widehat{GC}(\mathbb{G})$ decomposes as $\bigoplus_{d,s\in\mathbb{Z}} \widehat{GC}_d(\mathbb{G},s)$, and the differential $\widehat{\partial}$ drops Maslov grading by one and preserves Alexander grading, i.e.

$$\widehat{\partial}\colon \widehat{GC}_d(\mathbb{G},s) \to \widehat{GC}_{d-1}(\mathbb{G},s).$$

Thus, both gradings descend to give the stated bigrading on the homology $\widehat{GH}(K)$. The chain complex itself depends on the grid presentation of the knot, but its homology is independent of this choice.

As a simple example, consider the unknot $\mathcal{O}$. It turns out that the grid homology $\widehat{GH}(\mathcal{O})$ has total dimension one, supported in Alexander and Maslov gradings equal to zero; i.e.

$$\widehat{GH}_d(\mathcal{O},s) = \begin{cases} \mathbb{F} & \text{if } d = s = 0 \\ 0 & \text{otherwise.} \end{cases} \tag{1.1}$$

Since grid homology has two gradings, its Poincaré polynomial is naturally a Laurent polynomial in two formal variables q and t:

$$P_K(q,t) = \sum_{d,s\in\mathbb{Z}} \dim_{\mathbb{F}} \widehat{GH}_d(K,s) \cdot t^s q^d. \tag{1.2}$$

The natural *graded Euler characteristic* in this case is a Laurent polynomial in a single indeterminate t, obtained by setting $q = -1$:

$$\chi(\widehat{GH}(K)) = P_K(-1,t) = \sum_{d,s\in\mathbb{Z}} (-1)^d \dim_{\mathbb{F}} \widehat{GH}_d(K,s) \cdot t^s \in \mathbb{Z}[t,t^{-1}].$$

One of the key properties of grid homology is its relationship with a classical polynomial invariant for knots, the *Alexander polynomial*, see [**1**]. This relationship is expressed in the following:

THEOREM 1.1.1. *The graded Euler characteristic of the grid homology of a knot* $K \subset S^3$ *coincides with the (symmetrized) Alexander polynomial* $\Delta_K(t)$ *of* K*:*

$$\chi(\widehat{GH}(K)) = \Delta_K(t).$$

The Alexander polynomial of a knot satisfies a number of basic properties. We recall some of these below:

(A-1) It is symmetric in t; i.e. $\Delta_K(t) = \Delta_K(t^{-1})$.
(A-2) It is invariant under taking the mirror image of the knot; i.e. if $m(K)$ denotes the mirror of K, then $\Delta_K(t) = \Delta_{m(K)}(t)$.
(A-3) Evaluating the Alexander polynomial at $t = 1$ gives the value 1.
(A-4) The Alexander polynomial satisfies a "skein relation", which relates the Alexander polynomials of two knots K_+ and K_- that differ in a single crossing change, and the Alexander polynomial of the two-component link K_0 obtained by taking the oriented resolution at the crossing:

$$\Delta_{K_+}(t) - \Delta_{K_-}(t) = (t^{\frac{1}{2}} - t^{-\frac{1}{2}})\Delta_{K_0}(t).$$

(The third term here uses a natural extension of the Alexander polynomial to oriented links in S^3. Notice that our convention for the Alexander polynomial differs from the convention used in [**119**] by a multiplicative factor of $(-1)^{|L|-1}$, where $|L|$ denotes the number of components of the link L.)

These results have analogues in grid homology. For example, there are symmetries between grid homology groups generalizing Properties (A-1) and (A-2) (see Propositions 7.1.1 and 7.1.2 respectively). Property (A-3) has a manifestation in grid homology (see Proposition 6.1.4) which is used in the construction of an integer-valued invariant $\tau(K)$, which we will discuss in the next section. Perhaps the most interesting of the above properties is the skein relation, Property (A-4). The analogue of this result is a "skein exact sequence" which is a long exact sequence relating the grid homology groups $\widehat{GH}$ of K_+, K_-, and K_0, where the third term uses a version of grid homology for links. For the exact sequence of grid homologies, see Theorems 9.1.1 and 9.1.2.

1.2. Applications of grid homology

Grid homology is useful, for example, in the study of three numerical invariants of knots: the Seifert genus, the slice genus, and the unknotting number. A knot K in S^3 bounds embedded, oriented, compact surfaces (such a surface is called a *Seifert surface* of K), and the minimal genus of such a surface is the *Seifert genus* $g(K)$ of K. A smoothly embedded, oriented, compact surface $(F, \partial F) \to (D^4, \partial D^4 = S^3)$ in the four-ball D^4 with the property that $\partial F = K$ is called a *slice surface* for K. By taking the minimum of the genera of slice surfaces of K, we arrive at an inherently four-dimensional invariant, the *slice genus* (or *four-ball genus*) $g_s(K)$ of K. The *unknotting number* of a knot K is the minimum number of times the knot needs to pass through itself to obtain the unknot. The following inequalities are easily verified: $g_s(K) \leq g(K)$ and $g_s(K) \leq u(K)$.

Common features of the slice genus and the unknotting number are that both can be easily bounded from above, but both are difficult to compute. For example, for the (p, q) torus knot, it is straightforward to find an unknotting with $\frac{(p-1)(q-1)}{2}$ crossing changes. John Milnor conjectured in 1968 [**144**] that this bound is sharp.

This conjecture was verified by Peter Kronheimer and Tomasz Mrowka in 1993 [**106**] using smooth four-manifold topology and, specifically, the four-manifold invariants introduced by Simon Donaldson [**34**]:

THEOREM 1.2.1 (Kronheimer-Mrowka, [**106**]). *Both the unknotting number and the slice genus of the* (p, q) *torus knot* $T_{p,q}$ *are equal to* $\frac{(p-1)(q-1)}{2}$. □

The original proof of the above theorem is based on a stronger result, the *generalized Thom conjecture*, that concerns the minimal genus problem in a smooth Kähler surface. After this breakthrough, a number of other proofs have emerged, using Seiberg-Witten theory [**107, 150**] and Heegaard Floer homology [**168**]. Other proofs of Theorem 1.2.1 have been found using knot Floer homology [**170, 191**] and Khovanov homology [**194**].

Following work of Sarkar [**204**], grid homology can also be used to give a self-contained proof of Theorem 1.2.1, as follows. First, by attaching further structures to grid homology, we obtain a variant $GH^-(K)$ of the construction, which is a bigraded module over the polynomial algebra $\mathbb{F}[U]$ in an indeterminate U. This bigraded module encodes an integer invariant $\tau(K)$ for knots. In Chapter 6, we present a simple proof of the inequality

$$|\tau(K)| \leq u(K), \tag{1.3}$$

which quickly leads to a proof of the Milnor conjecture for torus knots. In Chapter 8, this inequality is sharpened, to give

$$|\tau(K)| \leq g_s(K), \tag{1.4}$$

leading to a proof of Theorem 1.2.1.

The invariant τ can also be used to find knots with trivial Alexander polynomial that have $g_s(K) > 0$. Combining the existence of such knots with work of Michael Freedman [**59, 60**], exotic differentiable structures on $\mathbb{R}^4$ can be constructed; see Section 8.6. (The first examples of exotic differentiable structures on $\mathbb{R}^4$ were discovered by combining work of Donaldson [**33**] and Freedman [**59**]; cf. [**30, 76**].)

In a different direction, grid homology can be used to effectively study contact geometric properties of knots in the standard three-sphere. Recall that a *contact structure* on a three-manifold is a two-plane field ξ that is nowhere integrable. More formally, it is the kernel of a one-form α with the property that $\alpha \wedge d\alpha$ is a volume form. There is a canonical contact structure on $\mathbb{R}^3$, specified by the one-form $\alpha = dz - y\,dx$, which naturally extends to the one-point compactification S^3. The contact structure on $\mathbb{R}^3$ is canonical in the sense that every contact structure in dimension three is locally contactomorphic to this standard model.

There are two natural variations on knot theory one can study in the presence of a contact structure. One of these is *Legendrian knot theory*, where one considers *Legendrian knots*, which are smoothly embedded knots everywhere tangent to ξ. Two such knots are considered equivalent if they are isotopic via a one-parameter family of Legendrian knots. Another variation of knot theory in a contact manifold is provided by *transverse knots*, i.e. by knots that are everywhere transverse to the two-plane field ξ. These are then studied up to transverse isotopy. (For further background on these notions, see Chapter 12.)

Besides their smooth knot types, Legendrian knots come with two further "classical" numerical invariants, their Thurston-Bennequin and rotation numbers. A motivating problem of Legendrian knot theory is to understand to what extent these invariants determine the Legendrian knot type. A smooth knot type is said to be *(Legendrian) non-simple* if it contains Legendrian non-isotopic pairs of knots with identical classical invariants. Building on work of Yasha Eliashberg, Alexander Givental and Helmut Hofer [**40**], Yuri Chekanov [**23**] produced the first pair of such Legendrian knots.

Unlike Legendrian knots, transverse knots have only one classical numerical invariant, their self-linking number. The definition of Legendrian non-simplicity naturally adapts to the transverse setting: a knot type is *transversely non-simple* if it contains two transverse knots with equal self-linking numbers which are not transversely isotopic. Examples of transversely non-simple knot types were given by John Etnyre and Ko Honda using convex surface theory [**47**], see also [**13**].

The connection of Legendrian and transverse knot theory with grid homology is provided by the following observation: grid position naturally realizes the underlying knot as a Legendrian knot in the standard contact three-sphere. Moreover, the grid complex associated to a given grid diagram comes equipped with two canonical cycles, whose homology classes give invariants of the Legendrian knot type. Similarly, a grid diagram naturally defines a transverse knot, and one of the two cycles above provides a transverse invariant. With the use of these invariants, results of Chekanov and Etnyre-Honda can be reproved in a fairly simple manner, following [**185**] and [**25, 102, 157**] respectively. Specifically, in Sections 12.4 and 12.6, we will prove the following two theorems:

THEOREM 1.2.2 (Chekanov [**23**]). *The knot type $m(5_2)$ shown in Figure* 1.2 *is Legendrian non-simple.* □

THEOREM 1.2.3 (Etnyre-Honda [**47**]). *There are transversely non-simple knot types in S^3.* □

While Etnyre and Honda verified transverse non-simplicity for a certain cable of the trefoil knot [**47**], in this book we will show transverse non-simplicity of several other simpler knot types, including $m(10_{132})$ shown in Figure 1.3. This latter knot was shown to be transversely non-simple in [**157**]; our presentation will also draw on the argument from [**102**].

1.3. Knot Floer homology

Grid homology is a special case of a holomorphic construction of an earlier-defined invariant, *knot Floer homology* $\widehat{\mathrm{HFK}}$ and HFK^- [**172, 191**]. Grid diagrams

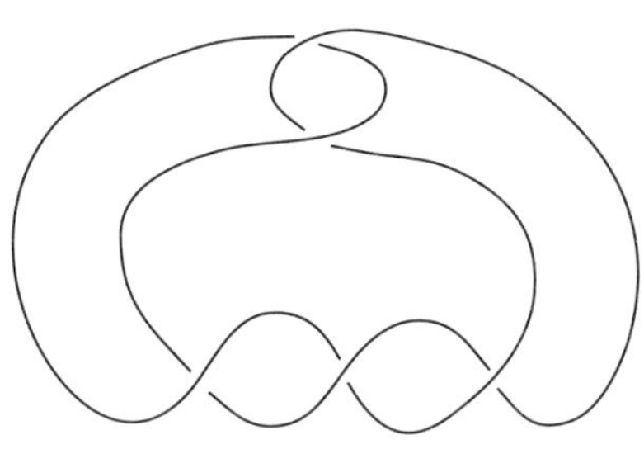

FIGURE 1.2. **The twist knot $m(5_2)$.** This is the knot with smallest crossing number which is Legendrian non-simple. It is the mirror image of the knot type 5_2 from [**198**].

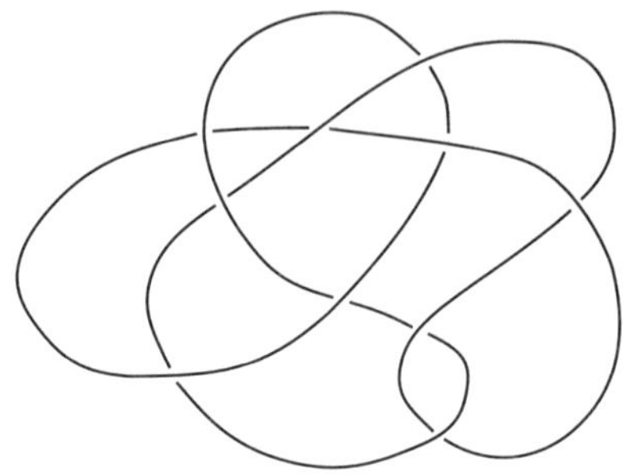

FIGURE 1.3. **The knot** $m(10_{132})$, the mirror image of the knot 10_{132} of Rolfsen's table [**198**]. This knot type contains transverse knots which are not transversely isotopic, although have equal self-linking numbers.

allow us to bypass the technically more involved theory of pseudo-holomorphic curves central to knot Floer homology. The price is that the present approach appears to be too rigid to verify some of the results that are accessible with the strength of the whole theory. To put grid homology in context, we collect some important results of knot Floer homology here. (Below we will refer to knot Floer homology with coefficients in $\mathbb{F} = \mathbb{Z}/2\mathbb{Z}$.)

A key feature of knot Floer homology, proved using the holomorphic theory, states that it recognizes the unknot, in the sense that:

THEOREM 1.3.1 ([**171**]). *The knot Floer homology* $\widehat{\mathrm{HFK}}(K)$ *of a knot* $K \subset S^3$ *has total dimension equal to one if and only if* K *is the unknot* $\mathcal{O}$. □

The above theorem follows from a more general result, which states that knot Floer homology detects the Seifert genus $g(K)$ of a knot K. To state it, note that knot Floer homology $\widehat{\mathrm{HFK}}$ is bigraded, i.e. it is given with a splitting

$$\widehat{\mathrm{HFK}}(K) = \bigoplus_{d,s\in\mathbb{Z}} \widehat{\mathrm{HFK}}_d(K,s),$$

so that its graded Euler characteristic is the Alexander polynomial.

THEOREM 1.3.2 ([**171**]). *If* $K \subset S^3$ *is a knot, then the Seifert genus of* K *is equal to the maximal integer* s *for which* $\widehat{\mathrm{HFK}}_*(K,s)$ *is non-zero.* □

(In [**171**], the above result was stated using $\widehat{\mathrm{HFK}}$ with $\mathbb{Z}$ coefficients; but the proof immediately adapts to show the same result with $\mathbb{Z}/2\mathbb{Z}$ coefficients, as well.)

A classical property of the Alexander polynomial is that it is monic if K is fibered. The corresponding property for knot Floer homology is fairly straightforward to establish [**175**]. The converse to this result, proved by Paolo Ghiggini in the case where $g(K) = 1$ [**71**] and in general by Yi Ni [**160**], is much deeper and gives the following:

THEOREM 1.3.3 ([**71, 160**]). *If* $K \subset S^3$ *is a knot with Seifert genus* g, *then* $\dim_{\mathbb{F}} \widehat{\mathrm{HFK}}_*(K,g) = 1$ *if and only if* K *is fibered.* □

András Juhász has given alternate proofs of both Theorems 1.3.2 and 1.3.3, using *sutured Floer homology* [**96, 97**].

A knot K is called *alternating* if it admits a diagram with the property that crossings alternate between over- and under-crossings, as we travel along the knot. The study of alternating knots is rather simpler than the general case. In particular, in the alternating case, the Alexander polynomial contains much more information about the knot. If K is alternating then the Alexander polynomial detects whether or not K is the unknot; more generally it encodes the Seifert genus of K; and it also encodes whether or not K is a fibered knot. (For details of this statement, see Theorem 2.4.13.) In fact Theorems 1.3.1, 1.3.2 and 1.3.3 can be viewed as extensions of these results to arbitrary knots, replacing the Alexander polynomial with the Poincaré polynomial of knot Floer homology. Indeed, for an alternating knot K the Poincaré polynomial of $\widehat{\mathrm{HFK}}(K)$ is determined by the Alexander polynomial $\Delta_K(t)$ and the signature $\sigma(K) \in \mathbb{Z}$ of K; we prove the analogous statement for grid homology in Theorem 10.3.1.

In addition to the above topological applications, the flexibility afforded by knot Floer homology makes it amenable to performing computations for certain infinite families of knots, for example, for torus knots (see Theorem 16.2.6).

The definition of knot Floer homology admits a natural extension to knots (and links) in arbitrary closed, oriented three-manifolds. This extension fits into the broader framework of Heegaard Floer homology for three-manifolds [**173, 174**]. Moreover, knot Floer homology is a useful device for computing the Heegaard Floer homology groups of three-manifolds obtained by Dehn surgery on K; see [**181**]. The details of these constructions are, however, beyond the scope of this book.

1.4. Comparison with Khovanov homology

It is interesting to compare the formal structure of knot Floer homology with a different kind of knot invariant defined by Mikhail Khovanov [**103**], and its various generalizations due to Khovanov and Lev Rozansky [**104, 105**]. For the sake of exposition, we focus on the simplest version, from [**103**]; see also [**8, 222**].

Like knot Floer homology, Khovanov homology is a bigraded vector space associated to a knot or link in S^3. Its graded Euler characteristic is another familiar knot polynomial: the *Jones polynomial* [**93**]. Khovanov homology can be enriched by a further differential, introduced by Eun-Soo Lee [**117**]; and Rasmussen [**194**] has used that differential to define a concordance invariant s similar in spirit to τ. These ideas led Rasmussen to give the first combinatorial proof of Theorem 1.2.1.

An analogue of Theorem 1.3.1 holds for Khovanov homology: Kronheimer and Mrowka [**112**] have shown that the unknot is the only knot whose Khovanov homology has dimension two over $\mathbb{Q}$. In fact, it is conjectured that the Jones polynomial also detects the unknot [**94**].

Intriguing though these similarities are, a precise mathematical relationship between the constructions of knot Floer and Khovanov homologies has yet to be discovered; compare [**36, 191**].

In a different direction, it is also interesting to compare knot Floer homology with analogous invariants coming from gauge theory, especially instanton homology for knots [**54, 113**].

1.5. On notational conventions

In the present book, we wish to distinguish properties of knot Floer homology that follow from the holomorphic theory (such as Theorem 1.3.1) from properties that can be derived purely within the combinatorial framework, and which we will proceed to prove in this book. To help emphasize this point, we make a semantic distinction between knot Floer homology and grid homology; and we make a corresponding notational distinction as well.

For the convenience of the experts, we have assembled a dictionary below connecting the traditional notation for knot Floer homology with the notation for grid homology we use here. The first column describes some construction and, in

The complex for knot/link homology (crossing no basepoints) *Fully blocked grid complex*	$\widetilde{CL}(G)$ **[136]**	$\widetilde{GC}(\mathbb{G})$
Link Floer homology (crossing no basepoints) *Fully blocked grid homology*	$\widetilde{HL}(\vec{L})$ **[136]**	$\widetilde{GH}(\vec{L})$
The knot/link Floer complex (crossing $\mathbb{O}$ basepoints) *Unblocked grid complex*	$\mathrm{CFK}^-(K)$ **[185]**; $\mathrm{CFL}^-(\vec{L})$ **[180]**; $CL^-(G)$ **[136]**	$GC^-(\mathbb{G})$
Knot/link Floer homology (crossing $\mathbb{O}$ basepoints) *Unblocked grid homology*	$\mathrm{HFK}^-(K)$ **[185]**; $\mathrm{HFL}^-(\vec{L})$ **[180]**	$GH^-(\vec{L})$
The complex for knot Floer homology with $U = 0$ *Simply blocked grid complex*	$\widehat{\mathrm{CFK}}(K)$ **[172]**	$\widehat{GC}(\mathbb{G})$
Knot Floer homology with $U = 0$ *Simply blocked grid homology*	$\widehat{\mathrm{HFK}}(K)$ **[172]**	$\widehat{GH}(K)$
Link Floer homology with collapsed grading and with $U = 0$ *Simply blocked, bigraded grid* *homology for links*	$\widehat{\mathrm{HFK}}(\vec{L})$ **[172]**	$\widehat{GH}(\vec{L})$
Link Floer homology *Simply blocked, multi-graded grid* *homology for links*	$\widehat{\mathrm{HFL}}(\vec{L})$ **[180]**	$\widehat{\mathbf{GH}}(\vec{L})$
Link Floer homology *Unblocked, multi-graded grid homology* *for links*	$\mathrm{HFL}^-(\vec{L})$ **[180]**	$\mathbf{GH}^-(\vec{L})$
The filtered knot complex $\widehat{\mathrm{CF}}(S^3)$ (crossing all $\mathbb{X}$ and all but one of the $\mathbb{O}$) *Simply blocked filtered grid complex*	$CFK^{0,*}(S^3, K)$ **[172]**	$\widehat{\mathcal{GC}}(\mathbb{G})$
The filtered knot complex $\mathrm{CF}^-(S^3)$ (crossing all basepoints) *(Unblocked) filtered grid complex*	$CFK^{-,*}(S^3, K)$ **[172]**	$\mathcal{GC}^-(\mathbb{G})$
The multi-filtered link complex (crossing all basepoints) *(Unblocked) filtered grid complex*	$CFL^{-,*}(\vec{L})$ **[180]**	$\boldsymbol{\mathcal{GC}}^-(\mathbb{G})$

italics, gives the name we call it in the present book. In the next column, we give the notation of the concept customary in knot Floer homology and indicate references where the notation is introduced or used extensively. In the final column, we include the notation used in this book.

Link Floer homology is typically thought of as endowed with an Alexander multi-grading. In the text we distinguish the multi-graded versions from their bigraded analogues by using boldface for the multi-graded ones ($\mathbf{GC}^-(\mathbb{G})$, $\widehat{\mathbf{GC}}(\mathbb{G})$, $\mathbf{GH}^-(\vec{L})$, $\widehat{\mathbf{GH}}(\vec{L})$), rather than the usual typeface for their bigraded analogues ($GC^-(\mathbb{G})$, $\widehat{GC}(\mathbb{G})$, $GH^-(\vec{L})$ and $\widehat{GH}(\vec{L})$).

There are two natural conventions for the Maslov grading of a link, which are different when the link has more than one component. In the present work, we made a choice which differs from the choices made elsewhere (e.g. in [**172**]). These two gradings differ by $\frac{\ell-1}{2}$, where ℓ denotes the number of components; see Equation (9.18). Using our present choice, for any link, the Maslov grading is always integer-valued.

1.6. Necessary background

This book is aimed at a wide audience, ranging from motivated undergraduates curious about modern methods in knot theory, to graduate students and researchers who want a leisurely introduction to the combinatorial aspects of Heegaard Floer homology. Some familiarity with knot theory would be helpful, say, on the level of [**28, 119, 198**], though in Chapter 2 this is recalled. The development of grid homology also uses some basic tools from homological algebra: chain complexes, chain maps, chain homotopies, and mapping cones. These concepts can be found in introductory books on algebraic topology such as [**83**]; see also [**226**]. For the reader's convenience, though, the relevant part of homological algebra is briefly summarized either before it is used or in Appendix A.

As we shall see in Chapter 12, Legendrian and transverse knots and links in the standard contact three-sphere lend themselves to study through grids and grid homology. Although we have attempted to make this chapter as self-contained as possible, at times we will refer the reader to basic texts in contact topology, such as [**45, 68**].

1.7. The organization of this book

We start our discussion in Chapter 2 with a short review of the classical theory of knots and links in the three-space. The concept of grid diagrams, their relation to the Alexander polynomial, and to Seifert surfaces are described in Chapter 3. In Chapter 4 we introduce the main object of the book, grid homologies of knots. We start with a simple version and then build up to the further, slightly more complicated variants. The invariance of these homology groups (i.e. independence of the presentation of a knot by a grid diagram) is discussed in Chapter 5.

After setting up the basics of the theory, we turn to the first applications. In Chapter 6 we give a short proof of the fact that the τ-invariant bounds the unknotting number. This result then leads to the verification of Milnor's conjecture on the unknotting numbers of torus knots. Further basic properties are discussed in Chapter 7.

The rest of the book is devoted to more advanced topics in grid homology. In Chapter 8 we strengthen the unknotting bound and show that $|\tau(K)| \leq g_s(K)$, completing the proof of Theorem 1.2.1. Chapter 9 provides an important computational tool, the skein exact sequence. We also show instances where this long exact sequence can be conveniently used in explicit computations. A variation of the skein exact sequence, presented in Chapter 10, then leads us to the computation of grid homology for all alternating knots. In Chapter 11 we give the details of the extension of grid homology from knots to links. To put this construction into perspective, we also recall some standard facts about the multi-variable Alexander polynomial of a link and the Thurston norm of a link complement. In Chapter 12 we give applications of grid homology for Legendrian and transverse knots. In particular, in Sections 12.4 and 12.6 we prove Theorems 1.2.2 and 1.2.3. Chapter 13 provides further algebraic background and describes a generalization of the invariant, which is the filtered quasi-isomorphism type of a filtered chain complex over an appropriate polynomial ring. In Chapter 14 further properties of the filtered chain complex are discussed. In Chapter 15, we explain the sign conventions required to develop grid homology with coefficients in $\mathbb{Z}$. In Chapter 16 we review some basic aspects of Heegaard Floer homology, a theory initiated in [**173, 174**] and further developed in [**172, 180, 191**] for knots and links in general three-manifolds. We also explain why the combinatorial theory discussed in the earlier chapters is, indeed, a special case of this more general theory. In Chapter 17 we collect some open problems, hopefully motivating further reading and research in the subject. For the sake of completeness, in Appendix A, we include elements of homological algebra which we use throughout the book. The text also uses some basic theorems of knot theory (such as the Reidemeister and the Reidemeister-Singer theorems and Cromwell's theorem on grids). Appendix B is devoted to the proofs of these results.

We have included a number of exercises throughout the text, with varying levels of difficulty. More challenging problems are marked with an asterisk *.

The material presented here is an exposition of results which have already appeared in the literature. Our exposition of basic knot theory was influenced by [**100, 101, 119, 189, 198**]; for grid diagrams we have drawn on [**28, 37**]. Our discussion of grid homology follows the account from [**136**] (see also [**135**]), with further topics coming from [**66, 85, 157, 167, 185, 204, 231**]. The proof of the unknotting bound from Chapter 6 is new, though it is inspired by [**204**]. The proof that grid homology gives bounds on the Seifert genus (Proposition 7.2.2) is new; but our treatment falls short of showing that these bounds are sharp, as we know from the holomorphic theory. The organization of the invariance proof of the grid invariants is a little different from that in [**136**]: we have chosen here to start from the simplest cases (invariance of the grid homology groups with coefficients mod 2) and build up to the general case (for the filtered quasi-isomorphism type), rather than going the other direction, as was done in [**136**]. As a by-product, the invariance proof presented here is somewhat simpler than the proof from [**136**].

This book could be used as a textbook for a semester-long course either at the graduate level or at an advanced undergraduate level. The background is reviewed in Chapter 2 and in Appendix A (especially Sections A.1–A.5). The material developed in Chapters 3-7 could serve as the core for this course. The topics covered in Chapters 8, 9, and the first three sections of Chapter 15 are essentially independent, and any of these could be added as additional topics.

A course with more advanced topics could include the computation of the grid homology for alternating knots from Chapter 10; and the extension of the theory to links from Chapter 11. Chapter 12 is a further essentially independent discussion on the role of grid homology in contact knot theory. The reader interested in pursuing Heegaard Floer homology further is encouraged to study Chapters 13 and 14 (and to read Chapter 16, as a preview).

1.8. Acknowledgements

The work presented here owes a great deal to the work of other researchers, both in this and in related subjects.

Knot Floer homology is built on Heegaard Floer homology, which in turn grew out of studying Seiberg-Witten gauge theory and Heegaard splittings.

The Seiberg-Witten equations on smooth four-manifolds were introduced by Nathan Seiberg and Edward Witten [**208, 209, 230**], see also [**149**]. Their introduction led to many important applications for smooth four-dimensional topology, and in particular, thanks to the fundamental contributions of Cliff Taubes [**215**], a link between four-manifold topology and Mikhail Gromov's invariants in symplectic geometry [**81**]. It also provided a framework for new three-dimensional invariants. Peter Kronheimer and Tom Mrowka were pioneers in the development of Seiberg-Witten theory for three-manifolds [**110, 111**]. Seiberg-Witten theory, in turn, was formulated as an alternative to Simon Donaldson's invariants for smooth four-manifolds [**35**], constructed using the anti-self-dual Yang-Mills equations. Donaldson's breakthrough formed the foundation of the modern understanding of smooth four-dimensional topology. The corresponding three-manifold invariant was defined and explored by Andreas Floer [**49**], whose work provided further guiding principles in Heegaard Floer homology.

Much of the work presented here is based on joint work with other collaborators. Grid homology was discovered in collaboration with Ciprian Manolescu and Sucharit Sarkar, and further developed in collaboration with Dylan Thurston. Hence our discussion here is heavily influenced by their ideas. Some of the applications of the transverse invariant were found in collaboration with Lenny Ng.

Moreover, the exposition of the solution for the Milnor conjecture we give here was motivated by Sucharit Sarkar's proof [**204**]. Our treatment of grid homology for alternating knots follows the work of Mike Wong [**231**], which in turn rests on earlier work of Ciprian Manolescu [**132**]. Our discussion of the theory over $\mathbb{Z}$ uses ideas of Étienne Gallais [**66**]. Some examples of transversally non-simple knots were taken from work of Lenny Ng and Tirasan Khandhawit [**102**].

We would like to thank the following people for providing valuable input on early drafts of this work: Daniel Copeland, Stefan Friedl, Marco Golla, Robert Lipshitz, Lilya Lyubich, András Szűcs and Mike Wong. We are also very grateful to Robert Lipshitz and Sucharit Sarkar for discussions pertinent to Chapter 17. Many of the figures were created by Boldizsár Kalmár; some others were drawn with the help of Mathematica.

Finally, all three of us owe a great deal to our advisor, John Morgan, who introduced us patiently to modern developments in low dimensional topology when we were graduate students, and has supplied both guidance and friendship in the years since.

During the course of this work Peter Ozsváth was partially supported by NSF DMS-1405114. András Stipsicz was partially supported by the *Lendület program ADT*, ERC Advanced Grant LDTBud and OTKA NK81203, K112735; he would also like to thank the Institute for Advanced Study in Princeton for its hospitality during much of this work. Zoltán Szabó was partially supported by NSF DMS-1006006 and NSF DMS-1309152.

CHAPTER 2

Knots and links in S^3

In this chapter we collect the notions and results from classical knot theory most relevant to our subsequent discussions. In Section 2.1 we provide some basic definitions and describe some families of knots that will serve as guiding examples in the further chapters. We discuss Seifert surfaces in Section 2.2, and we define the Seifert form in Section 2.3. Based on this notion, we define the signature of a knot and use it to bound the unknotting number. We devote Section 2.4 to the definition and basic properties of the Alexander polynomial (returning to its multi-variable generalization in Section 11.5). Extending ideas from Section 2.3, in Section 2.6 we give a proof of the lower bound of the slice genus provided by the signature. Finally, in Section 2.7 we use the Goeritz matrix associated to a diagram to derive a simple formula for the signature of a knot. This material is standard; for a more detailed treatment see [**28, 119, 198**]. Further basic theorems of knot theory are collected in Appendix B.

2.1. Knots and links

DEFINITION 2.1.1. An ℓ-component ***link*** L in S^3 is a collection of ℓ disjoint smoothly embedded simple closed curves. A 1-component link K is a ***knot***. The links we consider in this book will typically be oriented. If we want to emphasize the choice of an orientation, we write $\vec{L}$ for a link, equipped with its orientation. The links $\vec{L}_1, \vec{L}_2$ are ***equivalent*** if they are ***ambiently isotopic***, that is, there is a smooth map $H\colon S^3 \times [0,1] \to S^3$ such that $H_t = H|_{S^3\times\{t\}}$ is a diffeomorphism for each $t \in [0,1]$, $H_0 = \mathrm{id}_{S^3}$, $H_1(\vec{L}_1) = \vec{L}_2$ and H_1 preserves the orientation on the components. An equivalence class of links under this equivalence relation is called a ***link*** (or ***knot***) ***type***.

The above definition can be made with $\mathbb{R}^3 = S^3 \setminus \{p\}$ instead of S^3, but the theory is the same: two knots in $\mathbb{R}^3$ are equivalent if and only if they are equivalent when viewed in S^3. For this reason, we think of links as embedded in $\mathbb{R}^3$ or S^3 interchangeably.

Two ℓ-component links $\vec{L}_i$ ($i = 0, 1$) are *isotopic* if the two smooth maps $f_i\colon \cup_{j=1}^{\ell} S^1 \to S^3$ defining the links are isotopic, that is, there is a smooth map $F\colon (\cup_{j=1}^{\ell} S^1) \times [0,1] \to S^3$ which has the property that $F_t = F|_{(\cup_{j=1}^{\ell} S^1)\times\{t\}}$ are ℓ-component links with $F_i = f_i$ ($i = 0, 1$). By the isotopy extension theorem [**87**, Section 8, Theorem 1.6], two links are ambiently isotopic if and only if they are isotopic.

Reflecting L through a plane in $\mathbb{R}^3$ gives the *mirror image* $m(L)$ of L. Reversing orientations of all the components of $\vec{L}$ gives $-\vec{L}$.

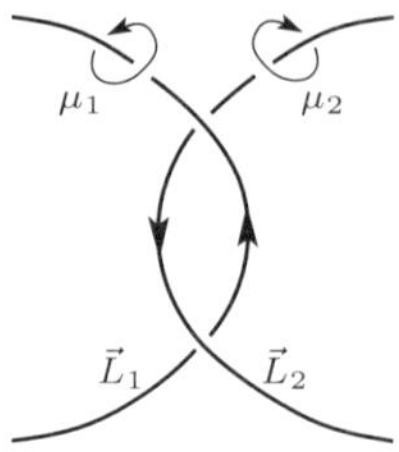

FIGURE 2.1. **Meridians of the components of a link.** The oriented link $\vec{L}$ of the diagram has two components L_1, L_2, with oriented meridians μ_1 and μ_2.

REMARK 2.1.2. Another way to define equivalence of links is to say that $\vec{L}_1$ and $\vec{L}_2$ are equivalent if there is an orientation-preserving diffeomorphism $f\colon S^3 \to S^3$ so that $f(\vec{L}_1) = \vec{L}_2$. In fact, this gives the same equivalence relation since the group of orientation-preserving diffeomorphisms of S^3 is connected [**22**].

REMARK 2.1.3. It is not hard to see that the smoothness condition in the above definition can be replaced by requiring the maps to be PL (piecewise linear). For basic notions of PL topology, see [**201**]. The PL condition provides an equivalent theory of knots and links, cf. [**18**]. (Assuming only continuity would allow wild knots, which we want to avoid.)

The complements of equivalent links are homeomorphic; therefore the fundamental group of the complement, also called the *link group* (or the *knot group* for a knot), is an invariant of the link type. The first homology group of an ℓ-component link $L = (L_1, \ldots, L_\ell)$ is given by

$$H_1(S^3 \setminus L; \mathbb{Z}) \cong \mathbb{Z}^\ell. \tag{2.1}$$

An isomorphism $\phi\colon H_1(S^3 \setminus L; \mathbb{Z}) \to \mathbb{Z}^\ell$ is specified by an orientation and a labeling of the components of L: ϕ sends the homology class of the positively oriented meridian $\mu_i \in H_1(S^3 \setminus L; \mathbb{Z})$ of the ith component L_i to the vector $(0, \ldots, 0, 1, 0, \ldots, 0) \in \mathbb{Z}^\ell$ (where 1 occurs at the ith position). For the orientation convention on the meridian, see Figure 2.1. Suppose that L is a link in $\mathbb{R}^3$ and $pr_P\colon \mathbb{R}^3 \to P$ is the orthogonal projection to an oriented plane $P \subset \mathbb{R}^3$. For a generic choice of P the projection pr_P restricted to L is an immersion with finitely many double points. At the double points, we illustrate the strand passing under as an interrupted curve segment. If L is oriented, the orientation is specified by placing an arrow on the diagram tangent to each component of L. The resulting diagram $\mathcal{D}$ is called a *knot* or *link diagram* of $\vec{L}$. Obviously, a link diagram determines a link type.

The local modifications of a link diagram indicated in Figure 2.2 are the *Reidemeister moves*; there are three types of these moves, denoted R_1, R_2 and R_3. When thinking of oriented link diagrams, the strands in the local picture can be oriented in any way. The figures indicate changes to the diagram within a small disk; the rest of the diagram is left alone. The Reidemeister moves obviously preserve the link type. The importance of the Reidemeister moves is underscored by the following theorem. (For a proof of this fundamental result, see Section B.1.)

THEOREM 2.1.4 (Reidemeister, [**196**]). *The link diagrams $\mathcal{D}_1$ and $\mathcal{D}_2$ correspond to equivalent links if and only if these diagrams can be transformed into each other by a finite sequence of Reidemeister moves and planar isotopies.* □

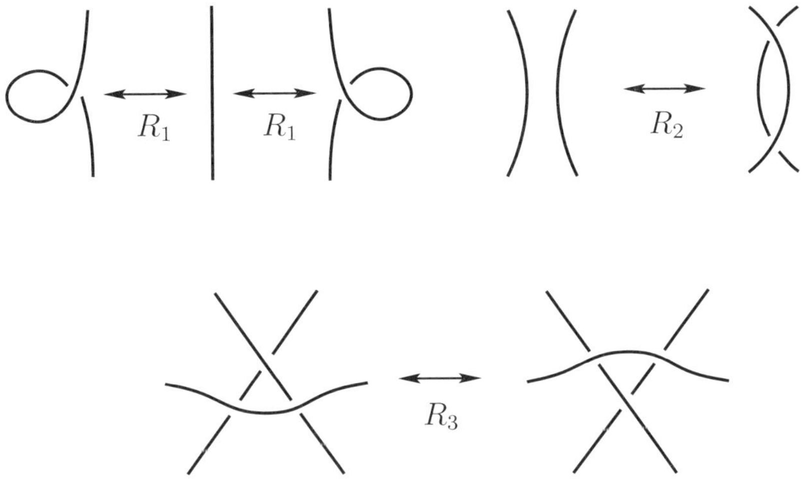

FIGURE 2.2. **The Reidemeister moves R_1, R_2, R_3.**

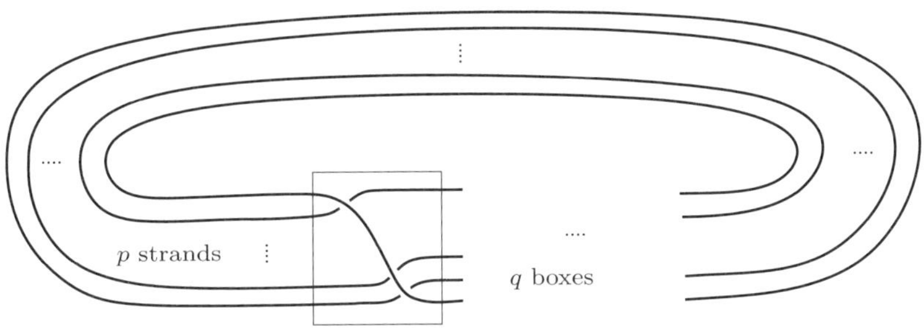

FIGURE 2.3. **Diagram of the torus link $T_{p,q}$.** The result is a knot if $\gcd(p,q) = 1$; in general the torus link has $\gcd(p,q)$ components.

The following examples will appear throughout the text.

EXAMPLES 2.1.5.
- Let $p, q > 1$ be relatively prime integers. The (p,q) ***torus knot*** $T_{p,q}$ is defined as the set of points

$$\{(z_1, z_2) \in \mathbb{C}^2 \mid z_1\overline{z}_1 + z_2\overline{z}_2 = 1,\ z_1^p + z_2^q = 0\} \subset S^3. \tag{2.2}$$

  This knot can be drawn on a standard, unknotted torus in three-space, so that it meets a longitudinal curve q times (each with local intersection number $+1$) and a meridional curve p times (again, each with local intersection number $+1$). A diagram for $T_{p,q}$ is shown in Figure 2.3. It is easy to see that $T_{p,q}$ and $T_{q,p}$ are isotopic knots. The mirror image $m(T_{p,q})$ of $T_{p,q}$ is called the ***negative torus knot*** $T_{-p,q}$. For general choices of p and q, the definition of Equation 2.2 produces a link, the ***torus link*** $T_{p,q}$, a link with $\gcd(p,q)$ components. $T_{2,3}$ is the ***right-handed trefoil*** knot, and $T_{-2,3}$ is the ***left-handed trefoil*** knot.
- For $a_1, \ldots, a_n \in \mathbb{Z}$ the diagram of Figure 2.5 defines the $(a_1, \ldots, a_n)$ ***pretzel knot*** (or ***pretzel link***) $P(a_1, \ldots a_n)$ on n strands. Informally,

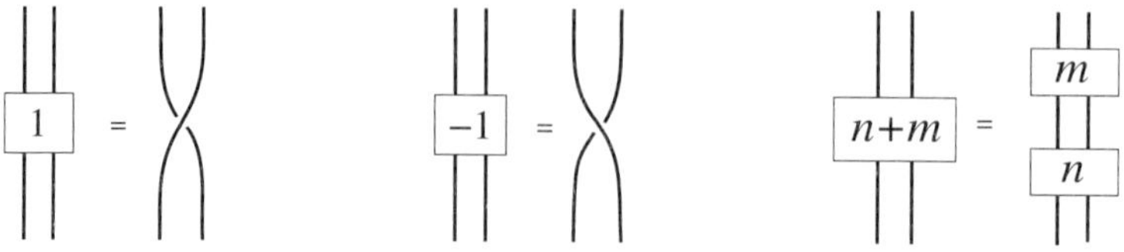

FIGURE 2.4. **Convention for twists.**

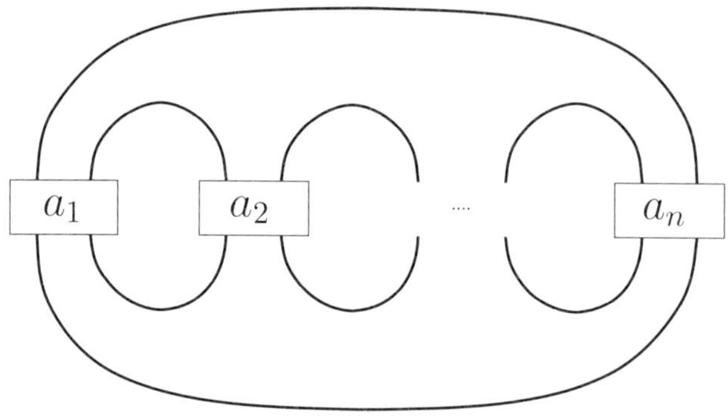

FIGURE 2.5. **Diagram of the pretzel link $P(a_1, \ldots, a_n)$.** A box with an integer a_i means a_i right half-twists for $a_i \geq 0$ and $-a_i$ left half-twists for $a_i < 0$, cf. Figure 2.4.

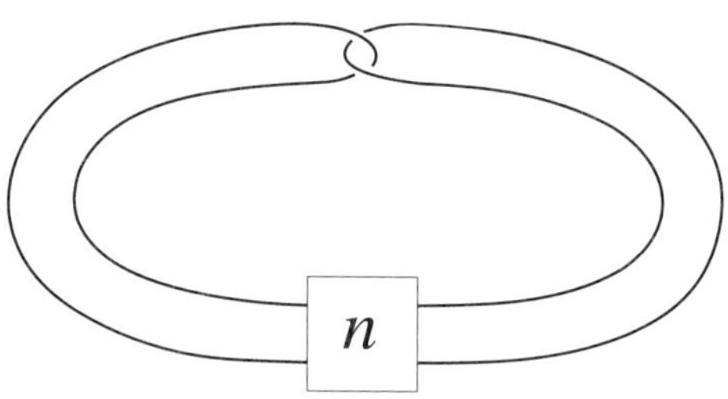

FIGURE 2.6. **Diagram of the twist knot W_n.** Clearly, W_{-1} and W_0 are both unknots, W_{-2} is the right-handed trefoil knot $T_{2,3}$, and W_1 is the left-handed trefoil knot $m(T_{2,3}) = T_{-2,3}$. The knot W_2 is also called the *figure-eight knot.*

the pretzel link is constructed by taking $2n$ strands $s_1, s_2, \ldots, s_{2n-1}, s_{2n}$, introducing $|a_i|$ half-twists (right half-twists when $a_i \geq 0$ and left half-twists when $a_i < 0$) on the two strands s_{2i-1}, s_{2i} and then closing up the strands as shown in Figure 2.5. (The conventions on the half-twists are indicated in Figure 2.4.)

- For $n \in \mathbb{Z}$ we define the ***twist knot*** W_n by Figure 2.6. Notice that we fix the clasp, and allow the twist in the box to have arbitrary sign and parity. Informally, a twist knot is constructed by considering two strands, adding $|n|$ half-twists (right if $n \geq 0$ and left if $n < 0$) to them and then closing up with the clasp shown by Figure 2.6.

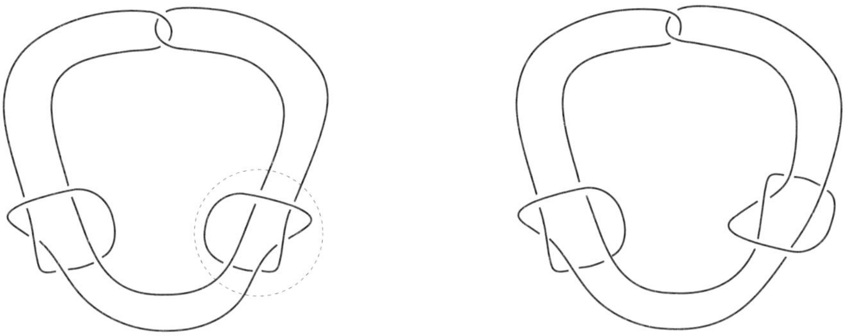

FIGURE 2.7. **The Kinoshita-Terasaka knot KT (on the left) and its Conway mutant, the Conway knot C (on the right).** The two knots are mutants of each other, as the dashed circle on the Kinoshita-Terasaka knot shows.

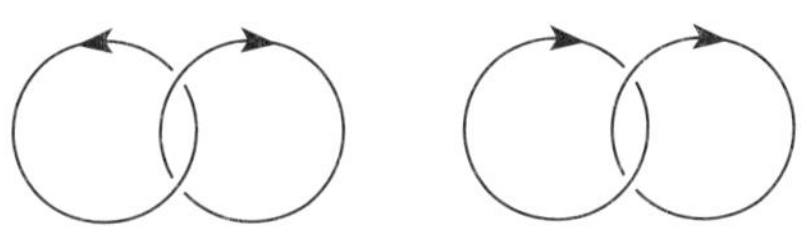

FIGURE 2.8. **The two Hopf links H_+ and H_-.**

- A diagram of the ***Kinoshita-Terasaka knot*** KT is shown on the left of Figure 2.7; the knot diagram on the right of Figure 2.7 represents the ***Conway knot*** C. These two knots are ***mutants*** of each other, that is, if we cut out the dashed disk from the left diagram of Figure 2.7 and glue it back after a 180° rotation, we get the Conway knot.

REMARK 2.1.6. If we equip the torus knot $T_{p,q}$ with its two possible orientations, we get isotopic knots. Similarly, the twist knots W_n are isotopic when equipped with the two possible orientations. When p and q are not relatively prime, we define the oriented link $\vec{T}_{p,q}$ by orienting all the parallel strands in Figure 2.3 in the same direction.

EXERCISE 2.1.7. **(a)** The above families are not disjoint. Find knots that appear in more than one family.
(b) Using the Seifert-Van Kampen theorem [**83**, Theorem 1.20], show that for $(p, q) = 1$, the knot group of the torus knot $T_{p,q}$ is isomorphic to $\langle x, y \mid x^p = y^q \rangle$.
(c) Compute the link group of $T_{3,6}$.
(d) Verify the claim of Remark 2.1.6 for the right-handed trefoil and for the figure-eight knots.
(e) Show that the figure-eight knot W_2 and its mirror $m(W_2)$ are isotopic.

The oriented link $\vec{T}_{2,2}$ is also called the *positive Hopf link* H_+. Reversing the orientation on one component of $\vec{T}_{2,2}$, we get the *negative Hopf link* H_-; see Figure 2.8. A simple three-component link is the *Borromean rings*; see Figure 2.9.

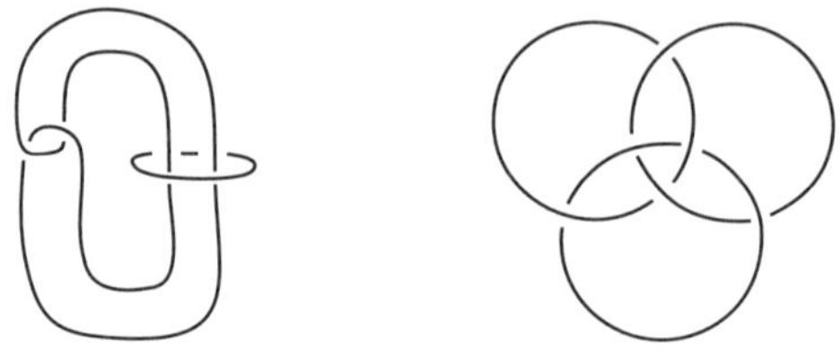

FIGURE 2.9. **The Whitehead link (on the left) and the Borromean rings (on the right).**

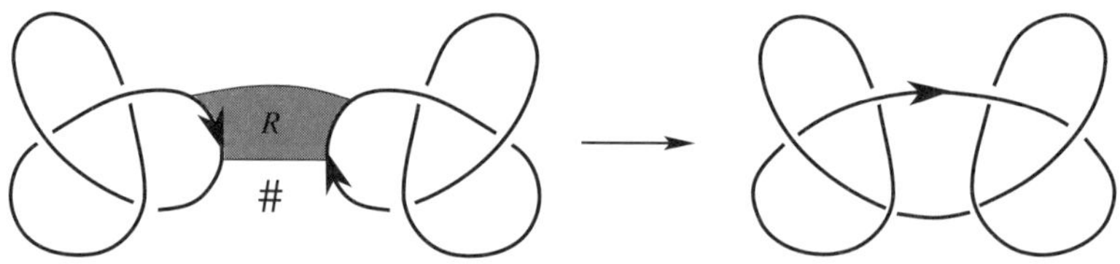

FIGURE 2.10. **The connected sum operation.** The band R is shown by the shaded rectangle.

An interesting property of knots and links is related to the existence of a fibration on their complement.

DEFINITION 2.1.8. A link $\vec{L}$ is ***fibered*** if the complement $S^3 \setminus \vec{L}$ admits a fibration $\varphi\colon S^3 \setminus \vec{L} \to S^1$ over the circle with the property that for each $t \in S^1$ the closure $\overline{\varphi^{-1}(t)}$ of the fiber $\varphi^{-1}(t)$ is equal to $\varphi^{-1}(t) \cup \vec{L}$ and is a compact, oriented surface with oriented boundary $\vec{L}$. (For more on fibered knots, see [**18**, Chapter 5].)

EXERCISE 2.1.9. Verify that the torus knot $T_{p,q}$ is fibered. (*Hint:* Refer to Example 2.1.5 and consider the map $f/|f|$ for $f(z_1, z_2) = z_1^p + z_2^q$.)

A well-studied and interesting class of knots are defined as follows.

DEFINITION 2.1.10. A link diagram $\mathcal{D}$ is called ***alternating*** if the crossings alternate between over-crossings and under-crossings, as we traverse each component of the link. A link admitting an alternating diagram is called an ***alternating link***.

EXAMPLES 2.1.11. The twist knots W_n are alternating for all n. More generally, consider the pretzel links $P(a_1, \dots, a_n)$ where the signs of the a_i are all the same; these pretzel links are also alternating. The Borromean rings is an alternating link.

Suppose that $\vec{K}_1, \vec{K}_2$ are two oriented knots in S^3 that are separated by an embedded sphere. Form the *connected sum* $\vec{K}_1 \# \vec{K}_2$ of $\vec{K}_1$ and $\vec{K}_2$ as follows. First choose an oriented rectangular disk R with boundary ∂R composed of four oriented arcs $\{e_1, e_2, e_3, e_4\}$ such that $\vec{K}_1 \cap R = -e_1 \subset \vec{K}_1$ and $\vec{K}_2 \cap R = -e_3 \subset \vec{K}_2$, and the separating sphere intersects R in a single arc and intersects e_2 and e_4 in a single point each. Then define $\vec{K}_1 \# \vec{K}_2$ as

$$\vec{K}_1 \# \vec{K}_2 = (\vec{K}_1 \setminus e_1) \cup e_2 \cup e_4 \cup (\vec{K}_2 \setminus e_3).$$

The resulting knot type is independent of the chosen band R. For a pictorial presentation of the connected sum of two knots, see Figure 2.10.

FIGURE 2.11. **Signs of crossings.** The crossing shown on the left is positive, while the one on the right is negative.

The connected sum operation for knots is reminiscent to the product of integers: every knot decomposes (in an essentially unique way) as the connected sum of basic knots (called *prime knots*). For more on prime decompositions see [**119**, Theorem 2.12]. As it turns out, fiberedness of the connected sum is determined by the same property of the components: by a result of Gabai [**64**], the connected sum of two knots is fibered if and only if the two knots are both fibered.

We define now a numerical obstruction to pulling arbitrarily far apart two disjoint, oriented knots $\vec{K}_1$ and $\vec{K}_2$.

DEFINITION 2.1.12. Suppose that $\vec{K}_1, \vec{K}_2 \subset S^3$ are two disjoint, oriented knots. Let $\mathcal{D}$ be a diagram for the oriented link $\vec{K}_1 \cup \vec{K}_2$. The ***linking number*** $\ell k(\vec{K}_1, \vec{K}_2)$ of $\vec{K}_1$ with $\vec{K}_2$ is half the sum of the signs of those crossings (in the sense of Figure 2.11) where one strand comes from $\vec{K}_1$ and the other from $\vec{K}_2$.

PROPOSITION 2.1.13. *The linking number $\ell k(\vec{K}_1, \vec{K}_2)$ has the following properties:*

- *it is independent of the diagram used in its definition;*
- *if $\vec{K}_1$ and $\vec{K}_2$ can be separated by a two-sphere, then $\ell k(\vec{K}_1, \vec{K}_2) = 0$;*
- *it is integral valued;*
- *it is symmetric; i.e. $\ell k(\vec{K}_1, \vec{K}_2) = \ell k(\vec{K}_2, \vec{K}_1)$.*

Proof. The fact that $\ell k(\vec{K}_1, \vec{K}_2)$ is independent of the diagram is a straightforward verification using the Reidemeister moves. It follows immediately that if $\vec{K}_1$ and $\vec{K}_2$ can be separated by a two-sphere, then $\ell k(\vec{K}_1, \vec{K}_2) = 0$.

Let $\vec{K}_1'$ be obtained from $\vec{K}_1$ by changing a single crossing with $\vec{K}_2$ (with respect to some fixed diagram $\mathcal{D}$). It is straightforward to see that $\ell k(\vec{K}_1', \vec{K}_2)$ differs from $\ell k(\vec{K}_1, \vec{K}_2)$ by ± 1. Continue to change crossings of $\vec{K}_1$ with $\vec{K}_2$ to obtain a new link $\vec{K}_1'' \cup \vec{K}_2$ (and a diagram of $\vec{K}_1'' \cup \vec{K}_2$) with the property that at any crossings between $\vec{K}_1''$ and $\vec{K}_2$, the strand in $\vec{K}_1''$ is above the strand in $\vec{K}_2$. It follows that the difference between $\ell k(\vec{K}_1'', \vec{K}_2)$ and $\ell k(\vec{K}_1, \vec{K}_2)$ is an integer. Since $\vec{K}_1''$ can be lifted above $\vec{K}_2$, and then separated from it by a two-sphere, $\ell k(\vec{K}_1'', \vec{K}_2) = 0$. We conclude that $\ell k(\vec{K}_1, \vec{K}_2)$ is integral valued. Finally, the definition of linking number is manifestly symmetric in the roles of $\vec{K}_1$ and $\vec{K}_2$. □

The linking number has a straightforward generalization to pairs $\vec{L}_1$ and $\vec{L}_2$ of oriented links. Clearly, the linking number is not the only obstruction to pulling apart the link $\vec{L}_1 \cup \vec{L}_2$. For instance, the two components of the Whitehead link of Figure 2.9 have zero linking number, but cannot be separated by a sphere (cf.

Exercise 2.4.12(f)). Similarly, for any two components of the Borromean rings the linking number is zero; but no component can be separated from the other two.

DEFINITION 2.1.14. Let $\mathcal{D}$ be a diagram of the link $\vec{L}$. The ***writhe*** $\mathrm{wr}(\mathcal{D})$ of the diagram $\mathcal{D}$ is defined to be the number of positive crossings in $\mathcal{D}$ minus the number of negative ones. Notice that if $\mathcal{D}$ is the diagram of a knot, then the chosen orientation does not influence the value of the writhe $\mathrm{wr}(\mathcal{D})$.

EXERCISE 2.1.15. **(a)** Consider the homology element $[\vec{K}_2] \in H_1(S^3 \setminus \vec{K}_1; \mathbb{Z}) \cong \mathbb{Z}$ given by $\vec{K}_2 \subset S^3 \setminus \vec{K}_1$. If μ_1 is the homology class of an oriented normal circle of $\vec{K}_1$, then show that $[\vec{K}_2] = \ell k(\vec{K}_1, \vec{K}_2) \cdot \mu_1$.
(b) Show that the Reidemeister moves R_2 and R_3 do not change the writhe of a projection. Determine the change of the writhe under the two versions of R_1.
(c) Suppose that $\mathcal{D}$ is the diagram of the link $\vec{L} = \vec{L}_1 \cup \vec{L}_2$. Reverse the orientation on all components of $\vec{L}_2$ (while keeping the orientations of the components of $\vec{L}_1$ fixed). Let $\mathcal{D}'$ denote the resulting diagram. Show that

$$\mathrm{wr}(\mathcal{D}) = \mathrm{wr}(\mathcal{D}') + 4\ell k(\vec{L}_1, \vec{L}_2).$$

2.2. Seifert surfaces

Knots and links can be studied via the surfaces they bound. More formally:

DEFINITION 2.2.1. A smoothly embedded, compact, connected, oriented surface-with-boundary in $\mathbb{R}^3$ is a ***Seifert surface*** of the oriented link $\vec{L}$ if $\partial\Sigma = L$, and the orientation induced on $\partial\Sigma$ agrees with the orientation specified by $\vec{L}$.

Recall that a connected, compact, orientable surface Σ is classified (up to diffeomorphism) by its number of boundary components $b(\Sigma)$, and an additional numerical invariant g, called the *genus*; see [**137**, Theorem 11.1]. This quantity can be most conveniently described through the Euler characteristic $\chi(\Sigma)$ of the surface, since

$$\chi(\Sigma) = 2 - 2g(\Sigma) - b(\Sigma).$$

From a given Seifert surface Σ of $\vec{L}$ further Seifert surfaces can be obtained by *stabilizing* (or *tubing*) Σ: connect two distinct points $p, q \in \mathrm{Int}\,\Sigma$ by an arc γ in $S^3 \setminus \Sigma$ that approaches Σ at p and q from the same side of Σ. Deleting small disk neighborhoods of p and q from Σ and adding an annulus around γ, we get a new surface, which (by our assumption on how γ approaches Σ) inherits a natural orientation from Σ, and has genus $g(\Sigma) + 1$, cf. Figure 2.12. According to the following result, any two Seifert surfaces of a given link can be transformed into each other by this operation (and isotopy). For a proof of the following result, see [**9**] or Section B.3.

THEOREM 2.2.2 (Reidemeister-Singer, [**212**]). *Any two Seifert surfaces Σ_1 and Σ_2 of a fixed oriented link $\vec{L}$ can be stabilized sufficiently many times to obtain Seifert surfaces Σ_1' and Σ_2' that are ambient isotopic.* □

EXERCISE 2.2.3. **(a)** Show that any knot or link in S^3 admits a Seifert surface. (*Hint:* Using the orientation, resolve all crossings in a diagram to get a disjoint union of oriented circles in the plane, and consider disks bounded by the resulting unknots. Move these disks appropriately to different heights and restore the crossings by adding bands to the disks. Connectedness can be achieved by tubing

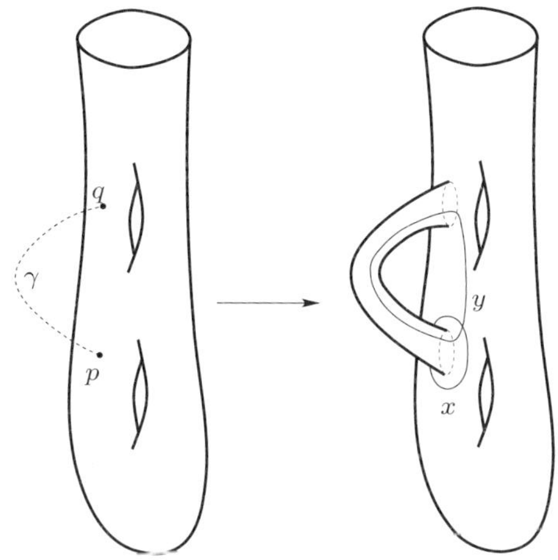

FIGURE 2.12. **Schematic picture of a stabilization of a Seifert surface.** The arc γ in the complement of the surface is assumed to approach Σ from the same side at p and q, so the result of the stabilization admits a natural orientation. Although the diagram shows an unknotted arc, γ is allowed to be knotted.

together various components. For further details, see Section B.3 or [**119**, Chapter 8].)
(b) Find a genus one Seifert surface for W_n.
(c) Find a genus one Seifert surface for the 3-stranded pretzel knot $P(a_1, a_2, a_3)$ with a_i odd for $i = 1, 2, 3$.
(d) Find a Seifert surface of genus n for the $(2, 2n+1)$ torus knot $T_{2,2n+1}$.

DEFINITION 2.2.4. The ***genus*** (or Seifert genus) $g(\vec{L})$ of a link $\vec{L}$ is the minimal genus of any Seifert surface for $\vec{L}$.

EXERCISE 2.2.5. Show that the unique knot with $g(K) = 0$ is the unknot $\mathcal{O}$.

REMARK 2.2.6. The linking number from Definition 2.1.12 has the following alternative definition using Seifert surfaces: $\ell k(\vec{K}_1, \vec{K}_2)$ is the algebraic intersection number of a Seifert surface for $\vec{K}_1$ with the oriented knot $\vec{K}_2$; see [**198**, Chapter 5].

Unlike the case of knots, the Seifert genus of a link in general depends on the orientations of the various components of L.

EXAMPLE 2.2.7. Let $\vec{L}_1$ denote the torus link $\vec{T}_{2,4}$, and let $\vec{L}_2$ be the same link with the orientation reversed on one component. It is easy to see that $\vec{L}_2$ bounds an annulus, hence $g(\vec{L}_2) = 0$, while $g(\vec{L}_1) = 1$.

It is proved in [**119**, Theorem 2.4] that the Seifert genus is additive under connected sum of oriented knots.

2.3. Signature and the unknotting number

A Seifert surface Σ for an oriented link $\vec{L}$ determines a bilinear form on $H_1(\Sigma; \mathbb{Z})$ as follows. Consider two elements $x, y \in H_1(\Sigma; \mathbb{Z})$ and represent them by oriented, embedded one-manifolds. More precisely, x can be represented by a collection γ_x

of pairwise disjoint, oriented, simple closed curves, and y can be represented by a similar γ_y. (Note that γ_x and γ_y might have non-empty intersection, though.) Let γ_y^+ denote the push-off of γ_y in the positive normal direction of Σ.

DEFINITION 2.3.1. The ***Seifert form*** S for the Seifert surface Σ of the link $\vec{L}$ is defined for $x, y \in H_1(\Sigma;\mathbb{Z})$ by

$$S(x,y) = \ell k(\gamma_x, \gamma_y^+).$$

It is easy to see that the resulting form is independent of the chosen representatives of the homology classes, and it is a bilinear form on $H_1(\Sigma;\mathbb{Z})$. By choosing a basis $\{a_1,\dots,a_n\}$ of $H_1(\Sigma;\mathbb{Z})$ (represented by embedded circles $\alpha_1,\dots,\alpha_n$), the form is represented by a ***Seifert matrix*** $(S_{i,j}) = (\ell k(\alpha_i, \alpha_j^+))$.

The Seifert form gives rise to various invariants of knots and links. In the following we will concentrate on the *signature* and the *Alexander polynomial* (in Section 2.4). The reason for this choice is that these two invariants have analogues in grid homology: the τ-invariant (to be defined in Chapter 6 and further explored in Chapters 7 and 8) shares a number of formal properties with the signature, while the Poincaré polynomial of grid homology can be regarded as a generalization of the Alexander polynomial.

Before proceeding with these definitions, we recall some simple facts from linear algebra. The signature of a symmetric, bilinear form Q on a finite dimensional real vector space V is defined as follows. Let V^+ resp. V^- be any maximal positive resp. negative definite subspace of V. The dimensions of V^+ and V^- are invariants of Q, and the signature $\sigma(V)$ of V is given by $\sigma(V) = \dim(V^+) - \dim(V^-)$. We define the signature of a symmetric $n \times n$ matrix M as the signature of the corresponding symmetric bilinear form Q_M on $\mathbb{R}^n$.

EXERCISE 2.3.2. **(a)** Let V be a vector space equipped with a symmetric, bilinear form Q. Let $W \subset V$ be a codimension one subspace. Show that

$$|\sigma(Q|_W) - \sigma(Q)| \leq 1.$$

(b) Suppose that Q on V is specified by a symmetric matrix M. Let Q' be represented by a matrix M' which differs from M by adding 1 to one of the diagonal entries. Show that $\sigma(Q) \leq \sigma(Q') \leq \sigma(Q) + 2$.

DEFINITION 2.3.3. Suppose that Σ is a Seifert surface for the oriented link $\vec{L}$ and S is a Seifert matrix of Σ. The ***signature*** $\sigma(\vec{L})$ of $\vec{L}$ is defined as the signature of the symmetrized Seifert matrix $S + S^T$. The ***determinant*** $\det(\vec{L})$ of the link $\vec{L}$ is $|\det(S + S^T)|$. The ***unnormalized determinant*** $\mathrm{Det}(\vec{L})$ of $\vec{L}$ is defined as $i^n \cdot \det(S + S^T) = \det(iS + iS^T)$, where $S + S^T$ is an $n \times n$ matrix. Note that if $\vec{L}$ has an odd number of components (hence n is even) then $\mathrm{Det}(\vec{L}) \in \mathbb{Z}$.

We wish to show that $\sigma(\vec{L})$, $\det(\vec{L})$, and $\mathrm{Det}(\vec{L})$ are independent of the chosen Seifert matrix of $\vec{L}$. A key step is the following:

LEMMA 2.3.4. *If Σ is a Seifert surface for $\vec{L}$ and Σ' is a stabilization of Σ, then there is a basis for $H_1(\Sigma';\mathbb{Z})$ whose Seifert matrix has the form*

$$\begin{pmatrix} S & \xi & 0 \\ 0 & 0 & 1 \\ 0 & 0 & 0 \end{pmatrix} \quad \text{or} \quad \begin{pmatrix} S & 0 & 0 \\ \xi^T & 0 & 0 \\ 0 & 1 & 0 \end{pmatrix}$$

where S is a Seifert matrix for Σ and ξ is some vector.

Proof. Suppose that $\{a_1, \ldots, a_n\}$ is a basis for $H_1(\Sigma; \mathbb{Z})$, giving the Seifert matrix S. Adding the two new homology classes y and x of the stabilized surface Σ' (as shown by Figure 2.12), we add two columns and two rows to the Seifert matrix. Clearly, $\ell k(a_i, x^+) = \ell k(x, a_i^+) = 0$ for all $i = 1, \ldots, n$, and $\ell k(x, x^+) = 0$. Furthermore, according to which side the stabilizing curve γ approaches Σ, either $\ell k(x, y^+) = 0$ and $\ell k(x^+, y) = 1$ or $\ell k(x, y^+) = 1$ and $\ell k(x^+, y) = 0$ (after replacing y by $-y$, if needed). Now, changing the basis by adding multiples of x if necessary to the a_i's and y, we get a Seifert matrix of the desired form. □

THEOREM 2.3.5. *The quantities $\sigma(\vec{L})$, $\det(\vec{L})$ and $\mathrm{Det}(\vec{L})$ are independent of the chosen Seifert matrix of $\vec{L}$ giving invariants of the link $\vec{L}$.*

Proof. This follows immediately from Theorem 2.2.2 and Lemma 2.3.4. □

The signature, the determinant, and the unnormalized determinant are constrained by the following identity:

PROPOSITION 2.3.6. *For an oriented link $\vec{L}$,*

$$\mathrm{Det}(\vec{L}) = i^{\sigma(\vec{L})} \det(\vec{L}).$$

Proof. If A is a symmetric matrix over $\mathbb{R}$, then it is elementary to verify that $\det(iA) = i^{\mathrm{sgn}(A)} |\det(A)|$, where $\mathrm{sgn}(A)$ denotes the signature of A. This is obvious if A is singular. If A is a non-singular $n \times n$ matrix, and n_+ and n_- are the dimensions of the maximal positive definite resp. negative definite subspaces of A, then $n = n_+ + n_-$ and

$$\det(iA) = i^{n-2n_-} |\det(A)| = i^{n_+ - n_-} |\det(A)| = i^{\mathrm{sgn}(A)} |\det(A)|.$$

Applying this to the symmetric matrix $S + S^T$, where S is a Seifert matrix for the link, we get the desired statement. □

EXERCISE 2.3.7. **(a)** Show that for any knot, $\det(S - S^T) = 1$, and for a link with more than one component $\det(S - S^T) = 0$ holds.
(b) Show that the signature of a knot is an even integer, and for the unknot $\mathcal{O}$ we have $\sigma(\mathcal{O}) = 0$. Compute $\det(\mathcal{O})$ using a genus one Seifert surface.
(c) Prove that $\sigma(m(K)) = -\sigma(K)$ and $\sigma(-K) = \sigma(K)$.
(d) Show that for $n \geq 0$, the signature of $T_{2,2n+1}$ is $-2n$.
(e) Compute the signature of the three-stranded pretzel knots $P(a_1, a_2, a_3)$ with a_1, a_2, and a_3 odd.
(f) Show that the signature is additive under connected sum, that is, $\sigma(K_1 \# K_2) = \sigma(K_1) + \sigma(K_2)$.
(g) Suppose that L is a *split link*, that is, L can be written as $L = L_1 \cup L_2$ with L_i non-empty in such a way that there is an embedded sphere $S^2 \subset S^3 \setminus L$ separating L_1 and L_2. Show that for any orientation $\vec{L}$ on L, $\det(\vec{L}) = 0$.
(h) Compute $\sigma(T_{3,4})$ and $\sigma(T_{3,7})$. (Cf. Exercise 2.7.9.)

Imagine modifying a knot in the following manner: allow the knot to move around in three-space, so that at one moment, two different strands are allowed to pass through one another transversely. These two knots are said to be related

FIGURE 2.13. **A strand passes through another one in unknotting a knot.**

FIGURE 2.14. **Changing the knot at a crossing.**

by a *crossing change*. Alternatively, take a suitable diagram of the initial knot, and modify it at exactly one crossing, as indicated in Figure 2.13. Any knot can be turned into the unknot after a finite sequence of such crossing changes. The minimal number of crossing changes required to unknot K is called the *unknotting number* or *Gordian number* $u(K)$ of the knot. Clearly, $u(K) = u(m(K))$.

EXERCISE 2.3.8. **(a)** Suppose that the diagram $\mathcal{D}$ of a knot K has the following property: there is a point p on $\mathcal{D}$ such that starting from p and traversing the knot, we reach each crossing for the first time on the overcrossing strand. Show that in this case K is the unknot.
(b) Verify that for any diagram of a knot K half of the number of crossings provides an upper bound for $u(K)$.
(c) Suppose that $\mathcal{D}$ is a diagram of the knot K with $c(\mathcal{D})$ crossings, and it contains an arc α with $c(\alpha)$ overcrossings and no undercrossings. Improving the result of (b) above, show that $u(K) \leq \frac{1}{2}(c(\mathcal{D}) - c(\alpha))$. Using the diagram of Example 2.1.5, show that $u(T_{p,q}) \leq \frac{1}{2}(p-1)(q-1)$.

Computing the unknotting number of a knot is a difficult task. There is no general algorithm to determine $u(K)$, since $u(K)$ is difficult to bound from below effectively. The signature provides a lower bound for $u(K)$, as we shall see below. (See Chapter 6 for an analogous bound using grid homology.)

PROPOSITION 2.3.9. ([**155**, Theorem 6.4.7]) *Let K_+ and K_- be two knots before and after a crossing change, as shown in Figure 2.14. Then, the signatures of K_+ and K_- are related by the following:*

$$-2 \leq \sigma(K_+) - \sigma(K_-) \leq 0.$$

Proof. Consider the oriented resolution K_0 of K_+ at its distinguished crossing. This is a two-component oriented link, where the crossing is locally removed, in a manner consistent with the orientation on K_+ (compare Figure 2.15). Fix a Seifert surface Σ_0 for K_0. Adding a band B to Σ_0 gives a Seifert surface Σ_+ for K_+, while adding B after introducing an appropriate twist, we get a Seifert surface Σ_- for K_-. Fix a basis for $H_1(\Sigma_0;\mathbb{Z})$ and extend it to a basis for $H_1(\Sigma_\pm;\mathbb{Z})$ by adding the homology element $\gamma_\pm$, obtained as the union of some fixed path in Σ_0 and an arc which passes through the band B. The two resulting Seifert matrices differ only at the diagonal entry corresponding to $\gamma_\pm$, which is the linking number $\ell k(\gamma_\pm, (\gamma_\pm)^+)$. When we change the band from the Seifert surface of K_+ to the Seifert surface of K_- this linking number increases by one. When relating the symmetrized Seifert matrices of the two surfaces, this fact implies that the signature of the symmetrized Seifert matrix either does not change or it increases by two (cf. Exercise 2.3.2(b)), proving the lemma. □

COROLLARY 2.3.10. *For any knot $K \subset S^3$ we have the inequality $\frac{1}{2}|\sigma(K)| \leq u(K)$.*

Proof. This follows immediately from Proposition 2.3.9 and the fact that the unknot $\mathcal{O}$ has vanishing signature. □

EXERCISE 2.3.11. Prove that for $n \geq 0$ the unknotting number of $T_{2,2n+1}$ is n.

REMARK 2.3.12. By Proposition 2.3.6, the parity of half the signature is determined by the sign of $\mathrm{Det}(\vec{L})$. Knowing this parity alone leads to the following method for determining the signature of an arbitrary knot K. Start from an unknotting sequence for K, and look at it in reverse order; i.e. starting at the unknot, which has vanishing signature. Observe that at each step in the sequence, $\frac{1}{2}\sigma$ can change by zero or ± 1. The parity of half the signature determines whether or not the change is non-zero, and in that case, Proposition 2.3.9 shows that the change in signature is determined by the type of the crossing change.

Note that, Proposition 2.3.9 gives a bound on $u(K)$ which is slightly stronger than the one stated in Corollary 2.3.10: if the signature of the knot K is positive, then in any unknotting sequence for K at least $\frac{1}{2}\sigma(K)$ moves must change a negative crossing to a positive one. Sometimes this stronger bound is referred to as a *signed unknotting bound*.

2.4. The Alexander polynomial

Beyond the signature and the determinant, further knot and link invariants can be derived from the Seifert matrix. Suppose that $\vec{L} \subset S^3$ is a given link in S^3 with a Seifert surface Σ and a corresponding Seifert matrix S. Consider the matrix $t^{-\frac{1}{2}}S - t^{\frac{1}{2}}S^T$ and define the *(symmetrized) Alexander polynomial* $\Delta_{\vec{L}}(t)$ by

$$\Delta_{\vec{L}}(t) = \det(t^{-\frac{1}{2}}S - t^{\frac{1}{2}}S^T). \tag{2.3}$$

Although the Seifert matrix S in the formula depends on certain choices, the above determinant (as the notation suggests) is an invariant of $\vec{L}$:

THEOREM 2.4.1. *The Laurent polynomial $\Delta_{\vec{L}}(t) \in \mathbb{Z}[t^{\frac{1}{2}}, t^{-\frac{1}{2}}]$ is independent from the chosen Seifert surface and Seifert matrix of $\vec{L}$ and hence is an invariant of the oriented link $\vec{L}$.*

Proof. The independence of $\Delta_{\vec{L}}(t)$ from the chosen basis of $H_1(\Sigma;\mathbb{Z})$ is a simple exercise in linear algebra. Indeed, a base change replaces S with PSP^T for a matrix with $\det P = \pm 1$, hence the Alexander polynomial is the same for the two bases. The theorem now follows from Theorem 2.2.2 and Lemma 2.3.4. □

EXAMPLE 2.4.2. The Alexander polynomial of the torus knot $T_{p,q}$ is equal to

$$\Delta_{T_{p,q}}(t) = t^k \frac{(t^{pq}-1)(t-1)}{(t^p-1)(t^q-1)} \tag{2.4}$$

with $k = -\frac{(p-1)(q-1)}{2}$, cf. Exercise 2.4.15.

It follows immediately from the definitions that

$$\mathrm{Det}(\vec{L}) = \Delta_{\vec{L}}(-1),$$

where the value of $\Delta_{\vec{L}}$ at -1 is to be interpreted as substituting $-i$ for $t^{\frac{1}{2}}$.

LEMMA 2.4.3. *For a knot K the Alexander polynomial $\Delta_K(t)$ is a symmetric Laurent polynomial, that is,*

$$\Delta_K(t^{-1}) = \Delta_K(t). \tag{2.5}$$

Proof. Let Σ be a genus g Seifert surface for K. Since $H_1(\Sigma;\mathbb{Z}) \cong \mathbb{Z}^{2g}$, S is a $2g \times 2g$ matrix, hence we have $\Delta_K(t^{-1}) = (-1)^{2g}\det(t^{-\frac{1}{2}}S^T - t^{\frac{1}{2}}S) = \Delta_K(t)$, concluding the proof. □

More generally, if $\vec{L}$ is an oriented link, then $\Delta_{\vec{L}}(t^{-1}) = (-1)^{|L|-1}\Delta_{\vec{L}}(t)$, where $|L|$ denotes the number of components of L.

EXERCISE 2.4.4. **(a)** Show that for a knot K the Alexander polynomial is in $\mathbb{Z}[t,t^{-1}]$. Verify the same for any link with an odd number of components.
(b) Show that for a knot K the Alexander polynomials of K, $-K$, and $m(K)$ are all equal.
(c) Show that the Alexander polynomial of the twist knot W_k is given by the formulas

$$\Delta_{W_{2n}}(t) = -nt + (2n+1) - nt^{-1}$$
$$\Delta_{W_{2n-1}}(t) = nt - (2n-1) + nt^{-1}.$$

(d) Compute the Alexander polynomial of the $(2, 2n+1)$ torus knot $T_{2,2n+1}$.
(e) Let P denote the 3-stranded pretzel knot $P(2b_1+1, 2b_2+1, 2b_3+1)$ with integers b_i $(i = 1, 2, 3)$. Compute the Seifert form corresponding to a Seifert surface of genus equal to one. Show that the Alexander polynomial of P is

$$\Delta_P(t) = Bt + (1-2B) + Bt^{-1},$$

where $B = b_1b_2 + b_1b_3 + b_2b_3 + b_1 + b_2 + b_3 + 1$.

Note that there are infinitely many pretzel knots with Alexander polynomial $\Delta_K(t) \equiv 1$; the smallest non-trivial one is the pretzel knot $P(-3, 5, 7)$.

The following exercise demonstrates that the Alexander polynomial depends on the orientation of a link:

EXERCISE 2.4.5. Consider the $(2, 2n)$ torus link $T_{2,2n}$ for $n \geq 1$. Orient the two strands so that the linking number of the two components is n, and compute the Alexander polynomial. Now reverse the orientation on one of the components, and compute the Alexander polynomial of this new oriented link.

Some important properties of the Alexander polynomial $\Delta_K(t)$ for knots are collected in the next result. Since the Alexander polynomial $\Delta_K(t)$ of the knot $K \subset S^3$ is symmetric in t, we can write it as

$$\Delta_K(t) = a_0 + \sum_{i=1}^{n} a_i(t^i + t^{-i}). \tag{2.6}$$

We define the *degree* $d(K)$ of $\Delta_K(t)$ as the maximal d for which $a_d \neq 0$.

THEOREM 2.4.6 ([**119, 198**]). *Suppose that the knot $K \subset S^3$ has Alexander polynomial $\Delta_K(t)$ of degree $d(K)$. Then*

(1) *The Seifert genus $g(K)$ of K satisfies $g(K) \geq d(K)$.*
(2) *For any two knots K_1 and K_2, $\Delta_{K_1 \# K_2}(t) = \Delta_{K_1}(t) \cdot \Delta_{K_2}(t)$.*
(3) *For any knot K, $\Delta_K(1) = 1$.*

Proof. For the first claim, choose a Seifert surface for K with genus $g(K)$, and use its associated Seifert form to compute the Alexander polynomial. The inequality $g(K) \geq d(K)$ follows at once.

The second property is clear by choosing Seifert surfaces Σ_1 and Σ_2 for K_1 and K_2 and taking their boundary connected sum.

Given any two curves γ_1 and γ_2 in Σ, $\ell k(\gamma_1^+, \gamma_2) - \ell k(\gamma_2^+, \gamma_1)$ is the algebraic intersection number of γ_1 and γ_2. To prove the third property, choose a basis $\{\alpha_i, \beta_j\}_{i,j=1}^{g}$ for $H_1(\Sigma)$ so that $\#(\alpha_i \cap \beta_i) = 1$ and all other pairs of curves are disjoint. If S is the Seifert matrix with respect to this basis, then the matrix $S^T - S$ decomposes as blocks of $\begin{pmatrix} 0 & 1 \\ -1 & 0 \end{pmatrix}$; and since this matrix has determinant 1, the claim follows at once. □

An argument using a $\mathbb{Z}$-fold covering of $S^3 \setminus K$ shows that the Alexander polynomial provides an obstruction for a knot being fibered.

THEOREM 2.4.7. ([**198**, page 326]) *If K is fibered, then $g(K) = d(K)$ and $a_{d(K)} = \pm 1$.* □

EXAMPLE 2.4.8. The computation of the Alexander polynomials for twist knots (given in Exercise 2.4.4(c)) together with the above result shows that W_{2n} and W_{2n-1} are not fibered once $|n| > 1$.

An important computational tool for the symmetrized Alexander polynomial $\Delta_{\vec{L}}$ is provided by the *skein relation.*

DEFINITION 2.4.9. Three oriented links $(\vec{L}_+, \vec{L}_-, \vec{L}_0)$ are said to form an ***oriented skein triple*** if they can be specified by diagrams $\mathcal{D}_+$, $\mathcal{D}_-$, $\mathcal{D}_0$ that are identical outside of a small disk, in which they are as illustrated in Figure 2.15. In this case, $\mathcal{D}_0$ is called the *oriented resolution* of $\mathcal{D}_+$ (or $\mathcal{D}_-$) at the distinguished crossing.

THEOREM 2.4.10. *Let $(\vec{L}_+, \vec{L}_-, \vec{L}_0)$ be an oriented skein triple. Then,*

$$\Delta_{\vec{L}_+}(t) - \Delta_{\vec{L}_-}(t) = (t^{\frac{1}{2}} - t^{-\frac{1}{2}})\Delta_{\vec{L}_0}(t). \tag{2.7}$$

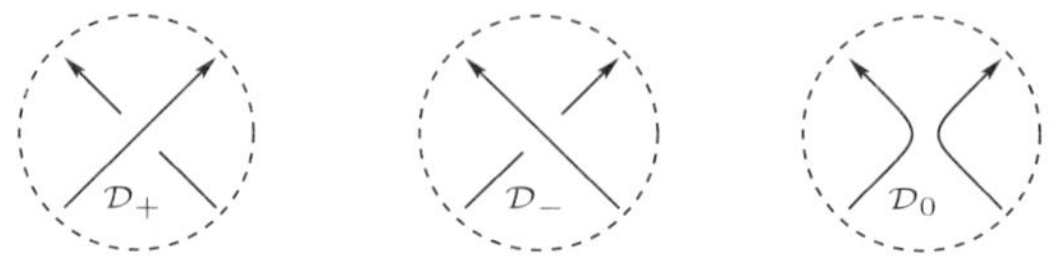

FIGURE 2.15. **Diagrams for the skein relation.** Three diagrams differ only inside the indicated disk.

Proof. Fix a Seifert surface Σ_0 for $\vec{L}_0$ as in the proof of Proposition 2.3.9, and consider the Seifert surfaces Σ_+ and Σ_- for $\vec{L}_+$ and $\vec{L}_-$ obtained from Σ_0 by adding the appropriate bands around the crossing. Let S_0 denote the Seifert form corresponding to a chosen basis of $H_1(\Sigma_0;\mathbb{Z})$. As in the proof of Proposition 2.3.9, such a basis can be extended to bases of $H_1(\Sigma_\pm;\mathbb{Z})$ by adding one further basis element $\gamma_\pm$ that passes through the band.

When computing the determinants defining the terms in the skein relation (2.7), on the left-hand-side all terms cancel except the ones involving the diagonal entries given by $\ell k(\gamma_\pm,(\gamma_\pm)^+)$ in the Seifert form. In the computation of the determinant, this entry gives rise to a factor $(t^{\frac{1}{2}}-t^{-\frac{1}{2}})$, which is multiplied with the determinant of the corresponding minor. Since that minor is $t^{-\frac{1}{2}}S_0-t^{\frac{1}{2}}S_0^T$, whose determinant is $\Delta_{\vec{L}_0}(t)$, the statement of the theorem follows at once. □

EXAMPLE 2.4.11. Using the skein relation, it follows immediately that the Alexander polynomial of the Hopf link $H_\pm$ is equal to $\pm(t^{\frac{1}{2}}-t^{-\frac{1}{2}})$. A slightly longer computation shows that the Alexander polynomial $\Delta_B(t)$ of the Borromean rings B is equal to $(t^{\frac{1}{2}}-t^{-\frac{1}{2}})^4$.

EXERCISE 2.4.12. **(a)** Show that for a split link $\vec{L}$ we have $\Delta_{\vec{L}}(t)\equiv 0$.
(b) Show that the skein relation, together with the normalization $\Delta_{\mathcal{O}}(t)=1$ on the unknot $\mathcal{O}$, determines the Alexander polynomial for all oriented links.
(c) Using the skein relation, determine the Alexander polynomial of W_n for all n. Determine the Seifert genus of W_n.
(d) Verify that the Kinoshita-Terasaka and the Conway knots both have Alexander polynomial equal to 1.
(e) Given a knot K, consider the 2-component link L we get by adding a meridian to K. Depending on the orientation of the meridian we get $L(+)$ and $L(-)$ (in the first case the linking number of the two components is 1, while in the second case it is -1). Show that $\Delta_{L(\pm)}(t)=\pm(t^{\frac{1}{2}}-t^{-\frac{1}{2}})\Delta_K$.
(f) Determine the Alexander polynomial of the Whitehead link of Figure 2.9.

The Alexander polynomial is an effective tool for studying alternating knots. (Compare the results below with Theorems 2.4.6 and 2.4.7.)

THEOREM 2.4.13 ([**152, 155**]). *Suppose that K is an alternating knot with Alexander polynomial $\Delta_K(t)=a_0+\sum_{i=1}^d a_i(t^i+t^{-i})$ and with degree $d(K)$.*

- *The genus $g(K)$ of the knot K is equal to $d(K)$. In particular, if the Alexander polynomial of K is trivial, then K is the unknot.*
- *For $i=0,\dots,d(K)-1$ the product a_ia_{i+1} is negative, that is, none of the coefficients (of index between 0 and $d(K)$) of the Alexander polynomial of K vanish, and these coefficients alternate in sign.*
- *The knot K is fibered if and only if $a_{d(K)}=\pm 1$.* □

EXERCISE 2.4.14. Identify the torus knots that are alternating.

2.4.1. The Alexander polynomial via Fox calculus. There is an algebraic way to compute the Alexander polynomial of a link through *Fox's free differential calculus*. For this construction, fix a presentation of the fundamental group of the link complement

$$\pi_1(S^3 \setminus L) = \langle g_1, \ldots, g_n \mid r_1, \ldots, r_m \rangle.$$

(By possibly adding trivial relations, we can always assume that $m \geq n-1$.) We associate to the presentation its $n \times m$ *Jacobi matrix* $J = (J_{i,j})$ over $\mathbb{Z}[t, t^{-1}]$, which is defined as follows. The presentation gives a surjective homomorphism of groups $F_n \to \pi_1(S^3 \setminus L)$, where F_n denotes the free group generated by the letters $g_1, \ldots, g_n$. Consider the induced map $\mathbb{Z}[F_n] \to \mathbb{Z}[\pi_1(S^3 \setminus L)]$ on the group rings. Composing this map with the abelianization, we get a map $\mathbb{Z}[F_n] \to \mathbb{Z}[H_1(S^3 \setminus L; \mathbb{Z})]$. Recall that the orientation of the link L provides a further map $H_1(S^3 \setminus L; \mathbb{Z}) \to \mathbb{Z}$, sending the oriented meridians of the components to 1. Hence, after identifying the group ring $\mathbb{Z}[\mathbb{Z}]$ with $\mathbb{Z}[t, t^{-1}]$, we get a map

$$\phi\colon \mathbb{Z}[F_n] \to \mathbb{Z}[t, t^{-1}]. \tag{2.8}$$

For a word $w \in F_n$ define the *free derivative*

$$\frac{\partial w}{\partial g_j} \in \mathbb{Z}[F_n]$$

by the rules

$$\frac{\partial uv}{\partial x} = \frac{\partial u}{\partial x} + u\frac{\partial v}{\partial x}, \qquad \frac{\partial g_i}{\partial g_i} = 1, \qquad \frac{\partial g_i}{\partial g_j} = 0 \quad (i \neq j).$$

EXERCISE 2.4.15. **(a)** Show that for $n \in \mathbb{N}$

$$\frac{\partial x^n}{\partial x} = \frac{x^n - 1}{x - 1} \quad \text{and} \quad \frac{\partial x^{-n}}{\partial x} = -x^{-1}\frac{x^{-n} - 1}{x^{-1} - 1}.$$

(b) Suppose that for $p, q \in \mathbb{N}$ relatively prime integers the group G is presented as $\langle x, y \mid x^p y^{-q} \rangle$. Determine $\frac{\partial (x^p y^{-q})}{\partial x}$ and $\frac{\partial (x^p y^{-q})}{\partial y}$.

Applying the map ϕ of Equation (2.8) to the free derivative $\frac{\partial r_i}{\partial g_j}$ we get a polynomial $J_{i,j}$, the (i,j)-term of the Jacobi matrix J of the presentation. Consider the ideal ϵ_1 generated by the determinants of the $(n-1) \times (n-1)$-minors of the Jacobian J. For the proof of the following theorem, see [**119**, Chapters 6 and 11].

THEOREM 2.4.16. *The ideal ϵ_1 is a principal ideal, and its generator $P(t)$ is $\pm t^{\frac{k}{2}}$ times the Alexander polynomial $\Delta_{\vec{L}}(t)$, for some $k \in \mathbb{Z}$.* □

EXERCISE 2.4.17. **(a)** Using Fox calculus, verify Equation (2.4), and compute the Alexander polynomial of the (p,q) torus knot. (*Hint:* Recall Exercise 2.1.7(b) and apply Exercise 2.4.15(b)).
(b) Using the Alexander polynomial (and the result of Theorem 2.4.6), show that the Seifert genus of the torus knot $T_{p,q}$ is given by $g(T_{p,q}) = \frac{1}{2}(p-1)(q-1)$.
(c) Find a presentation of $\pi_1(S^3 \setminus B)$ for the Borromean rings B, and compute $\pm\Delta_B(t)$ with the aid of Fox calculus.

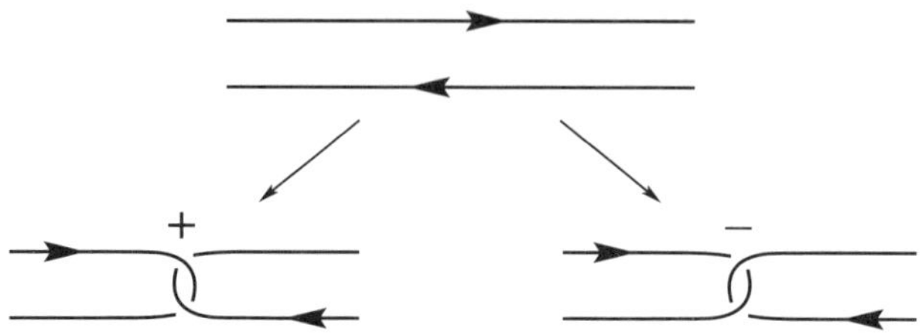

FIGURE 2.16. **By introducing a clasp we construct the Whitehead double of K.** Notice that there are two different ways for introducing the clasp, providing a further parameter $\pm$ besides the chosen framing.

2.5. Further constructions of knots and links

The normal bundle $\nu(K) \to K$ of a knot $K \subset S^3$ is an oriented D^2-bundle over S^1, hence it is trivial. A trivialization of this bundle is called a *framing* of K. Thought of as a complex line bundle, the normal bundle can be trivialized by a single (nowhere zero) section, hence by a push-off K' of K. The linking number $\ell k(K, K')$ of the knot K with the push-off K' determines the framing up to isotopy. With this identification, the push-off along a Seifert surface, providing the *Seifert framing*, corresponds to 0.

EXERCISE 2.5.1. Suppose that the knot K is given by the diagram $\mathcal{D}$. The diagram provides a framing by pushing off the arcs of $\mathcal{D}$ within the plane. Show that the resulting framing corresponds to the writhe $\mathrm{wr}(\mathcal{D}) \in \mathbb{Z}$.

Knots with interesting properties can be constructed as follows. For a given knot K consider the push-off K' of K corresponding to the framing $k \in \mathbb{Z}$, and orient K' opposite to K. Then the resulting two-component link $L_k(K)$ bounds an annulus between K and K', and it is easy to see from the definition that for the given framing k, the link will have Alexander polynomial equal to $\Delta_{L_k(K)}(t) = k(t^{-\frac{1}{2}} - t^{\frac{1}{2}})$. (The annulus provides a Seifert surface with corresponding 1×1 Seifert matrix (k).) In particular, for $k = 0$ the resulting link $L_0(K)$ has vanishing Alexander polynomial.

Modify now the link $L_k(K)$ constructed above by replacing the two close parallel segments near a chosen point p with a clasp as shown in Figure 2.16. The resulting knot is called a *Whitehead double* of K. Notice that since the clasp can be positive or negative, for each framing k we actually have two doubles, $W_k^+(K)$ and $W_k^-(K)$; the *k-framed positive resp. negative Whitehead double of* K. Observe that the k-framed Whitehead double of the unknot is a twist knot; more precisely, $W_k^+(\mathcal{O}) = W_{2k}$ and $W_k^-(\mathcal{O}) = W_{2k-1}$.

LEMMA 2.5.2. *The 0-framed Whitehead doubles $W_0^{\pm}(K)$ for any knot K have Alexander polynomial $\Delta_{W_0^{\pm}(K)}(t) = 1$.*

Proof. Use the skein relation at a crossing of the clasp, and note that the oriented resolution has vanishing Alexander polynomial, as shown above, while the knot obtained by a crossing change is the unknot. □

EXERCISE 2.5.3. Compute $\mathrm{Det}(W_0^{\pm}(K))$ and show that $\sigma(W_0^{\pm}(K)) = 0$.

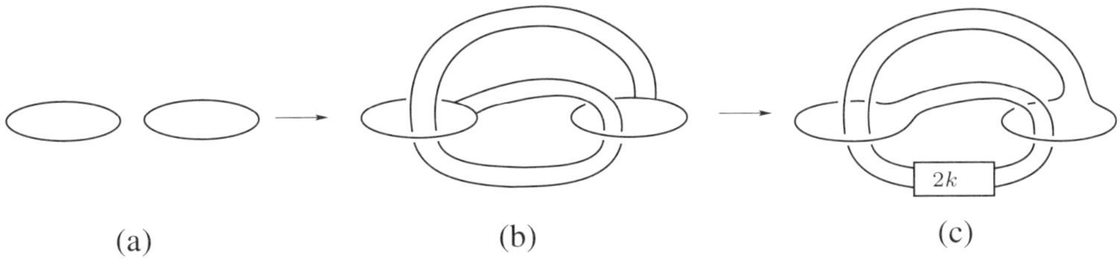

FIGURE 2.17. **Adding a band.** Start from the two-component unlink of (a) and add the band B (an example shown in (b)) to get the knot $K(B)$. Adding k full twists to B we get the family $K(B,k)$ of knots, shown by (c).

Another class of examples is provided by the two-component unlink, equipped with an embedded band B added to the unlink which turns it into a knot $K(B)$, cf. Figure 2.17. In this construction B can be any band whose interior is disjoint from the unlink, and whose ends are contained in different components of the unlink. A band B gives rise to further bands by adding twists to it: by adding k full twists to B, we get $K(B,k)$.

LEMMA 2.5.4. *The Alexander polynomial of $K(B,k)$ is independent of k:*

$$\Delta_{K(B,k)}(t) = \Delta_{K(B)}(t).$$

Proof. Applying the skein relation to a crossing coming from the twist on the band B, the three links in the skein triple are $K(B,k)$, $K(B,k-1)$ and the two-component unlink. Since the two-component unlink has vanishing Alexander polynomial, induction on k verifies the statement of the lemma. □

REMARK 2.5.5. Using other knot invariants, it is not hard to see that $K(B,k)$ for various k can be distinct. For example, if $K(B)$ has non-trivial Jones polynomial (cf. [**119**]), then the Jones polynomials distinguish the $K(B,k)$ for various values of k.

For a variation on this theme, consider the *Kanenobu knots* $K(p,q)$ shown in Figure 2.18. These knots are constructed from the two-component unlink by a similar procedure as our previous examples $K(B,k)$ in two different ways: we could regard the region with the p full twists as a band added to the two-component unlink (cf. Figure 2.19(a)), or we can do the same with the region of the q full twists, as shown in Figure 2.19(b). It follows that all of them have the same Alexander polynomial.

If we allow both parameters to change so that $p+q$ stays fixed, then not only the Alexander polynomials, but also the HOMFLY (and hence the Jones) polynomials and Khovanov (and Khovanov-Rozansky) homologies of the resulting knots stay equal. For these latter computations see [**224**], cf. also [**85**].

REMARK 2.5.6. The definition of the Alexander polynomial through Fox calculus provides further invariants by considering the kth elementary ideals ϵ_k generated by the determinants of the $(n-k)\times(n-k)$ minors of a Jacobi matrix J for $k>1$. Indeed, the Kanenobu knots (of Figure 2.18) can be distinguished by the Jones

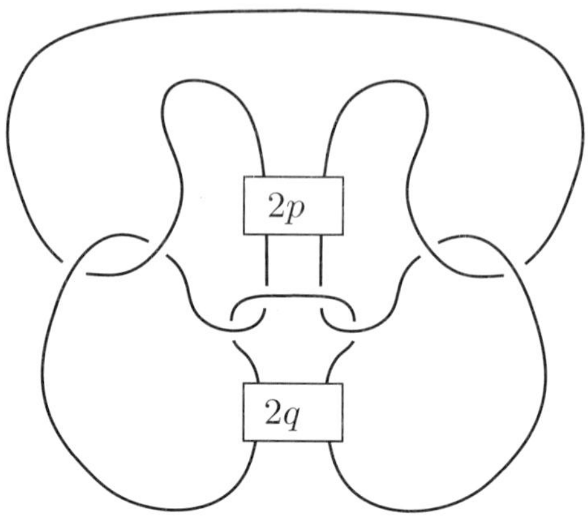

FIGURE 2.18. **The Kanenobu knot $K(p, q)$.** The boxes represent $2p$ and $2q$ half twists; that is p and q full twists.

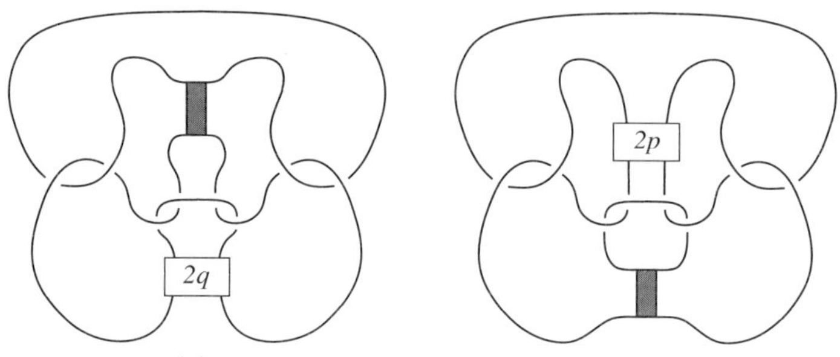

FIGURE 2.19. **Two ribbon representations of $K(p, q)$.**

polynomial together with the second elementary ideal ϵ_2: for $K(p, q)$ it is generated by the two polynomials $t^2 - 3t + 1$ and $p - q$, hence for fixed $p + q$ this ideal determines p and q. For this computation and further related results see [**99**].

EXERCISE 2.5.7. Determine $\Delta_{K(p,q)}(t)$. (*Hint:* Pick $p = q = 0$ and identify $K(0, 0)$ with the connected sum of two copies of the figure-eight knot.)

The construction of the knots $K(B)$ naturally generalizes by considering the n-component unlink and adding $(n - 1)$ disjoint bands to it in such a way that the result is connected. A knot presented in this way is called a *ribbon knot.*

EXERCISE 2.5.8. Show that $K \subset S^3$ is ribbon if and only if it bounds an immersed disk in S^3, where the double points of the immersion, that is, the self-intersections of the disk locally look like the picture of Figure 2.20.

2.6. The slice genus

A further basic knot invariant is the *(smooth) slice genus* (or *four-ball genus*) $g_s(K)$ of a knot K, defined as follows. An oriented, smoothly embedded surface $(F, \partial F) \subset (D^4, \partial D^4 = S^3)$ with $\partial F = K$ is called a *slice surface* of K.

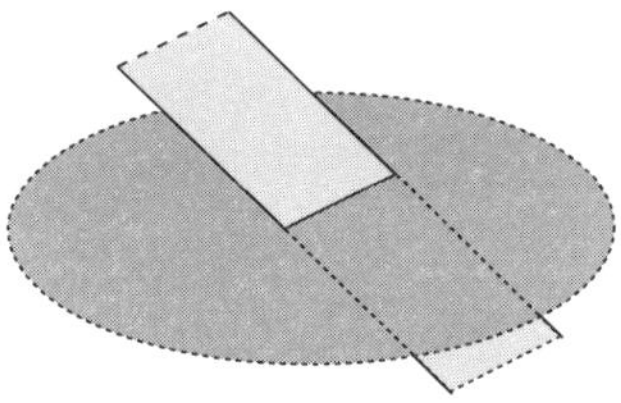

FIGURE 2.20. **Ribbon singularity.**

DEFINITION 2.6.1. The integer

$$g_s(K) = \min\{g(F) \mid (F, \partial F) \subset (D^4, S^3) \text{ is a slice surface for } K\}$$

is the ***slice genus*** (or four-ball genus) of the knot K. A knot K is a ***slice knot*** if $g_s(K) = 0$, that is, if it admits a slice disk.

The invariant g_s provides a connection between knot theory and 4-dimensional topology; see also Section 8.6. The slice genus is related to the Seifert genus and the unknotting number by the inequalities:

$$g_s(K) \leq g(K), \qquad g_s(K) \leq u(K). \tag{2.9}$$

The first is immediate: just push the interior of a Seifert surface into the interior of D^4. For the second, note that a d-step unknotting of K (followed by capping off the unknot at the end) can be viewed as an immersed disk in D^4 with d double points. Resolving the double points gives a smoothly embedded genus d surface which meets $\partial D^4 = S^3$ at K. In more detail, this resolution is done by removing two small disks at each double point of the immersed disk, and replacing them with an embedded annulus. Clearly, for each double point, this operation drops the Euler characteristic by two and hence increases the genus by one. One can find knots K for which the differences $g(K) - g_s(K)$ and $u(K) - g_s(K)$ are arbitrarily large. (See for instance Exercise 2.6.2(b) and Example 8.7.1.)

EXERCISE 2.6.2. **(a)** Show that a ribbon knot is slice. In particular, verify that the knots $K(B, k)$ from Lemma 2.5.4 are slice.
(b) Show that for any knot K, $K\#m(-K)$ is a slice knot. Show that $K\#m(-K)$ is, indeed, ribbon.

REMARK 2.6.3. There is no known example of a slice knot which is not ribbon. Indeed, the ***slice-ribbon conjecture*** of Fox [**57**] asserts that any slice knot is ribbon. The conjecture has been verified for 2-bridge knots [**124**] and for certain Montesinos knots [**116**], but it is open in general.

A further property, the *Fox-Milnor condition*, of the Alexander polynomial can be used to show that certain knots are not slice.

THEOREM 2.6.4 (Fox-Milnor, [**56, 58**]). *If K is a slice knot, then there is a polynomial f with the property that $\Delta_K(t) = f(t) \cdot f(t^{-1})$.* □

EXERCISE 2.6.5. Compute the slice genus of the figure-eight knot W_2 (cf. Figure 2.6) and of the right-handed trefoil knot $T_{2,3}$.

Like the unknotting number $u(K)$, the slice genus $g_s(K)$ is poorly understood; in fact it is unknown even for some small-crossing knots. However, there are some classical lower bounds on the slice genus; we review here one coming from the signature (generalizing Corollary 2.3.10):

THEOREM 2.6.6. *For a knot $K \subset S^3$, $\frac{1}{2}|\sigma(K)| \leq g_s(K)$.*

We return to a proof of Theorem 2.6.6 after some further discussion.

The bound in Theorem 2.6.6 is typically not sharp. For example, as we will see, the slice genus of the (p, q) torus knot $T_{p,q}$ is $\frac{1}{2}(p-1)(q-1)$, while the signature can be significantly smaller. (For a recursive formula for $\sigma(T_{p,q})$ see [**155**].) For example, $\frac{1}{2}\sigma(T_{3,7}) = -4$ (cf. Exercise 2.3.7(h)), while $g_s(T_{3,7}) = u(T_{3,7}) = 6$. Similarly, in Chapter 8 (see Remark 8.6.5) we will show that $g_s(W_0^-(T_{-2,3})) = 1$, while (according to Exercise 2.5.3) it has vanishing signature.

The conclusion of Theorem 2.6.4 holds even when the hypothesis that K is slice is replaced by the following weaker condition:

DEFINITION 2.6.7. A knot K is called ***topologically slice*** if there is a continuous embedding $\phi\colon (D^2 \times D^2, (\partial D^2) \times D^2) \to (D^4, \partial D^4 = S^3)$ such that $\phi(\partial D^2 \times \{0\})$ is K.

Note that the "normal" D^2-direction (required by the above definition) automatically exists for smooth embeddings of D^2 in D^4. The topologically slice condition on K is strictly weaker than the (smoothly) slice condition: for example the Whitehead double of any knot (with respect to the Seifert framing) is a topologically slice knot, but in many cases (for example, for the negatively clasped Whitehead double of the left-handed trefoil knot) it is not smoothly slice. The fact that these knots are topologically slice follows from a famous result of Freedman [**59**] (see also [**67**]), showing that any knot whose Alexander polynomial $\Delta_K(t) = 1$ is topologically slice. The fact that $W_0^-(T_{-2,3})$ is not smoothly slice will be demonstrated using the τ invariant in grid homology, cf. Lemma 8.6.4. In particular, the condition that $\Delta_K = 1$ is not sufficient for a knot to admit a smooth slice disk. Recall that both the Kinoshita-Terasaka knot and the Conway knot have $\Delta_K = 1$. The Kinoshita-Terasaka knot is smoothly slice, while the (smooth) slice genus of the Conway knot is unknown. Note that the distinction between smooth and topological does not appear for the Seifert genus, cf. [**2**, Section 1.1].

EXERCISE 2.6.8. Find a slice disk for the Kinoshita-Terasaka knot.

The rest of this section is devoted to the proof of Theorem 2.6.6. During the course of the proof, we give some preparatory material which will also be used in Chapter 8, where we present an analogous bound coming from grid homology.

We prefer to recast Theorem 2.6.6 in terms of knot cobordisms, defined as follows. Given two oriented links $\vec{L}_0, \vec{L}_1 \subset S^3$, a *cobordism* between them is a smoothly embedded, compact, oriented surface-with-boundary $W \subset S^3 \times [0,1]$ such that $W \cap (S^3 \times \{i\})$ is $\vec{L}_i$ for $i = 0, 1$, and the orientation of W induces the orientation of $\vec{L}_1$ and the negative of the orientation of $\vec{L}_0$ on the two ends.

We will prove the following variant of Theorem 2.6.6. (The proof we describe here is similar to the one given by Murasugi [**153**].)

THEOREM 2.6.9. *Suppose that W is a smooth genus g cobordism between the knots K_1 and K_2. Then*

$$|\sigma(K_1) - \sigma(K_2)| \leq 2g.$$

Before we provide the details of the proof, we need a definition.

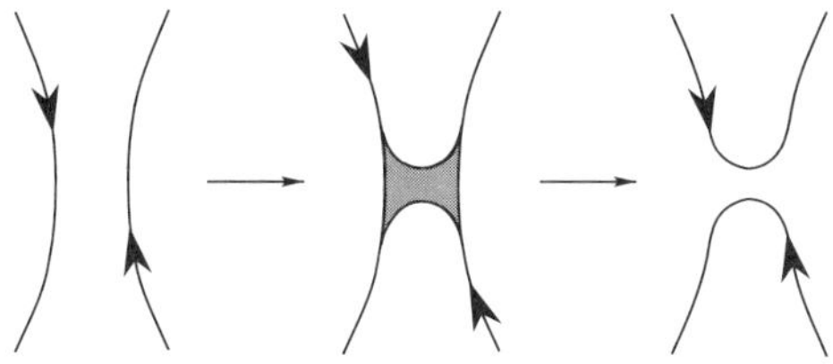

FIGURE 2.21. **A saddle move.** Adding a band to $\vec{L}$ (on the left) we get the link $\vec{L}'$ (on the right), and the two links are related by a saddle move.

DEFINITION 2.6.10. Given two oriented links $\vec{L}$ and $\vec{L}'$, we say that $\vec{L}$ and $\vec{L}'$ are related by a ***saddle move*** if there is a smoothly embedded, oriented rectangle R with oriented edges $e_1, \dots, e_4$, whose interior is disjoint from $\vec{L}$, with the property that $\vec{L} \cap R = (-e_1) \cup (-e_3)$, and $\vec{L}'$ is obtained by removing e_1 and e_3 from $\vec{L}$ and attaching e_2 and e_4 with the given orientations (and smoothing the corners). This relation is clearly symmetric in $\vec{L}$ and $\vec{L}'$, see Figure 2.21. (Notice that the connected sum of two knots is a special case of this operation.) If we have k disjoint rectangles between $\vec{L}$ and $\vec{L}'$ as above, we say that $\vec{L}$ and $\vec{L}'$ are related by k ***simultaneous saddle moves***.

In the course of the verification of the inequality of Theorem 2.6.9, we use the following standard result. (See also Section B.5.) For the statement, we introduce the following notational convention: given a knot K and an integer n, let $\mathcal{U}_n(K)$ denote the link obtained by adding n unknotted, unlinked components to K.

PROPOSITION 2.6.11 (cf. Section B.5). *Suppose that two knots K_1 and K_2 can be connected by a smooth, oriented, genus g cobordism W. Then, there are knots K_1' and K_2' and integers b and d with the following properties:*

(1) $\mathcal{U}_b(K_1)$ can be obtained from K_1' by b simultaneous saddle moves.
(2) K_1' and K_2' can be connected by a sequence of $2g$ saddle moves.
(3) $\mathcal{U}_d(K_2)$ can be obtained from K_2' by d simultaneous saddle moves. □

The proof of this proposition relies on the concept of *normal forms* of cobordisms between knots, as explained in Section B.5.

With Proposition 2.6.11 at our disposal, the proof of Theorem 2.6.9 will easily follow from the two lemmas below:

LEMMA 2.6.12. *If $\vec{L}$ and $\vec{L}'$ are oriented links that differ by a saddle move, then* $|\sigma(\vec{L}) - \sigma(\vec{L}')| \leq 1$.

Proof. A Seifert surface for $\vec{L}$ can be obtained from one for $\vec{L}'$ by adding a band to it, cf. the proof of Proposition 2.3.9. Thus, a Seifert matrix for $\vec{L}$ is obtained by adding one row and one column to a Seifert matrix of $\vec{L}'$. This fact immediately implies that the signature can change by at most one (cf. Exercise 2.3.2(a)). □

LEMMA 2.6.13. *Let K_1 and K_2 be knots with the property that K_2 can be obtained from $\mathcal{U}_d(K_1)$ by d simultaneous saddle moves. Then, $\sigma(K_1) = \sigma(K_2)$.*

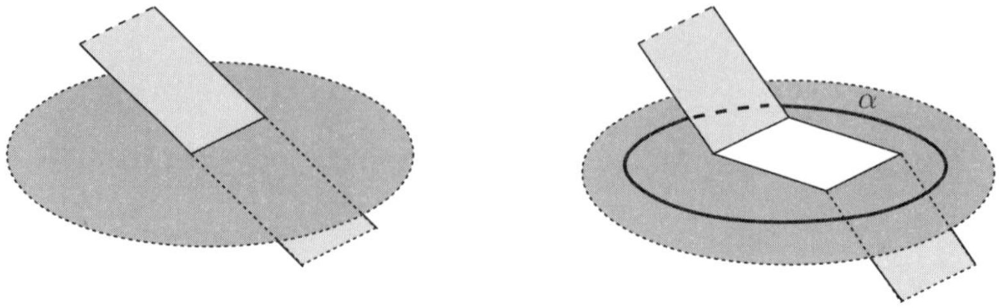

FIGURE 2.22. **Resolution of a ribbon singularity.** By pulling apart the bands slightly, we get a Seifert surface for the knot we got by attaching the bands to $\mathcal{U}_d(K)$. The curve α is indicated in the picture on the right.

Proof. Fix a Seifert surface Σ for K_1 and the spanning disks for the d unknot components in $\mathcal{U}_d(K_1)$. By making the bands sufficiently thin, we can arrange that the intersections of the saddle bands with the spanning disks or Σ are ribbon singularities, as shown in Figure 2.20 (or on the left in Figure 2.22). A Seifert surface Σ' for K_2 can be constructed by pulling the bands slightly apart at the ribbon singularities, as shown on the right in Figure 2.22. Each time we apply this operation, we increase the genus of the surface Σ by one, hence we increase the number of rows (and columns) of the Seifert matrix by two. One of the two new homology elements, called α_p at the ribbon singularity p, can be visualized on the picture: it encircles the square we opened up. The linking number of α_p with α_p^+ vanishes. Furthermore, the linking number of α_p with all homology elements on the Seifert surface Σ, and with the other α_q also vanish.

The surfaces Σ and Σ' give Seifert forms S and S'. We wish to compare the signatures of the bilinear forms Q and Q' represented by $S + S^T$ and $S' + (S')^T$. There is a natural embedding $H_1(\Sigma) \hookrightarrow H_1(\Sigma')$, and the restriction of Q' to $H_1(\Sigma)$ is Q. Since the determinant of a knot is always non-trivial (cf. Exercise 2.3.7(a)), it follows that Q and Q' are both non-degenerate; so there is a perpendicular splitting (with respect to Q')

$$H_1(\Sigma'; \mathbb{R}) \cong H_1(\Sigma; \mathbb{R}) \oplus V.$$

The curves α_p are linearly independent, since surgery along them gives a connected surface. Thus, the α_p span a half-dimensional subspace of V, moreover Q' vanishes on their span. Since Q' is non-degenerate on V, it follows that the signature of V vanishes; and hence the signature of Q equals the signature of Q'. □

Proof of Theorem 2.6.9. The theorem is now a direct consequence of Proposition 2.6.11, Lemma 2.6.12 and Lemma 2.6.13. □

It follows that the signature bounds the slice genus:

Proof of Theorem 2.6.6. Removing a small ball around some point on a smooth slice surface gives a smooth genus g cobordism from K to the unknot. Applying Theorem 2.6.9 and the fact that the signature of the unknot vanishes, the result follows at once. □

REMARK 2.6.14. The above proof of Theorem 2.6.6 rests on the normal form for cobordisms (Proposition 2.6.11), whose hypothesis is that the surface is smoothly embedded. With different methods it can be shown that $\frac{1}{2}|\sigma(K)| \leq g_{top}(K)$, for the

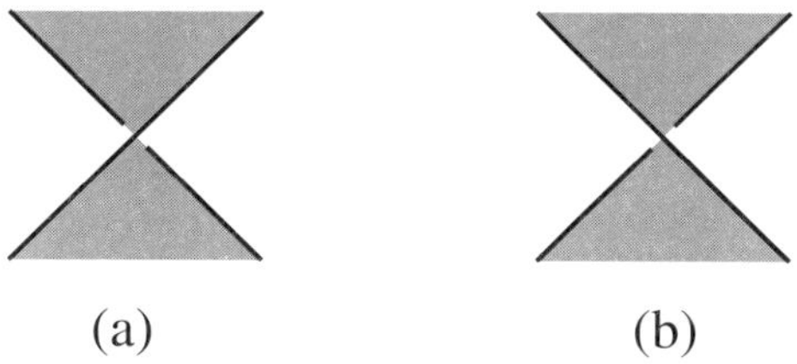

FIGURE 2.23. **The sign $\epsilon = \pm 1$ associated to a crossing in the diagram.** If the crossing is positioned as (a) with respect to the black regions, we associate $+1$ to it, while if the crossing has the shape of (b), then we associate -1 to it.

topological slice genus $g_{top}(K)$, the minimal genus of a locally flat embedded surface in D^4 bounding K [**119**, Theorem 8.19]. Consequently, the signature $\sigma(K)$ vanishes for any topologically slice knot, and therefore it cannot be used to distinguish topological and smooth sliceness.

2.7. The Goeritz matrix and the signature

We include here a handy formula, due to Gordon and Litherland [**78**], for computing the signature of a link in terms of its diagram. (This material will be needed in Section 10.3, where we compute the grid homology for alternating knots.)

Let $\mathcal{D}$ be a diagram of a link. The diagram admits a chessboard coloring: the components of the complement of the diagram in the plane can be colored black and white in such a manner that domains with the same color do not share sides. Indeed, the diagram $\mathcal{D}$ admits two such colorings; choose the one where the unbounded region is white and call this unbounded region d_0. Let the other white regions be denoted by $d_1, \ldots, d_n$.

DEFINITION 2.7.1. The black regions can be glued together to form a compact surface, the ***black surface*** $F_b \subset \mathbb{R}^3$ with the given link as its boundary ∂F_b: at each crossing glue the domains together with a twisted band to restore the crossing in the diagram.

EXERCISE 2.7.2. Consider the alternating diagram of the $(2, 2n+1)$ torus knot $T_{2,2n+1}$ given by Figure 2.3. Show that the surface F_b is homeomorphic to the Möbius band, so F_b is not a Seifert surface.

The chessboard coloring gives rise to a matrix defined as follows. First associate to each crossing p of the diagram $\mathcal{D}$ a sign $\epsilon(p) \in \{\pm 1\}$ shown in Figure 2.23. (Conventions on ϵ are not uniform in the literature; we are using the one from [**18**].)

DEFINITION 2.7.3. Define the ***unreduced Goeritz matrix*** $G' = (g_{i,j})_{i,j=0}^n$ as follows. For $i \neq j$, let

$$g_{i,j} = -\sum_p \epsilon(p),$$

where the sum is taken over all crossings p shared by the white domains d_i and d_j; for $i = j$, let

$$g_{i,i} = -\sum_{k \neq i} g_{i,k}.$$

FIGURE 2.24. **Types of crossings in an oriented diagram.**

The ***reduced Goeritz matrix*** $G = G(\mathcal{D})$ is obtained from G' by considering the n rows and columns corresponding to $i, j > 0$.

Recall that the link (and hence its projection $\mathcal{D}$) is equipped with an orientation. Classify a crossing p of $\mathcal{D}$ as type I or type II according to whether at p the positive quadrant is white or black; see Figure 2.24. The type of a crossing is insensitive to which of the two strands passes over the other one; but it takes the orientation of the link into account. Define $\mu(\mathcal{D})$ as $\sum \epsilon(p)$, where the summation is for all crossings in $\mathcal{D}$ of type II. The Goeritz matrix (together with the correction term $\mu(\mathcal{D})$ above) can be used to give an explicit formula for computing the signature of a link $\vec{L}$ from a diagram $\mathcal{D}$. This formula is often more convenient than the original definition using the Seifert form of a Seifert surface. (The proof given below follows [**189**].)

THEOREM 2.7.4 (Gordon-Litherland formula, [**78**]). *Suppose that $\mathcal{D}$ is a diagram of a link $\vec{L}$ with reduced Goeritz matrix $G = G(\mathcal{D})$. Let $\sigma(G)$ denote the signature of the symmetric matrix G. Then, $\sigma(\vec{L})$ is equal to $\sigma(G) - \mu(\mathcal{D})$.*

In the course of the proof of this theorem we will need the following definition and lemma (the proof of which will be given in Appendix B).

DEFINITION 2.7.5. The diagram $\mathcal{D}$ of a link $\vec{L}$ is ***special*** if it is connected, and the associated black surface F_b (from Definition 2.7.1) is a Seifert surface for $\vec{L}$.

LEMMA 2.7.6 (see Proposition B.3.3). *Any oriented link admits a special diagram.* □

Using the above result, the proof of Theorem 2.7.4 will be done in two steps. First we assume that $\mathcal{D}$ is a special diagram, and check the validity of the formula for the signature in this case. In the second step we show that $\sigma(G) - \mu(\mathcal{D})$ is a link invariant; i.e., it is independent of the chosen projection.

LEMMA 2.7.7. *Suppose that $\mathcal{D}$ is a special diagram of the oriented link $\vec{L}$. Then the signature of the reduced Goeritz matrix $G(\mathcal{D})$ is equal to $\sigma(\vec{L})$, and $\mu(\mathcal{D}) = 0$.*

Proof. The contour of any bounded white domain gives a closed, embedded curve, hence a one-dimensional homology class in the Seifert surface F_b. We claim these classes give a basis for $H_1(F_b; \mathbb{R})$. To show linear independence, for each crossing c consider the relative first homology class p_c in $H_1(F_b, \partial F_b; \mathbb{R})$ represented by the arc in F_b that is the pre-image of the crossing. For a crossing adjacent to the unbounded domain the corresponding arc is intersected by a single contour, and working our way towards the inner domains, an inductive argument establishes linear independence.

Let B denote the number of black regions, W the number of white regions, and C the number of crossings. Thinking of the connected projection as giving a cell decomposition of S^2, we see that $W + B - C = 2$. By definition we have $\chi(F_b) = B - C$. It follows that the first homology of F_b has dimension $W - 1$; thus the contours give a basis for $H_1(F_b; \mathbb{R})$.

A local computation shows that the reduced Goeritz matrix is equal to $S + S^T$, where S is the Seifert matrix of F_b for the above basis. The definition immediately provides the identity $\sigma(G(\mathcal{D})) = \sigma(\vec{L})$.

If F_b is a Seifert surface for $\vec{L}$, the diagram $\mathcal{D}$ has no type II crossing, hence for a special diagram $\mathcal{D}$ the correction term $\mu(\mathcal{D})$ is equal to zero. □

LEMMA 2.7.8. *Fix an oriented link $\vec{L}$ and consider a diagram $\mathcal{D}$ of it. The difference $\sigma(G) - \mu(\mathcal{D})$ is independent of the chosen diagram $\mathcal{D}$, and is an invariant of $\vec{L}$.*

Proof. By the Reidemeister Theorem 2.1.4, the claim follows once we show that $\sigma(G) - \mu(\mathcal{D})$ remains unchanged if we perform a Reidemeister move.

The first Reidemeister move creates one new domain, which in the chessboard coloring is either black or white. If it is black, the new crossing is of type I and the Goeritz matrix remains unchanged, hence the quantity $\sigma(G) - \mu(\mathcal{D})$ remains unchanged, as well. If the new domain is white, then the new crossing p is of type II, hence μ changes by $\epsilon(p)$. The Goeritz matrix also changes, as follows. Let d_{new} denote the new white domain, and d_{next} the domain sharing a crossing with d_{new}. The new Goeritz matrix, written in the basis provided by the domains with the exception of taking $d_{next} + d_{new}$ instead of d_{next}, is the direct sum of the Goeritz matrix we had before the move, and the 1×1 matrix $(\epsilon(p))$. The invariance follows again.

For the second Reidemeister move, we have two cases again, depending on whether the bigon enclosed by the two arcs is black or white in the chessboard coloring. If it is black, then the matrix G does not change, and the two new intersections have the same type, with opposite ϵ-values, hence μ does not change either. If the new domain is white, then μ does not change under the move by the same reasoning as before. Now, however, the matrix changes, but (as a simple computation shows) its signature remains the same, verifying the independence.

The invariance under the Reidemeister move R_3 needs a somewhat longer case-analysis, corresponding to the various orientations of the three strands involved. Interpret the move as pushing a strand over a crossing p, and suppose that a black region disappears and a new white region is created. Inspecting each case, one sees that μ changes to $\mu - \epsilon(p)$, while G acquires a new row and column, and after an appropriate change of basis this row and column contains only zeros except one term -$\epsilon(p)$ in the diagonal; hence the invariance follows as before. □

After these preparations we are ready to prove the Gordon-Litherland formula:

Proof of Theorem 2.7.4. Consider the given diagram $\mathcal{D}$ and a special diagram $\mathcal{D}'$ for the fixed oriented link $\vec{L}$. By Lemma 2.7.8

$$\sigma(G(\mathcal{D})) - \mu(\mathcal{D}) = \sigma(G(\mathcal{D}')) - \mu(\mathcal{D}'),$$

while for the special diagram $\mathcal{D}'$ (by Lemma 2.7.7) we have $\sigma(G(\mathcal{D}')) = \sigma(\vec{L})$ and $\mu(\mathcal{D}') = 0$, verifying the identity of the theorem. □

EXERCISE 2.7.9. Compute the signature $\sigma(T_{3,3n+1})$ as a function of n.

If Σ is a Seifert surface for $\vec{L}$, then $-\Sigma$ is a Seifert surface for $-\vec{L}$. Thus, if S is the Seifert matrix for $\vec{L}$, then S^T is the Seifert matrix for $-\vec{L}$. The signature, determinant and the Alexander polynomial of an oriented link therefore remains unchanged if we reverse the orientations of all its components.

If we reverse the orientation of only some components of a link, however, the situation is different. As Exercise 2.4.5(b) shows, the Alexander polynomial changes in general. As it will be explained in Chapter 10, the determinant of the link will stay unchanged under such reversal of orientations.

The signature of a link depends on the orientation of the link, and it changes in a predictable way if we reverse the orientations of some of its components. Later we will need the exact description of this change, which was first established by Murasugi [**154**]. Here we follow the elegant derivation of [**78**], using the Gordon-Litherland formula of Theorem 2.7.4.

COROLLARY 2.7.10. *Let $\vec{L}_1$ and $\vec{L}_2$ be two disjoint, oriented links. Then,*

$$\sigma(\vec{L}_1 \cup \vec{L}_2) = \sigma((-\vec{L}_1) \cup \vec{L}_2) - 2\ell k(\vec{L}_1, \vec{L}_2).$$

Proof. Let $\vec{L} = \vec{L}_1 \cup \vec{L}_2$. Fix a diagram $\mathcal{D}$ for the oriented link $\vec{L}_1 \cup \vec{L}_2$, and let $\mathcal{D}'$ be the induced diagram for $(-\vec{L}_1) \cup \vec{L}_2$, obtained by changing the orientations on L_1. Since reversing the orientation on some of the components of $\mathcal{D}$ leaves the Goeritz matrix unchanged, the Gordon-Litherland formula gives

$$\sigma(\vec{L}_1 \cup \vec{L}_2) - \sigma((-\vec{L}_1) \cup \vec{L}_2) = \mu(\mathcal{D}') - \mu(\mathcal{D}).$$

The identification

$$\mu(\mathcal{D}') - \mu(\mathcal{D}) = -2\ell k(\vec{L}_1, \vec{L}_2)$$

is now a straightforward matter: at those crossings where both strands belong either to $\vec{L}_1$ or to $\vec{L}_2$ the same quantity appears in $\mu(\mathcal{D}')$ and in $\mu(\mathcal{D})$. At crossings of strands from $\vec{L}_1$ and $\vec{L}_2$ the orientation reversal changes the type, but leaves the quantity $\epsilon(p)$ unchanged. Summing up these contributions (as required in the sum given by $\mu(\mathcal{D}') - \mu(\mathcal{D})$) we get $-2\ell k(\vec{L}_1, \vec{L}_2)$ (cf. Definition 2.1.12), concluding the proof. □

The Gordon-Litherland formula has an interesting consequence for alternating links. To describe this corollary, we introduce the following notion. The *compatible coloring* for an alternating link arranges for each crossing to have the coloring shown in Figure 2.23(b). It is easy to see that a connected alternating diagram always has a unique compatible coloring.

Suppose now that $\mathcal{D}$ is a connected alternating diagram of a link L. Let $\mathrm{Neg}(\mathcal{D})$ (and similarly, $\mathrm{Pos}(\mathcal{D})$) denote the number of negative (resp. positive) crossings in $\mathcal{D}$, and let $\mathrm{White}(\mathcal{D})$ (and $\mathrm{Black}(\mathcal{D})$) denote the number of white (resp. black) regions, for the compatible coloring.

COROLLARY 2.7.11. *Let $\vec{L}$ be a link which admits a connected, alternating diagram $\mathcal{D}$. Equip $\mathcal{D}$ with a compatible coloring. Then,*

$$\sigma(\vec{L}) = \mathrm{Neg}(\mathcal{D}) - \mathrm{White}(\mathcal{D}) + 1 \quad \textit{and} \quad \sigma(\vec{L}) = \mathrm{Black}(\mathcal{D}) - \mathrm{Pos}(\mathcal{D}) - 1. \tag{2.10}$$

Proof. Thinking of the knot projection as giving a cell decomposition of S^2, it follows that

$$\mathrm{White}(\mathcal{D}) + \mathrm{Black}(\mathcal{D}) = \mathrm{Pos}(\mathcal{D}) + \mathrm{Neg}(\mathcal{D}) + 2;$$

so it suffices to prove only one of the two formulas in Equation (2.10).

Suppose first that for the compatible coloring of $\mathcal{D}$ we have that the unbounded domain is white. From our coloring conventions on the alternating projection it is clear that $\epsilon(p) = -1$ for all crossings, and furthermore positive crossings are of type I and negative crossings are of type II. Therefore, $\mu(\mathcal{D}) = -\mathrm{Neg}(\mathcal{D})$.

Next we claim that the Goeritz matrix of a compatibly colored, connected, alternating link diagram is negative definite. We see this as follows. By the alternating property it follows that $\epsilon(p) = -1$ for all crossings. Let m denote the number of crossings in the diagram. Consider the negative definite lattice $\mathbb{Z}^m$, equipped with a basis $\{e_p\}_{p=1}^m$ so that $\langle e_p, e_q\rangle = -\delta_{pq}$ (with δ_{pq} being the Kronecker delta). Think of the basis vectors e_p as being in one-to-one correspondence with the crossings in the projection. Consider next the vector space whose basis vectors correspond to the bounded white regions $\{d_i\}_{i=1}^n$ in the diagram. At each crossing, label one of the white quadrants with $+1$ and the other with -1. For $i = 1, \ldots, n$ and $p = 1, \ldots, m$, let $c_{i,p}$ be zero if the pth crossing does not appear on the boundary of d_i or if it appears twice on the boundary of d_i; otherwise, let $c_{i,p}$ be ± 1, depending on the sign of the quadrant at the pth crossing in d_i. Consider the linear map sending d_i to $\sum_p c_{i,p} \cdot e_p$. Since $\mathcal{D}$ is connected, this map is injective. (This follows from the inductive argument used in the proof of Lemma 2.7.7.) It is now straightforward to check that this linear map realizes an embedding of the lattice specified by the Goeritz matrix $G(\mathcal{D})$ into the standard, negative definite lattice. It follows at once that the Goeritz matrix is negative definite, as claimed.

The above argument shows that $\sigma(G(\mathcal{D}))$ is equal to $-(\mathrm{White}(\mathcal{D}) - 1)$, and so the Gordon-Litherland formula immediately implies the corollary.

Assume now that the compatible coloring on $\mathcal{D}$ provides a black unbounded domain. Reverse all crossings of $\mathcal{D}$ to get the *mirror diagram* $m(\mathcal{D})$, which represents the mirror link $m(\vec{L})$. Since the reversal also reverses the colors of the domains in the compatible coloring, the unbounded domain of the compatible coloring on $m(\mathcal{D})$ is white. For this diagram the previous argument then shows

$$\sigma(m(\vec{L})) = \mathrm{Neg}(m(\mathcal{D})) - \mathrm{White}(m(\mathcal{D})) + 1.$$

Since $\mathrm{Neg}(m(\mathcal{D})) = \mathrm{Pos}(\mathcal{D})$, $\mathrm{White}(m(\mathcal{D})) = \mathrm{Black}(\mathcal{D})$ and $\sigma(m(\vec{L})) = -\sigma(\vec{L})$, we get

$$\sigma(\vec{L}) = \mathrm{Black}(\mathcal{D}) - \mathrm{Pos}(\mathcal{D}) - 1. \qquad \square$$

CHAPTER 3

Grid diagrams

In this chapter we introduce the concept of a *grid diagram*, giving a convenient combinatorial way to represent knots and links in S^3. Grid diagrams will play an essential role in the rest of the book. These diagrams, as a tool for studying knots and links, made their first appearance in the work of Brunn in the late 19th century [**17**]. Other variants have been used since then, for example, in *bridge positions* [**127**], or in *arc presentations* of Cromwell and Dynnikov [**27, 37**]. Dynnikov used grid diagrams in his algorithm for recognizing the unknot [**37**]; see also [**12**]. Our presentation rests on Cromwell's theorem which describes the moves connecting different grid presentations of a given link type.

In Section 3.1 we introduce planar grid diagrams and their grid moves. Planar grid diagrams can be naturally transferred to the torus, to obtain *toroidal grid diagrams*, used in the definition of grid homology. Toroidal grid diagrams are discussed in Section 3.2. In Section 3.3 we show how grid diagrams can be used to compute the Alexander polynomial, while in Section 3.4 we introduce a method which provides Seifert surfaces for knots and links in grid position. Finally, in Section 3.5 we describe a presentation of the fundamental group of a link complement that is naturally associated to a grid diagram.

3.1. Planar grid diagrams

The present section will concern the following object:

DEFINITION 3.1.1. A ***planar grid diagram*** $\mathbb{G}$ is an $n \times n$ grid on the plane; that is, a square with n rows and n columns of small squares. Furthermore, n of these small squares are marked with an X and n of them are marked with an O; and these markings are distributed subject to the following rules:

(G-1) Each row has a single square marked with an X and a single square marked with an O.
(G-2) Each column has a single square marked with an X and a single square marked with an O.
(G-3) No square is marked with both an X and an O.

The number n is called the ***grid number*** of $\mathbb{G}$.

We denote the set of squares marked with an X by $\mathbb{X}$ and the set of squares marked with an O by $\mathbb{O}$. Sometimes, we will find it convenient to label the O-markings $\{O_i\}_{i=1}^n$. A grid diagram can be described by two permutations $\sigma_{\mathbb{O}}$ and $\sigma_{\mathbb{X}}$. If there is an O-marking in the intersection of the ith column and the jth row, then the permutation $\sigma_{\mathbb{O}}$ maps i to j. We will indicate this permutation as an n-tuple, $(\sigma_{\mathbb{O}}(1), \ldots, \sigma_{\mathbb{O}}(n))$. (By convention, we regard the left-most column and the bottom-most row as first.) The permutation $\sigma_{\mathbb{X}}$ is defined analogously, using the X-markings in place of the O-markings.

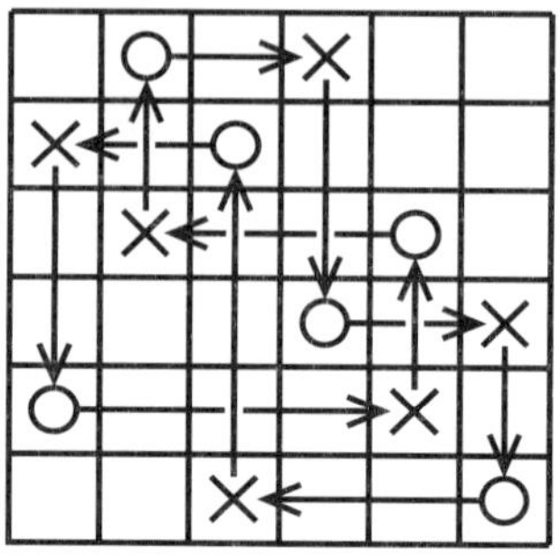

FIGURE 3.1. **The knot associated to the pictured grid diagram, with orientation and crossing conventions.** The diagram can be described by the two permutations $\sigma_{\mathbb{O}}$ and $\sigma_{\mathbb{X}}$, specifying the locations of the O's and X's. Above, the two permutations are $\sigma_{\mathbb{O}} = (2, 6, 5, 3, 4, 1)$ and $\sigma_{\mathbb{X}} = (5, 4, 1, 6, 2, 3)$.

In this section, we will use the terms "planar grid diagram" and simply "grid diagram" interchangeably. Care must be taken once we introduce the notion of a "toroidal grid diagram", later in the chapter.

3.1.1. Specifying links via planar grid diagrams. A grid diagram $\mathbb{G}$ specifies an oriented link $\vec{L}$ via the following procedure. Draw oriented segments connecting the X-marked squares to the O-marked squares in each column; then draw oriented segments connecting the O-marked squares to the X-marked squares in each row, with the convention that the vertical segments always cross above the horizontal ones. See Figure 3.1 for an example. In this case, we say that $\mathbb{G}$ is a *grid diagram for* $\vec{L}$.

REMARK 3.1.2. The permutation $\sigma_{\mathbb{X}} \cdot \sigma_{\mathbb{O}}^{-1}$ can be decomposed as a product of ℓ disjoint cycles for some ℓ. This number is equal to the number of components of the link specified by the grid diagram.

LEMMA 3.1.3. *Every oriented link in S^3 can be represented by a grid diagram.*

Proof. Approximate the link $\vec{L}$ by a PL-embedding with the property that the projection admits only horizontal and vertical segments. At a crossing for which the horizontal segment is an over-crossing, apply the modification indicated in Figure 3.2. Finally, move the link into general position, so that different horizontal (or vertical) segments are not collinear. Mark the turns by X's and O's, chosen so that vertical segments point from X to O, while horizontal segments point from O to X. The result is a grid diagram representing $\vec{L}$. □

EXAMPLES 3.1.4. **(a)** Given $p, q > 1$, define a $(p+q) \times (p+q)$ grid $\mathbb{G}(p, q)$ by $\sigma_{\mathbb{O}} = (p+q, p+q-1, \ldots, 2, 1)$ and $\sigma_{\mathbb{X}} = (p, p-1, \ldots, 1, p+q, p+q-1, \ldots, p+2, p+1)$. For $\mathbb{G}(2, 3)$ see the right picture in Figure 3.3. When $(p, q) = 1$, $\mathbb{G}(p, q)$ represents the torus knot $T_{p,q}$, cf. Exercise 3.1.5(a). More generally, $\mathbb{G}(p, q)$ represents the torus link $T_{p,q}$.
(b) Figure 3.4 provides grid presentations of the Kinoshita-Terasaka and Conway knots.

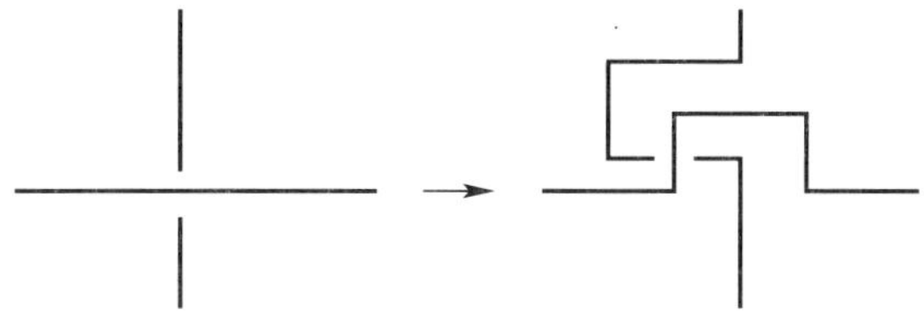

FIGURE 3.2. **The local modification for correcting crossings.**

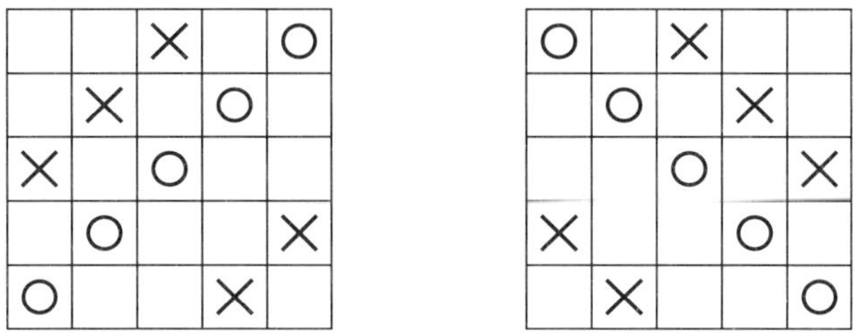

FIGURE 3.3. **Grid diagrams for the trefoils.** The left-handed trefoil $T_{-2,3}$ is on the left; the right-handed trefoil $T_{2,3}$ is on the right.

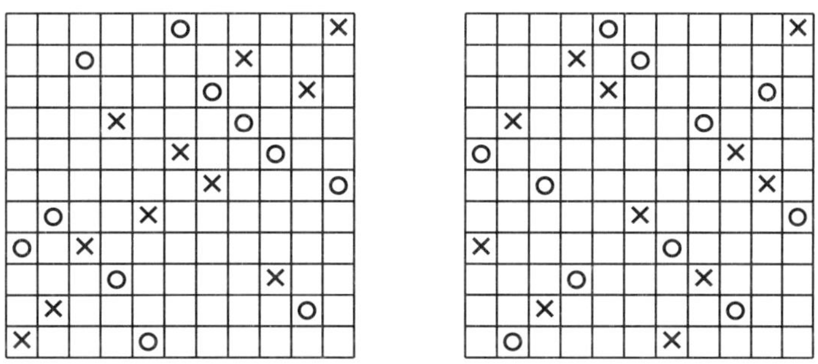

FIGURE 3.4. **Grid diagrams for the Conway knot (left) and the Kinoshita-Terasaka knot (right).**

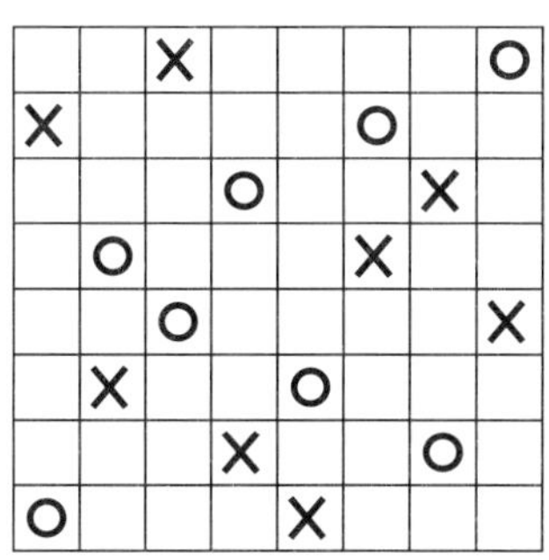

FIGURE 3.5. **Grid diagram for the Borromean rings.**

(c) The diagram of Figure 3.5 is a grid diagram for the Borromean rings.

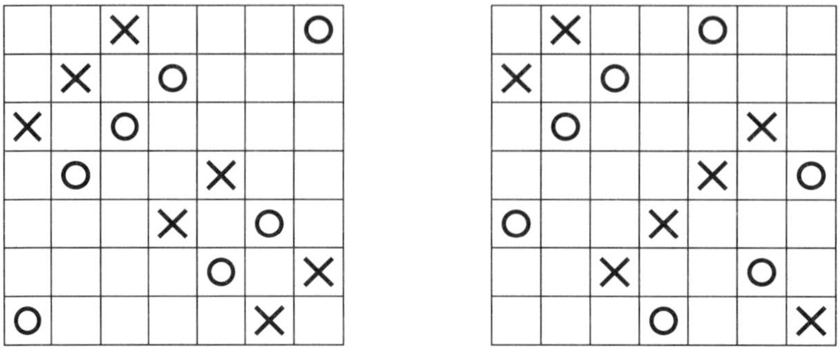

FIGURE 3.6. **Two grid diagrams of the 5_2 knot, isotopic to the twist knot W_3.**

EXERCISE 3.1.5. **(a)** Show that when $(p, q) = 1$, $\mathbb{G}(p, q)$ represents the (p, q) torus knot $T_{p,q}$.
(b) Find a grid presentation of the twist knot W_n from Example 2.1.5. (*Hint*: For the special case $n = 3$ consult Figure 3.6. Notice that both diagrams present the same knot, which is 5_2 in the knot tables.)
(c) Consider the permutations $\sigma_{\mathbb{X}} = (p + 1, p + 2, \dots, p + q, 1, \dots, p)$ and $\sigma_{\mathbb{O}} = (1, 2, \dots, p + q)$. Show that the resulting $(p + q) \times (p + q)$ grid diagram represents $T_{-p,q}$, the mirror of the torus knot $T_{p,q}$.
(d) Show that by reversing the roles of $\mathbb{X}$ and $\mathbb{O}$, the resulting diagram represents the same link with the opposite orientation.
(e) Similarly, by reflecting a given grid $\mathbb{G}$ across the diagonal, the resulting grid $\mathbb{G}'$ represents the same link as $\mathbb{G}$, but with the opposite orientation.
(f) Suppose that the grid $\mathbb{G}$ represents the knot K. Reflect $\mathbb{G}$ through the horizontal symmetry axis of the grid square and show that the resulting grid diagram $\mathbb{G}^*$ represents $m(K)$, the mirror image of K.
(g) Find diagrams for the Hopf links $H_{\pm}$.

3.1.2. Grid moves. Following Cromwell [**27**] (compare also Dynnikov [**37**]), we define two moves on planar grid diagrams.

DEFINITION 3.1.6. Each column in a grid diagram determines a closed interval of real numbers that connects the height of its O-marking with the height of its X-marking. Consider a pair of consecutive columns in a grid diagram $\mathbb{G}$. Suppose that the two intervals associated to the consecutive columns are either disjoint, or one is contained in the interior of the other. Interchanging these two columns gives rise to a new grid diagram $\mathbb{G}'$. We say that the two grid diagrams $\mathbb{G}$ and $\mathbb{G}'$ differ by a ***column commutation***. A ***row commutation*** is defined analogously, using rows in place of columns. A column or a row commutation is called a ***commutation move***, cf. Figure 3.7.

The second move on grid diagrams will change the grid number.

DEFINITION 3.1.7. Suppose that $\mathbb{G}$ is an $n \times n$ grid diagram. A grid diagram $\mathbb{G}'$ is called a ***stabilization*** of $\mathbb{G}$ if it is an $(n+1) \times (n+1)$ grid diagram obtained by splitting a row and column in $\mathbb{G}$ in two, as follows. Choose some marked square in $\mathbb{G}$, and erase the marking in that square, in the other marked square in its row,

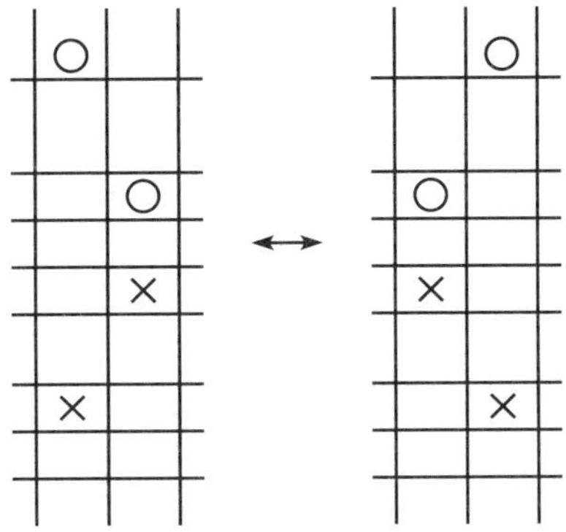

FIGURE 3.7. **The commutation move of two consecutive columns in a grid diagram.** Rotation by $90°$ gives an example of a row commutation.

and in the other marked square in its column. Now, split the row and column in two (i.e. add a new horizontal and a new vertical line). There are four ways to insert markings in the two new columns and rows in the $(n+1)\times(n+1)$ grid to obtain a grid diagram; see Figure 3.8 in the case where the initial square in $\mathbb{G}$ was marked with an X. Let $\mathbb{G}'$ be any of these four new grid diagrams. The inverse of a stabilization is called a ***destabilization***.

We will find it useful to classify the various types of stabilizations in a grid diagram. To this end, observe that for any stabilization, the original marked square gets subdivided into four squares, arranged in a 2×2 block. Exactly three of these new squares will be marked. The type of a stabilization is encoded by a letter X or an O, according to the marking on the original square chosen for stabilization (or equivalently, which letter appears twice in the newly-introduced 2×2 block), and by the position of the square in the 2×2 block which remains empty, which we indicate by a direction: *northwest* NW, *southwest* SW, *southeast* SE, or *northeast* NE. It is easy to see that a stabilization changes the projection either by a planar isotopy or by a Reidemeister move R_1. For example, in the diagrams of Figure 3.8 the stabilizations $X{:}NW$, $X{:}NE$, $X{:}SE$ give isotopies while $X{:}SW$ corresponds to the Reidemeister move R_1.

DEFINITION 3.1.8. We call commutations, stabilizations, and destabilizations ***grid moves*** collectively.

Grid diagrams are an effective tool for constructing knot invariants, thanks to the following theorem of Cromwell [**27**], see also [**37**] and Section B.4:

THEOREM 3.1.9 (Cromwell, [**27**]). *Two planar grid diagrams represent equivalent links if and only if there is a finite sequence of grid moves that transform one into the other.* □

3.1.3. Other moves between grid diagrams. Interchanging two consecutive rows or columns can be a commutation move; there are two other possibilities:

DEFINITION 3.1.10. Consider two consecutive columns in a grid diagram. These columns are called ***special*** if the X-marking in one of the columns occurs in the

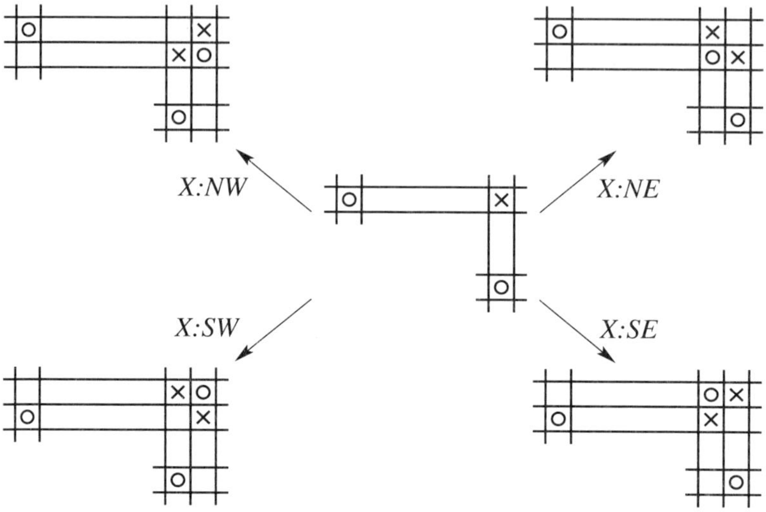

FIGURE 3.8. **The stabilization at an X-marking.** There are four different stabilizations which can occur at a given X-marking: *X:NW*, *X:NE*, *X:SE*, and *X:SW*. The further four types of stabilizations (i.e. at O-markings) can be derived from these diagrams by interchanging all X- and O-markings.

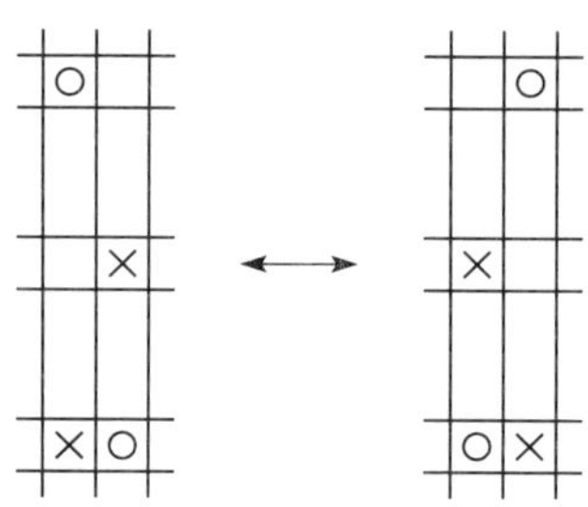

FIGURE 3.9. **A switch of two special columns.** Rotate both diagrams by 90° to get an example of a switch of two special rows.

same row as the O-marking in the other column. If $\mathbb{G}'$ is obtained from $\mathbb{G}$ by interchanging a pair of special columns, then we say that $\mathbb{G}$ and $\mathbb{G}'$ are related by a ***switch***. Similarly, if two consecutive rows have an X- and an O-marking in the same column, interchanging them is also called a switch. See Figure 3.9.

Clearly, grid diagrams that differ by a switch determine the same link type.

EXERCISE 3.1.11. Express a switch as a sequence of commutations, stabilizations, and destabilizations.

DEFINITION 3.1.12. Fix two consecutive columns (or rows) in a grid diagram $\mathbb{G}$, and let $\mathbb{G}'$ be obtained by interchanging those two columns (or rows). Suppose that the interiors of their corresponding intervals intersect non-trivially, but neither is contained in the other; then we say that the grid diagrams $\mathbb{G}$ and $\mathbb{G}'$ are related by a ***cross-commutation***.

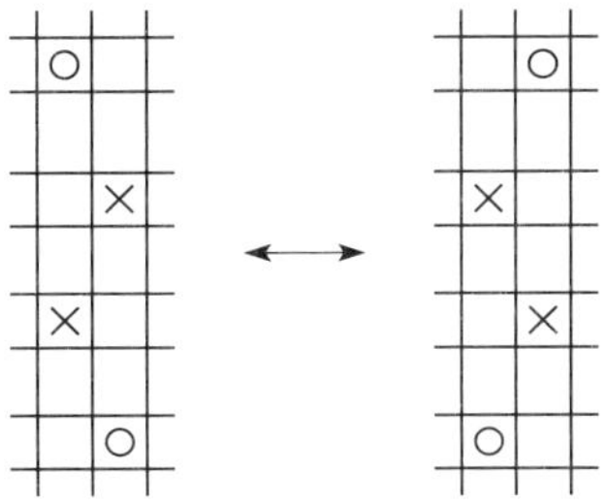

FIGURE 3.10. **A cross-commutation move.**

The proof of the following proposition is straightforward:

PROPOSITION 3.1.13. *If $\mathbb{G}$ and $\mathbb{G}'$ are two grid diagrams that are related by a cross-commutation, then their associated oriented links $\vec{L}$ and $\vec{L}'$ are related by a crossing change.* □

Grid diagrams can be used to show that any knot can be untied by a finite sequence of crossing changes (compare Exercise 2.3.8). Pick an X-marking and move the row containing the O-marking sharing the column with the chosen X-marking until these two markings occupy neighbouring squares. These moves are either commutations, switches (as such, leaving the link type unchanged), or cross-commutations, causing crossing changes. Then commute the column of the chosen X-marking until it reaches the O-marking in its row (compare Figure 3.12) and destabilize. This procedure reduces the grid number of the diagram. Repeatedly applying the procedure we turn the initial grid diagram into a 2×2 grid diagram representing the unknot, while changing the knot diagram by planar isotopies, Reidemeister moves and crossing changes only.

3.2. Toroidal grid diagrams

We find it convenient to transfer our planar grid diagrams to the torus $\mathbb{T}$ obtained by identifying the top boundary segment with the bottom one, and the left boundary segment with the right one. In the torus, the horizontal and vertical segments (which separate the rows and columns of squares) become horizontal and vertical circles. The torus inherits its orientation from the plane. We call the resulting object a *toroidal grid diagram*.

Conversely, a toroidal grid diagram can be cut up to give a planar grid diagram in n^2 different ways. We call these *planar realizations* of the given toroidal grid diagram. It is straightforward to see that two different planar realizations of the same grid diagram represent isotopic links. The relationship between these different planar realizations can be formalized with the help of the following:

DEFINITION 3.2.1. Let $\mathbb{G}$ be a planar grid diagram, and let $\mathbb{G}'$ be a new planar diagram obtained by cyclically permuting the rows or the columns of $\mathbb{G}$. (Notice that this move has no effect on the induced toroidal grid diagram.) In this case, we say that $\mathbb{G}'$ is obtained from $\mathbb{G}$ by a ***cyclic permutation***. See Figure 3.11 for an example.

Clearly, two different planar realizations of a toroidal grid diagram can be connected by a sequence of cyclic permutations.

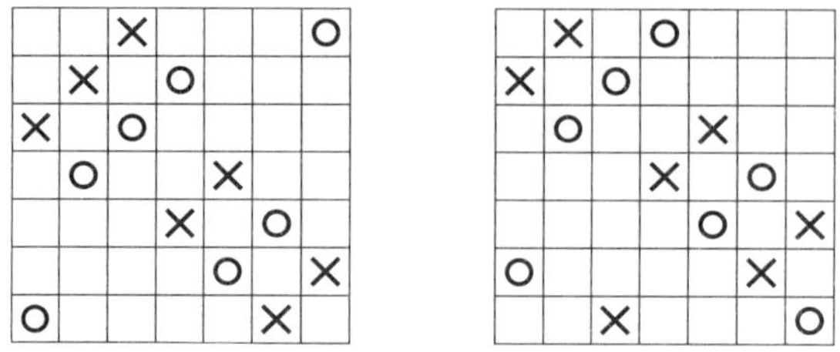

FIGURE 3.11. **Cyclic permutation.** By moving the top row of the left grid to the bottom, we get the grid on the right. In a cyclic permutation we can move several consecutive rows from top to bottom (or from bottom to top), and there is a corresponding move for columns as well.

The toroidal grid diagram inherits a little extra structure from the planar diagram. Thinking of the coordinate axes on the plane as oriented, there are induced orientations on the horizontal and vertical circles: explicitly, the grid torus is expressed as a product of two circles $\mathbb{T} = S^1 \times S^1$, where $S^1 \times \{p\}$ is a horizontal circle and $\{p\} \times S^1$ is a vertical circle; and both the horizontal and vertical circles are oriented. At each point in the torus, there are four preferred directions, which we think of as *North*, *South*, *East*, and *West*. (More formally, "North" refers to the oriented tangent vector of the circles $\{p\} \times S^1$; "South" to the opposite direction; "East" refers to the positive tangent vector of the circles $S^1 \times \{p\}$; and "West" to its opposite.) Correspondingly, each of the squares in the toroidal grid has a northern edge, an eastern edge, a southern edge, and a western edge.

Commutation and stabilization moves have natural adaptations to the toroidal case. For example, we say that two toroidal grid diagrams differ by a commutation move if they have planar realizations which differ by a commutation move. Stabilization moves on toroidal grids are defined analogously. The classification of the types of stabilizations carries over to the toroidal case.

The grid chain complex (introduced in the next chapter) is associated to a toroidal grid diagram for a knot K. The resulting homology, however, depends on K only. The proof of this statement will hinge on Theorem 3.1.9: we will check that grid homology is invariant under grid moves. In the course of the proof it will be helpful to express certain grid moves in terms of others.

LEMMA 3.2.2. *A stabilization of type O:NE (respectively O:SE, O:NW, or O:SW) can be realized by a stabilization of type X:SW (respectively X:NW, X:SE, or X:NE), followed by a sequence of commutation moves on the torus.*

Proof. Let $\mathbb{G}$ be a grid diagram and $\mathbb{G}_1$ be the stabilization of $\mathbb{G}$ at an O-marking. Commute the new length one vertical segment repeatedly until it meets another X-marking, and let $\mathbb{G}_2$ denote the resulting grid diagram. A type X destabilization on $\mathbb{G}_2$ gives $\mathbb{G}$ back. See Figure 3.12 for an illustration. □

COROLLARY 3.2.3. *Any stabilization can be expressed as a stabilization of type X:SW followed by sequence of switches and commutations.*

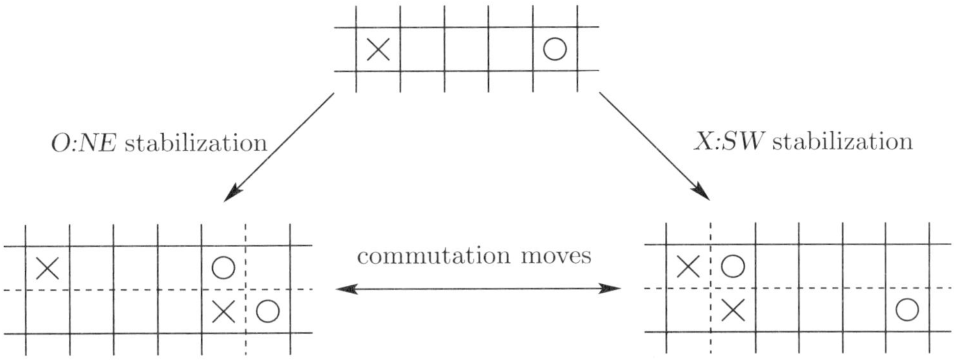

FIGURE 3.12. **A stabilization of type** O:NE **is equivalent to an** X:SW **stabilization and a sequence of commutations.**

Proof. First use Lemma 3.2.2 to express stabilizations of type O in terms of stabilizations of type X (and commutations). Next note that stabilizations *X:SE*, *X:SW*, *X:NE*, *X:NW* differ from each other by one or two switches. □

In a similar spirit, we have the following lemma, which will be used in Chapter 12:

LEMMA 3.2.4. *A cyclic permutation is equivalent to a sequence of commutations in the plane, stabilizations, and destabilizations of types* X:NW, X:SE, O:NW, *and* O:SE.

Proof. Consider the case of moving a horizontal segment from the top to the bottom, and suppose moreover that the left end of that segment is marked X_1, and the right end is marked O_2. Let O_1 (respectively X_2) be the other marking in the column containing X_1 (respectively O_2). Apply a stabilization of type *X:NW* at X_2, and commute the resulting horizontal segment of length 1 to the bottom of the diagram. We now have a vertical segment stretching the height of the diagram; apply commutation moves until it is just to the right of the column containing X_1. Now the horizontal segment starting at X_1 is of length 1, and so can be commuted down until it is just above O_1, where we can apply a destabilization of type *O:SE* to get the desired cyclic permutation. See Figure 3.13 for a picture of the sequence of moves we just performed. The other cases are handled similarly. □

3.3. Grids and the Alexander polynomial

In this section grids will refer to planar grids, unless explicitly stated otherwise. Let $\mathbb{G}$ be an $n \times n$ planar grid diagram for a link placed in the $[0, n] \times [0, n]$ square on the plane. (The horizontal segments of the grid now have integral y-coordinates, while the vertical ones have integral x-coordinates. The O- and X-markings have half-integer coordinates.) Remember that the grid can be specified by the two permutations $\sigma_{\mathbb{O}}$ and $\sigma_{\mathbb{X}}$ describing the locations of the two sets of markings.

Recall the following construction from elementary topology:

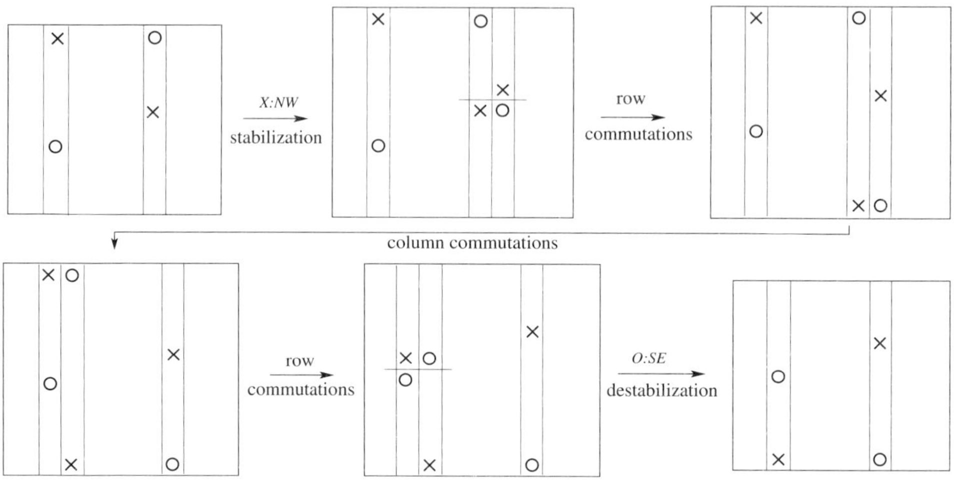

FIGURE 3.13. **The steps of the proof of Lemma 3.2.4.**

DEFINITION 3.3.1. Let γ be a closed, piecewise linear, oriented (not necessarily embedded and possibly disconnected) curve γ in the plane and a point $p \in \mathbb{R}^2 \setminus \gamma$. The ***winding number*** $w_\gamma(p)$ of γ around the point p is defined as follows. Draw a ray ρ from p to ∞, and let $w_\gamma(p)$ be the algebraic intersection of ρ with γ. The winding number is independent of the choice of the ray ρ.

With this terminology in place, we associate a matrix to the grid $\mathbb{G}$ as follows.

DEFINITION 3.3.2. Fix a grid diagram $\mathbb{G}$ representing the link $\vec{L}$. Form the $n \times n$ matrix whose $(i,j)^{th}$ entry (the element in the ith row and j^{th} column) is obtained by raising the formal variable t to the power given by (-1)-times the winding number of the link diagram given by $\mathbb{G}$ around the $(j-1, n-i)^{th}$ lattice point with $1 \leq i,j \leq n$. Call this matrix the ***grid matrix***, and denote it by $\mathbf{M}(\mathbb{G})$.

Notice that the left-most column and the bottom-most row of $\mathbf{M}(\mathbb{G})$ consist of 1's only. To explain our above convention, note that the $(1,1)$ entry of a matrix is in the upper left corner, while in our convention for grids the bottom-most row is the first. As an example, for the grid diagram in Figure 3.1 (compare Figure 3.14) the grid matrix is

$$\begin{pmatrix} 1 & 1 & t & t & 1 & 1 \\ 1 & t^{-1} & 1 & t & 1 & 1 \\ 1 & t^{-1} & t^{-1} & 1 & t^{-1} & 1 \\ 1 & t^{-1} & t^{-1} & 1 & 1 & t \\ 1 & 1 & 1 & t & t & t \\ 1 & 1 & 1 & 1 & 1 & 1 \end{pmatrix}.$$

EXERCISE 3.3.3. Consider the 2×2 and 3×3 grid diagrams for the unknot, given by Figure 3.15. Compute the determinants of the associated matrices.

Consider the function $\det(\mathbf{M}(\mathbb{G}))$ associated to the diagram. According to Exercise 3.3.3, one immediately realizes that this determinant is not a link invariant: it does depend on the choice of the grid diagram representing the given link. However,

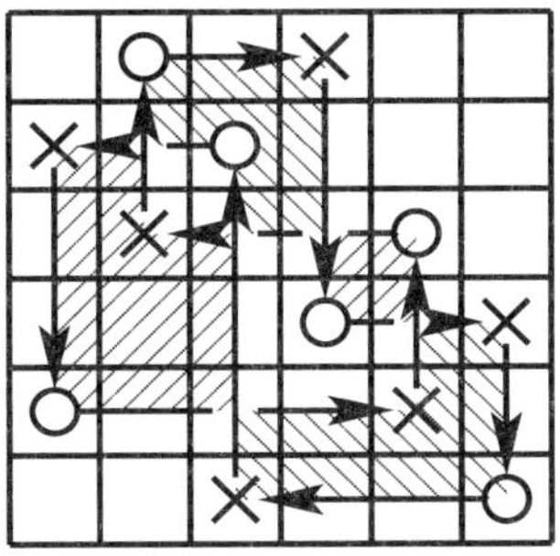

FIGURE 3.14. For the grid diagram illustrated in Figure 3.1, we shade regions according to the winding number of the knot: diagonal hatchings from lower left to upper right indicate regions with winding number $+1$, the other hatchings indicate winding number -1, and no hatchings indicate winding number 0.

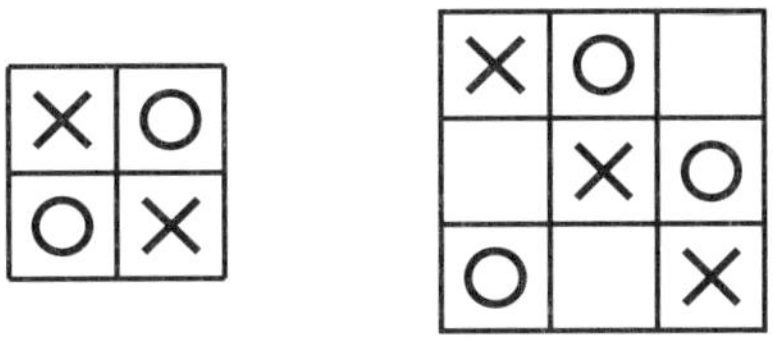

FIGURE 3.15. **Two grids representing the unknot.** It is easy to see that the associated determinants $\det(\mathbf{M}(\mathbb{G}))$ are different.

as we will see, after a suitable normalization of this quantity, we obtain the Alexander polynomial of the link represented by the grid. To describe the normalization, we consider the following quantity $a(\mathbb{G})$ associated to the grid. For an O and an X consider the four corners of the square in the grid occupied by the marking and sum up the winding numbers in these corners. By summing these contributions for all O's and X's and dividing the result by 8, we get a number $a(\mathbb{G})$ associated to the $n \times n$ grid. Finally, let $\epsilon(\mathbb{G}) \in \{\pm 1\}$ be the sign of the permutation that connects $\sigma_{\mathbb{O}}$ and $(n, n-1, ..., 1)$.

DEFINITION 3.3.4. Suppose that $\mathbb{G}$ is an $n \times n$ grid. Define the function $D_{\mathbb{G}}(t)$ to be the product

$$\epsilon(\mathbb{G}) \cdot \det(\mathbf{M}(\mathbb{G})) \cdot (t^{-\frac{1}{2}} - t^{\frac{1}{2}})^{1-n} t^{a(\mathbb{G})}.$$

EXERCISE 3.3.5. Compute $D_{\mathbb{G}}(t)$ for the two grids in Figure 3.15.

The next theorem relates the function $D_{\mathbb{G}}(t)$ with the Alexander polynomial.

THEOREM 3.3.6. *Let $\mathbb{G}$ be a grid diagram that represents $\vec{L}$. Then, the function $D_{\mathbb{G}}(t)$ is a link invariant and it coincides with the symmetrized Alexander polynomial $\Delta_{\vec{L}}(t)$ of the link $\vec{L}$ (as it is defined in Equation* (2.3)*).*

To prove Theorem 3.3.6, we establish some invariance properties of $D_{\mathbb{G}}(t)$.

LEMMA 3.3.7. *The function $D_{\mathbb{G}}(t)$ is invariant under commutation moves.*

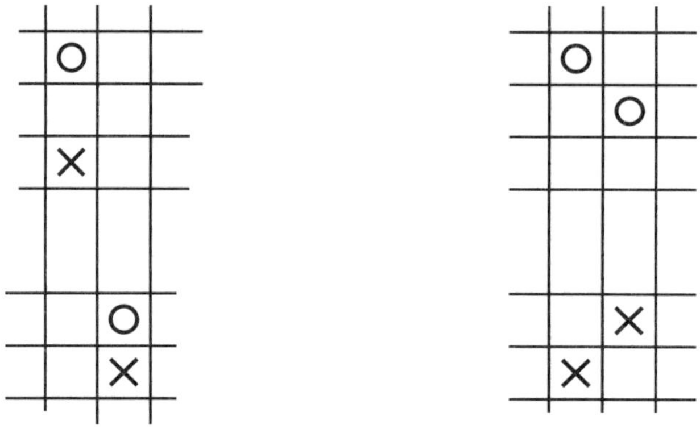

FIGURE 3.16. **The two cases in Lemma 3.3.7.** On the left the y-projections of the $O - X$ intervals are disjoint, while on the right the projections are nested. The diagrams represent two consecutive columns of the grid (say the ith and $(i+1)^{st}$), and therefore the vertical lines correspond to the ith, $(i+1)^{st}$ and $(i+2)^{nd}$ columns of the grid matrix.

Proof. Suppose that the grid $\mathbb{G}'$ is derived from $\mathbb{G}$ by commuting the ith and $(i+1)^{st}$ columns. Then the matrices $\mathbf{M}(\mathbb{G})$ and $\mathbf{M}(\mathbb{G}')$ differ in the $(i+1)^{st}$ column only.

We distinguish two cases, depending on whether the two intervals we are about to commute project disjointly to the y-axis, or one projection contains the other one (the two possibilities are shown by the left and right diagrams of Figure 3.16). In the first case, subtract the ith column from the $(i+1)^{st}$ in $\mathbf{M}(\mathbb{G})$ and the $(i+2)^{nd}$ from the $(i+1)^{st}$ in $\mathbf{M}(\mathbb{G}')$. The resulting matrices will differ only in the sign of the $(i+1)^{st}$ column, hence their determinants are opposites of each other. Since neither the size of the grid nor the quantity $a(\mathbb{G})$ changes, while $\epsilon(\mathbb{G}) = -\epsilon(\mathbb{G}')$, the invariance of $D_{\mathbb{G}}(t)$ under such commutation follows at once.

In the second case, we distinguish further subcases, depending on the relative positions of the O- and X-markings in the two columns. In the right diagram of Figure 3.16 we show the case when in both columns the O-marking is over the X-marking; the further three cases can be given by switching one or both pairs within their columns. In the following we will give the details of the argument only for the configuration shown by Figure 3.16; the verifications for the other cases proceed along similar lines.

As before, we subtract one column from the other one in each matrix. The choice of the columns in this case is important. In the case shown by Figure 3.16 we subtract the $(i+2)^{nd}$ column from the $(i+1)^{st}$; in general we subtract the column on the side of the shorter O-X-interval (that is, on the side where the two markings in the column are closer to each other). After performing the subtraction in both matrices $\mathbf{M}(\mathbb{G})$ and $\mathbf{M}(\mathbb{G}')$, we realize that the $(i+1)^{st}$ columns of the two matrices differ not only by a sign, but also by a multiple of t. A simple calculation shows that this difference is compensated by the difference in the terms originating from $a(\mathbb{G})$ and $a(\mathbb{G}')$, while the sign difference is absorbed by the change of ϵ. This results that $D_{\mathbb{G}}(t)$ remains unchanged under commuting columns. A similar argument verifies the result when we commute rows, completing the argument. □

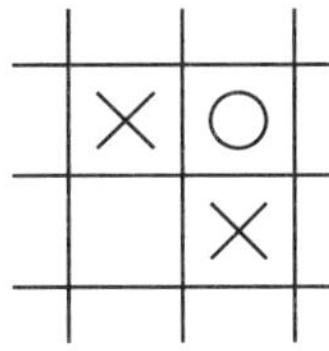

FIGURE 3.17. **The convention used in the proof of Lemma 3.3.8.**

LEMMA 3.3.8. *The function $D_{\mathbb{G}}(t)$ is invariant under stabilization moves.*

Proof. Consider the case where the stabilization is of type *X:SW*. In the matrix of the stabilized diagram, if we subtract the $(i+2)^{nd}$ row of Figure 3.17 from the $(i+1)^{st}$ row (passing between the two X's in the stabilization), then we get a matrix which has a single non-zero term in this row. The determinant of the minor corresponding to this single element is, up to sign, the determinant of the matrix we had before the stabilization. The sign change is compensated by the introduction of $\epsilon(\mathbb{G})$, while the t-power in front of the determinant of the minor is absorbed by the change of the quantity $a(\mathbb{G})$ and the change of the size of the diagram, showing that $D_{\mathbb{G}}(t)$ remains unchanged. Other stabilizations work similarly. □

Combining the above lemmas with Cromwell's Theorem 3.1.9, the function $D_{\mathbb{G}}(t)$ is a link invariant. Therefore, if $\mathbb{G}$ represents the link type $\vec{L}$ then $D_{\mathbb{G}}(t)$ will be denoted by $D_{\vec{L}}(t)$.

The proof of Theorem 3.3.6 will use the fact that $D_{\vec{L}}$ satisfies the skein relation. We start with a definition adapting the notion of an oriented skein triple to the grid context.

DEFINITION 3.3.9. Let $(\vec{L}_+, \vec{L}_-, \vec{L}_0)$ be an oriented skein triple, as in Definition 2.4.9. A ***grid realization of the oriented skein triple*** consists of four grid diagrams $\mathbb{G}_+$, $\mathbb{G}_-$, $\mathbb{G}_0$, and $\mathbb{G}'_0$, representing the links $\vec{L}_+$, $\vec{L}_-$, $\vec{L}_0$, and $\vec{L}_0$ respectively. These diagrams are further related as follows: $\mathbb{G}_+$ and $\mathbb{G}_-$ differ by a cross-commutation, $\mathbb{G}_0$ and $\mathbb{G}'_0$ differ by a commutation, and $\mathbb{G}_+$ and $\mathbb{G}_0$ differ in the placement of their X-markings. See Figure 3.18 for a picture.

LEMMA 3.3.10. *Any oriented skein triple has a grid realization.*

Proof. Consider the diagrams $(\mathcal{D}_+, \mathcal{D}_-, \mathcal{D}_0)$ given by the skein triple. Approximate the diagrams by horizontal and vertical segments as explained in the proof of Lemma 3.1.3, with the additional property that in the small disk where the diagrams differ, the approximation is as given by $\mathbb{G}_+, \mathbb{G}_-$ and $\mathbb{G}'_0$ of Figure 3.18, while outside of the disk the three approximations are identical. Applying the commutation move on the first two columns of the grid $\mathbb{G}'_0$ we get $\mathbb{G}_0$, concluding the argument. □

PROPOSITION 3.3.11. *The invariant $D_{\vec{L}}(t)$ satisfies the skein relation, that is, for an oriented skein triple $(\vec{L}_+, \vec{L}_-, \vec{L}_0)$ we have*

$$D_{\vec{L}_+}(t) - D_{\vec{L}_-}(t) = (t^{\frac{1}{2}} - t^{-\frac{1}{2}}) D_{\vec{L}_0}(t).$$

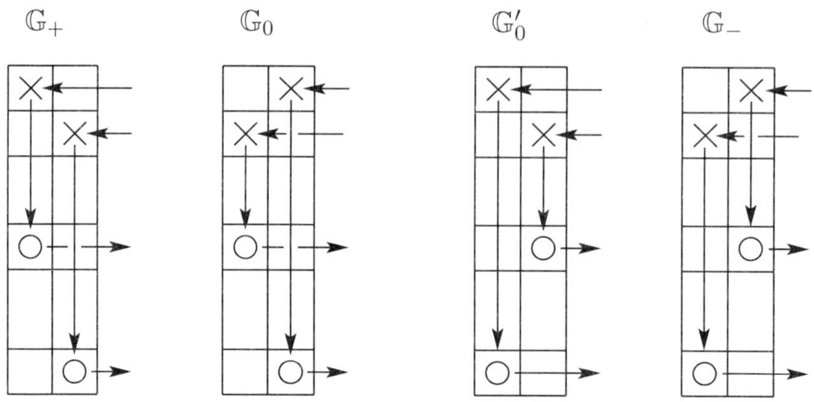

FIGURE 3.18. **The four grid diagrams which show up in the skein relation.**

Proof. Let $(\vec{L}_+, \vec{L}_-, \vec{L}_0)$ be an oriented skein triple, and let $(\mathbb{G}_+, \mathbb{G}_-, \mathbb{G}_0, \mathbb{G}'_0)$ be its grid realization, provided by Lemma 3.3.10. These grid diagrams agree in the placements of their $\mathbb{X}$- and $\mathbb{O}$-markings in all but two consecutive columns, which we think of as left-most. In these two columns of $\mathbb{G}_+$ we either move the two X-markings (transforming $\mathbb{G}_+$ to $\mathbb{G}_0$), or the two O-markings (giving $\mathbb{G}'_0$ from $\mathbb{G}_+$), or both (realizing a cross-commutation, transforming $\mathbb{G}_+$ to $\mathbb{G}_-$). In Figure 3.18 we depict the left-most two columns of these grids.

Now, the four associated grid matrices differ only in their second columns; and in fact, we have the relation

$$\det(\mathbf{M}(\mathbb{G}_+)) + \det(\mathbf{M}(\mathbb{G}_-)) = \det(\mathbf{M}(\mathbb{G}_0)) + \det(\mathbf{M}(\mathbb{G}'_0)). \tag{3.1}$$

It is straightforward to verify that

$$\begin{aligned} a(\mathbb{G}_-) = a(\mathbb{G}_+) = a(\mathbb{G}_0) + \frac{1}{2} = a(\mathbb{G}'_0) - \frac{1}{2}, \\ \epsilon(\mathbb{G}_+) = -\epsilon(\mathbb{G}_-) = \epsilon(\mathbb{G}_0) = -\epsilon(\mathbb{G}'_0). \end{aligned}$$

Combining these with Equation (3.1) gives the skein relation for $D_{\vec{L}}(t)$. □

Proof of Theorem 3.3.6. The Alexander polynomial for an oriented link satisfies the skein relation (Theorem 2.4.10). In fact, it is not hard to see that the Alexander polynomial is characterized by this relation, and its normalization for the unknot. Since $D_{\vec{L}}(t)$ satisfies this skein relation (Proposition 3.3.11), and $D_{\mathcal{O}}(t) = 1$ (as can be seen by checking in a 2×2 grid diagram), the result follows. □

The determinant of the grid matrix can be thought of as a weighted count of permutations, where the weight is obtained as a monomial in t, with exponent given by a winding number. In Chapter 4, grid homology will be defined as the homology of a bigraded chain complex whose generators correspond to these permutations, equipped with two gradings.

3.4. Grid diagrams and Seifert surfaces

It turns out that planar grid diagrams can also be applied to study Seifert surfaces of knots and links. Suppose that the $n \times n$ grid diagram $\mathbb{G}$ represents the knot $K \subset S^3$. Consider the *winding matrix* $\mathbf{W}(\mathbb{G})$ associated to $\mathbb{G}$ in the following way: the $(i,j)^{th}$ entry is the winding number of the projection of K given by the grid $\mathbb{G}$ around the $(j-1,n-i)^{th}$ lattice point with $1 \le i,j \le n$. In the following we describe a method which produces a Seifert surface for K based on $\mathbf{W}(\mathbb{G})$.

Let R_i (and similarly C_j) denote the $n \times n$ matrix with 1's in the ith row (j^{th} column), and 0 everywhere else. Obviously, by adding sufficiently many R_i's or C_j's, or both, any integral matrix can be turned into one which has only non-negative entries.

DEFINITION 3.4.1. Define the ***complexity*** $c(A)$ of a non-negative matrix A to be the sum of all its entries: $c(A) = \sum_{i,j} a_{i,j}$. An integral matrix $A \in M_n(\mathbb{Z})$ with non-negative entries is called ***minimal*** if its complexity is minimal among those non-negative integral matrices which can be given by repeatedly adding/subtracting C_i's and R_j's to A.

The following lemma gives a criterion for minimality:

LEMMA 3.4.2. *The matrix $A = (a_{i,j}) \in M_n(\mathbb{Z})$ with non-negative entries is minimal if and only if there is a permutation $\sigma \in \mathfrak{S}_n$ such that $a_{i,\sigma(i)} = 0$ for all $i \in \{1, \dots, n\}$.*

Proof. Suppose that there is a permutation σ with the property that $a_{i,\sigma(i)} = 0$ for all i. Consider integers m_i and n_i for $i = 1, \dots, n$ so that

$$A' = A + \sum_i n_i R_i + \sum_i m_i C_i$$

is a matrix with non-negative entries. Since $a_{i,\sigma(i)} = 0$, we conclude that $n_i + m_{\sigma(i)} \ge 0$ for all $i = 1, \dots, n$. Clearly,

$$c(A') = c(A) + \sum_{i=1}^{n} n(n_i + m_{\sigma(i)}) \ge c(A),$$

so the complexity of A is minimal, as claimed.

For the converse direction, let $\mathcal{C}$ denote the set of columns, while $\mathcal{R}$ the set of rows of the given non-negative matrix A. Connect $c_j \in \mathcal{C}$ with $r_i \in \mathcal{R}$ if and only if the $(i,j)^{th}$ entry $a_{i,j}$ of A is equal to zero. Let G denote the resulting bipartite graph on $2n$ vertices. According to Hall's Theorem [**82**] (a standard result in graph theory) either there is a perfect matching in G, providing the desired permutation, or there is a subset $C \subset \mathcal{C}$ such that the cardinality of the set R formed by those elements in $\mathcal{R}$ which are connected to C is smaller than $|C|$. Now if we add the R_j's with $j \in R$ to A, the columns corresponding to elements of C become positive, hence can be subtracted while keeping the matrix non-negative. Since $|C| > |R|$, we reduced the complexity of A, hence it was non-minimal. □

Returning to the matrix $\mathbf{W}(\mathbb{G})$, add and subtract appropriate R_i's and C_j's until it becomes a minimal, non-negative integral matrix, and denote the result by H. (Notice that this matrix is not uniquely associated to $\mathbf{W}(\mathbb{G})$ — it depends on the way we turned our starting matrix into a minimal, non-negative one.)

LEMMA 3.4.3. *Adjacent entries of H differ by at most one; i.e.*

$$|h_{i,j} - h_{i,j+1}| \le 1, \qquad |h_{i+1,j} - h_{i,j}| \le 1, \tag{3.2}$$

where i and j are taken modulo n; e.g. we are viewing the last column as adjacent to the first one. More generally, for each 2×2 block of entries in the matrix H (viewed on the torus), there is a non-negative integer a so that the block has one of the following five possible shapes, up to rotation by multiples of $90°$. *In cases where the center of the 2×2 block is unmarked, the possibilities are:* $\begin{array}{c|c} a & a \\ \hline a & a \end{array}$, $\begin{array}{c|c} a & a+1 \\ \hline a & a+1 \end{array}$ *and* $\begin{array}{c|c} a & a+1 \\ \hline a+1 & a+2 \end{array}$. *In cases where the center is marked with an O or an X, the possibilities are* $\begin{array}{c|c} a & a \\ \hline a & a+1 \end{array}$ *and* $\begin{array}{c|c} a & a+1 \\ \hline a+1 & a+1 \end{array}$.

Proof. Consider a 2×2 block in $\mathbf{W}(\mathbb{G})$, with entries $\begin{array}{c|c} a_{i,j} & a_{i,j+1} \\ \hline a_{i+1,j} & a_{i+1,j+1} \end{array}$. By thinking about winding numbers, it is clear that $a_{i,j} + a_{i+1,j+1} - a_{i,j+1} - a_{i+1,j} = 0$ unless the corner point corresponds to an O- or X-marking, in which case $a_{i,j} + a_{i+1,j+1} - a_{i,j+1} - a_{i+1,j} = \pm 1$. Since the expression $a_{i,j} + a_{i+1,j+1} - a_{i,j+1} - a_{i+1,j}$ is unchanged after the addition of C_i or R_j, we conclude that for each i and j,

$$|h_{i,j} - h_{i,j+1} - h_{i+1,j} + h_{i+1,j+1}| \leq 1; \tag{3.3}$$

and for each fixed j there are at most two i for which equality holds. Moreover, for fixed j, we can find k and ℓ so that $h_{k,j} = h_{\ell,j+1} = 0$. Since the entries of H are all non-negative, it follows that $|h_{i,j} - h_{i,j+1}| \leq 1$ for all i. The same reasoning gives the other bound.

Combining the bounds from Equations (3.2) and (3.3), we arrive at the five possibilities for the 2×2 blocks listed above. □

Next we associate a surface $F_H \subset S^3$ to H. Consider first a disjoint union of squares $s^k_{i,j}$ with $i, j \in \{1, \dots, n\}$ and $k \in \{1, \dots h_{i,j}\}$. Glue the right edge of $s^k_{i,j}$ to the left edge of $s^k_{i,j+1}$ for $k \leq \min(h_{i,j}, h_{i,j+1})$, and the bottom edge of $s^{h_{i,j}-k}_{i,j}$ to the top edge of $s^{h_{i+1,j}-k}_{i+1,j}$ for $0 \leq k \leq \min(h_{i,j}, h_{i+1,j}) - 1$. The result F_H is an oriented two-manifold with boundary, equipped with an orientation-preserving map to the torus. It is connected, since H vanishes somewhere.

Construct an embedding of F_H into S^3, as follows. View the grid torus as standardly embedded in S^3, and view $\{s^k_{i,j}\}_{k=1}^{h_{i,j}}$ as a collection of disjoint squares, stacked above the $(i,j)^{th}$ square in the grid torus, so that $s^{k+1}_{i,j}$ is above $s^k_{i,j}$ in the pile; see Figure 3.19. Instead of the edge identifications described earlier, we glue the various squares together by attaching strips; see Figure 3.20. The result is an embedding of the surface F_H constructed above into S^3.

PROPOSITION 3.4.4. *Suppose that the knot $K \subset S^3$ is represented by the grid diagram $\mathbb{G}$. Assume that the minimal, non-negative matrix H is given by adding and subtracting R_i's and C_j's to the matrix $\mathbf{W}(\mathbb{G})$ associated to $\mathbb{G}$. Then, the above embedding of the 2-complex F_H is a Seifert surface of K.*

Proof. We have seen that F_H is a 2-dimensional connected, oriented manifold embedded in S^3. By analyzing the local behavior from Lemma 3.4.3, it follows that the boundary of F_H is isotopic to K. See Figure 3.21 for an example. □

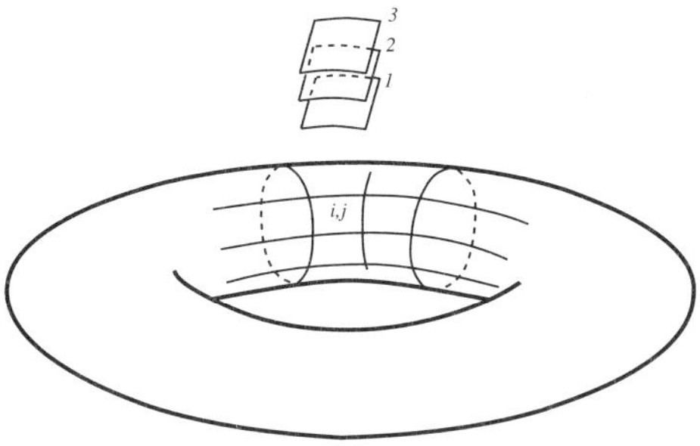

FIGURE 3.19. **Squares over the grid.** In this picture, $h_{i,j} = 3$.

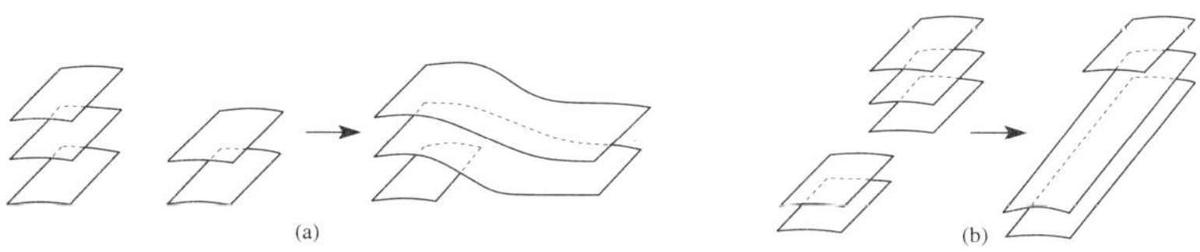

FIGURE 3.20. **Gluing squares to construct the embedding of F_H.** Neighbouring stacks of squares are glued together either from the top (as in (a)) or from the bottom (as in (b)).

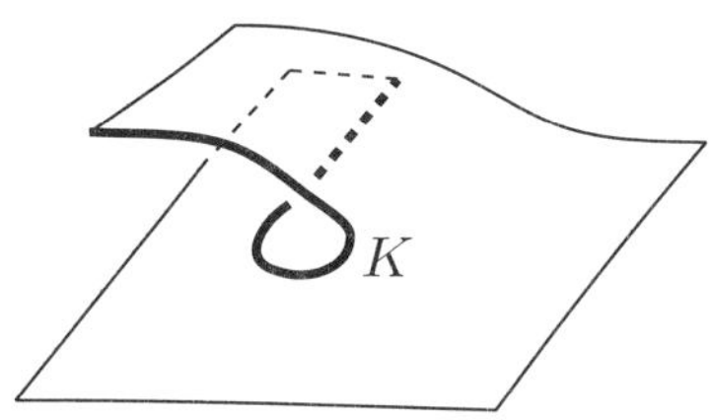

FIGURE 3.21. **A portion of F_H.** We have illustrated here the portion of F_H over a 2×2 block, one square with multiplicity 2 and the others with multiplicity 1. The the knot K is drawn thicker.

EXERCISE 3.4.5. Draw the local picture of the embedding of F_H over a 2×2 block for the five possibilities listed in Lemma 3.4.3, with $a = 2$.

We will compute the Euler characteristic of F_H via a formula for any surface-with-boundary obtained by gluing squares, in the following sense:

DEFINITION 3.4.6. A ***nearly flattened surface*** is a topological space F which is obtained as a disjoint union of oriented squares, which are identified along certain pairs of edges via orientation-reversing maps; and only edges of different squares are identified. The resulting space F is naturally a CW complex, with 0-cells corresponding to the corners in the squares (modulo identifications), 1-cells corresponding to edges of the squares (possibly identified in pairs), and 2-cells corresponding to squares. A ***flattened surface*** is a nearly flattened surface with the property that every 0-cell which is not contained on the boundary is a corner for exactly four rectangles.

A flattened surface is homeomorphic to a compact two-manifold with boundary.

LEMMA 3.4.7. *Let F be a flattened surface. At each corner $p \in F$ (i.e. each point of F coming from some corner of some square), let n_p denote the number of squares which meet at p. Let $C_{\partial F}$ denote the set of corner points in ∂F. Then, the Euler characteristic of F is computed as*

$$\chi(F) = \sum_{p \in C_{\partial F}} \left(\frac{1}{2} - \frac{n_p}{4}\right). \tag{3.4}$$

Proof. Take a sum over all the squares in F with the following weights: each square is counted with weight 1, each edge on each square with weight $-\frac{1}{2}$, and each corner (on each square) is counted with weight $\frac{1}{4}$. Adding up these weights, each 2-cell is counted with weight 1 (which is the contribution of each 2-cell to $\chi(F)$), each interior edge with total weight -1 (the contribution of the corresponding 1-cell to $\chi(F)$), and each boundary edge with weight $-\frac{1}{2}$ (which is $\frac{1}{2}$ greater than the contribution to $\chi(F)$), and each corner point with weight $\frac{n_p}{4}$. Since the total contribution of each square vanishes, we conclude that

$$\chi(F) = \sum_{p \in C_{\partial F}} \left(1 - \frac{n_p}{4}\right) - \frac{1}{2}\#\{e \subset \partial F\}.$$

Since the Euler characteristic of the boundary vanishes, and it is computed by $\#\{p \in C_{\partial F}\} - \#\{e \subset \partial F\}$, we can subtract half this Euler characteristic to deduce the claimed formula. $\square$

DEFINITION 3.4.8. Given a square Q in the grid marked with an X or an O, let $\theta(Q, H)$ denote the average of the four $h_{i,j}$ adjoining Q. Given a matrix H, let $\theta(H) = \sum_{X \in \mathbb{X}} \theta(X, H) + \sum_{O \in \mathbb{O}} \theta(O, H)$.

PROPOSITION 3.4.9. *Fix a grid diagram $\mathbb{G}$ for a knot K with grid number n, and let H be any minimal matrix with non-negative entries associated to $\mathbf{W}(\mathbb{G})$. The Euler characteristic of F_H is given by $n - \theta(H)$; so the genus of F_H is given by*

$$g(F_H) = \frac{1}{2}\theta(H) - \frac{n-1}{2}. \tag{3.5}$$

Proof. Clearly, F_H is a nearly flattened surface. We can check that it is a flattened surface by analyzing the local picture above each 2×2 block in H. When all four local multiplicities equal to a, the center point lifts to a different 0-cells, none of which is contained in the boundary, and each of which appears as the corner of exactly four rectangles. When three of the local multiplicities equal one another, the center point lifts to a single corner point contained on the boundary of the surface. When two of the local multiplicities are a and the other two are $a+1$, the center point lifts to a different interior 0-cells, and a single 0-cell on the boundary, which is contained in two edges. Finally, when there are three different local multiplicities a, $a+1$, and $a+2$, the center point lifts to a interior 0-cells, and two 0-cells appearing on the boundary, and each is contained in two 1-cells.

Consider Equation (3.4), which, according to Lemma 3.4.7, computes the Euler characteristic of F_H. The boundary points for F_H for which $n_p \neq 2$ (i.e. for

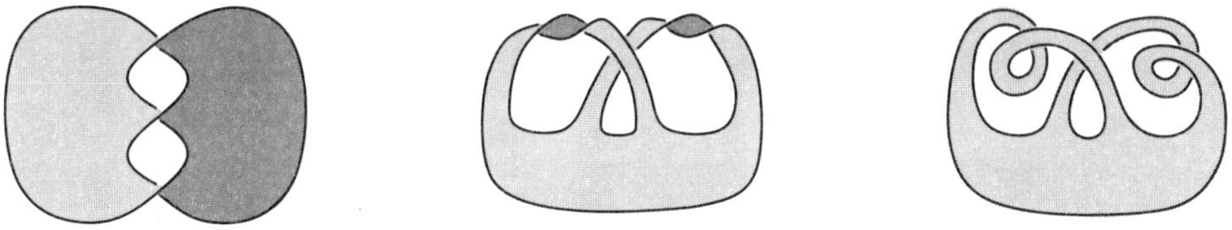

FIGURE 3.22. **Isotopy of a Seifert surface.** A Seifert surface of the right-handed trefoil knot (shown on the left) is isotoped to a disk with 1-handles attached to it (in the middle). In the final figure, further isotopies are applied so that the projection is an orientation preserving immersion.

which the contribution to the right-hand-side of Equation (3.4) does not vanish) are exactly those $2n$ points which are marked with an O or an X; i.e. those which lie over the center point of the 2×2 blocks where exactly three of the local multiplicities are equal to one another. For these points, n_p is the sum of the local multiplicities at the four adjacent entries. Lemma 3.4.7 gives the stated result. □

By considering various grid presentations of the fixed knot K, and various ways to turn $\mathbf{W}(\mathbb{G})$ into a minimal, non-negative matrix, the above algorithm provides a large collection of Seifert surfaces.

COROLLARY 3.4.10. *For a fixed grid diagram $\mathbb{G}$ the Euler characteristic of F_H is independent of the choice of the minimal, non-negative matrix H (obtained by adding and subtracting rows to $\mathbf{W}(\mathbb{G})$) used in its construction.*

Proof. Observe that $\theta(M)$ grows by 2 whenever we add a row or a column to the matrix M, and also the complexity increases by n. It follows at once from Lemma 3.4.2 that for two minimal complexity, non-negative matrices H and H' derived from $\mathbf{W}(\mathbb{G})$, $\theta(H) = \theta(H')$. By Proposition 3.4.9 the Euler characteristic of F_H can be computed from $\theta(H)$ and the grid number n, implying the claim. □

The above corollary shows that each grid diagram $\mathbb{G}$ representing a knot K determines an integer $g(\mathbb{G})$, the *associated genus* of $\mathbb{G}$, which is the genus of any Seifert surface of K constructed from $\mathbb{G}$ by the above procedure.

PROPOSITION 3.4.11. *If K is a knot with Seifert genus g, then there is a grid diagram $\mathbb{G}$ for K whose associated genus is g.*

Proof. Any Seifert surface F for K can be thought of as obtained from a disk by adding handles. After isotopy, we can think of these handles as very thin bands. After further isotopies, we can assume that the Seifert surface immerses orientation preservingly onto the plane; see Figure 3.22, for example. Approximate the cores of the one-handles so that their projections consist of horizontal and vertical segments only. Performing a further local move as in Figure 3.2, we can arrange that for all crossings, the vertical segments are overcrossings. Approximate the result to get a grid diagram $\mathbb{G}$, and a surface F_0, isotopic to the original F, which projects onto $\mathbb{G}$. The projection of the Seifert surface produces a matrix H_0 all of whose coefficients are 0, 1, and 2; the coefficients of 2 correspond to the intersections of the projections of the bands. We claim that the genus of the Seifert surface F is greater than or equal to the genus associated to the grid. Indeed, the genus of F_0 is computed by the same formula as in Equation (3.5):

$$g(F_0) = \frac{1}{2}\theta(H_0) - \frac{n-1}{2}.$$

Lemma 3.4.2 gives a minimal complexity non-negative matrix H with

$$H + \sum_i m_i C_i + \sum_j n_j R_j = H_0,$$

and $\sum_i m_i + n_i \geq 0$. Since $\theta(H) + 2(\sum m_i + n_i) = \theta(H_0)$, it follows that $\theta(H) \leq \theta(H_0)$, and (by Proposition 3.4.9), $g(F_H) \leq g(F_0)$. Applying this reasoning to a surface F with minimal genus (among Seifert surfaces for K), we conclude that $g(F_H) = g(F_0)$. □

EXERCISE 3.4.12. Find a Seifert surface for the trefoil knot $T_{2,3}$ with the method above, using the grid diagram of Figure 3.3. Do the same for the figure-eight knot, using the grid of Figure 3.1.

EXAMPLES 3.4.13. We demonstrate the above construction by two other examples. Let us first consider the Conway knot (of Figure 2.7), represented by the grid diagram of Figure 3.4. The winding matrix $\mathbf{W}(\mathbb{G})$ is now equal to

$$\begin{pmatrix}
0 & 0 & 0 & 0 & 0 & 0 & -1 & -1 & -1 & -1 & -1 \\
0 & 0 & 0 & -1 & -1 & -1 & -2 & -2 & -1 & -1 & -1 \\
0 & 0 & 0 & -1 & -1 & -1 & -2 & -3 & -2 & -2 & -1 \\
0 & 0 & 0 & -1 & 0 & 0 & -1 & -2 & -2 & -2 & -1 \\
0 & 0 & 0 & -1 & 0 & 0 & 0 & -1 & -1 & -2 & -1 \\
0 & 0 & 0 & -1 & 0 & 0 & 0 & 0 & 0 & -1 & 0 \\
0 & 0 & -1 & -2 & -1 & 0 & 0 & 0 & 0 & -1 & 0 \\
0 & -1 & -2 & -2 & -1 & 0 & 0 & 0 & 0 & -1 & 0 \\
0 & -1 & -2 & -2 & -2 & -1 & -1 & -1 & -1 & -1 & 0 \\
0 & -1 & -1 & -1 & -1 & 0 & 0 & 0 & 0 & 0 & 0 \\
0 & 0 & 0 & 0 & 0 & 0 & 0 & 0 & 0 & 0 & 0
\end{pmatrix}.$$

Adding C_i's to the columns with multiplicities $(0, 1, 2, 2, 2, 1, 2, 3, 2, 2, 1)$ we get a non-negative matrix, and then adding C_1 and subtracting R_6 and R_{11} we get a minimal non-negative matrix. Using this matrix the construction provides a Seifert surface of genus three for the Conway knot.

In a similar manner, we consider the Kinoshita-Terasaka knot of Figure 2.7, and represent it by the grid diagram we get from the grid of Figure 3.4 after commuting the first two columns. Then, after taking the winding matrix, and adding C_i's to the columns with multiplicities $(0, 0, 0, 1, 0, 1, 1, 1, 2, 1, 1)$, then adding C_1 and subtracting R_8 and R_{10} we get the non-negative minimal matrix

$$\begin{pmatrix} 1 & 0 & 0 & 1 & 0 & 0 & 0 & 0 & 1 & 0 & 0 \\ 1 & 0 & 0 & 1 & 1 & 1 & 0 & 0 & 1 & 0 & 0 \\ 1 & 0 & 0 & 1 & 1 & 2 & 1 & 1 & 2 & 1 & 0 \\ 1 & 1 & 1 & 2 & 2 & 3 & 2 & 2 & 2 & 1 & 0 \\ 1 & 1 & 0 & 1 & 1 & 2 & 1 & 1 & 1 & 1 & 0 \\ 1 & 1 & 0 & 0 & 0 & 1 & 0 & 0 & 0 & 0 & 0 \\ 1 & 1 & 0 & 0 & 0 & 1 & 1 & 1 & 1 & 1 & 1 \\ 0 & 0 & 0 & 0 & 0 & 1 & 1 & 0 & 0 & 0 & 0 \\ 1 & 1 & 1 & 1 & 0 & 1 & 1 & 0 & 1 & 1 & 1 \\ 0 & 0 & 0 & 1 & 0 & 1 & 1 & 0 & 1 & 0 & 0 \\ 1 & 0 & 0 & 1 & 0 & 1 & 1 & 1 & 2 & 1 & 1 \end{pmatrix}.$$

A simple calculation shows that the corresponding Seifert surface has genus two.

3.5. Grid diagrams and the fundamental group

A planar grid diagram determines a simple presentation of the link group $\pi_1(S^3 \setminus L)$ of the underlying link as follows.

The generators $\{x_1, \dots, x_n\}$ correspond to the vertical segments in the grid diagram (connecting the O- and the X-markings). The relations $\{r_1, \dots, r_{n-1}\}$ correspond to the horizontal lines separating the rows. The relation r_j is the product of the generators corresponding to those vertical segments of the link diagram which meet the jth horizontal line, in the order they are encountered, from left to right. See Figure 3.23.

LEMMA 3.5.1. *The presentation*

$$\langle x_1, \dots, x_n \mid r_1, \dots, r_{n-1} \rangle \tag{3.6}$$

described above is a presentation of the link group $\pi_1(S^3 \setminus L)$ of L.

Proof. The result follows from the Seifert-Van Kampen theorem for a suitable decomposition of the link complement into two open subsets. (See for instance [**137**].) To visualize this decomposition, consider the planar grid $\mathbb{G}$ and assume that the link is isotoped into the following position. In the usual coordinates (x, y, z) of $\mathbb{R}^3$ (with the understanding that the planar grid lies in the plane $\{z = 0\}$) we assume that the horizontal segments of the grid presenting L are in the plane $\{z = 0\}$, while the vertical segments are in the plane $\{z = 1\}$. Over the markings $\mathbb{X}$ and $\mathbb{O}$, these segments are joined by segments parallel to the z-axis. The resulting polygon in $\mathbb{R}^3$ is a PL representative of L.

Take $X_1 = \{(x, y, z) \in \mathbb{R}^3 \setminus L \mid z > 0\}$ and $X_2 = \{(x, y, z) \in \mathbb{R}^3 \setminus L \mid z < 1\}$, decomposing the knot complement into two path-connected open subsets, in such a way that $X_1 \cap X_2$ is also path-connected. Fix the basepoint x_0 on the plane $\{z = \frac{1}{2}\}$. By choosing convenient generators of the free groups $\pi_1(X_1, x_0)$ and $\pi_1(X_2, x_0)$, the Seifert-Van Kampen theorem provides the desired presentation of the link group $\pi_1(\mathbb{R}^3 \setminus L, x_0) = \pi_1(S^3 \setminus L, x_0)$. □

EXAMPLE 3.5.2. Consider the planar grid diagram of Figure 3.3, representing the right-handed trefoil knot $T_{2,3}$. The knot group G has the presentation

$$\langle x_1, x_2, x_3, x_4, x_5 \mid x_1x_3, \quad x_1x_2x_3x_4, \quad x_1x_2x_4x_5, \quad x_2x_5 \rangle.$$

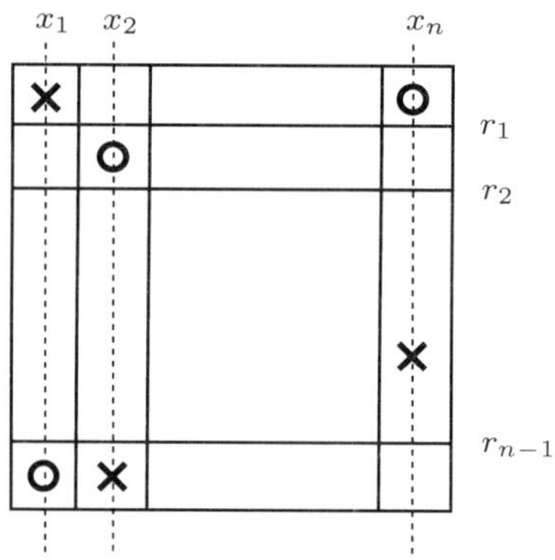

FIGURE 3.23. **The presentation of the fundamental group of the knot complement from a grid diagram.**

Since $x_3 = x_1^{-1}$, $x_5 = x_2^{-1}$, and $x_4 = x_1 x_2^{-1} x_1^{-1}$, G has the simpler presentation $\langle x_1, x_2 \mid x_1 x_2 x_1 = x_2 x_1 x_2 \rangle$. Taking $u = x_1 x_2$ and $v = x_2 x_1 x_2$, the above presentation is equivalent to $\langle u, v \mid u^3 = v^2 \rangle$.

EXERCISE 3.5.3. Using the grid diagram of Figure 3.5, find a presentation of the link group of the Borromean rings.

CHAPTER 4

Grid homology

The aim of the present chapter is to define the chain complexes for computing grid homology, following [**135, 136**]. We define three versions: $\widetilde{GC}(\mathbb{G})$, $\widehat{GC}(\mathbb{G})$, and $GC^-(\mathbb{G})$. The first of these is the simplest, and the first two are both specializations of the last one, which in turn is a specialization of a more complicated algebraic object $\mathcal{GC}^-(\mathbb{G})$ that we will meet in Chapter 13. In this chapter, and in fact, all the way until Chapter 8, we will consider primarily the case of knots.

This chapter is organized as follows. Section 4.1 introduces *grid states*, the generators of the grid chain complexes. Differentials count rectangles in the torus, and in Section 4.2 we describe how rectangles can connect grid states. In Section 4.3, we define two functions, the Maslov function and the Alexander function on grid states; these functions will induce the bigradings on the grid complexes. In Section 4.4 we define the grid complex $\widetilde{GC}$, the variant with the minimal amount of algebraic structure. In Section 4.5, we give a quick overview of some of the basic constructions from homological algebra (chain complexes, chain homotopies, quotient complexes) which will be of immediate use. (For more, see Appendix A.) In Section 4.6, we define further versions of the grid complex GC^- and $\widehat{GC}$. In Section 4.7, we interpret the Alexander function in terms of the winding number, leading to an expression of the Euler characteristic of $\widetilde{GH}$ and $\widehat{GH}$ in terms of the Alexander polynomial. Section 4.8 gives some concrete calculations of grid homology. In Section 4.9, we conclude with some remarks relating the combinatorial constructions with analogous holomorphic constructions.

4.1. Grid states

Consider a toroidal grid diagram for a knot K with grid number n, as described in Section 3.1. Think of each square in the grid as bounded by two horizontal and two vertical arcs. The horizontal arcs can be assembled to form n horizontal circles in the torus, denoted $\boldsymbol{\alpha} = \{\alpha_i\}_{i=1}^n$, and the vertical ones can be assembled to form n vertical circles, denoted $\boldsymbol{\beta} = \{\beta_i\}_{i=1}^n$.

DEFINITION 4.1.1. A ***grid state*** for a grid diagram $\mathbb{G}$ with grid number n is a one-to-one correspondence between the horizontal and vertical circles. More geometrically, a grid state is an n-tuple of points $\mathbf{x} = \{x_1, \dots, x_n\}$ in the torus, with the property that each horizontal circle contains exactly one of the elements of $\mathbf{x}$ and each vertical circle contains exactly one of the elements of $\mathbf{x}$. The set of grid states for $\mathbb{G}$ is denoted $\mathbf{S}(\mathbb{G})$.

A grid state $\mathbf{x}$ can be thought of as a graph of a permutation; i.e. if $\mathbf{x} = \{x_1, \dots, x_n\}$, then $\sigma = \sigma_{\mathbf{x}}$ is specified by the property that $x_i = \alpha_{\sigma(i)} \cap \beta_i$. The correspondence between grid states and permutations is, of course, not canonical: it depends on a numbering of the horizontal and vertical circles.

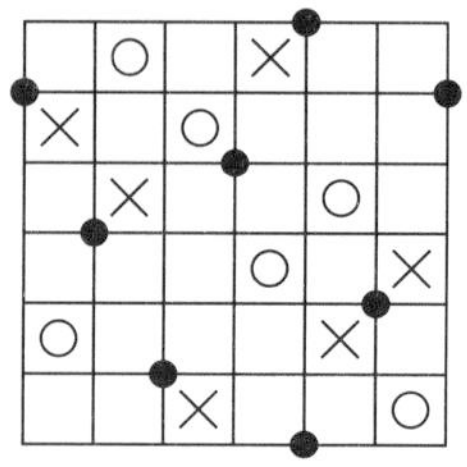

FIGURE 4.1. **A grid state in $\mathbf{S}(\mathbb{G})$.** Labelling the circles from left to right and bottom to top in this picture, the grid state corresponds to the permutation $(1,2,3,4,5,6) \mapsto (6,4,2,5,1,3)$.

When illustrating the diagrams and the states, we use planar grids; that is, cut the toroidal grid along a vertical and a horizontal circle. The square obtained by cutting up the torus is a *fundamental domain* for the torus, and the induced planar grid diagram is the planar realization of the grid diagram. Figure 4.1 illustrates a grid state in a grid diagram for the figure-eight knot. To emphasize the side identifications used in going from the planar to the toroidal grid, we repeat components of the grid state on the left and the right edge, and the top and the bottom edge.

4.2. Rectangles connecting grid states

The chain complexes associated to a grid diagram are generated by grid states, and their differentials count rectangles connecting states. The various versions of the grid complex differ in how they count rectangles. We formalize the concept of connecting rectangles, as follows.

Fix two grid states $\mathbf{x}, \mathbf{y} \in \mathbf{S}(\mathbb{G})$, and an embedded rectangle r in the torus whose boundary lies in the union of the horizontal and vertical circles, satisfying the following relationship. The sets $\mathbf{x}$ and $\mathbf{y}$ overlap in $n-2$ points in the torus, and the four points left out are the four corners of r. There is a further condition stated in terms of the orientation r inherits from the torus. The oriented boundary of r consists of four oriented segments, two of which are vertical and two of which are horizontal. The rectangle r *goes from* $\mathbf{x}$ *to* $\mathbf{y}$ if the horizontal segments in ∂r point from the components of $\mathbf{x}$ to the components of $\mathbf{y}$, while the vertical segments point from the components of $\mathbf{y}$ to the components of $\mathbf{x}$.

More formally, if r is a rectangle, let $\partial_\alpha r$ denote the portion of the boundary of r in the horizontal circles $\alpha_1 \cup \ldots \cup \alpha_n$, and let $\partial_\beta r$ denote the portion of the boundary of r in the vertical ones. The boundary inherits an orientation from r. The rectangle r goes from $\mathbf{x}$ to $\mathbf{y}$ if

$$\partial(\partial_\alpha r) = \mathbf{y} - \mathbf{x} \qquad \text{and} \qquad \partial(\partial_\beta r) = \mathbf{x} - \mathbf{y},$$

where $\mathbf{x} - \mathbf{y}$ is thought of as a formal sum of points; e.g. at points in $p \in \mathbf{x} \cap \mathbf{y}$, the difference cancels.

If there is a rectangle from $\mathbf{x}$ to $\mathbf{y}$, then all but two points in $\mathbf{x}$ must also be in $\mathbf{y}$. This is equivalent to the condition that the associated permutations $\sigma_{\mathbf{x}}$ and $\sigma_{\mathbf{y}}$ satisfy the property that $\sigma_{\mathbf{x}}^{-1} \cdot \sigma_{\mathbf{y}}$.

ξ associated to $\mathbf{x}$ and the permutation η associated to $\mathbf{y}$ are related by the property that $\xi \cdot \eta^{-1}$ is a transposition.

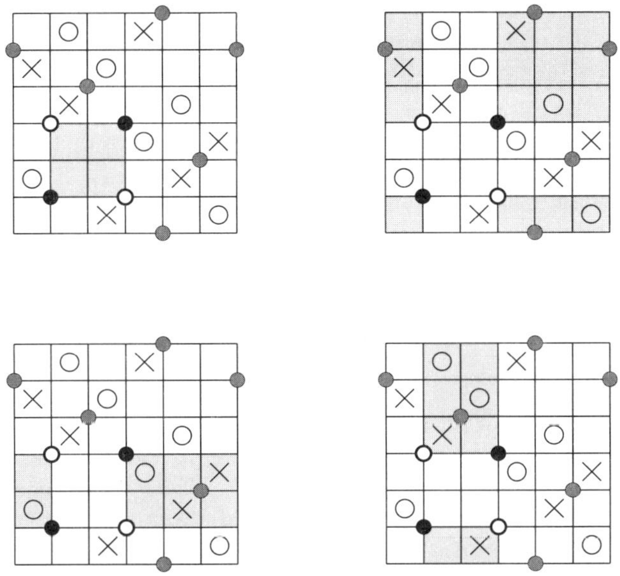

FIGURE 4.2. **Two grid states and the four rectangles connecting them.** Black dots appear in only one, call it $\mathbf{x}$; white dots appear in only the other one, call it $\mathbf{y}$; and gray dots appear in both. The two rectangles on the top row go from $\mathbf{x}$ to $\mathbf{y}$, and the other two go from $\mathbf{y}$ to $\mathbf{x}$. The top left rectangle is empty, and the other three are not.

For $\mathbf{x}, \mathbf{y} \in \mathbf{S}(\mathbb{G})$, let $\mathrm{Rect}(\mathbf{x}, \mathbf{y})$ denote the set of rectangles from $\mathbf{x}$ to $\mathbf{y}$. The set $\mathrm{Rect}(\mathbf{x}, \mathbf{y})$ is either empty, or it consists of exactly two rectangles, in which case $\mathrm{Rect}(\mathbf{y}, \mathbf{x})$ also consists of two rectangles. See Figure 4.2 for an illustration.

When we speak of a "rectangle", we will think of it as the geometric subset of the torus, together with the initial and the terminal grid states $\mathbf{x}$ and $\mathbf{y}$. Thus, if $\mathbf{x} \neq \mathbf{x}'$, a rectangle from $\mathbf{x}$ to $\mathbf{y}$ is always thought of as different from a rectangle from $\mathbf{x}'$ to $\mathbf{y}'$, even if their underlying polygons in the torus are the same. The underlying polygon is called the *support* of the rectangle.

Label the four corners of any rectangle as *northeast*, *southeast*, *northwest*, and *southwest*. This can be done, for example, by lifting r to the universal cover, which inherits a preferred pair of coordinates: the horizontal direction which is oriented eastward, and the vertical direction, which is oriented northward, following the conventions of Section 3.2. Sometimes we refer to the northwest corner as the upper left one, and the southeast corner as the lower right one.

If r is a rectangle from $\mathbf{x}$ to $\mathbf{y}$, then r contains elements of $\mathbf{x}$ and $\mathbf{y}$ on its boundary. The northeast and southwest corners of r are elements of $\mathbf{x}$, called *initial corners*, and the southeast and northwest corners of r are elements of $\mathbf{y}$, called *terminal corners*. The rectangle r might in addition contain elements of $\mathbf{x}$ in its interior $\mathrm{Int}(r)$. Note that $\mathbf{x} \cap \mathrm{Int}(r) = \mathbf{y} \cap \mathrm{Int}(r)$.

The following rectangles will play a special role in our subsequent constructions:

DEFINITION 4.2.1. A rectangle $r \in \mathrm{Rect}(\mathbf{x}, \mathbf{y})$ is called an ***empty rectangle*** if

$$\mathbf{x} \cap \mathrm{Int}(r) = \mathbf{y} \cap \mathrm{Int}(r) = \emptyset.$$

The set of empty rectangles from $\mathbf{x}$ to $\mathbf{y}$ is denoted $\mathrm{Rect}^\circ(\mathbf{x},\mathbf{y})$.

4.3. The bigrading on grid states

The grid complexes are equipped with two gradings called the *Maslov grading* and the *Alexander grading*, both induced by integral-valued functions on grid states for a toroidal grid diagram. The aim of this section is to construct these functions. Key properties of both functions are stated in the next two propositions, whose proofs will occupy the rest of the section. We start with the Maslov function.

PROPOSITION 4.3.1. *For any toroidal grid diagram* $\mathbb{G}$, *there is a function*

$$M_{\mathbb{O}}\colon \mathbf{S}(\mathbb{G}) \to \mathbb{Z},$$

called the ***Maslov function on grid states***, *which is uniquely characterized by the following two properties:*

(M-1) Let $\mathbf{x}^{NWO}$ *be the grid state whose components are the upper left corners of the squares marked with* O. *Then,*

$$M_{\mathbb{O}}(\mathbf{x}^{NWO}) = 0. \tag{4.1}$$

(M-2) If $\mathbf{x}$ *and* $\mathbf{y}$ *are two grid states that can be connected by some rectangle* $r \in \mathrm{Rect}(\mathbf{x},\mathbf{y})$, *then*

$$M_{\mathbb{O}}(\mathbf{x}) - M_{\mathbb{O}}(\mathbf{y}) = 1 - 2\#(r\cap\mathbb{O}) + 2\#(\mathbf{x}\cap \mathrm{Int}(r)). \tag{4.2}$$

Note that $M_{\mathbb{O}}$ is independent of the placement of the X-markings. There is another function, $M_{\mathbb{X}}$ defined as in Proposition 4.3.1, only using the X-markings in place of the O-markings. Unless explicitly stated otherwise, the Maslov function on states refers to $M_{\mathbb{O}}$; and we will usually drop $\mathbb{O}$ from its notation.

DEFINITION 4.3.2. The ***Alexander function on grid states*** is defined in terms of the Maslov functions by the formula

$$A(\mathbf{x}) = \frac{1}{2}(M_{\mathbb{O}}(\mathbf{x}) - M_{\mathbb{X}}(\mathbf{x})) - \left(\frac{n-1}{2}\right). \tag{4.3}$$

Key properties of the Alexander function are captured in the following:

PROPOSITION 4.3.3. *Let* $\mathbb{G}$ *be a toroidal grid diagram for a knot. The function* A *is characterized, up to an overall additive constant, by the following property. For any rectangle* $r\in\mathrm{Rect}(\mathbf{x},\mathbf{y})$ *connecting two grid states* $\mathbf{x}$ *and* $\mathbf{y}$,

$$A(\mathbf{x}) - A(\mathbf{y}) = \#(r\cap\mathbb{X}) - \#(r\cap\mathbb{O}). \tag{4.4}$$

Furthermore, A *is integral valued.*

We prove Proposition 4.3.1 first. This is done by constructing a candidate function for $M_{\mathbb{O}}$, and verifying that it has the properties specified in Proposition 4.3.1. The candidate function is defined in terms of a planar realization of the toroidal grid, using the following construction.

DEFINITION 4.3.4. Consider the partial ordering on points in the plane $\mathbb{R}^2$ specified by $(p_1,p_2) < (q_1,q_2)$ if $p_1<q_1$ and $p_2<q_2$. If P and Q are sets of finitely many points in the plane, let $\mathcal{I}(P,Q)$ denote the number of pairs $p\in P$ and $q\in Q$ with $p<q$. We symmetrize this function, defining

$$\mathcal{J}(P,Q) = \frac{\mathcal{I}(P,Q)+\mathcal{I}(Q,P)}{2}.$$

Consider a fundamental domain $[0, n) \times [0, n)$ for the torus in the plane, with its left and bottom edges included. A grid state $\mathbf{x} \in \mathbf{S}(\mathbb{G})$ can be viewed as a collection of points with integer coordinates in this fundamental domain. Similarly, $\mathbb{O} = \{O_i\}_{i=1}^n$ can be viewed as a collection of points in the plane with half-integer coordinates in the fundamental domain.

During the course of our proof, we will find that $M_{\mathbb{O}}$ is given by the formula

$$M_{\mathbb{O}}(\mathbf{x}) = \mathcal{J}(\mathbf{x}, \mathbf{x}) - 2\mathcal{J}(\mathbf{x}, \mathbb{O}) + \mathcal{J}(\mathbb{O}, \mathbb{O}) + 1, \tag{4.5}$$

which we write more succinctly as

$$M_{\mathbb{O}}(\mathbf{x}) = \mathcal{J}(\mathbf{x} - \mathbb{O}, \mathbf{x} - \mathbb{O}) + 1,$$

thinking of $\mathcal{J}$ as extended bilinearly over formal sums and formal differences of subsets of the plane. Correspondingly, $M_{\mathbb{X}}$ is given by

$$M_{\mathbb{X}}(\mathbf{x}) = \mathcal{J}(\mathbf{x} - \mathbb{X}, \mathbf{x} - \mathbb{X}) + 1.$$

LEMMA 4.3.5. *Fix a planar realization of a toroidal grid diagram. The function* $M(\mathbf{x}) = \mathcal{J}(\mathbf{x} - \mathbb{O}, \mathbf{x} - \mathbb{O}) + 1$ *satisfies Properties* (M-1) *and* (M-2).

Proof. Let $NW(O_i)$ denote the northwest corner of the square marked with O_i, and then project it to the fundamental domain. Clearly,

$$\begin{aligned} M(\mathbf{x}^{NWO}) =& \#\{(i,j) | NW(O_i) < NW(O_j)\} - \#\{(i,j) | NW(O_i) < O_j\} \\ &- \#\{(i,j) | O_i < NW(O_j)\} + \#\{(i,j) | O_i < O_j\} + 1. \end{aligned} \tag{4.6}$$

To verify Equation (4.1), we count the number of times each pair (i, j) appears in the four sets on the right of Equation (4.6). We break this analysis into the following cases:

- If $i \neq j$ and neither O_i nor O_j is in the top row, then the following four inequalities are equivalent: $O_i < O_j$, $NW(O_i) < O_j$, $O_i < NW(O_j)$, and $NW(O_i) < NW(O_j)$.
- If O_j is in the top row and $i \neq j$, then $O_i < O_j$ is equivalent to $NW(O_i) < O_j$; while neither of $O_i < NW(O_j)$ nor $NW(O_i) < NW(O_j)$ can hold (since $NW(O_j)$ is in the bottom segment).
- If O_i is in the top row and $i \neq j$, neither $O_i < O_j$ nor $O_i < NW(O_j)$ can hold, while $NW(O_i) < O_j$ is equivalent to $NW(O_i) < NW(O_j)$.
- When $i = j$, there is exactly one O_i-marking for which $NW(O_i) < O_i$, when the O_i is in the top row. Note also that the three other inequalities $O_i < O_i$, $O_i < NW(O_i)$ and $NW(O_i) < NW(O_i)$ are never satisfied.

The total to the right-hand-side of Equation (4.6) from the first three cases are all 0, while the last case contributes -1. It follows that $M(\mathbf{x}^{NWO}) = 0$; i.e. M satisfies Property (M-1), as stated.

We verify that M satisfies Property (M-2), starting with the case where the rectangle r is contained in the fundamental domain for the torus used to define M. Label the southwest, northeast, northwest, and southeast corners of r by x_1, x_2, y_1, and y_2 respectively. Clearly,

$$\mathbf{x} = \{x_1, x_2\} \cup \mathbf{p} \text{ and } \mathbf{y} = \{y_1, y_2\} \cup \mathbf{p},$$

where $\mathbf{p} = \mathbf{x} \cap \mathbf{y}$. It is easy to see that

$$\begin{aligned}\mathcal{J}(\mathbf{x},\mathbf{x}) - \mathcal{J}(\mathbf{y},\mathbf{y}) = 1 &+ \#\{x \in \mathbf{p} \big| x > x_1\} + \#\{x \in \mathbf{p} \big| x > x_2\} \\ &+ \#\{x \in \mathbf{p} \big| x < x_1\} + \#\{x \in \mathbf{p} \big| x < x_2\} - \#\{x \in \mathbf{p} \big| x > y_1\} \\ &- \#\{x \in \mathbf{p} \big| x > y_2\} - \#\{x \in \mathbf{p} \big| x < y_1\} - \#\{x \in \mathbf{p} \big| x < y_2\} \\ = 1 &+ 2\{x \in \mathbf{p} \big| x_1 < x < x_2\} = 1 + 2\#(\mathbf{x} \cap \operatorname{Int}(r)).\end{aligned}$$

Above, the contribution of 1 comes from the pair $x_1 < x_2$. Similarly,

$$\begin{aligned}2\mathcal{J}(\mathbf{x},\mathbb{O}) - 2\mathcal{J}(\mathbf{y},\mathbb{O}) &= \#\{O_i \in \mathbb{O} \big| O_i > x_1\} + \#\{O_i \in \mathbb{O} \big| O_i > x_2\} \\ &+ \#\{O_i \in \mathbb{O} \big| O_i < x_1\} + \#\{O_i \in \mathbb{O} \big| O_i < x_2\} - \#\{O_i \in \mathbb{O} \big| O_i > y_1\} \\ &- \#\{O_i \in \mathbb{O} \big| O_i > y_2\} - \#\{O_i \in \mathbb{O} \big| O_i < y_1\} - \#\{O_i \in \mathbb{O} \big| O_i < y_2\} \\ &= 2\#\{O_i \in \mathbb{O} \big| x_1 < O_i < x_2\} = 2\#(\mathbb{O} \cap r).\end{aligned}$$

The above two equations imply that Equation (4.2) holds when r is contained in the fundamental domain used to define Equation (4.5).

Next suppose that r satisfies Equation (4.2). Suppose that $r' \in \operatorname{Rect}(\mathbf{y},\mathbf{x})$ is the rectangle with the property that r and r' meet along both horizontal edges, so that, in particular, both have the same width v. Then, since every column contains an O, and every vertical circle contains a component of $\mathbf{x}$, it follows that

$$\begin{aligned}\#(r' \cap \mathbb{O}) + \#(r \cap \mathbb{O}) &= v \\ \#(\mathbf{x} \cap \operatorname{Int}(r')) + \#(\mathbf{x} \cap \operatorname{Int}(r)) &= v - 1.\end{aligned}$$

These two equations, together with Equation (4.2) (for r), show that

$$M(\mathbf{y}) - M(\mathbf{x}) = 1 - 2\#(r' \cap \mathbb{O}) + 2\#(\mathbf{x} \cap \operatorname{Int}(r')),$$

which is the analogue of Equation (4.2) for r'.

In exactly the same manner, Equation (4.2) for r implies the same property for the rectangle that shares two vertical edges with r.

It follows that if Equation (4.2) holds for any rectangle $r \in \operatorname{Rect}(\mathbf{x},\mathbf{y})$, then the same holds for any other rectangle in $\operatorname{Rect}(\mathbf{x},\mathbf{y}) \cup \operatorname{Rect}(\mathbf{y},\mathbf{x})$. It is easy to see that at least one of the four rectangles in $\operatorname{Rect}(\mathbf{x},\mathbf{y}) \cup \operatorname{Rect}(\mathbf{y},\mathbf{x})$ is contained in the fundamental domain, for which we have already verified Equation (4.2); and hence the function defined in Equation (4.5) satisfies Property (M-2). □

Proof of Proposition 4.3.1. Lemma 4.3.5 verifies the existence of a function that satisfies Properties (M-1) and (M-2). To see that the function is uniquely characterized by these properties, observe that for any two grid states $\mathbf{x}$ and $\mathbf{y}$, there is a sequence of grid states $\mathbf{x} = \mathbf{x}_1, \mathbf{x}_2, \dots, \mathbf{x}_k = \mathbf{y}$ and rectangles $r_i \in \operatorname{Rect}(\mathbf{x}_i, \mathbf{x}_{i+1})$. This follows from the fact that the symmetric group is generated by transpositions. Thus the function $M(\mathbf{x})$ is uniquely determined up to an overall additive constant by Equation (4.2). Equation (4.1) specifies this constant. □

Note that Equation (4.5) specifies $M_{\mathbb{O}}$ using a fundamental domain; but the properties from Proposition 4.3.1 that uniquely characterize $M_{\mathbb{O}}$ make no reference to this choice. It follows that $M_{\mathbb{O}}$ is independent of the fundamental domain.

Next, we verify Equation (4.4), characterizing the Alexander function A.

Proof of Proposition 4.3.3. By Equation (4.2), if $r \in \mathrm{Rect}(\mathbf{x}, \mathbf{y})$ is any rectangle connecting the two grid states $\mathbf{x}$ and $\mathbf{y}$, then

$$M_{\mathbb{O}}(\mathbf{x}) - M_{\mathbb{O}}(\mathbf{y}) = 1 - 2\#(r \cap \mathbb{O}) + 2\#(\mathbf{x} \cap \mathrm{Int}(r))$$
$$M_{\mathbb{X}}(\mathbf{x}) - M_{\mathbb{X}}(\mathbf{y}) = 1 - 2\#(r \cap \mathbb{X}) + 2\#(\mathbf{x} \cap \mathrm{Int}(r))$$

Taking the difference of these two equations, and applying Equation (4.3), we conclude that Equation (4.4) holds. The function A is characterized up to an additive constant by Equation (4.4), since we can connect any two grid states by a sequence of rectangles.

The fact that M takes values in $\mathbb{Z}$ implies only that A takes values in $\frac{1}{2}\mathbb{Z}$. In view of Equation (4.4), to see that the Alexander function is integral, it suffices to show that there is one grid state $\mathbf{x}$ for which $A(\mathbf{x})$ is integral. Taking $\mathbf{x} = \mathbf{x}^{NWO}$, and using Equation (4.1), it suffices to show that

$$M_{\mathbb{X}}(\mathbf{x}^{NWO}) \equiv n - 1 \pmod 2. \tag{4.7}$$

To this end, we find a sequence of grid states $\mathbf{x}_i \in \mathbf{S}(\mathbb{G})$ for $i = 1, \ldots, n$, with the following properties:

- $\mathbf{x}_1 = \mathbf{x}^{NWX}$ is the grid state whose components are the northwest corners of the squares marked with X,
- $\mathbf{x}_n = \mathbf{x}^{NWO}$,
- there is a (not necessarily empty) rectangle connecting $\mathbf{x}_i$ to $\mathbf{x}_{i+1}$.

This sequence can be found, since the permutation that connects $\mathbf{x}^{NWO}$ to $\mathbf{x}^{NWX}$ is a cycle of length n (since the grid represents a knot), and such a cycle can be written as a product of $n-1$ transpositions. By Equation (4.1), $M_{\mathbb{X}}(\mathbf{x}_1) = 0$; so Equation (4.7) now follows from the mod 2 reduction of Equation (4.2). □

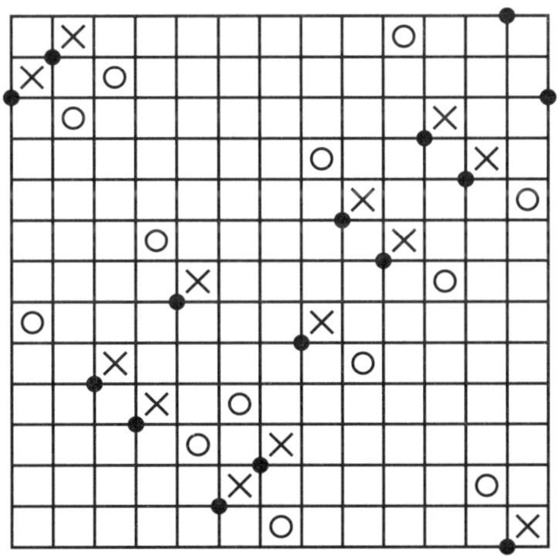

FIGURE 4.3. **Grid diagram for the 0-framed, negative Whitehead double of the left-handed trefoil knot.** The grid state $\mathbf{x}$ depicted in the diagram has Maslov grading 2.

EXERCISE 4.3.6. Consider the grid diagram $\mathbb{G}$ of Figure 4.3. Show that $\mathbb{G}$ represents $W_0^-(T_{-2,3})$, the 0-framed, negative Whitehead double of the left-handed trefoil knot. Determine the Maslov and Alexander gradings of the grid state $\mathbf{x}$ indicated in the diagram. This example will play a crucial role in Section 8.6.

The following result will be useful later:

PROPOSITION 4.3.7. *Let* $\mathbf{x}^{SWO}$ *be the grid state whose components are the lower left (SW) corners of the squares marked with* O. *Then,* $M(\mathbf{x}^{SWO}) = 1 - n$ *for any* $n \times n$ *grid.*

Proof. Using the formula $M(\mathbf{x}^{SWO}) = \mathcal{J}(\mathbf{x}^{SWO} - \mathbb{O}, \mathbf{x}^{SWO} - \mathbb{O}) + 1$, we see that almost all terms cancel in pairs, except for the n pairs O_i and their corresponding $SW(O_i)$. The result follows. □

EXERCISE 4.3.8. Let $\mathbb{G}$ be any grid diagram, and let $\mathbf{x}^{SEO}$ and $\mathbf{x}^{NEO}$, respectively, be the grid state whose components are the lower resp. the upper right corners of the squares marked with O. Compute $M(\mathbf{x}^{SEO})$ and $M(\mathbf{x}^{NEO})$.

4.4. The simplest version of grid homology

In the various grid complexes studied in the present book, the boundary maps count certain empty rectangles. The various constructions differ in how the empty rectangles are counted, in terms of how they interact with the $\mathbb{X}$- and $\mathbb{O}$-markings. (Compare also Section 5.5, where a different construction is outlined.) The simplest version of the grid complex is the following:

DEFINITION 4.4.1. The ***fully blocked grid chain complex*** associated to the grid diagram $\mathbb{G}$ is the chain complex $\widetilde{GC}(\mathbb{G})$, whose underlying vector space over $\mathbb{F} = \mathbb{Z}/2\mathbb{Z}$ has a basis corresponding to the set of grid states $\mathbf{S}(\mathbb{G})$, and whose differential is specified by

$$\tilde{\partial}_{\mathbb{O},\mathbb{X}}(\mathbf{x}) = \sum_{\mathbf{y}\in\mathbf{S}(\mathbb{G})} \#\{r \in \mathrm{Rect}^\circ(\mathbf{x},\mathbf{y}) \big| r \cap \mathbb{O} = r \cap \mathbb{X} = \emptyset\} \cdot \mathbf{y}. \tag{4.8}$$

Here $\#\{\cdot\}$ denotes the number of elements in the set modulo 2. (The subscript on $\tilde{\partial}_{\mathbb{O},\mathbb{X}}$ indicates the fact that the map counts rectangles that are disjoint from $\mathbb{O}$ and $\mathbb{X}$.)

The reader is invited to show that $\tilde{\partial}^2_{\mathbb{O},\mathbb{X}} = 0$. This is verified by interpreting the terms in $\tilde{\partial}^2_{\mathbb{O},\mathbb{X}}$ as counts of regions in the grid diagram that are compositions of two rectangles, and then showing that such regions have exactly two different decompositions into two rectangles, giving rise to pairwise cancelling terms in $\tilde{\partial}^2_{\mathbb{O},\mathbb{X}}$. A more general fact will be proved in Lemma 4.6.7 below.

The Maslov and Alexander functions on $\mathbf{S}(\mathbb{G})$ induce two gradings on $\widetilde{GC}(\mathbb{G})$: we define $\widetilde{GC}_d(\mathbb{G}, s)$ to be the $\mathbb{F}$-vector space generated by grid states $\mathbf{x}$ with $M(\mathbf{x}) = d$ and $A(\mathbf{x}) = s$. It follows quickly from Equations (4.2) and (4.4) that the restriction of $\tilde{\partial}_{\mathbb{O},\mathbb{X}}$ to $\widetilde{GC}_d(\mathbb{G}, s)$ maps into $\widetilde{GC}_{d-1}(\mathbb{G}, s)$. Thus, the bigrading descends to a bigrading on the homology groups of $\widetilde{GC}(\mathbb{G})$. Explicitly, letting

$$\widetilde{GH}_d(\mathbb{G}, s) = \frac{\mathrm{Ker}(\tilde{\partial}_{\mathbb{O},\mathbb{X}}) \cap \widetilde{GC}_d(\mathbb{G}, s)}{\mathrm{Im}(\tilde{\partial}_{\mathbb{O},\mathbb{X}}) \cap \widetilde{GC}_d(\mathbb{G}, s)},$$

then

$$\widetilde{GH}(\mathbb{G}) = \bigoplus_{d,s\in\mathbb{Z}} \widetilde{GH}_d(\mathbb{G}, s).$$

A *bigraded vector space* X is a vector space equipped with a splitting indexed by a pair of integers: $X = \bigoplus_{d,s\in\mathbb{Z}} X_{d,s}$. In this language, the Maslov and Alexander functions give $\widetilde{GH}(\mathbb{G})$ the structure of a bigraded vector space.

DEFINITION 4.4.2. The ***fully blocked grid homology*** of $\mathbb{G}$, denoted $\widetilde{GH}(\mathbb{G})$, is the homology of the chain complex $(\widetilde{GC}(\mathbb{G}), \tilde{\partial}_{\mathbb{O},\mathbb{X}})$, thought of as a bigraded vector space.

EXERCISE 4.4.3. Let $\mathcal{O}$ denote the unknot. Compute $\widetilde{GH}(\mathbb{G})$ for a 2×2 grid for $\mathcal{O}$. Compute $\widetilde{GH}(\mathbb{G})$ for a 3×3 grid for $\mathcal{O}$.

The above exercise demonstrates the fact that the total dimension of the homology $\widetilde{GH}(\mathbb{G})$ depends on the grid presentation of the knot. In fact, the following will be proved in Section 5.3:

THEOREM 4.4.4. *If $\mathbb{G}$ is a grid diagram with grid number n representing a knot K, then the renormalized dimension $\dim_{\mathbb{F}}(\widetilde{GH}(\mathbb{G}))/2^{n-1}$ is an integer-valued knot invariant; in particular, it is independent of the chosen grid presentation of K.*

4.5. Background on chain complexes

Theorem 4.4.4 might seem mysterious at this point. Indeed, even the fact that the dimension of $\widetilde{GH}(\mathbb{G})$ is divisible by 2^{n-1} is surprising. To verify this latter fact, it is helpful to enrich our coefficient ring to a polynomial algebra and to define a version of the grid complex over this algebra.

In the present section, we recall the necessary tools from homological algebra needed to study this enrichment. This material is essentially standard, with small modifications needed to accommodate the natural gradings arising in grid homology. More details, and proofs of some of these results, are provided in Appendix A.

Fix a commutative ring $\mathbb{K}$ with unit, which in our applications will be either $\mathbb{Z}$, the finite field $\mathbb{Z}/p\mathbb{Z}$ for some prime p, or $\mathbb{Q}$. In fact, through most of this text, we will take $\mathbb{K} = \mathbb{Z}/2\mathbb{Z} = \mathbb{F}$. Consider the polynomial ring $\mathcal{R} = \mathbb{K}[V_1, \ldots, V_n]$ in n formal variables $V_1, \ldots, V_n$. (We also allow $n = 0$, so that $\mathcal{R} = \mathbb{K}$.)

DEFINITION 4.5.1. A ***bigraded $\mathcal{R}$-module*** M is an $\mathcal{R}$-module, together with a splitting $M = \bigoplus_{d,s\in\mathbb{Z}} M_{d,s}$ as a $\mathbb{K}$-module, so that for each $i = 1, \ldots, n$, V_i maps $M_{d,s}$ into $M_{d-2,s-1}$. A ***bigraded $\mathcal{R}$-module homomorphism*** is a homomorphism $f\colon M \to M'$ between two bigraded $\mathcal{R}$-modules that sends $M_{d,s}$ to $M'_{d,s}$ for all $d, s \in \mathbb{Z}$. More generally, an $\mathcal{R}$-module homomorphism from $f\colon M \to M'$ is said to be ***homogeneous of degree*** (m, t) if it sends $M_{d,s}$ to $M'_{d+m,s+t}$ for all $d, s \in \mathbb{Z}$.

A ***bigraded chain complex over*** $\mathcal{R} = \mathbb{K}[V_1, \ldots, V_n]$ is a bigraded $\mathcal{R}$-module C, equipped with an $\mathcal{R}$-module homomorphism $\partial\colon C \to C$ with $\partial \circ \partial = 0$ that maps $C_{d,s}$ into $C_{d-1,s}$; in particular, ∂ is a homomorphism of $\mathcal{R}$-modules that is homogeneous of degree $(-1, 0)$.

The case where $n = 1$ will be of particular relevance to us. In this case, we write the algebra $\mathcal{R}$ simply as $\mathbb{K}[U]$. When $n = 0$ and $\mathbb{K} = \mathbb{F}$, the bigraded modules are bigraded vector spaces, the structures we encountered in Section 4.4.

DEFINITION 4.5.2. Let (C,∂) and (C',∂') be two bigraded chain complexes over $\mathcal{R}=\mathbb{K}[V_1,\dots,V_n]$. A ***chain map*** $f\colon(C,\partial)\to(C',\partial')$ is a homomorphism of $\mathcal{R}$-modules, satisfying the property that $\partial'\circ f=f\circ\partial$. The chain map f is called a ***bigraded chain map*** if it is also a bigraded homomorphism. More generally, a chain map is called ***homogeneous of degree*** (m,t) if the underlying homomorphism is bigraded of degree (m,t). An ***isomorphism*** of bigraded chain complexes is a bigraded chain map $f\colon(C,\partial)\to(C',\partial')$ for which there is another bigraded chain map $g\colon(C',\partial')\to(C,\partial)$ with $f\circ g=\mathrm{Id}_{C'}$ and $g\circ f=\mathrm{Id}_C$. If there is an isomorphism from (C,∂) to (C',∂'), we say that they are ***isomorphic bigraded chain complexes***, and write $(C,\partial)\cong(C',\partial')$.

A bigraded chain map $f\colon C\to C'$ between two bigraded chain complexes over $\mathcal{R}$ induces a well-defined bigraded map on homology, denoted $H(f)\colon H(C)\to H(C')$.

If (C,∂) and (C',∂') are bigraded chain complexes over $\mathcal{R}$, and $f\colon(C,\partial)\to(C',\partial')$ is a chain map, we can form the quotient complex $(C',\partial')/\mathrm{Im}(f)$, which is also a chain complex over $\mathcal{R}$. When f is homogeneous of degree (m,t), the quotient complex is also a bigraded chain complex of $\mathcal{R}$-modules.

For example, if (C,∂) is a bigraded chain complex over $\mathcal{R}=\mathbb{K}[V_1,\dots,V_n]$ with $n\geq 1$, then multiplication by V_i $i\in\{1,\dots,n\}$ is a chain map $V_i\colon(C,\partial)\to(C,\partial)$. In this case, the quotient complex is denoted $\frac{C}{V_i}$; or more suggestively $\frac{C}{V_i=0}$. This construction can be iterated; e.g. we can take the quotient of the chain complex by the map $V_j\colon\frac{C}{V_i}\to\frac{C}{V_i}$; the corresponding quotient will be denoted $\frac{C}{V_i=V_j=0}$.

A short exact sequence of chain complexes induces a long exact sequence on homology, according to the following:

LEMMA 4.5.3. *Let (C,∂), (C',∂'), and (C'',∂'') be three bigraded chain complexes over $\mathcal{R}=\mathbb{K}[V_1,\dots,V_n]$. Suppose that $f\colon C\to C'$ is a chain map which is homogeneous of degree (m,t), and $g\colon C'\to C''$ is a bigraded chain map, both of which fit into a short exact sequence*

$$0\longrightarrow C\xrightarrow{\ f\ }C'\xrightarrow{\ g\ }C''\longrightarrow 0.$$

Then, there is a homomorphism of $\mathcal{R}$-modules $\delta\colon H(C'')\to H(C)$ that is homogeneous of degree $(-m-1,-t)$, which fits into a long exact sequence

$$\cdots\longrightarrow H_{d-m,s-t}(C)\xrightarrow{H(f)}H_{d,s}(C')\xrightarrow{H(g)}H_{d,s}(C'')\xrightarrow{\ \delta\ }H_{d-m-1,s-t}(C)\longrightarrow\cdots$$

The proof of the above standard result is recalled in Appendix A; see Lemma A.2.1.

DEFINITION 4.5.4. Suppose that $f,g\colon(C,\partial)\to(C',\partial')$ are two bigraded chain maps between two bigraded chain complexes over $\mathcal{R}$. The maps f and g are said to be ***chain homotopic*** if there is an $\mathcal{R}$-module homomorphism $h\colon C\to C'$ that is homogeneous of degree $(1,0)$, and that satisfies the formula

$$f-g=\partial'\circ h+h\circ\partial. \tag{4.9}$$

In this case, h is called a ***chain homotopy from*** g ***to*** f. More generally, if $f,g\colon(C,\partial)\to(C',\partial')$ are two chain maps that are homogeneous of degree (m,t), they are called ***chain homotopic*** if there is a map $h\colon C\to C'$ that is an $\mathcal{R}$-module homomorphism homogeneous of degree $(m+1,t)$ and satisfies Equation (4.9).

It is easy to verify that chain homotopic maps induce the same map on homology.

DEFINITION 4.5.5. A chain map $f\colon C \to C'$ is a ***chain homotopy equivalence*** if there is a chain map $\phi\colon C' \to C$, called a ***chain homotopy inverse to*** f, with the property that $f \circ \phi$ and $\phi \circ f$ are both chain homotopic to the respective identity maps. If there is a chain homotopy equivalence from C to C', then C and C' are said to be ***chain homotopy equivalent*** complexes.

PROPOSITION 4.5.6. *Let C and C' be two bigraded chain complexes of $\mathcal{R} = \mathbb{K}[V_1, \dots, V_n]$-modules. A chain map $f\colon C \to C'$, homogeneous of degree (m,t), naturally induces a chain map $\overline{f}\colon \frac{C}{V_i} \to \frac{C}{V_i}$ that is also homogeneous of degree (m,t). Moreover, if g is another chain map that is homogeneous of degree (m,t), a chain homotopy h from f to g induces a chain homotopy $\overline{h}$ from $\overline{f}$ to $\overline{g}$.*

Proof. Any $\mathcal{R}$-module homomorphism $\phi\colon C \to C'$ induces an $\mathcal{R}$-module homomorphism $\overline{\phi}\colon \frac{C}{V_i} \to \frac{C'}{V_i}$. In this notation, the differential ∂ on C induces the differential $\overline{\partial}$ on $\frac{C}{V_i}$. Also, the chain maps f, g, and the chain homotopy h induce maps $\overline{f}$, $\overline{g}$, and $\overline{h}\colon \frac{C}{V_i} \to \frac{C'}{V_i}$. The relation $\overline{\partial}' \circ \overline{h} + \overline{h} \circ \overline{\partial} = \overline{f} - \overline{g}$ is a consequence of the relation $\partial' \circ h + h \circ \partial = f - g$. □

4.6. The grid chain complex GC^-

We now enrich the grid complex to a bigraded chain complex over the ring $\mathcal{R} = \mathbb{F}[V_1, \dots, V_n]$. Various specializations of this complex give rise to different versions of grid homology.

To define the enrichment, it is useful to enumerate the set $\mathbb{O} = \{O_i\}_{i=1}^n$. This puts the O-markings in one-to-one correspondence with the generators V_i of the polynomial algebra $\mathcal{R}$. Informally, the unblocked grid complex is the $\mathcal{R}$-module generated by grid states, equipped with a differential $\partial_{\mathbb{X}}^-$ counting empty rectangles that may cross the O- but not the X-markings. The *multiplicity* $O_i(r)$ of the rectangle r at the marking O_i is defined to be either 1 or 0, depending on whether or not r contains O_i. This multiplicity is recorded as the exponent of the formal variable V_i. More explicitly:

DEFINITION 4.6.1. The ***(unblocked) grid complex*** $GC^-(\mathbb{G})$ is the free module over $\mathcal{R}$ generated by $\mathbf{S}(\mathbb{G})$, equipped with the $\mathcal{R}$-module endomorphism whose value on any $\mathbf{x} \in \mathbf{S}(\mathbb{G})$ is given by

$$\partial_{\mathbb{X}}^- \mathbf{x} = \sum_{\mathbf{y} \in \mathbf{S}(\mathbb{G})} \sum_{\{r \in \mathrm{Rect}^\circ(\mathbf{x},\mathbf{y}) \,\big|\, r \cap \mathbb{X} = \emptyset\}} V_1^{O_1(r)} \cdots V_n^{O_n(r)} \cdot \mathbf{y}. \tag{4.10}$$

The elements $V_1^{k_1} \cdots V_n^{k_n} \cdot \mathbf{x}$ where $\mathbf{x} \in \mathbf{S}(\mathbb{G})$ and $k_1, \dots, k_n$ are arbitrary, non-negative integers form a basis for the $\mathbb{F}$-vector space $GC^-(\mathbb{G})$. Extend the Maslov and Alexander functions (Proposition 4.3.1 and Definition 4.3.2) to this basis by

$$M(V_1^{k_1} \dots V_n^{k_n} \cdot \mathbf{x}) = M(\mathbf{x}) - 2k_1 - \dots - 2k_n, \tag{4.11}$$

$$A(V_1^{k_1} \dots V_n^{k_n} \cdot \mathbf{x}) = A(\mathbf{x}) - k_1 - \dots - k_n. \tag{4.12}$$

These extensions equip $GC^-(\mathbb{G})$ with a bigrading: let $GC^-_d(\mathbb{G},s)$ denote the vector subspace spanned by basis vectors $V_1^{k_1}\cdots V_n^{k_n}\cdot\mathbf{x}$ with $M(V_1^{k_1}\cdots V_n^{k_n}\cdot\mathbf{x})=d$, and $A(V_1^{k_1}\cdots V_n^{k_n}\cdot\mathbf{x})=s$. If $x\in GC^-(\mathbb{G})$ lies in some $GC^-_d(\mathbb{G},s)$ for some d and s, we say that x is *homogeneous with bigrading* (d,s) or simply *homogeneous*. (Note that the element 0 is homogeneous with any bigrading.)

REMARK 4.6.2. In Chapter 13, we will study another variant of the grid complex, $\mathcal{GC}^-(\mathbb{G})$, which has the same underlying $\mathcal{R}$-module as $GC^-(\mathbb{G})$, a grading induced by M, and a differential specified by

$$\partial^-\mathbf{x}=\sum_{\mathbf{y}\in\mathbf{S}(\mathbb{G})}\sum_{r\in\mathrm{Rect}^\circ(\mathbf{x},\mathbf{y})}V_1^{O_1(r)}\cdots V_n^{O_n(r)}\cdot\mathbf{y}. \tag{4.13}$$

This complex has a filtration which is a knot invariant, and its total homology is isomorphic to $\mathbb{F}[U]$. The normalization of M specified by Equation (4.1) was chosen so that the generator of this homology module has grading equal to zero.

THEOREM 4.6.3. *The object $(GC^-(\mathbb{G}),\partial^-_{\mathbb{X}})$ is a bigraded chain complex over the ring $\mathbb{F}[V_1,\dots,V_n]$, in the sense of Definition* 4.5.1.

We break the proof of Theorem 4.6.3 into pieces, starting with the verification that $\partial^-_{\mathbb{X}}\circ\partial^-_{\mathbb{X}}=0$. To this end, it is convenient to generalize the notion of rectangles.

Recall that the circles $\alpha_1,\dots,\alpha_g,\beta_1,\dots,\beta_g$ divide the torus into oriented squares $S_1,\dots,S_{n^2}$. A formal linear combination of the closures of these squares, $\mathcal{D}=\sum a_i\cdot\overline{S_i}$ with $a_i\in\mathbb{Z}$, has a boundary, which is a formal linear combination of intervals contained inside $\alpha_1\cup\dots\cup\alpha_n\cup\beta_1\cup\dots\cup\beta_n$. Let $\partial_\alpha\mathcal{D}$ be the portion of the boundary contained in $\alpha_1\cup\dots\cup\alpha_n$ and $\partial_\beta\mathcal{D}$ be the portion in $\beta_1\cup\dots\cup\beta_n$.

DEFINITION 4.6.4. Fix $\mathbf{x},\mathbf{y}\in\mathbf{S}(\mathbb{G})$. A ***domain*** ψ from $\mathbf{x}$ to $\mathbf{y}$ is a formal linear combination of the closures of the squares in $\mathbb{G}\setminus(\boldsymbol{\alpha}\cup\boldsymbol{\beta})$, with the property that $\partial(\partial_\alpha\psi)=\mathbf{y}-\mathbf{x}$ and hence $\partial(\partial_\beta\psi)=\mathbf{x}-\mathbf{y}$. In these equations, the two sides represent a formal linear combinations of points; e.g. if $\mathbf{x}=\{x_1,\dots,x_n\}$ and $\mathbf{y}=\{y_1,\dots,y_n\}$, then $\mathbf{x}-\mathbf{y}=\sum_{i=1}^n(x_i-y_i)$. Denote the set of domains from $\mathbf{x}$ to $\mathbf{y}$ by $\pi(\mathbf{x},\mathbf{y})$. A domain ψ is called ***positive*** if each square in the torus (with its orientation inherited from the torus) appears in the expression for ψ with non-negative multiplicity.

REMARK 4.6.5. The grid diagram $\mathbb{G}$ equips the torus with a CW-decomposition, whose 0-cells are the n^2 intersection points of the horizontal and the vertical circles; its 1-cells are the $2n^2$ intervals on the horizontal and vertical circles between consecutive intersections of these circles, and its 2-cells are the n^2 small squares of the grid diagram. A formal sum ψ of rectangles is a 2-chain in this CW-complex structure. The group of 1-chains splits as the sum of the span of the horizontal intervals and vertical intervals. The 1-chain $\partial_\alpha\psi$ is the part of $\partial\psi$ in the span of the horizontal intervals, so the relation $\partial(\partial_\alpha\psi)=\mathbf{y}-\mathbf{x}$ is an equation of 0-chains.

Domains can be composed: if $\phi\in\pi(\mathbf{x},\mathbf{y})$ and $\psi\in\pi(\mathbf{y},\mathbf{z})$, then by adding the two underlying 2-chains we get a new domain, written $\phi*\psi\in\pi(\mathbf{x},\mathbf{z})$.

EXERCISE 4.6.6. **(a)** Show that any two $\mathbf{x},\mathbf{y}\in\mathbf{S}(\mathbb{G})$ can be connected by a domain $\psi\in\pi(\mathbf{x},\mathbf{y})$.
(b) Show that any two $\mathbf{x},\mathbf{y}\in\mathbf{S}(\mathbb{G})$ can be connected by a domain $\psi\in\pi(\mathbf{x},\mathbf{y})$ with $\mathbb{X}\cap\psi=\emptyset$.

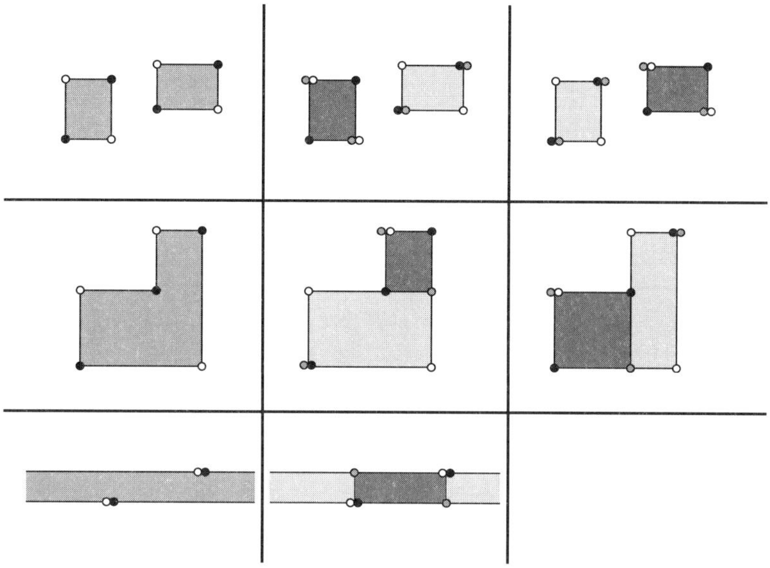

FIGURE 4.4. **Cases in the proof of Lemma 4.6.7.** The left column illustrates the three basic types of domains ψ with $N(\psi) > 0$ (Cases (R-1), (R-2), and (R-3), respectively). The initial grid state is indicated by black dots; the terminal one by the white dots. The second and third columns show the decompositions of the domain in the first column. The first rectangle in the decomposition is darker than the second. The intermediate grid state is indicated by gray dots. In the first row we consider the case of two disjoint rectangles; these rectangles can also overlap as in Figure 4.5.

(c) If $\mathbb{G}$ represents a knot, show that any domain $\psi \in \pi(\mathbf{x}, \mathbf{y})$ with $\mathbb{X} \cap \psi = \emptyset$ is uniquely determined by its multiplicities at the $\mathbb{O}$. What if $\mathbb{G}$ represents a link?

The next lemma will be used to establish Theorem 4.6.3. Its proof will serve as a prototype for many of the proofs from Chapter 5.

LEMMA 4.6.7. *The operator* $\partial_{\mathbb{X}}^- : GC^-(\mathbb{G}) \to GC^-(\mathbb{G})$ *satisfies* $\partial_{\mathbb{X}}^- \circ \partial_{\mathbb{X}}^- = 0$.

Proof. For grid states $\mathbf{x}$ and $\mathbf{z}$ fix $\psi \in \pi(\mathbf{x}, \mathbf{z})$ and (for the purposes of this proof) let $N(\psi)$ denote the number of ways we can decompose ψ as a composite of two empty rectangles $r_1 * r_2$. Observe that if $\psi = r_1 * r_2$ for some $r_1 \in \mathrm{Rect}(\mathbf{x}, \mathbf{y})$ and $r_2 \in \mathrm{Rect}(\mathbf{y}, \mathbf{z})$, the following statements hold:

- $\psi \cap \mathbb{X}$ is empty if and only if $r_i \cap \mathbb{X}$ is empty for both $i = 1, 2$.
- The local multiplicities of ψ, r_1, and r_2 at any $O_i \in \mathbb{O}$ are related by
$$O_i(\psi) = O_i(r_1) + O_i(r_2).$$

It follows that for any $\mathbf{x} \in \mathbf{S}(\mathbb{G})$,

$$\partial_{\mathbb{X}}^- \circ \partial_{\mathbb{X}}^-(\mathbf{x}) = \sum_{\mathbf{z} \in \mathbf{S}(\mathbb{G})} \sum_{\left\{\psi \in \pi(\mathbf{x}, \mathbf{z}) \mid \psi \cap \mathbb{X} = \emptyset\right\}} N(\psi) \cdot V_1^{O_1(\psi)} \cdots V_n^{O_n(\psi)} \cdot \mathbf{z}.$$

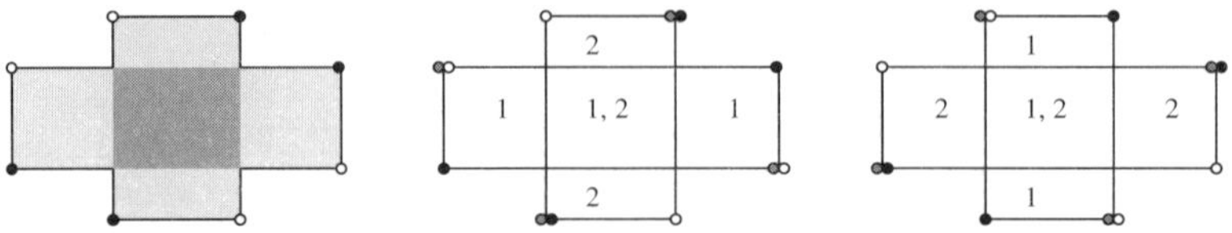

FIGURE 4.5. **Overlapping domains counted in $\partial_{\mathbb{X}}^- \circ \partial_{\mathbb{X}}^- = 0$.** Part of the domain on the left has local multiplicity two (indicated by the darker shading). The next two pictures show the two decompositions of this domain as a juxtaposition of two rectangles. The rectangles are labeled by integers 1 and 2, indicating their order in the decomposition.

Consider a pair of empty rectangles $r_1 \in \mathrm{Rect}^\circ(\mathbf{x},\mathbf{y})$ and $r_2 \in \mathrm{Rect}^\circ(\mathbf{y},\mathbf{z})$, so that $r_1 * r_2 = \psi$ is a domain with $N(\psi) > 0$. There are three basic cases (see also Figure 4.4 for an illustration):

(R-1) $\mathbf{x} \setminus (\mathbf{x} \cap \mathbf{z})$ consists of 4 elements. In this case, the corners of r_1 and r_2 are all distinct. There is a unique $\mathbf{y}' \in \mathbf{S}(\mathbb{G})$ and rectangles $r_1' \in \mathrm{Rect}^\circ(\mathbf{x},\mathbf{y}')$ and $r_2' \in \mathrm{Rect}^\circ(\mathbf{y}',\mathbf{z})$ so that r_1 and r_2' have the same support and r_2 and r_1' have the same support. See the top row of Figure 4.4 (and also Figure 4.5). Then, $r_1 * r_2 = r_1' * r_2'$ and in fact $N(\psi) = 2$.

(R-2) $\mathbf{x} \setminus (\mathbf{x} \cap \mathbf{z})$ consists of 3 elements. In this case, the local multiplicities of ψ are all 0 or 1 and the corresponding region in the torus has six corners, five of which are 90°, and one of which is 270°. Cutting at the 270° corner in two different directions gives the two decompositions of ψ as a juxtaposition of empty rectangles $\psi = r_1 * r_2 = r_1' * r_2'$, where $r_1 \in \pi(\mathbf{x},\mathbf{y})$, $r_2 \in \pi(\mathbf{y},\mathbf{z})$, $r_1' \in \pi(\mathbf{x},\mathbf{y}')$, and $r_2' \in \pi(\mathbf{y}',\mathbf{z})$ (with $\mathbf{y} \neq \mathbf{y}'$). In particular, $N(\psi) = 2$ in this case, as well. See the middle row of Figure 4.4.

(R-3) $\mathbf{x} = \mathbf{z}$. In this case, $\psi = r_1 * r_2$, where r_1 and r_2 intersect along two edges and therefore ψ is an annulus. Since r_1 and r_2 are empty, this annulus has height or width equal to 1. Such an annulus is called a *thin annulus*; see the bottom row of Figure 4.4. Thin annuli have $N(\psi) = 1$.

Contributions from Cases (R-1) and (R-2) cancel in pairs, since we are working modulo 2. There are no contributions from Case (R-3), since every thin annulus contains one X-marking in it, concluding the proof of the lemma. □

LEMMA 4.6.8. *The differential $\partial_{\mathbb{X}}^-$ is homogeneous of degree $(-1,0)$.*

Proof. If $V_1^{k_1} \cdots V_n^{k_n} \cdot \mathbf{y}$ appears in $\partial_{\mathbb{X}}^- \mathbf{x}$, then there is a rectangle $r \in \mathrm{Rect}^\circ(\mathbf{x},\mathbf{y})$ with $r \cap \mathbb{X} = \emptyset$, and $O_i(r) = k_i$ for $i = 1,\dots,n$. By Equations (4.2) and (4.11),

$$M(V_1^{k_1} \cdots V_n^{k_n} \cdot \mathbf{y}) = M(\mathbf{y}) - 2\#(r \cap \mathbb{O}) = M(\mathbf{x}) - 1,$$

so the Maslov grading drops by one under the differential. Similarly, Equations (4.4) and (4.12) give

$$A(V_1^{k_1} \cdots V_n^{k_n} \cdot \mathbf{y}) = A(\mathbf{y}) - \#(r \cap \mathbb{O}) = A(\mathbf{x}) - \#(r \cap \mathbb{X}). \tag{4.14}$$

Since $r \cap \mathbb{X} = \emptyset$, it follows that $A(V_1^{k_1} \cdots V_n^{k_n} \cdot \mathbf{y}) = A(\mathbf{x})$; i.e. $\partial_{\mathbb{X}}^-$ preserves the Alexander grading. □

Proof of Theorem 4.6.3. Equations (4.11) and (4.12) ensure that multiplication by V_i is a homogeneous map of degree $(-2,-1)$; i.e. $GC^-(\mathbb{G})$ is a bigraded module over $\mathbb{F}[V_1,\dots,V_n]$. The differential is defined to be an $\mathcal{R}$-module homomorphism; Lemma 4.6.8 ensures that it is homogeneous of degree $(-1,0)$. The theorem now follows from Lemma 4.6.7. □

The complex $GC^-(\mathbb{G})$ generalizes $\widetilde{GC}(\mathbb{G})$, since

$$\frac{GC^-(\mathbb{G})}{V_1=\dots=V_n=0}\cong\widetilde{GC}(\mathbb{G}). \tag{4.15}$$

We study now further properties of $GC^-(\mathbb{G})$.

LEMMA 4.6.9. *For any pair of integers $i,j\in\{1,\dots,n\}$ multiplication by V_i is chain homotopic to multiplication by V_j, when thought of as homogeneous maps from $GC^-(\mathbb{G})$ to itself of degree $(-2,-1)$.*

Proof. Variables V_i and V_j are called *consecutive* if there is a square marked by X in the same row as O_i and in the same column as O_j. Suppose that V_i and V_j are consecutive, and let X_i denote the X-marking in the same row as O_i and in the same column as O_j. Define a corresponding homotopy operator that counts rectangles that contain X_i in their interior:

$$\mathcal{H}_i(\mathbf{x})=\mathcal{H}_{X_i}(\mathbf{x})=\sum_{\mathbf{y}\in\mathbf{S}(\mathbb{G})}\ \sum_{\{r\in\operatorname{Rect}^\circ(\mathbf{x},\mathbf{y})\,\big|\,\operatorname{Int}(r)\cap\mathbb{X}=X_i\}} V_1^{O_1(r)}\cdots V_n^{O_n(r)}\cdot\mathbf{y}. \tag{4.16}$$

It follows immediately from Proposition 4.3.1 and Equation (4.4) that $\mathcal{H}_i$ is homogeneous of degree $(-1,-1)$. The proof of Lemma 4.6.7 shows that [1]

$$\partial^-_{\mathbb{X}}\circ\mathcal{H}_i+\mathcal{H}_i\circ\partial^-_{\mathbb{X}}=V_i-V_j.$$

In this adaptation, count decompositions of domains ψ with $N(\psi)>0$ and which contain X_i (and no other $X\in\mathbb{X}$) with multiplicity one in their interior. In addition to the types of pairs appearing in Cases (R-1) and (R-2) of Lemma 4.6.7, there are two thin annuli that contribute to $\partial^-_{\mathbb{X}}\circ\mathcal{H}_i+\mathcal{H}_i\circ\partial^-_{\mathbb{X}}$, and those are the two annuli (horizontal and vertical) through X_i. The contributions of these two annuli are multiplication by V_i and multiplication by V_j.

For general V_i and V_j, since K is a knot there is a sequence of variables $V_i=V_{n_1},\dots,V_{n_m}=V_j$ where V_{n_k} and $V_{n_{k+1}}$ are consecutive. Adding the chain homotopies, we deduce that V_i is homotopic to V_j. □

REMARK 4.6.10. Lemma 4.6.9 uses the fact that the grid diagram $\mathbb{G}$ represents a knot, rather than a link: in general, the actions of variables corresponding to different link components are not chain homotopic; cf. also Lemma 8.2.3. For more on the case of links, see Section 9.1 and Chapter 11.

DEFINITION 4.6.11. Fix some $i\in\{1,\dots,n\}$. The ***unblocked grid homology of*** $\mathbb{G}$, denoted $GH^-(\mathbb{G})$, is the homology of $(GC^-(\mathbb{G}),\partial^-_{\mathbb{X}})$, viewed as a bigraded module over $\mathbb{F}[U]$, where the action of U is induced by multiplication by V_i.

[1] Note that $V_i-V_j=V_i+V_j$ in $\mathcal{R}$. We write V_i-V_j, since that expression is what shows up when we work with $\mathbb{Z}$ coefficients, as in Chapter 15.

Lemma 4.6.9 shows that the grid homology groups, thought of as bigraded modules over $\mathbb{F}[U]$, are independent of the choice of i. Lemma 4.6.9 also inspires the following further construction:

DEFINITION 4.6.12. Fix some $i = 1, \dots, n$. The quotient complex $GC^-(\mathbb{G})/V_i$ is called the ***simply blocked grid complex***, and it is denoted $\widehat{GC}(\mathbb{G})$. The ***simply blocked grid homology of*** $\mathbb{G}$, $\widehat{GH}(\mathbb{G})$, is the bigraded vector space obtained as the homology of $\widehat{GC}(\mathbb{G}) = (GC^-(\mathbb{G})/V_i, \partial_{\mathbb{X}}^-)$.

REMARK 4.6.13. Explicitly, $GC^-(\mathbb{G})/V_n$ is the bigraded $\mathbb{F}$-vector space with basis $V_1^{k_1} \cdots V_{n-1}^{k_{n-1}} \cdot \mathbf{x}$, where $k_1, \dots, k_{n-1}$ are arbitrary non-negative integers and $\mathbf{x} \in \mathbf{S}(\mathbb{G})$; equipped with a differential $\widehat{\partial}_{\mathbb{X},O_n}$ specified by $\widehat{\partial}_{\mathbb{X},O_n} \circ V_j = V_j \circ \widehat{\partial}_{\mathbb{X},O_n}$ for $j = 1, \dots, n-1$, and for any $\mathbf{x} \in \mathbf{S}(\mathbb{G})$,

$$\widehat{\partial}_{\mathbb{X},O_n}(\mathbf{x}) = \sum_{\mathbf{y} \in \mathbf{S}(\mathbb{G})} \sum_{\{r \in \mathrm{Rect}^\circ(\mathbf{x},\mathbf{y}) \big| r \cap \mathbb{X} = \emptyset, O_n(r) = 0\}} V_1^{O_1(r)} \cdots V_{n-1}^{O_{n-1}(r)} \cdot \mathbf{y}.$$

We shall see that $\widehat{GH}$ is a finite dimensional vector space that is independent of the choice of i, in Corollaries 4.6.16 and 4.6.17 below. But first, we explain how to extract the vector space $\widehat{GH}(\mathbb{G})$ from $\widetilde{GH}(\mathbb{G})$, in terms of the following notation. Let X and Y be two bigraded vector spaces

$$X = \bigoplus_{d,s \in \mathbb{Z}} X_{d,s} \qquad \text{and} \qquad Y = \bigoplus_{d,s \in \mathbb{Z}} Y_{d,s}.$$

Their tensor product $X \otimes Y = \bigoplus_{d,s \in \mathbb{Z}} (X \otimes Y)_{d,s}$ is the bigraded vector space, with

$$(X \otimes Y)_{d,s} = \bigoplus_{\substack{d_1 + d_2 = d \\ s_1 + s_2 = s}} X_{d_1,s_1} \otimes Y_{d_2,s_2}. \tag{4.17}$$

DEFINITION 4.6.14. Let X be a bigraded vector space, and fix integers a and b. The corresponding ***shift of*** X, denoted $X[\![a,b]\!]$, is the bigraded vector space that is isomorphic to X as a vector space and given the bigrading $X[\![a,b]\!]_{d,s} = X_{d+a,s+b}$.

Let W be the two-dimensional bigraded vector space with one generator in bigrading $(0,0)$ and another in bigrading $(-1,-1)$, and let X be any other bigraded vector space, then the tensor product $X \otimes W$ is identified with two copies of X, one of which is equipped with a shift in degree:

$$X \otimes W \cong X \oplus X[\![1,1]\!]. \tag{4.18}$$

This can be iterated; for example, $X \otimes W^{\otimes 2} \cong X \oplus X[\![1,1]\!] \oplus X[\![1,1]\!] \oplus X[\![2,2]\!]$.

PROPOSITION 4.6.15. *Let $\mathbb{G}$ be a grid diagram representing a knot. Let W be the two-dimensional bigraded vector space, with one generator in bigrading $(0,0)$ and the other in bigrading $(-1,-1)$. Then, there is an isomorphism*

$$\widetilde{GH}(\mathbb{G}) \cong \widehat{GH}(\mathbb{G}) \otimes W^{\otimes(n-1)} \tag{4.19}$$

of bigraded vector spaces, where $\widehat{GH}(\mathbb{G}) = H(\frac{GC^-(\mathbb{G})}{V_i})$ for any $i = 1, \dots, n$.

Proof. We will prove by induction on j that

$$H\left(\frac{GC^-(\mathbb{G})}{V_1 = \cdots = V_j = 0}\right) \cong H\left(\frac{GC^-(\mathbb{G})}{V_1 = 0}\right) \otimes W^{\otimes(j-1)}. \tag{4.20}$$

We interpret W^0 as a one-dimensional vector space in bigrading $(0,0)$, so that the isomorphism $W^{\otimes a} \otimes W \cong W^{\otimes(a+1)}$ holds for all $a \geq 0$. In the basic case where $j = 1$, Equation (4.20) is a tautology.

For the inductive step, for $j > 1$ consider the short exact sequence

(4.21)
$$0 \longrightarrow \frac{GC^-(\mathbb{G})}{V_1=\cdots=V_{j-1}=0} \xrightarrow{V_j} \frac{GC^-(\mathbb{G})}{V_1=\cdots=V_{j-1}=0} \longrightarrow \frac{GC^-(\mathbb{G})}{V_1=\cdots=V_j=0} \longrightarrow 0.$$

Using Proposition 4.5.6, the chain homotopy between V_j and V_1 provided by Lemma 4.6.9, induces a chain homotopy between the chain map

$$V_j \colon \frac{GC^-(\mathbb{G})}{V_1 = \cdots = V_{j-1} = 0} \to \frac{GC^-(\mathbb{G})}{V_1 = \cdots = V_{j-1} = 0}$$

and the 0 map, so the long exact sequence on homology associated to the short exact squence from Equation (4.21) (cf. Lemma 4.5.3) becomes a short exact sequence

$$0 \longrightarrow H(\frac{GC^-(\mathbb{G})}{V_1=\cdots=V_{j-1}=0}) \longrightarrow H(\frac{GC^-(\mathbb{G})}{V_1=\cdots=V_j=0}) \longrightarrow H(\frac{GC^-(\mathbb{G})}{V_1=\cdots=V_{j-1}=0}) \longrightarrow 0,$$

where the second arrow preserves bigradings, and the third is homogeneous of degree $(1,1)$. Thus, this short exact sequence of vector spaces translates into the first isomorphism of bigraded vector spaces appearing in the following:

$$H\left(\frac{GC^-(\mathbb{G})}{V_1 = \cdots = V_j = 0}\right) \cong H\left(\frac{GC^-(\mathbb{G})}{V_1 = \cdots = V_{j-1} = 0}\right) \otimes W \cong \widehat{GH}(\mathbb{G}) \otimes W^{\otimes(j-1)},$$

and the second isomorphism follows from the inductive hypothesis. This completes the inductive step, verifying Equation (4.20) for all $j = 1, \ldots, n$.

In view of Equation (4.15), when $j = n$, Equation (4.20) gives Equation (4.19) for $i = 1$. Numbering our formal variables differently, we conclude that Equation (4.19) holds for arbitrary i. $\square$

The chain complex $\widetilde{GC}(\mathbb{G})$ is finite dimensional over $\mathbb{F}$, so its homology $\widetilde{GH}(\mathbb{G})$ is also finite dimensional. Although $\widehat{GC}(\mathbb{G})$ is infinite dimensional over $\mathbb{F}$, Proposition 4.6.15 has the following immediate consequence:

COROLLARY 4.6.16. *For a grid diagram $\mathbb{G}$ with grid number n, the vector space $\widehat{GH}(\mathbb{G})$ is finite dimensional, the dimension of $\widehat{GH}(\mathbb{G})$ is divisible by 2^{n-1}, and in fact $2^{n-1} \cdot \dim_{\mathbb{F}} \widehat{GH}(\mathbb{G}) = \dim_{\mathbb{F}} \widetilde{GH}(\mathbb{G})$.* $\square$

COROLLARY 4.6.17. *The simply blocked grid homology*

$$\widehat{GH}(\mathbb{G}) = H(GC^-(\mathbb{G})/V_i)$$

is independent of the choice of $i = 1, \ldots, n$.

Proof. From Proposition 4.6.15, it follows that for i, j,

$$H\left(\frac{GC^-(\mathbb{G})}{V_i}\right) \otimes W^{(n-1)} \cong H\left(\frac{GC^-(\mathbb{G})}{V_j}\right) \otimes W^{(n-1)} \tag{4.22}$$

as bigraded vector spaces.

Just as a finite dimensional vector space is determined up to isomorphism by its dimension, a finite dimensional bigraded vector space Y is determined up to isomorphism by its *Poincaré polynomial* P_Y, the Laurent polynomial in q and t:

$$P_Y(q,t) = \sum_{d,s\in\mathbb{Z}} \dim Y_{d,s} \cdot q^d t^s. \tag{4.23}$$

Letting $Y_i = H(\frac{GC^-(\mathbb{G}))}{V_i})$, Equation (4.22) translates into the equation

$$(1+q^{-1}t^{-1})^{n-1} \cdot P_{Y_i}(q,t) = (1+q^{-1}t^{-1})^{n-1} \cdot P_{Y_j}(q,t),$$

so $P_{Y_i} = P_{Y_j}$, and hence $H(\frac{GC^-(\mathbb{G})}{V_i}) \cong H(\frac{GC^-(\mathbb{G})}{V_j})$ as bigraded vector spaces. □

Another relation among the grid homology groups is given by the following:

PROPOSITION 4.6.18. *There is a long exact sequence relating $\widehat{GH}(\mathbb{G})$ and $GH^-(\mathbb{G})$:*

$$\dots \to GH^-_{d+2}(\mathbb{G}, s+1) \xrightarrow{U} GH^-_d(\mathbb{G}, s) \to \widehat{GH}_d(\mathbb{G}, s) \to GH^-_{d+1}(\mathbb{G}, s+1) \to \dots$$

Proof. Consider the short exact sequence

$$0 \to GC^-(\mathbb{G}) \xrightarrow{V_i} GC^-(\mathbb{G}) \to \widehat{GC}(\mathbb{G}) \to 0$$

of bigraded chain complexes of $\mathbb{F}[V_i]$-modules, where the first map is, of course, homogeneous of degree $(-2,-1)$. The associated long exact sequence in homology (Lemma 4.5.3) gives the statement of the proposition. □

A key feature of the grid homology groups $\widehat{GH}(\mathbb{G})$ and $GH^-(\mathbb{G})$ is that they are knot invariants, in the following sense.

THEOREM 4.6.19. *The homologies $\widehat{GH}(\mathbb{G})$ and $GH^-(\mathbb{G})$ (the former thought of as a bigraded $\mathbb{F}$-vector space, the latter thought of as a bigraded $\mathbb{F}[U]$-module) depend on the grid $\mathbb{G}$ only through its underlying (unoriented) knot K.*

The proof of the above theorem will be given in Chapter 5.

4.7. The Alexander grading as a winding number

The aim of the present section is to give geometric insight into the bigrading from Section 4.3. Byproducts include a practical formula for computing A and a relationship between grid homology and the Alexander polynomial. The geometric interpretation of the Alexander grading rests on the following formula, which expresses the winding number about a knot projection in terms of planar grid diagrams.

LEMMA 4.7.1. *Let $\mathbb{G}$ be a planar grid diagram of a knot K, let $\mathcal{D} = \mathcal{D}(\mathbb{G})$ be the corresponding knot projection in the plane, and let p be any point not on $\mathcal{D}$. Then, the winding number $w_{\mathcal{D}}(p)$ of $\mathcal{D}$ around p is computed by the formula*

$$w_{\mathcal{D}}(p) = \mathcal{J}(p, \mathbb{O} - \mathbb{X}). \tag{4.24}$$

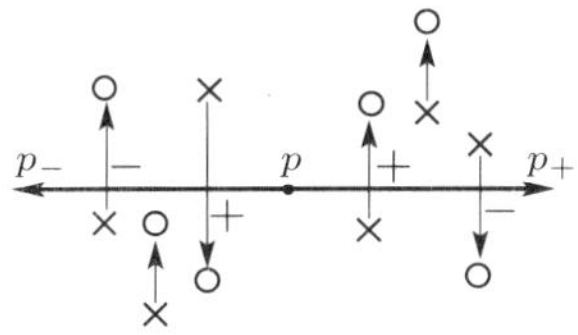

FIGURE 4.6. **Winding numbers**. The diagram illustrates the equality $w_{\mathcal{D}}(p) = \mathcal{I}(p, \mathbb{O} - \mathbb{X})$ (interpreting the winding number as intersection of the knot projection with the ray p_+), and at the same time the equality $w_{\mathcal{D}}(p) = \mathcal{I}(\mathbb{O} - \mathbb{X}, p)$ (interpreting the winding number as intersection with the ray p_-).

Proof. If $p = (x, y)$ is any point not contained in $\mathcal{D}$, then $\mathcal{I}(p, \mathbb{O} - \mathbb{X})$ is the (signed) intersection number of the ray p_+ from p to $(+\infty, y)$ with $\mathcal{D}$: the vertical arc connecting some O with X contributes $+1$ if the O lies in this upper right quadrant and the X does not, and it contributes -1 if the X lies in this upper right quadrant and the O does not, and it contributes 0 otherwise; i.e.

$$\#(p_+ \cap \mathcal{D}) = \mathcal{I}(p, \mathbb{O} - \mathbb{X}).$$

Similarly, the intersection number of the ray p_- from p to $(-\infty, y)$ with $\mathcal{D}$ is

$$\#(p_- \cap \mathcal{D}) = \mathcal{I}(\mathbb{O} - \mathbb{X}, p).$$

Clearly, $w_{\mathcal{D}}(p) = \#(p_+ \cap \mathcal{D}) = \#(p_- \cap \mathcal{D})$. Average the above two equations to get Equation (4.24). □

Fix a planar realization of a toroidal grid diagram, and consider the function A' on the grid state $\mathbf{x} \in \mathbf{S}(\mathbb{G})$ defined by

$$A'(\mathbf{x}) = -\sum_{x \in \mathbf{x}} w_{\mathcal{D}}(x). \tag{4.25}$$

As we shall see shortly, A and A' differ by a constant (independent of the grid state). We express this constant in terms of quantities which we have met already in Section 3.3. To this end, recall that each of the $2n$ squares marked with an X or O has 4 corners, giving us a total of $8n$ lattice points on the grid (possibly counted with multiplicity, when the marked squares share a corner), which we denote $p_1, \dots, p_{8n}$. The sum of the winding numbers at these points, divided by 8, was denoted by $a(\mathbb{G})$ in Section 3.3. The precise relationship between A and A' can now be stated as follows:

PROPOSITION 4.7.2. *Choose a planar realization of a toroidal grid diagram $\mathbb{G}$ representing a knot K. Let $\mathcal{D}$ be the corresponding diagram of K. The Alexander function A can be expressed in terms of the winding numbers $w_{\mathcal{D}}$ by the following formula:*

$$A(\mathbf{x}) = -\sum_{x \in \mathbf{x}} w_{\mathcal{D}}(x) + \frac{1}{8}\sum_{j=1}^{8n} w_{\mathcal{D}}(p_j) - \left(\frac{n-1}{2}\right) = A'(\mathbf{x}) + a(\mathbb{G}) - \frac{n-1}{2}. \tag{4.26}$$

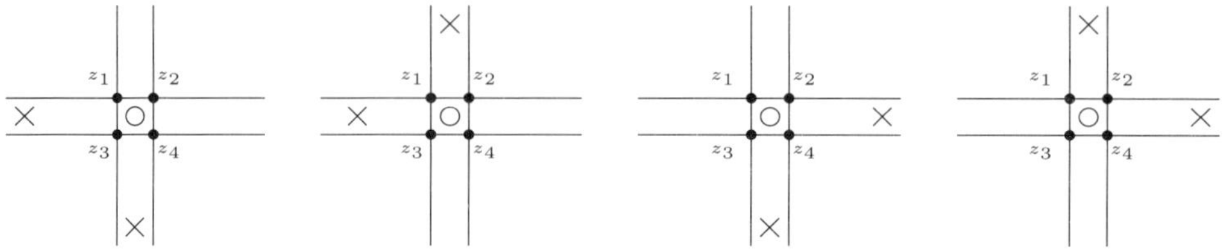

FIGURE 4.7. **Verification of Equation** (4.28). Here are the four cases where the distinguished square z is marked with an O. To verify the equation, find the pairs contributing to $\mathcal{J}(z_1 + \cdots + z_4, X)$, where X is in the same row or column as z, and to $\mathcal{J}(z_1 + \cdots + z_4, O)$, where the O-marking is at z.

Proof. Summing Equation (4.24) over all the components $x \in \mathbf{x}$, gives $A'(\mathbf{x}) = -\mathcal{J}(\mathbf{x}, \mathbb{O} - \mathbb{X})$; so

$$\begin{aligned} A(\mathbf{x}) &= \frac{1}{2}(M_{\mathbb{O}}(\mathbf{x}) - M_{\mathbb{X}}(\mathbf{x})) - \left(\frac{n-1}{2}\right) \\ &= -\mathcal{J}(\mathbf{x}, \mathbb{O} - \mathbb{X}) + \frac{1}{2}(\mathcal{J}(\mathbb{O}, \mathbb{O}) - \mathcal{J}(\mathbb{X}, \mathbb{X})) - \left(\frac{n-1}{2}\right) \\ &= A'(\mathbf{x}) + \frac{1}{2}\mathcal{J}(\mathbb{O} + \mathbb{X}, \mathbb{O} - \mathbb{X}) - \left(\frac{n-1}{2}\right). \end{aligned}$$

Thus, Equation (4.26) now follows once we show that

$$\frac{1}{2}\mathcal{J}(\mathbb{O} + \mathbb{X}, \mathbb{O} - \mathbb{X}) = \frac{1}{8}\sum_{i=1}^{8n} w_{\mathcal{D}}(p_i). \tag{4.27}$$

To check Equation (4.27), we first verify the following: given any small square (in a planar grid) whose center z is marked with an O or an X, if $z_1, \dots, z_4$ denote its four corner points in the plane, then
(4.28)

$$\mathcal{J}(z, \mathbb{O} - \mathbb{X}) = \frac{1}{4}\mathcal{J}(z_1 + z_2 + z_3 + z_4, \mathbb{O} - \mathbb{X}) + \begin{cases} -\frac{1}{4} & \text{if } z \text{ is marked with an } O \\ \frac{1}{4} & \text{if } z \text{ is marked with an } X. \end{cases}$$

Suppose for definiteness that z is marked with an O. Then, for any marking $O' \in \mathbb{O}$ with $O \neq O'$,

$$\mathcal{J}(z, O') = \frac{1}{4}\mathcal{J}(z_1 + z_2 + z_3 + z_4, O').$$

Also, for any X-marking not in the same row or column as z,

$$\mathcal{J}(z, X) = \frac{1}{4}\mathcal{J}(z_1 + z_2 + z_3 + z_4, X).$$

The correction of $-\frac{1}{4}$ comes from the pairing of the X-markings in the same row and column as z with the formal sum $z_1 + \cdots + z_4$, combined with the pairing of the O-marking on z with $z_1 + \cdots + z_4$; see Figure 4.7. A similar reasoning gives Equation (4.28) when z is marked with an X.

Equation (4.27) follows from summing up Equation (4.28) over all O- and X-marked squares, and using Lemma 4.7.1. □

LEMMA 4.7.3. *The sign of the permutation that connects* $\mathbf{x}$ *with* $\mathbf{x}^{NWO}$ *is* $(-1)^{M(\mathbf{x})}$.

Proof. This is an immediate consequence of Proposition 4.3.1, combined with the mod 2 reductions of Equations (4.1) and (4.2). □

DEFINITION 4.7.4. Let $X = \bigoplus_{d,s} X_{d,s}$ be a bigraded vector space. Define the ***graded Euler characteristic of*** X to be the Laurent polynomial in t given by

$$\chi(X) = \sum_{d,s} (-1)^d \dim X_{d,s} \cdot t^s.$$

The Euler characteristic of grid homology is related to the Alexander polynomial, according to the following:

PROPOSITION 4.7.5. *Let* $\mathbb{G}$ *be a grid diagram for a knot* K *with grid number* n*. The graded Euler characteristic of the bigraded vector space* $\widetilde{GH}(\mathbb{G})$ *is given by*

$$\chi(\widetilde{GH}(\mathbb{G})) = (1 - t^{-1})^{n-1} \cdot \Delta_K(t), \tag{4.29}$$

where $\Delta_K(t)$ *is the symmetrized Alexander polynomial of Equation* (2.3).

Proof. It is a standard fact that the Euler characteristic of a chain complex agrees with that of its homology (and this fact remains true in the bigraded case). Thus,

$$\chi(\widetilde{GH}(\mathbb{G})) = \chi(\widetilde{GC}(\mathbb{G})) = \sum_{\mathbf{x} \in \mathbf{S}(\mathbb{G})} (-1)^{M(\mathbf{x})} t^{A(\mathbf{x})}.$$

By Proposition 4.7.2 (for the t-power) and Lemma 4.7.3 together with Proposition 4.3.7 (for the sign), it follows that this graded Euler characteristic agrees with

$$\sum_{\mathbf{x}} (-1)^{M(\mathbf{x})} t^{A(\mathbf{x})} = (-1)^{n-1} \epsilon(\mathbb{G}) \cdot \det(\mathbf{M}(\mathbb{G})) \cdot t^{a(\mathbb{G})} \cdot t^{\frac{1-n}{2}}.$$

The result now follows from Theorem 3.3.6. □

Proposition 4.7.5 relates the Euler characteristic of $\widetilde{GH}(\mathbb{G})$ and the Alexander polynomial of the underlying knot. This leads quickly to the following relationship between the Alexander polynomial and the graded Euler characteristic of $\widehat{GH}$

$$\chi(\widehat{GH}(K)) = \sum_{d,s} (-1)^d \dim\ \widehat{GH}_d(K, s) \cdot t^s \in \mathbb{Z}[t, t^{-1}]. \tag{4.30}$$

THEOREM 4.7.6 ([**172, 191**]). *The graded Euler characteristic of the simply blocked grid homology is equal to the (symmetrized) Alexander polynomial* $\Delta_K(t)$*:*

$$\chi(\widehat{GH}(K)) = \Delta_K(t).$$

Proof. The graded Euler characteristic of the bigraded vector space W from Lemma 4.6.15 is $\chi(W) = 1 - t^{-1}$, so the identity follows immediately from Propositions 4.7.5 and 4.6.15. □

4.8. Computations

Assuming Theorem 4.6.19, we can directly compute some of the homology groups defined earlier in this chapter. See also Chapter 10 for more computations.

PROPOSITION 4.8.1. *For the unknot* $\mathcal{O}$, $\widehat{GH}(\mathcal{O}) \cong \mathbb{F}$ *is supported in bigrading* $(0,0)$; *and* $GH^-(\mathcal{O}) \cong \mathbb{F}[U]$, *and its generator has bigrading* $(0,0)$.

Proof. In the 2×2 grid diagram $\mathbb{G}$ representing the unknot, there are exactly two generators; call them $\mathbf{p}$ and $\mathbf{q}$, with $A(\mathbf{p})=0$, $M(\mathbf{p})=0$, $A(\mathbf{q})=-1$, $M(\mathbf{q})=-1$. The complex $GC^-(\mathbb{G})$ is generated over $\mathbb{F}[V_1,V_2]$ by these two generators, and its boundary map is specified by

$$\partial^-_{\mathbb{X}}(\mathbf{p})=0, \qquad \partial^-_{\mathbb{X}}(\mathbf{q})=(V_1+V_2)\cdot\mathbf{p}.$$

The homology of this complex is clearly isomorphic to $\mathbb{F}[U]$, generated by the cycle $\mathbf{p}$; this completes the computation of $GH^-(\mathcal{O})$.

For $\widehat{GH}(\mathcal{O})$, we can set $V_2=0$, to obtain the complex over $\mathbb{F}[V_1]$ with generators $\mathbf{p}$ and $\mathbf{q}$, and boundary specified by

$$\partial^-_{\mathbb{X}}(\mathbf{p})=0, \qquad \partial^-_{\mathbb{X}}(\mathbf{q})=V_1\cdot\mathbf{p},$$

whose homology is $\mathbb{F}$, generated by the cycle $\mathbf{p}$. $\square$

With more work, one can show that the grid homology groups of the right-handed trefoil knot $K=T_{2,3}$ are given by:

$$\widehat{GH}_d(K,s)=\begin{cases}\mathbb{F} & \text{if } (d,s)\in\{(0,1),(-1,0),(-2,-1)\}\\ 0 & \text{otherwise.}\end{cases} \tag{4.31}$$

$$GH^-_d(K,s)=\begin{cases}\mathbb{F} & \text{if } (d,s)=(0,1) \text{ or } (d,s)=(-2k,-k) \text{ for } k\geq 1\\ 0 & \text{otherwise.}\end{cases} \tag{4.32}$$

In the second case, the $\mathbb{F}[U]$-module structure is determined by the property that $U\colon GH^-_{-2k}(K,-k)\to GH^-_{-2k-2}(K,-k-1)$ is an isomorphism for all $k\geq 1$. More succinctly, we write

$$GH^-(K)\cong(\mathbb{F}[U]/U)_{(0,1)}\oplus(\mathbb{F}[U])_{(-2,-1)},$$

where the subscripts on the cyclic $\mathbb{F}[U]$-modules denote the bigradings of their generators.

EXERCISE 4.8.2. Let K denote the right-handed trefoil knot.
(a) Use Figure 3.3 to verify Equation (4.31). (*Hint:* Show first that there are no generators for $\widetilde{GC}(\mathbb{G})$ in Alexander grading greater than 1. Next, find generators of $\widetilde{GC}(\mathbb{G})$ in Alexander gradings 0, 1, and -5, and apply Proposition 4.6.15.)
(b) Verify Equation (4.32). (*Hint:* Proposition 4.6.18 might be helpful.)
(c) Let K denote the left-handed trefoil knot. Compute $\widehat{GH}(K)$ and $GH^-(K)$.

REMARK 4.8.3. The result of Exercise 4.8.2 shows that grid homology distinguishes the right-handed trefoil $T_{2,3}$ from its mirror $T_{-2,3}$. See Proposition 7.1.2 for a general description of how homology behaves under reflection.

Restricting attention to a carefully chosen Alexander grading, we can give a more general computation valid for all torus knots.

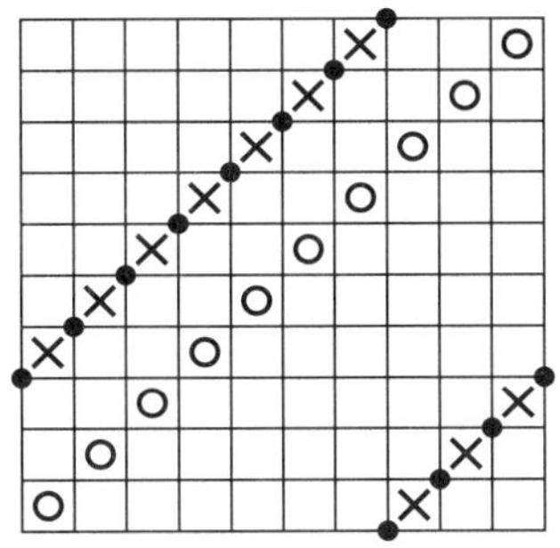

FIGURE 4.8. **Grid diagram for $T_{-3,7}$.** This is the diagram for $T_{-p,q}$ from Exercise 3.1.5(c), when $p = 3$ and $q = 7$. The grid state $\mathbf{x}^+$ is indicated by the heavy dots in the grid.

LEMMA 4.8.4. *Let $p, q > 1$ be relatively prime integers. There is a grid diagram $\mathbb{G}$ for $T_{-p,q}$ with the following property. If $\mathbf{x}^+ = \mathbf{x}^{NEX}$ is the grid state which occupies the upper right corner of each square marked with X, then this grid state is the unique generator with maximal Alexander grading among all generators, and*

$$A(\mathbf{x}^+) = \frac{(p-1)(q-1)}{2}.$$

Proof. Let $\mathbb{G}$ be the $(p+q) \times (p+q)$ grid diagram with $\sigma_{\mathbb{O}} = (1, \dots, p+q)$ and $\sigma_{\mathbb{X}} = (p+1, p+2, \dots, p)$; see Figure 4.8. (Compare also Exercise 3.1.5(c)).

Consider the associated winding matrix $M_{p,q} = \mathbf{W}(\mathbb{G})$. In the jth row, the winding numbers start out zero for a while, they increase by 1's until they reach their maximum, then they stay constant, and then eventually they drop by 1's. More precisely: the left column and bottom row vanish; for $j = 1, \dots, q$, in the jth row (from the top), the first $q - j + 1$ entries are 0 and all others are positive; while for $j = q+1, \dots, p+q-1$, the last $j - q$ entries and the first entry are 0 and all others are positive.

For example, for the torus knot $T_{-3,7}$ from Figure 4.8, this matrix is

$$M_{3,7} = \begin{pmatrix} 0 & 0 & 0 & 0 & 0 & 0 & 0 & 1 & 1 & 1 \\ 0 & 0 & 0 & 0 & 0 & 0 & 1 & 2 & 2 & 1 \\ 0 & 0 & 0 & 0 & 0 & 1 & 2 & 3 & 2 & 1 \\ 0 & 0 & 0 & 0 & 1 & 2 & 3 & 3 & 2 & 1 \\ 0 & 0 & 0 & 1 & 2 & 3 & 3 & 3 & 2 & 1 \\ 0 & 0 & 1 & 2 & 3 & 3 & 3 & 3 & 2 & 1 \\ 0 & 1 & 2 & 3 & 3 & 3 & 3 & 3 & 2 & 1 \\ 0 & 1 & 2 & 2 & 2 & 2 & 2 & 2 & 1 & 0 \\ 0 & 1 & 1 & 1 & 1 & 1 & 1 & 1 & 0 & 0 \\ 0 & 0 & 0 & 0 & 0 & 0 & 0 & 0 & 0 & 0 \end{pmatrix}.$$

It follows at once that for $\mathbf{x}^+$ all the winding numbers are zero, and for all other grid states $\mathbf{x}$, the sum of the winding numbers $-A'(\mathbf{x})$ is positive; so by Proposition 4.7.2, $\mathbf{x}^+$ is the unique grid state with maximal Alexander grading.

Elementary computation shows that

$$\mathcal{J}(\mathbf{x}^+,\mathbf{x}^+) = \frac{p(p-1)+q(q-1)}{2} \qquad \mathcal{J}(\mathbb{O},\mathbb{O}) = \frac{(p+q)(p+q-1)}{2}$$

$$\mathcal{J}(\mathbf{x}^+,\mathbb{O}) = \frac{p^2+q^2}{2} \qquad M_{\mathbb{X}}(\mathbf{x}^+) = 1-p-q;$$

so using Definition 4.3.2 we find that

$$A(\mathbf{x}^+) = \frac{1}{2}(M_{\mathbb{O}}(\mathbf{x}^+) - M_{\mathbb{X}}(\mathbf{x}^+)) - \frac{p+q-1}{2} = \frac{(p-1)(q-1)}{2}. \qquad \square$$

PROPOSITION 4.8.5. *Fix relatively prime, positive integers p and q with $p, q > 1$. Some of the grid homology groups $\widehat{GH}(T_{-p,q})$ are given by the following:*

$$\widehat{GH}_d(T_{-p,q}, s) = \begin{cases} \mathbb{F} & \text{if } s = \frac{(p-1)(q-1)}{2} \text{ and } d = (p-1)(q-1) \\ 0 & \text{if } s = \frac{(p-1)(q-1)}{2} \text{ and } d \neq (p-1)(q-1) \\ 0 & \text{if } s > \frac{(p-1)(q-1)}{2}. \end{cases}$$

Proof. According to Lemma 4.8.4, $\widehat{GC}(T_{-p,q})$ has no generators with Alexander grading greater than $\frac{(p-1)(q-1)}{2}$; and it has a single one with Alexander grading equal to $\frac{(p-1)(q-1)}{2}$. The formulas in the proof of Lemma 4.8.4 also show that $M(\mathbf{x}^+) - (p-1)(q-1)$. $\square$

To give further examples, we find it convenient to encode the grid homology by its Poincaré polynomial $P_K(q,t) = \sum_{d,s} \dim\ \widehat{GH}_d(K,s)t^s q^d$ (introduced in Equation (1.2)). Using a direct computer computation, Baldwin and Gillam [**5**] computed the grid homology of all knots with at most twelve crossings. In particular, for the eleven crossing Kinoshita-Terasaka knot KT and for its Conway mutant C of Figure 2.7 (compare also [**182**, Section 5.4] and [**179**, Section 3]) they found that:

$$P_{KT}(q,t) = (q^{-3}+q^{-2})t^{-2}+4(q^{-2}+q^{-1})t^{-1}+6q^{-1}+7+4(1+q)t+(q+q^2)t^2, \tag{4.33}$$

$$\begin{aligned} P_C(q,t) = {} & (q^{-4}+q^{-3})t^{-3} + 3(q^{-3}+q^{-2})t^{-2} + 3(q^{-2}+q^{-1})t^{-1} + 2q^{-1} \\ & +3 + 3(1+q)t + 3(q+q^2)t^2 + (q^2+q^3)t^3. \end{aligned} \tag{4.34}$$

(Among non-trivial knots with at most eleven crossings, these are the two knots with Alexander polynomial equal to 1.)

Although these are not computations one would wish to perform by hand, there are pieces which can be verified directly. For example:

EXERCISE 4.8.6. Consider Figure 4.9, a grid diagram for the Conway knot.
(a) Show that the grid states pictured on the figure are the only two grid states in Alexander grading 3, and that there are no grid states in greater Alexander grading.
(b) Show that there are no empty rectangles connecting the two grid states. Use this to verify that the coefficient in front of the t^3 term in the Poincaré polynomial $P_C(q,t)$ of the Conway knot is, indeed, (q^2+q^3), as stated in Equation (4.34), and that all higher t-powers have vanishing coefficients.

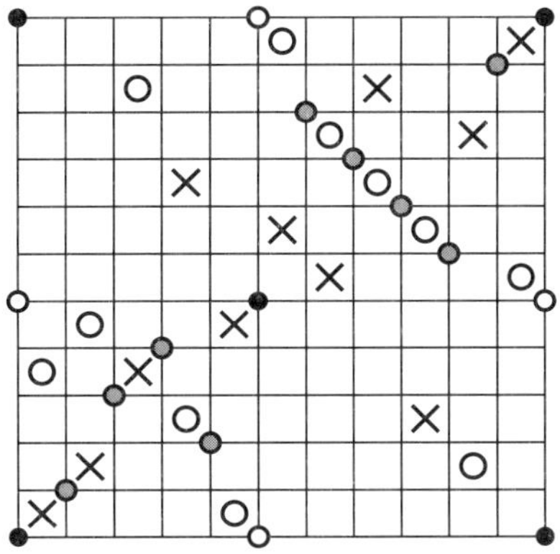

FIGURE 4.9. **Two grid states for the Conway knot.** The white ones appear only in one of the grid states, the black ones appear only in the other, and the gray dots appear in both.

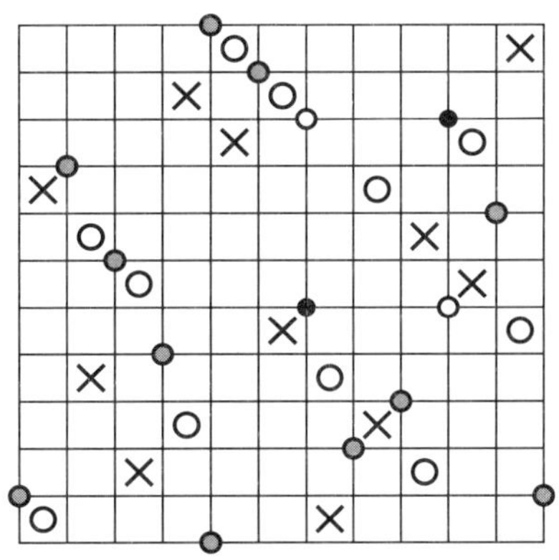

FIGURE 4.10. **Two grid states for the Kinoshita-Terasaka knot.** We have exhibited here two grid states on the same grid, with the same conventions as in Figure 4.9.

EXERCISE 4.8.7. Consider the grid diagram of the Kinoshita-Terasaka knot KT from Figure 4.10. (Notice that the diagram for the Kinoshita-Terasaka knot we gave in Figure 3.4 differs from this diagram by a single commutation.)
(a) Show that there are exactly four grid states in Alexander grading 2, and none with Alexander grading greater than 2. (*Hint:* Two of the grid states in Alexander grading 2 are pictured in Figure 4.10. Find the other two.)
(b) Show that the homology of the resulting chain complex in Alexander grading 2 has dimension 2. Use this to verify that the coefficient in front of t^2 in $P_{KT}(q,t)$ is $(q^2 + q^1)$, and all coefficients with higher t-powers vanish, as stated in Equation (4.33).

REMARK 4.8.8. For a typical grid diagram of the Kinoshita-Terasaka knot with grid number 11, the number of generators in Alexander grading 2 is rather large. For the choice we gave here, there are four generators, and this makes the computation of the grid homology in this Alexander grading easy.

4.9. Further remarks

The argument from Lemma 4.6.7 is a combinatorial analogue of the proof, in Lagrangian Floer homology, that the Lagrangian Floer complex is, in fact, a chain complex. The proof in that case hinges on Gromov's compactness theorem, together with gluing results for solutions of the relevant non-linear Cauchy-Riemann operator. These results are key ingredients in the development of Lagrangian Floer homology. Likewise, the combinatorial arguments from Lemma 4.6.7, although they are much simpler, also lie at the core of grid homology. Arguments of this type will appear throughout the text. (See for example Lemma 4.6.9 and Lemma 5.1.4.)

CHAPTER 5

The invariance of grid homology

This chapter gives a self-contained proof that the grid homology from Chapter 4 is a knot invariant. The discussion here is closely modeled on [**136**], with some simplifications. Most importantly, we consider invariance for grid homology, rather than the more general filtered version of the grid complex; we will return to the more general setting in Chapter 13. The present version, though, is sufficient for many topological applications, such as those given in Chapters 6, 8, and 12.

In view of Cromwell's theorem (Theorem 3.1.9), to prove invariance, it suffices to verify invariance under the grid moves: commutations and stabilizations. Section 5.1 is devoted to the proof of commutation invariance, and Section 5.2 to stabilization invariance. In Section 5.3, we put together the pieces and prove the invariance statement for grid homology. We also show that grid homology is independent of the orientation on K. In Section 5.4, we give a more detailed analysis of the maps induced by stabilizations; this material will be used in Chapter 6. In Section 5.5 we sketch a further variant of grid homology (due to Lipshitz); in Section 5.6, we indicate the relationship of the present chapter with the pseudo-holomorphic theory; and in Section 5.7 we give the reader some exercises, to construct chain homotopy equivalences for stabilizations, that induce the isomorphisms studied earlier.

5.1. Commutation invariance

In this section, we define isomorphisms on grid homology induced by commutation moves. To define these maps, we introduce some notation. As in Definition 3.1.6, let $\mathbb{G}$ be a toroidal grid diagram, and $\mathbb{G}'$ be a different grid diagram obtained by commuting two columns of $\mathbb{G}$. It will be helpful to view both diagrams on one torus: regard all the X- and O-markings as fixed for both $\mathbb{G}$ and $\mathbb{G}'$, and draw two of the vertical circles (one for $\mathbb{G}$, and another for $\mathbb{G}'$) curved, as shown in Figure 5.1. Let $\boldsymbol{\alpha} = \{\alpha_1, \dots, \alpha_n\}$ and $\boldsymbol{\beta} = \{\beta_1, \dots, \beta_n\}$ denote the horizontal and vertical circles for $\mathbb{G}$. The set of horizontal circles for $\mathbb{G}'$ is $\boldsymbol{\alpha} = \{\alpha_1, \dots, \alpha_n\}$ as before, but its vertical circles are $\boldsymbol{\gamma} = \{\beta_1, \dots, \beta_{i-1}, \gamma_i, \beta_{i+1}, \dots, \beta_n\}$. Label the vertical circles compatibly with the cyclic ordering inherited from the toroidal grid; i.e. for $k = 1, \dots, n-1$, β_{k+1} is the vertical circle immediately to the east of β_k.

Draw the vertical circles so that β_i and γ_i meet perpendicularly in two points, denoted a and b, both of which miss the horizontal circles. Distinguish a and b, as follows. The complement of $\beta_i \cup \gamma_i$ in the grid torus contains two bigons, which intersect in $\{a, b\}$. One of these bigons has a portion of β_i on its western boundary and a portion of γ_i on its eastern boundary. For this bigon, the intersection point a occurs at its southern tip. We will define chain maps between the grid complexes for $\mathbb{G}$ and $\mathbb{G}'$ that count suitable pentagons, defined as follows.

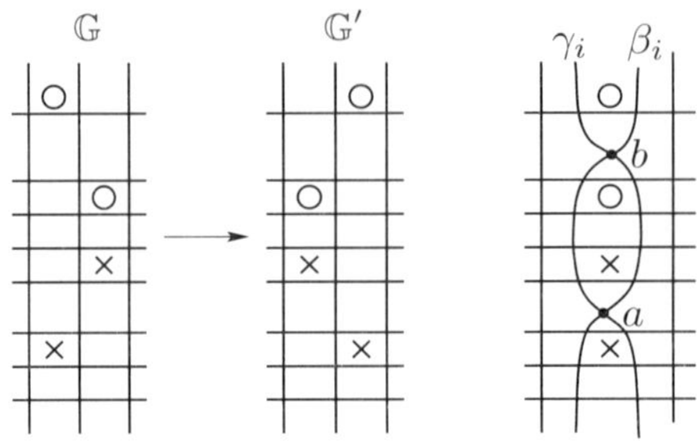

FIGURE 5.1. **Commutation diagram.** The commutation move indicated in the left and middle pictures is encoded in the diagram on the right, which contains both diagrams simultaneously.

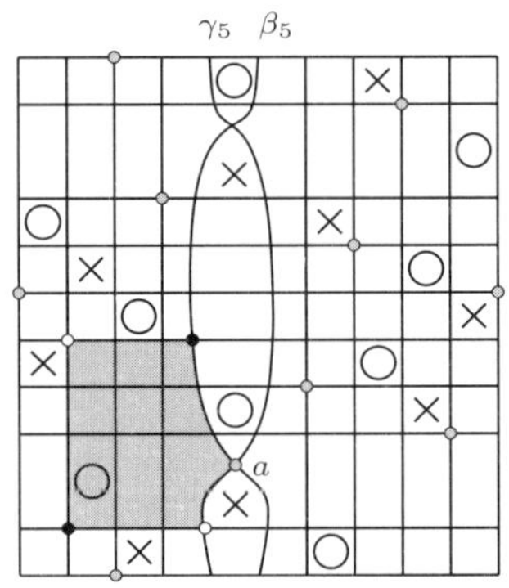

FIGURE 5.2. **Empty pentagon.** The shaded region is a pentagon counted in the commutation map.

DEFINITION 5.1.1. Fix grid states $\mathbf{x} \in \mathbf{S}(\mathbb{G})$ and $\mathbf{y}' \in \mathbf{S}(\mathbb{G}')$. An embedded disk p in the torus whose boundary is the union of five arcs, each of which lies in some α_j, β_j, or γ_i, is called a ***pentagon from*** $\mathbf{x}$ ***to*** $\mathbf{y}'$ if it satisfies the following conditions:

- Four of the corners of p are in $\mathbf{x} \cup \mathbf{y}'$.
- Note that each corner point x of p is an intersection of two of the curves taken from the $\{\alpha_j, \beta_j, \gamma_i\}_{j=1}^n$; and a small disk centered at x is divided into four quadrants by these two curves. The pentagon p contains exactly one of the four quadrants.
- Letting $\partial_\alpha p$ denote the portion of the boundary of p in $\alpha_1 \cup \cdots \cup \alpha_n$,

$$\partial(\partial_\alpha p) = \mathbf{y}' - \mathbf{x}. \tag{5.1}$$

The set of such pentagons is denoted $\text{Pent}(\mathbf{x}, \mathbf{y}')$. (See Figure 5.2 for a picture.)

The set $\text{Pent}(\mathbf{x}, \mathbf{y}')$ is empty unless $\mathbf{x}$ and $\mathbf{y}'$ share exactly $n-2$ elements. Moreover, $\text{Pent}(\mathbf{x}, \mathbf{y}')$ consists of at most one element. The conditions on a pentagon ensure that its fifth corner point is at the distinguished point a.

We will also need pentagons from $\mathbf{y}' \in \mathbb{G}'$ to $\mathbf{x} \in \mathbb{G}$, that satisfy all the properties from Definition 5.1.1, except that Equation (5.1) is replaced by the condition that $\partial(\partial_\alpha p) = \mathbf{x} - \mathbf{y}'$. For such pentagons, the fifth corner point is at b.

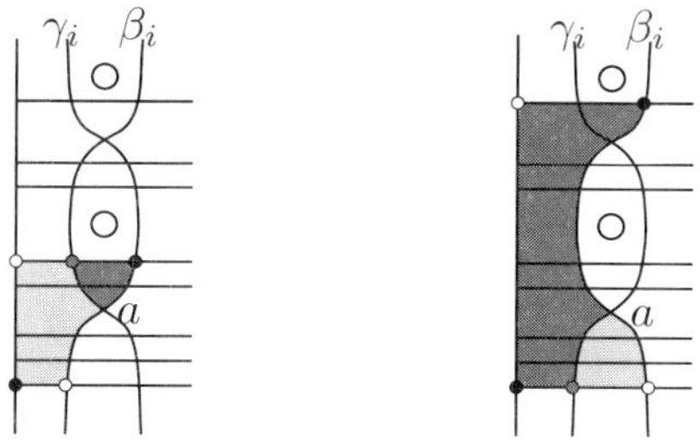

FIGURE 5.3. **Composing pentagons and triangles.** On the left, the pentagon can be pre-composed with a triangle to get a rectangle; on the right, the pentagon can be post-composed with a triangle to get a rectangle.

DEFINITION 5.1.2. The pentagon $p \in \text{Pent}(\mathbf{x}, \mathbf{y}')$ is said to be ***empty*** if

$$\text{Int}(p) \cap \mathbf{x} = \emptyset = \text{Int}(p) \cap \mathbf{y}'.$$

The set of empty pentagons from $\mathbf{x}$ to $\mathbf{y}'$ is denoted $\text{Pent}^\circ(\mathbf{x}, \mathbf{y}')$.

Define the $\mathbb{F}[V_1, \dots, V_n]$-module map $P\colon GC^-(\mathbb{G}) \to GC^-(\mathbb{G}')$ by the formula:

$$P(\mathbf{x}) = \sum_{\mathbf{y}' \in \mathbf{S}(\mathbb{G}')} \sum_{\{p \in \text{Pent}^\circ(\mathbf{x}, \mathbf{y}') \big| p \cap \mathbb{X} = \emptyset\}} V_1^{O_1(p)} \cdots V_n^{O_n(p)} \cdot \mathbf{y}'. \tag{5.2}$$

LEMMA 5.1.3. *The above map P is a bigraded map.*

Proof. There is a one-to-one correspondence

$$I\colon \mathbf{S}(\mathbb{G}') \to \mathbf{S}(\mathbb{G}), \tag{5.3}$$

the *nearest point map*, sending a grid state $\mathbf{x}' \in \mathbf{S}(\mathbb{G}')$ to the unique grid state $\mathbf{x} = I(\mathbf{x}')$ that agrees with $\mathbf{x}'$ in all but one component (i.e., the elements of $\mathbf{x}'$ on the generic vertical circles coincide with the corresponding elements of $I(\mathbf{x}')$).

There is a canonical small (positive) triangle from $\mathbf{x}'$ to $I(\mathbf{x}') = \mathbf{x}$, written $t_{\mathbf{x}}$. These triangles can be composed with pentagons to obtain rectangles, via a juxtaposition that will be still denoted by $*$. If $p \in \text{Pent}(\mathbf{x}, \mathbf{y}')$ then at least one of the following holds: $t_{\mathbf{x}} * p \in \text{Rect}(\mathbf{x}', \mathbf{y}')$ or $p * t_{\mathbf{y}} \in \text{Rect}(\mathbf{x}, \mathbf{y})$; see Figure 5.3.

We claim that if $t_{\mathbf{x}}$ is the triangle from $\mathbf{x}'$ to $\mathbf{x} = I(\mathbf{x}')$, then

$$M(\mathbf{x}) - M(\mathbf{x}') = -1 + 2\#(t_{\mathbf{x}} \cap \mathbb{O}). \tag{5.4}$$

To see this, choose the fundamental domain so that the triangle has the form of one of the four triangles illustrated in Figure 5.4. In all four cases $\mathcal{J}(\mathbf{x}, \mathbf{x}) - \mathcal{J}(\mathbf{x}', \mathbf{x}') = 0$, $\mathcal{J}(\mathbb{O}, \mathbb{O}) - \mathcal{J}(\mathbb{O}', \mathbb{O}') = -1$, while $\mathcal{J}(\mathbb{O}, \mathbf{x}) - \mathcal{J}(\mathbb{O}', \mathbf{x}')$ can be 0 or -1; and all four cases are consistent with Equation (5.4).

Equations (5.4) and (4.2) show that if $p \in \text{Pent}(\mathbf{x}, \mathbf{y}')$, then

$$M(\mathbf{x}) - M(\mathbf{y}') = -2\#(p \cap \mathbb{O}) + 2\#(\mathbf{x} \cap \text{Int}(p)),$$

bearing in mind that $\mathbf{y} \cap \text{Int}(t_{\mathbf{x}}) = \emptyset$ for any triangle $t_{\mathbf{x}}$ and any grid state $\mathbf{y}$. This implies that P preserves Maslov grading. Switching the roles of $\mathbb{X}$ and $\mathbb{O}$, and using Equation (4.3) we see that for any $p \in \text{Pent}(\mathbf{x}, \mathbf{y}')$

$$A(\mathbf{x}) - A(\mathbf{y}') = \#(p \cap \mathbb{X}) - \#(p \cap \mathbb{O}). \tag{5.5}$$

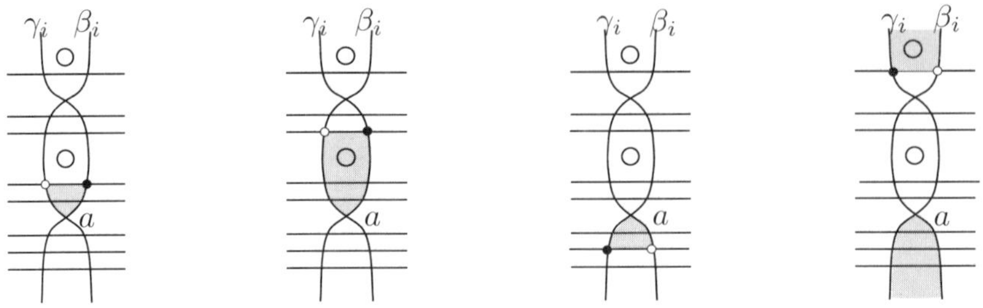

FIGURE 5.4. **Triangles shift Maslov grading.** The four possible types of small triangles with initial vertex at a.

Since all the pentagons counted in P have $p \cap \mathbb{X} = \emptyset$, it follows that P preserves the Alexander grading. □

LEMMA 5.1.4. *The map P defined in Equation* (5.2) *is a chain map.*

Proof. We verify the identity $\partial_{\mathbb{X}}^- \circ P = P \circ \partial_{\mathbb{X}}^-$ along the lines of Lemma 4.6.7, in this case by analyzing how pentagons and rectangles can interact.

We start by generalizing the notion of a domain from Definition 4.6.4. Cut the grid torus along each α_j, β_j, and γ_i, to obtain a union of triangles, rectangles, and pentagons. For $\mathbf{x} \in \mathbf{S}(\mathbb{G})$ and $\mathbf{y}' \in \mathbf{S}(\mathbb{G}')$, a domain ψ from $\mathbf{x}$ to $\mathbf{y}'$ is a formal sum of the closures of these elementary regions, taken with integral multiplicities, with the further property that $\partial_\alpha \psi$, the portion of the boundary in $\alpha_1 \cup \dots \cup \alpha_n$ (thought of as a formal sum of intervals), satisfies $\partial(\partial_\alpha \psi) = \mathbf{y}' - \mathbf{x}$.

Next, let $N(\psi)$ denote the number of decompositions of ψ as a composite of an empty rectangle (for $\mathbb{G}$) followed by an empty pentagon or as an empty pentagon followed by an empty rectangle (for $\mathbb{G}'$); compare the proof of Lemma 4.6.7. Then,

$$(5.6)\quad (\partial_{\mathbb{X}}^- \circ P + P \circ \partial_{\mathbb{X}}^-)(\mathbf{x}) = \sum_{\mathbf{z}' \in \mathbf{S}(\mathbb{G}')} \sum_{\left\{\psi \in \pi(\mathbf{x},\mathbf{z}') \middle| \mathbb{X} \cap \psi = \emptyset\right\}} N(\psi) \cdot V_1^{O_1(\psi)} \cdots V_n^{O_n(\psi)} \cdot \mathbf{z}'.$$

When $N(\psi) > 0$, then ψ satisfies one of the following three conditions:

(P-1) $\mathbf{x} \setminus (\mathbf{x} \cap \mathbf{z}')$ consists of 4 points. In this case, there are exactly two decompositions of ψ, as $r * p$ or $p' * r'$, where p and p' resp. r and r' have the same underlying pentagon resp. rectangle, i.e. they differ only in the grid states they connect. This shows that $N(\psi) = 2$.

(P-2) $\mathbf{x} \setminus (\mathbf{x} \cap \mathbf{z}')$ consists of 3 points. There are two cases: either all of the local multiplicities of ψ are 0 and 1, or some local multiplicity is 2. In the first case, ψ has seven corners, one of them is a $270°$ corner. Cutting at this $270°$ corner in two different directions, gives two different decompositions of ψ. Both are juxtapositions of a rectangle and a pentagon, but the precise order depends on the geometry of ψ. In the second case, ψ has a $270°$ corner at a, and cutting in both ways there gives the two decompositions of ψ. In all cases $N(\psi) = 2$. Both Cases (P-1) and (P-2) are illustrated in Figure 5.5.

(P-3) $\mathbf{x} \setminus (\mathbf{x} \cap \mathbf{z}')$ consists of 1 point. This happens when the domain ψ goes around the torus, which in turn can happen in two ways: the composition is either a vertical thin annulus together with a small triangle (shown

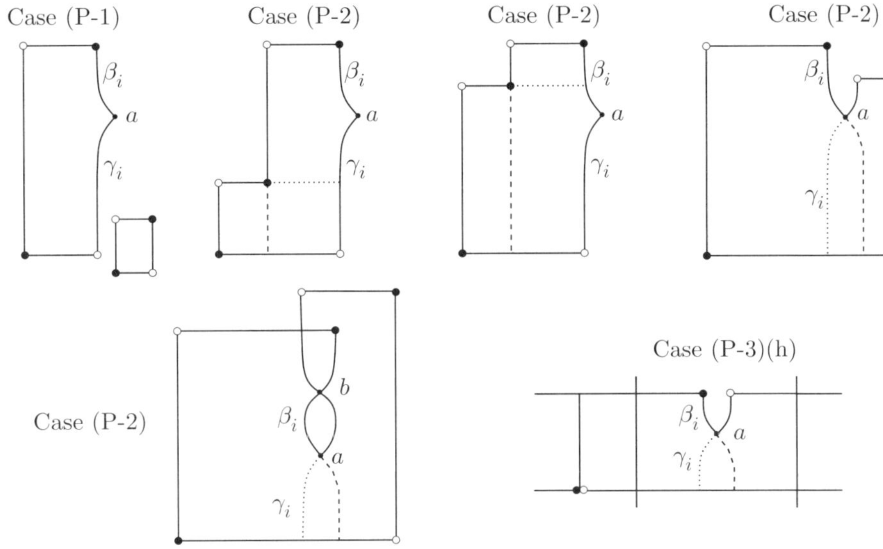

FIGURE 5.5. **Cases from the proof of Lemma 5.1.4.** The lower left shows a domain with some local multiplicity 2.

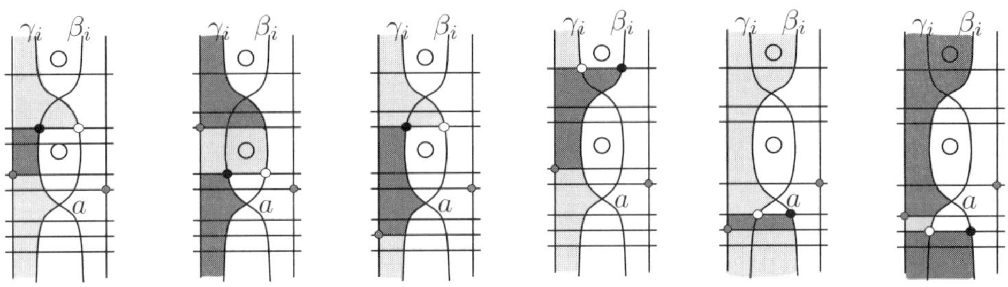

FIGURE 5.6. **Case (P-3)(v).** The diagram shows the six types of decompositions of the region ψ_1 from the proof of Lemma 5.1.4.

in Figure 5.6) or a horizontal thin annulus minus a small triangle, as in Figure 5.5. We denote these two cases Case (P-3)(v) and Case (P-3)(h) respectively. In Case (P-3)(v), ψ has a unique decomposition as a thin pentagon and a thin rectangle; in Case (P-3)(h), ψ has $N(\psi) = 2$.

Verifying that $\partial_{\mathbb{X}}^- \circ P(\mathbf{x}) + P \circ \partial_{\mathbb{X}}^-(\mathbf{x}) = 0$ now reduces to showing that the contributions coming from Case (P-3)(v) cancel in pairs, which we see as follows. Clearly, any configuration ψ as in Case (P-3)(v) is supported in the annulus between β_{i-1} and β_{i+1}, since the rectangle and the pentagon are assumed to be empty. In fact, there are exactly two possible domains ψ_1 and ψ_2 as in Case (P-3)(v): one has multiplicity 1 immediately to the east of β_{i-1} (and multiplicity 0 immediately to the west of β_{i+1}) and the other has multiplicity 1 to the west of β_{i+1} (and multiplicity 0 to the east of β_{i-1}). In the decomposition of ψ_1 we use the component of $\mathbf{x}$ on β_{i-1}, and when decomposing ψ_2 we use the component of $\mathbf{x}$ on β_{i+1}, cf. the diagrams on Figure 5.6 for ψ_1.

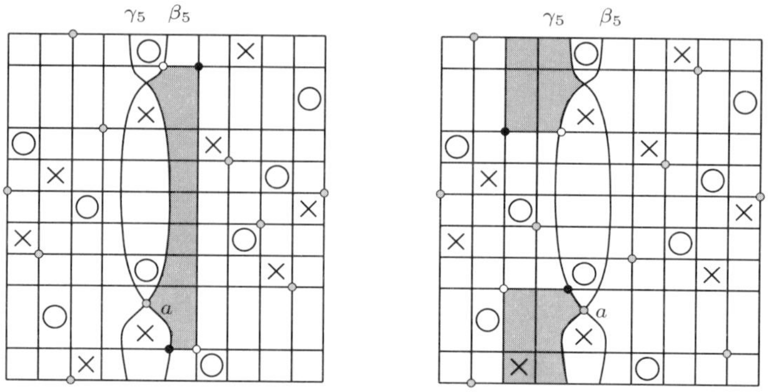

FIGURE 5.7. **Examples of hexagons.** The hexagon can be on either side of the commutation region; an example of each sides is shown by the diagram. Both hexagons are empty.

To express the relationship between ψ_1 and ψ_2, we set up some notation. The vertical circles β_{i-1} and β_{i+1} bound an annulus which lies to the east of β_{i-1} and to the west of β_{i+1}. The complement of $\beta_i \cup \gamma_i$ in this annulus consists of two bigons and two annuli; the annuli are denoted χ_1 and χ_2, numbered so that β_{i-1} meets χ_1 and β_{i+1} meets χ_2. If $t'_{\mathbf{x}}$ denotes the triangle connecting $\mathbf{x}$ and $\mathbf{z}'$ (with $I(\mathbf{z}') = \mathbf{x}$, in the notation of Lemma 5.1.3) with third corner b, then the support of ψ_1 is the union $\chi_1 \cup t'_{\mathbf{x}}$, while the support of ψ_2 is equal to $\chi_2 \cup t'_{\mathbf{x}}$. Since χ_1 and χ_2 are disjoint from $\mathbb{O}$ and $\mathbb{X}$, and $\psi_1 - \psi_2 = \chi_1 - \chi_2$, it follows that ψ_1 and ψ_2 have the same contribution in the right-hand-side of Equation (5.6). $\square$

Define an analogous map $P' \colon GC^-(\mathbb{G}') \to GC^-(\mathbb{G})$ by counting empty pentagons from grid states of $\mathbb{G}'$ to grid states of $\mathbb{G}$; i.e. for $\mathbf{x}' \in \mathbf{S}(\mathbb{G}')$, let

$$P'(\mathbf{x}') = \sum_{\mathbf{y}\in\mathbf{S}(\mathbb{G})} \sum_{\{p\in\mathrm{Pent}^\circ(\mathbf{x}',\mathbf{y}) \big| p\cap\mathbb{X}=\emptyset\}} V_1^{O_1(p)} \cdots V_n^{O_n(p)} \cdot \mathbf{y}, \tag{5.7}$$

counting pentagons with one vertex at $b \in \beta_i \cap \gamma_i$. Next we will show that the two maps P and P' are homotopy inverses of each other, via a homotopy operator that counts suitable hexagons, defined below.

DEFINITION 5.1.5. Fix grid states $\mathbf{x}, \mathbf{y} \in \mathbf{S}(\mathbb{G})$. An embedded disk h in the torus whose boundary is in the union of the α_j, β_j (for $j = 1, \dots, n$) and γ_i is called a ***hexagon from* $\mathbf{x}$ *to* $\mathbf{y}$** if it satisfies the following conditions:

- At any of the six corner points x of h, the hexagon contains exactly one of the four quadrants determined by the two intersecting curves at x.
- Four of the corner points of h are in $\mathbf{x} \cup \mathbf{y}$, and the two other corners are a and b.
- $\partial(\partial_\alpha h) = \mathbf{y} - \mathbf{x}$.

Denote the set of hexagons from $\mathbf{x}$ to $\mathbf{y}$ by $\mathrm{Hex}(\mathbf{x}, \mathbf{y})$. A hexagon $h \in \mathrm{Hex}(\mathbf{x}, \mathbf{y})$ is ***empty*** if $\mathrm{Int}(h) \cap \mathbf{x} = \emptyset = \mathrm{Int}(h) \cap \mathbf{y}$. The set of empty hexagons from $\mathbf{x}$ to $\mathbf{y}$ is denoted $\mathrm{Hex}^\circ(\mathbf{x}, \mathbf{y})$. (See Figure 5.7 for an example.)

Consider the $\mathbb{F}[V_1, \dots, V_n]$-module homomorphism $H\colon GC^-(\mathbb{G}) \to GC^-(\mathbb{G})$, counting certain empty hexagons: for each $\mathbf{x} \in \mathbf{S}(\mathbb{G})$,

$$H(\mathbf{x}) = \sum_{\mathbf{y}\in\mathbf{S}(\mathbb{G})} \sum_{\{h\in\mathrm{Hex}^\circ(\mathbf{x},\mathbf{y})\,\big|\,h\cap\mathbb{X}=\emptyset\}} V_1^{O_1(h)} \cdots V_n^{O_n(h)} \cdot \mathbf{y}. \tag{5.8}$$

An analogous map $H'\colon GC^-(\mathbb{G}') \to GC^-(\mathbb{G}')$, is defined by counting empty hexagons from $\mathbb{G}'$ to itself.

LEMMA 5.1.6. *The map $H\colon GC^-(\mathbb{G}) \to GC^-(\mathbb{G})$ provides a homotopy from the bigraded chain map $P' \circ P$ to the identity map on $GC^-(\mathbb{G})$.*

Proof. First, we show that H is homogeneous of degree $(1,0)$. Suppose that $V_1^{k_1} \cdots V_n^{k_n} \cdot \mathbf{y}$ occurs with non-zero coefficient in $H(\mathbf{x})$. The corresponding hexagon $h \in \mathrm{Hex}^\circ(\mathbf{x},\mathbf{y})$ can be naturally extended to a rectangle $r \in \mathrm{Rect}(\mathbf{x},\mathbf{y})$ that differs from h by the addition of a bigon next to β_i containing exactly one X and exactly one O. Combining with Equations (4.2) and (4.4), we conclude that

$$\begin{aligned} M(\mathbf{x}) - M(V_1^{O_1(h)} \cdots V_n^{O_n(h)} \cdot \mathbf{y}) &= -1 + 2\#(\mathbf{x} \cap \mathrm{Int}(h)) \\ A(\mathbf{x}) - A(V_1^{O_1(h)} \cdots V_n^{O_n(h)} \cdot \mathbf{y}) &= \#(h \cap \mathbb{X}). \end{aligned}$$

Since h is empty and $h \cap \mathbb{X} = \emptyset$, the claim for the grading shift follows.

The proof of the identity

$$\partial_{\mathbb{X}}^- \circ H + H \circ \partial_{\mathbb{X}}^- = \mathrm{Id} + P' \circ P. \tag{5.9}$$

is similar to the proof of Lemmas 4.6.7 and 5.1.4, analyzing how empty rectangles and empty hexagons can interact. Specifically, for $\psi \in \pi(\mathbf{x},\mathbf{z})$ let $N(\psi)$ denote the number of ways of decomposing ψ as either:

- $\psi = r * h$, where r is an empty rectangle and h is an empty hexagon;
- $\psi = h * r$, where h is an empty hexagon and r is an empty rectangle;
- $\psi = p * p'$, where p is an empty pentagon from $\mathbb{G}$ to $\mathbb{G}'$ and p' is an empty pentagon from $\mathbb{G}'$ back to $\mathbb{G}$.

Clearly,

$$(\partial_{\mathbb{X}}^- \circ H + H \circ \partial_{\mathbb{X}}^- + P' \circ P)(\mathbf{x}) = \sum_{\mathbf{z}\in\mathbf{S}(\mathbb{G})} \sum_{\{\psi\in\pi(\mathbf{x},\mathbf{z})\,\big|\,\psi\cap\mathbb{X}=\emptyset\}} N(\psi) \cdot V_1^{O_1(\psi)} \cdots V_n^{O_n(\psi)} \cdot \mathbf{z}.$$

Again, there are three cases of $\psi \in \pi(\mathbf{x},\mathbf{z})$ with $N(\psi) > 0$:

(H-1) $\mathbf{x} \setminus (\mathbf{x} \cap \mathbf{z})$ consists of 4 elements. In this case $N(\psi) = 2$, and the two decompositions are as $r_1 * h_1$ and $h_2 * r_2$, where r_1 and r_2 are rectangles for $\mathbb{G}$, and h_1 and h_2 are hexagons for $\mathbb{G}$ with the same support.

(H-2) $\mathbf{x} \setminus (\mathbf{x} \cap \mathbf{z})$ consists of 3 elements. In this case, ψ has eight corners. Either seven of these corners are 90°, and one is 270°, or five are 90°, and three are 270°. In the first case, cutting at the 270° corner in two different directions gives the two decompositions of ψ. In the second case, the three corners with 270° include both a and b. Again, $N(\psi) = 2$: at one of the three corners there is a choice of two different directions for cutting, while at the other two corners, the direction for cutting is uniquely determined. Both Cases (H-1) and (H-2) are illustrated in Figure 5.8.

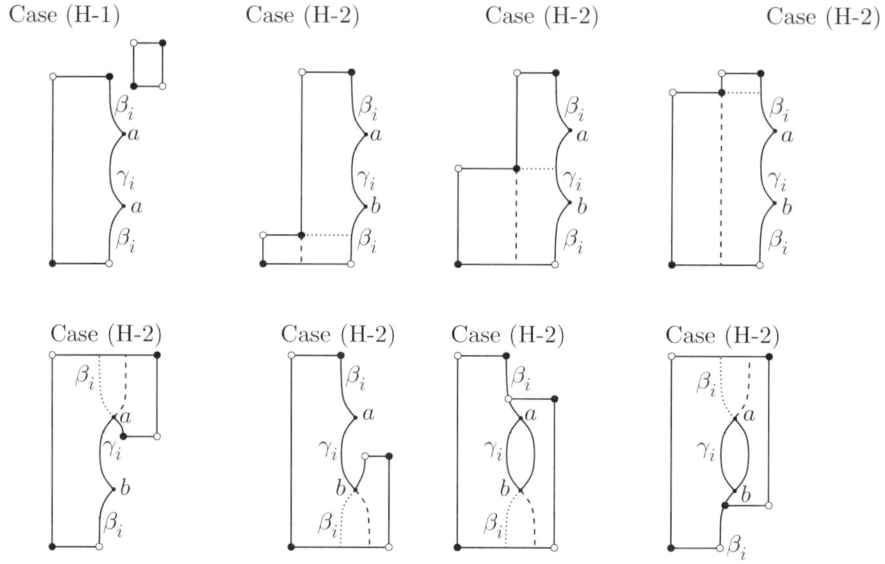

FIGURE 5.8. **Cases (H-1) and (H-2) from Lemma 5.1.6.**

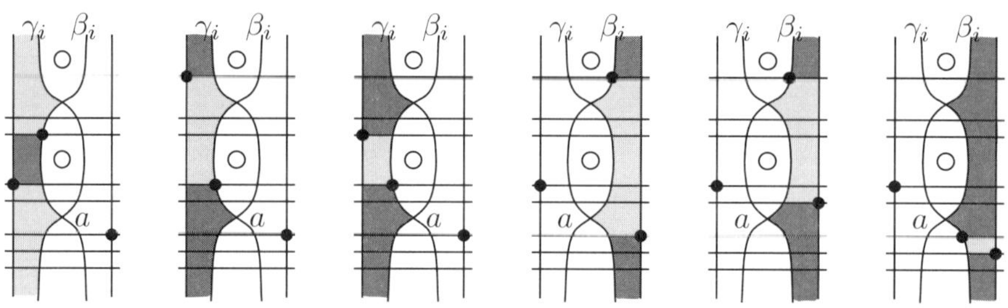

FIGURE 5.9. **Various decompositions of Case (H-3) from Lemma 5.1.6.** The initial grid state is labeled with a black dot; it is also the terminal grid state.

(H-3) $\mathbf{x} = \mathbf{z}$. In this case, ψ is supported inside an annulus between β_i and one of the other consecutive vertical circles. In this case $N(\psi) = 1$, and ψ could have decompositions of any of the three types (that is, rectangle-hexagon, hexagon-rectangle or pentagon-pentagon), depending on the exact placement of $\mathbf{x}$ on β_{i-1}, β_i, and β_{i+1}; see Figure 5.9.

For any given $\mathbf{x}$, the domain ψ appearing in Case (H-3) is unique, and it contains no markings (i.e. $\mathbb{X} \cap \psi = \mathbb{O} \cap \psi = \emptyset$). Hence, it contributes the identity map, as illustrated in Figure 5.9, verifying the identity from Equation (5.9). □

Now we are ready to prove the main result of the section:

PROPOSITION 5.1.7. *If $\mathbb{G}$ and $\mathbb{G}'$ are two grid diagrams which differ by a commutation move, then there is an isomorphsm of bigraded vector spaces $\widehat{GH}(\mathbb{G}) \cong \widehat{GH}(\mathbb{G}')$ and an isomorphism of bigraded $\mathbb{F}[U]$-modules $GH^-(\mathbb{G}) \cong GH^-(\mathbb{G}')$.*

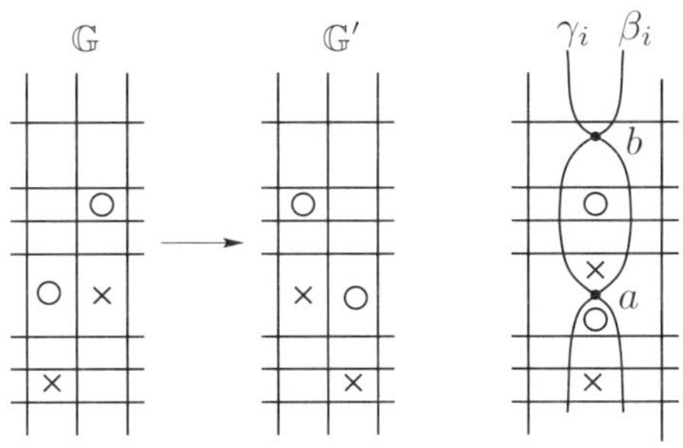

FIGURE 5.10. **Common diagram for a switch.**

Proof. Suppose that $\mathbb{G}$ and $\mathbb{G}'$ differ by commuting two columns. The map $P\colon GC^-(\mathbb{G}) \to GC^-(\mathbb{G}')$ is a chain map by Lemma 5.1.4 that, by Lemma 5.1.3, respects the gradings; and by applying Lemma 5.1.6 twice (for $P' \circ P$ and H as stated, and for $P \circ P'$ and H' with the necessary minor modifications), it is a bigraded chain homotopy equivalence.

Furthermore, a bigraded chain homotopy equivalence between $GC^-(\mathbb{G})$ and $GC^-(\mathbb{G}')$ induces a bigraded chain homotopy equivalence on the $V_1 = 0$ specialization $\widehat{GC}(\mathbb{G}) = \frac{GC^-(\mathbb{G})}{V_1} \to \widehat{GC}(\mathbb{G}') = \frac{GC^-(\mathbb{G}')}{V_1}$.

Since a bigraded chain homotopy equivalence induces isomorphisms on homology in both cases ($\widehat{GH}$ and GH^-), the result follows when columns are commuted.

When rows are commuted, the same result is proved by a slight modification of the earlier proof: one has to simply rotate all pictures (such as those appearing in the definition of Definition 5.1.1) by 90°. □

5.1.1. Invariance under switches. Recall that Definition 3.1.6 and Definition 3.1.10 define similar moves: commutations and switches. The above proof in fact establishes invariance under both kinds of moves. As before, we draw the two grid diagrams $\mathbb{G}$ and $\mathbb{G}'$ connected by a switch on one torus, so that two of the vertical circles are curved, rather than straight. When switching two columns, the O- and the X-markings sharing a row now will be in the same square determined by the straight lines, separated only by the curved ones, as illustrated in Figure 5.10. With this detail understood, the pentagon counting maps as in the proof of Proposition 5.1.7 induce isomorphisms on homologies, proving the following:

PROPOSITION 5.1.8. *Suppose that the two toroidal grid diagrams $\mathbb{G}$ and $\mathbb{G}'$ differ by a switch. Then, there is an isomorphism of bigraded vector spaces $\widehat{GH}(\mathbb{G}) \cong \widehat{GH}(\mathbb{G}')$ and an isomorphism of bigraded $\mathbb{F}[U]$-modules $GH^-(\mathbb{G}) \cong GH^-(\mathbb{G}')$.* □

The need to distinguish commutations and switches will arise later, in the proof of the commutation invariance of a particular grid state, cf. Lemma 6.4.4.

5.2. Stabilization invariance

We now turn to the stabilization invariance of grid homology:

PROPOSITION 5.2.1. *If $\mathbb{G}'$ is obtained from $\mathbb{G}$ by stabilization, then there are isomorphisms:*

$$\widehat{GH}(\mathbb{G}) \cong \widehat{GH}(\mathbb{G}'), \tag{5.10}$$

$$GH^-(\mathbb{G}) \cong GH^-(\mathbb{G}'), \tag{5.11}$$

the first of which is an isomorphism of bigraded vector spaces, and the second of which is an isomorphism of bigraded $\mathbb{F}[U]$-modules.

5.2.1. Stabilization in the fully and simply blocked theories. As a warm-up, we will start with the easier case of the effect of stabilization on the fully blocked homology group $\widetilde{GH}$, from which invariance for $\widehat{GH}$ follows quickly. We return to the more general case of GH^- in Section 5.2.3.

PROPOSITION 5.2.2. *Suppose that $\mathbb{G}'$ is a stabilization of $\mathbb{G}$. Then, there is an isomorphism of bigraded vector spaces*

$$\widetilde{GH}(\mathbb{G}') \cong \widetilde{GH}(\mathbb{G}) \oplus \widetilde{GH}(\mathbb{G})[\![1,1]\!]. \tag{5.12}$$

COROLLARY 5.2.3. *The simply blocked grid homology is invariant under stabilization; that is, if the grid diagram $\mathbb{G}'$ is given as a stabilization of $\mathbb{G}$, then Equation* (5.10) *holds.*

Proof. This is an immediate consequence of Propositions 5.2.2 and 4.6.15. □

We prove Proposition 5.2.2 after introducing some notation. Assume that $\mathbb{G}'$ is obtained from $\mathbb{G}$ by a stabilization of type *X:SW*, in the terminology from Section 3.1; i.e. $\mathbb{G}$ is gotten from $\mathbb{G}'$ by destabilizing at a two-by-two square $\begin{array}{c|c} X & O \\ \hline & X \end{array}$. Number the O-markings so that O_1 is the newly-introduced one, and O_2 is the O-marking in the row just below O_1. Let c denote the intersection point of the new horizontal and vertical circles in $\mathbb{G}'$. We will find it also useful to label the two X-markings as $\begin{array}{c|c} X_1 & O_1 \\ \hline & X_2 \end{array}$.

Decompose the set of grid states $\mathbf{S}(\mathbb{G}')$ of the stabilized diagram $\mathbb{G}'$ as the disjoint union $\mathbf{I}(\mathbb{G}') \cup \mathbf{N}(\mathbb{G}')$, where $\mathbf{I}(\mathbb{G}')$ is the set of grid states $\mathbf{x} \in \mathbf{S}(\mathbb{G}')$ with $c \in \mathbf{x}$. This decomposition splits the vector space $\widetilde{GC}(\mathbb{G}') \cong \widetilde{\mathbf{I}} \oplus \widetilde{\mathbf{N}}$, where $\widetilde{\mathbf{I}}$ and $\widetilde{\mathbf{N}}$ denote the spans of the grid states from $\mathbf{I}(\mathbb{G}')$ and $\mathbf{N}(\mathbb{G}')$ respectively. Note that $\widetilde{\mathbf{N}}$ is a subcomplex: any rectangle $\operatorname{Rect}(\mathbf{x}, \mathbf{y})$ with $\mathbf{x} \in \mathbf{N}(\mathbb{G}')$ and $\mathbf{y} \in \mathbf{I}(\mathbb{G}')$ must contain one of X_1 or X_2. With respect to the splitting $\widetilde{GC}(\mathbb{G}') \cong \widetilde{\mathbf{I}} \oplus \widetilde{\mathbf{N}}$, the map $\tilde{\partial}_{\mathbb{O},\mathbb{X}}$ can be written in the matrix form as

$$\tilde{\partial}_{\mathbb{O},\mathbb{X}} = \begin{pmatrix} \widetilde{\partial}_{\mathbf{I}}^{\mathbf{I}} & 0 \\ \widetilde{\partial}_{\mathbf{I}}^{\mathbf{N}} & \widetilde{\partial}_{\mathbf{N}}^{\mathbf{N}} \end{pmatrix},$$

so that $\widetilde{\partial}_{\mathbf{I}}^{\mathbf{N}} \colon (\widetilde{\mathbf{I}}, \widetilde{\partial}_{\mathbf{I}}^{\mathbf{I}}) \to (\widetilde{\mathbf{N}}, \widetilde{\partial}_{\mathbf{N}}^{\mathbf{N}})$ is a chain map. Proposition 5.2.2 will follow quickly from the three statements, which we prove in the subsequent lemmas:

- The homology of the chain complex $(\widetilde{\mathbf{I}}, \widetilde{\partial}_{\mathbf{I}}^{\mathbf{I}})$ is isomorphic to $\widetilde{GH}(\mathbb{G})[\![1,1]\!]$, as a bigraded vector space (Lemma 5.2.5).
- The homology of the chain complex $(\widetilde{\mathbf{N}}, \widetilde{\partial}_{\mathbf{N}}^{\mathbf{N}})$ is isomorphic to $\widetilde{GH}(\mathbb{G})$ as a bigraded vector space (Lemma 5.2.6, together with Lemma 5.2.5).

- The chain map $\widetilde{\partial}_{\mathbf{I}}^{\mathbf{N}}$ induces the zero map in homology (Lemma 5.2.8).

As a preliminary to the first step, note that there is a one-to-one correspondence between grid states in $\mathbf{S}(\mathbb{G})$ and those in $\mathbf{I}(\mathbb{G}')$, which takes $\mathbf{x} \in \mathbf{S}(\mathbb{G})$ to the state $\mathbf{x} \cup \{c\} \in \mathbf{I}(\mathbb{G}')$.

LEMMA 5.2.4. *For* $\mathbf{x} \in \mathbf{S}(\mathbb{G})$, *let* $\mathbf{x}' = \mathbf{x} \cup \{c\} \in \mathbf{I}(\mathbb{G}')$; *then*

$$M(\mathbf{x}') = M(\mathbf{x}) - 1 \qquad \text{and} \qquad A(\mathbf{x}') = A(\mathbf{x}) - 1.$$

Proof. Consider a fundamental domain, where the stabilization is in the upper right corner. Using the formula of Equations (4.5), a simple calculation shows that $M_{\mathbb{O}'}(\mathbf{x}') = M_{\mathbb{O}}(\mathbf{x}) - 1$, and $M_{\mathbb{X}'}(\mathbf{x}') = M_{\mathbb{X}}(\mathbf{x})$, implying $A(\mathbf{x}') = A(\mathbf{x}) - 1$. □

LEMMA 5.2.5. *The map* $\widetilde{e}\colon \widetilde{\mathbf{I}} \to \widetilde{GC}(\mathbb{G})$ *induced by the correspondence* $\mathbf{x} \cup \{c\} \mapsto \mathbf{x}$ *gives an isomorphism between the bigraded chain complexes* $(\widetilde{\mathbf{I}}, \widetilde{\partial}_{\mathbf{I}}^{\mathbf{I}})$ *and* $\widetilde{GH}(\mathbb{G})[\![1,1]\!]$.

Proof. The map $\widetilde{e}$ is a bijection on the corresponding grid states, so it induces an isomorphism between the vector spaces $\widetilde{\mathbf{I}}$ and $\widetilde{GC}(\mathbb{G})$. Clearly, empty rectangles disjoint from $\mathbb{X} \cup \mathbb{O}$ in $\mathbb{G}$ correspond to empty rectangles disjoint from $\mathbb{X}' \cup \mathbb{O}'$ in $\mathbb{G}'$. This shows that $\widetilde{e}$ is an isomorphism of chain complexes; the grading shift was verified in Lemma 5.2.4. □

To relate the homology of $\widetilde{\mathbf{I}}$ and $\widetilde{\mathbf{N}}$, consider the linear map $\widetilde{\mathcal{H}}_{X_2}^{\mathbf{I}} : \widetilde{\mathbf{N}} \to \widetilde{\mathbf{I}}$ whose value on any $\mathbf{x} \in \mathbf{N}(\mathbb{G}')$ is given by

$$\widetilde{\mathcal{H}}_{X_2}^{\mathbf{I}}(\mathbf{x}) = \sum_{\mathbf{y} \in \mathbf{I}(\mathbb{G}')} \#\{r \in \operatorname{Rect}^\circ(\mathbf{x}, \mathbf{y}) \big| \operatorname{Int}(r) \cap \mathbb{O} = \emptyset, \operatorname{Int}(r) \cap \mathbb{X} = X_2\} \cdot \mathbf{y}, \tag{5.13}$$

and consider the linear map $\widetilde{\mathcal{H}}_{O_1} : \widetilde{\mathbf{I}} \to \widetilde{\mathbf{N}}$ whose value on any $\mathbf{x} \in \mathbf{I}(\mathbb{G}')$ is

$$\widetilde{\mathcal{H}}_{O_1}(\mathbf{x}) = \sum_{\mathbf{y} \in \mathbf{N}(\mathbb{G}')} \#\{r \in \operatorname{Rect}^\circ(\mathbf{x}, \mathbf{y}) \big| \operatorname{Int}(r) \cap \mathbb{X} = \emptyset, \operatorname{Int}(r) \cap \mathbb{O} = O_1\} \cdot \mathbf{y}.$$

LEMMA 5.2.6. *The map* $\widetilde{\mathcal{H}}_{X_2}^{\mathbf{I}}$ *drops both the Maslov and Alexander gradings by one, while* $\widetilde{\mathcal{H}}_{O_1}$ *increases both gradings by one. Moreover the maps* $\widetilde{\mathcal{H}}_{X_2}^{\mathbf{I}}$ *and* $\widetilde{\mathcal{H}}_{O_1}$ *are chain maps, and induce isomorphisms on homology.*

Proof. The grading changes of $\widetilde{\mathcal{H}}_{X_2}^{\mathbf{I}}$ and $\widetilde{\mathcal{H}}_{O_1}$ follow readily from Equations (4.2) and (4.4); and the methods in the proof of Lemma 4.6.7 shows that both maps are chain maps.

The value of the composition $\widetilde{\mathcal{H}}_{X_2}^{\mathbf{I}} \circ \widetilde{\mathcal{H}}_{O_1}$ on a generator $\mathbf{x} \in \mathbf{I}(\mathbb{G}')$ counts juxtapositions $r_1 * r_2$ of pairs of empty rectangles, where r_1 starts at $\mathbf{x} \in \mathbf{I}(\mathbb{G}')$ and contains O_1; and r_2 contains X_2 and terminates in some state in $\mathbf{I}(\mathbb{G}')$. By simple geometric reasoning, for any $\mathbf{x} \in \mathbf{I}(\mathbb{G}')$, there exists a unique such juxtaposition, forming a vertical annulus of width one through O_1 and X_2. It follows that

$$\widetilde{\mathcal{H}}_{X_2}^{\mathbf{I}} \circ \widetilde{\mathcal{H}}_{O_1} = \operatorname{Id}_{\widetilde{\mathbf{I}}}. \tag{5.14}$$

Consider the linear map $\widetilde{\mathcal{H}}_{O_1,X_2}\colon \widetilde{\mathbf{N}} \to \widetilde{\mathbf{N}}$ whose value on $\mathbf{x} \in \mathbf{N}(\mathbb{G}')$ is given by

(5.15)
$$\widetilde{\mathcal{H}}_{O_1,X_2}(\mathbf{x}) = \sum_{\mathbf{y}\in\mathbf{N}(\mathbb{G}')} \#\{r \in \mathrm{Rect}^\circ(\mathbf{x},\mathbf{y}) \big| \mathrm{Int}(r)\cap\mathbb{O} = O_1, \mathrm{Int}(r)\cap\mathbb{X} = X_2\}\cdot\mathbf{y}.$$

It is easy to see that this map is homogeneous of degree $(1,0)$. A minor adaptation of the earlier argument shows that

$$\widetilde{\mathcal{H}}_{O_1} \circ \widetilde{\mathcal{H}}^{\mathbf{I}}_{X_2} + \widetilde{\mathcal{H}}_{O_1,X_2} \circ \widetilde{\partial}^{\mathbf{N}}_{\mathbf{N}} + \widetilde{\partial}^{\mathbf{N}}_{\mathbf{N}} \circ \widetilde{\mathcal{H}}_{O_1,X_2} = \mathrm{Id}_{\widetilde{\mathbf{N}}}. \tag{5.16}$$

Indeed, the left-hand-side counts domains starting and ending at states in $\mathbf{N}(\mathbb{G}')$, that can be given as composition of two empty rectangles, with the property that the union intersects $\mathbb{X}\cup\mathbb{O}$ in $\{X_2, O_1\}$. For most such domains ψ we have $N(\psi) = 2$ (as defined in the proof of Lemma 4.6.7), and hence these contributions cancel. The exceptional cases are those domains ψ that are thin annuli containing X_2 and O_1. These domains identify the left-hand side of Equation (5.16) with $\mathrm{Id}_{\widetilde{\mathbf{N}}}$.

Equations (5.14) and (5.16) show that $\widetilde{\mathcal{H}}^{\mathbf{I}}_{X_2}$ and $\widetilde{\mathcal{H}}_{O_1}$ are chain homotopy equivalences, verifying the lemma. □

REMARK 5.2.7. The identities used to prove Lemma 5.2.6 can also be derived from the relation $\partial^-_{\mathbb{X}} \circ \partial^-_{\mathbb{X}} = 0$ (Lemma 4.6.7), and the fact that $\mathcal{H}_{X_2}$ (from Equation (4.16)) is a chain homotopy between multiplication by V_1 and V_2.

LEMMA 5.2.8. *The chain map* $\widetilde{\partial}^{\mathbf{N}}_{\mathbf{I}}\colon (\widetilde{\mathbf{I}}, \widetilde{\partial}^{\mathbf{I}}_{\mathbf{I}}) \to (\widetilde{\mathbf{N}}, \widetilde{\partial}^{\mathbf{N}}_{\mathbf{N}})$ *induces thc trivial map on homology.*

Proof. If $\mathbf{x},\mathbf{z} \in \mathbf{I}(\mathbb{G}')$, $\mathbf{y} \in \mathbf{N}(\mathbb{G}')$, $r_1 \in \mathrm{Rect}^\circ(\mathbf{x},\mathbf{y})$ and $r_2 \in \mathrm{Rect}^\circ(\mathbf{y},\mathbf{z})$, it is easy to see that $\mathbf{x} = \mathbf{z}$ and $r_1 * r_2$ is a thin annulus; in particular, one of r_1 or r_2 contains an O-marking. It follows immediately that $\widetilde{\mathcal{H}}^{\mathbf{I}}_{X_2} \circ \widetilde{\partial}^{\mathbf{N}}_{\mathbf{I}} = 0$. Since $\widetilde{\mathcal{H}}^{\mathbf{I}}_{X_2}$ induces an isomorphism in homology (Lemma 5.2.6), the lemma follows. □

Proof of Proposition 5.2.2. Start with the case of destabilizations of type *X:SW*, where $\widetilde{\mathbf{N}}$ is a subcomplex of $\widetilde{GC}$ with quotient complex $\widetilde{\mathbf{I}}$. Recalling the construction of the long exact sequence in homology associated to the short exact sequence

$$0 \longrightarrow \widetilde{\mathbf{N}} \longrightarrow \widetilde{GC}(\mathbb{G}') \longrightarrow \widetilde{\mathbf{I}} \longrightarrow 0,$$

we find that the connecting homomophism is the map on homology induced by $\widetilde{\partial}^{\mathbf{N}}_{\mathbf{I}}\colon \widetilde{\mathbf{I}} \to \widetilde{\mathbf{N}}$ (cf. Lemma A.2.1 from Appendix A). Since $\widetilde{\partial}^{\mathbf{N}}_{\mathbf{I}}$ vanishes on homology (Lemma 5.2.8), the long exact sequence reduces to the short exact sequence

$$0 \longrightarrow H(\widetilde{\mathbf{N}}) \longrightarrow \widetilde{GH}(\mathbb{G}') \longrightarrow H(\widetilde{\mathbf{I}}) \longrightarrow 0.$$

Replace $H(\widetilde{\mathbf{I}})$ and $H(\widetilde{\mathbf{N}})$ by $\widetilde{GH}(\mathbb{G})$ (with suitable grading shifts, as needed), following Lemmas 5.2.5 and 5.2.6, respectively. We obtain the short exact sequence

$$0 \longrightarrow \widetilde{GH}(\mathbb{G}) \longrightarrow \widetilde{GH}(\mathbb{G}') \longrightarrow \widetilde{GH}(\mathbb{G})[\![1,1]\!] \longrightarrow 0.$$

Since we are we working over a field, this short exact sequence implies Equation (5.12) for stabilizations of type *X:SW*. Invariance under the other types of stabilizations follows from this case, together with invariance under commutations and switches (Propositions 5.1.7 and 5.1.8), thanks to Corollary 3.2.3. □

5.2.2. Mapping cones: an algebraic interlude. Before moving on to stabilization invariance for the unblocked theory, we find it convenient to collect a few further tools from homological algebra that will be helpful. Again, we will state the results over some commutative base ring $\mathbb{K}$ with a unit, which we will choose to be $\mathbb{F} = \mathbb{Z}/2\mathbb{Z}$ in immediate applications.

DEFINITION 5.2.9. Let (C, ∂) and (C', ∂') be two bigraded chain complexes over $\mathcal{R} = \mathbb{K}[V_1, \dots, V_n]$. A map $f \colon C \to C'$ is a ***quasi-isomorphism*** if it is a bigraded chain map which induces an isomorphism in homology.

Note that every chain homotopy equivalence is a quasi-isomorphism.

DEFINITION 5.2.10. Fix chain complexes (C, ∂) and (C', ∂') over $\mathcal{R}$ and let $f \colon C \to C'$ be a chain map. The ***mapping cone*** $\mathrm{Cone}(f \colon C \to C')$ is the chain complex whose underlying $\mathcal{R}$-module is $C \oplus C'$, equipped with the differential

$$D(c, c') = (-\partial(c), \partial(c') + f(c)). \tag{5.17}$$

When C and C' are bigraded chain complexes over $\mathcal{R}$, and the chain map $f \colon C \to C'$ is bigraded of degree (m, t), the mapping cone $\mathrm{Cone}(f \colon C \to C')$ is a bigraded chain complex over $\mathcal{R}$, with bigrading specified by

$$\mathrm{Cone}(f \colon C \to C')_{d,s} = C_{d-m-1, s-t} \oplus C'_{d,s}. \tag{5.18}$$

For example, when f is multiplication by V_i, for all $d, s \in \mathbb{Z}$ we have an identification of $\mathbb{K}$-modules:

$$\mathrm{Cone}(V_i \colon C \to C)_{d,s} = C_{d+1, s+1} \oplus C_{d,s}. \tag{5.19}$$

We will need the following two basic properties of mapping cones. (Both are restated and proved in Appendix A; see Lemmas A.3.2 and A.3.8 respectively.)

LEMMA 5.2.11. *Let C and C' be two bigraded chain complexes of $\mathcal{R}$-modules, and $f \colon C \to C'$ is a chain map that is homogeneous of degree (m, t). Then, there is a bigraded map of $\mathcal{R}$-modules $H(i) \colon H(C') \to H(\mathrm{Cone}(f))$ and a map of $\mathcal{R}$-modules $H(p) \colon H(\mathrm{Cone}(f)) \to H(C)$ that is homogeneous with degree $(-m-1, -t)$, all of which fit into the long exact sequence*

$$\cdots \to H_{d+1,s}(C') \xrightarrow{H(i)} H_{d+1,s}(\mathrm{Cone}(f)) \xrightarrow{H(p)} H_{d-m,s-t}(C) \xrightarrow{H(f)} H_{d,s}(C') \to \cdots$$

LEMMA 5.2.12. *A commutative diagram of chain complexes of bigraded $\mathcal{R}$-modules*

$$\begin{array}{ccc} C & \xrightarrow{f} & C' \\ {\scriptstyle \phi}\downarrow & & \downarrow{\scriptstyle \phi'} \\ E & \xrightarrow{g} & E', \end{array}$$

where the horizontal maps are homogeneous of degree (m, t) and the vertical ones are bigraded, induces a bigraded chain map $\Phi \colon \mathrm{Cone}(f) \to \mathrm{Cone}(g)$, compatible with the long exact sequence from Lemma 5.2.11, in the sense that the following diagram *commutes:*

$$\begin{array}{ccccccccc}
\cdots \longrightarrow & H_{d+1,s}(C') & \longrightarrow & H_{d+1,s}(\mathrm{Cone}(f)) & \longrightarrow & H_{d-m,s-t}(C) & \xrightarrow{H(f)} & H_{d,s}(C') & \rightarrow \cdots \\
& \Big\downarrow H(\phi') & & \Big\downarrow H(\Phi) & & \Big\downarrow H(\phi) & & \Big\downarrow H(\phi') & \\
\cdots \longrightarrow & H_{d+1,s}(E') & \longrightarrow & H_{d+1,s}(\mathrm{Cone}(g)) & \longrightarrow & H_{d-m,s-t}(E) & \xrightarrow{H(g)} & H_{d,s}(E') & \rightarrow \cdots
\end{array}$$

Supposing moreover that the vertical maps ϕ and ϕ' are quasi-isomorphisms, then it follows that Φ is a quasi-isomorphism, as well. □

It is suggestive to think of the mapping cone as a kind of quotient complex. To this end, there is the following result which we use first in Chapter 7:

LEMMA 5.2.13. *If $f\colon C \to C'$ is an injective chain map, then there is a quasi-isomorphism from* $\mathrm{Cone}(f)$ *to the quotient $C'/f(C)$.* □

Proofs of the above, and the next result, which we also include for future reference, can be found in Appendix A. (See Lemmas A.3.9 and A.3.7 respectively.)

LEMMA 5.2.14. *If C, C' are two bigraded chain complexes over $\mathcal{R}$, and $f, g\colon C \to C'$ are two chain homotopic chain maps that are homogeneous of degree (m,t), then there is an isomorphism of bigraded chain complexes* $\mathrm{Cone}(f) \simeq \mathrm{Cone}(g)$. □

5.2.3. Stabilization in the unblocked theory. Now we study the effect of a stabilization on the unblocked theory GH^-; our goal is to verify Equation (5.11). As in Section 5.2.1, we will consider stabilizations of type *X:SW*.

A little complication arises when comparing $GC^-(\mathbb{G})$ and $GC^-(\mathbb{G}')$: they are defined over different polynomial rings. Numbering the variables suitably, we can think of $GC^-(\mathbb{G})$ as defined over $\mathbb{F}[V_2,\dots,V_n]$, while $GC^-(\mathbb{G}')$ is defined over $\mathbb{F}[V_1,\dots,V_n]$.

As a first step, we promote $GC^-(\mathbb{G})$ to a complex, denoted $GC^-(\mathbb{G})[V_1]$, defined over the base ring $\mathbb{F}[V_1,\dots,V_n]$. This promotion is done as follows:

DEFINITION 5.2.15. If M is a bigraded module over $\mathbb{F}[V_2,\dots,V_n]$, let $M[V_1]$ denote the module over $\mathbb{F}[V_1,\dots,V_n]$ given by

$$M[V_1] = M \otimes_{\mathbb{F}[V_2,\dots,V_n]} \mathbb{F}[V_1,\dots,V_n];$$

i.e. $M[V_1]$ consists of finite sums $\sum_k m_k \otimes V_1^k$ with $m_k \in M$ and $k \geq 0$, thought of as an $\mathbb{F}[V_1,\dots,V_n]$-module, where V_1 multiplies on the second factor and $V_2,\dots,V_n$ act on the first. The bigrading is specified by declaring $m \otimes V_1^k$ to be homogeneous of degree $(d-2k, s-k)$ if $m \in M$ is homogeneous of degree (d,s).

If (C,∂_C) is a bigraded chain complex over $\mathbb{F}[V_2,\dots,V_n]$, let $(C,\partial_C)[V_1]$ be the chain complex over $\mathbb{F}[V_1,\dots,V_n]$, whose underlying $\mathbb{F}[V_1,\dots,V_n]$-module is $C[V_1]$, and whose differential is specified by $\partial_{C[V_1]}(m \otimes V_1^{\otimes k}) = (\partial_C m) \otimes V_1^{\otimes k}$.

The homology of $GC^-(\mathbb{G})[V_1]$ is much larger than the homology of $GC^-(\mathbb{G})$. To rectify this, we divide out in a suitable sense by $V_1 - V_2$. More precisely, consider the (degree-shifting) chain map $V_1 - V_2\colon GC^-(\mathbb{G})[V_1] \to GC^-(\mathbb{G})[V_1]$, defined as multiplication by $V_1 - V_2 \in \mathbb{F}[V_1,\dots,V_n]$, and take its mapping cone.

According to the following lemma, choosing U to be any V_i for $i > 1$, there is an isomorphism of bigraded $\mathbb{F}[U]$-modules

$$H(\mathrm{Cone}(V_1 - V_2 \colon GC^-(\mathbb{G})[V_1] \to GH^-(\mathbb{G})[V_1])) \cong GH^-(\mathbb{G}).$$

LEMMA 5.2.16. *Let C be a bigraded chain complex over $\mathbb{F}[V_2, \ldots, V_n]$. Then, there is an isomorphism of bigraded $\mathbb{F}[V_2, \ldots, V_n]$-modules*

$$H(\mathrm{Cone}(V_1 - V_2 \colon C[V_1] \to C[V_1])) \cong H(C).$$

Proof. Observe that $H(C[V_1]) \cong H(C)[V_1]$, since the chain complex underlying $C[V_1]$ is a direct sum of infinitely many copies of C, where the ith summand consists of elements $c \cdot V_1^i$ with $c \in C$. It follows that the injective map $V_1 - V_2 \colon C[V_1] \to C[V_1]$ induces the injective map $V_1 - V_2 \colon H(C)[V_1] \to H(C)[V_1]$ on homology. Abbreviate $\mathrm{Cone}(V_1 - V_2 \colon C[V_1] \to C[V_1])$ as $\mathrm{Cone}(V_1 - V_2)$. The long exact sequence of a mapping cone (Lemma 5.2.11) now specializes to the short exact sequence of bigraded $\mathbb{F}[V_1, \ldots, V_n]$-modules

$$0 \longrightarrow H(C)[V_1] \xrightarrow{V_1 - V_2} H(C)[V_1] \longrightarrow H(\mathrm{Cone}(V_1 - V_2)) \longrightarrow 0;$$

and hence gives an isomorphism of bigraded $\mathbb{F}[V_1, \ldots, V_n]$-modules,

$$H(\mathrm{Cone}(V_1 - V_2)) \cong \frac{H(C)[V_1]}{V_1 - V_2}.$$

The lemma follows now from the observation that for any bigraded $\mathbb{F}[V_2, \ldots, V_n]$-module M, such as $M = H(C)$, the inclusion $M \to M[V_1]$ induces an isomorphism $M \to \frac{M[V_1]}{V_1 - V_2}$ of bigraded $\mathbb{F}[V_2, \ldots, V_n]$-modules. □

Thus, the key step in establishing stabilization invariance is the following:

PROPOSITION 5.2.17. *Suppose that $\mathbb{G}'$ is obtained from $\mathbb{G}$ by a stabilization of type X:SW. Then, there is a quasi-isomorphism of bigraded chain complexes over $\mathbb{F}[V_1, \ldots, V_n]$ from $GC^-(\mathbb{G}')$ to* $\mathrm{Cone}(V_1 - V_2)$.

We continue notation from Subsection 5.2.1: the grid states in $\mathbf{S}(\mathbb{G}')$ are partitioned into $\mathbf{I}(\mathbb{G}') \cup \mathbf{N}(\mathbb{G}')$ according to whether or not the grid state contains c. Consider the corresponding $\mathbb{F}[V_1, \ldots, V_n]$-module splitting of $GC^-(\mathbb{G}') \cong \mathbf{I} \oplus \mathbf{N}$, where $\mathbf{I}$, resp. $\mathbf{N}$, is the submodule generated by grid states in $\mathbf{I}(\mathbb{G}')$, resp. $\mathbf{N}(\mathbb{G}')$. The submodule $\mathbf{N}$ is in fact a subcomplex, since any rectangle $\mathrm{Rect}(\mathbf{x}, \mathbf{y})$ with $\mathbf{x} \in \mathbf{N}(\mathbb{G}')$ and $\mathbf{y} \in \mathbf{I}(\mathbb{G}')$ must contain one of X_1 or X_2. Thus, we can write the differential on $GC^-(\mathbb{G}')$ as a 2×2 matrix

$$\partial_{\mathbb{X}}^- = \begin{pmatrix} \partial_{\mathbf{I}}^{\mathbf{I}} & 0 \\ \partial_{\mathbf{I}}^{\mathbf{N}} & \partial_{\mathbf{N}}^{\mathbf{N}} \end{pmatrix};$$

i.e. $GC^-(\mathbb{G}')$ is the mapping cone of the chain map $\partial_{\mathbf{I}}^{\mathbf{N}} \colon (\mathbf{I}, \partial_{\mathbf{I}}^{\mathbf{I}}) \to (\mathbf{N}, \partial_{\mathbf{N}}^{\mathbf{N}})$.

LEMMA 5.2.18. *The natural identification of $\mathbf{S}(\mathbb{G})$ with $\mathbf{I}(\mathbb{G}')$ induces an isomorphism $e \colon (\mathbf{I}, \partial_{\mathbf{I}}^{\mathbf{I}}) \to GC^-(\mathbb{G})[V_1][\![1,1]\!]$ of bigraded chain complexes over $\mathbb{F}[V_1, \ldots, V_n]$.*

Proof. The fact that e respects gradings follows immediately from Lemma 5.2.4. Since the stabilization is of type X, the correspondence also induces a one-to-one correspondence between empty rectangles disjoint from $\mathbb{X}$ in $\mathbb{G}$ and empty rectangles disjoint from $\mathbb{X}'$ in $\mathbb{G}'$, written $r \mapsto r'$. Moreover, for $r' \in \mathrm{Rect}^\circ(\mathbf{x},\mathbf{y})$ with $\mathbf{x},\mathbf{y} \in \mathbf{I}(\mathbb{G}')$ we have $O_1(r') = 0$ and for all $i = 2,\dots,n$, $O_i(r') = O_i(r)$, so e is a chain map. It follows that e is an isomorphism, since it is induced by a one-to-one correspondence of generators. □

Define $\mathcal{H}^{\mathbf{I}}_{X_2}\colon \mathbf{N} \to \mathbf{I}$ to be the $\mathbb{F}[V_1,\dots,V_n]$-module homomorphism with

$$\mathcal{H}^{\mathbf{I}}_{X_2}(\mathbf{x}) = \sum_{\mathbf{y}\in\mathbf{I}(\mathbb{G}')} \sum_{\{r\in\mathrm{Rect}^\circ(\mathbf{x},\mathbf{y})\big|\mathrm{Int}(r)\cap\mathbb{X}=X_2\}} V_1^{O_1(r)}\cdots V_n^{O_n(r)}\cdot\mathbf{y}, \tag{5.20}$$

for all $\mathbf{x} \in \mathbf{N}(\mathbb{G}')$. Note that $\mathcal{H}^{\mathbf{I}}_{X_2}$ is a component of the homotopy operator defined in Equation (4.16).

We have the following analogue of Lemma 5.2.6:

LEMMA 5.2.19. *The map $\mathcal{H}^{\mathbf{I}}_{X_2}\colon (\mathbf{N},\partial^{\mathbf{N}}_{\mathbf{N}}) \to (\mathbf{I},\partial^{\mathbf{I}}_{\mathbf{I}})[\![-1,-1]\!]$ is a chain homotopy equivalence of bigraded chain complexes.*

Proof. First, we verify that $\mathcal{H}^{\mathbf{I}}_{X_2}$ is a chain map. Consider the homotopy operator $\mathcal{H}_{X_2}$ appearing in the proof of Lemma 4.6.9. In that proof, we verified that the operator satisfies the equation

$$\partial^-_{\mathbb{X}} \circ \mathcal{H}_{X_2} + \mathcal{H}_{X_2} \circ \partial^-_{\mathbb{X}} = (V_1 - V_2). \tag{5.21}$$

The fact that $\mathcal{H}^{\mathbf{I}}_{X_2}$ is a chain map follows quickly from the above equation, and the observation that $\mathcal{H}^{\mathbf{I}}_{X_2}$ is the portion of $\mathcal{H}_{X_2}$ mapping $\mathbf{N}$ to $\mathbf{I}$. The grading shift follows as in Lemma 4.6.9.

To see that $\mathcal{H}^{\mathbf{I}}_{X_2}$ is an isomorphism on homology, consider the $\mathbb{F}[V_1,\dots,V_n]$-module map $\mathcal{H}^-_{O_1}\colon \mathbf{I} \to \mathbf{N}$ determined by

$$\mathcal{H}^-_{O_1}(\mathbf{x}) = \sum_{\mathbf{y}\in\mathbf{N}(\mathbb{G}')} \sum_{\{r\in\mathrm{Rect}^\circ(\mathbf{x},\mathbf{y})\big|\mathrm{Int}(r)\cap\mathbb{X}=\emptyset,O_1\in r\}} V_2^{O_2(r)}\cdots V_n^{O_n(r)}\cdot\mathbf{y},$$

for any $\mathbf{x} \in \mathbf{I}(\mathbb{G}')$; furthermore let $\mathcal{H}^-_{O_1,X_2}\colon \mathbf{N} \to \mathbf{N}$ be determined by

$$\mathcal{H}^-_{O_1,X_2}(\mathbf{x}) = \sum_{\mathbf{y}\in\mathbf{N}(\mathbb{G}')} \sum_{\{r\in\mathrm{Rect}^\circ(\mathbf{x},\mathbf{y})\big|\mathrm{Int}(r)\cap\mathbb{X}=X_2,O_1\in r\}} V_2^{O_2(r)}\cdots V_n^{O_n(r)}\cdot\mathbf{y}$$

for any $\mathbf{x} \in \mathbf{N}(\mathbb{G}')$. Note that $\mathcal{H}^-_{O_1}$ and $\mathcal{H}^-_{O_1,X_2}$ are components of the maps $\partial^-_{\mathbb{X}}$ and $\mathcal{H}_{X_2}$, appearing as multiples of V_1. These satisfy the formulas

$$\mathcal{H}^{\mathbf{I}}_{X_2} \circ \mathcal{H}^-_{O_1} = \mathrm{Id}_{\mathbf{I}}$$
$$\mathcal{H}^-_{O_1} \circ \mathcal{H}^{\mathbf{I}}_{X_2} + \partial^{\mathbf{N}}_{\mathbf{N}} \circ \mathcal{H}^-_{O_1,X_2} + \mathcal{H}^-_{O_1,X_2} \circ \partial^{\mathbf{N}}_{\mathbf{N}} = \mathrm{Id}_{\mathbf{N}},$$

which in turn follow from Equation (5.21). These identities then verify the statement of the lemma. □

Proof of Proposition 5.2.17. Equation (5.21) implies the commutativity of the square of chain maps:

$$\begin{array}{ccc} \mathbf{I} & \xrightarrow{\quad \partial_{\mathbf{I}}^{\mathbf{N}} \quad} & \mathbf{N} \\ {\scriptstyle e}\big\downarrow & & \big\downarrow{\scriptstyle e\circ \mathcal{H}^{\mathbf{I}}_{X_2}} \\ GC^-(\mathbb{G})[V_1][\![1,1]\!] & \xrightarrow{\quad V_1 - V_2 \quad} & GC^-(\mathbb{G})[V_1] \end{array} \tag{5.22}$$

where the vertical maps are bigraded quasi-isomorphisms by Lemmas 5.2.18 and 5.2.19, and the horizontal maps are of bidegree $(-1,0)$. Using Lemma 5.2.12, this commutative square induces the claimed quasi-isomorphism from $\mathrm{Cone}(\partial_{\mathbf{I}}^{\mathbf{N}}) = GC^-(\mathbb{G}')$ to $\mathrm{Cone}(V_1 - V_2)$. ☐

Proof of Proposition 5.2.1. For stabilizations of type *X:SW*, this follows from Proposition 5.2.17 and Lemma 5.2.16. Other stabilizations can be reduced to this case using Corollary 3.2.3, together with the invariance under commutations and switches (Propositions 5.1.7 and 5.1.8). ☐

5.3. Completion of the invariance proof for grid homology

We can now assemble the ingredients of the proof of invariance. We start the argument for oriented knots.

THEOREM 5.3.1. *The bigraded vector space $\widehat{GH}(\mathbb{G})$ and the bigraded $\mathbb{F}[U]$-module $GH^-(\mathbb{G})$ depend on the grid $\mathbb{G}$ only through its underlying oriented knot $\vec{K}$.*

Proof. For a fixed toroidal grid diagram $\mathbb{G}$, the vector space $\widehat{GC}(\mathbb{G})$ and the $\mathbb{F}[U]$-module structure on $GC^-(\mathbb{G})$ depends on a choice of V_i. By Corollary 4.6.17 and Lemma 4.6.9 respectively, $\widehat{GH}(\mathbb{G})$ and $GH^-(\mathbb{G})$ are independent of this choice, and depend only on the toroidal grid diagram $\mathbb{G}$. Independence of the choice of the grid diagram follows from Cromwell's Theorem 3.1.9, together with commutation invariance (Proposition 5.1.7) and stabilization invariance (Proposition 5.2.1). ☐

PROPOSITION 5.3.2. *The bigraded vector space $\widehat{GH}(K)$ and the bigraded $\mathbb{F}[U]$-module $GH^-(K)$ are independent of the choice of orientation on K.*

Proof. If $\mathbb{G}$ represents $\vec{K}$, then the grid diagram $\mathbb{G}'$ obtained by reflecting $\mathbb{G}$ across the diagonal represents $-\vec{K}$. Reflection also induces a one-to-one correspondence $\phi\colon \mathbf{S}(\mathbb{G}) \to \mathbf{S}(\mathbb{G}')$. This reflection preserves the fundamental domain $[0,n) \times [0,n)$. In addition, given any two sets P and Q in the fundamental domain, we have $\mathcal{J}(P,Q) = \mathcal{J}(\phi(P),\phi(Q))$. From these observations, it follows immediately that $M(\mathbf{x}) = M(\phi(\mathbf{x}))$ and $A(\mathbf{x}) = A(\phi(\mathbf{x}))$. The reflection also identifies rectangles of $\mathrm{Rect}^\circ(\mathbf{x},\mathbf{y})$ with rectangles of $\mathrm{Rect}^\circ(\phi(\mathbf{x}),\phi(\mathbf{y}))$, so it follows immediately that ϕ extends to an isomorphism $GC^-(\mathbb{G}) \to GC^-(\mathbb{G}')$ of bigraded chain complexes. The result now follows. ☐

Proof of Theorem 4.6.19. Combine Theorem 5.3.1 and Proposition 5.3.2. □

Proof of Theorem 4.4.4. Combine Theorem 5.3.1 (for the independence from the grid diagram of $\widehat{GH}$), Proposition 5.3.2 (for the independence from the orientation of K), and Corollary 4.6.16 (for the relationship between $\widehat{GH}$ and $\widetilde{GH}$). □

5.4. The destabilization maps, revisited

In some applications (see Chapter 6), it will be useful to track more precisely the isomorphisms between grid homology modules induced by the four types of X-destabilizations, without using switch moves.

To this end, suppose that $\mathbb{G}'$ is obtained from $\mathbb{G}$ by a single stabilization of type X, and let c be the corresponding intersection point between the two new curves. As we did for stabilizations of type *X:SW*, we partition grid states $\mathbf{S}(\mathbb{G}') = \mathbf{I}(\mathbb{G}') \cup \mathbf{N}(\mathbb{G}')$, according to whether or not they contain c.

Label the markings in the stabilization region O_1, X_1, and X_2, as follows:

$$\begin{array}{c|c} X_1 & O_1 \\ \hline & X_2 \end{array} \qquad \begin{array}{c|c} X_2 & \\ \hline O_1 & X_1 \end{array} \qquad \begin{array}{c|c} & X_2 \\ \hline X_1 & O_1 \end{array} \qquad \begin{array}{c|c} O_1 & X_1 \\ \hline X_2 & \end{array},$$

and let O_2 be the O-marking in the same row as X_2.

In all cases, there is an identification between $\mathbf{S}(\mathbb{G})$ and $\mathbf{I}(\mathbb{G}')$. Letting $\mathbf{I}$ be the free $\mathbb{F}[V_1, \dots, V_n]$-module generated by $\mathbf{I}(\mathbb{G}')$, the above one-to-one correspondence extends to give an isomorphism of $\mathbb{F}[V_1, \dots, V_n]$-modules

$$e \colon \mathbf{I} \to GC^-(\mathbb{G})[V_1]. \tag{5.23}$$

If the stabilization is of type *X:SW* or *X:NE*, then e is a homogeneous map of degree $(1,1)$, and if the stabilization is of type *X:NW* or *X:SE*, then e is bigraded. We will also use the natural projection

$$\pi \colon H(GC^-(\mathbb{G})[V_1]) \cong GH^-(\mathbb{G})[V_1] \to \frac{GH^-(\mathbb{G})[V_1]}{V_1 - V_2} \cong GH^-(\mathbb{G}); \tag{5.24}$$

between bigraded $\mathbb{F}[U]$-modules, where U acts as multiplication by any V_i with $2 \leq i \leq n$.

For stabilizations of type *X:SW*, we defined a map $\mathcal{H}^{\mathbf{I}}_{X_2}$ in Equation (5.20). This definition extends immediately to stabilizations of type *X:NE*, as well.

PROPOSITION 5.4.1. *If $\mathbb{G}'$ is obtained from $\mathbb{G}$ by a stabilization, then there is an isomorphism of bigraded $\mathbb{F}[U]$-modules from $GH^-(\mathbb{G}')$ to $GH^-(\mathbb{G})$. More concretely*

- *If the stabilization is of type X:SW or X:NE, then $\mathbf{N}(\mathbb{G}')$ generates a subcomplex $\mathbf{N}$ of $GC^-(\mathbb{G}')$, whose quotient complex is $\mathbf{I}$. There is an isomorphism from $GH^-(\mathbb{G}')$ to $GH^-(\mathbb{G})$ whose value on the homology classes of cycles from the subcomplex $\mathbf{N} \subset GC^-(\mathbb{G}')$ is the composition of the map on homology induced by $e \circ \mathcal{H}^{\mathbf{I}}_{X_2}$ (where $\mathcal{H}^{\mathbf{I}}_{X_2}$ is defined as in Equation (5.20)) followed by the projection map π.*
- *If the stabilization is of type X:SE or X:NW, then $\mathbf{I}(\mathbb{G}')$ generates a subcomplex $\mathbf{I} \subset GC^-(\mathbb{G}')$. There is an isomorphism from $GH^-(\mathbb{G}')$ to $GH^-(\mathbb{G})$ whose restriction to cycles from $\mathbf{I}$ is the map on homology induced by e, followed by the projection map π.*

Proof. For stabilizations of type *X:SW*, the above was verified in the proof of Proposition 5.2.17, combined with Lemma 5.2.16. Restricted to the subcomplex $\mathbf{N}$ of $GC^-(\mathbb{G}')$, Proposition 5.2.17 shows that the isomorphism $GH^-(\mathbb{G}') \to GH^-(\mathbb{G})$ is the map on homology induced by the chain map $e \circ \mathcal{H}^{\mathbf{I}}_{X_2}\colon \mathbf{N} \to GC^-(\mathbb{G})[V_1]$, followed by the map on homology induced by the inclusion $i\colon GC^-(\mathbb{G})[V_1] \to \mathrm{Cone}(V_1 - V_2)$, followed by the isomorphism

$$q\colon H(\mathrm{Cone}(V_1 - V_2)) \to \frac{GH^-(\mathbb{G})[V_1]}{V_1 - V_2} \cong GH^-(\mathbb{G}),$$

coming from Lemma 5.2.16. Since $q \circ H(i) = \pi$, this case is complete.

The proof of Proposition 5.2.17 can be readily adapted to the case of stabilizations of type *X:NE*. With our labeling conventions, all the statements in the proof of Proposition 5.2.17 remain true.

For stabilizations of type *X:SE* or *X:NW*, the modifications are more extensive. In these two cases, the differential has the form

$$\partial^-_{\mathbb{X}} = \begin{pmatrix} \partial^{\mathbf{I}}_{\mathbf{I}} & \partial^{\mathbf{I}}_{\mathbf{N}} \\ 0 & \partial^{\mathbf{N}}_{\mathbf{N}} \end{pmatrix}.$$

Lemma 5.2.18 is replaced by the statement that $e\colon (\mathbf{I}, \partial^{\mathbf{I}}_{\mathbf{I}}) \to GC^-(\mathbb{G})[V_1]$ is an isomorphism of bigraded chain complexes. We will use its inverse e'. Define $\mathcal{H}^{\mathbf{N}}_{X_2}\colon \mathbf{I} \to \mathbf{N}$ by

$$\mathcal{H}^{\mathbf{N}}_{X_2}(\mathbf{x}) = \sum_{\mathbf{y}\in\mathbf{N}(\mathbb{G}')} \sum_{\{r\in\mathrm{Rect}^\circ(\mathbf{x},\mathbf{y}) \,\big|\, \mathrm{Int}(r)\cap\mathbb{X}=X_2\}} V_1^{O_1(r)} \cdots V_n^{O_n(r)} \cdot \mathbf{y}. \tag{5.25}$$

Adapting the proof of Lemma 5.2.19, we see that $\mathcal{H}^{\mathbf{N}}_{X_2}\colon (\mathbf{I}, \partial^{\mathbf{I}}_{\mathbf{I}})[\![1,1]\!] \to (\mathbf{N}, \partial^{\mathbf{N}}_{\mathbf{N}})$ is a chain homotopy equivalence of bigraded chain complexes. Adapting the proof of Proposition 5.2.17, the commuting square replacing Equation (5.22) is

$$\begin{array}{ccc} \mathbf{N} & \xrightarrow{\quad\partial^{\mathbf{I}}_{\mathbf{N}}\quad} & \mathbf{I} \\ {\scriptstyle \mathcal{H}^{\mathbf{N}}_{X_2}\circ e'}\big\uparrow & & \big\uparrow{\scriptstyle e'} \\ GC^-(\mathbb{G})[V_1][\![1,1]\!] & \xrightarrow{\quad V_1 - V_2\quad} & GC^-(\mathbb{G})[V_1] \end{array} \tag{5.26}$$

where the vertical maps are quasi-isomorphisms. According to Lemma 5.2.12, this square induces a quasi-isomorphism $\Phi\colon \mathrm{Cone}(V_1 - V_2) \to GC^-(\mathbb{G}')$, whose induced map on homology fits into the commutative square:

$$\begin{array}{ccc} H(\mathbf{I}) & \xrightarrow{\qquad\qquad} & GH^-(\mathbb{G}') \\ {\scriptstyle H(e')}\big\uparrow & & \big\uparrow{\scriptstyle H(\Phi)} \\ GH^-(\mathbb{G})[V_1] & \xrightarrow{\quad\pi\quad} & H(\mathrm{Cone}(V_1 - V_2)) \cong GH^-(\mathbb{G}) \end{array}$$

Using the isomorphism $H(\Phi)^{-1}$, and observing that $H(e')^{-1} = H(e)$, the claims of the proposition follow in this case, as well. □

5.5. Other variants of the grid complex

There are other versions of the grid complex. We will meet one in Chapter 13, where we drop the condition in the differential (in Equation (4.10)) that the rectangles are disjoint from the X-markings.

There is a different version, dropping the condition that the rectangles are empty. This construction was proposed and studied by Lipshitz [**121**]; see also [**120**] for Lipshitz's corresponding holomorphic construction. Below we describe the boundary map in this theory, and return to the construction briefly in Section 17.2.

For a fixed integer m, modify the differential on $GC^-(\mathbb{G})$ to give a function

$$\partial_m^-(\mathbf{x}) = \sum_{\mathbf{y}} \sum_{\{r \in \mathrm{Rect}(\mathbf{x},\mathbf{y}) \big| \mathrm{Int}(r) \cap \mathbb{X} = \emptyset,\ |\mathbf{x} \cap \mathrm{Int}(r)| = m\}} V_1^{O_1(r)} \cdots V_n^{O_n(r)} \mathbf{y};$$

of course, when $m = 0$, this is the differential $\partial_{\mathbb{X}}^-$ on $GC^-(\mathbb{G})$. Enhance the ground ring to $\mathcal{R}_v = \mathbb{F}[V_1, \ldots, V_n, v]$, including a new formal variable v, and consider the module $GC_{big}(\mathbb{G}) = GC^-(\mathbb{G})[v]$, equipped with the differential

$$\partial_{big}(\mathbf{x}) = \sum_{m=0}^{\infty} v^m \cdot \partial_m^-(\mathbf{x}).$$

We will call this the *double-point enhanced grid complex*; and its homology the *double-point enhanced grid homology*. The Maslov and Alexander functions induce a bigrading on the double-point enhanced grid complex, so that multiplication by v increases Maslov grading by two and preserves Alexander grading.

EXERCISE 5.5.1. **(a)** Verify that ∂_{big} is a differential.
(b)* Verify that the double-point enhanced grid homology is a knot invariant.

5.6. On the holomorphic theory

The proof of invariance described in Sections 5.1 and 5.2 is inspired by the (holomorphic) proof of the invariance of knot Floer homology. For example, in the commutation invariance proof, the pentagon counting map P from Equation (5.2) is a combinatorial realization of the map obtained by counting "pseudo-holomorphic triangles" that appears in the handle-slide invariance of knot Floer homology; see for example [**172**]. The holomorphic triangle map induces a chain homotopy equivalence, and the chain homotopy is constructed by counting "pseudo-holomorphic quadrilaterals". The combinatorial realization of this quadrilateral-counting map is supplied by the hexagon counting map H of Equation (5.8). Similar remarks hold for the stabilization invariance proof. The mapping cone $\mathrm{Cone}(V_1 - V_2)$ is the knot complex of an intermediate Heegaard diagram. The isomorphisms induced by destabilizations count holomorphic triangles in the corresponding Heegaard triples.

5.7. Further remarks on stabilization maps

In Section 5.1, we defined homotopy equivalences between grid complexes of grids which differ by commutation moves. In Section 5.2, we defined quasi-isomorphisms for destabilizations. With a little extra work, we can find geometrically defined homotopy inverses to these destabilization maps. Although these maps are not necessary for the development of the theory, the reader may find it rewarding to work them out, after some hints.

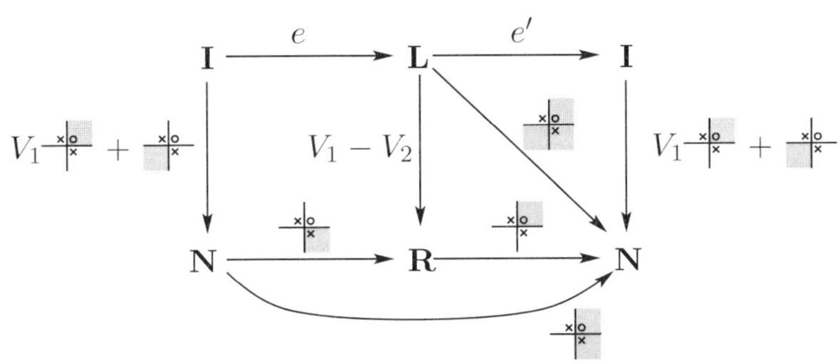

FIGURE 5.11. **Stabilization and destabilization maps for type X:SW.** The first and third column represent the complex $GC^-(\mathbb{G}')$, and the second represents $\mathrm{Cone}(V_1 - V_2)$. The arrows from the first to the middle column represent the destabilization map D; the arrows from the middle to the right column represent the stabilization map S. The long horizontal arrow represents a homotopy operator. The diagonal map counts concave hexagons. The rectangle counting maps are indicated by shaded regions (where the rectangles meet the destabilization region).

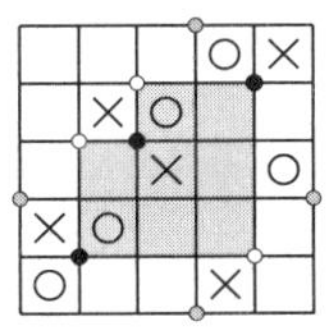

FIGURE 5.12. **A concave hexagon.**

We start with the case of stabilizations of type X:SW, writing the summand $GC^-(\mathbb{G})[V_1][\![1,1]\!] \subset \mathrm{Cone}(V_1 - V_2)$ (i.e. the domain of $V_1 - V_2$) as $\mathbf{L}$ and the summand $GC^-(\mathbb{G})[V_1] \subset \mathrm{Cone}(V_1 - V_2)$ (i.e. the range of $V_1 - V_2$) as $\mathbf{R}$.

The quasi-isomorphism from the proof of Proposition 5.2.17 can be represented as the left square in Figure 5.11: the two columns represent $GC^-(\mathbb{G}')$ and $\mathrm{Cone}(V_1 - V_2)$ respectively, and the horizontal maps are induced by the identification e and the count of rectangles crossing X_2. The two maps in this left square of Figure 5.11 provide the map $D\colon GC^-(\mathbb{G}') \to \mathrm{Cone}(V_1 - V_2)$.

To define a stabilization map $S\colon \mathrm{Cone}(V_1 - V_2) \to GC^-(\mathbb{G}')$, it is useful to have the following notion. (See Figure 5.12.)

DEFINITION 5.7.1. An ***empty, concave hexagon*** is an element $h \in \pi(\mathbf{x}, \mathbf{y})$ with $\mathbf{x} \in \mathbf{I}(\mathbb{G}')$ and $\mathbf{y} \in \mathbf{N}(\mathbb{G}')$ satisfying the following properties:

- the local multiplicities of h are all 0 or 1,
- h has six corners, five $90°$ corners and one $270°$ corner which is at c,
- h contains O_1 and $h \cap \mathbb{X} = X_2$,
- h is empty, in the sense that $\mathbf{x} \cap \mathrm{Int}(h) = \emptyset$.

EXERCISE 5.7.2. Verify that $GC^-(\mathbb{G}')$ and $\mathrm{Cone}(V_1 - V_2)$ are (bigraded) chain homotopy equivalent chain complexes, in the following steps:

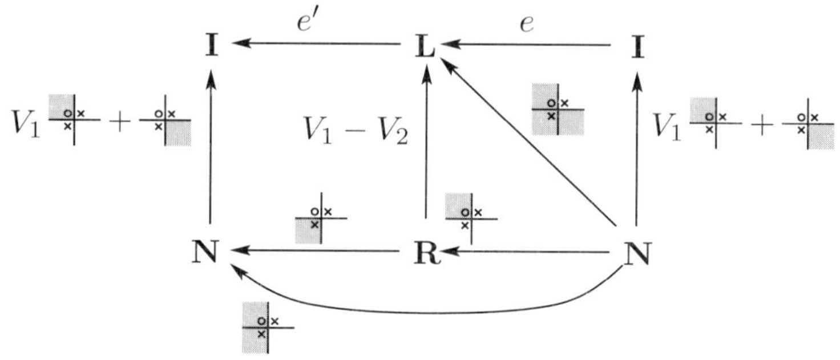

FIGURE 5.13. **Stabilization and destabilization maps in type** $X{:}SE$**.** This figure is obtained from Figure 5.11 by reflecting all the pictures through the vertical circle containing c, and reversing directions of all the arrows.

(a) Construct a stabilization map S, which is a chain map from $\mathrm{Cone}(V_1 - V_2)$ to $GC^-(\mathbb{G}')$, as indicated by the second square in Figure 5.11 (where the diagonal arrow counts empty concave hexagons).
(b) Show that the stabilization map preserves bigrading.
(c) Using the destabilization map D indicated by the left square of Figure 5.11, show that $D \circ S$ induces the identity map on $\mathrm{Cone}(V_1 - V_2)$.
(d) Show that

$$S \circ D + \partial^-_{\mathbb{X}} \circ \mathcal{H}^-_{O_1,X_2} + \mathcal{H}^-_{O_1,X_2} \circ \partial^-_{\mathbb{X}} = \mathrm{Id}_{GC^-(\mathbb{G}')},$$

where, as usual, $\mathcal{H}^-_{O_1,X_2} \colon \mathbf{N} \to \mathbf{N}$ is the map counting empty rectangles containing O_1 and X_2; on Figure 5.11 this map is indicated by the lower long arrow.

For other stabilizations, the local pictures from Figure 5.11 can be rotated or reflected. Special care must be taken, since a $90°$ rotation reverses the roles of horizontal and vertical circles, and hence reverses the directions of the rectangles.

EXERCISE 5.7.3. Modify the stabilization and destabilization maps to construct chain homotopy equivalences for stabilizations of the remaining three types $X{:}NE$, $X{:}SE$, and $X{:}NW$. (*Hint:* For $X{:}SE$ stabilizations, see Figure 5.13.)

CHAPTER 6

The unknotting number and τ

In the previous two chapters, we constructed knot invariants $\widehat{GH}(K)$ and $GH^-(K)$, and verified their invariance properties. The first invariant, $\widehat{GH}(K)$ is a bigraded vector space; the invariant $GH^-(K)$ comes with some extra algebraic structure. As a vector space, GH^- splits as $GH^-(K) = \bigoplus_{d,s\in\mathbb{Z}} GH^-_d(K,s)$, and it is equipped with an endomorphism U with the property that

$$U\colon GH^-_d(K,s) \to GH^-_{d-2}(K,s-1). \tag{6.1}$$

Thus, we can think of $GH^-(K)$ as an unconventionally graded $\mathbb{F}[U]$-module.

In this chapter, we turn to our first application of grid homology, showing that the bigraded $\mathbb{F}[U]$-module structure of $GH^-(K)$ can be used to extract a numerical invariant $\tau(K)$ for knots, which in turn gives the following lower bound on the unknotting number $u(K)$ of $K : |\tau(K)| \leq u(K)$.

The invariant τ is defined in Section 6.1, where the unknotting bound is also verified. Both the definition and the estimate rely on certain maps associated to cross-commutations, which are constructed in Section 6.2. In Section 6.3, τ is computed for torus knots, leading quickly to a proof of the Milnor conjecture [**144**] (first proved in [**106**]). Our proof is inspired by Sarkar's combinatorial proof of this result [**204**]. In Section 6.4 we give a bound on τ that can be arrived at with very little computation. This estimate is used in Section 8.6; the material in this section is also fundamental to the constructions in Chapter 12.

6.1. The definition of τ and its unknotting estimate

Let K_+ be a knot with a distinguished positive crossing and K_- be the knot with the crossing changed.

PROPOSITION 6.1.1. *There are $\mathbb{F}[U]$-module maps*

$$C_-\colon GH^-(K_+) \to GH^-(K_-) \qquad \textit{and} \qquad C_+\colon GH^-(K_-) \to GH^-(K_+),$$

where C_- is bigraded and C_+ is homogeneous of degree $(-2,-1)$, with

$$C_-\circ C_+ = U \qquad \textit{and} \qquad C_+\circ C_- = U, \tag{6.2}$$

where U denotes the endomorphism induced by the algebra action.

We postpone the proof of Proposition 6.1.1 to Section 6.2, and give some of its consequences now. The first consequence is a computation of the rank of $GH^-(K)$. We start with some algebraic preliminaries.

DEFINITION 6.1.2. Let M be a module over $\mathbb{F}[U]$. The ***torsion submodule*** $\mathrm{Tors} = \mathrm{Tors}(M)$ of M is

$$\mathrm{Tors} = \{m \in M \mid \text{there is a non-zero } p \in \mathbb{F}[U] \text{ with } p\cdot m = 0\}.$$

When M is bigraded, the quotient M/Tors inherits a bigrading, as well. When $M/\text{Tors} \cong \mathbb{F}[U]^r$ for some integer r (for example, when M is finitely generated), r is called the ***rank*** of M, rk $M = \text{rk}_{\mathbb{F}[U]} M$.

Clearly, any $\mathbb{F}[U]$-module homomorphism $\phi\colon M \to N$ maps $\text{Tors}(M)$ into $\text{Tors}(N)$.

LEMMA 6.1.3. *Let M and N be two modules over $\mathbb{F}[U]$. If $\phi\colon M \to N$ and $\psi\colon N \to M$ are two module maps with the property that $\psi \circ \phi = U$, then ϕ induces an injective map from $M/\text{Tors}(M)$ into $N/\text{Tors}(N)$.*

Proof. If $\phi(m) \in \text{Tors}(N)$, then there is a k with $U^k \cdot \phi(m) = 0$, so $\psi(U^k \cdot \phi(m)) = U^{k+1} \cdot m = 0$, so $m \in \text{Tors}(M)$. □

PROPOSITION 6.1.4. *For any knot K, $GH^-(K)$ has rank 1; in fact,*

$$GH^-(K)/\text{Tors} \cong \mathbb{F}[U]$$

is supported in bigradings (d, s) satisfying $d - 2s = 0$.

Proof. For the unknot $\mathcal{O}$, Proposition 4.8.1 shows that $GH^-(\mathcal{O}) = \mathbb{F}[U]$, whose torsion submodule is trivial. Thus, $GH^-(\mathcal{O}) = GH^-(\mathcal{O})/\text{Tors} \cong \mathbb{F}[U]$, supported in bigradings $d - 2s = 0$. This verifies the theorem for the unknot.

Since any knot K can be connected to the unknot by a sequence of unknotting operations, Proposition 6.1.1 and Lemma 6.1.3 gives an injective module map from $\mathbb{F}[U]$ into $GH^-(K)/\text{Tors}$, showing that $GH^-(K)/\text{Tors}$ is non-trivial; and it also gives an injective module map from $GH^-(K)/\text{Tors}$ to $\mathbb{F}[U]$. Since every non-trivial submodule of $\mathbb{F}[U]$ is isomorphic to $\mathbb{F}[U]$, it follows that $GH^-(K)/\text{Tors} \cong \mathbb{F}[U]$. Since the maps preserve the grading $M - 2A$, the grading statement follows. □

Proposition 6.1.4 implies that there are homogeneous, non-torsion elements in $GH^-(K)$ for any knot. Thus, the following definition makes sense:

DEFINITION 6.1.5. For any knot K, $\tau(K)$ is -1 times the maximal integer i for which there is a homogeneous, non-torsion element in $GH^-(K)$ with Alexander grading equal to i.

REMARK 6.1.6. The sign in the above definition of τ is chosen to make it agree with the original definition of τ (defined in [**170**] using filtrations; see also Section 14.1).

By Proposition 6.1.1, τ cannot change much under a crossing change:

THEOREM 6.1.7. *If K_+ is a knot with a distinguished positive crossing and K_- is the knot with the crossing changed, then $0 \le \tau(K_+) - \tau(K_-) \le 1$.*

Proof. Let $\xi \in GH^-(K_+)$ be a homogeneous, non-torsion element with maximal Alexander grading; i.e. the Alexander grading of ξ is $-\tau(K_+)$. By Proposition 6.1.1 its image $C_-(\xi) \in GH^-(K_-)$ has Alexander grading $-\tau(K_+)$. Equation (6.2) ensures that $C_+(C_-(\xi)) = U \cdot \xi$ is non-torsion; it follows that $C_-(\xi)$ is non-torsion, as C_+ is a module map. The fact that $C_-(\xi)$ is a homogeneous, non-torsion element of degree $-\tau(K_+)$ immediately implies that $-\tau(K_+) \le -\tau(K_-)$.

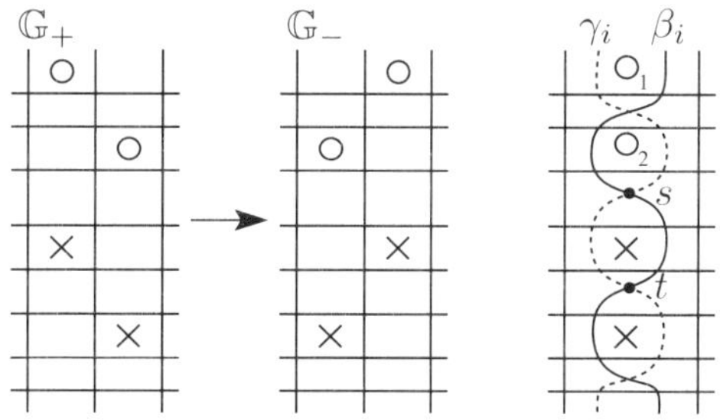

FIGURE 6.1. **Grid diagrams for crossing changes.** The common diagram is shown on the right. (Compare Figure 5.1.)

Let $\eta \in GH^-(K_-)$ be a homogeneous, non-torsion element with Alexander grading $-\tau(K_-)$. By Proposition 6.1.1, $C_+(\eta)$ is a homogeneous non-torsion element of $GH^-(K_+)$ with Alexander grading $-\tau(K_-)-1$; so $-\tau(K_-)-1 \leq -\tau(K_+)$. □

Notice the similarity of this theorem with Proposition 2.3.9. Like the result on signature (Corollary 2.3.10), Theorem 6.1.7 has the following consequence (which will also be generalized from the unknotting number to the slice genus in Chapter 8):

COROLLARY 6.1.8. *For any knot K, $|\tau(K)| \leq u(K)$.*

Proof. This follows from the fact that $\tau(K)$ changes by at most 1 under each unknotting operation (Theorem 6.1.7) and $\tau(\mathcal{O}) = 0$ (Proposition 4.8.1). □

We will deduce the Milnor conjecture from the above, and a computation of τ for torus knots in Section 6.3. First, we prove Proposition 6.1.1.

6.2. Construction of the crossing change maps

Our aim here is to construct the maps C_- and C_+ appearing in Proposition 6.1.1, similar to the commutation maps from Section 5.1.

Let K_+ be a knot with a distinguished positive crossing and K_- be the knot with the crossing changed. Represent these two knots by the grid diagrams $\mathbb{G}_+$ and $\mathbb{G}_-$ that differ by a cross-commutation of columns as in Definition 3.1.12. (See Figure 6.1, and also Figure 3.18).

Like in Section 5.1, we draw both grids $\mathbb{G}_+$ and $\mathbb{G}_-$ on the same torus, thinking of the $\mathbb{O}$ and $\mathbb{X}$ as fixed; see the right-most diagram of Figure 6.1. Start with n horizontal circles, $n-1$ vertical circles, and two additional candidate vertical circles β_i and γ_i. The diagram $\mathbb{G}_+$ includes β_i among its vertical circles, while $\mathbb{G}_-$ includes γ_i. Of course, we cannot draw β_i and γ_i simultaneously as straight, since the O- and X-markings are fixed. We draw them so that they intersect in four points. We will need to specify two of these points in order to define the maps appearing in Proposition 6.1.1. To this end, it will help to introduce some notation. The complement of $\beta_i \cup \gamma_i$ has five components, four of which are bigons, and each of these bigons is marked by a single X or a single O. Since we have a cross-commutation, the two X-marked bigons share a vertex on $\beta_i \cap \gamma_i$; call it t. The two O-marked bigons also share a vertex. Label the O-markings so that O_1 is above

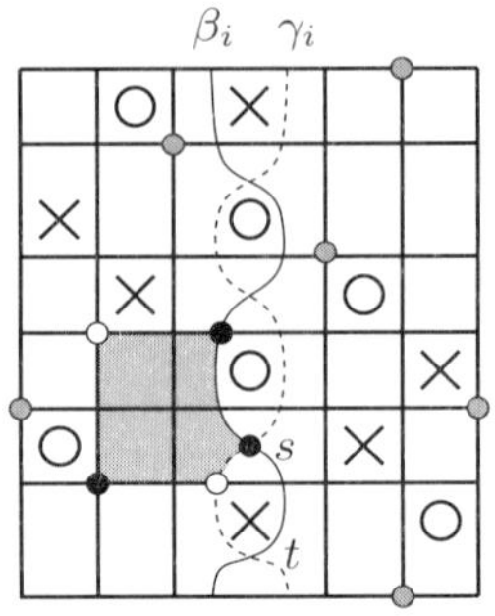

FIGURE 6.2. **A pentagon counted in the unknotting maps.** The shaded pentagon goes from the black state to the white one.

O_2. The bigon labelled by O_2 and one of the X-labelled bigons share a vertex; call it s. These notational choices are indicated in Figure 6.1.

We will define maps that count pentagons containing either the vertex s or the vertex t. Specifically, lifting the definition of empty pentagons from Definition 5.1.1 and fixing arbitrary grid states $\mathbf{x}_+ \in \mathbf{S}(\mathbb{G}_+)$ and $\mathbf{x}_- \in \mathbf{S}(\mathbb{G}_-)$, let $\mathrm{Pent}_s^\circ(\mathbf{x}_+, \mathbf{x}_-)$ denote the space of empty pentagons from $\mathbf{x}_+$ to $\mathbf{x}_-$ with one vertex at s, and let $\mathrm{Pent}_t^\circ(\mathbf{x}_-, \mathbf{x}_+)$ denote the space of empty pentagons from $\mathbf{x}_-$ to $\mathbf{x}_+$ with one vertex at t. (See Figure 6.2 for an example.) Correspondingly, define the $\mathbb{F}[V_1, \dots, V_n]$-module maps $c_- : GC^-(\mathbb{G}_+) \to GC^-(\mathbb{G}_-)$ and $c_+ : GC^-(\mathbb{G}_-) \to GC^-(\mathbb{G}_+)$ specified on $\mathbf{x}_\pm \in \mathbf{S}(\mathbb{G}_\pm)$ by

$$c_-(\mathbf{x}_+) = \sum_{\mathbf{y}_- \in \mathbf{S}(\mathbb{G}_-)} \sum_{\{p \in \mathrm{Pent}_s^\circ(\mathbf{x}_+, \mathbf{y}_-) \,|\, p \cap \mathbb{X} = \emptyset\}} V_1^{O_1(p)} \cdots V_n^{O_n(p)} \cdot \mathbf{y}_- \tag{6.3}$$

$$c_+(\mathbf{x}_-) = \sum_{\mathbf{y}_+ \in \mathbf{S}(\mathbb{G}_+)} \sum_{\{p \in \mathrm{Pent}_t^\circ(\mathbf{x}_-, \mathbf{y}_+) \,|\, p \cap \mathbb{X} = \emptyset\}} V_1^{O_1(p)} \cdots V_n^{O_n(p)} \cdot \mathbf{y}_+. \tag{6.4}$$

LEMMA 6.2.1. *The map c_- preserves bigradings, while c_+ is homogeneous of degree $(-2, -1)$.*

Proof. We verify the grading shifts by relating the pentagons counted in c_- and c_+ with rectangles, as follows. The four special markings divide β_i into four segments, which we denote by $\mathbf{A}$, $\mathbf{B}$, $\mathbf{C}$, and $\mathbf{D}$, as shown in Figure 6.3. There is a natural one-to-one correspondence between $\mathbf{S}(\mathbb{G}_+)$ and $\mathbf{S}(\mathbb{G}_-)$, associating to $\mathbf{x}_+ \in \mathbf{S}(\mathbb{G}_+)$ the grid state $\mathbf{x}_- \in \mathbf{S}(\mathbb{G}_-)$ that agrees with $\mathbf{x}_+$ in $n-1$ components, cf. the nearest point map of Equation (5.3). By a straightforward local computation,

$$M(\mathbf{x}_-) = M(\mathbf{x}_+) + \begin{cases} 1 & \text{if } \mathbf{x}_+ \cap \beta_i \in \mathbf{B} \cup \mathbf{C} \cup \mathbf{D} \\ -1 & \text{if } \mathbf{x}_+ \cap \beta_i \in \mathbf{A}. \end{cases} \tag{6.5}$$

Letting $x = \mathbf{x}_+ \cap \beta_i$, the difference can be seen by computing the difference between the "local Maslov contributions" $\mathcal{J}(\{O_1, O_2\} - \{x\}, \{O_1, O_2\} - \{x\})$, computed for some planar realizations of $\mathbb{G}_+$ and $\mathbb{G}_-$. See Figure 6.3 for these computations.

We classify pentagons as *left* and *right*, depending on whether they lie to the left or to the right of the cross-commutation. Associate to each pentagon $p \in \mathrm{Pent}^\circ(\mathbf{x}_+, \mathbf{y}_-)$ the rectangle $r \in \mathrm{Rect}^\circ(\mathbf{x}_+, \mathbf{y}_+)$ with the same local multiplicities as p away from the four bigons between β_i and γ_i. This sets up a correspondence

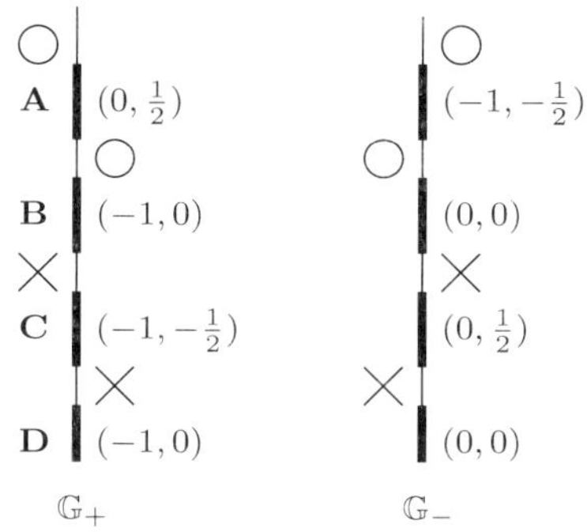

FIGURE 6.3. **Grading shifts for the identification of grid states of $\mathbb{G}_+$ and $\mathbb{G}_-$.** The four markings near β_i for $\mathbb{G}_+$ and γ_i for $\mathbb{G}_-$ divide the circle into the intervals **A**, **B**, **C**, and **D** shown here. Local Maslov and Alexander contributions for $\mathbb{G}_+$ and $\mathbb{G}_-$ are displayed next to the four intervals.

between left pentagons and certain left rectangles in $\mathbb{G}_+$ that contain the point s in the boundary. The left pentagons p that are not blocked by an X-marking correspond to rectangles r whose terminal generators are of types **B** or **C**. Moreover, the local multiplicities at all the O-markings of p correspond to the local multiplicities of r at all the O-markings. It follows now that each non-zero term in $c_-(\mathbf{x}_+)$ corresponding to a left pentagon has the same Maslov grading as $\mathbf{x}_+$, since the terminal corner of a pentagon cannot be of type **A**; cf. Equation (6.5).

To study the right pentagons, it is simplest to compare with the right rectangles in $\mathbb{G}_-$. The verification that Maslov gradings are preserved in this case follows from a similar reasoning to the case considered before, since a right pentagon cannot correspond to a rectangle in $\mathbb{G}_-$ whose initial corner is of type **A**. The Maslov grading shift for c_+ is computed similarly.

For the Alexander grading shifts, a similar argument works. Note that the square associated to a pentagon might cross X-markings (and hence shift Alexander gradings), but this is compensated by the local Alexander grading changes for the types **A**, **B**, **C**, and **D**. These local computations are also displayed in Figure 6.3. □

LEMMA 6.2.2. *The maps c_- and c_+ are chain maps.*

Proof. The proof follows the same logic as the proof of Lemma 5.1.4. For example, consider the equation

$$\partial_{\mathbb{X}}^- \circ c_-(\mathbf{x}_+) + c_- \circ \partial_{\mathbb{X}}^-(\mathbf{x}_+) = 0. \tag{6.6}$$

Most of the domains contributing to the equation contribute in pairs, but there might be two exceptional domains, each of which admit a unique decomposition. These domains are associated to pairs $\mathbf{x}_+$ and $\mathbf{x}_-$ of grid states for $\mathbb{G}_+$ and $\mathbb{G}_-$ respectively, that agree in all but one component. There are two thin annular regions A_1 and A_2, both of which have exactly three corners: one corner is at s; another corner is at the β_i component of $\mathbf{x}_+$, and the third corner is at the γ_i component of $\mathbf{x}_-$. (See Figure 6.4.) Both A_1 and A_2 have a unique decomposition

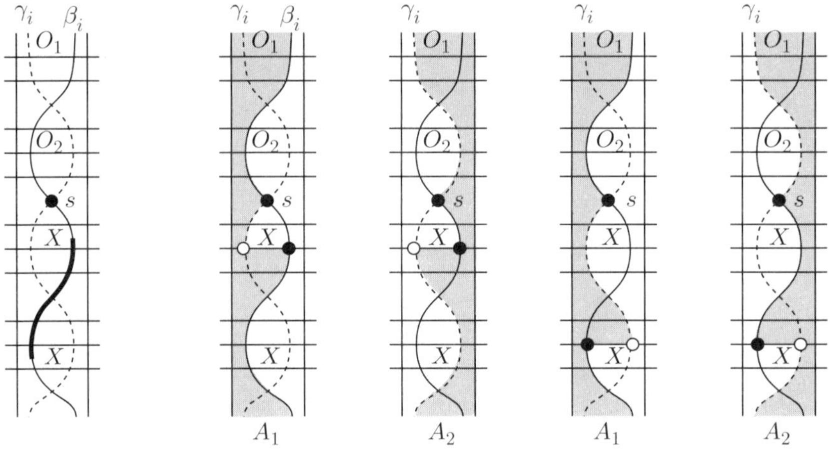

FIGURE 6.4. **Thin regions in the verification of Equation (6.6).** The annuli A_1 and A_2 cover none of the X-markings precisely when the component x of $\mathbf{x}_+$ on β_i is in the darkened arc. There are two combinatorial types of A_1 in this case, depending on which X-labelled bigon the component x is contained in; both are illustrated above, along with the corresponding A_2.

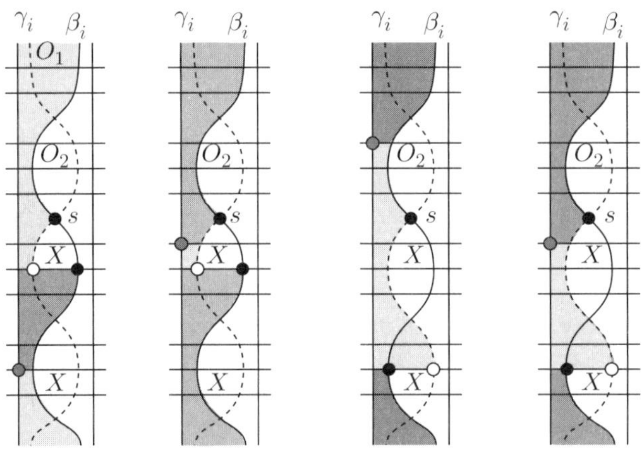

FIGURE 6.5. **Decomposing** A_1. The annulus A_1 can be decomposed as a rectangle and a pentagon, in an order dictated by the position of $\mathbf{x}_+ \cap \beta_{i-1}$ relative to $\mathbf{x}_+ \cap \beta_i$. This decomposition is exhibited for both types of A_1 from Figure 6.4.

as either a juxtaposition of an empty pentagon with a vertex at s followed by an empty rectangle in $\mathbb{G}_-$ or as a juxtaposition of an empty rectangle in $\mathbb{G}_+$ followed by an empty pentagon with a vertex at s. Moreover, A_1 and A_2 cross exactly the same X- and O-markings, so their contributions to $\partial^-_{\mathbb{X}} \circ c_-(\mathbf{x}_+) + c_- \circ \partial^-_{\mathbb{X}}(\mathbf{x}_+)$ cancel; see Figure 6.5. (Compare the proof of Lemma 5.1.4.)

A similar argument verifies that c_+ is a chain map. □

The above chain maps c_- and c_+ induce the maps C_- and C_+ on homology appearing in the statement of Proposition 6.1.1. In order to verify Equation (6.2), we construct chain homotopies between the composites $c_- \circ c_+$ resp. $c_+ \circ c_-$ and multiplication by V_1 that count hexagons in the diagram. These hexagons are required to have two consecutive vertices at s and t; or at t and s. Correspondingly, for $\mathbf{x}_-, \mathbf{y}_- \in \mathbf{S}(\mathbb{G}_-)$, let $\mathrm{Hex}^\circ_{s,t}(\mathbf{x}_-, \mathbf{y}_-)$ denote the set of empty hexagons with two consecutive corners, one at s and the next at t in the order specified by the orientation of the hexagon; and for $\mathbf{x}_+, \mathbf{y}_+ \in \mathbf{S}(\mathbb{G}_+)$ let $\mathrm{Hex}^\circ_{t,s}(\mathbf{x}_+, \mathbf{y}_+)$ be the analogous set with the order of s and t reversed. Let $H_- : GC^-(\mathbb{G}_-) \to GC^-(\mathbb{G}_-)$ be the $\mathbb{F}[V_1, \dots, V_n]$-module map whose value on any $\mathbf{x}_- \in \mathbf{S}(\mathbb{G}_-)$ is

$$H_-(\mathbf{x}_-) = \sum_{\mathbf{y}_- \in \mathbf{S}(\mathbb{G}_-)} \sum_{\{h \in \mathrm{Hex}^\circ_{s,t}(\mathbf{x}_-, \mathbf{y}_-) \mid h \cap \mathbb{X} = \emptyset\}} V_1^{O_1(h)} \cdots V_n^{O_n(h)} \cdot \mathbf{y}_-.$$

Define another map $H_+ : GC^-(\mathbb{G}_+) \to GC^-(\mathbb{G}_+)$, using $\mathrm{Hex}^\circ_{t,s}(\mathbf{x}_+, \mathbf{y}_+)$ instead.

Proof of Proposition 6.1.1. Let c_- and c_+ be the $\mathbb{F}[V_1, \dots, V_n]$-module maps defined in Equations (6.3) and (6.4). According to Lemma 6.2.2, these maps are chain maps. By Lemma 6.2.1, their induced maps on homology C_- and C_+ shift bigradings as stated in the proposition. Hence we are left with the proof of Equation (6.2). Consider the maps H_+ and H_- defined above. We claim first that

$$H_+ : GC^-_d(\mathbb{G}_+, s) \to GC^-_{d-1}(\mathbb{G}_+, s-1) \tag{6.7}$$

$$H_- : GC^-_d(\mathbb{G}_-, s) \to GC^-_{d-1}(\mathbb{G}_-, s-1). \tag{6.8}$$

This is easy to see: for instance, if a hexagon h from $\mathbf{x}_+$ and $\mathbf{y}_+$ is counted in H_+, then there is a corresponding empty rectangle r from $\mathbf{x}_+$ and $\mathbf{y}_+$ that contains exactly one X-marking, and the same number of O-markings as h.

Furthermore, these maps satisfy the homotopy formulas

$$\partial^-_{\mathbb{X}} \circ H_+ + H_+ \circ \partial^-_{\mathbb{X}} = c_+ \circ c_- + V_1 \tag{6.9}$$

$$\partial^-_{\mathbb{X}} \circ H_- + H_- \circ \partial^-_{\mathbb{X}} = c_- \circ c_+ + V_1. \tag{6.10}$$

The proof of Equations (6.9) and (6.10) is analogous to the proof of Lemma 5.1.6 (see especially Equation (5.9)). Again, we analyze domains given by juxtaposing empty rectangles and hexagons (in any order) or pairs of empty pentagons. These cancel in pairs, except for one special thin domain which wraps around the torus. This domain is forced to cover O_1.

In a little more detail, consider Equation (6.9). There are two thin annular regions, both containing O_1, and with corners at s and t: one to the left (i.e. meeting β_{i-1}), and the other to the right (i.e. meeting β_{i+1}), see Figure 6.6. Consider a grid state $\mathbf{x}_+ \in \mathbf{S}(\mathbb{G}_+)$. There are two cases, according to whether the component of $\mathbf{x}_+$ on β_i is on the short arc (adjacent to the X-marked bigon) connecting s and t or not. In the first case, consider the annular region to the right. (See the left picture in Figure 6.6.) That annulus has a unique decomposition into two domains which could contribute to one of $\partial^-_{\mathbb{X}} \circ H_+$, $H_+ \circ \partial^-_{\mathbb{X}}$, or $c_+ \circ c_-$, and the other annulus has no such decomposition. (See Figure 6.7.) In the second case (indicated in the right picture of Figure 6.6), it is the other annulus that has a unique decomposition. The verification of Equation (6.10) works similarly.

Equation (6.2) follows from Equations (6.9) and (6.10), by considering the corresponding actions on homology. □

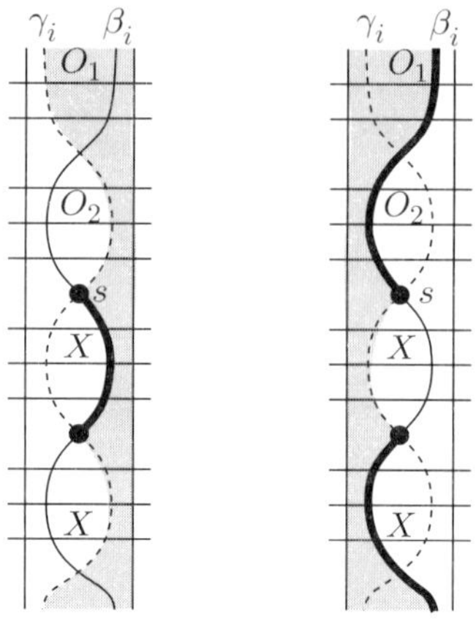

FIGURE 6.6. **Two cases in the verification of Equation** (6.9). These are distinguished by the component of $\mathbf{x}_+$ on β_i: if it is the shorter arc darkened on the left picture, we consider the domain shaded in that diagram; if it is on the longer arc darkened on the right, we consider the domain shaded there.

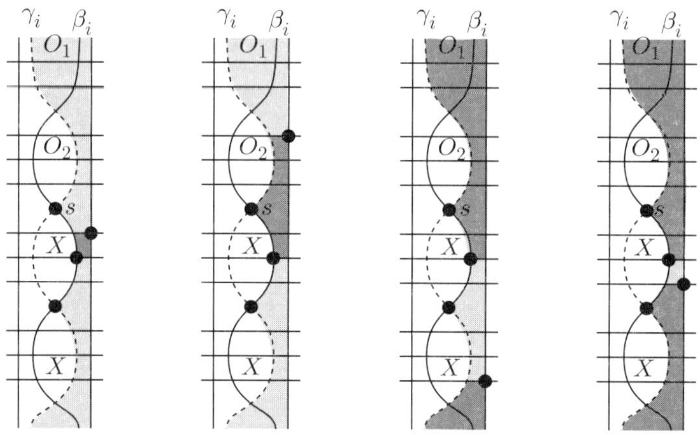

FIGURE 6.7. **Decomposing the annulus to the right of β_i, for Equation** (6.9). When $\mathbf{x}_+ \cap \beta_i$ is on the short arc between s and t, the annular region to the right of β_i has a unique decomposition, in which t is encountered before s.

6.3. The Milnor conjecture for torus knots

We turn now to a computation of τ for torus knots.

PROPOSITION 6.3.1. *For the negative (p, q) torus knot $T_{-p,q}$,*

$$\tau(T_{-p,q}) = -\frac{(p-1)(q-1)}{2}.$$

Proof. Use the grid diagram for $T_{-p,q}$ from the proof of Lemma 4.8.4. In this diagram the X-markings all line up on a diagonal in the torus; see Figure 4.8.

Consider the grid state $\mathbf{x}^+$ that occupies the upper right corner of each square marked with X. According to Lemma 4.8.4, its Alexander grading, $\frac{(p-1)(q-1)}{2}$ is maximal among all grid states. We establish the following further two properties:

(1) $\mathbf{x}^+$ is a cycle in $GC^-(\mathbb{G})$, and
(2) the homology class $[\mathbf{x}^+]$ represented by this cycle is non-torsion.

The placement of $\mathbf{x}^+$ ensures that for any $\mathbf{y} \in \mathbf{S}(\mathbb{G})$ and any rectangle $r \in \operatorname{Rect}(\mathbf{x}^+, \mathbf{y})$, r contains some X-marking in its interior; Property (1) follows.

To verify Property (2), we use another specialization $C'(\mathbb{G})$ of the grid complex, the quotient of $GC^-(\mathbb{G})$ by the sums of multiples of the $V_i - 1$:

$$C'(\mathbb{G}) = \frac{GC^-(\mathbb{G})}{V_1 = \cdots = V_n = 1}.$$

Equivalently, $C'(\mathbb{G})$ is the chain complex generated over $\mathbb{F}$ by all grid states, equipped with the differential

$$\partial'_{\mathbb{X}}\mathbf{x} = \sum_{\mathbf{y}\in\mathbf{S}(\mathbb{G})} \#\{r \in \operatorname{Rect}^\circ(\mathbf{x}, \mathbf{y}) \big| r \cap \mathbb{X} = \emptyset\} \cdot \mathbf{y},$$

thought of as a module over $\mathbb{F}[V_1, \dots, V_n]$, with $V_i \cdot \mathbf{x} = \mathbf{x}$ for $i = 1, \dots, n$. Note that $C'(\mathbb{G})$ does not inherit a bigrading. (The $\mathbb{Z}$-grading on $GC^-(\mathbb{G})$ specified by $M - 2A$ does descend to $C'(\mathbb{G})$, but this is not important for our present purposes.)

There is a quotient map $\Phi\colon GC^-(\mathbb{G}) \to C'(\mathbb{G})$, which is a homomorphism of $\mathbb{F}[V_1, \dots, V_n]$-modules and a chain map; denote its induced map on homology by $\phi\colon GH^-(\mathbb{G}) \to H(C'(\mathbb{G}))$. If $\phi(\xi) \neq 0$, then for all $k \geq 0$, $\phi(U^k \cdot \xi) = U^k\phi(\xi) = \phi(\xi) \neq 0$; thus, ξ is a non-torsion homology class in $GH^-(\mathbb{G})$.

If $\mathbf{y}$ is any grid state for which there is an empty rectangle $r \in \operatorname{Rect}^\circ(\mathbf{y}, \mathbf{x}^+)$ with $r \cap \mathbb{X} = \emptyset$ (i.e. r contributes $\mathbf{x}^+$ to $\partial'_{\mathbb{X}}\mathbf{y}$) then the other rectangle $r' \in \operatorname{Rect}(\mathbf{y}, \mathbf{x}^+)$ also satisfies $\operatorname{Int}(r') \cap \mathbf{x}^+ = \emptyset$ and $r' \cap \mathbb{X} = \emptyset$. This follows readily from the fact (visible from inspecting our grid diagram) that the complement of $r \cup r'$ is a union of two squares, which together contain all the X-markings and all the components of $\mathbf{x}^+$. Since the differential $\partial'_{\mathbb{X}}$ makes no reference to the O-markings, the contributions of these rectangles r and r' cancel.

It follows that $\mathbf{x}^+$, thought of as an element of $C'(\mathbb{G})$, is homologically nontrivial; and hence Property (2) follows.

By Lemma 4.8.4, $\mathbf{x}^+$ has maximal Alexander grading among all grid states, so of course $[\mathbf{x}^+]$ has maximal Alexander grading among all non-torsion homology classes, i.e. $\tau(T_{-p,q}) = -A(\mathbf{x}^+)$. Also by Lemma 4.8.4, $A(\mathbf{x}^+) = \frac{(p-1)(q-1)}{2}$. □

REMARK 6.3.2. As we shall see in Section 6.4, the definition of $\mathbf{x}^+$ and Properties (1) and (2) hold in an arbitrary grid diagram for a knot.

Proposition 6.3.1, combined with the inequality of Corollary 6.1.8 gives the following result (conjectured by Milnor [**144**] and first proved by Kronheimer and Mrowka [**106**] using gauge theory; compare Theorem 1.2.1):

COROLLARY 6.3.3. *The unknotting number of the (p, q) torus knot is $\frac{(p-1)(q-1)}{2}$.*

We deduce the corollary from Corollary 6.1.8, together with the following:

LEMMA 6.3.4. *The torus knot $T_{p,q}$ can be unknotted in $\frac{(p-1)(q-1)}{2}$ crossing changes.*

FIGURE 6.8. **Part of $T_{p,q}$.** We have illustrated the case where $p = 4$. Observe that s_i and s_j cross each other in this region twice.

Proof. We prove this by induction on $\max(p,q)$. Assume without loss of generality that $q > p$. If $p = 1$, $T_{p,q}$ is the unknot, where the result is clear. When $p > 1$, consider a region in the standard diagram for the (p,q) torus knot with p strands, containing one full twist. This region contains $p(p-1)$ crossings, and any two strands cross in this region twice (once over, and once under). Let s_i denote the ith strand from the top, as in Figure 6.8. Change half the crossings in the region so that for all $i < j$, in both crossings of s_i with s_j, the strand s_i is the overcrossing. Eliminate these crossings in pairs (by isotopies), to give a diagram for the torus knot $T_{p,q-p}$. The lemma now follows from the inductive hypothesis. □

Proof of Corollary 6.3.3. Lemma 6.3.4 shows that the unknotting number $u(T_{p,q})$ of the (p,q) torus knot satisfies $u(T_{p,q}) \leq \frac{(p-1)(q-1)}{2}$. The result of Corollary 6.1.8, giving $|\tau(T_{-p,q})| \leq u(T_{-p,q}) = u(T_{p,q})$, together with $|\tau(T_{-p,q})| = \frac{(p-1)(q-1)}{2}$ (Proposition 6.3.1) concludes the argument. □

Some special cases of Corollary 6.3.3 can be proved by more traditional means. For example, when $p = 2$, the corollary follows using the signature, cf. Exercise 2.3.11. The bounds from the signature are, in general, weaker: $\frac{1}{2}\sigma(T_{3,7}) = -4$, while $|\tau(T_{-3,7})| = 6 = u(T_{3,7})$.

A little extra information can be gleaned from the sign of τ: a more careful look at the proof of Corollary 6.1.8 shows that any unknotting of the positive torus knot $T_{p,q}$ involves at least $\frac{(p-1)(q-1)}{2}$ crossing changes which change positive crossings to negative ones. Compare this to the similar signed unknotting bound given by the knot signature, discussed at the end of Section 2.3.

6.4. Canonical grid cycles and estimates on τ

We explain how to estimate $\tau(K)$ given a grid diagram, without computing grid homology. This is done with the help of two distinguished grid states, which are cycles in the grid complex, representing non-torsion classes in GH^- (Proposition 6.4.8).

DEFINITION 6.4.1. Let $\mathbb{G}$ be a grid diagram for a knot K, and consider the grid chain complex $GC^-(\mathbb{G})$. The ***canonical grid states*** $\mathbf{x}^+ = \mathbf{x}^+(\mathbb{G})$ and $\mathbf{x}^- = \mathbf{x}^-(\mathbb{G}) \in \mathbf{S}(\mathbb{G})$ are defined as follows. Each component of the grid state $\mathbf{x}^+$ is the northeast corner of a square decorated with an $X \in \mathbb{X}$, while each component of $\mathbf{x}^-$ is the southwest corner of a square decorated with an $X \in \mathbb{X}$.

LEMMA 6.4.2. *For any grid diagram $\mathbb{G}$, $\mathbf{x}^+(\mathbb{G})$ and $\mathbf{x}^-(\mathbb{G})$ are cycles in $GC^-(\mathbb{G})$.*

Proof. Suppose that $\mathbf{y} \in \mathbf{S}(\mathbb{G})$ is another grid state and $r \in \mathrm{Rect}(\mathbf{x}^+, \mathbf{y})$. Let $x_1 \in \mathbf{x}^+ = \mathbf{x}^+(\mathbb{G})$ denote the upper right corner of the rectangle r. By the definition of $\mathbf{x}^+$, there is an X_i in the square immediately to the lower left of x_1, which is therefore also in r. Thus, r cannot contribute a term in $\partial^-_{\mathbb{X}}(\mathbf{x}^+)$. Since r was arbitrary, it follows that $\partial^-_{\mathbb{X}}(\mathbf{x}^+) = 0$. An analogous argument applies to $\mathbf{x}^- = \mathbf{x}^-(\mathbb{G})$ (now by taking the lower left corner of r). $\square$

REMARK 6.4.3. The homology classes in $GH^-(K)$ represented by $\mathbf{x}^+(\mathbb{G})$ and $\mathbf{x}^-(\mathbb{G})$ will play a central role in Chapter 12: these homology classes are the "Legendrian grid invariants" studied there.

Next we examine some invariance properties of these homology classes:

LEMMA 6.4.4. *Let $\mathbb{G}$ and $\mathbb{G}'$ be two grid diagrams that differ by a commutation move. Then, there is a quasi-isomorphism $P\colon GC^-(\mathbb{G}) \to GC^-(\mathbb{G}')$ sending the cycle $\mathbf{x}^+(\mathbb{G})$ to $\mathbf{x}^+(\mathbb{G}')$ and $\mathbf{x}^-(\mathbb{G})$ to $\mathbf{x}^-(\mathbb{G}')$.*

Proof. We start with the case of $\mathbf{x}^+$, assuming that the two grids $\mathbb{G}$ and $\mathbb{G}'$ differ by a column commutation. We continue to use notation from Section 5.1. Notice that there is some freedom in choosing where the two curves β_i and γ_i of Figure 5.1 intersect each other. In the present proof we will assume that they intersect each other outside of the squares occupied by the X-markings. For a commutation we can always choose the distinguished curves in this way. With this choice of curves, consider the pentagon count map $P\colon GC^-(\mathbb{G}) \to GC^-(\mathbb{G}')$, as defined in Equation (5.2).

Choose any $p \in \mathrm{Pent}^\circ(\mathbf{x}^+(\mathbb{G}), \mathbf{y}')$ for arbitrary $\mathbf{y}' \in \mathbf{S}(\mathbb{G}')$. By the definition of $\mathbf{x}^+(\mathbb{G})$, p contains an X-marking in its upper right corner, unless the pentagon is the unique thin pentagon from $\mathbf{x}^+(\mathbb{G})$ to $\mathbf{y}' = \mathbf{x}^+(\mathbb{G}')$ as in Figure 6.9. It follows that $P(\mathbf{x}^+(\mathbb{G})) = \mathbf{x}^+(\mathbb{G}')$. A similar argument shows $P(\mathbf{x}^-(\mathbb{G})) = \mathbf{x}^-(\mathbb{G}')$.

The case of row commutations works similarly. $\square$

REMARK 6.4.5. The above proof does not work for switches (cf. Definition 3.1.6) in place of commutations. For a switch, the curves β_i and γ_i intersect in a square occupied by one of the X-markings, and hence either $\mathrm{Rect}^\circ(\mathbf{x}^+(\mathbb{G}), \mathbf{x}^+(\mathbb{G}'))$ or $\mathrm{Rect}^\circ(\mathbf{x}^-(\mathbb{G}), \mathbf{x}^-(\mathbb{G}'))$ is empty.

We consider how the homology classes of $\mathbf{x}^+$ and $\mathbf{x}^-$ transform under stabilization.

LEMMA 6.4.6. *Let $\mathbb{G}$ be a grid diagram and $\mathbb{G}'$ a stabilization of $\mathbb{G}$.*

(S-1) If $\mathbb{G}'$ is obtained from $\mathbb{G}$ by a stabilization of type X:NW or X:SE, then there is an isomorphism $GH^-(\mathbb{G}') \cong GH^-(\mathbb{G})$ of bigraded $\mathbb{F}[U]$-modules that sends the homology class $[\mathbf{x}^+(\mathbb{G}')]$ to $[\mathbf{x}^+(\mathbb{G})]$ and the homology class $[\mathbf{x}^-(\mathbb{G}')]$ to $[\mathbf{x}^-(\mathbb{G})]$.

(S-2) If $\mathbb{G}'$ is obtained from $\mathbb{G}$ by a stabilization of type X:SW, then there is an isomorphism $GH^-(\mathbb{G}') \cong GH^-(\mathbb{G})$ of bigraded $\mathbb{F}[U]$-modules that sends $[\mathbf{x}^+(\mathbb{G}')]$ to $[\mathbf{x}^+(\mathbb{G})]$ and $[\mathbf{x}^-(\mathbb{G}')]$ to $U \cdot [\mathbf{x}^-(\mathbb{G})]$.

(S-3) If $\mathbb{G}'$ is obtained from $\mathbb{G}$ by a stabilization of type X:NE, then there is an isomorphism $GH^-(\mathbb{G}') \cong GH^-(\mathbb{G})$ that sends $[\mathbf{x}^+(\mathbb{G}')]$ to $U \cdot [\mathbf{x}^+(\mathbb{G})]$ and $[\mathbf{x}^-(\mathbb{G}')]$ to $[\mathbf{x}^-(\mathbb{G})]$.

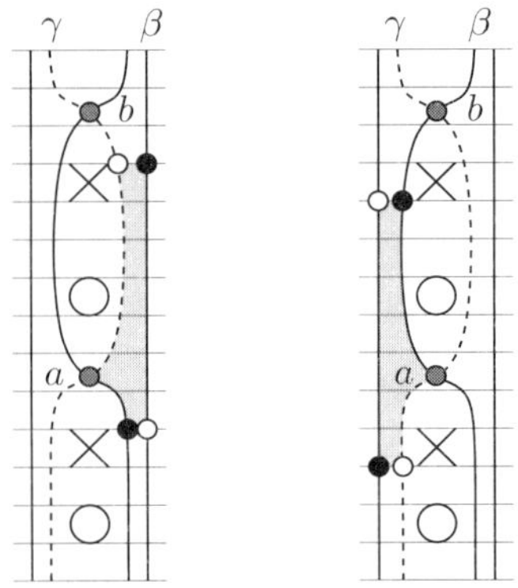

FIGURE 6.9. **Thin pentagons.** The dark circles represent the canonical grid states $\mathbf{x}^+$ and $\mathbf{x}^-$ for the diagram $\mathbb{G}$ involving β, while the white ones represent the corresponding states for the diagram $\mathbb{G}'$ involving γ. The shaded pentagons show that $\mathbf{x}^{\pm}(\mathbb{G})$ are mapped to $\mathbf{x}^{\pm}(\mathbb{G}')$.

Proof. For Part (S-1), Proposition 5.4.1 gives an isomorphism $GH^-(\mathbb{G}') \cong GH^-(\mathbb{G})$ that takes the homology class represented by a cycle ξ from the subcomplex $(\mathbf{I}, \partial_{\mathbf{I}}^{\mathbf{I}})$ to the corresponding element $e(\xi) \in GC^-(\mathbb{G})[V_1]$ (from Equation (5.23)), and projects it to $\frac{GC^-(\mathbb{G})[V_1]}{V_1 - V_2} \cong GC^-(\mathbb{G})$. Since the image of $\mathbf{x}^{\pm}(\mathbb{G}') \in \mathbf{I} \subset GC^-(\mathbb{G}')$ under e is $\mathbf{x}^{\pm}(\mathbb{G})$, Part (S-1) follows.

Consider now Part (S-2); that is, a stabilization of type *X:SW*, and continue with the labeling conventions from Subsection 5.2.1. According to Proposition 5.4.1, there is an isomorphism $GH^-(\mathbb{G}') \cong GH^-(\mathbb{G})$ induced by the chain map that sends $\mathbf{x} \in \mathbf{N}(\mathbb{G}')$ to $\mathcal{H}^{\mathbf{I}}_{X_2}(\mathbf{x}) \in \mathbf{I}$ (Equation (5.20)), followed by the isomorphism $e\colon \mathbf{I} \to GC^-(\mathbb{G})[V_1]$ induced by the identification $\mathbf{I}(\mathbb{G}') \cong \mathbf{S}(\mathbb{G})$, composed with the quotient map to $\frac{GH^-(\mathbb{G})[V_1]}{V_1 - V_2} \cong GH^-(\mathbb{G})$.

Observe that in this case $\mathbf{x}^+(\mathbb{G}') \in \mathbf{N}(\mathbb{G}')$, and there is a single $\mathbf{y} \in \mathbf{I}(\mathbb{G}')$ for which there is some rectangle $r \in \mathrm{Rect}^\circ(\mathbf{x}^+(\mathbb{G}'), \mathbf{y})$ with $\mathrm{Int}(r) \cap \mathbb{X} = X_2$. Indeed, such a rectangle r is unique: it is a width one rectangle that wraps vertically nearly around the torus, crossing no O-markings, and it terminates in the grid state $\mathbf{y}$ with $e(\mathbf{y}) = \mathbf{x}^+(\mathbb{G})$; see the upper right diagram in Figure 6.10. This verifies that $[\mathbf{x}^+(\mathbb{G}')]$ is mapped to $[\mathbf{x}^+(\mathbb{G})]$ under the identification of homologies.

Similarly, continuing the case of stabilizations of type *X:SW*, note that $\mathbf{x}^-(\mathbb{G}') \in \mathbf{N}(\mathbb{G}')$. There is only one rectangle in $\mathrm{Rect}^\circ(\mathbf{x}^-(\mathbb{G}'), \mathbf{y})$ with $\mathbf{y} \in \mathbf{I}(\mathbb{G}')$ and $\mathrm{Int}(r) \cap \mathbb{X} = X_2$, and that is the height one rectangle which wraps horizontally nearly around the torus, crossing the O-marking in the same row as X_2; see the lower right picture in Figure 6.10. This verifies that under the identification of homologies, $[\mathbf{x}^-(\mathbb{G}')]$ is mapped to $U \cdot [\mathbf{x}^-(\mathbb{G})]$.

The case of stabilizations of type *X:NE* (in Part (S-3)) follows similarly; see the first column of Figure 6.10. □

Lemmas 6.4.4 and 6.4.6 allow us to compare the cycles $\mathbf{x}^+$ and $\mathbf{x}^-$ for two grid diagrams representing isotopic knots. For crossing changes, we have the following:

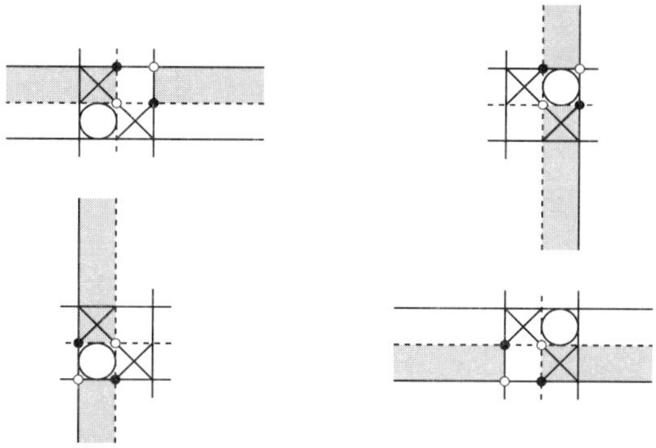

FIGURE 6.10. **Behaviour of $\mathbf{x}^+$ and $\mathbf{x}^-$ under stabilizations of type *X:NE* and *X:SW*.** The black and white states represent canonical elements ($\mathbf{x}^+$ on the top row and $\mathbf{x}^-$ on the bottom) for the stabilized and destabilized diagrams.

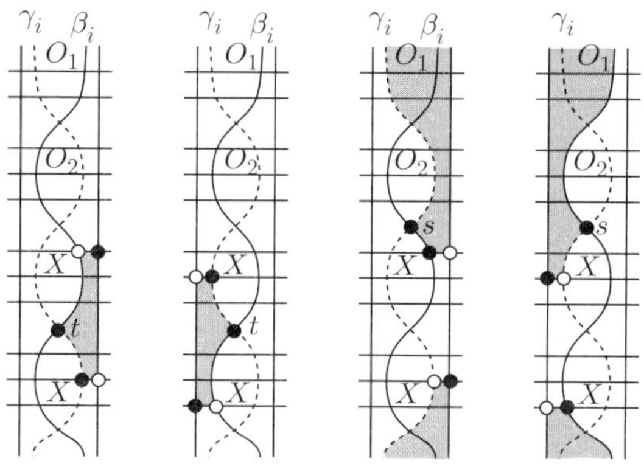

FIGURE 6.11. **Canonical cycles and cross-commutations.**

PROPOSITION 6.4.7. *Suppose that the knots K_+ and K_- differ by a crossing change. Then, there are grid diagrams $\mathbb{G}_+$ and $\mathbb{G}_-$ whose canonical states $\mathbf{x}^\pm$ are related by $C_+([\mathbf{x}^+(\mathbb{G}_-)]) = [\mathbf{x}^+(\mathbb{G}_+)]$ and $C_+([\mathbf{x}^-(\mathbb{G}_-)]) = [\mathbf{x}^-(\mathbb{G}_+)]$, while $C_-([\mathbf{x}^+(\mathbb{G}_+)]) = U[\mathbf{x}^+(\mathbb{G}_-)]$ and $C_-([\mathbf{x}^-(\mathbb{G}_+)]) = U[\mathbf{x}^-(\mathbb{G}_-)]$, where C_+ and C_- are the crossing change maps from Proposition* 6.1.1.

Proof. Consider the maps c_+ and c_- for a cross-commutation, as defined in Equations (6.3) and (6.4) respectively. There is only one empty pentagon $p \in \mathrm{Pent}_t^\circ(\mathbf{x}^+(\mathbb{G}_-), \mathbf{y})$ with no X-marking, and that has $\mathbf{y} = \mathbf{x}^+(\mathbb{G}_+)$ and no O-markings inside it. It follows that $c_+(\mathbf{x}^+(\mathbb{G}_-)) = \mathbf{x}^+(\mathbb{G}_+)$. The same argument shows that $c_+(\mathbf{x}^-(\mathbb{G}_-)) = \mathbf{x}^-(\mathbb{G}_+)$. For c_-, note that the thin pentagons connecting the canonical states of $\mathbb{G}_+$ to the ones of $\mathbb{G}_-$ cover O_1. (Figure 6.11 illustrates, from left to right, $c_+(\mathbf{x}^+(\mathbb{G}_-)) = \mathbf{x}^+(\mathbb{G}_+)$, $c_+(\mathbf{x}^-(\mathbb{G}_-)) = \mathbf{x}^-(\mathbb{G}_+)$, $c_-(\mathbf{x}^+(\mathbb{G}_+)) = V_1 \cdot \mathbf{x}^+(\mathbb{G}_-)$, and $c_-(\mathbf{x}^-(\mathbb{G}_+)) = V_1 \cdot \mathbf{x}^-(\mathbb{G}_-)$ respectively). $\square$

PROPOSITION 6.4.8. *The states* $\mathbf{x}^+ = \mathbf{x}^+(\mathbb{G})$ *and* $\mathbf{x}^- = \mathbf{x}^-(\mathbb{G})$ *are cycles in* $GC^-(\mathbb{G})$*, representing non-torsion homology classes in* $GH^-(\mathbb{G})$*, with*

$$M(\mathbf{x}^+) = 2A(\mathbf{x}^+), \qquad M(\mathbf{x}^-) = 2A(\mathbf{x}^-). \tag{6.11}$$

Proof. By Lemma 6.4.2 the grid states $\mathbf{x}^+$ and $\mathbf{x}^-$ are cycles.

Suppose that there is a grid diagram $\mathbb{G}$ for a knot K so that $\mathbf{x}^+$ or $\mathbf{x}^-$ is torsion. Then, by Lemmas 6.4.4 and 6.4.6, together with Theorem 3.1.9, the same will be true for all grid diagrams for K. In fact, by Proposition 6.4.7, this would hold for any grid diagram for the unknot, contradicting a straightforward computation in a 2×2 grid diagram. Thus, these classes are always non-torsion. Equation (6.11) follows from Proposition 6.1.4. □

Proposition 6.4.8 has the following immediate consequence:

COROLLARY 6.4.9. *Given any grid diagram* $\mathbb{G}$ *representing* K, $-A(\mathbf{x}^+(\mathbb{G})) \geq \tau(K)$ *and* $-A(\mathbf{x}^-(\mathbb{G})) \geq \tau(K)$. □

For a more conceptual interpretation of the above bounds, see Proposition 12.4.2.

EXERCISE 6.4.10. Consider the grid diagram $\mathbb{G}$ of Figure 4.3 of the negative, 0-framed Whitehead double $W_0^-(T_{-2,3})$ of the left-handed trefoil knot. Show that $\tau(W_0^-(T_{-2,3})) = -1$. (*Hint:* Note that any Whitehead double can be unknotted with one crossing change.)

CHAPTER 7

Basic properties of grid homology

This chapter describes some of the basic properties of grid homology. Section 7.1 verifies some symmetries of the simply blocked grid homology groups $\widehat{GH}$, and Section 7.2 relates $\widehat{GH}$ with the Seifert genus. Section 7.3 explores the algebraic structure of $GH^-(K)$, building on work from Chapter 6. Finally, in Section 7.4 symmetries of $GH^-(K)$ are discussed.

7.1. Symmetries of the simply blocked grid homology

Many of the familiar properties of the Alexander polynomial have analogues in grid homology. The Alexander polynomial for a knot is a symmetric function of t. This has the following manifestation in grid homology:

PROPOSITION 7.1.1. *If K is a knot, then for all $d, s \in \mathbb{Z}$,*

$$\widehat{GH}_d(K, s) \cong \widehat{GH}_{d-2s}(K, -s).$$

Proof. Let $\mathbb{G}_1$ be a grid diagram for K, and let $\mathbb{G}_2$ be the new grid diagram obtained by reversing the roles of $\mathbb{X}$ and $\mathbb{O}$. Both $\mathbb{G}_1$ and $\mathbb{G}_2$ represent K, but with opposite orientations. There is an isomorphism of chain complexes between $\widetilde{GC}(\mathbb{G}_1)$ and $\widetilde{GC}(\mathbb{G}_2)$, induced by the natural identification of the grid states, which does not respect the bigradings. Letting M_i and A_i denote Maslov and the Alexander gradings calculated with respect to the diagram $\mathbb{G}_i$ for $i = 1, 2$, Equation (4.3), gives

$$M_1(\mathbf{x}) - M_2(\mathbf{x}) = 2A_1(\mathbf{x}) + n - 1,$$
$$A_1(\mathbf{x}) + A_2(\mathbf{x}) = 1 - n.$$

It follows that

$$\widetilde{GH}_d(\mathbb{G}_1, s) \cong \widetilde{GH}_{d-2s+1-n}(\mathbb{G}_2, -s+1-n). \tag{7.1}$$

This symmetry, together with the relationship between $\widetilde{GH}$ and $\widehat{GH}$ (given in Proposition 4.6.15), implies the stated symmetry of $\widehat{GH}$. To see how, it is convenient to express the symmetries in terms of the Poincaré polynomials

$$\widetilde{P}_{\mathbb{G}}(q,t) = \sum_{d,s} \dim\ \widetilde{GH}_d(\mathbb{G}, s) \cdot q^d t^s \quad \text{and} \qquad \widehat{P}_{\mathbb{G}}(q,t) = \sum_{d,s} \dim\ \widehat{GH}_d(\mathbb{G}, s) \cdot q^d t^s.$$

In this notation, Equation (7.1) can be expressed as the relation

$$\widetilde{P}_{\mathbb{G}_1}(q,t) = (qt)^{1-n} \cdot \widetilde{P}_{\mathbb{G}_2}(q, q^{-2}t^{-1}).$$

The isomorphism of Proposition 4.6.15 (applied to $\mathbb{G}_1$ and $\mathbb{G}_2$) gives the relations

$$(1 + q^{-1}t^{-1})^{n-1}\widehat{P}_{\mathbb{G}_1}(q,t) = \widetilde{P}_{\mathbb{G}_1}(q,t) \quad \text{and} \quad (1 + q^{-1}t^{-1})^{n-1}\widehat{P}_{\mathbb{G}_2}(q,t) = \widetilde{P}_{\mathbb{G}_2}(q,t).$$

It follows from these equations that $\widehat{P}_{\mathbb{G}_1}(q,t) = \widehat{P}_{\mathbb{G}_2}(q, q^{-2}t^{-1})$, which in turn gives the stated isomorphism. $\square$

The Alexander polynomial is invariant under mirroring, whereas grid homology, by contrast, is not, as was illustrated by the computations of the grid homology groups of the two trefoils in Exercise 4.8.2. More generally, we have the following:

PROPOSITION 7.1.2. *If K is a knot and $m(K)$ is its mirror, then for all $d, s \in \mathbb{Z}$, $\widehat{GH}_d(K,s) \cong \widehat{GH}_{2s-d}(m(K), s)$.*

Proof. Let $\mathbb{G}$ be a grid diagram (with markings $\mathbb{O}$ and $\mathbb{X}$) for K, and let $\mathbb{G}^*$ be the diagram (with markings $\mathbb{O}^*$ and $\mathbb{X}^*$) obtained by reflecting $\mathbb{G}$ through a horizontal axis. The diagram $\mathbb{G}^*$ represents $m(K)$. Reflection induces a bijection $\mathbf{x} \mapsto \mathbf{x}^*$ between grid states for $\mathbb{G}$ and those for $\mathbb{G}^*$, inducing a bijection between empty rectangles in $\mathrm{Rect}^\circ(\mathbf{x},\mathbf{y})$ and empty rectangles in $\mathrm{Rect}^\circ(\mathbf{y}^*,\mathbf{x}^*)$. Thus, reflection induces an isomorphism of chain complexes $\widetilde{GC}(\mathbb{G}^*) \cong \mathrm{Hom}(\widetilde{GC}(\mathbb{G}), \mathbb{F})$. Since we are working over the field $\mathbb{F} = \mathbb{Z}/2\mathbb{Z}$, the Universal Coefficient Theorem in cohomology (Theorem A.5.2) gives an isomorphism $\widetilde{GH}(\mathbb{G}) \cong \widetilde{GH}(\mathbb{G}^*)$. By Proposition 4.6.15, we get that $\widehat{GH}(K) \cong \widehat{GH}(m(K))$.

It remains to keep track of gradings in the above isomorphism. We verify first that

$$M_{\mathbb{O}}(\mathbf{x}) + M_{\mathbb{O}^*}(\mathbf{x}^*) = 1 - n. \tag{7.2}$$

Choosing $\mathbf{x} = \mathbf{x}^{NWO}$, $M_{\mathbb{O}}(\mathbf{x}) = 0$ by Equation (4.1), and $M_{\mathbb{O}^*}(\mathbf{x}^*) = 1 - n$ by Proposition 4.3.7, so Equation (7.2) follows for this choice. For arbitrary grid states $\mathbf{x}, \mathbf{y}$, if $r \in \mathrm{Rect}(\mathbf{x},\mathbf{y})$ then $r^* \in \mathrm{Rect}(\mathbf{y}^*,\mathbf{x}^*)$ and Equation (4.2) gives

$$M_{\mathbb{O}}(\mathbf{x}) + M_{\mathbb{O}^*}(\mathbf{x}^*) = M_{\mathbb{O}}(\mathbf{y}) + M_{\mathbb{O}^*}(\mathbf{y}^*).$$

Since any two grid states can be connected by a sequence of rectangles, Equation (7.2) follows for all grid states.

It follows from Equation (7.2) (along with its analogue, replacing $\mathbb{O}$ by $\mathbb{X}$) that

$$A_{\mathbb{G}}(\mathbf{x}) + A_{\mathbb{G}^*}(\mathbf{x}^*) = 1 - n, \tag{7.3}$$

which, along with Equation (7.2), identifies $\widetilde{GH}_d(\mathbb{G}, s) \cong \widetilde{GH}_{1-n-d}(\mathbb{G}^*, 1-n-s)$. Combining this with Equation (7.1), gives an isomorphism

$$\widetilde{GH}_d(\mathbb{G}, s) \cong \widetilde{GH}_{2s-d}(\mathbb{G}', s), \tag{7.4}$$

where $\mathbb{G}'$ is obtained from $\mathbb{G}^*$ by switching the roles of $\mathbb{X}$ and $\mathbb{O}$. The above isomorphism can be combined with Proposition 4.6.15, to give the desired symmetry: Equation (7.4) gives the relation between the Poincaré polynomials

$$\widetilde{P}_{\mathbb{G}}(q,t) = \widetilde{P}_{\mathbb{G}'}(q^{-1}, q^2 t),$$

Proposition 4.6.15 gives the relations

$$(1+q^{-1}t^{-1})^{n-1}\widehat{P}_{\mathbb{G}}(q,t) = \widetilde{P}_{\mathbb{G}}(q,t) \qquad \text{and} \qquad (1+q^{-1}t^{-1})^{n-1}\widehat{P}_{\mathbb{G}'}(q,t) = \widetilde{P}_{\mathbb{G}'}(q,t);$$

so $\widehat{P}_{\mathbb{G}}(q,t) = \widehat{P}_{\mathbb{G}'}(q^{-1}, q^2 t)$, which gives the stated symmetry. $\square$

7.2. Genus bounds

Write the Alexander polynomial of K as $\Delta_K(t) = a_0 + \sum_{i=1}^{d} a_i(t^i + t^{-i})$. By Theorem 2.4.6, the Seifert genus of K is bounded below by the maximal i for which a_i is non-zero. In this section, we generalize this result to grid homology.

Before giving this generalization, we give a result concerning grid diagrams. In Section 3.4, we associated to any grid diagram $\mathbb{G}$ a collection of Seifert surfaces. The genus of any of these surfaces depends only on the diagram, and was called the *associated genus* of $\mathbb{G}$. The associated genus has the following reformulation in terms of grid states.

PROPOSITION 7.2.1. *Let $\mathbb{G}$ be any grid diagram for a knot K. Its associated genus $g(\mathbb{G})$ is the maximum value of the Alexander function over all grid states for $\mathbb{G}$.*

Proof. According to Proposition 4.7.2, the matrix $W = \mathbf{W}(\mathbb{G})$ computes the Alexander grading by the formula

$$A_W(\mathbf{x}) = -\sum_{x\in\mathbf{x}} w(x) + \frac{1}{8}\sum_{k=1}^{8n} w(p_k) - \left(\frac{n-1}{2}\right).$$

Using a different $n\times n$ matrix W' in place of $\mathbf{W}(\mathbb{G})$, the right hand side still makes sense, inducing a function $A_{W'}$ on grid states. The function A_W is unchanged if we add a row or column to W. For columns, this is true because

$$A_{W+C_i}(\mathbf{x}) - A_W(\mathbf{x}) = -\#(\mathbf{x}\cap C_i) + \frac{1}{8}\#(\{p_k\}\cap C_i),$$

and $\mathbf{x}$ has one component in C_i, and eight of the p_k are in C_i; the case of rows works the same.

Let H be a minimal complexity matrix obtained by adding and subtracting rows or columns to $\mathbf{W}(\mathbb{G})$ (in the sense of Definition 3.4.1). According to Lemma 3.4.2, the grid state $\mathbf{x}$ that maximizes A_H has $h(x) = 0$ for all $x \in \mathbf{x}$. Thus, the maximum of the Alexander function (or equivalently, A_H) over all grid states is given by

$$\frac{1}{8}\sum_{k=1}^{8n} h(p_k) - \left(\frac{n-1}{2}\right);$$

which, by Proposition 3.4.9, computes the genus of F_H. □

We can now give the grid homology bound on the Seifert genus. We use the shorthand $\widehat{GH}(K,s) = \bigoplus_d \widehat{GH}_d(K,s)$.

PROPOSITION 7.2.2. *For any knot $K \subset S^3$, $\max\{s \mid \widehat{GH}(K,s) \neq 0\} \leq g(K)$.*

Proof. For any grid diagram $\mathbb{G}$, Proposition 4.6.15 and 7.2.1 give

$$\max\{s\mid \widehat{GH}(K,s)\neq 0\} = \max\{s\mid \widetilde{GH}(\mathbb{G},s)\neq 0\} \leq \max\{s\mid \widetilde{GC}(\mathbb{G},s)\neq 0\} = g(\mathbb{G}).$$

By Proposition 3.4.11, we can choose $\mathbb{G}$ so that $g(K) = g(\mathbb{G})$. □

The following Seifert genera were first computed by Gabai [**65**]:

LEMMA 7.2.3. *The Seifert genus of the Kinoshita-Tersasaka knot KT is 2 and, the Seifert genus of the Conway knot C is 3. (See Figure 2.7 for these knots.)*

Proof. Proposition 7.2.2, combined with calculations from Exercise 4.8.7 and 4.8.6 give bounds $2 \leq g(KT)$ and $3 \leq g(C)$ respectively; and Seifert surfaces of genus 2 and 3 for KT and C were described in Example 3.4.13. $\square$

Unlike the bound coming from the Alexander polynomial, the bound on the Seifert genus $g(K)$ of a knot K coming from grid homology in Proposition 7.2.2 is sharp; see Theorem 1.3.2. The proof of this result, however, remains outside the combinatorial framework described in this book; see also Problem 17.1.2.

7.3. General properties of unblocked grid homology

We turn to the structure of the unblocked grid homology $GH^-(K)$, which is a module over $\mathbb{F}[U]$. Proposition 4.6.18 relates $GH^-(K)$ and $\widehat{GH}(K)$ by a long exact sequence
(7.5)

$$\cdots \to GH^-_{d+2}(K, s+1) \xrightarrow{U} GH^-_d(K, s) \to \widehat{GH}_d(K, s) \to GH^-_{d+1}(K, s+1) \to \dots.$$

PROPOSITION 7.3.1. *For each $d, s \in \mathbb{Z}$, $GH^-_d(K,s)$ is a finite dimensional vector space that vanishes if either d or s is sufficiently large. Moreover, $GH^-(K)$ is a finitely generated bigraded $\mathbb{F}[U]$-module.*

Proof. Fix a grid diagram $\mathbb{G}$ representing K. The chain complex $GC^-(\mathbb{G})$ is, by construction, a finitely generated module over $\mathbb{F}[V_1, \dots, V_n]$, that is bigraded. Since multiplication by V_i shifts both Maslov and Alexander gradings down, it follows that for each $d, s \in \mathbb{Z}$, $GC^-_d(\mathbb{G}, s)$ is a finite-dimensional vector space that vanishes if either d or s are sufficiently large.

The homology $GH^-(K)$ inherits these properties; in particular, $GH^-(K)$ is finitely generated as a $\mathbb{F}[V_1, \dots, V_n]$-module because $\mathbb{F}[V_1, \dots, V_n]$ is a Noetherian ring (see Proposition A.4.1). Since each V_i acts on $GH^-(K)$ the same way (Lemma 4.6.9), we conclude that $GH^-(K)$ is a finitely generated module over $\mathbb{F}[U]$. $\square$

We aim first to study the Euler characteristic of the unblocked invariant, and then we return to further properties of its algebraic structure.

Let $\mathbb{Z}[t, t^{-1}]\!]$ denote the ring of formal Laurent series in t; that is an element of $\mathbb{Z}[t, t^{-1}]\!]$ is a formal sum $\sum_{s \in \mathbb{Z}} a_s t^s$, where $a_s \in \mathbb{Z}$, and $a_s = 0$ for all sufficiently large s. By Proposition 7.3.1 we can consider the *graded Euler characteristic*

$$\chi(GH^-(K)) = \sum_{d,s} (-1)^d \dim_{\mathbb{F}} GH^-_d(K, s) \cdot t^s,$$

which is a formal Laurent series. As noted in the proof of Proposition 7.3.1, for each $d, s \in \mathbb{Z}$, $GC^-_d(\mathbb{G}, s)$ is finite dimensional, and $GC^-_d(\mathbb{G}, s) = 0$ if d or s is sufficiently large. It follows that the graded Euler characteristic can be taken at the chain level:

$$\chi(GC^-(\mathbb{G})) = \sum_{d,s} (-1)^d \dim_{\mathbb{F}} GC^-_d(K, s) \cdot t^s \in \mathbb{Z}[t, t^{-1}]\!];$$

and general properties of Euler characteristics ensure $\chi(GH^-(K)) = \chi(GC^-(\mathbb{G}))$.

For the next statement, recall the expansion $\frac{1}{1-t^{-1}} = \sum_{i\in\mathbb{Z}^{\geq 0}} t^{-i} \in \mathbb{Z}[t, t^{-1}]\!]$.

PROPOSITION 7.3.2. *The graded Euler characteristic of $GH^-(K)$ is given by*

$$\chi(GH^-(K)) = \frac{\Delta_K(t)}{1-t^{-1}}.$$

Proof. The short exact sequence

$$0 \longrightarrow GC^-(\mathbb{G})[\![2,1]\!] \xrightarrow{V_i} GC^-(\mathbb{G}) \longrightarrow \widehat{GC}(\mathbb{G}) \longrightarrow 0$$

(used in the verification of Equation (7.5); cf. the proof of Proposition 4.6.18) implies the relation on Euler characteristics:

$$\begin{aligned}\chi(GC^-(\mathbb{G})) &= \chi(GC^-(\mathbb{G})[\![2,1]\!]) + \chi(\widehat{GC}(\mathbb{G})) \\ &= \chi(GC^-(\mathbb{G}))t^{-1} + \chi(\widehat{GC}(\mathbb{G})).\end{aligned}$$

Since the graded Euler characteristic is unchanged under homology, Theorem 4.7.6 identifies $\chi(\widehat{GC}(K)) = \Delta_K(t)$, and the result follows. □

Combining Propositions 7.3.1 and 6.1.4 with the classification of finitely generated modules over a principal ideal domain, we can bring the $\mathbb{F}[U]$-module $GH^-(K)$ to a standard form. To describe this, we introduce some notation. Let $\mathbb{F}[U]/U^n_{(d,s)}$ denote the bigraded cyclic torsion module whose generator g has bigrading given by (d,s) and satisfies the relation $U^n g = 0$ (while $U^{n-1}g \neq 0$); that is, $\mathbb{F}[U]/U^n_{(d,s)}$ is isomorphic to $\mathbb{F}$ in bigrading $(d-2i, s-i)$ for $i = 0, \dots, n-1$, and U times the non-zero element in bigrading $(d-2i, s-i)$ is the one in bigrading $(d-2i-2, s-i-1)$, provided that $0 \leq i \leq n-2$. Similarly, let $\mathbb{F}[U]_{(d,s)}$ denote the free, rank one $\mathbb{F}[U]$-module whose generator has bigrading (d,s).

PROPOSITION 7.3.3. *For each knot K, we can find $k \geq 0$ and a set of triples of integers $\{(d_i, s_i, n_i)\}_{i=1}^k$ with $n_i > 0$, so that*

$$GH^-(K) \cong \left(\bigoplus_{i=1}^k \mathbb{F}[U]/U^{n_i}_{(d_i,s_i)}\right) \oplus \mathbb{F}[U]_{(-2\tau,-\tau)} \tag{7.6}$$

as $\mathbb{F}[U]$-modules, where $\tau = \tau(K)$ from Definition 6.1.5. The bigraded vector space $\widehat{GH}(K)$ is determined from the above decomposition by

$$\widehat{GH}(K) \cong \left(\bigoplus_{i=1}^k \mathbb{F}_{(d_i,s_i)} \oplus \mathbb{F}_{(d_i-2n_i+1, s_i-n_i)}\right) \oplus \mathbb{F}_{(-2\tau,-\tau)}. \tag{7.7}$$

Proof. According to Proposition 7.3.1, $GH^-(K)$ is a finitely generated module over $\mathbb{F}[U]$. By the classification of finitely generated modules over a principal ideal domain (see for example [**92**, Chapter 3.8]), $GH^-(K)$ splits as a direct sum of finitely many cyclic modules, which are of the form $\mathbb{F}[U]/p$, for some polynomial $p \in \mathbb{F}[U]$. The gradings force these summands to be either of the form $\mathbb{F}[U]$ (i.e. free summands), or of the form $\mathbb{F}[U]/U^n$ for some $n > 0$. (For more details, and a proof of the classification theorem in the form we need it, see Proposition A.4.3.)

The free summand in $GH^-(K)$ is given by $GH^-(K)/\mathrm{Tors}$, which we showed in Proposition 6.1.4 to be isomorphic to $\mathbb{F}[U]_{(2a,a)}$; so $a = -\tau(K)$. This establishes Equation (7.6). The structure of $\widehat{GH}(K)$ now follows from Equation (7.5). □

REMARK 7.3.4. According to Theorem 2.4.6, the value of the Alexander polynomial $\Delta_K(t)$ of a knot K at $t = 1$ equals 1. This is equivalent to the property that the formal power series $\Delta_K(t)/(1 - t^{-1}) \in \mathbb{Z}[t, t^{-1}]]$ has coefficient 1 in front of each t^{-k} for k sufficiently large. The fact that $GH^-(K)$ has rank one can be thought of as the grid homology manifestation of this property.

7.4. Symmetries of the unblocked theory

We saw that simply blocked grid homology satisfies two symmetries: it is symmetric in the Alexander grading (Proposition 7.1.1), and it transforms in a predictable manner under mirroring (Proposition 7.1.2).

Proposition 7.1.1 does not hold in the case of GH^-. In fact, it fails already on the level of its Euler characteristic: the Euler characteristic of GH^- as stated in Proposition 7.3.2 is manifestly not symmetric under the map $t \mapsto t^{-1}$.

Proposition 7.1.2 admits a generalization in terms of the dual complex. To study this property, we introduce some more algebraic constructions. It will be convenient to work with a stabilized version of $GC^-(\mathbb{G})$, where the various V_i variables are set equal to one another, $\frac{GC^-(\mathbb{G})}{V_1=\cdots=V_n}$, i.e. the quotient of $GC^-(\mathbb{G})$, divided by sums of multiples of $V_i - V_{i+1}$ for $i = 1, \dots, n-1$. Equivalently, the complex $\frac{GC^-(\mathbb{G})}{V_1=\cdots=V_n}$ is the free module over $\mathbb{F}[U]$, generated by the grid states $\mathbf{S}(\mathbb{G})$ and equipped with the differential

$$\partial^{c}_{\mathbb{X}}(\mathbf{x}) = \sum_{\mathbf{y}\in\mathbf{S}(\mathbb{G})} \sum_{\{r\in\text{Rect}^\circ(\mathbf{x},\mathbf{y})\,|\,r\cap\mathbb{X}=\emptyset\}} U^{\#(r\cap\mathbb{O})} \cdot \mathbf{y}. \tag{7.8}$$

To state the relationship between $H(\frac{GC^-(\mathbb{G})}{V_1=\cdots=V_n})$ and $GH^-(\mathbb{G})$, we use the two-dimensional bigraded vector space $W \cong \mathbb{F}_{(0,0)} \oplus \mathbb{F}_{(-1,-1)}$. Given a bigraded $\mathbb{F}[U]$-module X, $X \otimes W$ is the $\mathbb{F}[U]$-module, where U acts on the X factor; i.e. there is an isomorphism of bigraded modules over $\mathbb{F}[U]$

$$X \otimes W \cong X \oplus X[\![1, 1]\!].$$

LEMMA 7.4.1. *Let C be a free chain complex over $\mathbb{F}[V_1, \dots, V_n]$, and suppose that multiplication by V_1 is chain homotopic to V_2. Then, there is a quasi-isomorphism from $C \otimes W$ to $\frac{C}{V_1 - V_2}$ of chain complexes over $\mathbb{F}[V_1, \dots, V_n]$.*

Proof. Consider $\text{Cone}(V_1 - V_2 \colon C \to C)$. Since $V_1 - V_2$ is chain homotopic to the 0 map, Lemma 5.2.14 gives an isomorphism $C \otimes W \cong \text{Cone}(V_1 - V_2)$. Since C is free, multiplication by $V_1 - V_2$ is an injective endomorphism of C; so Lemma 5.2.13 gives a quasi-isomorphism from $\text{Cone}(V_1 - V_2)$ to $\frac{C}{V_1 - V_2}$. Composing with the previous isomorphism gives the claimed quasi-isomorphism. □

Iterating this gives the following:

LEMMA 7.4.2. *Suppose that $\mathbb{G}$ is a grid diagram with grid number n representing a knot. Then, there is an isomorphism of bigraded $\mathbb{F}[U]$-modules*

$$H\left(\frac{GC^-(\mathbb{G})}{V_1 = \ldots = V_n}\right) \cong GH^-(\mathbb{G}) \otimes W^{\otimes(n-1)}.$$

Proof. For each $i = 1, \dots, n$, we construct inductively a quasi-isomorphism of $\mathbb{F}[V_i, \dots V_n]$-modules

$$\Phi_i \colon GC^-(\mathbb{G}) \otimes W^{\otimes(i-1)} \to \frac{GC^-(\mathbb{G})}{V_1 = \dots = V_i}, \tag{7.9}$$

where the basic case is a tautology. For the inductive step, since $GC^-(\mathbb{G})$ is a free module over $\mathbb{F}[V_1, \dots, V_n]$, the quotient $\frac{GC^-(\mathbb{G})}{V_1=\dots=V_i}$ is a free module over $\mathbb{F}[V_i, \dots, V_n]$, Lemma 7.4.1 gives a quasi-isomorphism

$$\frac{GC^-(\mathbb{G})}{V_1 = \dots = V_i} \otimes W \to \frac{GC^-(\mathbb{G})}{V_1 = \dots = V_{i+1}}.$$

Precomposing this with the isomorphism from Equation (7.9) (which holds by the induction hypothesis) tensored with W, gives the quasi-isomorphism Φ_{i+1} needed to establish Equation (7.9) for all $i - 1, \dots, n$.

Equation (7.9) in the case where $i = n$ gives the statement of the lemma. □

We will need another algebraic construction. Suppose that C is a bigraded chain complex over $\mathbb{F}[U]$. The *dual complex* of C is the complex over $\mathbb{F}[U]$ whose elements are $\mathbb{F}[U]$-module homomorphisms $\phi\colon C \to \mathbb{F}[U]$; i.e. $\phi(U \cdot \xi) = U \cdot \phi(\xi)$, endowed with the differential ∂^* satisfying $\partial^*\phi(c) = \phi(\partial c)$. We write this complex as $\mathrm{Hom}(C, \mathbb{F}[U])$. This is like the usual construction of cochain complex in the definition of cohomology, except that we will take a different grading convention.

Equip $\mathbb{F}[U]$ with the bigrading for which $U^k \in \mathbb{F}[U]$ has Maslov grading $-2k$ and Alexander grading $-k$. Endow the dual complex $\mathrm{Hom}(C, \mathbb{F}[U])$ with the bigrading for which a homomorphism $\phi\colon C \to \mathbb{F}[U]$ has bigrading (d, s) if it is a homogeneous map of degree (d, s); i.e. for any homogeneous element $c \in C$ with bigrading (m, t), the image $\phi(c)$ has bigrading $(m + d, t + s)$. The homology groups of a dual complex are related to the homology groups of the original complex, via a Universal Coefficient Theorem; see Theorem A.5.6, whose statement uses the bigraded Ext functor described in Section A.5.

Proposition 7.1.2 has the following analogue for GH^-:

PROPOSITION 7.4.3. *If K is a knot, and $m(K)$ is its mirror, then the homology of the grid complex of $m(K)$ is isomorphic to the homology of the dual complex of K; i.e. there is an isomorphism of bigraded $\mathbb{F}[U]$-modules:*

$$GH^-(m(K)) \cong \mathrm{Hom}(GH^-(K), \mathbb{F}[U]) \oplus \mathrm{Ext}(GH^-(K), \mathbb{F}[U])[\![1, 0]\!], \tag{7.10}$$

where the right-hand-side contains the bigraded Ext *modules. More explicitly, if*

$$GH^-(K) \cong \left(\bigoplus_{i=1}^{k} \mathbb{F}[U]/U^{n_i}_{(d_i, s_i)}\right) \oplus \mathbb{F}[U]_{(-2\tau, -\tau)},$$

then

$$GH^-(m(K)) \cong \left(\bigoplus_{i=1}^{k} \mathbb{F}[U]/U^{n_i}_{(2n_i - d_i - 1, n_i - s_i)}\right) \oplus \mathbb{F}[U]_{(2\tau, \tau)}, \tag{7.11}$$

where $\tau = \tau(K)$.

Proof. Let $\mathbb{G}$ be a grid complex for K, and $\mathbb{G}^*$ be its horizontal reflection (as in the proof of Proposition 7.1.2). A grid state $\mathbf{x}$ in $\mathbb{G}$ determines a map from $\mathbf{S}(\mathbb{G}^*)$

to $\mathbb{F}[U]$, defined as follows. The reflection of the grid state $\mathbf{x}$ (which is a grid state for $\mathbb{G}^*$) is mapped to 1, and all other grid states for $\mathbb{G}^*$ are mapped to 0. It is a straightforward exercise to verify that this map induces an isomorphism of bigraded chain complexes:

$$\frac{GC^-(\mathbb{G})}{V_1 = \cdots = V_n} \to \mathrm{Hom}\left(\frac{GC^-(\mathbb{G}^*)}{V_1 = \cdots = V_n}, \mathbb{F}[U]\right)[\![n-1, n-1]\!]. \tag{7.12}$$

Abbreviate $C = \frac{GC^-(\mathbb{G})}{V_1=\cdots=V_n}$ and $C' = \frac{GC^-(\mathbb{G}^*)}{V_1=\cdots=V_n}$. Lemma 7.4.2 gives

$$H(C) \cong GH^-(\mathbb{G}) \otimes W^{\otimes(n-1)} \qquad \text{and} \qquad H(C') \cong GH^-(\mathbb{G}^*) \otimes W^{\otimes(n-1)}. \tag{7.13}$$

The Universal Coefficient Theorem (Theorem A.5.6) gives a bigraded isomorphism which, when combined with Equation (7.13), gives

$$\begin{aligned} H(\mathrm{Hom}(C', \mathbb{F}[U])) &\cong \mathrm{Hom}(H(C'), \mathbb{F}[U]) \oplus \mathrm{Ext}(H(C'), \mathbb{F}[U])[\![1, 0]\!] \\ &\cong \Big(\mathrm{Hom}(GH^-(\mathbb{G}^*), \mathbb{F}[U]) \oplus \mathrm{Ext}(GH^-(\mathbb{G}^*), \mathbb{F}[U])[\![1, 0]\!]\Big) \otimes (W^*)^{\otimes(n-1)}, \end{aligned}$$

where W^* it is the the two-dimensional vector space spanned by vectors in bigrading $(0,0)$ and $(1,1)$. This isomorphism, together with Equations (7.12) and (7.13) and the observation that $(W^*)^{\otimes(n-1)}[\![n-1, n-1]\!] \cong W^{\otimes(n-1)}$ gives

$$\begin{aligned} &GH^-(\mathbb{G}) \otimes W^{\otimes(n-1)} \cong \\ &\qquad \Big(\mathrm{Hom}(GH^-(\mathbb{G}^*), \mathbb{F}[U]) \oplus \mathrm{Ext}(GH^-(\mathbb{G}^*), \mathbb{F}[U])[\![1, 0]\!]\Big) \otimes W^{\otimes(n-1)}. \end{aligned}$$

Equation (7.10) then follows.

For Equation (7.11), the free summand $\mathbb{F}[U]_{(-2a,-a)}$ of $GH^-(m(K))$ comes from $\mathrm{Hom}(GH^-(K), \mathbb{F}[U])$, while all the torsion summands of $GH^-(m(K))$ come from the isomorphism $\mathrm{Ext}(\mathbb{F}[U]/U^n_{(d,s)}, \mathbb{F}[U]) \cong \mathbb{F}[U]/U^n_{(2n-d,n-s)}$. □

EXAMPLE 7.4.4. We saw in Equation (4.32) that for the right-handed trefoil knot $GH^-(T_{2,3}) \cong (\mathbb{F}[U]/U)_{(0,1)} \oplus \mathbb{F}[U]_{(-2,-1)}$. It follows from Proposition 7.4.3 that for the left-handed trefoil knot $GH^-(T_{-2,3}) \cong (\mathbb{F}[U]/U)_{(1,0)} \oplus \mathbb{F}[U]_{(2,1)}$.

Proposition 7.4.3 specializes to the following:

COROLLARY 7.4.5. *Let K be knot and $m(K)$ its mirror, then $\tau(m(K)) = -\tau(K)$.* □

EXAMPLE 7.4.6. By Proposition 6.3.1 and Corollary 7.4.5, for the positive (p,q) torus knot, $\tau(T_{p,q}) = \frac{(p-1)(q-1)}{2}$.

CHAPTER 8

The slice genus and τ

In Chapter 6, we saw that τ gives a lower bound on the unknotting number of a knot. The aim of this chapter is to strengthen this bound from the unknotting number to the slice genus. These bounds allow us to compute the slice genus of torus knots, completing the proof of Theorem 1.2.1.

In view of the normal form for knot cobordisms (Proposition 2.6.11), the proof of these genus bounds depends on two key ingredients:

- an estimate on how τ differs between two knots that can be connected by a sequence of saddle moves; and
- an identification $\tau(K) = \tau(K')$, in the case where $\mathcal{U}_n(K)$ can be connected to K' by exactly n saddle moves. (Recall that $\mathcal{U}_n(K)$ denotes the disjoint union of the knot K with n unknotted, unlinked components.)

The first of these ingredients is obtained by methods similar to those from Chapter 6. To state the estimate, though, we need first to give a (fairly straightforward) generalization of grid homology to links, which we do in Section 8.2. In Section 8.3, we associate maps to saddle moves between the grid homologies of the corresponding links. With the help of these saddle maps, we generalize τ to links, and derive an estimate on how these numerical link invariants change under saddle moves.

For the second ingredient, in Section 8.4 we give a description of the grid homology of $\mathcal{U}_n(K)$ in terms of the grid homology of K. This identification is obtained in a manner reminiscent of the stabilization invariance proof for grid homology. With all the pieces in place, we prove the slice genus bound in Section 8.5.

In Section 8.6, we give an example of an *exotic* $\mathbb{R}^4$, a smooth four-manifold homeomorphic but not diffeomorphic to $\mathbb{R}^4$. This construction uses a knot that is topologically slice, according a deep result of Freedman, but not smoothly slice, which we deduce using τ. We conclude with a brief comparison of slice genus and the unknotting number in Section 8.7.

8.1. Slice genus bounds from τ and their consequences

We start by stating the main result of this chapter, and then give some of its consequences. Unless explicitly stated otherwise, all of the surfaces in D^4 or in $[0,1] \times S^3$ we consider are smoothly embedded.

THEOREM 8.1.1. *If K_0 and K_1 are two knots that can be connected by a genus g cobordism in $[0,1] \times S^3$, then $|\tau(K_0) - \tau(K_1)| \leq g$.*

COROLLARY 8.1.2. *For any knot K, $|\tau(K)| \leq g_s(K)$; in particular, if K is slice, then $\tau(K) = 0$.*

Proof. Let F be a genus-minimizing surface in D^4 with K as its boundary. Removing a small four-ball from D^4 centered at a point in F gives a genus g

cobordism from K to the unknot. The corollary now follows from Theorem 8.1.1 and the fact that $\tau(\mathcal{O}) = 0$. $\square$

In Corollary 6.3.3, we computed the unknotting number of the (p,q) torus knot. This same quantity computes both the Seifert and the four-ball genus of this knot, according to the following:

COROLLARY 8.1.3. *For the (p,q) torus knot $T_{p,q}$ the Seifert genus and the four-ball genus are both equal to* $\frac{(p-1)(q-1)}{2}$.

Proof. Seifert's algorithm (recalled in Section B.3) applied to a standard projection of $T_{p,q}$ (see Figure 2.3) gives a Seifert surface of genus $\frac{(p-1)(q-1)}{2}$. Combine Proposition 6.3.1, Corollary 8.1.2, and the fact that $g_s(K) \leq g(K)$ to get

$$\frac{(p-1)(q-1)}{2} = |\tau(T_{-p,q})| \leq g_s(T_{-p,q}) = g_s(T_{p,q}) \leq g(T_{p,q}) \leq \frac{(p-1)(q-1)}{2},$$

which, taken together, imply the corollary. $\square$

Together, Corollaries 6.3.3 and 8.1.3 give Theorem 1.2.1. To prove Theorem 8.1.1, we use the normal form decomposition from Proposition 2.6.11, which brings links into play.

8.2. A version of grid homology for links

We introduce a generalization of grid homology to oriented links, the *collapsed grid homology for links*, which can be thought of as a grid homology analogue of the one-variable Alexander polynomial. A more elaborate version, the grid homology analogue of the multi-variable Alexander polynomial, will be discussed in Chapter 11 (along with an intermediate version, the *uncollapsed grid homology.*)

Let $\mathbb{G}$ be a grid diagram representing an oriented link $\vec{L}$. Consider the chain complex $GC^-(\mathbb{G})$ generated freely over $\mathbb{F}[V_1,\dots,V_n]$ by grid states, with differential

$$\partial^-_{\mathbb{X}}\mathbf{x} = \sum_{\mathbf{y}\in\mathbf{S}(\mathbb{G})} \sum_{\{r\in\mathrm{Rect}^\circ(\mathbf{x},\mathbf{y})\,\big|\,r\cap\mathbb{X}=\emptyset\}} V_1^{O_1(r)}\cdots V_n^{O_n(r)}\cdot\mathbf{y}. \tag{8.1}$$

So far this looks exactly like the definition of the unblocked grid complex of a knot; see Definition 4.6.1. Equip the complex with the Maslov grading defined in Equation (4.5), and normalize the Alexander function to take into account the number of components ℓ of $\vec{L}$, as follows:

$$A(\mathbf{x}) = \frac{1}{2}(M_{\mathbb{O}}(\mathbf{x}) - M_{\mathbb{X}}(\mathbf{x})) - \left(\frac{n-\ell}{2}\right). \tag{8.2}$$

LEMMA 8.2.1. *For any grid state* $\mathbf{x}$ *the Alexander function* $A(\mathbf{x})$ *is an integer.*

Proof. For a link with ℓ components, we claim that

$$M_{\mathbb{X}}(\mathbf{x}^{NWO}) \equiv n-\ell \pmod 2. \tag{8.3}$$

Since the permutation taking $\mathbf{x}^{NWO}$ to $\mathbf{x}^{NWX}$ is a product of ℓ disjoint cycles in the permutation group on n letters (see Remark 3.1.2), Equation (8.3) follows from the fact that such a permutation can be written as a product of $n-\ell$ transpositions. The rest of the argument is exactly as in Proposition 4.3.3. $\square$

DEFINITION 8.2.2. Let $\mathbb{G}$ be a grid diagram representing an oriented link $\vec{L}$. The ***uncollapsed, bigraded grid complex*** $GC^-(\mathbb{G})$ of the grid diagram $\mathbb{G}$ is the chain complex defined as above; i.e. with differential specified by Equation (8.1), and bigrading specified by the Maslov and Alexander functions from Equations (4.5) and (8.2).

Most of the definitions from Chapter 4 adapt in a straightforward way. The proof of Lemma 4.6.9 adapts to links, to give the following:

LEMMA 8.2.3. *Let $\mathbb{G}$ be a grid diagram for an oriented link $\vec{L}$, and let O_i and O_j be markings on the same component of the link $\vec{L}$. Then, multiplication by V_i, thought of as an endomorphism of the bigraded grid complex $GC^-(\mathbb{G})$, is chain homotopic to multiplication by V_j.* □

To get a homology module associated to $\vec{L}$ that is a module over a polynomial algebra in one variable, we set the V-variables corresponding to different link components equal to each other, as follows:

DEFINITION 8.2.4. Let $\mathbb{G}$ be a grid diagram representing an oriented link $\vec{L}$ with ℓ components. Choose O-markings $O_{j_1},\dots,O_{j_\ell}$ lying on the ℓ different components of $\vec{L}$. The ***collapsed grid complex*** of the link $\vec{L}$ is the chain complex $cGC^-(\mathbb{G})$, formed as the quotient of the bigraded grid complex $GC^-(\mathbb{G})$:

$$cGC^-(\mathbb{G}) = \frac{GC^-(\mathbb{G})}{V_{j_1} = \dots = V_{j_\ell}}. \tag{8.4}$$

The homology of $cGC^-(\mathbb{G})$, written $cGH^-(\mathbb{G})$, is called the ***collapsed grid homology of*** $\mathbb{G}$. It can be viewed as a module over $\mathbb{F}[U]$, where the variable U acts as multiplication by any V_i for $i \in \{1,\dots,n\}$.

THEOREM 8.2.5. *The collapsed grid homology $cGH^-(\vec{L})$ of $\vec{L}$, thought of as a bigraded $\mathbb{F}[U]$-module, is an invariant of the oriented link $\vec{L}$.*

Proof. Recall that $cGC^-(\mathbb{G})$ depends on a choice of a sequence $O_{j_1},\dots,O_{j_\ell}$; and its $\mathbb{F}[U]$-module structure is defined using a choice of V_i for $i \in \{1,\dots,n\}$.

To see that the $\mathbb{F}[U]$-module structure on the homology is independent of the choice of V_i, Lemma 8.2.3 shows that multiplication by any V_i is chain homotopic to multiplication by V_{j_k}, where O_{j_k} is chosen to be on the same component as O_i; and this remains true in the quotient complex $cGC^-(\mathbb{G})$. Since all the different V_{j_k} are identified in $cGC^-(\mathbb{G})$, it follows that all the V_i's induce chain homotopic endomorphisms of $cGC^-(\mathbb{G})$.

We claim next that the homology module is also independent of the choice of the distinguished sequence of O-markings, $O_{j_1},\dots,O_{j_\ell}$. Lemma 5.2.13 gives a quasi-isomorphism from the iterated mapping cone of $V_{j_k} - V_{j_{k+1}}$ acting on $GC^-(\mathbb{G})$, where $k = 1,\dots,\ell-1$, to the quotient complex $cGC^-(\mathbb{G})$; Lemma 8.2.3 gives homotopies between $V_{j_k} - V_{j_{k+1}}$ and $V_{j'_k} - V_{j'_{k+1}}$ (as we vary the sequence $O_{j_1},\dots,O_{j_\ell}$); and Lemma 5.2.14 gives an isomorphism from the iterated mapping cone for $V_{j_k} - V_{j_{k+1}}$ with that for $V_{j'_k} - V_{j'_{k+1}}$.

Having established that the $\mathbb{F}[U]$-module $cGH^-(\mathbb{G})$ depends on the grid diagram only, it remains to see that it is independent of this choice, as well. This follows from the observation that grid moves for links induce isomorphisms on collapsed grid homology by the same formulas as they did for knots in Chapter 5. □

REMARK 8.2.6. Another approach to get a link invariant of the ℓ-component oriented link $\vec{L}$ is to view the homology of the bigraded grid complex $GC^-(\mathbb{G})$ as a module over the polynomial algebra $\mathbb{F}[U_1, \dots, U_\ell]$, whose variables correspond to the various components of $\vec{L}$; see Definition 11.1.1. The above discussion can be adapted to show that the resulting bigraded homology is also an invariant of the oriented link $\vec{L}$. This bigraded grid homology will be further studied in Chapter 11.

8.2.1. Simply blocked grid homology for links. Using the collapsed grid homology of links, the notion of simply blocked grid homology (of Definition 4.6.12) extends to links. Although we will not need this version for the rest of the present chapter, we give the construction now and compute its Euler characteristic.

DEFINITION 8.2.7. Suppose that $\mathbb{G}$ is a grid diagram representing an oriented ℓ-component link $\vec{L}$, and let $GC^-(\mathbb{G})$ be its grid complex. Choose a sequence $O_{j_1}, \dots, O_{j_\ell}$ of O-markings, one on each component of $\vec{L}$, and let

$$\widehat{GC}(\mathbb{G}) = \frac{cGC^-(\mathbb{G})}{V_{j_1} = 0} = \frac{GC^-(\mathbb{G})}{V_{j_1} = \cdots = V_{j_\ell} = 0}.$$

The chain complex $\widehat{GC}(\mathbb{G})$ is the ***simply blocked, bigraded grid complex*** for the grid diagram $\mathbb{G}$ of $\vec{L}$, and its homology $\widehat{GH}(\mathbb{G})$, is called the ***simply blocked, bigraded grid homology*** for the link $\vec{L}$. We can also form the fully blocked complex $\widetilde{GC}(\mathbb{G})$ (with Alexander grading as given in Equation (8.2)), counting only rectangles disjoint from $\mathbb{O}, \mathbb{X}$; i.e. equivalently, this is the quotient

$$\widetilde{GC}(\mathbb{G}) = \frac{GC^-(\mathbb{G})}{V_1 = \cdots = V_n = 0}.$$

The homology of this chain complex is the ***fully blocked grid homology*** of $\mathbb{G}$.

The fully blocked grid homology is in practice easier to compute. It is related to the simply blocked homology just as in the case of knots. In fact, the proof of Proposition 4.6.15 immediately adapts to give the following:

PROPOSITION 8.2.8. *Let W be the two-dimensional bigraded vector space, with one generator in bigrading $(0,0)$ and the other in bigrading $(-1,-1)$. Then, there is an isomorphism $\widetilde{GH}(\mathbb{G}) \cong \widehat{GH}(\mathbb{G}) \otimes W^{\otimes(n-\ell)}$ of bigraded vector spaces.* □

PROPOSITION 8.2.9. *The bigraded vector space $\widehat{GH}(\mathbb{G})$ is an invariant of the oriented link $\vec{L}$ specified by $\mathbb{G}$.*

Proof. The statement follows exactly as the invariance of the collapsed grid homology $cGH^-(\vec{L})$. (See Theorem 8.2.5.) □

PROPOSITION 8.2.10. *Let $\vec{L}$ be an oriented link with ℓ components. The Euler characteristics of the various versions of grid homology are related to the symmetrized Alexander polynomial $\Delta_{\vec{L}}(t)$ by the formulas:*

$$\chi(\widetilde{GH}(\vec{L})) = \Delta_{\vec{L}}(t) \cdot (1 - t^{-1})^{n-1} \cdot t^{\frac{\ell-1}{2}} \tag{8.5}$$

$$\chi(\widehat{GH}(\vec{L})) = \Delta_{\vec{L}}(t) \cdot (t^{\frac{1}{2}} - t^{-\frac{1}{2}})^{\ell-1} \tag{8.6}$$

$$\chi(cGH^-(\vec{L})) = \Delta_{\vec{L}}(t) \cdot t^{\frac{1}{2}} \cdot (t^{\frac{1}{2}} - t^{-\frac{1}{2}})^{\ell-2}. \tag{8.7}$$

Proof. We have the following analogue of Proposition 4.7.2 (Equation (4.26)):

$$A(\mathbf{x}) = -\sum_{x\in\mathbf{x}} w_{\mathcal{D}}(x) + \frac{1}{8}\sum_{j=1}^{8n} w_{\mathcal{D}}(p_j) - \frac{n-\ell}{2} = A'(\mathbf{x}) + a(\mathbb{G}) - \frac{n-\ell}{2}, \tag{8.8}$$

where $\mathcal{D}$ is the oriented link diagram defined by the grid $\mathbb{G}$ presenting $\vec{L}$. The novelty here is the appearance of the number ℓ of components in the formula; its proof, though, is straightforward. It follows that

$$\begin{aligned}\chi(\widetilde{GH}(\vec{L})) &= \chi(\widetilde{GC}(\mathbb{G}))\\ &= \sum_{\mathbf{x}\in\mathbf{S}(\mathbb{G})} (-1)^{M(\mathbf{x})} t^{A(\mathbf{x})} = (-1)^{n-1}\epsilon(\mathbb{G})\det(\mathbf{M}(\mathbb{G}))\cdot t^{a(\mathbb{G})-\frac{n-\ell}{2}}\\ &= D_{\mathbb{G}}(t)(t^{\frac{1}{2}}-t^{-\frac{1}{2}})^{n-1}t^{\frac{\ell-n}{2}} = \Delta_{\vec{L}}(t)(1-t^{-1})^{n-1}t^{\frac{\ell-1}{2}},\end{aligned}$$

verifying Equation (8.5). The first step uses the fact that the Euler characteristic of a complex agrees with that of its homology, the second is straightforward, the third uses Equation (8.8), the fourth is the definition of $D_{\mathbb{G}}(t)$, and the fifth uses Theorem 3.3.6. The other two equations follow from Equation (8.5), since $\chi(\widetilde{GH}(\vec{L})) = \chi(\widehat{GH}(\vec{L}))(1-t^{-1})^{n-\ell}$ and $\chi(\widehat{GH}(\vec{L})) = \chi(cGH^-(\vec{L}))(1-t^{-1})$. □

8.3. Grid homology and saddle moves

We define and study maps on collapsed grid homology associated to saddle moves. If $\vec{L}'$ is obtained from $\vec{L}$ by a saddle move, and the number of components of $\vec{L}'$ is one more than the number of components of $\vec{L}$, then we say that $\vec{L}'$ is obtained from $\vec{L}$ by a *split move* and $\vec{L}$ is obtained from $\vec{L}'$ by a *merge move.*

PROPOSITION 8.3.1. *As usual, let* $W = \mathbb{F}_{(0,0)} \oplus \mathbb{F}_{(-1,-1)}$. *If* $\vec{L}'$ *is obtained from* $\vec{L}$ *by a split move, then there are* $\mathbb{F}[U]$*-module maps*

$$\sigma\colon cGH^-(\vec{L})\otimes W \to cGH^-(\vec{L}')$$
$$\mu\colon cGH^-(\vec{L}') \to cGH^-(\vec{L})\otimes W$$

with the following properties:

- σ *is homogeneous of degree* $(-1,0)$,
- μ *is homogeneous of degree* $(-1,-1)$,
- $\mu\circ\sigma$ *is multiplication by* U,
- $\sigma\circ\mu$ *is multiplication by* U.

We postpone the proof of the above proposition, after giving some of its immediate consequences. The first application is the following generalization of Proposition 6.1.4:

PROPOSITION 8.3.2. *Let* $\vec{L}$ *be an* ℓ*-component link,* $\mathrm{Tors} = \mathrm{Tors}(cGH^-(\vec{L}))$, *and* $r = 2^{\ell-1}$. *Then there is an isomorphism*

$$cGH^-(\vec{L})/\mathrm{Tors} \cong \mathbb{F}[U]^r. \tag{8.9}$$

Proof. We will use the fact that any submodule of $\mathbb{F}[U]^m$ is of the form $\mathbb{F}[U]^k$ with $k\leq m$. Equation (8.9) is proved by induction on the number of components ℓ of $\vec{L}$, with the basic case $\ell=1$ supplied by Proposition 6.1.4. Assume that Equation (8.9) holds for all links with ℓ components. If $\vec{L}'$ has $\ell+1$ components, there is a saddle move connecting $\vec{L}'$ to $\vec{L}$, where $\vec{L}$ has ℓ components. Proposition 8.3.1 and Lemma 6.1.3 gives an inclusion of $cGH^-(\vec{L}')/\mathrm{Tors}$ into $cGH^-(\vec{L})\otimes W/\mathrm{Tors}\cong \mathbb{F}[U]^{2r}$ (by the induction hypothesis), showing that $cGH^-(\vec{L}')/\mathrm{Tors}\cong\mathbb{F}[U]^{r'}$ with $r'\leq 2^\ell$. Proposition 8.3.1 and Lemma 6.1.3 give an inclusion of $cGH^-(\vec{L})\otimes W/\mathrm{Tors}\cong\mathbb{F}[U]^{2r}$ into $cGH^-(\vec{L}')/\mathrm{Tors}$, showing that $r'=2^\ell$, as claimed. □

The above proposition can be used to generalize τ to links, as follows:

DEFINITION 8.3.3. Let $\vec{L}$ be an oriented link with ℓ components. The ***τ-set of*** $\vec{L}$ is the sequence of integers $\tau_{min}(\vec{L})=\tau_1\leq\tau_2\leq\dots\leq\tau_{2^{\ell-1}}=\tau_{max}(\vec{L})$, defined as follows. Choose a generating set of $2^{\ell-1}$ elements for $cGH^-(\vec{L})/\mathrm{Tors}$, with the property that each element is homogeneous with respect to the Alexander grading. The τ-set obtained by multiplying the Alexander gradings of these generators by (-1), and placing the result in increasing order.

COROLLARY 8.3.4. *The τ-set of $\vec{L}$ is an oriented link invariant.*

Proof. Definition 8.3.3 appears to depend on a choice of basis; to show it does not, we give an alternative, basis-independent formulation. Consider the non-torsion quotient $M=cGH^-(\vec{L})/\mathrm{Tors}$, thought of as a graded $\mathbb{F}[U]$-module induced by the Alexander grading. Decompose $M=\bigoplus_{s\in\mathbb{Z}}M_s$, where for each s, M_s is the finite dimensional vector space spanned by elements with Alexander grading s. Form the Poincaré series of M, $P_M(t)=\sum_{s\in\mathbb{Z}}\dim M_s\cdot t^s\in\mathbb{Z}[t,t^{-1}]]$, and notice that $P_M(t)$ is a sum, over each generator ξ in $cGH^-(\vec{L})/\mathrm{Tors}$ of $\sum_{s=0}^\infty t^{-s}$ times $t^{A(\xi)}$. By Proposition 8.3.2, the τ-set of $\vec{L}$ is the ordered list of (-1) times the exponents appearing in the polynomial $(1-t^{-1})\cdot P_M(t)$, taken with multiplicity specified by their respective coefficients. The fact that the τ-set has $r=2^{\ell-1}$ elements in it follows immediately from Proposition 8.3.2.

Since $P_M(t)$ is naturally associated to the bigraded $\mathbb{F}[U]$-module $cGH^-(\vec{L})$, which is a link invariant by Theorem 8.2.5, it follows that the τ-set is also an invariant. □

REMARK 8.3.5. Grid homology for knots, and hence also τ, is independent of the orientation on the knot (Proposition 5.3.2). The same is not true for links.

We will need only $\tau_{max}(\vec{L})$ and $\tau_{min}(\vec{L})$, which can be characterized as follows: $\tau_{min}(\vec{L})$ is (-1) times the maximal Alexander grading of any homogeneous, non-torsion element in $cGH^-(\vec{L})$; $\tau_{max}(\vec{L})$ is (-1) times the minimal Alexander grading of any homogeneous element of $cGH^-(\vec{L})/\mathrm{Tors}$ that is not contained in $U\cdot(cGH^-(\vec{L})/\mathrm{Tors})$.

THEOREM 8.3.6. *Let $\vec{L}$ and $\vec{L}'$ be two oriented links that differ by a saddle move, and suppose that $\vec{L}'$ has one more component than $\vec{L}$. Then,*

$$\tau_{min}(\vec{L})-1\leq\tau_{min}(\vec{L}')\leq\tau_{min}(\vec{L}) \tag{8.10}$$

$$\tau_{max}(\vec{L})\leq\tau_{max}(\vec{L}')\leq\tau_{max}(\vec{L})+1. \tag{8.11}$$

Proof. Consider a non-torsion element $\xi \in cGH^-(\vec{L}) \otimes W$ with maximal Alexander grading, i.e. the Alexander grading of ξ is $-\tau_{min}(\vec{L})$. By Proposition 8.3.1, its image $\sigma(\xi)$ is non-torsion, and the Alexander grading of $\sigma(\xi)$ is $-\tau_{min}(\vec{L})$, hence $-\tau_{min}(\vec{L}) \leq -\tau_{min}(\vec{L}')$. Similarly, if $\eta \in cGH^-(\vec{L}')$ is a non-torsion element with maximal Alexander grading $-\tau_{min}(\vec{L}')$, then its image $\mu(\eta)$ has Alexander grading $-\tau_{min}(\vec{L}') - 1$, and it is non-torsion, so $-\tau_{min}(\vec{L}') - 1 \leq -\tau_{min}(\vec{L})$. Rearranging this gives Inequality (8.10).

To verify Inequality (8.11), consider the free $\mathbb{F}[U]$-modules

$$(cGH^-(\vec{L}) \otimes W)/\text{Tors}$$

and

$$cGH^-(\vec{L}')/\text{Tors}.$$

The split and merge maps induce $\mathbb{F}[U]$-module maps on these quotients. Pick a homogeneous element $g \in cGH^-(\vec{L}')/\text{Tors}$ with Alexander grading $-\tau_{max}(\vec{L}')$ that generates a free summand. Then $\mu(g)$ has Alexander grading $-\tau_{max}(\vec{L}') - 1$; since $\sigma \circ \mu = U$, either $\mu(g)$ generates a free summand in $cGH^-(\vec{L}) \otimes W/\text{Tors}$, or it is U-times such an element. Since the minimal Alexander grading of any generator of a free summand in $cGH^-(\vec{L}) \otimes W/\text{Tors}$ is $-\tau_{max}(\vec{L}) - 1$, it follows that $-\tau_{max}(\vec{L}') - 1 \geq -\tau_{max}(\vec{L}) - 2$, giving the upper bound on $\tau_{max}(\vec{L}')$ from Equation (8.11). The lower bound follows similarly. □

EXERCISE 8.3.7. **(a)** The positive Hopf link H_+ differs by a single saddle move from both the unknot (which has $\tau = 0$) and the right-handed trefoil (which has $\tau = 1$). Use Theorem 8.3.6 to determine $\tau_{max}(H_+)$ and $\tau_{min}(H_+)$.
(b) Let $\vec{L}$ be an oriented link and $-\vec{L}$ be the same link with the opposite orientation on all of its components. Show that the τ-sets of $\vec{L}$ and $-\vec{L}$ coincide.
(c) Find a link with two different orientations whose associated τ-sets are different.

Having drawn consequences of Proposition 8.3.1, we turn to its proof. Let $\mathbb{G}$ and $\mathbb{G}'$ be two grid diagrams that differ only in the placement of their O-markings in two distinguished columns. Assume that in those two distinguished columns, the circular ordering of the O- and the X-markings for $\mathbb{G}$ and $\mathbb{G}'$ are as specified in Figure 8.1; and label the distinguished O-markings O_1 and O_2 for $\mathbb{G}$ and O_1' and O_2' for $\mathbb{G}'$ as in the figure. As pictured there, the distinguished vertical circle β_i that separates the two markings is divided into two disjoint arcs $\mathbf{A}$ and $\mathbf{B}$, so that $\mathbf{B}$ is adjacent to the X-markings, but $\mathbf{A}$ is not; O_1 is at the lower left end of $\mathbf{A}$ while O_2 is at the upper right; O_1' is at the lower right and O_2' is at the upper left. If $\vec{L}$ and $\vec{L}'$ differ by a saddle move, then we can find grid diagrams $\mathbb{G}$ and $\mathbb{G}'$ as above for $\vec{L}$ and $\vec{L}'$ respectively. We need the following:

LEMMA 8.3.8. *If O_1 and O_2 are on the same component of the link specified by $\mathbb{G}$, then there is an isomorphism of bigraded $\mathbb{F}[U]$-modules*

$$H\left(\frac{cGC^-(\mathbb{G})}{V_1 = V_2}\right) \cong cGH^-(\mathbb{G}) \otimes W.$$

Proof. Since O_1 and O_2 are on the same component of the link, $cGC^-(\mathbb{G})$ can be viewed as a free module over $\mathbb{F}[V_1, V_2]$. By Lemma 7.4.2, multiplication

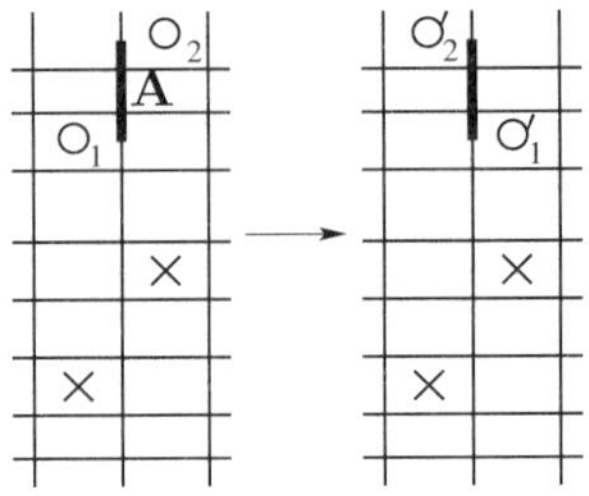

FIGURE 8.1. **Grid diagrams for a saddle move.** The arc **A** is indicated by the thicker segment. The complement of **A** in the circle is **B**.

by V_1, thought of as an endomorphism of $cGC^-(\mathbb{G})$, is chain homotopic to V_2. Lemma 7.4.1 now applies, and provides the stated isomorphism. □

Proof of Proposition 8.3.1. Identify the grid states of $\mathbb{G}$ and $\mathbb{G}'$. The grid states in $\mathbb{G}$ (and likewise in $\mathbb{G}'$) are classified in two types: denote by $\mathcal{A}$ the set of grid states whose component in the distinguished vertical circle β_i lies in the arc **A** between O_1 and O_2; and denote by $\mathcal{B}$ the set of grid states whose β_i-component is on the other arc **B**.

Since O_1 and O_2 are in different components in $\mathbb{G}'$, we can choose them to be part of the sequence of O_{j_k} defining $cGC^-(\mathbb{G}')$. With this choice, multiplication by V_1 equals multiplication by V_2, when thought of as endomorphisms of either complexes $\frac{cGC^-(\mathbb{G})}{V_1=V_2}$ or $cGC^-(\mathbb{G}')$. The resulting endomorphism is denoted by an undecorated U.

Now we define the chain maps inducing the maps σ and μ on homology promised in Proposition 8.3.1. These chain maps will be denoted by the same symbols σ and μ. Define

$$\sigma\colon \frac{cGC^-(\mathbb{G})}{V_1 = V_2} \to cGC^-(\mathbb{G}') \qquad \text{and} \qquad \mu\colon cGC^-(\mathbb{G}') \to \frac{cGC^-(\mathbb{G})}{V_1 = V_2}$$

by

$$\sigma(\mathbf{x}) = \begin{cases} U\cdot\mathbf{x} & \text{if } \mathbf{x}\in\mathcal{A} \\ \mathbf{x} & \text{if } \mathbf{x}\in\mathcal{B}. \end{cases} \qquad \text{and} \qquad \mu(\mathbf{x}) = \begin{cases} \mathbf{x} & \text{if } \mathbf{x}\in\mathcal{A} \\ U\cdot\mathbf{x} & \text{if } \mathbf{x}\in\mathcal{B}. \end{cases}$$

Obviously, both composite maps $\sigma\circ\mu$ and $\mu\circ\sigma$ are multiplications by U.

The maps appearing in the statement of the proposition are the maps induced by these chain maps on homology, composed with the identification from Lemma 8.3.8. The proposition is proved once we show that the maps defined on the chain level are, indeed, chain maps which have the required behaviour on bi-gradings.

To prove that σ is a chain map, consider a rectangle $r\in\mathrm{Rect}(\mathbf{x},\mathbf{y})$ connecting the grid states $\mathbf{x}$ and $\mathbf{y}$. Since the $\mathbb{X}$-markings are at the same positions in both diagrams, r is disjoint from $\mathbb{X}$ in $\mathbb{G}$ exactly when it is disjoint from $\mathbb{X}$ in $\mathbb{G}'$. It is easy to see that if both grid states $\mathbf{x},\mathbf{y}$ are in $\mathcal{A}$ or both are in $\mathcal{B}$, then the rectangle r intersects $\{O_1,O_2\}$ with the same multiplicity as it intersects $\{O_1',O_2'\}$, viewed as a rectangle in either $\mathbb{G}$ or $\mathbb{G}'$. If $\mathbf{x}\in\mathcal{A}$ and $\mathbf{y}\in\mathcal{B}$, then r, thought of as a rectangle in $\mathbb{G}$, contains exactly one of O_1 or O_2, but it does not contain either of O_1' or O_2'. Similarly, if $\mathbf{x}\in\mathcal{B}$ and $\mathbf{y}\in\mathcal{A}$, r contains neither O_1 nor O_2, but it contains exactly one of O'_1 or O'_2. It follows from these observations that σ is a chain map. The map μ is a chain map by the same logic.

In comparing the Maslov gradings of $\mathbf{x}$ for $\mathbb{G}$ and $\mathbb{G}'$, note that for an element $\mathbf{x} \in \mathcal{A}$ we have $M_{\mathbb{O}'}(\mathbf{x}) = M_{\mathbb{O}}(\mathbf{x}) + 1$, while for $\mathbf{x} \in \mathcal{B}$, $M_{\mathbb{O}'}(\mathbf{x}) = M_{\mathbb{O}}(\mathbf{x}) - 1$. The Alexander gradings are given by

$$A(\mathbf{x}) = \frac{1}{2}(M_{\mathbb{O}}(\mathbf{x}) - M_{\mathbb{X}}(\mathbf{x})) - \frac{n-\ell}{2} \quad \text{and}$$
$$A'(\mathbf{x}) = \frac{1}{2}(M_{\mathbb{O}'}(\mathbf{x}) - M_{\mathbb{X}}(\mathbf{x})) - \frac{n-\ell-1}{2},$$

where ℓ denotes the number of components of $\vec{L}$. It follows that σ and μ have the stated behaviour on the bigradings. □

8.4. Adding unknots to a link

To establish Theorem 8.1.1, we will need to understand links of the form $\mathcal{U}_d(K)$. For such links, the τ-set has a very simple structure.

LEMMA 8.4.1. *If L is a link of the form $L = \mathcal{U}_d(K)$ for some knot K, then, for any orientation $\vec{L}$ on L, $\tau_{min}(\vec{L}) = \tau_{max}(\vec{L}) = \tau(K)$.*

This lemma is a quick consequence of the following more general description of how collapsed grid homology changes when we add an unknotted, unlinked component:

LEMMA 8.4.2. *Let $\vec{L}$ be an oriented link, and let $\vec{L}' = \mathcal{U}_1(\vec{L})$. Then, there is an isomorphism of bigraded $\mathbb{F}[U]$-modules*

$$cGH^-(\vec{L}') \cong cGH^-(\vec{L})[\![1,0]\!] \oplus cGH^-(\vec{L}). \tag{8.12}$$

Before proving this lemma, we give a slight extension of grid diagrams which will be useful for studying the links $\mathcal{U}_d(\vec{L})$.

DEFINITION 8.4.3. An ***extended grid diagram*** is a generalization of grid diagrams, where now squares can be marked by both an X and an O; i.e. an extended grid diagram satisfies Conditions (G-1) and (G-2) in Definition 3.1.1. A square marked by both an X and O is called a ***doubly-marked*** square.

An extended grid diagram specifies an oriented link as in the case of usual grid diagrams, with the understanding that each doubly-marked square corresponds to an unknotted, unlinked component, oriented arbitrarily. Let $\overline{\mathbb{G}}$ be a given extended grid diagram. The notions of grid states, the grid complexes GC^- and cGC^-, and the collapsed homology cGH^- generalize immediately to the case of extended grids.

EXAMPLE 8.4.4. The 1×1 extended grid $\overline{\mathbb{G}}$ (with an X and an O in the single square) represents the unknot, has a single generator and $\partial^-_{\mathbb{X}} = 0$. It follows at once that $cGH^-(\overline{\mathbb{G}}) \cong \mathbb{F}[U]$. In particular, this homology is isomorphic to $GH^-(\mathcal{O})$.

EXAMPLE 8.4.5. The two-component unlink can be represented by a 2×2 extended grid diagram $\overline{\mathbb{G}}$, with two squares which are simultaneously marked with an O and an X. This grid diagram has two grid states $\mathbf{p}$ and $\mathbf{q}$, both with Alexander grading zero. Moreover, $M(\mathbf{p}) = 0$ and $M(\mathbf{q}) = -1$, and the differential $\partial^-_{\mathbb{X}}$ vanishes identically. It follows that $cGH^-(\overline{\mathbb{G}})$ is isomorphic to $\mathbb{F}[U]_{(0,0)} \oplus \mathbb{F}[U]_{(-1,0)}$.

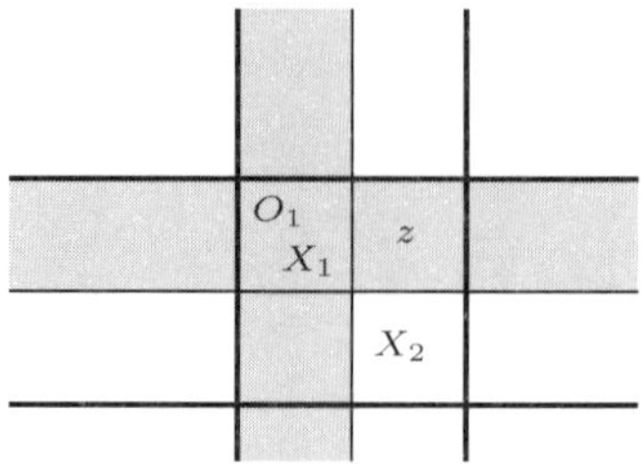

FIGURE 8.2. **Destabilizing doubly-marked squares.** Destabilize the 2×2 grid containing a doubly-marked square. We have shaded the row and the column whose contributions cancel in $\mathcal{H}^{\mathbf{I}}_{O_1,X_1} \circ \partial^{\mathbf{I}}_{\mathbf{N}}$.

An $n \times n$ extended grid diagram $\overline{\mathbb{G}}$ is called an *extended destabilization* of a grid diagram $\mathbb{G}$ with grid number $n+1$ if the markings in the complement of the row and column through a doubly-marked square in $\overline{\mathbb{G}}$ agree with the markings in the complement of a pair of consecutive rows and columns in $\mathbb{G}$. In this case, there is a distinguished 2×2 block in $\mathbb{G}$, with all four squares marked. Adapting the stabilization invariance proof in Proposition 5.2.1 gives the following:

LEMMA 8.4.6. *Let $\overline{\mathbb{G}}$ be an extended grid diagram with a single doubly-marked square, and suppose that $\overline{\mathbb{G}}$ is an extended destabilization of $\mathbb{G}$. Then, there is an isomorphism of bigraded $\mathbb{F}[U]$-modules $cGH^-(\mathbb{G}) \cong cGH^-(\overline{\mathbb{G}})$.*

Proof. Given $\overline{\mathbb{G}}$, there are two choices of $\mathbb{G}$, corresponding to the two choices of markings in the distinguished 2×2 block. Since these two grid diagrams can be connected by a switch, it suffices to consider the case where, in the distinguished 2×2 square, the two O-markings are diagonal. We would like to use the destabilization argument from Chapter 5 in this case. To this end, notice that proof of Proposition 5.2.1 uses the placement of the three special markings in the distinguished 2×2 square (of type *X:SW*), but does not use significantly the fact that the fourth square is unmarked. Thus, the proof adapts, to give a quasi-isomorphism

$$cGC^-(\mathbb{G}) \to \mathrm{Cone}(V_1 - V_2\colon cGC^-(\overline{\mathbb{G}})[V_1] \to cGC^-(\overline{\mathbb{G}})[V_1]).$$

Compose the induced map on homology with the $\mathbb{F}[U]$-module isomorphism

$$H(\mathrm{Cone}(V_1 - V_2\colon cGC^-(\overline{\mathbb{G}})[V_1] \to cGC^-(\overline{\mathbb{G}})[V_1]) \cong cGH^-(\overline{\mathbb{G}})$$

from Lemma 5.2.16 to get the stated isomorphism. □

Assume that $\vec{L}$ is a link of $\ell - 1$ components and $\vec{L}' = \mathcal{U}_1(\vec{L})$. Suppose that $\overline{\mathbb{G}}$ is an extended grid diagram representing $\mathcal{U}_1(\vec{L})$, and the unlink component is represented by a square marked simultaneously by X_1 and O_1. Assume moreover that there is a 2×2 box of squares, the upper left of which is the doubly-marked square, and the lower right of which is marked X_2. Number the variables so that the V-variable corresponding to this lower row is V_2. Let $\mathbb{G}$ be the grid diagram obtained by deleting the row and column of the doubly-marked square; see Figure 8.2.

LEMMA 8.4.7. *For $\mathbb{G}$ and $\overline{\mathbb{G}}$ as above, there is an isomorphism of bigraded $\mathbb{F}[U]$-modules $cGH^-(\overline{\mathbb{G}}) \cong cGH^-(\mathbb{G})[\![1,0]\!] \oplus cGH^-(\mathbb{G})$.*

Proof. The proof of Proposition 5.2.17 applies, after suitable modifications. Let $\mathbf{I}(\overline{\mathbb{G}})$ denote the set of those generators in $\mathbf{S}(\overline{\mathbb{G}})$ which have a component at the lower right corner of the O_1X_1-marked square, and let $\mathbf{N}(\overline{\mathbb{G}})$ be the complement of $\mathbf{I}(\overline{\mathbb{G}})$ in $\mathbf{S}(\overline{\mathbb{G}})$. From the placement of the X-markings, we see that $\mathbf{N}(\overline{\mathbb{G}})$ spans a subcomplex $\mathbf{N}$ in $GC^-(\overline{\mathbb{G}})$. Write the differential on $GC^-(\overline{\mathbb{G}})$ as a 2×2 matrix

$$\partial_{\mathbb{X}}^- = \begin{pmatrix} \partial_{\mathbf{I}}^{\mathbf{I}} & 0 \\ \partial_{\mathbf{I}}^{\mathbf{N}} & \partial_{\mathbf{N}}^{\mathbf{N}} \end{pmatrix}.$$

There is a one-to-one correspondence between elements of $\mathbf{I}(\overline{\mathbb{G}})$ and grid states in $\mathbf{S}(\mathbb{G})$ for the destabilized diagram, inducing an isomorphism of bigraded chain complexes $e\colon (\mathbf{I},\partial_{\mathbf{I}}^{\mathbf{I}}) \to GC^-(\mathbb{G})[V_1]$. (Compare Lemma 5.2.18.) Observe that the grading shifts here are different from the case of Lemma 5.2.18 because of two reasons: the link represented by $\mathbb{G}$ has one fewer component, and the O-marking in the destablized 2×2 block is placed differently.

Define a map $\mathcal{H}^{\mathbf{I}}_{O_1,X_1}\colon \mathbf{N}\to\mathbf{I}$ by

$$\mathcal{H}^{\mathbf{I}}_{O_1,X_1}(\mathbf{x}) = \sum_{\mathbf{y}\in\mathbf{I}(\overline{\mathbb{G}})} \; \sum_{\{r\in\mathrm{Rect}^\circ(\mathbf{x},\mathbf{y})\,\big|\,O_1\in r,\, r\cap\mathbb{X}=\{X_1\}\}} V_2^{O_2(r)}\cdots V_n^{O_n(r)}\cdot\mathbf{y}.$$

We claim that $\mathcal{H}^{\mathbf{I}}_{O_1,X_1}$ is a chain homotopy equivalence between the chain complexes $(\mathbf{N},\partial_{\mathbf{N}}^{\mathbf{N}})$ and $(\mathbf{I},\partial_{\mathbf{I}}^{\mathbf{I}})$, which is homogeneous of degree $(1,0)$. To see this, we mark the upper right square in the 2×2 block with a new marking z, and consider the corresponding operator $\mathcal{H}_z^-\colon\mathbf{I}\to\mathbf{N}$ defined by

$$\mathcal{H}_z^-(\mathbf{x}) = \sum_{\mathbf{y}\in\mathbf{N}(\overline{\mathbb{G}})} \; \sum_{\{r\in\mathrm{Rect}^\circ(\mathbf{x},\mathbf{y})\,\big|\,z\in r,\, r\cap\mathbb{X}=\emptyset\}} V_2^{O_2(r)}\cdots V_n^{O_n(r)}\cdot\mathbf{y}.$$

The usual arguments show that $\mathcal{H}^{\mathbf{I}}_{O_1,X_1}$ and $\mathcal{H}_z^-$ are homotopy inverses to one another (using a homotopy counting rectangles crossing O_1 and X_1 and z).

We claim that $\mathcal{H}^{\mathbf{I}}_{O_1,X_1}\circ\partial_{\mathbf{I}}^{\mathbf{N}}=0$. This follows from the fact that there are two domains which contribute to the composite $\mathcal{H}^{\mathbf{I}}_{O_1,X_1}\circ\partial_{\mathbf{I}}^{\mathbf{N}}$: the horizontal and the vertical annuli through O_1X_1. Both of these contribute 1; so their contributions cancel. This relation can be expressed by the commutative diagram:

$$\begin{array}{ccc}
\mathbf{I} & \xrightarrow{\;\partial_{\mathbf{I}}^{\mathbf{N}}\;} & \mathbf{N} \\
{\scriptstyle e}\big\downarrow & & \big\downarrow{\scriptstyle e\circ\mathcal{H}^{\mathbf{I}}_{O_1,X_1}} \\
GC^-(\mathbb{G})[V_1] & \xrightarrow{\;0\;} & GC^-(\mathbb{G})[V_1]\llbracket 1,0\rrbracket
\end{array} \tag{8.13}$$

(Compare this diagram with Equation (5.22): the grading shifts are different, and rather than multiplication by V_1-V_2 along the bottom horizontal, we have the zero map; cf. Figure 8.2.)

According to Lemma A.3.8, Equation (8.13) naturally induces a chain map

$$\Phi\colon \mathrm{Cone}(\partial_{\mathbf{I}}^{\mathbf{N}}\colon\mathbf{I}\to\mathbf{N}) \to GC^-(\mathbb{G})[V_1]\oplus GC^-(\mathbb{G})[V_1]\llbracket 1,0\rrbracket.$$

Since the vertical maps from Equation (8.13) are quasi-isomorphisms, we conclude that Φ is a quasi-isomorphism. Specialize Φ to get a quasi-isomorphism between the quotient complexes by setting $V_{j_1} = \cdots = V_{j_\ell}$. Note that

$$\frac{GC^-(\mathbb{G})[V_1]}{V_{j_1} = \cdots = V_{j_\ell}} = cGC^-(\mathbb{G}),$$

since $V_1 = V_{j_\ell}$; so the stated identification follows. □

Proof of Lemma 8.4.2. Suppose that the unknotted, unlinked component is represented in a 2×2 box of squares. After a sequence of commutation moves, we can arrange that the 2×2 box of squares shares a corner with an X-marking. Use Lemma 8.4.6 to replace the 2×2 box with a doubly-marked square, without changing the homology. Lemma 8.4.7 completes the argument. □

Lemma 8.4.2 illustrates the strength of grid homology over the Alexander polynomial: the Alexander polynomial of $\mathcal{U}_1(\vec{L})$ vanishes, since $\mathcal{U}_1(\vec{L})$ is a split link cf. Exercise 2.4.12(a), while grid homology of $\mathcal{U}_1(\vec{L})$ remembers the grid homology of $\vec{L}$.

REMARK 8.4.8. Notice that we did not use any invariance property of the grid homology of an extended grid: invariance follows from the identification of Lemma 8.4.6, but it is not needed for the proof of Lemma 8.4.2.

Proof of Lemma 8.4.1. Apply Lemma 8.4.2 d times. □

EXERCISE 8.4.9. **(a)** Let $\vec{L}$ be a link with ℓ components. Show that

$$0 \leq \tau_{max}(\vec{L}) - \tau_{min}(\vec{L}) \leq \ell - 1.$$

(b) For any $0 \leq k \leq \ell - 1$, find an ℓ-component link for which $\tau_{max}(\vec{L}) - \tau_{min}(\vec{L}) = k$. (*Hint:* To realize $\tau_{max}(\vec{L}) - \tau_{min}(\vec{L}) = \ell - 1$, consider an unknotting sequence for the (p, q) torus knot, and break up each unknotting operation into a pair of saddles. Rearrange the saddles and use Theorem 8.3.6.)

8.5. Assembling the pieces: τ bounds the slice genus

According to Proposition 2.6.11, to prove Theorem 8.1.1, there are two key properties of τ we must verify, which we do in the following two propositions.

PROPOSITION 8.5.1. *Suppose that K_1 and K_2 are two knots so that K_2 is constructed from K_1 by a sequence of $2g$ saddle moves. Then $|\tau(K_1) - \tau(K_2)| \leq g$.*

Proof. Let $\vec{L}_1$ and $\vec{L}_2$ be two links that are connected by a sequence of saddle moves; and let m denote the number of merge moves and s the number of split moves in this sequence. Iterating Inequalities (8.10) and (8.11) from Theorem 8.3.6 gives $\tau_{max}(\vec{L}_1) \leq \tau_{max}(\vec{L}_2) + m$ and $\tau_{min}(\vec{L}_2) - m \leq \tau_{min}(\vec{L}_1)$. When the starting and ending links are knots, $m = s = g$ and of course $\tau_{max}(K_1) = \tau_{min}(K_1) = \tau(K_1)$ and $\tau_{max}(K_2) = \tau_{min}(K_2) = \tau(K_2)$; so the above inequalities give the claimed bound. □

REMARK 8.5.2. Since a crossing change can be factored as a sequence of two saddle moves, Proposition 8.5.1 gives another proof of the bound $u(K) \leq |\tau(K)|$ of Corollary 6.1.8; but it does not give the signed refinement proved in Theorem 6.1.7.

The second ingredient in the proof of Theorem 8.1.1 is the following:

PROPOSITION 8.5.3. *Suppose that K_1 and K_2 are two knots, and K_2 is obtained from $\mathcal{U}_d(K_1)$ by exactly d saddle moves. Then $\tau(K_1) = \tau(K_2)$.*

Proposition 8.5.3 will be proved by induction on the number of saddles. The following lemma makes the induction possible:

LEMMA 8.5.4. *Suppose that $\vec{L}'$ is a link with the property that $\tau_{max}(\vec{L}') = \tau_{min}(\vec{L}')$; and suppose that $\vec{L}$ is obtained from $\vec{L}'$ by a merge move. Then, $\tau_{max}(\vec{L}) = \tau_{min}(\vec{L}) = \tau_{max}(\vec{L}')$.*

Proof. Tautologically, $\tau_{min}(\vec{L}) \leq \tau_{max}(\vec{L})$. Combine this with Inequalities (8.10) and (8.11) from Theorem 8.3.6 to get $\tau_{min}(\vec{L}') \leq \tau_{min}(\vec{L}) \leq \tau_{max}(\vec{L}) \leq \tau_{max}(\vec{L}')$. Since $\tau_{min}(\vec{L}') = \tau_{max}(\vec{L}')$, it follows that $\tau_{min}(\vec{L}) = \tau_{max}(\vec{L}) = \tau_{max}(\vec{L}')$. □

Proof of Proposition 8.5.3. Taking saddle moves one at a time, we obtain a sequence of links $\vec{L}_1 = \mathcal{U}_d(K_1), \vec{L}_2, \dots, \vec{L}_d, \vec{L}_{d+1} = K_2$. We prove $\tau_{max}(\vec{L}_k) = \tau_{min}(\vec{L}_k) = \tau(K_1)$ by induction on k. The case where $k = 1$ is given by Lemma 8.4.1, and the inductive step is by Lemma 8.5.4. The case where $k = d + 1$ gives $\tau(K_1) = \tau(K_2)$. □

Proof of Theorem 8.1.1. Fix a genus g cobordism from K_1 to K_2. Let K_1' and K_2' be as in the statement of Proposition 2.6.11. By Proposition 8.5.3, $\tau(K_1) = \tau(K_1')$ and $\tau(K_2) = \tau(K_2')$; by Proposition 8.5.1 $|\tau(K_1') - \tau(K_2')| \leq g$. It follows that $|\tau(K_1) - \tau(K_2)| \leq g$. □

8.6. The existence of an exotic structure on $\mathbb{R}^4$

Corollary 8.1.2 can be used to show the existence of an exotic structure on the 4-dimensional Euclidean space $\mathbb{R}^4$. The argument uses notions from four-dimensional topology, including the concept of smooth handlebodies (cf. [**77**, Chapter 4]), and three further facts which we list below. (For related discussion see also [**77**, Section 9.4].)

- By [**146**], any closed topological 3-manifold admits a smooth structure which is unique up to diffeomorphism; indeed, any homeomorphism between two closed smooth 3-manifolds can be isotoped to a diffeomorphism.
- According to a result of Freedman [**59, 67**], any knot $K \subset S^3$ with $\Delta_K(t) = 1$ is topologically slice (as in Definition 2.6.7).
- A theorem of Freedman and Quinn [**60**] provides that any connected, non-compact topological four-manifold (possibly with non-empty boundary) admits a smooth structure.

Let X_K be the smooth, compact four-manifold-with-boundary obtained by attaching a 4-dimensional 2-handle to D^4 along K with framing 0. This means that X_K is obtained by gluing together D^4 and $D^2 \times D^2$, so that $\mathrm{nd}(K) \cong S^1 \times D^2 \subset S^3 = \partial D^4$ is identified with $(\partial D^2) \times D^2 = S^1 \times D^2$ using the Seifert framing along K.

LEMMA 8.6.1. *For any topologically slice knot $K \subset S^3$, there is a smooth 4-manifold $R = R_K$ homeomorphic to $\mathbb{R}^4$ so that X_K smoothly embeds into R.*

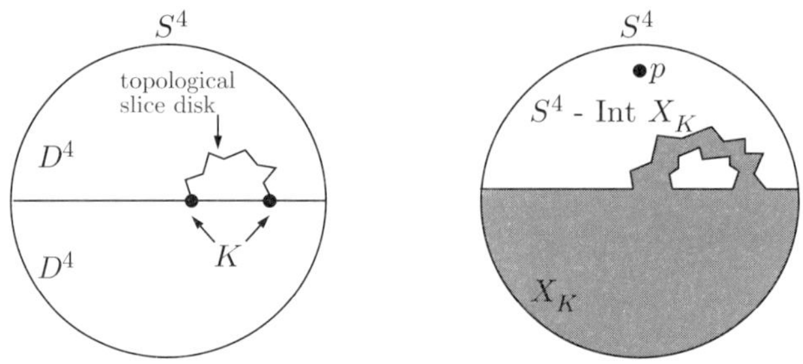

FIGURE 8.3. **Drilling out the topological slice disk from one of the two copies of D^4 in S^4.** The manifold-with-boundary X_K topologically embeds into $\mathbb{R}^4 = S^4 - \{p\}$.

Proof. View S^4 as the union of two D^4's, drill out a tubular neighborhood of the topological slice disk of K from one of them, and glue this neighbourhood to the other D^4. The resulting manifold is homeomorphic to X_K, and it is equipped with a topological embedding into $\mathbb{R}^4 = S^4 \setminus \{p\} \subset S^4$; see Figure 8.3. By [**60**] any connected, non-compact topological four-manifold (with possibly non-empty boundary) admits a smooth structure, so we can equip $\mathbb{R}^4 \setminus \operatorname{Int} X_K$ with a smooth structure. Since 3-manifolds admit unique smooth structures and any homeomorphism is isotopic to a diffeomorphism [**146**], the identification of ∂X_K with $\partial(\mathbb{R}^4 \setminus \operatorname{Int} X_K)$ can be isotoped to a diffeomorphism. Gluing back X_K with this diffeomorphism gives a smooth manifold R. By the choice of the gluing diffeomorphism, it is obviously homeomorphic to $\mathbb{R}^4$; and by construction, X_K is smoothly embedded into R. □

LEMMA 8.6.2. *Let $K \subset S^3$ be a knot. If X_K embeds smoothly into $\mathbb{R}^4$, then K is smoothly slice.*

Proof. As a preliminary point, let $D_\epsilon \subset D^4$ denote the ball of radius ϵ around the origin in D^4. It is easy to see that for any smooth map $\phi\colon D^4 \to \mathbb{R}^4$ with non-zero differential at the origin, for all sufficiently small $\epsilon > 0$, the complement of $\phi(D_\epsilon)$ is diffeomorphic to $\mathbb{R}^4 \setminus D^4$.

Recall that X_K is the union of D^4 and the 4-dimensional 2-handle $D^2 \times D^2$, the latter attached to D^4 by identifying the tubular neighbourhood of $K \subset \partial D^4$ with $(\partial D^2) \times D^2$ (using the Seifert framing). For all $\epsilon > 0$, there is a copy of K in $S^3 \cong \partial D_\epsilon \subset D^4 \subset X_K$, and there is a smoothly embedded, two-dimensional disk in $X_K \setminus \operatorname{Int}(D_\epsilon)$ whose boundary is K. Thus, the image of this disk under the embedding of X_K in $\mathbb{R}^4$ is a slice disk for K. □

COROLLARY 8.6.3. *Suppose that $K \subset S^3$ is a knot with $\Delta_K(t) = 1$ and $\tau(K) \neq 0$. Then the manifold R_K from Lemma* 8.6.1 *is homeomorphic but not diffeomorphic to the standard 4-dimensional Euclidean space $\mathbb{R}^4$.*

Proof. Since $\Delta_K(t) = 1$, Freedman's theorem [**59, 67**] ensures that K is topologically slice, so Lemma 8.6.1 gives a smooth four-manifold R_K, homeomorphic to $\mathbb{R}^4$, which contains a smoothly embedded copy of X_K. Since $\tau(K) \neq 0$, Corollary 8.1.2

ensures that the knot K is not smoothly slice, so by Lemma 8.6.2 the four-manifold X_K does not embed smoothly into $\mathbb{R}^4$. □

COROLLARY 8.6.4. *There is an exotic smooth structure on* $\mathbb{R}^4$.

Proof. Consider the 0-framed negative Whitehead double $W_0^-(T_{-2,3})$ of the left-handed trefoil knot $T_{-2,3}$. The Alexander polynomial of this knot is equal to 1. In Exercise 4.3.6, we considered a grid diagram $\mathbb{G}$ of $W_0^-(T_{-2,3})$ together with a generator $\mathbf{x} = \mathbf{x}^-(\mathbb{G})$; a straightforward computation shows that $A(\mathbf{x}) = 1$. By Corollary 6.4.9, $\tau(W_0^-(T_{-2,3})) \leq -1$; in particular, $\tau(W_0^-(T_{-2,3}))$ is non-zero. Applying Corollary 8.6.3 concludes the proof. □

REMARK 8.6.5. Since the Seifert genus of $W_0^-(T_{-2,3})$ is equal to 1, it follows that $\tau(W_0^-(T_{-2,3})) = -1$. By Corollary 7.4.5 this equation implies that $\tau(W_0^+(T_{2,3})) = g_s(W_0^+(T_{2,3}) = 1$. (See also Section 12.4 for a more conceptual interpretation of the above computation.) The slice genus of $W_0^-(T_{2,3})$ (or equivalently, of $W_0^+(T_{-2,3})$), on the other hand, is still unknown.

REMARK 8.6.6. There are many other topologically slice, but not smoothly slice knots, which can also be used to construct exotic $\mathbb{R}^4$'s. The methods applied here, however, fall short of reproving the result that there are infinitely many (in fact, uncountably many) distinct exotic $\mathbb{R}^4$'s. For further examples, see [**77**, Section 9.4].

8.7. Slice bounds vs. unknotting bounds

In Chapter 6 we discussed bounds on the unknotting number coming from τ. In fact, τ should really be thought of as a tool for studying the four-ball genus; the fact that it says something about the unknotting number is a consequence of the familiar inequality $g_s(K) \leq u(K)$.

There are other methods for studying the unknotting number, independently from the slice genus. One particularly effective tool is Montesinos' trick [**147**]: if $K \subset S^3$ has unknotting number equal to n, then the three-manifold $\Sigma(K)$ obtained as the branched double cover of S^3 branched along K can be constructed by Dehn surgery on an n-component link in S^3. In particular, the abelian group $H_1(\Sigma(K);\mathbb{Z})$ can be generated by n elements. This obstruction can be made explicit: the Goeritz matrix for a knot gives a presentation of $H_1(\Sigma(K);\mathbb{Z})$. (See for example [**18**, Theorem 8.20].)

Montesinos' trick can be used to show that the four-ball genus and the unknotting number can be arbitrarily large, as follows:

EXAMPLE 8.7.1. Consider the family of slice knots $K_n = \#^n(T_{2,3}\#T_{-2,3})$ indexed by $n \in \mathbb{Z}^{\geq 0}$. Since the branched double cover of S^3 along $T_{2,3}$ has first homology $\mathbb{Z}/3\mathbb{Z}$, it follows that $H_1(\Sigma(K_n);\mathbb{Z}) \cong (\mathbb{Z}/3\mathbb{Z})^{2n}$. Montesinos's trick gives the inequality that $u(K_n) \geq 2n$. It follows easily that $u(K_n) = 2n$.

Montesinos' trick can be used effectively as a tool for bounding the unknotting number, especially when it is combined with further obstructions to realizing $\Sigma(K)$ as surgery on an n-component link. A nice geometric construction for ruling out knots with unknotting number equal to one is given in [**79**]. Other such surgery obstructions are provided by Heegaard Floer homology, see for example [**162, 176**].

CHAPTER 9

The oriented skein exact sequence

The skein relation (Theorem 2.4.10) is an effective device for computing the Alexander polynomial. The analogue of this tool for grid homology is the *(oriented) skein exact sequence*, which is the subject of the present chapter.

In Section 9.1, we state the skein exact sequence, and in Section 9.2, we prove a chain level analogue. The versions stated in Section 9.1 are deduced in Section 9.3. In Section 9.4, we give some simple computations using the skein exact sequence. Section 9.5 gives a further application of this tool: we describe infinite collections of knots with the same grid homology, generalizing our earlier discussion for the Alexander polynomial. (Compare Lemma 2.5.4.) Examples of this kind were first described in [**84**]. In Section 9.6, we relate a map appearing in the skein exact sequence with the map associated to a crossing change from Chapter 6. This relationship will be used in Chapter 10. We conclude in Section 9.7 with a short discussion of skein sequences in related contexts.

A skein exact sequence for knot Floer homology was first proved in [**172**]; we follow here the treatment from [**167**].

9.1. The skein exact sequence

Recall the skein relation for the Alexander polynomial from Theorem 2.4.10; see also Section 3.3. To set up notation, let $\vec{L}_+$, $\vec{L}_-$, and $\vec{L}_0$ be three oriented links which form a skein triple in the sense of Definition 2.4.9; see Figure 9.1. As explained in Section 2.4, the Alexander polynomials of these three oriented links are related by the skein relation $\Delta_{\vec{L}_+}(t) - \Delta_{\vec{L}_-}(t) = (t^{\frac{1}{2}} - t^{-\frac{1}{2}}) \cdot \Delta_{\vec{L}_0}(t)$.

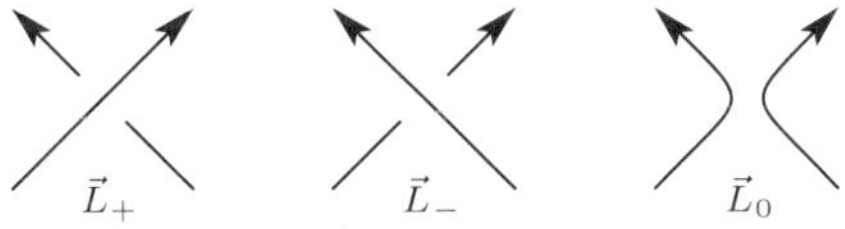

FIGURE 9.1. **Crossing conventions for the skein exact sequence.**

Let ℓ_+, ℓ_-, and ℓ_0 denote the number of components of $\vec{L}_+$, $\vec{L}_-$, and $\vec{L}_0$ respectively. Clearly, $\ell_+ = \ell_-$. If the two strands at the crossing in $\vec{L}_+$ belong to the same component, then $\ell_0 = \ell_+ + 1$; otherwise $\ell_0 = \ell_+ - 1$.

In grid homology the above skein relation is replaced by a *skein exact sequence*. Like the skein relation, this exact sequence involves oriented link invariants. There are several versions of the skein sequence; the first variant we state (Theorem 9.1.1) is expressed in terms of the collapsed grid homology of a link cGH^- introduced in Section 8.2. When the link has one component, this invariant is the grid homology of that knot in the sense of Chapter 4.

We will be taking the tensor product (over $\mathbb{F}$) with a four-dimensional bigraded vector space $J \cong \mathbb{F}_{(0,1)} \oplus \mathbb{F}_{(-1,0)} \oplus \mathbb{F}_{(-1,0)} \oplus \mathbb{F}_{(-2,-1)}$. If X is a bigraded $\mathbb{F}[U]$-module, then there is an isomorphism of bigraded $\mathbb{F}[U]$-modules

$$X \otimes J \cong X[\![0,-1]\!] \oplus X[\![1,0]\!] \oplus X[\![1,0]\!] \oplus X[\![2,1]\!]. \tag{9.1}$$

THEOREM 9.1.1. *Let $(\vec{L}_+, \vec{L}_-, \vec{L}_0)$ be an oriented skein triple. Let ℓ and ℓ_0 denote the number of components of $\vec{L}_+$ and $\vec{L}_0$ respectively. If $\ell_0 = \ell + 1$, then there is a long exact sequence*

$$\to cGH^-_m(\vec{L}_+, s) \xrightarrow{f^-} cGH^-_m(\vec{L}_-, s) \xrightarrow{g^-} cGH^-_{m-1}(\vec{L}_0, s) \xrightarrow{h^-} cGH^-_{m-1}(\vec{L}_+, s) \to \tag{9.2}$$

If $\ell_0 = \ell - 1$, there is an exact sequence

$$\to cGH^-_m(\vec{L}_+, s) \xrightarrow{f^-} cGH^-_m(\vec{L}_-, s) \xrightarrow{g^-} (cGH^-(\vec{L}_0) \otimes J)_{m-1,s} \xrightarrow{h^-} cGH^-_{m-1}(\vec{L}_+, s) \to \tag{9.3}$$

where J is as in Equation (9.1). In both cases, the maps f^-, g^-, and h^- fit together to give homomorphisms of $\mathbb{F}[U]$-modules.

For example, if $\vec{L}_+ = K_+$ and $\vec{L}_- = K_-$ are knots, so that $\vec{L}_0 = \vec{K}_0$ is a two-component link, then Equation (9.2) specializes to the exact sequence

$$\longrightarrow GH^-_m(K_+, s) \xrightarrow{f^-} GH^-_m(K_-, s) \xrightarrow{g^-} cGH^-_{m-1}(\vec{K}_0, s) \xrightarrow{h^-} GH^-_{m-1}(K_+, s) \longrightarrow$$

Theorem 9.1.1 has an analogue using simply blocked grid homology.

THEOREM 9.1.2. *Let $(\vec{L}_+, \vec{L}_-, \vec{L}_0)$ be an oriented skein triple, with ℓ, ℓ_0, and J as above. If $\ell_0 = \ell + 1$, then there is an exact sequence*

$$\longrightarrow \widehat{GH}_m(\vec{L}_+, s) \xrightarrow{\widehat{f}} \widehat{GH}_m(\vec{L}_-, s) \xrightarrow{\widehat{g}} \widehat{GH}_{m-1}(\vec{L}_0, s) \xrightarrow{\widehat{h}} \widehat{GH}_{m-1}(\vec{L}_+, s) \longrightarrow$$

If $\ell_0 = \ell - 1$, there is an exact sequence

$$\to \widehat{GH}_m(\vec{L}_+, s) \xrightarrow{\widehat{f}} \widehat{GH}_m(\vec{L}_-, s) \xrightarrow{\widehat{g}} (\widehat{GH}(\vec{L}_0) \otimes J)_{m-1,s} \xrightarrow{\widehat{h}} \widehat{GH}_{m-1}(\vec{L}_+, s) \to$$

It is convenient to express three-periodic long exact sequences as *exact triangles*. (See Section A.2.) In this notation, the long exact sequence appearing in Equation (9.2) takes the following form:

$$\begin{array}{ccc} cGH^-(\vec{L}_+) & \xrightarrow{\qquad f^- \qquad} & cGH^-(\vec{L}_-) \\ {\scriptstyle h^-}\nwarrow & & \swarrow{\scriptstyle g^-} \\ & cGH^-(\vec{L}_0) & \end{array}$$

where f^- and h^- are bigraded $\mathbb{F}[U]$-module maps, and g^- is an $\mathbb{F}[U]$-module map that is homogeneous of degree $(-1, 0)$.

Taking the graded Euler characteristics of the skein exact sequences (in Theorems 9.1.1 and 9.1.2) gives back the skein relation for the Alexander polynomial.

Whereas the skein relation computes the Alexander polynomial of any link in a skein triple in terms of the other two, the skein exact sequence gives only a relationship between the grid homologies of these three links.

In the proof of the skein exact sequence, we use two closely related grid diagrams $\mathbb{G}_+$ and $\mathbb{G}_-$, representing $\vec{L}_+$ and $\vec{L}_-$ respectively; see Figure 9.2 or 3.18. By changing the placements of the X-markings locally on $\mathbb{G}_-$ and $\mathbb{G}_+$, we obtain two different grid diagrams $\mathbb{G}_0$ and $\mathbb{G}_0'$, both representing $\vec{L}_0$, see Proposition 3.3.11.

We find it convenient to work with the uncollapsed, bigraded grid complex from Definition 8.2.2. We construct a chain map $P_{+,-}\colon GC^-(\mathbb{G}_+) \to GC^-(\mathbb{G}_-)$ between these complexes, whose mapping cone, with a suitable grading shift, is quasi-isomorphic to the mapping cone of $V_2 - V_1\colon GC^-(\mathbb{G}_0) \to GC^-(\mathbb{G}_0)$, where V_1 and V_2 are two formal variables corresponding to the two strands meeting in $\vec{L}_+$; see Theorem 9.2.1. The quasi-isomorphism between these mapping cones gives rise to all of the exact sequences appearing in Theorems 9.1.1 and 9.1.2. The different variations result from different algebraic specializations of the grid complexes.

9.2. The skein relation on the chain level

Let $(\vec{L}_+, \vec{L}_-, \vec{L}_0)$ be an oriented skein triple (as in Definition 2.4.9) and choose a grid realization $(\mathbb{G}_+, \mathbb{G}_-, \mathbb{G}_0, \mathbb{G}_0')$ of this triple (as in Definition 3.3.9). The main result of this section is the following chain level analogue of the skein exact sequence:

THEOREM 9.2.1. *Let ℓ denote the number of components of $\vec{L}_+$ (and $\vec{L}_-$), and let ℓ_0 denote the number of components of $\vec{L}_0$ (in particular, $|\ell - \ell_0| = 1$). There is a bigraded chain map $P_{+,-}\colon GC^-(\mathbb{G}_+) \to GC^-(\mathbb{G}_-)$ of $\mathbb{F}[V_1, \dots, V_n]$-modules, between the (uncollapsed) grid complexes, and a bigraded quasi-isomorphism from the mapping cone of $P_{+,-}$ to*

$$\mathrm{Cone}(V_2 - V_1\colon GC^-(\mathbb{G}_0) \to GC^-(\mathbb{G}_0))[\![-1, \frac{\ell_0 - \ell - 1}{2}]\!].$$

The proof of Theorem 9.2.1, will occupy the rest of this section.

The four grid diagrams, forming the grid realization of the skein triple, that come into play in the proof are illustrated in Figure 9.2. The existence of such grid realizations was verified in Lemma 3.3.10. We will need the additional property that the two distinguished X-markings occur in consecutive rows; this can be arranged by adapting the proof of Lemma 3.3.10 in a straightforward way. In two of the diagrams, we label the two distinguished X-markings by "Y", as in Figure 9.2.

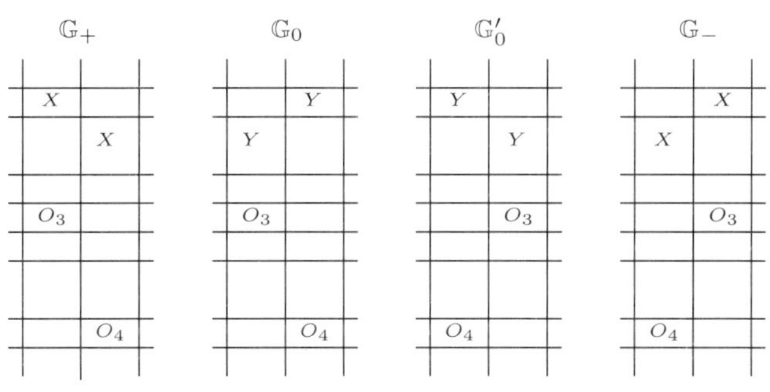

FIGURE 9.2. **Grid diagrams for the skein exact sequence.**

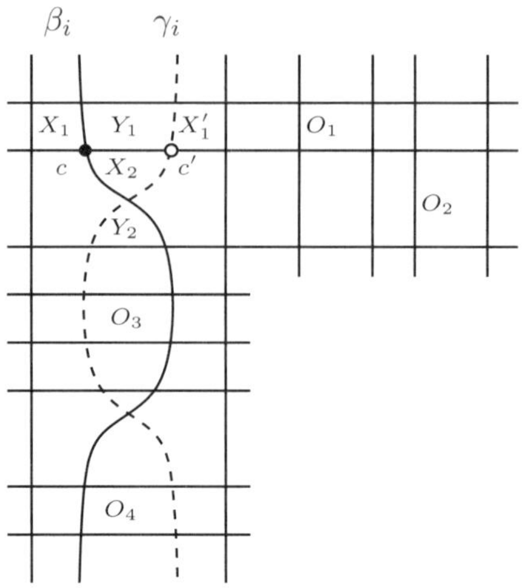

FIGURE 9.3. **Simultaneous grid diagram for a skein triple.**

Draw all four diagrams simultaneously on one grid, as shown in Figure 9.3. That figure also indicates labels on the various X- and Y-markings. As in the proof of commutation invariance, instead of moving the O-markings (which we think of as fixed for all four diagrams), we vary the choice of one distinguished vertical circle, β_i and γ_i. That is, all four grid diagrams $\mathbb{G}_+$, $\mathbb{G}_-$, $\mathbb{G}_0$, and $\mathbb{G}'_0$ use the same horizontal circles and O-markings; the grid diagram $\mathbb{G}_+$ uses X-markings $\mathbb{X} = \{X_1, \ldots, X_n\}$ and vertical circles $\boldsymbol{\beta} = \{\beta_1, \ldots, \beta_n\}$; while $\mathbb{G}_-$ uses the X-markings $\mathbb{X}' = \{X'_1, X_2, \ldots, X_n\}$, and vertical circles $\boldsymbol{\gamma} = \{\beta_1, \ldots, \beta_{i-1}, \gamma_i, \beta_{i+1}, \ldots \beta_n\}$. Letting $\mathbb{Y} = \{Y_1, Y_2, X_3, \ldots, X_n\}$, the grid diagram $\mathbb{G}_0$ uses the vertical circles $\boldsymbol{\beta}$ and X-markings $\mathbb{Y}$, and $\mathbb{G}'_0$ uses the vertical circles $\boldsymbol{\gamma}$ and X-markings $\mathbb{Y}$.

The two distinguished X-marked squares in $\mathbb{G}_+$ (marked by X_1 and X_2) meet at a common corner, denoted c. The two distinguished X-marked squares in $\mathbb{G}_0$ (marked by Y_1 and Y_2) also meet at c. Similarly, the two distinguished X-marked squares in $\mathbb{G}_-$ (marked by X'_1 and X_2) share a corner denoted c'; and the two distinguished X-marked squares in $\mathbb{G}'_0$ (marked by Y_1 and Y_2) also meet at c'.

As in the stabilization invariance proof, partition the grid states according to whether or not they contain the distinguished corner. Specifically, partition $\mathbf{S}(\mathbb{G}_+) = \mathbf{I}(\mathbb{G}_+) \cup \mathbf{N}(\mathbb{G}_+)$ according to whether or not the given grid state $\mathbf{x}$ contains the corner point c; and partition $\mathbf{S}(\mathbb{G}_0) = \mathbf{I}(\mathbb{G}_0) \cup \mathbf{N}(\mathbb{G}_0)$ according to the same criterion. Similarly, partition $\mathbf{S}(\mathbb{G}_-) = \mathbf{I}'(\mathbb{G}_-) \cup \mathbf{N}'(\mathbb{G}_-)$ and $\mathbf{S}(\mathbb{G}'_0) = \mathbf{I}'(\mathbb{G}'_0) \cup \mathbf{N}'(\mathbb{G}'_0)$ according to whether or not $\mathbf{x}$ contains the corner point c'. Think of $GC^-(\mathbb{G}_+)$ as the mapping cone of a map $\partial_{\mathbf{I}}^{\mathbf{N}} \colon (\mathbf{I}, \partial_{\mathbf{I}}^{\mathbf{I}}) \to (\mathbf{N}, \partial_{\mathbf{N}}^{\mathbf{N}})$ that counts rectangles crossing exactly one of the two distinguished Y-markings; and $GC^-(\mathbb{G}_0)$ as the mapping cone of $\partial_{\mathbf{N}}^{\mathbf{I}} \colon (\mathbf{N}, \partial_{\mathbf{N}}^{\mathbf{N}}) \to (\mathbf{I}, \partial_{\mathbf{I}}^{\mathbf{I}})$ that counts rectangles crossing exactly one of the two distinguished X-markings. Although the roles of $\mathbf{I}$ and $\mathbf{N}$ as sub- or quotient-complexes are reversed for $\mathbb{G}_+$ and $\mathbb{G}_0$, the complexes themselves are naturally identified. Similarly, think of $GC^-(\mathbb{G}'_0)$ as the mapping cone of $\partial_{\mathbf{I}'}^{\mathbf{N}'} \colon (\mathbf{I}', \partial_{\mathbf{I}'}^{\mathbf{I}'}) \to (\mathbf{N}', \partial_{\mathbf{N}'}^{\mathbf{N}'})$ (counting rectangles crossing exactly one of the two distinguished$X$-markings), and $GC^-(\mathbb{G}_-)$ as the mapping cone of $\partial_{\mathbf{N}'}^{\mathbf{I}'} \colon (\mathbf{N}', \partial_{\mathbf{N}'}^{\mathbf{N}'}) \to (\mathbf{I}', \partial_{\mathbf{I}'}^{\mathbf{I}'})$ (counting rectangles crossing exactly one of the two distinguished Y-markings).

Under the natural identification of grid states for $\mathbb{G}_+$ and $\mathbb{G}_0$, the Alexander functions for $\mathbb{G}_+$ and $\mathbb{G}_0$ can be thought of as giving two different functions A_+ and A_0 on the same set of grid states. Similarly, identify grid states for $\mathbb{G}_-$ and $\mathbb{G}'_0$, equipped with two different Alexander functions A_- and A'_0. Consider also the one-to-one correspondence $T\colon \mathbf{I}'(\mathbb{G}'_0) \to \mathbf{I}(\mathbb{G}_+)$, uniquely characterized by the property that $T(\mathbf{x}) \setminus (T(\mathbf{x}) \cap \beta_i) = \mathbf{x} \setminus (\mathbf{x} \cap \gamma_i)$.

LEMMA 9.2.2. *For* $\mathbf{x} \in \mathbf{N}(\mathbb{G}_+) = \mathbf{N}(\mathbb{G}_0)$,

$$M_+(\mathbf{x}) = M_0(\mathbf{x}), \qquad A_+(\mathbf{x}) - \frac{1}{2} - \frac{\ell}{2} = A_0(\mathbf{x}) - \frac{\ell_0}{2};$$

for $\mathbf{x} \in \mathbf{N}'(\mathbb{G}_-) = \mathbf{N}'(\mathbb{G}'_0)$,

$$M_-(\mathbf{x}) = M'_0(\mathbf{x}), \qquad A_-(\mathbf{x}) + \frac{1}{2} - \frac{\ell}{2} = A'_0(\mathbf{x}) - \frac{\ell_0}{2};$$

for $\mathbf{x} \in \mathbf{I}'(\mathbb{G}'_0)$ *and* $T(\mathbf{x}) \in \mathbf{I}(\mathbb{G}_+)$,

$$M'_0(\mathbf{x}) = M_+(T(\mathbf{x})) + 1, \qquad A'_0(\mathbf{x}) - \frac{\ell_0}{2} = A_+(T(\mathbf{x})) + \frac{1}{2} - \frac{\ell}{2};$$

while for $\mathbf{x} \in \mathbf{I}'(\mathbb{G}_-)$ *and* $T(\mathbf{x}) \in \mathbf{I}(\mathbb{G}_0)$

$$M_-(\mathbf{x}) = M_0(T(\mathbf{x})) + 1, \qquad A_-(\mathbf{x}) - \frac{\ell}{2} = A_0(T(\mathbf{x})) + \frac{1}{2} - \frac{\ell_0}{2}.$$

Proof. This is a straightforward computation in a fundamental domain. □

LEMMA 9.2.3. *The above identification* $T\colon \mathbf{I}'(\mathbb{G}'_0) \to \mathbf{I}(\mathbb{G}_+)$ *extends to an isomorphism* $T\colon (\mathbf{I}', \partial_{\mathbf{I}'}^{\mathbf{I}'}) \to (\mathbf{I}, \partial_{\mathbf{I}}^{\mathbf{I}})$ *of chain complexes over* $\mathbb{F}[V_1, \dots, V_n]$.

Proof. Since $\mathbb{G}'_0$ and $\mathbb{G}_+$ differ only in a neighborhood of β_i or γ_i, there is a one-to-one correspondence between rectangles connecting generators in $\mathbb{G}'_0$ and those connecting corresponding generators in $\mathbb{G}_+$. Under this correspondence, the local multiplicities of the rectangles at the O- and X-markings coincide, except when the rectangle in $\mathbb{G}_+$ has an edge on β_i, so that its corresponding rectangle in $\mathbb{G}'_0$ has an edge on γ_i. Since rectangles connecting $\mathbf{x}', \mathbf{y}' \in \mathbf{I}'(\mathbb{G}'_0)$ or $\mathbf{x}, \mathbf{y} \in \mathbf{I}(\mathbb{G}_+)$ do not have edges along these distinguished curves, the contribution of a rectangle to $\partial_{\mathbf{I}'}^{\mathbf{I}'}(\mathbf{x}')$ is the same as the contribution of the corresponding rectangle to $\partial_{\mathbf{I}}^{\mathbf{I}}(T(\mathbf{x}'))$; it follows that T induces an isomorphism of chain complexes. □

In the next lemma, we will verify that the following diagram commutes:

(9.4)
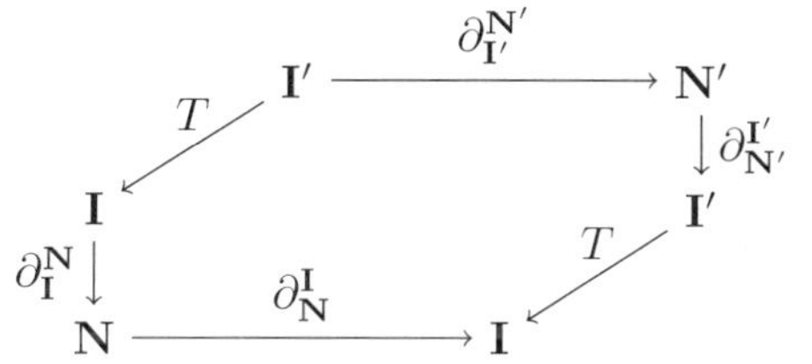

LEMMA 9.2.4. *Consider the composite maps* $\partial_{\mathbf{N}}^{\mathbf{I}} \circ \partial_{\mathbf{I}}^{\mathbf{N}} : \mathbf{I} \to \mathbf{I}$ *and* $\partial_{\mathbf{N}'}^{\mathbf{I}'} \circ \partial_{\mathbf{I}'}^{\mathbf{N}'} : \mathbf{I}' \to \mathbf{I}'$. *Both these maps are multiplication by* $V_1 + V_2 - V_3 - V_4$.

Proof. Consider the composition $\partial_{\mathbf{N}}^{\mathbf{I}} \circ \partial_{\mathbf{I}}^{\mathbf{N}} \colon \mathbf{I} \to \mathbf{I}$. This counts domains that can be written as a juxtaposition of two empty rectangles, the first of which goes from some $\mathbf{x} \in \mathbf{I}(\mathbb{G}_+)$ to some $\mathbf{y} \in \mathbf{N}(\mathbb{G}_+)$ and the second of which goes from $\mathbf{y} \in \mathbf{N}(\mathbb{G}_+)$ to some element $\mathbf{z} \in \mathbf{I}(\mathbb{G}_+)$. For fixed initial state $\mathbf{x}$, it is easy to see that there are exactly four such domains: the two vertical annuli through the preferred two X-markings, and the two horizontal annuli through the preferred two X-markings. It follows that $\partial_{\mathbf{N}}^{\mathbf{I}} \circ \partial_{\mathbf{I}}^{\mathbf{N}}$ is multiplication by $V_1 + V_2 - V_3 - V_4$. (We are working over $\mathbb{F}$; the signs become relevant when working over $\mathbb{Z}$; see Chapter 15.)

The other composition $\partial_{\mathbf{N}'}^{\mathbf{I}'} \circ \partial_{\mathbf{I}'}^{\mathbf{N}'} \colon \mathbf{I}' \to \mathbf{I}'$ works the same way. $\square$

Identifying $\mathbf{I}$ and $\mathbf{I}'$ using T, Equation (9.4) can be viewed as the chain complex

$$\begin{array}{ccc} \mathbf{I}' & \xrightarrow{\partial_{\mathbf{I}'}^{\mathbf{N}'}} & \mathbf{N}' \\ {\scriptstyle \partial_{\mathbf{I}}^{\mathbf{N}} \circ T}\big\downarrow & & \big\downarrow{\scriptstyle T \circ \partial_{\mathbf{N}'}^{\mathbf{I}'}} \\ \mathbf{N} & \xrightarrow{\partial_{\mathbf{N}}^{\mathbf{I}}} & \mathbf{I} \end{array} \tag{9.5}$$

LEMMA 9.2.5. *There are bigradings for* $\mathbf{I}'$, $\mathbf{N}'$, $\mathbf{N}$, *and* $\mathbf{I}$ *so that, in Equation* (9.5)*:*

- *Each edge map is homogeneous of degree* $(-1, 0)$.
- *The left column, as a bigraded chain complex over* $\mathbb{F}[V_1, \dots, V_n]$, *is identified with* $GC^-(\mathbb{G}_+)[\![-1, 0]\!]$.
- *The right column, as a bigraded chain complex over* $\mathbb{F}[V_1, \dots, V_n]$, *is identified with* $GC^-(\mathbb{G}_-)$.
- *The top row is identified with* $GC^-(\mathbb{G}_0')[\![0, \frac{\ell_0 - \ell + 1}{2}]\!]$.
- *The bottom row is identified with* $GC^-(\mathbb{G}_0)[\![-1, \frac{\ell_0 - \ell - 1}{2}]\!]$.

Proof. Consider the bigrading on $\mathbf{I}$ and $\mathbf{N}$ given by $M_0 + 1$ and $A_0 + \frac{\ell - \ell_0 + 1}{2}$; and the bigrading on $\mathbf{I}'$ and $\mathbf{N}'$ given by M_0' and $A_0' + \frac{\ell - \ell_0 - 1}{2}$. The top row is clearly isomorphic to $GC^-(\mathbb{G}_0')[\![0, \frac{\ell_0 - \ell + 1}{2}]\!]$, while the bottom row gives $GC^-(\mathbb{G}_0)[\![-1, \frac{\ell_0 - \ell - 1}{2}]\!]$; in particular, the two horizontal maps in Equation (9.5) are homogeneous of degree $(-1, 0)$. Corresponding statements about the vertical maps follow from Lemma 9.2.3 together with Lemma 9.2.2. $\square$

Lemmas 9.2.4 and 9.2.5 together imply that the map

$$\partial_{\mathbf{I}}^{\mathbf{N}} \circ T + T \circ \partial_{\mathbf{N}'}^{\mathbf{I}'} \colon GC^-(\mathbb{G}_0') \to GC^-(\mathbb{G}_0)$$

is a chain map of $\mathbb{F}[V_1, \dots, V_n]$-modules that is homogeneous of degree $(-2, -1)$. We compute this map, after post-composing with a commutation map, in the following two lemmas. This computation uses the following further maps:

Let $h_{X_2} \colon GC^-(\mathbb{G}_0') \to GC^-(\mathbb{G}_0')$ be the $\mathbb{F}[V_1, \dots, V_n]$-module map specified by

$$h_{X_2}(\mathbf{x}) = \sum_{\mathbf{y} \in \mathbf{S}(\mathbb{G}_0')} \sum_{\{r \in \mathrm{Rect}^\circ(\mathbf{x}, \mathbf{y}) \big| (\mathbb{X} \cup \mathbb{Y}) \cap r = \{X_2\}\}} V_1^{O_1(r)} \cdots V_n^{O_n(r)} \cdot \mathbf{y};$$

and for $i = 1, 2$, let $h_{Y_i}\colon GC^-(\mathbb{G}'_0) \to GC^-(\mathbb{G}'_0)$ be the $\mathbb{F}[V_1, \ldots, V_n]$-module map specified by

$$h_{Y_i}(\mathbf{x}) = \sum_{\mathbf{y}\in\mathbf{S}(\mathbb{G}'_0)} \sum_{\{r\in\mathrm{Rect}^\circ(\mathbf{x},\mathbf{y})\,|\,\mathbb{Y}\cap r=\{Y_i\}\}} V_1^{O_1(r)} \cdots V_n^{O_n(r)} \cdot \mathbf{y}.$$

Finally, let $h_Y = h_{Y_1} + h_{Y_2}$.

LEMMA 9.2.6. *The map h_{X_2} is homogeneous of degree $(-1, 0)$; and for $i = 1, 2$, h_{Y_i} is homogeneous of degree $(-1, -1)$. Moreover, h_{X_2} vanishes on $\mathbf{N}'$ and sends $\mathbf{I}'$ to $\mathbf{N}'$; while h_{Y_i} vanishes on $\mathbf{I}'$.*

Proof. The grading shift properties follow quickly from Equations (4.2) and (4.4). The other properties are clear from the local picture near the corner point c'. □

LEMMA 9.2.7. *There is a quasi-isomorphism $P\colon GC^-(\mathbb{G}_0) \to GC^-(\mathbb{G}'_0)$ with the property that the following identity holds:*

$$P \circ \left(\partial_{\mathbf{I}}^{\mathbf{N}} \circ T + T \circ \partial_{\mathbf{N}'}^{\mathbf{I}'}\right) = h_{X_2} \circ h_Y + h_Y \circ h_{X_2}, \tag{9.6}$$

so $h_{X_2} \circ h_Y + h_Y \circ h_{X_2}$ is a chain map. Furthermore, there is a quasi-isomorphism

$$\begin{aligned}\mathrm{Cone}(\partial_{\mathbf{I}}^{\mathbf{N}} \circ T + T \circ \partial_{\mathbf{N}'}^{\mathbf{I}'}\colon GC^-(\mathbb{G}'_0) \to GC^-(\mathbb{G}_0))&\\ \longrightarrow \mathrm{Cone}(h_{X_2} \circ h_Y + h_Y \circ h_{X_2}\colon GC^-(\mathbb{G}'_0) \to GC^-(\mathbb{G}'_0)).&\end{aligned}$$

Proof. The quasi-isomorphism P is the map induced by commuting the first two columns, and is defined by counting pentagons, as in Equation (5.2). It is a quasi-isomorphism, as we saw in Proposition 5.1.7. We verify the identity

$$P \circ T \circ \partial_{\mathbf{N}'}^{\mathbf{I}'} = h_{X_2} \circ h_Y \tag{9.7}$$

by setting up a suitable one-to-one correspondence between domains contributing to both sides; see Figure 9.4. To this end, it is helpful to think of T as the map counting the small triangle containing X_2 in Figure 9.3.

Consider a domain contributing to $P \circ T \circ \partial_{\mathbf{N}'}^{\mathbf{I}'}$. Let r_1 be the rectangle contributing to $\partial_{\mathbf{N}'}^{\mathbf{I}'}$. Juxtaposing the pentagon (contributing to P) and the triangle (contributing to T) gives a rectangle, call it r_2, that contains X_2 but not $Y = \{Y_1, Y_2\}$. Thus, the juxtaposition $r_1 * r_2$ represents a term in $h_{X_2} \circ h_Y$. Conversely, if $r_1 * r_2$ contributes to $h_{X_2} \circ h_Y$, remove the small triangle containing X_2 from r_2, to obtain a pentagon p. Under this correspondence, the domain contributing to $P \circ T \circ \partial_{\mathbf{N}'}^{\mathbf{I}'}$ coincides with $r_1 * r_2$, so both contribute the same monomial in Equation (9.7).

Next, we verify the identity

$$P \circ \partial_{\mathbf{I}}^{\mathbf{N}} \circ T = h_Y \circ h_{X_2}. \tag{9.8}$$

This is also verified by setting up a one-to-one correspondence between domains contributing to both sides. (See again Figure 9.4.) For a domain contributing to $P \circ \partial_{\mathbf{I}}^{\mathbf{N}} \circ T$, the distinguished corner c' must be part of the initial generator. Since the domain contains a rectangle contributing to $\partial_{\mathbf{I}}^{\mathbf{N}}$, at least one of the Y-labelled regions adjacent to c' must have non-zero multiplicity. There are three possibilities: either c' is one of the $270°$ corners; or c' is a $180°$ corner; or c' is contained in the interior of the domain.

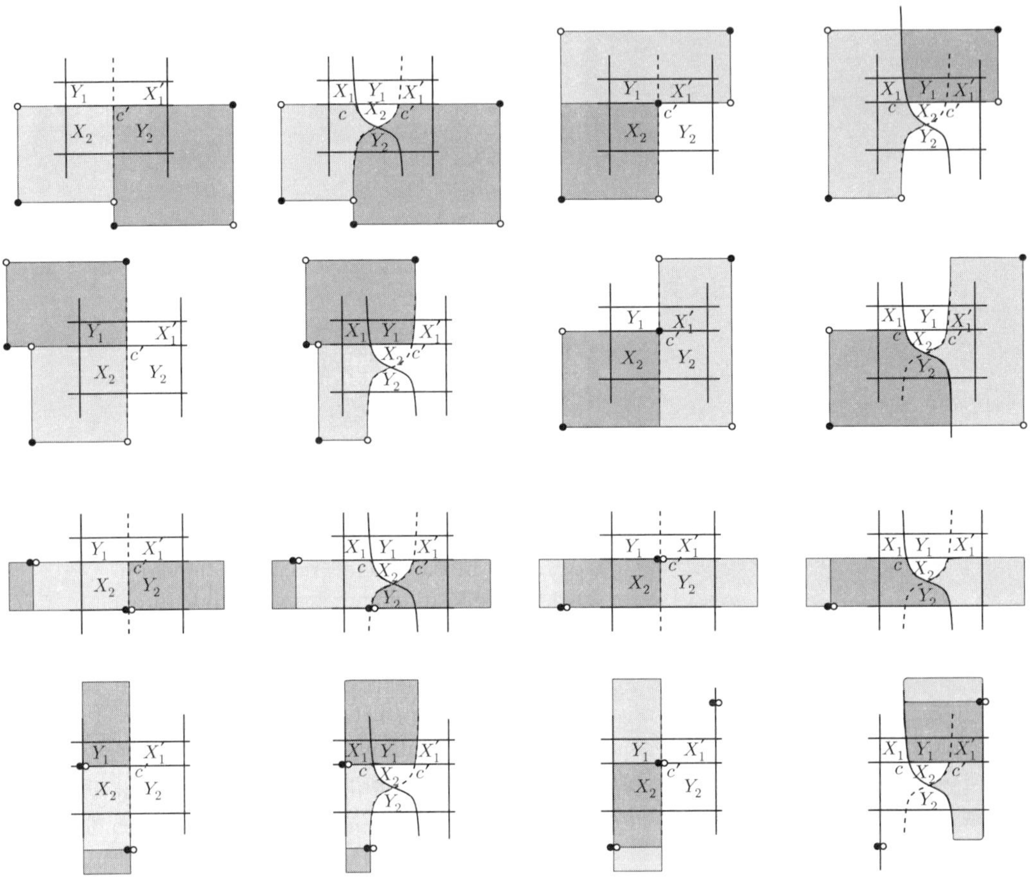

FIGURE 9.4. **Proof of Lemma 9.2.7.** In the first column, we have domains contributing to $h_{X_2} \circ h_Y$; in the second we have the corresponding domain contributing to $P \circ T \circ \partial^{\mathbf{I}'}_{\mathbf{N}'}$ (though the triangle representing T is left unshaded); in the third column we have domains contributing to $h_Y \circ h_{X_2}$; in the fourth column, we have domains contributing to $P \circ \partial^{\mathbf{N}}_{\mathbf{I}} \circ T$. As usual, the heavy shaded domain is applied first.

In the case where c' is one of the $270°$ corners, cut the juxtaposition into two rectangles $r_1 * r_2$, where r_1 contains X_2 and r_2 contains the Y-markings. This decomposition gives a contribution to $h_Y \circ h_{X_2}$. Conversely, if $r_1 * r_2$ contributes to $h_Y \circ h_{X_2}$, remove the small triangle from $r_1 * r_2$, and decompose the remainder uniquely as a pentagon and a rectangle (representing $\partial^{\mathbf{N}}_{\mathbf{I}}$). Under this correspondence, the domain contributing to $P \circ \partial^{\mathbf{N}}_{\mathbf{I}} \circ T$ is the same as the domain $r_1 * r_2$, so the two contributions to Equation (9.8) cancel.

Consider the degenerate case where the angle around c' is $180°$. In this case, the domain contributing to $P \circ \partial^{\mathbf{N}}_{\mathbf{I}} \circ T$ is an annulus of height one that passes through X_2. Let $r_1 * r_2$ be the corresponding decomposition of the annulus into rectangles. Once again, the contributions to Equation (9.8) cancel.

There is a final case where the angle around c' is 360°. In this case, the domain contributing to $P \circ \partial_{\mathbf{I}}^{\mathbf{N}} \circ T$ is an annulus of width one, which contains c' in its interior. Let the corresponding domain $r_1 * r_2$ be the thin, vertical annulus that contains X_2. This is the only case where the domain contributing to the left-hand-side of Equation (9.6) is different from the corresponding one contributing to the right-hand-side; however, both regions cover only O_4, as can be seen by looking at the last two pictures in Figure 9.4 (and comparing Figure 9.3), so their contributions cancel.

Adding up Equations (9.8) and (9.7) gives Equation (9.6), according to which the following square commutes:

$$\begin{array}{ccc} GC^-(\mathbb{G}'_0) & \xrightarrow{\partial_{\mathbf{I}}^{\mathbf{N}} \circ T + T \circ \partial_{\mathbf{N}'}^{\mathbf{I}'}} & GC^-(\mathbb{G}_0) \\ \Big\downarrow = & & \Big\downarrow P \\ GC^-(\mathbb{G}'_0) & \xrightarrow{h_{X_2} \circ h_Y + h_Y \circ h_{X_2}} & GC^-(\mathbb{G}'_0) \end{array}$$

Since P is a quasi-isomorphism, by Lemma 5.2.12 the above square induces the stated quasi-isomorphism between mapping cones. □

LEMMA 9.2.8. *The map $h_{X_2} \circ h_Y + h_Y \circ h_{X_2} \colon GC^-(\mathbb{G}'_0) \to GC^-(\mathbb{G}'_0)$ is chain homotopic, as homogeneous chain maps of degree $(-2, -1)$, to multiplication by $V_2 - V_1$; so there is an isomorphism*

$$\begin{aligned} &\operatorname{Cone}(h_{X_2} \circ h_Y + h_Y \circ h_{X_2} \colon GC^-(\mathbb{G}'_0) \to GC^-(\mathbb{G}'_0)) \\ &\qquad\qquad \cong \operatorname{Cone}(V_2 - V_1 \colon GC^-(\mathbb{G}'_0) \to GC^-(\mathbb{G}'_0)). \end{aligned}$$

Proof. For $i = 1, 2$, let $h_{X_2,Y_i} \colon GC^-(\mathbb{G}'_0) \to GC^-(\mathbb{G}'_0)$ be the map defined by

$$h_{X_2,Y_i}(\mathbf{x}) = \sum_{\mathbf{y} \in \mathbf{N}(\mathbb{G}'_0)} \sum_{\{r \in \operatorname{Rect}^\circ(\mathbf{x},\mathbf{y}) \big| r \cap \mathbb{Y} = Y_i, X_2 \in r\}} V_1^{O_1(r)} \cdots V_n^{O_n(r)} \cdot \mathbf{y},$$

and let $h_{X_2,Y} = h_{X_2,Y_1} + h_{X_2,Y_2}$. The following identity holds:

$$h_{X_2} \circ h_Y + h_Y \circ h_{X_2} + \partial_{\mathbf{N}'}^{\mathbf{N}'} \circ h_{X_2,Y} + h_{X_2,Y} \circ \partial_{\mathbf{I}'}^{\mathbf{N}'} + h_{X_2,Y} \circ \partial_{\mathbf{N}'}^{\mathbf{N}'} = (V_2 - V_4). \tag{9.9}$$

This can be seen by considering juxtapositions of rectangles that can contribute, and noting that the two terms that do not appear twice are the row through X_2 (responsible for the V_2) and the column through X_2 (responsible for the V_4).

Note that h_{X_2,Y_i} vanishes on $\mathbf{I}'$, maps $\mathbf{N}'$ to itself, and by Equations (4.2) and (4.4)

$$h_{X_2,Y_i} \colon GC_d^-(\mathbb{G}'_0, s) \to GC_{d-1}^-(\mathbb{G}'_0, s - 1);$$

so Equation (9.9) says that $h_{X_2,Y}$ gives a chain homotopy between $h_{X_2} \circ h_Y + h_Y \circ h_{X_2}$ and multiplication by $V_2 - V_4$. Since V_4 is chain homotopic to V_1 in $GC^-(\mathbb{G}'_0)$, it follows that $h_{X_2} \circ h_Y + h_Y \circ h_{X_2}$ is chain homotopic to $V_2 - V_1$. Chain homotopic maps have isomorphic mapping cones (Lemma 5.2.14), so the lemma follows. □

Proof of Theorem 9.2.1. By Lemma 9.2.5, the two horizontal maps from Equation (9.5) together induce a bigraded chain map $P_{+,-}\colon GC^-(\mathbb{G}_+) \to GC^-(\mathbb{G}_-)$. Applying in succession Lemmas 9.2.5, 9.2.7, and 9.2.8 gives a quasi-isomorphism of bigraded complexes over $\mathbb{F}[V_1,\dots,V_n]$ from $\mathrm{Cone}(P_{+,-}\colon GC^-(\mathbb{G}_+) \to GC^-(\mathbb{G}_-))$ to $\mathrm{Cone}(V_2 - V_1\colon GC^-(\mathbb{G}_0') \to GC^-(\mathbb{G}_0'))[\![-1, \frac{\ell_0-\ell-1}{2}]\!]$. The theorem is stated in terms of $GC^-(\mathbb{G}_0)$ instead of $GC^-(\mathbb{G}_0')$; but the proof of Proposition 5.1.7 gives a quasi-isomorphism from $GC^-(\mathbb{G}_0')$ and $GC^-(\mathbb{G}_0)$, and hence (for example, by appealing to Lemma 5.2.12) a quasi-isomorphism

$$\mathrm{Cone}(V_2 - V_1\colon GC^-(\mathbb{G}_0') \to GC^-(\mathbb{G}_0')) \to \mathrm{Cone}(V_2 - V_1\colon GC^-(\mathbb{G}_0) \to GC^-(\mathbb{G}_0)).$$

Composing these quasi-isomorphisms establishes the theorem. □

EXERCISE 9.2.9. **(a)** Show that $(t^{\frac{1}{2}} - t^{-\frac{1}{2}})\cdot \chi(GC^-(\mathbb{G})) = t^{\frac{\ell}{2}} \cdot \Delta_{\vec{L}}(t)$, for any grid diagram $\mathbb{G}$ representing $\vec{L}$.
(b) Show that the skein relation for the Alexander polynomial is a formal consequence of Theorem 9.2.1 together with part (a).

9.3. Proofs of the skein exact sequences

Having proved Theorem 9.2.1, we can quickly deduce the skein exact sequences, as stated in the beginning of the chapter. It will be useful to have the following algebraic observation:

LEMMA 9.3.1. *Let C and C' be two free, bigraded chain complexes over $\mathbb{F}[V_1,\dots,V_n]$. Fix any non-zero, homogeneous polynomial $r \in \mathbb{F}[V_1,\dots,V_n]$. A quasi-isomorphism $\phi\colon C \to C'$ induces a quasi-isomorphism $\overline{\phi}\colon \frac{C}{r} \to \frac{C'}{r}$, where $\frac{C}{r}$ is shorthand for the quotient complex $\frac{C}{rC}$.*

Proof. Since ϕ is an $\mathbb{F}[V_1,\dots,V_n]$-module map, and C and C' are both free, there is a map between the short exact sequences

$$\begin{array}{ccccccccc}
0 & \longrightarrow & C & \xrightarrow{r} & C & \longrightarrow & \frac{C}{r} & \longrightarrow & 0 \\
 & & \downarrow{\scriptstyle\phi} & & \downarrow{\scriptstyle\phi} & & \downarrow{\scriptstyle\overline{\phi}} & & \\
0 & \longrightarrow & C' & \xrightarrow{r} & C' & \longrightarrow & \frac{C'}{r} & \longrightarrow & 0.
\end{array}$$

Naturality of the associated long exact sequence (Lemma A.2.2) and the five lemma (Lemma A.2.3) now give the result. (Compare also Proposition A.3.5.) □

Proof of Theorem 9.1.1. Let ℓ be the number of components in $\vec{L}_+$. Label the O-markings so that $O_2,\dots,O_{\ell+1}$ belong to ℓ different components on $\vec{L}_+$, so

$$cGC^-(\mathbb{G}_+) = \frac{GC^-(\mathbb{G}_+)}{V_2 = \cdots = V_{\ell+1}} \qquad \text{and} \qquad cGC^-(\mathbb{G}_-) = \frac{GC^-(\mathbb{G}_-)}{V_2 = \cdots = V_{\ell+1}}.$$

The quasi-isomorphism from Theorem 9.2.1 induces a quasi-isomorphism

(9.10)

$$\begin{aligned}
&\mathrm{Cone}(\overline{P}_{+,-}\colon cGC^-(\mathbb{G}_+) \to cGC^-(\mathbb{G}_-)) \\
&\to \mathrm{Cone}\left(V_2 - V_1\colon \frac{GC^-(\mathbb{G}_0)}{V_2 = \cdots = V_{\ell+1}} \to \frac{GC^-(\mathbb{G}_0)}{V_2 = \cdots = V_{\ell+1}}\right)[\![-1, \frac{\ell_0 - \ell - 1}{2}]\!],
\end{aligned}$$

where $\overline{P}_{+,-}$ is the map induced by $P_{+,-}$ on the collapsed grid complex. This can be seen by applying Lemma 9.3.1 repeatedly, using

$$r = V_i - V_{i+1}$$
$$C = \mathrm{Cone}(\overline{P}_{+,-} : \frac{GC^-(\mathbb{G}_+)}{V_2 = \cdots = V_i} \to \frac{GC^-(\mathbb{G}_-)}{V_2 = \cdots = V_i})$$
$$C' = \mathrm{Cone}\left(V_2 - V_1 : \frac{GC^-(\mathbb{G}_0)}{V_2 = \cdots = V_i} \to \frac{GC^-(\mathbb{G}_0)}{V_2 = \cdots = V_i}\right) [\![-1, \frac{\ell_0 - \ell - 1}{2}]\!],$$

the latter two thought of as free chain complexes over $\mathbb{F}[V_i, \dots, V_n]$; and using the observation that for any bigraded chain map $f \colon C \to C'$ inducing $\overline{f} \colon \frac{C}{r} \to \frac{C'}{r}$, we have $\mathrm{Cone}(\overline{f}) \cong \frac{\mathrm{Cone}(f)}{r}$. This inductive procedure gives a chain map over $\mathbb{F}[V_1, \dots, V_n]$ of the form promised in Equation (9.10) which induces an isomorphism on homology; and so it is a quasi-isomorphism over $\mathbb{F}[V_1, \dots, V_n]$.

When $\vec{L}_0$ has $\ell_0 = \ell + 1$ components, Lemma 5.2.13 identifies the homology of the second mapping cone appearing in Equation (9.10) with the homology of

$$\frac{GC^-(\mathbb{G}_0)}{V_1 = \cdots = V_{\ell+1}} = cGC^-(\mathbb{G}_0). \tag{9.11}$$

Consider the long exact sequence on homology associated to the mapping cone $\mathrm{Cone}(\overline{P}_{+,-} \colon cGC^-(\mathbb{G}_+) \to GC^-(\mathbb{G}_-))$, and substitute $cGH^-(\vec{L}_0)$ (with grading shift as specified in Equation (9.10)) in place of the homology of the mapping cone. The resulting long exact sequence is Equation (9.2).

When $\vec{L}_0$ has $\ell_0 = \ell - 1$ components, for $\mathbb{G}_0$ exactly two of the variables $V_2, \dots, V_{\ell+1}$ belong to the same component; suppose those two variables are V_2 and V_3. Then, $\frac{GC^-(\mathbb{G}_0)}{V_3 = \cdots = V_{\ell+1}} = cGC^-(\mathbb{G}_0)$. By Lemma 8.2.3, V_2 is chain homotopic to V_3, so Lemma 7.4.1 gives a quasi-isomorphism from $cGC^-(\mathbb{G}_0) \otimes W$ to the quotient $\frac{GC^-(\mathbb{G}_0)}{V_2 = V_3 = \cdots = V_{\ell+1}}$. Instead of Equation (9.11), since V_1 and V_2 correspond to the same component on $\vec{L}_0$, Lemmas 8.2.3 and 7.4.1 give a quasi-isomorphism from the mapping cone $\mathrm{Cone}\left(V_2 - V_1 \colon cGC^-(\mathbb{G}_0) \otimes W \to cGC^-(\mathbb{G}_0) \otimes W\right)$ to $cGC^-(\mathbb{G}_0) \otimes W \otimes W$. In view of the bigraded isomorphism $W \otimes W[\![0, -1]\!] \cong J$, Equation (9.3) follows from the long exact sequence for the mapping cone for $\overline{P}_{+,-}$, and using $cGH^-(\vec{L}_0) \otimes J$ in place of the homology of the mapping cone. ☐

Proof of Theorem 9.1.2. Continue notation from the proof of Theorem 9.1.1, noting that $\widehat{GC}(\mathbb{G}) = \frac{cGC^-(\mathbb{G})}{V_1}$ for $\mathbb{G} = \mathbb{G}_+$, $\mathbb{G}_-$, and $\mathbb{G}_0$. Consider the quasi-isomorphism of Equation (9.10), composed with the quasi-isomorphism

$$\mathrm{Cone}\left(V_2 - V_1 \colon \frac{GC^-(\mathbb{G}_0)}{V_2 = \cdots = V_{\ell+1}} \to \frac{GC^-(\mathbb{G}_0)}{V_2 = \cdots = V_{\ell+1}}\right) \to \frac{GC^-(\mathbb{G}_0)}{V_1 = \cdots = V_{\ell+1}}$$

from Lemma 5.2.13. Divide by V_1 as in Lemma 9.3.1, to get a quasi-isomorphism

$$\mathrm{Cone}(\widehat{P}_{+,-} \colon \widehat{GC}(\mathbb{G}_+) \to \widehat{GC}(\mathbb{G}_-)) \to \frac{GC^-(\mathbb{G}_0)}{V_1 = \cdots = V_{\ell+1} = 0}[\![-1, \frac{\ell_0 - \ell - 1}{2}]\!].$$

When $\vec{L}_0$ has $\ell_0 = \ell + 1$ components, $\frac{GC^-(\mathbb{G}_0)}{V_1 = V_2 = \cdots = V_{\ell+1} = 0} = \widehat{GC}(\mathbb{G}_0)$, and when $\vec{L}_0$ has $\ell_0 = \ell - 1$ components, there is a quasi-isomorphism from $\widehat{GC}(\mathbb{G}_0) \otimes W \otimes W$ to

$\frac{GC^-(\mathbb{G}_0)}{V_1=V_2=\cdots=V_{\ell+1}=0}$. The long exact sequences of the theorem follow from the long exact sequence of the mapping cone of $\widehat{P}_{+,-}\colon \widehat{GC}(\mathbb{G}_+) \to \widehat{GC}(\mathbb{G}_-)$, the identification on homology from Equation (9.10), and the above observations. □

9.4. First computations using the skein sequence

We give now a few computations using the skein exact sequence. Use notation from Section 7.3: $\mathbb{F}[U]_{(d,s)}$ is the bigraded free $\mathbb{F}[U]$-module whose generator has bigrading (d,s) (so it is non-trivial in bigradings $(d-2i, s-i)$ for $i \geq 0$); and $\mathbb{F}_{(d,s)}$ is the one-dimensional $\mathbb{F}$-vector space supported in bigrading (d,s).

Recall (Proposition 4.8.1) that for the unknot $\mathcal{O}$ the grid homologies are

$$GH^-(\mathcal{O}) = \mathbb{F}[U]_{(0,0)} \text{ and } \widehat{GH}(\mathcal{O}) = \mathbb{F}_{(0,0)}.$$

Combining this with Lemma 8.4.2, this immediately gives the following:

LEMMA 9.4.1. *For the two-component unlink L,*

$$cGH^-(L) \cong \mathbb{F}[U]_{(0,0)} \oplus \mathbb{F}[U]_{(-1,0)} \qquad \textit{and} \qquad \widehat{GH}(L) \cong \mathbb{F}_{(0,0)} \oplus \mathbb{F}_{(-1,0)}.$$ □

EXERCISE 9.4.2. Give an alternate proof of Lemma 9.4.1 using the skein exact sequence and Proposition 8.3.2.

The skein exact sequence can be used to compute the grid homologies of the Hopf links H_+ and H_- from Figure 2.8.

To this end, consider the skein exact sequence associated to the skein triple where $\vec{L}_+ = H_+$, $\vec{L}_-$ is the two-component unlink, and $\vec{L}_0$ is the unknot. To compute the map $g^-\colon cGH^-(\vec{L}_-) \to cGH^-(\vec{L}_0) \otimes J$, we argue as follows. Let x denote the generator of $cGH^-(\vec{L}_0) \otimes J$ in bigrading $(0,1)$, and y the generator in bigrading $(-2,-1)$. Let a resp. b denote the elements of $GH^-(\vec{L}_-)$ in bigradings $(0,0)$ resp. $(-1,0)$. We claim that a and b map non-trivially into $cGH^-(\vec{L}_0) \otimes J$, for otherwise, the rank of $cGH^-(\vec{L}_+)$ (as an $\mathbb{F}[U]$-module) would be at least three, violating Proposition 8.3.2. Moreover, y and all its multiples (in the ring $\mathbb{F}[U]$) are in the cokernel of this map, since there is nothing in the appropriate bigrading in $cGH^-(\vec{L}_-)$ to map onto it (in view of the computation from Lemma 9.4.1). Also, x lies in the cokernel, for the same reason; but $U \cdot x = g^-(b)$ (again, this is the only possibility for grading reasons, since $g^-(b)$ is non-zero).

We conclude that

$$cGH^-(H_+) = \mathbb{F}_{(0,1)} \oplus \mathbb{F}[U]_{(-1,0)} \oplus \mathbb{F}[U]_{(-2,-1)};$$

the nonzero element of $\mathbb{F}_{(0,1)}$ is the image of x, the free summand starting at $(-1,0)$ comes from the cokernel of the map $\mathbb{F}[U] \cdot a$ into $(\mathbb{F}[U] \oplus \mathbb{F}[U])_{(-1,0)}$, and the remaining free summand is the image of y and its multiples.

It follows readily that $\widehat{GH}(H_+) \cong \mathbb{F}_{(0,1)} \oplus \mathbb{F}^2_{(-1,0)} \oplus \mathbb{F}_{(-2,-1)}$.

EXERCISE 9.4.3. Use the skein exact sequence to show that

$$cGH^-(H_-) \cong \mathbb{F}[U]_{(1,1)} \oplus \mathbb{F}[U]_{(0,0)} \oplus \mathbb{F}_{(0,0)}$$
$$\widehat{GH}(H_-) \cong \mathbb{F}_{(1,1)} \oplus \mathbb{F}^2_{(0,0)} \oplus \mathbb{F}_{(-1,-1)}.$$

Next, we compute GH^- for the right-handed trefoil knot $T_{2,3}$, by putting it into a skein triple with $\vec{L}_+ = T_{2,3}$, $\vec{L}_-$ the unknot $\mathcal{O}$, and $\vec{L}_0$ the Hopf link H_+

computed earlier. The map $g^-: cGH^-(\mathcal{O}) \to cGH^-(H_+)$ is non-trivial; for otherwise $GH^-(T_{2,3})$ would have rank three (as an $\mathbb{F}[U]$-module); but we know that its rank is one (Proposition 7.3.3). Indeed, the nontriviality of g^- implies that its image is $\mathbb{F}[U]_{(-1,0)}$, hence $GH^-(T_{2,3}) = \mathbb{F}_{(0,1)} \oplus \mathbb{F}[U]_{(-2,-1)}$; compare Section 4.8.

EXERCISE 9.4.4. **(a)** Use the skein exact sequence to compute $GH^-(T_{-2,3})$ for the left-handed trefoil knot $T_{-2,3}$.
(b) Compute GH^- and cGH^- for all $(2,n)$ torus knots and links.
(c) Compute GH^- for the twist knots W_n from Example 2.1.5.

9.5. Knots with identical grid homologies

As an application of the skein exact sequence, we give infinite families of distinct knots with the same grid homology [**84**]; compare also Lemma 2.5.4. Consider the family $K(B,k)$ of knots we get from the two-component unlink L by adding a k-fold twisted band to the two components, as in Lemma 2.5.4.

PROPOSITION 9.5.1. *Let $K(B,k)$ be a knot constructed from the two-component unlink by adding the band B with k full twists to the unlink. Then, the grid homologies GH^- and $\widehat{GH}$ of $K(B,k)$ are independent of k.*

Proof. By Proposition 7.3.3, for all k there is a splitting of $\mathbb{F}[U]$-modules

$$GH^-(K(B,k)) \cong M_k \oplus \mathbb{F}[U]_{(2a_k,a_k)}, \tag{9.12}$$

where $a_k = -\tau(K(B,k))$, and M_k is the torsion submodule of $GH^-(K(B,k))$. Let $x_k \in GH^-(K(B,k))$ be a generator of a non-torsion summand of $GH^-(K(B,k))$. Note that x_k is not canonical, since the splitting of Equation (9.12) is not canonical, but its Alexander grading a_k is.

Since there is a skein triple with $K_- = K(B,k)$, $K_+ = K(B,k-1)$, and $\vec{K}_0$, the 2-component unlink L, Theorem 9.1.1 gives the exact triangle:

$$\begin{array}{ccc} GH^-(K(B,k-1)) & \xrightarrow{\ f^-\ } & GH^-(K(B,k)) \\ & {\scriptstyle h^-}\nwarrow \quad \swarrow{\scriptstyle g^-} & \\ & \mathbb{F}[U]_{(0,0)} \oplus \mathbb{F}[U]_{(-1,0)} & \end{array} \tag{9.13}$$

where f^- and h^- are bigraded $\mathbb{F}[U]$-module maps, and g^- is homogeneous of degree $(-1,0)$. Let y_0 and y_{-1} be the two generators of $\mathbb{F}[U]_{(0,0)} \oplus \mathbb{F}[U]_{(-1,0)}$. Exactness of the triangle (together with the fact that $GH^-(K(B,j))$ for $j = k, k-1$ both have rank one) implies that the maps h^- and g^- are non-zero maps. Since g^- respects the $\mathbb{F}[U]$-module structure, it follows that g^- vanishes on the torsion part of $GH^-(K(B,k))$, so $\mathrm{Im}(g^-) = \mathbb{F}[U] \cdot g^-(x_k)$. Since a non-torsion summand for GH^- of a knot is in even Maslov grading, g^- drops Maslov grading by one, and g^- is non-trivial, it follows that for some $m = m_k \geq 0$,

$$g^-(x_k) = U^{m_k} \cdot y_{-1}. \tag{9.14}$$

Since g^- preserves Alexander grading and $A(y_{-1}) = 0$, it follows that

$$-\tau(K(B,k)) = A(x_k) = -m_k \leq 0. \tag{9.15}$$

Exactness ensures that $h^-(y_0)$ is non-torsion; so there is some $n = n_k \geq 0$ with

$$h^-(y_0) + U^{n_k} x_{k-1} \in M_{k-1}. \tag{9.16}$$

Since $A(y_0) = 0$ and h^- preserves Alexander grading,

$$-\tau(K(B, k-1)) = A(x_{k-1}) = n_k \geq 0. \tag{9.17}$$

Since both (9.15) and (9.17) hold for all k, we conclude that for all j (and in particular, for $j = k$ and $k-1$), $\tau(K(B,j)) = 0$.

Substituting $n_k = 0$ into Equation (9.16) shows $\mathrm{Coker}(h^-) = M_{k-1}$, and Equation (9.14) shows $\mathrm{Ker}(g^-) = M_k$. Exactness shows that f^- induces an identification $M_{k-1} \cong M_k$.

It follows that $GH^-(K(B,k-1)) \cong GH^-(K(B,k))$; so by Proposition 7.3.3, $\widehat{GH}(K(B,k-1)) \cong \widehat{GH}(K(B,k))$ as well. □

REMARK 9.5.2. Most of the effort in the above proof goes into showing that $\tau(K(B,k)) = 0$. We could have appealed to Corollary 8.1.2 (or indeed Proposition 8.5.3, with $K_1 = K(B,k)$ and K_2 the unknot), but we chose not to, in order to illustrate better the skein sequence.

Recall the Kanenobu knots $K(p,q)$ of Figure 2.18 from Section 2.4. These knots can be constructed from the two-component unlink $K(B,k)$ as in Proposition 9.5.1 in two different ways: either regard the region with the p full twists as a band added to the two-component unlink, or do the same with the region of the q full twists (cf. Figure 2.19). We compute grid homology for these knots Section 10.3; for the time being observe that Proposition 9.5.1 has the following immediate consequence:

LEMMA 9.5.3. *The grid homologies $GH^-(K(p,q))$ and $\widehat{GH}(K(p,q))$ are independent of p and q.* □

9.6. The skein exact sequence and the crossing change map

When $\mathbb{G}_+$ and $\mathbb{G}_-$ represent knots, we defined a map for the unknotting operation $C_-\colon GH^-(\mathbb{G}_+) \to GH^-(\mathbb{G}_-)$ in Chapter 6. In the skein exact sequence, there is another map $f^-\colon GH^-(\mathbb{G}_+) \to GH^-(\mathbb{G}_-)$, which is induced from the chain map $P_{+,-}\colon GC^-(\mathbb{G}_+) \to GC^-(\mathbb{G}_-)$ of Theorem 9.2.1. We describe here the relationship between these maps; this will be used in the proof of Theorem 10.2.4.

Assume still that the height difference between the two distinguished X-markings in Figure 9.3 is one. Draw the horizontal circles so that exactly one of them crosses the two X-marked bigons; see Figure 9.5. This can be done because of the hypothesis on the height difference between the two X-markings.

Recall that $P_{+,-}$ counts rectangles, as specified by the horizontal arrows in Equation (9.5). The map C_- is induced by the chain map from Lemma 6.2.2 $c_-\colon GC^-(\mathbb{G}_+) \to GC^-(\mathbb{G}_-)$, defined by counting pentagons with corner at s; i.e.

$$c_-(\mathbf{x}_+) = \sum_{\mathbf{y}_- \in \mathbf{S}(\mathbb{G}_-)} \sum_{\{p \in \mathrm{Pent}^\circ_s(\mathbf{x}_+,\mathbf{y}_-) \mid p \cap \mathbb{X} = \emptyset\}} V_1^{O_1(p)} \cdots V_n^{O_n(p)} \cdot \mathbf{y}_-,$$

where $\mathbf{x}_+ \in \mathbf{S}(\mathbb{G}_+)$ and $s \in \beta_i \cap \gamma_i$ is the northern corner point on the bigon in $\mathbb{T} \setminus \beta_i \cap \gamma_i$ containing X_1. Letting s' be the southern corner of the bigon containing X_2

(see Figure 9.5), consider the map $c'_- : GC^-(\mathbb{G}_+) \to GC^-(\mathbb{G}_-)$ counting pentagons with a corner at s'; i.e.

$$c'_-(\mathbf{x}_+) = \sum_{\mathbf{y}_- \in \mathbf{S}(\mathbb{G}_-)} \sum_{\{p \in \mathrm{Pent}^\circ_{s'}(\mathbf{x}_+, \mathbf{y}_-) \big| p \cap \mathbb{X} = \emptyset\}} V_1^{O_1(p)} \cdots V_n^{O_n(p)} \cdot \mathbf{y}_-.$$

The proof that this is a chain map follows exactly as in the proof of Lemma 6.2.2.

PROPOSITION 9.6.1. *If $\vec{L}_+$ is a knot, we can choose $P_{+,-} = c_- + c'_-$ for the map in Theorem* 9.2.1.

Proof. To compare the two maps, draw the two grid diagrams on one grid torus in two ways: fix the four distinguished markings and vary the vertical curve (illustrated on the left of Figure 9.5) as was done in Chapter 6; or alternatively, fix the vertical circle and vary all four distinguished markings (illustrated on the right of Figure 9.5). Here, we use X_1, X_2, O_3, and O_4 to represent $\mathbb{G}_+$ and X'_1, X'_2, O'_3, and O'_4 to represent $\mathbb{G}_-$. (Note that the pictures in Chapter 6 are labeled differently from the ones here.)

Let $\mathbf{x} \in \mathbf{I}$ and $\mathbf{y} \in \mathbf{N}$. There is a natural one-to-one correspondence between $r \in \mathrm{Rect}^\circ(\mathbf{x}, \mathbf{y})$ with $X'_1 \in r$ and $p \in \mathrm{Pent}^\circ_s(\mathbf{x}, \mathbf{y})$; and similarly, between $r \in \mathrm{Rect}^\circ(\mathbf{x}, \mathbf{y})$ with $X'_2 \in r$ and $p \in \mathrm{Pent}^\circ_{s'}(\mathbf{x}, \mathbf{y})$. Together, these rectangles are the ones counted in $\partial^{\mathbf{N}'}_{\mathbf{I}'} : \mathbf{I}' \to \mathbf{N}'$, the top horizontal map from Equation (9.5) (i.e. the restriction of $P_{+,-}$ to $\mathbf{I}'$). Similarly, there is a natural one-to-one correspondence between $r \in \mathrm{Rect}^\circ(\mathbf{y}, \mathbf{x})$ with $X_1 \in r$ and $p \in \mathrm{Pent}^\circ_s(\mathbf{y}, \mathbf{x})$; and $r \in \mathrm{Rect}^\circ(\mathbf{y}, \mathbf{x})$ with $X_2 \in r$ and $p \in \mathrm{Pent}^\circ_{s'}(\mathbf{y}, \mathbf{x})$. These are the rectangles counted in $\partial^{\mathbf{I}}_{\mathbf{N}} : \mathbf{N} \to \mathbf{I}$, the bottom horizontal map from Equation (9.5) (i.e. the restriction of $P_{+,-}$ to $\mathbf{N}$).

All pentagons not covered by the above correspondences contain X-markings in their interior. Under the above correspondence, some O_i is contained in the interior of the pentagon if and only if the corresponding O-marking is contained in the interior of the corresponding rectangle. It follows that the contributions to $c_- + c'_-$ and to $P_{+,-}$ coincide. $\square$

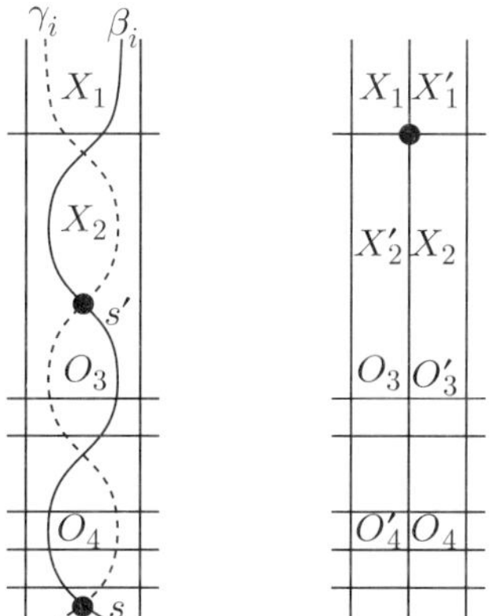

FIGURE 9.5. **Crossing change maps and the skein sequence.**

EXERCISE 9.6.2. Work out the commutation invariance of grid homology from Chapter 5, using rectangles instead of pentagons.

9.7. Further remarks

The skein exact triangle has a long history. In [**54**], Andreas Floer first proposed that a gauge-theoretic Floer homology group should satisfy a related exact triangle; see also [**16**]. Heegaard Floer homology was shown to satisfy an exact triangle soon after its discovery; see for example [**173**, Theorem 9.1], and compare [**114**] for an analogue in Seiberg-Witten theory. The exact triangle for Heegaard Floer homology also can be applied to knot Floer homology to give an oriented skein exact sequence, as described in [**172**, Theorem 10.2]. The present chapter follows the approach from [**167**]. The skein exact sequence can also be seen as a consequence of an exact triangle for singular knots; see [**165, 183**].

When comparing these results with the literature, the reader should be warned that there are several natural grading conventions. For example, in [**172**] Maslov gradings for links take possibly half-integer values; specifically, if M' is the Maslov grading appearing there and M denotes the Maslov grading we use in this book, then

$$M' = M + \frac{\ell - 1}{2}. \tag{9.18}$$

CHAPTER 10

Grid homologies of alternating knots

In Chapter 9, we studied an exact sequence relating the three links in an oriented skein triple. In this chapter we develop a different *unoriented skein exact sequence*, which relates grid homology groups of three links obtained from another kind of skein triple, described as follows. Fix an unoriented link L and a projection with a distinguished crossing, and form two more links by replacing the crossing by its two possible resolutions, as shown in Figure 10.1. With the help of this unoriented skein sequence, we will compute the grid homology for alternating knots. In fact, we will compute grid homology for a wider class of links, defined as follows.

The class $\mathcal{Q}$ of *quasi-alternating links* is the smallest set of link types with the following two properties:

- the unknot $\mathcal{O}$ is in $\mathcal{Q}$, and
- if L is a link that admits a diagram with a crossing whose two resolutions L_1 and L_2 (as in Figure 10.1) are in $\mathcal{Q}$, with $\det L_1 + \det L_2 = \det L$, then the link L is also in $\mathcal{Q}$.

Since the links appearing in the above definition do not have natural orientations, the definition of $\mathcal{Q}$ hinges on the fact that the determinant of a link L is independent of the choice of an orientation on L. (This claim will be proved in Proposition 10.1.8.)

The unoriented skein exact sequence can be thought of as a homological refinement of the additivity formula for the determinant of links, which we give in Section 10.1. In Section 10.2, we turn to the statement and proof of the unoriented skein exact sequence, using it to compute the grid homologies of alternating knots in Section 10.3 in terms of their Alexander polynomials and signatures. (See Theorem 10.3.1.)

10.1. Properties of the determinant of a link

The unoriented skein exact sequence we will see in Section 10.2 is an exact sequence satisfied by a certain variant of grid homology. As usual, this variant is generated by grid states; but it is given a single integral grading, called the *δ-grading*, which is the difference between the earlier defined Maslov and Alexander

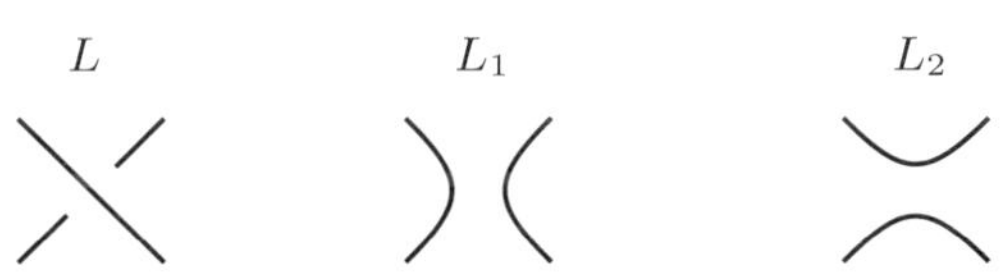

FIGURE 10.1. The two resolutions L_1 and L_2 of a crossing.

gradings:

$$\delta_{\mathbb{G}}(\mathbf{x}) = M(\mathbf{x}) - A(\mathbf{x}) = \frac{M_{\mathbb{O}}(\mathbf{x}) + M_{\mathbb{X}}(\mathbf{x}) + n - \ell}{2}. \tag{10.1}$$

Its Euler characteristic will turn out to be related to the unnormalized determinant of the link $\mathrm{Det}(\vec{L})$ (Definition 2.3.3). The aim of the present section is to study $\mathrm{Det}(\vec{L})$. We start by expressing it in terms of grid states (Lemma 10.1.2). This formulation has two consequences: first, it shows that the determinant of a link is independent of its orientation, and second, it leads to an additivity formula for the determinant for unoriented skein triples (Proposition 10.1.11); which in turn is used to verify that alternating knots are quasi-alternating (Theorem 10.1.13).

Although both of these statements have proofs without using grid states, we prefer the grid proofs, since these proofs will be useful for the grid homology analogues.

Recall that the unnormalized determinant $\mathrm{Det}(\vec{L})$ of an oriented link $\vec{L}$ is the value of the Alexander polynomial $\Delta_{\vec{L}}(t)$ at $t^{\frac{1}{2}} = -i$. The usual determinant $\det(\vec{L})$, which is the absolute value of $\mathrm{Det}(\vec{L})$, satisfies

$$\mathrm{Det}(\vec{L}) = i^{\sigma(\vec{L})} \det(\vec{L}), \tag{10.2}$$

where $\sigma(\vec{L})$ is the signature of $\vec{L}$; see Proposition 2.3.6. Note that if ℓ denotes the number of components of $\vec{L}$, then $i^{\ell-1}\mathrm{Det}(\vec{L})$ is always an integer.

EXAMPLE 10.1.1. For the two trefoils $\mathrm{Det}(T_{2,3}) = -3 = \mathrm{Det}(T_{-2,3})$. For the Hopf links of Figure 2.8, $\mathrm{Det}(H_{\pm}) = \mp 2i$.

LEMMA 10.1.2. *If $\mathbb{G}$ is a grid diagram representing an oriented link $\vec{L}$ with ℓ components, then*

$$\sum_{\mathbf{x}\in\mathbf{S}(\mathbb{G})} (-1)^{\delta_{\mathbb{G}}(\mathbf{x})} = 2^{n-1}(-i)^{\ell-1}\mathrm{Det}(\vec{L}). \tag{10.3}$$

Proof. We saw in the proof of Proposition 8.2.10 (Theorem 4.7.5 for the case of knots) that $\sum_{\mathbf{x}\in\mathbf{S}(\mathbb{G})}(-1)^{M(\mathbf{x})}t^{A(\mathbf{x})} = (1-t^{-1})^{n-1}t^{\frac{\ell-1}{2}} \cdot \Delta_{\vec{L}}(t)$. (In that proof, the left hand sum was thought of as the Euler characteristic of the fully blocked grid complex $\widetilde{GC}(\mathbb{G})$.) Setting $t^{\frac{1}{2}} = -i$ gives Equation (10.3). □

Using the above lemma, we will study the determinant of a link using grid diagrams. To this end, it is useful to have the following:

DEFINITION 10.1.3. For a planar grid diagram G, a marking (X or O), is called a ***local maximum*** (of the anti-diagonal height function) if the vertical arc in the projection containing the marking lies below the marking, and the horizontal arc lies to its right. A ***local minimum*** is defined analogously: it is a marking whose vertical arc lies above the marking and whose horizontal arc lies to the left. The ***bridge index*** $b(G)$ is the number of local maxima.

DEFINITION 10.1.4. The ***writhe*** of a planar grid diagram G, denoted $\mathrm{wr}(G)$, is the writhe of its underlying link diagram; i.e. it is the number of positive crossings minus the number of negative crossings.

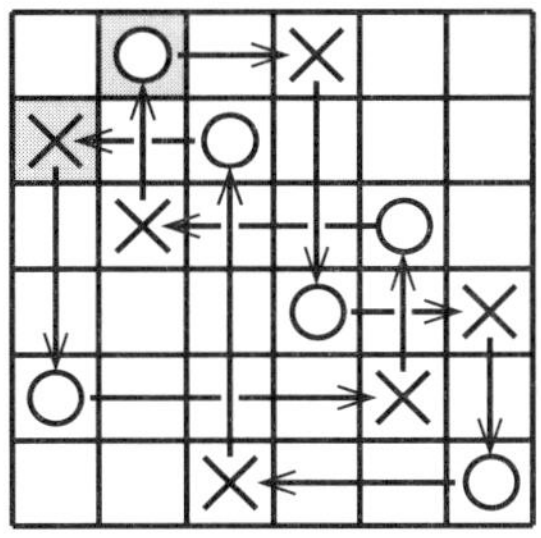

FIGURE 10.2. **Local maxima in a grid diagram.** The two squares that contain local maxima are shaded.

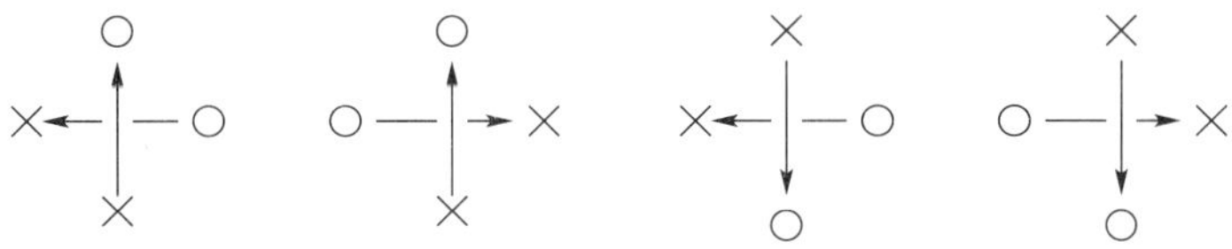

FIGURE 10.3. **Intersecting intervals.** There are four ways two intervals can intersect in their interior in a grid; for these four ways, $-z(a,b)$ is the sign of the crossing of a and b.

LEMMA 10.1.5. *Let $\mathbb{G}$ be a toroidal grid diagram for an oriented link $\vec{L}$, and let G be any planar realization of it. Then, $\mathcal{J}(\mathbb{O}-\mathbb{X},\mathbb{O}-\mathbb{X}) = b(G) - \mathrm{wr}(G)$.*

Proof. The link projection represented by G is composed of straight vertical and horizontal segments; let $\mathcal{C}_v$ and $\mathcal{C}_h$ denote the sets of vertical and horizontal segments. Each such segment $a \in \mathcal{C}_v \cup \mathcal{C}_h$ connects a pair of markings, $O(a)$ and $X(a)$. It is easy to see that

$$\mathcal{J}(\mathbb{O}-\mathbb{X},\mathbb{O}-\mathbb{X}) = \sum_{a\in\mathcal{C}_v, b\in\mathcal{C}_h} \mathcal{J}(\{O(a)\}-\{X(a)\},\{O(b)\}-\{X(b)\}).$$

Let $z(a,b) = \mathcal{J}(\{O(a)\}-\{X(a)\},\{O(b)\}-\{X(b)\})$. For $a \in \mathcal{C}_v$ and $b \in \mathcal{C}_h$, it is easy to see that $z(a,b)$ is computed as follows. If a and b are disjoint, then $z(a,b)=0$. If a and b meet in an endpoint that is a local maximum or a local minimum, then $z(a,b)=\frac{1}{2}$; if they meet in an endpoint which is neither, then $z(a,b)=0$. Finally, if they intersect in an interior point, then $z(a,b)$ is minus one times the sign of the crossing of a and b. (See Figure 10.3.)

The lemma follows from this local computation, together with the observation that the number of local maxima equals the number of local minima. □

EXAMPLE 10.1.6. For the picture in Figure 10.2, $\mathcal{J}(\mathbb{O},\mathbb{O})=5$, $\mathcal{J}(\mathbb{X},\mathbb{X})=6$, $\mathcal{J}(\mathbb{O},\mathbb{X})=4$, $\mathrm{wr}(G)=-1$, and $b(G)=2$.

LEMMA 10.1.7. *Fix a toroidal grid diagram $\mathbb{G}$, and let $\mathbb{G}'$ be obtained by exchanging some of the X- and O-markings, so that $\mathbb{G}$ and $\mathbb{G}'$ represent two different orientations on the same underlying link. View the two different δ-gradings $\delta_{\mathbb{G}}$ and $\delta_{\mathbb{G}'}$ as functions on the same set of grid states. For any grid state $\mathbf{x}$,*

$$\delta_{\mathbb{G}}(\mathbf{x}) + \frac{1}{4}\mathrm{wr}(G) = \delta_{\mathbb{G}'}(\mathbf{x}) + \frac{1}{4}\mathrm{wr}(G'), \tag{10.4}$$

where the planar realizations G and G' of $\mathbb{G}$ and $\mathbb{G}'$ are associated to the same fundamental domain in the torus. More invariantly, if L denotes the unoriented link represented by both $\mathbb{G}$ and $\mathbb{G}'$, and $L = L_1 \cup L_2$ denotes the splitting where $L_1 \subset L$ is the sublink where the two orientations agree, and $\vec{L}_1$ and $\vec{L}_2$ denote the orientations induced from $\mathbb{G}$, then

$$\delta_{\mathbb{G}}(\mathbf{x}) - \delta_{\mathbb{G}'}(\mathbf{x}) = -\ell k(\vec{L}_1, \vec{L}_2). \tag{10.5}$$

Proof. Let $\mathbb{O}$, $\mathbb{X}$ be the markings in $\mathbb{G}$ and $\mathbb{O}'$, $\mathbb{X}'$ be the markings in $\mathbb{G}'$. Fix a fundamental domain for the torus and let G and G' be the planar realizations of $\mathbb{G}$ and $\mathbb{G}'$. Combining Equation (10.1) with the bilinearity of $\mathcal{J}$, we have that

$$\begin{aligned}\delta_{\mathbb{G}}(\mathbf{x}) - \delta_{\mathbb{G}'}(\mathbf{x}) &= \frac{1}{2}(M_{\mathbb{O}}(\mathbf{x}) + M_{\mathbb{X}}(\mathbf{x}) - M_{\mathbb{O}'}(\mathbf{x}) - M_{\mathbb{X}'}(\mathbf{x})) \\ &= \frac{1}{2}(\mathcal{J}(\mathbb{O},\mathbb{O}) + \mathcal{J}(\mathbb{X},\mathbb{X}) - \mathcal{J}(\mathbb{O}',\mathbb{O}') - \mathcal{J}(\mathbb{X}',\mathbb{X}')).\end{aligned}$$

Partition $\mathbb{O} = \mathbb{O}_1 \cup \mathbb{O}_2$, $\mathbb{X} = \mathbb{X}_1 \cup \mathbb{X}_2$, $\mathbb{O}' = \mathbb{O}_1 \cup \mathbb{X}_2$, and $\mathbb{X}' = \mathbb{X}_1 \cup \mathbb{O}_2$; i.e. $\mathbb{X}_1$ and $\mathbb{O}_1$ are the markings that coincide for the two orientations, and $\mathbb{X}_2$ and $\mathbb{O}_2$ are the ones that switch.

Bilinearity of $\mathcal{J}$, Lemma 10.1.5, and the relation $b(G') = b(G)$ give

$$\begin{aligned}&2(\mathcal{J}(\mathbb{O},\mathbb{O}) + \mathcal{J}(\mathbb{X},\mathbb{X}) - \mathcal{J}(\mathbb{O}',\mathbb{O}') - \mathcal{J}(\mathbb{X}',\mathbb{X}')) \\ &\quad = \mathcal{J}(\mathbb{O}_1 + \mathbb{O}_2 - \mathbb{X}_1 - \mathbb{X}_2, \mathbb{O}_1 + \mathbb{O}_2 - \mathbb{X}_1 - \mathbb{X}_2) \\ &\qquad - \mathcal{J}(\mathbb{O}_1 + \mathbb{X}_2 - \mathbb{X}_1 - \mathbb{O}_2, \mathbb{O}_1 + \mathbb{X}_2 - \mathbb{X}_1 - \mathbb{O}_2) \\ &\quad = \mathcal{J}(\mathbb{O} - \mathbb{X}, \mathbb{O} - \mathbb{X}) - \mathcal{J}(\mathbb{O}' - \mathbb{X}', \mathbb{O}' - \mathbb{X}') \\ &\quad = \mathrm{wr}(G') - \mathrm{wr}(G). \end{aligned} \tag{10.6}$$

This completes the proof of Equation (10.4).

It is elementary to verify that $\frac{1}{4}(\mathrm{wr}(G) - \mathrm{wr}(G')) = \ell k(\vec{L}_1, \vec{L}_2)$. For example, if a positive crossing in G changes to a negative one in G', then that crossing contributes 2 to $\mathrm{wr}(G) - \mathrm{wr}(G')$, and it contributes $1/2$ to $\ell k(\vec{L}_1, \vec{L}_2)$. □

The following result is used implicitly in the definition of quasi-alternating links:

PROPOSITION 10.1.8. *Let $\vec{L}$ and $\vec{L}'$ be any two different orientations on the same underlying link L, and partition $L = L_1 \cup L_2$, where L_1 is the subset where the two orientations agree; let $\vec{L}_1$ and $\vec{L}_2$ be the orientations induced from $\vec{L}$. Then,* $\mathrm{Det}(\vec{L}) = (-1)^{\ell k(\vec{L}_1, \vec{L}_2)}\mathrm{Det}(\vec{L}')$. *In particular,* $\det(\vec{L}) = \det(\vec{L}')$.

Proof. Letting $\mathbb{G}$ be a grid diagram for $\vec{L}$, a grid diagram for $\vec{L}'$ is obtained by exchanging some of the X- and O-markings in $\mathbb{G}$, so Lemma 10.1.7 can be used to compare the two δ-gradings. Applying Equations (10.3) and (10.5), we find

$$\begin{aligned}2^{n-1}(-i)^{\ell-1}\mathrm{Det}(\vec{L}) &= \sum_{\mathbf{x}\in\mathbf{S}(\mathbb{G})} (-1)^{\delta_{\mathbb{G}}(\mathbf{x})} = \sum_{\mathbf{x}\in\mathbf{S}(\mathbb{G})} (-1)^{\delta_{\mathbb{G}'}(\mathbf{x}) - \ell k(\vec{L}_1, \vec{L}_2)} \\ &= 2^{n-1}(-i)^{\ell-1}\mathrm{Det}(\vec{L}')(-1)^{\ell k(\vec{L}_1, \vec{L}_2)}.\end{aligned}$$

□

Let L be a link, and fix a diagram $\mathcal{D} = \mathcal{D}_L$ for it. It follows immediately from Proposition 10.1.8 that for any orientation on L, $i^{\frac{\mathrm{wr}(\mathcal{D})}{2}} \cdot \mathrm{Det}(\vec{L})$ is independent of the choice of orientation on L (although it depends on $\mathcal{D}$).

The main goal of this section is to establish the unoriented skein relation for the unnormalized determinant, formulated with the help of the following:

DEFINITION 10.1.9. Consider a diagram for the link L and fix a crossing in that diagram. We can form two further links L_1 and L_2, represented by the two different resolutions of L at the chosen crossing, as specified in Figure 10.1. The resulting triple (L, L_1, L_2) is called an ***unoriented skein triple***.

REMARK 10.1.10. The ordering on the three links in an unoriented skein triple depends on the choice of a diagram. As we shall exploit in the proof of Proposition 10.1.11 below, if (L, L_1, L_2) form an unoriented skein triple, then we can find different diagrams where the roles of the three links are cyclically permuted.

We now give the promised *unoriented skein relation for* Det:

PROPOSITION 10.1.11. *Let (L, L_1, L_2) be an unoriented skein triple, and choose orientations $\vec{L}$, $\vec{L}_1$, and $\vec{L}_2$ arbitrarily on the three terms. Then, there is a relation*

$$i^{\frac{\mathrm{wr}(\vec{L})}{2}} \mathrm{Det}(\vec{L}) = i^{\frac{\mathrm{wr}(\vec{L}_1)+1}{2}} \mathrm{Det}(\vec{L}_1) + i^{\frac{\mathrm{wr}(\vec{L}_2)-1}{2}} \mathrm{Det}(\vec{L}_2), \tag{10.7}$$

where the writhes are computed using diagrams for $\vec{L}$, $\vec{L}_1$, and $\vec{L}_2$ that are identical outside of a small neighborhood of the chosen crossing in L.

Proof. It is helpful to note that for any oriented skein triple $(\vec{L}_+, \vec{L}_-, \vec{L}_0)$,

$$\mathrm{Det}(\vec{L}_+) - \mathrm{Det}(\vec{L}_-) = -2i \cdot \mathrm{Det}(\vec{L}_0). \tag{10.8}$$

This is seen by setting $t^{\frac{1}{2}} = -i$ in the skein relation for the Alexander polynomial.

Fix a diagram $\vec{L}$ for an oriented link with a preferred crossing, let $\vec{L}_0$ denote the oriented resolution, and let $\vec{L}_n$ be any orientation on the other resolution. If the preferred crossing is positive, write $\vec{L} = \vec{L}_+$ and if it is negative, write $\vec{L} = \vec{L}_-$. Equation (10.7) is equivalent to the following two equations:

$$\mathrm{Det}(\vec{L}_+) = -i\mathrm{Det}(\vec{L}_0) + i^{\frac{\mathrm{wr}(L_n)-\mathrm{wr}(L_0)}{2}} \mathrm{Det}(\vec{L}_n) \tag{10.9}$$

$$\mathrm{Det}(\vec{L}_-) = i\mathrm{Det}(\vec{L}_0) + i^{\frac{\mathrm{wr}(L_n)-\mathrm{wr}(L_0)}{2}} \mathrm{Det}(\vec{L}_n). \tag{10.10}$$

Observe that if $\vec{L} = \vec{L}_+$, then the unoriented skein triple is $(L, L_1 = L_n, L_2 = L_0)$; if $\vec{L} = \vec{L}_-$, the unoriented skein triple is $(L, L_1 = L_0, L_2 = L_n)$.

Distinguish two cases, according to whether or not the strands of L meeting at the distinguished crossing belong to two different components. We verify Equations (10.9) and (10.10) in both cases.

Case 1: The two strands in L at the distinguished crossing belong to two different components. By assumption, L can be given two different orientations, $\vec{L}_+$ and $\vec{R}_-$, where the given crossing is oriented positively and negatively, respectively. Performing the crossing change gives two further oriented links $\vec{L}_-$ and $\vec{R}_+$. The oriented resolution $\vec{R}_0$ of $\vec{R}_\pm$ at the distinguished crossing can be thought of as the unoriented resolution $\vec{L}_n$ of the preferred crossing in $\vec{L}_\pm$.

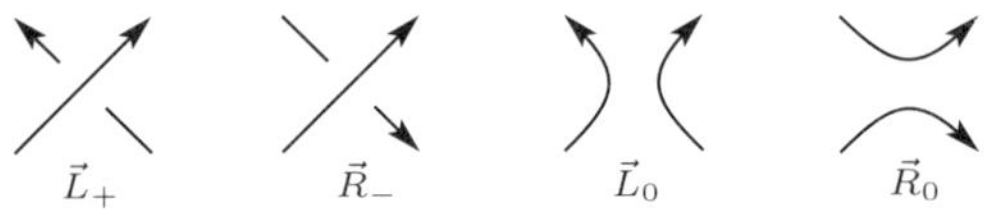

FIGURE 10.4. **The two strands belong to different components.** Reversing the orientation of one of the strands in the distinguished positive crossing of $\vec{L}_+$ gives $\vec{R}_-$, whose oriented resolution can be thought of as the unoriented resolution of $\vec{L}_+$.

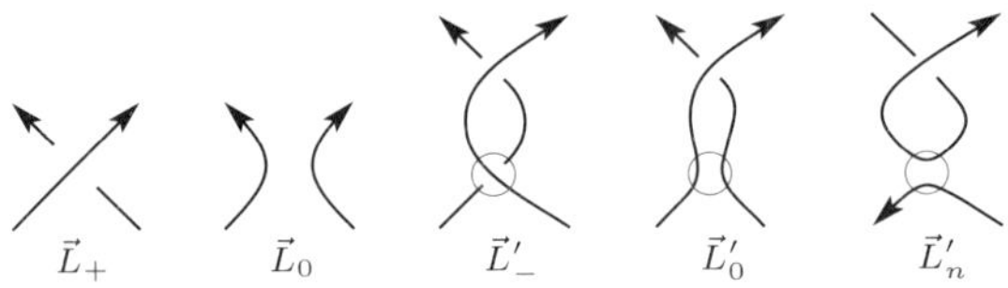

FIGURE 10.5. **The two strands belong to the same components.** Starting from $\vec{L}_+$, form its oriented resolution $\vec{L}_0$, introduce an extra pair of crossings and focus on the negative crossing, to get the diagram $\vec{L}'_-$. The oriented resolution $\vec{L}'_0$ clearly agrees with the original diagram $\vec{L}_+$.

Since $\mathrm{wr}(\vec{L}_\pm) = \mathrm{wr}(\vec{L}_0) \pm 1$ and $\mathrm{wr}(\vec{R}_\pm) = \mathrm{wr}(\vec{R}_0) \pm 1$, Proposition 10.1.8 gives

$$\begin{aligned}\mathrm{Det}(\vec{R}_-) &= i^{\frac{\mathrm{wr}(\vec{L}_+)-\mathrm{wr}(\vec{R}_-)}{2}}\mathrm{Det}(\vec{L}_+) = i^{\frac{\mathrm{wr}(\vec{L}_0)-\mathrm{wr}(\vec{R}_0)}{2}+1}\mathrm{Det}(\vec{L}_+)\\ \mathrm{Det}(\vec{R}_+) &= i^{\frac{\mathrm{wr}(\vec{L}_-)-\mathrm{wr}(\vec{R}_+)}{2}}\mathrm{Det}(\vec{L}_-) = i^{\frac{\mathrm{wr}(\vec{L}_0)-\mathrm{wr}(\vec{R}_0)}{2}-1}\mathrm{Det}(\vec{L}_-).\end{aligned}$$

Substituting these formulas back into Equation (10.8) (for the skein triple $(\vec{R}_+, \vec{R}_-, \vec{R}_0)$) gives

$$\mathrm{Det}(\vec{L}_+) + \mathrm{Det}(\vec{L}_-) = 2i^{\frac{\mathrm{wr}(\vec{R}_0)-\mathrm{wr}(\vec{L}_0)}{2}}\mathrm{Det}(\vec{R}_0), \tag{10.11}$$

which, when combined with Equation (10.8), verifies Equations (10.9) and (10.10).

Case 2: The two strands in L at the distinguished crossing belong to the same component. Use a Reidemeister 2 move on the diagram for $\vec{L}_0$, to get a new diagram for $\vec{L}_0 = \vec{L}'_-$, with an extra pair of crossings in it. Distinguish the newly-introduced negative crossing, and note that the two strands meeting there belong to different components. Observe that $\vec{L}'_0 = \vec{L}_+$. The link $\vec{L}'_n$ can be thought of as reversing the orientation on one of the strands in $\vec{L}'_-$, and forming the oriented resolution at the distinguished crossing. The resulting diagram for $\vec{L}'_n$ is a diagram for $\vec{L}_n$, with an extra negative crossing in it, cf. the right-most diagram of Figure 10.5. Equation (10.10) applied to $\vec{L}'_-$ gives

$$\begin{aligned}\mathrm{Det}(\vec{L}_0) &= i\mathrm{Det}(\vec{L}_+) + i^{\frac{\mathrm{wr}(\vec{L}_n)-1-\mathrm{wr}(\vec{L}_+)}{2}}\mathrm{Det}(\vec{L}_n)\\ &= i\mathrm{Det}(\vec{L}_+) + i^{\frac{\mathrm{wr}(\vec{L}_n)-\mathrm{wr}(\vec{L}_0)}{2}-1}\mathrm{Det}(\vec{L}_n),\end{aligned}$$

and Equation (10.9) in the present case follows. Equation (10.10) follows from this and Equation (10.8). $\square$

The following was formulated by Przytycki [**189**, Theorem 7.1]; see also [**133**]:

PROPOSITION 10.1.12. *Let* (L, L_1, L_2) *be an unoriented skein triple, consisting of three links all of whose determinants are non-zero. Then, the conditions* (D-1) *and* (D-2) *below are equivalent:*

(D-1) $\det(L) = \det(L_1) + \det(L_2)$
(D-2) The following two equations hold:

$$\sigma(\vec{L}) + \frac{\mathrm{wr}(\vec{L})}{2} = \sigma(\vec{L}_1) + \frac{\mathrm{wr}(\vec{L}_1)+1}{2} \tag{10.12}$$

$$\sigma(\vec{L}) + \frac{\mathrm{wr}(\vec{L})}{2} = \sigma(\vec{L}_2) + \frac{\mathrm{wr}(\vec{L}_2)-1}{2}, \tag{10.13}$$

where the writhes are computed using diagrams for L, L_1, *and* L_2 *that are identical outside of a neighborhood of a crossing in* L; *and the orientations are chosen arbitrarily.*

Proof. Note that, by the Gordon-Litherland formula (see Corollary 2.7.10), for any fixed projection of L, $\sigma(\vec{L}) + \frac{\mathrm{wr}(\vec{L})}{2}$ is indepedent of the orientation on L.

Substituting Equation (10.2) into Equation (10.7), we find that

$$i^{\sigma(\vec{L})+\frac{\mathrm{wr}(\vec{L})}{2}}|\mathrm{Det}(\vec{L})| = i^{\sigma(\vec{L}_1)+\frac{\mathrm{wr}(\vec{L}_1)+1}{2}}|\mathrm{Det}(\vec{L}_1)| + i^{\sigma(\vec{L}_2)+\frac{\mathrm{wr}(\vec{L}_2)-1}{2}}|\mathrm{Det}(\vec{L}_2)|. \tag{10.14}$$

Condition (D-2) now immediately implies Condition (D-1).

Conversely, assume Condition (D-1). It follows from Equation (10.14) that

$$\sigma(\vec{L}) + \frac{\mathrm{wr}(\vec{L})}{2} \equiv \sigma(\vec{L}_1) + \frac{\mathrm{wr}(\vec{L}_1)+1}{2} \pmod 4 \tag{10.15}$$

$$\sigma(\vec{L}) + \frac{\mathrm{wr}(\vec{L})}{2} \equiv \sigma(\vec{L}_2) + \frac{\mathrm{wr}(\vec{L}_2)-1}{2} \pmod 4. \tag{10.16}$$

There are once again two cases, according to whether or not the two strands meeting at the distinguished crossing of L belong to the same component.

Suppose that the two strands of L belong to different components. There are two different orientations on L, written $\vec{L}$ and $\vec{R}$, where the distinguished crossing is positive and negative respectively, whose oriented resolutions $\vec{L}_0$ and $\vec{R}_0$ are $\vec{L}_2$ and $\vec{L}_1$ respectively; see Figure 10.4. In particular, $|\mathrm{wr}(\vec{L}) - \mathrm{wr}(\vec{L}_2)| = 1$ and $|\sigma(\vec{L}) - \sigma(\vec{L}_2)| \leq 1$, according to Lemma 2.6.12. This, together with (10.16) implies (10.13). A similar argument shows that $|\mathrm{wr}(\vec{R}) - \mathrm{wr}(\vec{L}_1)| = 1$ and $|\sigma(\vec{R}) - \sigma(\vec{L}_1)| \leq 1$. Since

$$\sigma(\vec{R}) + \frac{\mathrm{wr}(\vec{R})}{2} = \sigma(\vec{L}) + \frac{\mathrm{wr}(\vec{L})}{2},$$

Equation (10.15) implies Equation (10.12).

Suppose that the two strands of L belong to the same component, and suppose that the crossing is positive (for any orientation on L). Choose an orientation $\vec{L}$ on L and let $\vec{L}_0$ be the oriented resolution on $\vec{L}$. Changing the orientation of one of the two strands in $\vec{L}_0$ gives a different orientation $\vec{L}_2$ with the property that $\vec{L}_1$ can be obtained by adding an oriented saddle to $\vec{L}_2$, and $\mathrm{wr}(\vec{L}_1) = \mathrm{wr}(\vec{L}_2)$; see Figure 10.6. Lemma 2.6.12 shows that $|\sigma(\vec{L}) - \sigma(\vec{L}_0)| \leq 1$; and clearly $|\mathrm{wr}(\vec{L}) - \mathrm{wr}(\vec{L}_0)| = 1$.

FIGURE 10.6. **Changing σ when the two strands belong to the same component.** This is the case where the crossing in $\vec{L} = \vec{L}_+$ is positive, so $\vec{L}_2$ is a different orientation on the oriented resolution $\vec{L}_0$. Pictures where the crossing is negative are similar, except then $\vec{L}_1$ and $\vec{L}_2$ are reversed.

Since

$$\sigma(\vec{L}_0) + \frac{\mathrm{wr}(\vec{L}_0)}{2} = \sigma(\vec{L}_2) + \frac{\mathrm{wr}(\vec{L}_2)}{2},$$

we conclude from Equation (10.16) that Equation (10.13) holds.

Again, by Lemma 2.6.12, $|\sigma(\vec{L}_1) - \sigma(\vec{L}_2)| \leq 1$; also, $\mathrm{wr}(\vec{L}_1) = \mathrm{wr}(\vec{L}_2)$. Thus,

$$\sigma(\vec{L}_1) + \frac{\mathrm{wr}(\vec{L}_1)+1}{2} \equiv \sigma(\vec{L}_2) + \frac{\mathrm{wr}(\vec{L}_2)-1}{2} \pmod 4$$

(which follows from Equations (10.15) and (10.16)) implies that

$$\sigma(\vec{L}_1) + \frac{\mathrm{wr}(\vec{L}_1)+1}{2} = \sigma(\vec{L}_2) + \frac{\mathrm{wr}(\vec{L}_2)-1}{2},$$

which, together with Equation (10.13), gives Equation (10.12).

The case where the distinguished crossing is negative follows similarly. □

THEOREM 10.1.13. *If L is link that admits a connected, alternating projection, then L is quasi-alternating.*

Proof. The proof proceeds by induction on the number of crossings in a connected, alternating diagram $\mathcal{D}$ of L. If $\mathcal{D}$ has no crossings, then it represents the unknot, which is quasi-alternating.

Consider a crossing in $\mathcal{D}$. Resolve it in two ways to get diagrams $\mathcal{D}_0$ and $\mathcal{D}_n$ for L_0 and L_n. Although both $\mathcal{D}_0$ and $\mathcal{D}_n$ are alternating, one of the projections might be disconnected. In this case, the chosen crossing is called *nugatory*: it can be eliminated by untwisting the knot projection to get a new projection $\mathcal{D}'$. (For an example of a nugatory crossing, see Figure 10.7.) It is straightforward to verify that the new projection of $\mathcal{D}'$ is also a connected, alternating projection, it has one fewer crossing, so the induction hypothesis applies.

When the chosen crossing is not nugatory, both $\mathcal{D}_0$ and $\mathcal{D}_n$ are connected, alternating projections, and have fewer crossings than $\mathcal{D}$. By the inductive hypothesis, $\mathcal{D}_0$ and $\mathcal{D}_n$ represent quasi-alternating links. It remains to verify that

$$\det(L) = \det(L_0) + \det(L_n). \tag{10.17}$$

Suppose that the crossing is positive for some orientation $\vec{L}_+$ on L. Fix a chessboard coloring compatible with the alternating projection, as illustrated in Figure 2.23(b).

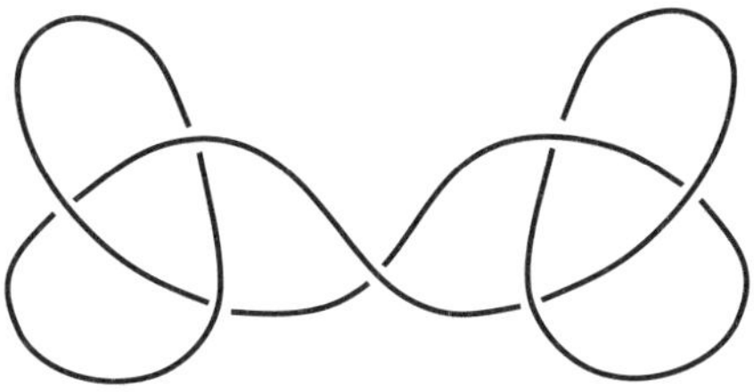

FIGURE 10.7. **A nugatory crossing.** The knot in the picture is the connected sum of two trefoils, and the diagram contains a crossing that can be eliminated by twisting one of the trefoils.

The Gordon-Litherland formula (Equation (2.10)) gives:

$$\sigma(\vec{L}_+) = \sigma(\vec{L}_0) - 1 \tag{10.18}$$

$$\sigma(\vec{L}_+) = \sigma(\vec{L}_n) - \frac{1}{2}(\mathrm{wr}(\vec{L}_0) - \mathrm{wr}(\vec{L}_n)), \tag{10.19}$$

where $\vec{L}_0$ is the oriented resolution of $\vec{L}_+$, and $\vec{L}_n$ is oriented arbitrarily. Since, $\mathrm{wr}(\vec{L}_+) = \mathrm{wr}(\vec{L}_0) + 1$, the above two computations of the signature are equivalent to Condition (D-2) from Proposition 10.1.12; so that proposition verifies Equation (10.17).

The case where the crossing is negative for some orientation $\vec{L}_-$ on L follows similarly, except that now the Gordon-Litherland formula gives

$$\sigma(\vec{L}_-) = \sigma(\vec{L}_0) + 1$$

$$\sigma(\vec{L}_-) = \sigma(\vec{L}_n) - \frac{1}{2}(\mathrm{wr}(\vec{L}_0) - \mathrm{wr}(\vec{L}_n)),$$

which also verifies Condition (D-2) from Proposition 10.1.12. □

EXERCISE 10.1.14. Let L be a link with a connected, alternating diagram $\mathcal{D}$. The "black graph" of $\mathcal{D}$ is the planar graph whose vertices correspond to the black regions in a chessboard coloring of $\mathcal{D}$, and whose edges correspond to crossings in $\mathcal{D}$: for each crossing in $\mathcal{D}$, the corresponding edge in the black graph connects the two vertices corresponding to the two black regions adjacent to the crossing. Show that $\det(L)$ is the number of spanning trees in the black graph.

By Theorem 10.1.13, all alternating knots are quasi-alternating. Figure 10.8 shows a non-alternating but quasi-alternating knot. Links L with disconnected projection have $\det(L) = 0$ and so are not quasi-alternating, according to the following:

LEMMA 10.1.15. *For any quasi-alternating link L,* $\det(L) \neq 0$. *Moreover* $\det(L) = 1$ *if and only if L is the unknot.*

Proof. We inductively define the *height* of a quasi-alternating link L, denoted $h(L)$, as follows. If L is the unknot, let $h(L) = 1$. Otherwise the height of L is the minimum of $\max(h(L_1), h(L_2)) + 1$, over all triples (L, L_1, L_2) where L_1 and L_2 are two quasi-alternating resolutions of L with $\det(L) = \det(L_1) + \det(L_2)$. By induction on the height, it follows readily that $|h(L)| \leq \det(L)$. The lemma is a simple consequence of this inequality. □

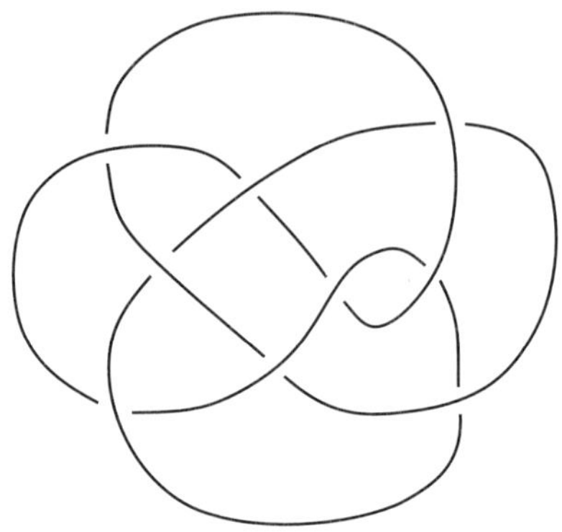

FIGURE 10.8. **A quasi-alternating, but non-alternating knot.** This is 9_{47} in the standard knot tables.

EXERCISE 10.1.16. Verify that the knot 9_{47} of Figure 10.8 is quasi-alternating.

10.2. The unoriented skein exact sequence

The aim of this section is to set up, state, and prove the promised unoriented skein exact sequence. Consider the chain complex $\widetilde{GC}(\mathbb{G})$, equipped with its single $\mathbb{Z}$-grading $\delta = M - A$ from Equation (10.1). Write the corresponding splitting

$$\widetilde{GC}(\mathbb{G}) = \bigoplus_{d\in\mathbb{Z}} \widetilde{GC}_d(\mathbb{G}),$$

where $\widetilde{GC}_d(\mathbb{G})$ is spanned by grid states with δ-grading equal to d. The differential

$$\tilde{\partial}_{\mathbb{O},\mathbb{X}}(\mathbf{x}) = \sum_{\mathbf{y}\in\mathbf{S}(\mathbb{G})} \#\{r \in \mathrm{Rect}^\circ(\mathbf{x},\mathbf{y}) \big| r \cap \mathbb{O} = r \cap \mathbb{X} = \emptyset\} \cdot \mathbf{y}$$

treats the two kinds of markings in the same way, so that the ungraded chain complex $\widetilde{GC}(\mathbb{G})$ depends only on the underlying unoriented link specified by $\mathbb{G}$. The δ-grading, however, depends on an orientation, according to the following:

PROPOSITION 10.2.1. *Let $\mathbb{G}$ be a toroidal grid diagram and $\mathbb{G}'$ another one obtained by switching some of the X- and O-markings in $\mathbb{G}$; let G and G' be planar realizations of $\mathbb{G}$ and $\mathbb{G}'$, corresponding to the same fundamental domain. Letting $c = \frac{1}{4}(\mathrm{wr}(G) - \mathrm{wr}(G'))$, there is an isomorphism*

$$\widetilde{GH}_\delta(\mathbb{G}) \cong \widetilde{GH}_{\delta+c}(\mathbb{G}').$$

Proof. The identification of grid states for $\mathbb{G}$ and for $\mathbb{G}'$ induces an isomorphism $\widetilde{GC}(\mathbb{G}) \to \widetilde{GC}(\mathbb{G}')$ of vector spaces. Since $\tilde{\partial}_{\mathbb{O},\mathbb{X}}$ treats both marking types the same way, this identification determines a chain map, inducing an isomorphism of chain complexes. To complete the proof, it remains to compare the δ-gradings of grid states induced from $\mathbb{G}$ and $\mathbb{G}'$; but this comparison was done in Lemma 10.1.7. □

The simply blocked grid complex for links $\widehat{GH}$ (see Definitions 4.6.12 and 8.2.7) can also be viewed as graded by $\delta = M - A$; correspondingly, its homology is a $\mathbb{Z}$-graded vector space. This vector space is related to $\widetilde{GH}$ with its δ-grading. To state the relationship, note that if $X = \bigoplus_{d\in\mathbb{Z}} X_d$ is a $\mathbb{Z}$-graded vector space and W is the two-dimensional $\mathbb{Z}$-graded vector space supported in grading 0, we can form the $\mathbb{Z}$-graded tensor product $X\otimes W = \bigoplus_{d\in\mathbb{Z}}(X\otimes W)_d$, where $(X\otimes W)_d = X_d \oplus X_d$.

PROPOSITION 10.2.2. *Let W be the two-dimensional vector space supported in $\mathbb{Z}$-grading equal to 0. If $\mathbb{G}$ is a grid diagram for $\vec{L}$, the δ-graded vector space $\widetilde{GH}(\mathbb{G})$ is related to the δ-graded link invariant by $\widetilde{GH}(\mathbb{G}) \cong \widehat{GH}(\vec{L}) \otimes W^{\otimes(n-\ell)}$.*

Proof. This follows from Proposition 4.6.15, since the two-dimensional bigraded vector space W appearing there is supported in δ-grading equal to zero. □

PROPOSITION 10.2.3. *Let $\mathbb{G}$ be a grid diagram with grid number n, representing the oriented link $\vec{L}$. Then, the Euler characteristic of $\widetilde{GH}(\mathbb{G})$, thought of as a $\mathbb{Z}$-graded vector space with grading δ, is given by $\chi(\widetilde{GH}(\mathbb{G})) = 2^{n-1}(-i)^{\ell-1}\mathrm{Det}(\vec{L})$.*

Proof. Since $\widetilde{GC}(\mathbb{G})$ is finite-dimensional,

$$\chi(\widetilde{GH}(\mathbb{G})) = \chi(\widetilde{GC}(\mathbb{G})) = \sum_{\mathbf{x}\in\mathbf{S}(\mathbb{G})} (-1)^{\delta(\mathbf{x})},$$

so the result follows from Lemma 10.1.2. □

THEOREM 10.2.4. *Let L, L_1, and L_2 be an unoriented skein triple, and let ℓ, ℓ_1, and ℓ_2 denote the number of components of L, L_1, and L_2 respectively. Then, for sufficiently large m, there is an exact triangle*

$$\begin{array}{ccc}
\widehat{GH}(L_1)\otimes W^{\otimes(m-\ell_1)} & \longrightarrow & \widehat{GH}(L_2)\otimes W^{\otimes(m-\ell_2)} \\
\nwarrow & & \swarrow \\
& \widehat{GH}(L)\otimes W^{\otimes(m-\ell)} &
\end{array} \tag{10.20}$$

where W is the two-dimensional $\mathbb{Z}$-graded vector space supported in grading zero, and $\widehat{GH}$ is thought of as $\mathbb{Z}$-graded, using δ. Let $L_3 = L$, fix projections of L, L_1, and L_2 that differ only around the fixed crossing of L, and fix orientations on L, L_1, and L_2 arbitrarily. Then, an arrow connecting the grid homology on L_i to L_j (where $j \equiv i+1 \pmod 3$) shifts grading by

$$\left(\frac{\ell_i - \ell_{i+1}}{2}\right) + \left(\frac{w_i - w_{i+1}}{4}\right) + \begin{cases} -\frac{1}{2} & \text{if } i = 1 \\ -\frac{1}{4} & \text{otherwise,} \end{cases} \tag{10.21}$$

where w_i is the writhe of the fixed projection of L_i ($i = 1,2,3$).

The grading shifts in Equation (10.20) can be alternatively expressed as follows:
(10.22)

$$\begin{array}{ccc} \widehat{GH}(L_1)\otimes W^{\otimes(m-\ell_1)}[\![-\frac{\ell_1}{2}-\frac{w_1+1}{4}]\!] & \xrightarrow{-1} & \widehat{GH}(L_2)\otimes W^{\otimes(m-\ell_2)}[\![-\frac{\ell_2}{2}-\frac{w_2-1}{4}]\!] \\ & \nwarrow \quad \swarrow & \\ & \widehat{GH}(L)\otimes W^{\otimes(m-\ell)}[\![-\frac{\ell}{2}-\frac{w}{4}]\!] & \end{array}$$

where the two unlabelled arrows preserve δ-grading, and the third arrow is homogeneous of degree -1. In the above triangle, we are using a shift operator for graded vector spaces defined analogously to the bigraded case: given a graded vector space $X=\bigoplus_d X_d$ and a number a, $X[\![a]\!]=\bigoplus_d X[\![a]\!]_d$ is the graded vector space with $X[\![a]\!]_d=X_{a+d}$. In the above triangle, the vector spaces are $\mathbb{Z}$-graded, but the shifts can be rational numbers.

REMARK 10.2.5. Let $\mathbb{G}$ be a grid diagram for $\vec{L}$, and let G be some planar realization of $\mathbb{G}$. It follows from Lemma 10.1.7 that the δ-graded vector space $\widetilde{GH}(\mathbb{G})[\![-\frac{\ell}{2}-\frac{\mathrm{wr}(G)}{4}]\!]$ (essentially the one appearing in the above triangle) is independent of the chosen orientation of $\vec{L}$.

The proof of Theorem 10.2.4 uses the following specialization of Theorem 9.2.1 (specialized to $\widetilde{GH}$ with its δ-grading):

THEOREM 10.2.6. *Let $\vec{L}_+$, $\vec{L}_-$, and $\vec{L}_0$ be an oriented skein triple, let ℓ denote the number of components in $\vec{L}_+$ (and $\vec{L}_-$); and let ℓ_0 denote the number of components of $\vec{L}_0$. Then, there is a chain map*

$$\widetilde{P}_{+,-}\colon \widetilde{GC}(\mathbb{G}_+)\to\widetilde{GC}(\mathbb{G}_-)$$

that preserves δ-gradings, and there is a quasi-isomorphism
(10.23)

$$\mathrm{Cone}\big(\widetilde{P}_{+,-}\colon \widetilde{GC}(\mathbb{G}_+)\to\widetilde{GC}(\mathbb{G}_-)\big)\to\Big(\widetilde{GC}(\mathbb{G}_0)\oplus\widetilde{GC}(\mathbb{G}_0)\Big)[\![\frac{\ell-\ell_0-1}{2}]\!].$$

Proof. Let $(\mathbb{G}_+,\mathbb{G}_-,\mathbb{G}_0)$ be a grid realization of the skein triple $(\vec{L}_+,\vec{L}_-,\vec{L}_0)$, with the property that the two distinguished X-markings are in consecutive rows. The map $\widetilde{P}_{+,-}\colon\widetilde{GC}(\mathbb{G}_+)\to\widetilde{GC}(\mathbb{G}_-)$ is obtained from the map $P_{+,-}$ of the proof of Theorem 9.2.1 by specializing $V_1=\cdots=V_n=0$. This is the map determined by the horizontal maps in Diagram (9.5); i.e., it is defined as a count of rectangles that cross none of the O- and X-markings. Theorem 9.2.1 states a quasi-isomorphism between two mapping cones. Specializing those mapping cones to $V_1=\cdots=V_n=0$ and collapsing the bigrading to $\delta=M-A$ gives the stated version. □

We separate two cases of Theorem 10.2.4, according to whether or not the strands crossing in L at the distinguished crossing belongs to two different components of L. Consider first the case where they belong to different components. Choose two different orientations $\vec{L}_+$ and $\vec{R}_-$ on L: for $\vec{L}_+$ the distinguished crossing is positive and for $\vec{R}_-$, the distinguished crossing is negative. Observe that $\vec{L}_1$ and $\vec{L}_2$ are the oriented resolutions of $\vec{R}_-$ and $\vec{L}_+$ respectively. We will also use the link $\vec{R}_+$, obtained from $\vec{R}_-$ by a crossing change. (See the top line of Figure 10.9.)

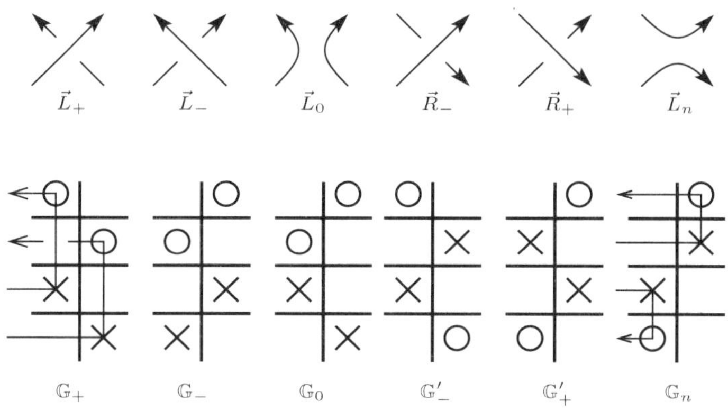

FIGURE 10.9. **Grid diagrams for an unoriented resolution.** The three grids in the left provide a grid realization $(\mathbb{G}_+, \mathbb{G}_-, \mathbb{G}_0)$ of the oriented skein triple $(\vec{L}_+, \vec{L}_-, \vec{L}_0)$. The markings in $\mathbb{G}_+$ are numbered by O_3 being the top, then O_4 and X_1, X_2. The three grids on the right describe a grid realization of the skein triple $(\vec{R}_-, \vec{R}_+, \vec{R}_0)$. The diagrams $\mathbb{G}'_\pm$ are derived from $\mathbb{G}_\mp$ by switching O_4 and X_2.

Let $(\mathbb{G}_+, \mathbb{G}_-, \mathbb{G}_0)$ be a grid realization of the skein triple $(\vec{L}_+, \vec{L}_-, \vec{L}_0)$. Arrange that the distinguished markings O_4, X_1, and X_2 are in consecutive rows. A grid realization $(\mathbb{G}'_+, \mathbb{G}'_-, \mathbb{G}_n)$ of the skein triple $(\vec{R}_+, \vec{R}_-, \vec{R}_0)$ can be given as follows: construct the grids $\mathbb{G}'_+$ and $\mathbb{G}'_-$ from $\mathbb{G}_-$ and $\mathbb{G}_+$ respectively by exchanging some X- and O-markings, as illustrated in Figure 10.9. In fact, these exchanges are done so that the X_2- and O_4-markings in both $\mathbb{G}_+$ and $\mathbb{G}_-$ are exchanged to O- and X-markings respectively in $\mathbb{G}'_-$ and $\mathbb{G}'_+$.

The fact that X_1 and X_2 are in consecutive rows ensures that the map

$$\widetilde{P}_{+,-}\colon \widetilde{GC}(\mathbb{G}_+) \to \widetilde{GC}(\mathbb{G}_-)$$

appearing in Theorem 10.2.6 is defined. Similarly, the fact that O_4 and X_1 are in consecutive rows ensures that the analogous map

$$\widetilde{P}'_{+,-}\colon \widetilde{GC}(\mathbb{G}'_+) \to \widetilde{GC}(\mathbb{G}'_-)$$

from Theorem 10.2.6, now for the skein triple $(\vec{R}_+, \vec{R}_-, \vec{R}_0)$, is defined.

Define

$$\widetilde{P}_{-,+}\colon \widetilde{GC}(\mathbb{G}_-) \to \widetilde{GC}(\mathbb{G}_+)$$

to be the composition

$$\widetilde{GC}(\mathbb{G}_-) \to \widetilde{GC}(\mathbb{G}'_+) \xrightarrow{\widetilde{P}'_{+,-}} \widetilde{GC}(\mathbb{G}'_-) \to \widetilde{GC}(\mathbb{G}_+), \tag{10.24}$$

where the unlabeled maps are isomorphisms induced by relabeling X- and O-markings, appearing in the proof of Proposition 10.2.1. By Proposition 10.2.1, $\widetilde{P}_{-,+}$ is a chain map that drops δ-grading by one.

LEMMA 10.2.7. *Let $(\vec{L}_+, \vec{L}_-, \vec{L}_0)$ be an oriented skein triple where the two strands meeting at the crossing belong to different components, and let $(\mathbb{G}_+, \mathbb{G}_-, \mathbb{G}_0)$ be a grid realization of the triple for which O_4, X_1, and X_2 are in consecutive rows.*

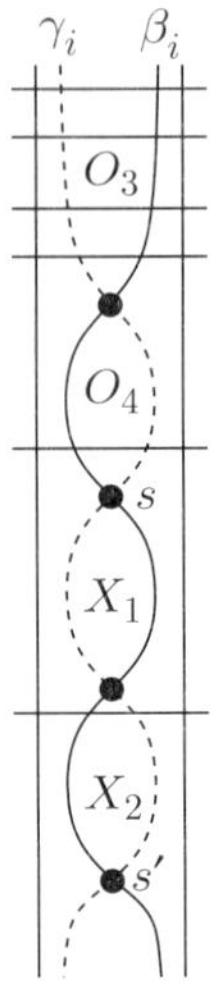

FIGURE 10.10. **Simultaneous realizations of $\mathbb{G}_+$, $\mathbb{G}_-$, $\mathbb{G}'_-$ and $\mathbb{G}'_+$.** A simultaneous representation of $\mathbb{G}_+$ and $\mathbb{G}_-$ is realized by using the markings X_1, X_2, O_3, and O_4, and using β_i and γ_i respectively. A simultaneous representation of $\mathbb{G}'_+$ and $\mathbb{G}'_-$ is obtained by relabeling X_2 and O_4 as O- and X-markings, and using γ_i and β_i respectively.

Let $\widetilde{P}_{+,-}\colon \widetilde{GC}(\mathbb{G}_+) \to \widetilde{GC}(\mathbb{G}_-)$ and $\widetilde{P}_{-,+}\colon \widetilde{GC}(\mathbb{G}_-) \to \widetilde{GC}(\mathbb{G}_+)$ be the maps as above. Then, $\widetilde{P}_{-,+} \circ \widetilde{P}_{+,-}$ and $\widetilde{P}_{+,-} \circ \widetilde{P}_{-,+}$ are null-homotopic.

Proof. Suppose that $\vec{L}_+$ is a knot. Recall that $\widetilde{P}_{+,-}$ and $\widetilde{P}_{-,+}$ are defined using rectangle counts, which can be identified with pentagon counts in a simultaneous drawing of $\mathbb{G}_+$ and $\mathbb{G}_-$ as in Chapter 6 (see also Proposition 9.6.1). In fact, we draw the grid diagrams $\mathbb{G}_+$ and $\mathbb{G}_-$ simultaneously on the same grid torus so that the bigons containing X_2 and O_4 each meet exactly one horizontal circle, and the bigon containing X_1 meets no horizontal circle; see Figure 10.10.

Let $\widetilde{c}_-$ be the count of pentagons based at s appearing in Equations (6.3) and (6.4) specialized to $\widetilde{GC}$ (i.e. setting $V_1 = \cdots = V_n = 0$), and let $\widetilde{c}'_-$ be the count of pentagons based at s'. Since the bigons containing X_1 and X_2 meet only one horizontal circle in Figure 10.10, we can apply Proposition 9.6.1 to conclude that $\widetilde{P}_{+,-} = \widetilde{c}_- + \widetilde{c}'_-$. Since the bigons containing X_2 and O_4 each meet only one horizontal circle, Proposition 9.6.1 also identifies $\widetilde{P}'_{+,-}$ with a count of pentagons that are based at t and t' (in Figure 10.10). Reformulated in terms of $\mathbb{G}_-$ and $\mathbb{G}_+$, this gives $\widetilde{P}_{-,+} = \widetilde{c}_+ + \widetilde{c}'_+$.

In the proof of Proposition 6.1.1, it is shown that $c_+ \circ c_-$ is homotopic to multiplication by some U variable. On the specialization to $\widetilde{GC}$, this shows that $\widetilde{c}_+ \circ \widetilde{c}_-$ is null-homotopic. It follows from the same argument that the other compositions $\widetilde{c}'_+ \circ \widetilde{c}_-$, $\widetilde{c}_+ \circ \widetilde{c}'_-$, and $\widetilde{c}'_+ \circ \widetilde{c}'_-$ are also null-homotopic. We conclude that $\widetilde{P}_{-,+} \circ \widetilde{P}_{+,-}$ is null-homotopic.

When $\vec{L}_+$ has more than one component, the same argument works. In particular, the maps $\widetilde{c}_-$, $\widetilde{c}_+$, $\widetilde{c}'_-$, and $\widetilde{c}'_+$, in cases where the grid diagrams represent links with more than one component, can be defined exactly as before. The identification $\widehat{P}_{-,+} = \widetilde{c}_+ + \widetilde{c}'_+$, from Proposition 9.6.1 is proved the same way, as is the verifications that the four maps $\widetilde{c}_+ \circ \widetilde{c}_-$, $\widetilde{c}'_+ \circ \widetilde{c}_-$, $\widetilde{c}_+ \circ \widetilde{c}'_-$, and $\widetilde{c}'_+ \circ \widetilde{c}'_-$ are null-homotopic.

The other composition $\widetilde{P}_{+,-} \circ \widetilde{P}_{-,+}$ follows similarly. □

The proof of Theorem 10.2.4 will use the following algebraic construction. Let $f\colon X \to Y$ and $g\colon Y \to X$ be two chain maps between the $\mathbb{Z}$-graded chain complexes X and Y, and suppose that f is homogeneous of degree a (i.e. f maps elements of degree d to elements of degree $d + a$) and g is homogeneous of degree b. Then, the mapping cones of f, g and $f \circ g$ fit into the following exact triangle (proved in Lemma A.3.10):

LEMMA 10.2.8. *Suppose that the maps f and g are given as above. Then, there is a chain map $\Phi\colon \mathrm{Cone}(f) \to \mathrm{Cone}(g)$ which is homogeneous of degree $-a-1$ and whose induced map on homology fits into an exact triangle*

$$
\begin{array}{ccc}
H(\mathrm{Cone}(f)) & \xrightarrow{\;-a-1\;} & H(\mathrm{Cone}(g)) \\
 & \nwarrow \qquad \swarrow_{a} & \\
 & H(\mathrm{Cone}(f \circ g)) &
\end{array}
\tag{10.25}
$$

where the integers next to the arrows indicate shifts in degree. □

Proof of Theorem 10.2.4. As usual, there are two cases, according to whether or not the two strands in L correspond to different components.

We consider first the case where they belong to different components. As noted earlier (Remark 10.2.5), the triangle is independent of the choice of orientation on L, so without loss of generality, we can assume that the crossing is positive; i.e. $\vec{L} = \vec{L}_+$. Take a grid realization $\mathbb{G}_+$, $\mathbb{G}_-$, $\mathbb{G}_0$ of the skein triple $(\vec{L}_+, \vec{L}_-, \vec{L}_0)$, and let $(\mathbb{G}'_+, \mathbb{G}'_-, \mathbb{G}_n)$ be the grid realization of the skein triple $(\vec{R}_+, \vec{R}_-, \vec{R}_0 = \vec{L}_n)$, where $\mathbb{G}'_+$ and $\mathbb{G}'_-$ are derived from $\mathbb{G}_-$ and $\mathbb{G}_+$ by relabeling some X- and O-marked squares.

Consider the maps $\widetilde{P}_{+,-}\colon \widetilde{GC}(\mathbb{G}_+) \to \widetilde{GC}(\mathbb{G}_-)$ and $\widetilde{P}_{-,+}\colon \widetilde{GC}(\mathbb{G}_-) \to \widetilde{GC}(\mathbb{G}_+)$ derived from Theorem 10.2.6 as above. Recall that the map $\widetilde{P}_{+,-}$ preserves δ-grading and $\widetilde{P}_{-,+}$ drops it by one. According to Lemma 10.2.8, the mapping cones of these maps fit into an exact triangle

$$
\begin{array}{ccc}
H(\mathrm{Cone}(\widetilde{P}_{-,+})) & \longrightarrow & H(\mathrm{Cone}(\widetilde{P}_{+,-})) \\
 & \nwarrow \qquad \swarrow_{-1} & \\
 & H(\mathrm{Cone}(\widetilde{P}_{-,+} \circ \widetilde{P}_{+,-})) &
\end{array}
\tag{10.26}
$$

(where the arrow marked with -1 shifts grading down by one).

The desired triangle is obtained by identifying the homology groups appearing in this triangle in terms of the grid homology groups of $\mathbb{G}_+$, $\mathbb{G}_n$, and $\mathbb{G}_0$, as follows.

Theorem 10.2.6 gives quasi-isomorphisms

$$\mathrm{Cone}(\widetilde{P}_{+,-}) \to \left(\widetilde{GC}(\mathbb{G}_0) \oplus \widetilde{GC}(\mathbb{G}_0)\right) \left[\!\!\left[\frac{\ell - \ell_0 - 1}{2}\right]\!\!\right],$$

$$\mathrm{Cone}(\widetilde{P}'_{+,-}) \to \left(\widetilde{GC}(\mathbb{G}_n) \oplus \widetilde{GC}(\mathbb{G}_n)\right) \left[\!\!\left[\frac{\ell - \ell_n - 1}{2}\right]\!\!\right].$$

Let $w_n = \mathrm{wr}(\mathbb{G}_n)$ and $w_0 = \mathrm{wr}(\mathbb{G}_0)$. Clearly, $\mathrm{wr}(\mathbb{G}_+) - \mathrm{wr}(\mathbb{G}'_-) = w_0 - w_n + 2$; so by Proposition 10.2.1 and Equation (10.24),

$$\mathrm{Cone}(\widetilde{P}_{-,+}) \cong \mathrm{Cone}(\widetilde{P}'_{+,-}) \left[\!\!\left[\frac{w_0 - w_n + 2}{4}\right]\!\!\right].$$

According to Lemma 10.2.7, the composition $\widetilde{P}_{-,+} \circ \widetilde{P}_{+,-}$ is null-homotopic; so Lemma 5.2.14 gives an isomorphism of chain complexes

$$\mathrm{Cone}\big(\widetilde{P}_{-,+} \circ \widetilde{P}_{+,-} \colon \widetilde{GC}(\mathbb{G}_+) \to \widetilde{GC}(\mathbb{G}_+)\big) \cong \widetilde{GC}(\mathbb{G}_+) \oplus \widetilde{GC}(\mathbb{G}_+).$$

Since $\widetilde{P}_{-,+}$ drops δ-grading by one and $\widetilde{P}_{+,-}$ preserves it, the above homotopy equivalence preserves the δ-grading.

Substituting the identifications on homology given by the above quasi-isomorphisms into the triangle from Equation (10.26), we get an exact triangle:

$$\begin{array}{ccc} \widetilde{GH}(\mathbb{G}_n) \otimes W[\![\frac{\ell-\ell_n}{2} + \frac{w_0 - w_n}{4}]\!] & \longrightarrow & \widetilde{GH}(\mathbb{G}_0) \otimes W[\![\frac{\ell-\ell_0-1}{2}]\!] \\ \nwarrow & & \swarrow {\scriptstyle -1} \\ & \widetilde{GH}(\mathbb{G}_+) \otimes W & \end{array}$$

Rearrange the degree shifts and use $w_+ - w_0 = 1$, to get the exact triangle

$$\begin{array}{ccc} \widetilde{GH}(\mathbb{G}_n) \otimes W[\![- \left(\frac{w_n+2\ell_n+1}{4}\right)]\!] & \longrightarrow & \widetilde{GH}(\mathbb{G}_0) \otimes W[\![- \left(\frac{w_0+2\ell_0+3}{4}\right)]\!] \\ \nwarrow & & \swarrow {\scriptstyle -1} \\ & \widetilde{GH}(\mathbb{G}_+) \otimes W[\![- \left(\frac{w_+ + 2\ell}{4}\right)]\!] & \end{array}$$

By Proposition 8.2.8, $\widetilde{GH}(\mathbb{G}) \cong \widehat{GH}(\mathbb{G}) \otimes W^{\otimes(n-\ell)}$ for any grid diagram $\mathbb{G}$. Substituting this back into the above triangle and readjusting the degree shifts gives the exact triangle from Equation (10.20) with its stated grading shifts (see especially Equation (10.22)), when the two strands of L belong to different components.

The case where the two strands in L belong to the same component can be reduced to the previous case, as follows. Suppose $\vec{L} = \vec{L}_+$, and let $\vec{L}'_-$ be obtained by adding an extra pair of crossings in the projection of $\vec{L}_0$ (see Figure 10.5), let $\vec{L}'_0$ be the oriented resolution of the newly formed negative intersection point, and let $\vec{L}'_n$ be be its unoriented resolution. The previous case gives a triangle

$$\begin{array}{ccc} \widehat{GH}(\vec{L}'_0) \otimes W^{\otimes(m-\ell'_0)}[\![-\frac{\ell'_0}{2} - \frac{w'_0+1}{4}]\!] & \xrightarrow{\;-1\;} & \widehat{GH}(\vec{L}'_n) \otimes W^{\otimes(m-\ell'_n)}[\![-\frac{\ell'_n}{2} - \frac{w'_n-1}{4}]\!] \\ \nwarrow & & \swarrow \\ & \widehat{GH}(\vec{L}'_-) \otimes W^{\otimes(m-\ell)}[\![-\frac{\ell'}{2} - \frac{w'}{4}]\!] & \end{array}$$

where $w' = \mathrm{wr}(\vec{L}'_-)$, $w'_0 = \mathrm{wr}(\vec{L}'_0)$, and $w'_n = \mathrm{wr}(\vec{L}'_n)$. Observe that $\vec{L}'_- = \vec{L}_0$, $\vec{L}'_0 = \vec{L}_+$, and $\vec{L}'_n = \vec{L}_n$, $w' = w_0$, $w'_0 = w$, and $w'_n = w_n - 1$. Substituting these back into the above triangle, we get the stated triangle with an overall shift. (Compare Remark 10.2.5.) The case where $\vec{L} = \vec{L}_-$ follows similarly. □

Equation (10.20) implies the following relation on the graded Euler characteristic:

$$i^{\ell+\frac{w}{2}}\chi(\widetilde{GH}(\mathbb{G})) = i^{\ell_1+\frac{w_1+1}{2}}\chi(\widetilde{GH}(\mathbb{G}_1)) + i^{\ell_2+\frac{w_2-1}{2}}\chi(\widetilde{GH}(\mathbb{G}_2)).$$

In view of Proposition 10.2.3, this formula is equivalent to the additivity formula for Det from Proposition 10.1.11.

The grading shifts in Theorem 10.2.4 can be interpreted in terms of signatures:

PROPOSITION 10.2.9. *Let (L, L_1, L_2) be an unoriented skein triple with* $\det(L) = \det(L_1) + \det(L_2)$. *Let L be denoted by L_3 and suppose that* $\det(L_1)$, $\det(L_2)$, *and* $\det(L_3)$ *are all non-zero. Then, in the exact triangle from Theorem 10.2.4, the arrow connecting the grid homology of L_i to L_{i+1} (when considering the indices mod 3) shifts grading by*

$$\left(\frac{\ell_i - \sigma(L_i)}{2}\right) - \left(\frac{\ell_{i+1} - \sigma(L_{i+1})}{2}\right) + \begin{cases} -1 & \text{if } i = 1 \\ 0 & \text{otherwise.} \end{cases} \tag{10.27}$$

Proof. Since Equation (D-1) from Proposition 10.1.12 is satisfied, we use that proposition to conclude that

$$\sigma(\vec{L}) + \frac{\mathrm{wr}(\vec{L})}{2} = \sigma(\vec{L}_1) + \frac{\mathrm{wr}(\vec{L}_1)+1}{2}$$
$$\sigma(\vec{L}) + \frac{\mathrm{wr}(\vec{L})}{2} = \sigma(\vec{L}_2) + \frac{\mathrm{wr}(\vec{L}_2)-1}{2}.$$

The stated equations now follow immediately from substituting the above equations into the formulas given in Equation (10.21). □

REMARK 10.2.10. Proposition 10.2.9 also suggests that we could consider the more invariantly formulated version of the δ-graded grid homology $\widehat{GH}(L)$, defined by $\widehat{GH}(\vec{L})[\![\frac{\sigma(\vec{L})-\ell+1}{2}]\!]$. In view of Lemma 10.1.7 and Corollary 2.7.10, this graded vector space is an invariant of the unoriented link L. Its gradings are either integers or half-integers, depending on the link.

10.3. Grid homology groups for alternating knots

We can now compute all version of grid homology for alternating knots.

THEOREM 10.3.1 ([**169**]). *If K is an alternating knot, then the simply blocked grid homology of K is determined by the signature $\sigma = \sigma(K)$ of K (from Definition 2.3.3) and the symmetrized Alexander polynomial $\Delta_K(t) = \sum_i a_i \cdot t^i$ by*

$$\widehat{GH}_d(K, s) = \begin{cases} \mathbb{F}^{|a_s|} & \text{if } d = s + \frac{\sigma}{2} \\ 0 & \text{otherwise.} \end{cases}$$

Theorem 10.3.1 is a special case of a more general result proved below. Before turning to the generalization, we give an immediate consequence. Theorem 10.3.1 in fact determines also GH^- for alternating knots. In the description, we abbreviate $\mathbb{F}_{(d,s)} = (\mathbb{F}[U]/U)_{(d,s)}$.

COROLLARY 10.3.2. *Let K be an alternating knot with with signature $\sigma = \sigma(K)$, and symmetrized Alexander polynomial $\Delta_K(t) = \sum_i a_i \cdot t^i$, and let $b_i = \sum_{j\geq 0} a_{i+j}$. The grid homology module $GH^-(K)$ is given by*

$$GH^-(K) \cong \left(\bigoplus_{i>\frac{\sigma}{2}} \mathbb{F}^{|b_i|}_{(i+\frac{\sigma}{2},i)}\right) \oplus \left(\bigoplus_{i\leq\frac{\sigma}{2}} \mathbb{F}^{|b_i-1|}_{(i+\frac{\sigma}{2},i)}\right) \oplus \mathbb{F}[U]_{(\sigma,\frac{\sigma}{2})}.$$

In particular, $\tau(K) = -\frac{\sigma(K)}{2}$.

Proof. By Equation (7.6), each summand $(\mathbb{F}[U]/U^{n_i})_{(d_i,s_i)}$ in $GH^-(K)$ contributes $\mathbb{F}_{(d_i,s_i)} \oplus \mathbb{F}_{(d_i-2n_i+1,s_i-n_i)}$ to $\widehat{GH}(K)$, and the difference in δ-gradings of these two elements is $n_i - 1$. Since by Theorem 10.3.1 the simply blocked grid homology $\widehat{GH}$ of an alternating knot is supported in a single δ-grading, it follows that all $n_i = 1$; i.e. $U \cdot \mathrm{Tors}(GH^-(K)) = 0$. In addition, according to Theorem 10.3.1 again, we have that $d_i - s_i = \frac{\sigma(K)}{2}$. The rest of the statement follows quickly from another look at Equation (7.6) and Proposition 7.3.2. □

THEOREM 10.3.3. *Let $\vec{L}$ be an oriented, quasi-alternating link with ℓ components and signature $\sigma = \sigma(\vec{L})$, and write $(t^{\frac{1}{2}} - t^{-\frac{1}{2}})^{\ell-1}\Delta_{\vec{L}}(t) = \sum_i \alpha_i \cdot t^i$. Then,*

$$\widehat{GH}_d(\vec{L}, s) = \begin{cases} \mathbb{F}^{|\alpha_s|} & \text{if } d = s + \frac{\sigma-\ell+1}{2} \\ 0 & \text{otherwise.} \end{cases} \tag{10.28}$$

Proof. We prove that for any quasi-alternating link $\vec{L}$, $\widehat{GH}(\vec{L})$ is supported in δ-grading equal to $\frac{\sigma(\vec{L})-\ell+1}{2}$, by induction on the determinant of $\vec{L}$. In the basic case (where $\det(\vec{L}) = 1$), Lemma 10.1.15 implies that $\vec{L}$ is the unknot, so the result is obviously true. For the inductive step, since $\vec{L}$ is quasi-alternating, there is an unoriented skein triple L, L_1, L_2, where $\det(L) = \det(L_1) + \det(L_2)$, $\det(L_1)$ and $\det(L_2)$ are non-zero, and L_1 and L_2 are quasi-alternating. By the inductive hypothesis, for $i = 1, 2$ the grid homology $\widehat{GH}(\vec{L}_i)$ is supported entirely in δ-grading

$$\delta_i = \frac{\sigma(\vec{L}_i) - \ell_i + 1}{2}. \tag{10.29}$$

Consider the unoriented skein exact sequence from Theorem 10.2.4, with grading shifts as interpreted in Proposition 10.2.9. In view of Equation (10.29), it follows that the map in the exact triangle from $\widehat{GH}(\vec{L}_1)\otimes W^{\otimes(m-\ell_1)}$ to $\widehat{GH}(\vec{L}_2)\otimes W^{\otimes(m-\ell_2)}$ is zero; so the long exact sequence gives a short exact sequence

$$0 \longrightarrow \widehat{GH}(\vec{L}_2)\otimes W^{\otimes(m-\ell_2)} \xrightarrow{f_2} \widehat{GH}(\vec{L})\otimes W^{\otimes(m-\ell)} \xrightarrow{f_3} \widehat{GH}(\vec{L}_1)\otimes W^{\otimes(m-\ell_1)} \longrightarrow 0.$$

Again, by the grading shift formulas from Proposition 10.2.9 and Equation (10.29) (for $i = 2$), the image of f_2 is contained in δ-grading $\frac{\sigma-\ell+1}{2}$; also, the cokernel of f_2 is contained in δ-grading $\frac{\sigma-\ell+1}{2}$. Thus, exactness allows us to conclude that $\widehat{GH}(\vec{L}) \otimes W^{\otimes(m-\ell)}$ is contained in δ-grading $\frac{\sigma-\ell+1}{2}$, as needed.

We have verified that there is a sequence of integers $\{c_s\}_{s\in\mathbb{Z}}$ so that there is an isomorphism of bigraded vector spaces $\widehat{GH}(\vec{L}) \cong \bigoplus_{s\in\mathbb{Z}} \mathbb{F}^{c_s}_{(s+\frac{\sigma-\ell+1}{2},s)}$. By Proposition 8.2.10, $c_s = |\alpha_s|$. □

Proof of Theorem 10.3.1. Apply Theorems 10.1.13 and 10.3.3. □

10.3.1. Some examples. We give a few special cases of the above theorems. The grid homologies of the $(2, 2n+1)$ torus knot (with $n > 0$) are given by

$$\widehat{GH}(T_{2,2n+1}) \cong \bigoplus_{s=-n}^{n} \mathbb{F}_{(s-n,s)}, \quad GH^-(T_{2,2n+1}) \cong \mathbb{F}[U]_{(-2n,-n)} \oplus \bigoplus_{i=0}^{n-1} \mathbb{F}_{(2i,n-2i)}.$$

For the $(2, 2n)$ torus link,

$$\widehat{GH}(T_{2,2n}) \cong \mathbb{F}_{(-2n,-n)} \oplus \left(\bigoplus_{s=-n+1}^{n-1} \mathbb{F}^2_{(s-n,s)} \right) \oplus \mathbb{F}_{(0,n)}.$$

For positive, even $n = 2k$ let C_n denote the two-component torus link $T_{2,2k}$, oriented so that the linking number of the strands is $-k$; that is, as the boundary of an unknotted k-times twisted embedded annulus. (The Alexander polynomial of this link was computed in Section 2.4: it is equal to $k(t^{-\frac{1}{2}} - t^{\frac{1}{2}})$.) Obviously C_0 is the two-component unlink, while C_2 is the negative Hopf link H_-. When $n = 2k > 0$

$$\widehat{GH}(C_n) \cong \mathbb{F}^k_{(-1,-1)} \oplus \mathbb{F}^{2k}_{(0,0)} \oplus \mathbb{F}^k_{(1,1)}.$$

For the twist knot W_n with $n = 2k > 0$ and the clasp given by Figure 2.6

$$\widehat{GH}(W_n) = \mathbb{F}^k_{(-1,-1)} \oplus \mathbb{F}^{2k+1}_{(0,0)} \oplus \mathbb{F}^k_{(1,1)} \qquad GH^-(W_{2k}) = \mathbb{F}[U]_{(0,0)} \oplus \mathbb{F}^k_{(1,1)} \oplus \mathbb{F}^k_{(0,0)}.$$

For the 3-strand pretzel knot $P = P(a_1, a_2, a_3)$ with $a_i = 2b_i + 1 > 0$ and $C = b_1b_2 + b_1b_3 + b_2b_3 + b_1 + b_2 + b_3 + 1$,

$$\widehat{GH}(P) \cong \mathbb{F}^C_{(0,-1)} \oplus \mathbb{F}^{2C-1}_{(1,0)} \oplus \mathbb{F}^{(C)}_{(2,1)}, \qquad GH^-(P) \cong \mathbb{F}[U]_{(2,1)} \oplus \mathbb{F}^{C-1}_{(2,1)} \oplus \mathbb{F}^C_{(1,0)}.$$

EXERCISE 10.3.4. **(a)** Determine $\widehat{GH}(W_n)$ for all n.
(b) Compute $cGH^-(T_{2,2n})$ and $cGH^-(C_n)$ for all $n > 0$.
(c) Express the τ-set of a quasi-alternating link $\vec{L}$ in terms of $\sigma(\vec{L})$ and ℓ.
(d) For an ℓ-component quasi-alternating link $\vec{L}$, express $cGH^-(\vec{L})$ in terms of σ, ℓ, and Δ.

Consider the Kanenobu knots $K(p, q)$ of Figure 2.18 (see also Section 9.5). When $p = q = 0$, the knot $K(0, 0) = W_2\#W_2$; in particular, it is alternating. An easy computation shows that $\Delta_{K(0,0)}(t) = t^2 - 6t + 11 - 6t^{-1} + t^{-2}$ and $\sigma(K_{0,0}) = 0$. By Theorem 10.3.1 and Lemma 9.5.3,

$$\widehat{GH}(K(p,q)) = \mathbb{F}_{(2,2)} \oplus \mathbb{F}^6_{(1,1)} \oplus \mathbb{F}^{11}_{(0,0)} \oplus \mathbb{F}^6_{(-1,-1)} \oplus \mathbb{F}_{(-2,-2)}$$
$$GH^-(K(p,q)) = \mathbb{F}_{(2,2)} \oplus \mathbb{F}^5_{(1,1)} \oplus \mathbb{F}^5_{(0,0)} \oplus \mathbb{F}_{(-1,-1)} \oplus \mathbb{F}[U]_{(0,0)}.$$

10.4. Further remarks

Knot Floer homology for alternating knots was computed in [**169**], generalizing a result of Rasmussen [**190**]. Quasi-alternating links were introduced in [**178**]. The definition was designed to compute a different invariant which also satisfies an unoriented skein exact sequence, the Heegaard Floer homology of the branched double cover of S^3 along a link. The unoriented skein exact sequence for knot Floer homology is due to Manolescu [**132**], and it leads quickly to the computation of knot Floer homology for quasi-alternating links [**134**]. A grid diagrammatic proof of the unoriented skein exact sequence was given by Wong [**231**].

The δ-grading we are using here differs by a multiplicative constant (-1), and an additive shift involving ℓ from the one used by Manolescu [**132**] and Wong [**231**]. Also, in the present set-up, the unoriented skein exact sequence is deduced from the oriented one. This forces an extra tensor factor of W in all three terms in the long exact sequence, which is not present in, for example, Wong's version [**231**]. In our applications, this extra tensor factor plays no role.

CHAPTER 11

Grid homology for links

Grid homology generalizes from the case of knots to the case of oriented links in a fairly straightforward manner. This adaptation can be made in various levels of generality, some of which we have already met. One version, the *collapsed grid homology* $cGH^-(\vec{L})$ of the oriented link $\vec{L}$, defined in Definition 8.2.4 has the structure of a bigraded $\mathbb{F}[U]$-module. The module is obtained from the bigraded chain complex associated to a grid representing the link (Definition 8.2.2), setting some of the variables equal to each other, and then taking homology. A further refinement, the *uncollapsed, bigraded link homology*, is simply the homology of the bigraded chain complex of a link; see Definition 11.1.1. In this version, the module structure reflects the fact that we are working with a link rather than a knot: $GH^-(\vec{L})$ is a bigraded $\mathbb{F}[U_1, \dots, U_\ell]$-module, where ℓ is the number of components of the link $\vec{L}$. Finally, the *simply blocked grid homology* $\widehat{GH}(\vec{L})$ of $\vec{L}$ was introduced in Definition 8.2.7. This version is merely a bigraded vector space over $\mathbb{F}$.

In the present chapter, we describe a further refinement (of both the simply blocked and of the uncollapsed grid homologies) whose Alexander grading is enhanced to take values in an ℓ-dimensional lattice. This enhancement reflects the fact that, for a link with ℓ components, the first homology of the link complement is ℓ-dimensional. For the oriented link $\vec{L}$ we define a *vector-valued Alexander function* on grid states, with values in the affine lattice $\mathbf{H}(L) \subset H_1(S^3 \setminus L; \mathbb{Q})$ defined in Definition 11.1.3. The Alexander function induces a grading by $\mathbf{H}(L)$ on the simply blocked and unblocked grid complexes, which, along with the $\mathbb{Z}$-valued Maslov grading, descend to the respective homologies, the former thought of as an $\mathbb{F}$-vector space and the latter as an $\mathbb{F}[U_1, \dots, U_\ell]$-module, with a correspondence between the variables and the components of $\vec{L}$. We denote these multi-graded grid homologies $\widehat{\mathbf{GH}}(\vec{L})$ and $\mathbf{GH}^-(\vec{L})$ respectively. The gradings give splittings

$$\widehat{\mathbf{GH}}(\vec{L}) = \bigoplus_{d\in\mathbb{Z}, h\in\mathbf{H}(L)} \widehat{\mathbf{GH}}_d(\vec{L}, h), \qquad \mathbf{GH}^-(\vec{L}) = \bigoplus_{d\in\mathbb{Z}, h\in\mathbf{H}(L)} \mathbf{GH}^-_d(\vec{L}, h),$$

and the action of U_j on $\mathbf{GH}^-(\vec{L})$ drops the jth component of the Alexander multi-grading by one and the Maslov grading by two.

Some classical notions from knot theory have similar enhancements arising from the fact that $H_1(S^3 \setminus L) \cong \mathbb{Z}^\ell$. The Alexander polynomial generalizes to the *multi-variable Alexander polynomial*, which is a Laurent polynomial in ℓ indeterminates $t_1, \dots, t_\ell$. The knot genus naturally generalizes to a function on the first cohomology of the link complement, called the *Thurston norm*.

Like in the case for knots, the graded Euler characteristic of the multi-graded grid homology of a link equals the multi-variable Alexander polynomial, up to an overall multiplicative constant. The relation between the grid homology of a link and the Thurston norm of its complement is again similar to the case of knots: grid homology determines the Thurston norm, as stated in Theorem 11.9.9. The proof of that theorem, however, goes beyond the scope of this book.

The aim of this chapter is to define the multi-graded grid homology for links and to study its properties. The generalization is stated in Section 11.1, and some necessary proofs are given in Section 11.2. In Section 11.3 we give a few easy computations. In Section 11.4, we study some of the symmetries of the resulting grid homology. In Section 11.5, we recall the construction of the multi-variable Alexander polynomial, and in Section 11.6 we relate that invariant and the Euler characteristic of grid homology. In Section 11.7, we connect the Seifert genus of a link with its grid homology. In Section 11.8, we give some more involved computations of multi-graded link homologies. Finally, in Section 11.9 we sketch the Thurston norm for links, and state its relationship with grid homology.

This chapter generalizes the constructions from Chapters 4 and 5; we also draw on Sections 8.2 and 10.1 (especially Lemma 10.1.7); but the material is otherwise independent of the previous two chapters.

11.1. The definition of grid homology for links

Fix an ℓ-component oriented link $\vec{L} = (\vec{L}_1, \dots, \vec{L}_\ell)$, and let $\mathbb{G}$ be a toroidal grid diagram representing $\vec{L}$. In Definition 8.2.2, we associated a bigraded chain complex $GC^-(\mathbb{G})$ to $\mathbb{G}$, generated by the grid states $\mathbf{S}(\mathbb{G})$ of $\mathbb{G}$.

DEFINITION 11.1.1. Let $\mathbb{G}$ be a grid diagram representing the ℓ-component oriented link $\vec{L}$. The ***(uncollapsed) bigraded grid homology*** $GH^-(\vec{L})$ of $\vec{L}$ is the homology of the bigraded grid complex $GC^-(\mathbb{G})$, thought of as bigraded module over the polynomial algebra $\mathbb{F}[U_1, \dots, U_\ell]$. The action by U_i is defined to be multiplication by V_{j_i}, where O_{j_i} is an O-marking on the ith component of $\vec{L}$.

A straightforward adaptation of the invariance proof from Chapter 5 gives the following result. (See also the proof of Theorem 8.2.5.)

THEOREM 11.1.2. *The bigraded $\mathbb{F}[U_1, \dots, U_\ell]$-module $GH^-(\vec{L})$ is an invariant of the oriented link $\vec{L}$.* □

The present chapter concerns a refinement of the above construction, enhanced to have an ℓ-dimensional Alexander multi-grading. First we clarify the Alexander grading set. This set is an affine set on the first homology $H_1(S^3 \setminus L; \mathbb{Z})$. An isomorphism $H_1(S^3 \setminus L; \mathbb{Z}) \cong \mathbb{Z}^\ell$ is specified by orienting the link $\vec{L}$, and ordering its components, since the oriented meridians $\mu_1, \dots, \mu_\ell$ give a basis for the free abelian group $H_1(S^3 \setminus L; \mathbb{Z})$.

DEFINITION 11.1.3. The ***Alexander grading set*** $\mathbf{H}(L)$ of a link L is the affine space for $H_1(S^3 \setminus L; \mathbb{Z})$ inside $H_1(S^3 \setminus L; \mathbb{Q})$ consisting of elements $\sum_{i=1}^{\ell} s_i \cdot \mu_i$, where $s_i \in \frac{1}{2}\mathbb{Z}$ satisfy the condition that $2s_i + \ell k(L_i, L \setminus L_i)$ is an even integer.

While the definition of $\mathbf{H}(L)$ appears to use an orientation on L (to compute linking numbers), it is easy to see that the grading set is independent of this choice. As we shall see, the grading set is a natural choice for expressing symmetries of link homology. (See Corollary 11.4.3 and Theorem 11.5.3.)

EXAMPLE 11.1.4. When L is the two-component unlink, $\mathbf{H}(L) = \mathbb{Z} \oplus \mathbb{Z} \subset \mathbb{Q} \oplus \mathbb{Q} = H_1(S^3 \setminus L; \mathbb{Q})$. When L is the Hopf link, $\mathbf{H}(L) = (\frac{1}{2} + \mathbb{Z}) \oplus (\frac{1}{2} + \mathbb{Z}) \subset \mathbb{Q} \oplus \mathbb{Q}$.

Next, we explain the Alexander grading, with values in the Alexander grading set.

DEFINITION 11.1.5. Let $\mathbb{G}$ be a grid diagram for the oriented link $\vec{L}$, let G be a planar realization of $\mathbb{G}$, and let $\vec{L}_i$ denote the ith component of $\vec{L}$. The ith ***Alexander component*** A_i is defined by the formula

$$A_i(\mathbf{x}) = \mathcal{J}(\mathbf{x} - \frac{1}{2}(\mathbb{X} + \mathbb{O}), \mathbb{X}_i - \mathbb{O}_i) - \frac{n_i - 1}{2}, \tag{11.1}$$

where $\mathbb{O}_i \subset \mathbb{O}$ and $\mathbb{X}_i \subset \mathbb{X}$ are the markings on $\vec{L}_i$, and n_i denotes the number of elements in $\mathbb{O}_i$. Collect the A_i into a vector-valued Alexander function

$$\mathbb{A} = (A_1, \dots, A_\ell)\colon \mathbf{S}(\mathbb{G}) \to \left(\frac{1}{2}\mathbb{Z}\right)^\ell \subset \mathbb{Q}^\ell$$

called the ***Alexander vector*** or ***Alexander multi-grading***.

PROPOSITION 11.1.6. *For a toroidal grid diagram $\mathbb{G}$, the function $\mathbb{A}$ is independent of the choice of the planar realization which appears in its definition. Furthermore, under the identification $\mathbb{Q}^\ell \cong H_1(S^3 \setminus L; \mathbb{Q})$ induced by the orientation and the ordering of the link components, the Alexander multi-grading takes values in the Alexander grading set $\mathbf{H}(L) \subset \mathbb{Q}^\ell$ for the link. If $\mathbf{x}$ and $\mathbf{y}$ are two grid states and $r \in \text{Rect}(\mathbf{x}, \mathbf{y})$, then the component A_i of $\mathbb{A}$ for $i = 1, \dots, \ell$ satisfies*

$$A_i(\mathbf{x}) - A_i(\mathbf{y}) = \#(r \cap \mathbb{X}_i) - \#(r \cap \mathbb{O}_i). \tag{11.2}$$

Finally, the Alexander vector is related to the Alexander function A (considered in Chapter 8; see Equation (8.2)) by the formula:

$$\sum_{i=1}^{\ell} A_i = A. \tag{11.3}$$

We construct multi-graded grid homology for links, assuming the above proposition, which is proved in Section 11.2. Let $\mathbb{G}$ be a toroidal grid diagram representing an ℓ-component link $\vec{L}$. Recall (Definition 8.2.2) that the uncollapsed grid complex $GC^-(\mathbb{G})$ of $\mathbb{G}$ is a free module over the ring $\mathbb{F}[V_1, \dots, V_n]$ generated by the grid states $\mathbf{S}(\mathbb{G})$ and equipped with the differential

$$\partial^-_{\mathbb{X}} \mathbf{x} = \sum_{\mathbf{y} \in \mathbf{S}(\mathbb{G})} \sum_{\{r \in \text{Rect}^\circ(\mathbf{x}, \mathbf{y}) \big| r \cap \mathbb{X} = \emptyset\}} V_1^{O_1(r)} \cdots V_n^{O_n(r)} \cdot \mathbf{y}.$$

The complex inherits a Maslov grading exactly as it did in the case of knots (see Equations (4.5) and (4.11)). To equip it with the Alexander multi-grading, partition the O-markings according to the link components, via a map $\pi\colon \{1, \dots, n\} \to \{1, \dots, \ell\}$ defined so that if O_j is on the ith link component, then $\pi(j) = i$. For $i \in \{1, \dots, \ell\}$ the ith component $A_i(V_1^{k_1} \cdots V_n^{k_n} \mathbf{x})$ of $\mathbb{A}(V_1^{k_1} \cdots V_n^{k_n} \mathbf{x})$ is

$$A_i(V_1^{k_1} \cdots V_n^{k_n} \mathbf{x}) = A_i(\mathbf{x}) - \sum_{j \in \pi^{-1}(i)} k_j, \tag{11.4}$$

extending the Alexander component on states from Definition 11.1.5. Let $\mathbf{GC}^-(\mathbb{G})$ be the resulting $\mathbb{Z} \oplus \mathbf{H}(L)$-graded chain complex. Theorem 4.6.3 has the following immediate generalization:

THEOREM 11.1.7. *Let $\mathbb{G}$ be a grid diagram representing an oriented link $\vec{L}$. The object $(\mathbf{GC}^-(\mathbb{G}), \partial^-_{\mathbb{X}})$ has the following structure:*

- *It is graded by $\mathbb{Z} \oplus \mathbf{H}(L)$ (the first component being the Maslov grading, and the rest being the Alexander multi-grading); i.e. as a vector space,*
$$\mathbf{GC}^-(\mathbb{G}) = \bigoplus_{d \in \mathbb{Z}, h \in \mathbf{H}(L)} \mathbf{GC}^-_d(\mathbb{G}, h),$$
where d is the Maslov grading and h is the Alexander vector.
- *It is a chain complex, i.e. $\partial^-_{\mathbb{X}} \circ \partial^-_{\mathbb{X}} = 0$.*
- *The differential drops Maslov grading by one, and it preserves Alexander multi-grading; i.e. for $d \in \mathbb{Z}$ and $h \in \mathbf{H}(L)$*
$$\partial^-_{\mathbb{X}} : \mathbf{GC}^-_d(\mathbb{G}, h) \to \mathbf{GC}^-_{d-1}(\mathbb{G}, h).$$
- *$\mathbf{GC}^-(\mathbb{G})$ is a graded module over $\mathbb{F}[V_1, \dots, V_n]$, so that multiplication by V_j drops Maslov grading by two and Alexander grading by $\mu_{\pi(j)}$; i.e.*
$$V_j \colon \mathbf{GC}^-_d(\mathbb{G}, h) \to \mathbf{GC}^-_{d-2}(\mathbb{G}, h - \mu_{\pi(j)}).$$ □

If $\mathbb{G}$ represents an ℓ-component oriented link $\vec{L}$, then the Maslov grading and the Alexander multi-grading on the chain complex $\mathbf{GC}^-(\mathbb{G})$ descends to the homology. Choose a subsequence $\{j_1, \dots, j_\ell\} \subset \{1, \dots, n\}$ so that O_{j_i} is on the ith component of the link; i.e. $\pi(j_i) = i$. Define the action of $\mathbb{F}[U_1, \dots, U_\ell]$ on the homology of $\mathbf{GC}^-(\mathbb{G})$, so that the action of U_i is induced by multiplication by V_{j_i}.

DEFINITION 11.1.8. The homology of $\mathbf{GC}^-(\mathbb{G})$, thought of as a $\mathbb{Z} \oplus \mathbf{H}(L)$-graded $\mathbb{F}[U_1, \dots, U_\ell]$-module, is denoted by $\mathbf{GH}^-(\mathbb{G})$ and is called the ***multi-graded, unblocked grid homology*** of the grid diagram $\mathbb{G}$. Similarly, we let $\widehat{\mathbf{GC}}(\mathbb{G})$ denote the quotient complex $\frac{\mathbf{GC}^-(\mathbb{G})}{V_{j_1} = \dots = V_{j_\ell} = 0}$. Its homology, thought of as a $\mathbb{Z} \oplus \mathbf{H}(L)$-graded vector space, is denoted $\widehat{\mathbf{GH}}(\mathbb{G})$ and is called the ***multi-graded, simply blocked grid homology*** of $\mathbb{G}$. Finally, homology of the chain complex $\widetilde{\mathbf{GC}}(\mathbb{G}) = \frac{\mathbf{GC}^-(\mathbb{G})}{V_1 = \dots = V_n = 0}$, thought of as a $\mathbb{Z} \oplus \mathbf{H}(L)$-graded vector space, is denoted $\widetilde{\mathbf{GH}}(\mathbb{G})$ and called the ***multi-graded, fully blocked grid homology*** of $\mathbb{G}$.

The multi-graded simply-blocked and the fully blocked grid homologies are related by a generalization of Proposition 4.6.15. To state this, let $\mathbf{W}_i$ denote the two-dimensional vector space graded by the set $\mathbb{Z} \oplus H_1(S^3 \setminus L)$, with one generator in grading $(0, 0)$, and the other generator in grading $(-1, -\mu_i)$. If X is a vector space graded by $\mathbb{Z} \oplus \mathbf{H}(L)$, we can form the tensor product $X \otimes \mathbf{W}_i$, which is the $\mathbb{Z} \oplus \mathbf{H}(L)$-graded vector space $X \otimes \mathbf{W}_i \cong X \oplus X[\![1, \mu_i]\!]$.

The proof of Proposition 4.6.15, keeping track of the multi-grading of the action of V_j (as in Proposition 8.2.8), adapts readily to give the following:

PROPOSITION 11.1.9. *Let $\mathbf{W}_i$ be the two-dimensional graded vector space defined above. Let $\mathbb{G}$ be a grid diagram representing an oriented link $\vec{L}$, and let n_i denote the number of O-markings in $\mathbb{G}$ on the ith component of $\vec{L}$. Then there is an isomorphism of $\mathbb{Z} \oplus \mathbf{H}(L)$-graded vector spaces $\widetilde{\mathbf{GH}}(\mathbb{G}) \cong \widehat{\mathbf{GH}}(\mathbb{G}) \otimes \bigotimes_{i=1}^{\ell} \mathbf{W}_i^{\otimes(n_i - 1)}$.* □

LEMMA 11.1.10. *The $\mathbb{Z} \oplus \mathbf{H}(L)$-graded homologies $\widehat{\mathbf{GH}}(\mathbb{G})$ and $\mathbf{GH}^-(\mathbb{G})$, (the former thought of as a vector space and the latter as a module over $\mathbb{F}[U_1, \dots, U_\ell]$) are* *independent of the choice* $V_{j_1}, \ldots, V_{j_\ell}$ *appearing in the definition of the* $\mathbb{F}[U_1, \ldots, U_\ell]$*-module action.*

Proof. This is a direct consequence of Lemma 8.2.3, with a little remark on gradings. Consider the proof of Lemma 4.6.9 (which is adapted to prove Lemma 8.2.3). This lemma is proved by constructing a chain homotopy $\mathcal{H}$ between multiplications by V_i and V_j, in the case where V_i and V_j are consecutive, by counting rectangles which contain a specific X-marking, see Equation (4.16). Both V_i and V_j drop the Alexander multi-grading by $\mu_{\pi(i)} = \mu_{\pi(j)}$, and $\mathcal{H}$ drops it by the same basis vector, so $V_i \colon \mathbf{GH}^-_d(\mathbb{G}, h) \to \mathbf{GH}^-_{d-2}(\mathbb{G}, h - \mu_{\pi(i)})$ is determined by $\pi(i)$. □

THEOREM 11.1.11. *The* $\mathbb{Z}\oplus\mathbf{H}(L)$*-graded grid homologies* $\widehat{\mathbf{GH}}(\mathbb{G})$ *and* $\mathbf{GH}^-(\mathbb{G})$ *(the former thought of as an* $\mathbb{F}$*-vector space, the latter thought of as a module over* $\mathbb{F}[U_1, \ldots, U_\ell]$*) depend on the grid* $\mathbb{G}$ *only through its underlying oriented link* $\vec{L}$.

Proof. We follow the proof of Theorem 4.6.19 from Chapter 5. Most of the adaptation is straightforward. For example, the commutation invariance follows directly as before, with the observation that the pentagon counting maps defined in Section 5.1 preserve Alexander multi-gradings, as do the hexagon counting homotopies.

To adapt the proof of stabilization invariance, adjust the grading shift formulas to take into account the new gradings. For instance, there is now an identification

$$\mathrm{Cone}(V_1 - V_2 \colon \mathbf{GC}^-(\mathbb{G}) \to \mathbf{GC}^-(\mathbb{G})) = \mathbf{GC}^-(\mathbb{G})[\![1, \mu_j]\!] \oplus \mathbf{GC}^-(\mathbb{G}),$$

as $\mathbb{Z} \oplus \mathbf{H}(L)$-graded modules over $\mathbb{F}[V_1, \ldots, V_n]$, where $\mu_j \in H_1(S^3 \setminus L; \mathbb{Z})$ is the meridian of the component $\vec{L}_j$ that is being stabilized (so $\pi(1) = \pi(2) = j$). (This is the generalization of Equation (5.19).) Correspondingly, consider the one-to-one correspondence between the set of grid states for the destabilized diagram $\mathbb{G}$ and the set $\mathbf{I}(\mathbb{G}')$ of those grid states for the stabilized diagram $\mathbb{G}'$ which contain the preferred point c. Lemma 5.2.4 is replaced by the statement that, under the one-to-one correspondence, $M(\mathbf{x}') = M(\mathbf{x}) - 1$ and $\mathbb{A}(\mathbf{x}') = \mathbb{A}(\mathbf{x}) - \mu_j$. With these details understood, the proof of Theorem 4.6.19 (given in Section 5.3) applies. □

According to the above theorem, we suppress the grid diagram from the notation of grid homology. Since $\widehat{\mathbf{GH}}(\vec{L})$ is finite dimensional, we have the following:

DEFINITION 11.1.12. The ***grid homology polytope*** $B_{\widehat{GH}}(\vec{L})$ is the convex hull of the set of elements $h \in \mathbf{H}(L) \subset H_1(S^3 \setminus \vec{L}; \mathbb{R}) \cong \mathbb{R}^\ell$ for which $\widehat{\mathbf{GH}}(\vec{L}, h) \neq 0$.

In practice, grid homology is often easier to compute for those $h \in \mathbf{H}(L)$ that lie on the boundary of the grid polytope.

The multi-graded grid homology groups refine their earlier defined bigraded analogues (from Definition 11.1.1), in the following sense.

PROPOSITION 11.1.13. *The multi-graded grid homology groups are related to the bigraded constructions by the formulas*

$$\widehat{GH}_d(\vec{L}, s) = \bigoplus_{\{h \in \mathbf{H}(L) \big| s_1 + \cdots + s_\ell = s\}} \widehat{\mathbf{GH}}_d(\vec{L}, h)$$

$$GH^-_d(\vec{L}, s) = \bigoplus_{\{h \in \mathbf{H}(L) \big| s_1 + \cdots + s_\ell = s\}} \mathbf{GH}^-_d(\vec{L}, h)$$

where $h = \sum_{i=1}^{\ell} s_i \cdot \mu_i$. *The second identification is compatible with the* $\mathbb{F}[U_1, \ldots, U_\ell]$*-module structure.*

Proof. As $\mathbb{F}[U_1, \ldots, U_\ell]$-modules, the definitions of $GC^-(\mathbb{G})$ and $\mathbf{GC}^-(\mathbb{G})$ are the same. They differ in their gradings, which are related according to Proposition 11.1.6. The same argument works for $\widehat{GH}$. □

Note that the Alexander multi-grading does not descend to a multi-grading on the collapsed grid homology cGH^- from Definition 8.2.4.

11.2. The Alexander multi-grading on grid homology

The construction of grid homology for links depends on Proposition 11.1.6, which can be thought of as a combination of four statements (see Lemmas 11.2.1, 11.2.2, 11.2.3 and 11.2.4 below). In this section, we establish these statements, and give a further handy formula for the Alexander multi-grading (Proposition 11.2.6).

LEMMA 11.2.1. *Different planar realizations of the same toroidal grid diagram* $\mathbb{G}$ *give the same Alexander vector* $\mathbb{A}$ *from Definition* 11.1.5.

Proof. Exactly as in the case of knots, Proposition 4.3.1 shows that the Maslov grading is independent of the planar realization. Thus, it suffices to express the functions A_i (defined in Equation (11.1)) in terms of certain Maslov gradings. To this end, let $\mathbb{X}' = (\mathbb{X} \setminus \mathbb{X}_i) \cup \mathbb{O}_i$ and $\mathbb{O}' = (\mathbb{O} \setminus \mathbb{O}_i) \cup \mathbb{X}_i$. Then, the identity

$$\mathcal{J}(\mathbf{x} - \frac{1}{2}(\mathbb{X} + \mathbb{O}), \mathbb{X}_i - \mathbb{O}_i) = \frac{1}{4}\left(M_{\mathbb{O}}(\mathbf{x}) - M_{\mathbb{O}'}(\mathbf{x}) - M_{\mathbb{X}}(\mathbf{x}) + M_{\mathbb{X}'}(\mathbf{x})\right)$$

follows immediately from Equation (4.5), together with the symmetry and bilinearity of $\mathcal{J}$. Thus,

$$A_i = \frac{1}{4}\left(M_{\mathbb{O}} - M_{\mathbb{O}'} - M_{\mathbb{X}} + M_{\mathbb{X}'}\right) - \frac{n_i - 1}{2}, \tag{11.5}$$

as needed. □

LEMMA 11.2.2. *If* $\mathbf{x}$ *and* $\mathbf{y}$ *are two grid states, and* $r \in \mathrm{Rect}(\mathbf{x}, \mathbf{y})$, *then*

$$A_i(\mathbf{x}) - A_i(\mathbf{y}) = \#(r \cap \mathbb{X}_i) - \#(r \cap \mathbb{O}_i).$$

Proof. This follows immediately from Equation (11.5) and Equation (4.2). □

LEMMA 11.2.3. *The function $\mathbb{A}$ takes values in the Alexander grading set $\mathbf{H}(L)$.*

Proof. Our goal is to show that $2A_i + \ell k(L_i, L \setminus L_i) \equiv 0 \pmod 2$. Let $\mathbb{O}$, $\mathbb{X}$, $\mathbb{O}'$, and $\mathbb{X}'$ be as in the proof of Lemma 11.2.1. Let $\mathbb{G}'$ be the grid diagram specified by markings $\mathbb{O}'$ and $\mathbb{X}'$. Equation (11.5), together with Lemma 10.1.7, gives:

$$\begin{aligned} 2A_i &= \delta_{\mathbb{G}'} - \delta_{\mathbb{G}} + M_{\mathbb{O}} - M_{\mathbb{O}'} - n_i + 1 \\ &= \ell k(L_i, L \setminus L_i) + M_{\mathbb{O}} - M_{\mathbb{O}'} - n_i + 1. \end{aligned} \tag{11.6}$$

It remains to show that

$$M_{\mathbb{O}}(\mathbf{x}) - M_{\mathbb{O}'}(\mathbf{x}) \equiv n_i - 1 \pmod 2. \tag{11.7}$$

This follows as in the proof of Proposition 4.3.3, since $NW(\mathbb{O})$ and $NW(\mathbb{O}')$ can be connected by $n_i - 1$ transpositions. □

LEMMA 11.2.4. *The integral Alexander grading is the sum of the components of the Alexander multi-grading, that is, Equation* (11.3) *holds.*

Proof. Using the definition of the A_i from Equation (11.1) and the definition of A from Equation (8.2), we get:

$$\begin{aligned} \sum_{i=1}^{\ell} A_i(\mathbf{x}) &= \mathcal{J}(\mathbf{x} - \frac{1}{2}(\mathbb{X} + \mathbb{O}), \mathbb{X} - \mathbb{O}) - \sum_{i=1}^{\ell} \frac{n_i - 1}{2} \\ &= \frac{1}{2}\left(\mathcal{J}(\mathbf{x} - \mathbb{O}, \mathbf{x} - \mathbb{O}) - \mathcal{J}(\mathbf{x} - \mathbb{X}, \mathbf{x} - \mathbb{X})\right) - \frac{n - \ell}{2} = A(\mathbf{x}). \end{aligned}$$

□

Proof of Proposition 11.1.6. The statements of the proposition are verified in Lemmas 11.2.1, 11.2.2, 11.2.3 and 11.2.4. □

We turn now to an explicit formula for the Alexander multi-grading. It will help to have the following expression for winding numbers. Let $\mathcal{D}$ denote the diagram of $\vec{L}$ provided by a fixed planar realization of the grid $\mathbb{G}$, and $\mathcal{D}_i$ the diagram of the component $\vec{L}_i$. The proof of Lemma 4.7.1 adapts readily, to prove the following:

LEMMA 11.2.5. *Fix a planar grid diagram for an oriented link $\vec{L}$, and let p be any point on the plane which is not on the projection of $\vec{L}$. Then, the winding number of the diagram $\mathcal{D}_i$ of $\vec{L}_i$ around p is computed by the formula*

$$\mathcal{J}(p, \mathbb{O}_i - \mathbb{X}_i) = w_{\mathcal{D}_i}(p) \tag{11.8}$$

□

Fix a planar realization G of a grid diagram $\mathbb{G}$ for $\vec{L}$, whose associated link diagram is $\mathcal{D} = \cup_{i=1}^{\ell} \mathcal{D}_i$. The winding numbers of $\mathcal{D}_i$ around the components of $\mathbf{x}$ are related to the Alexander vector, by the following adaptation of Proposition 4.7.2:

PROPOSITION 11.2.6. *The Alexander grading A_i can be expressed in terms of the winding numbers $w_{\mathcal{D}_i}$ by the following formula:*

$$A_i(\mathbf{x}) = -\sum_{x \in \mathbf{x}} w_{\mathcal{D}_i}(x) + \frac{1}{8}\sum_{j=1}^{8n} w_{\mathcal{D}_i}(p_j) - \left(\frac{n_i - 1}{2}\right), \tag{11.9}$$

where $\{p_j\}_{j=1}^{8n}$ are the corner points of the squares marked by $O \in \mathbb{O}$ or $X \in \mathbb{X}$.

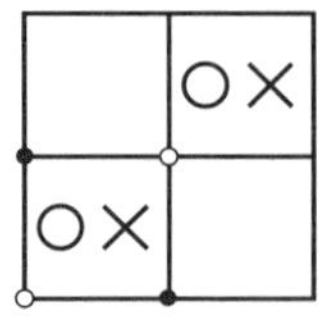

FIGURE 11.1. **Extended grid diagram for the two-component unlink.** The diagram admits two grid states $\mathbf{p}$ and $\mathbf{q}$, where $\mathbf{p}$ is denoted by full circles while $\mathbf{q}$ is denoted by hollow circles.

Proof. This follows as in the proof of Proposition 4.7.2. Specifically, use bilinearity of $\mathcal{J}$ and Lemma 11.2.5 to derive

$$\begin{aligned} A_i(\mathbf{x}) &= \mathcal{J}(\mathbf{x} - \frac{1}{2}(\mathbb{X}+\mathbb{O}), \mathbb{X}_i - \mathbb{O}_i) - \frac{n_i - 1}{2} \\ &= \mathcal{J}(\mathbf{x}, \mathbb{X}_i - \mathbb{O}_i) - \frac{1}{2}\mathcal{J}(\mathbb{X}+\mathbb{O}, \mathbb{X}_i - \mathbb{O}_i) - \frac{n_i - 1}{2} \\ &= -w_{\mathcal{D}_i}(\mathbf{x}) + \frac{1}{2}\mathcal{J}(\mathbb{X}+\mathbb{O}, \mathbb{O}_i - \mathbb{X}_i) - \frac{n_i - 1}{2}. \end{aligned}$$

It remains to show that $\frac{1}{8}\sum_{j=1}^{8n} w_{\mathcal{D}_i}(p_j) = \frac{1}{2}\mathcal{J}(\mathbb{X}+\mathbb{O}, \mathbb{O}_i - \mathbb{X}_i)$; and this follows in the same manner as Equation (4.27). $\square$

11.3. First examples

11.3.1. The two-component unlink. We compute the grid homology when $\vec{L}$ is the two-component unlink. Arguing as in Section 8.4, grid homology for $\vec{L}$ can be computed using the 2×2 extended grid diagram from Figure 11.1.

Gradings take values in $\mathbb{Z} \oplus \mathbf{H}(\vec{L}) \cong \mathbb{Z} \oplus (\mathbb{Z} \oplus \mathbb{Z})$, and the diagram has two grid states $\mathbf{p}$ and $\mathbf{q}$, satisfying

$$\mathbb{A}(\mathbf{p}) = \mathbb{A}(\mathbf{q}) = (0,0), \qquad M(\mathbf{p}) = 0, \qquad M(\mathbf{q}) = -1.$$

The differential vanishes identically, and the module $\mathbf{GH}^-(\vec{L})$ is isomorphic to the free module of rank two over the ring $\mathbb{F}[U_1, U_2]$, with one generator in grading $(0,(0,0))$ and another in $(-1,(0,0))$. Similarly, the vector space $\widehat{\mathbf{GH}}(\vec{L})$ is the two-dimensional vector space with generators in grading $(0,(0,0))$ and $(-1,(0,0))$. The grid homology polytope of the two-component unlink consists of a single point, the origin in $\mathbb{R}^2 \cong H_1(S^3 \setminus \vec{L}; \mathbb{R})$ see the left of Figure 11.2.

11.3.2. The Hopf links. Consider the grid diagram for the negative Hopf link H_- pictured in Figure 11.3.

PROPOSITION 11.3.1. *$\widehat{\mathbf{GH}}(H_-)$ is four-dimensional, spanned by vectors p, q, r, s with Maslov grading specified by*

$$M(p) = 1 \qquad M(q) = M(r) = 0 \qquad M(s) = -1,$$

and Alexander multi-grading specified by

$$\mathbb{A}(p) = (\frac{1}{2}, \frac{1}{2}) \qquad \mathbb{A}(q) = (-\frac{1}{2}, \frac{1}{2}) \qquad \mathbb{A}(r) = (\frac{1}{2}, -\frac{1}{2}) \qquad \mathbb{A}(s) = (-\frac{1}{2}, -\frac{1}{2}).$$

FIGURE 11.2. **Some grid homology polytopes.** The grid homology polytope for the two-component unlink (the origin) is on the left; the polytope for the Hopf link (the square with corners at $(\pm\frac{1}{2}, \pm\frac{1}{2})$) is on the right.

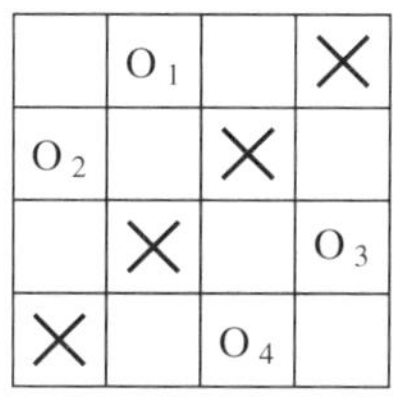

FIGURE 11.3. **Grid diagram for the negative Hopf link H_-.**

The grid homology $\mathbf{GH}^-(H_-)$ is the multi-graded $\mathbb{F}[U_1, U_2]$-module generated by p, q, and r with the above multi-gradings, modulo the relation $U_1 \cdot r + U_2 \cdot q = 0$.

Proof. We work with the toroidal grid diagram from Figure 11.3. Using Equation (11.9), for each grid state $\mathbf{x} \in \mathbf{S}(\mathbb{G})$ we have $A_i(\mathbf{x}) = -\sum_{x \in \mathbf{x}} w_{\mathcal{D}_i}(x) - \frac{3}{2}$ for $i = 1, 2$. Thus, looking at the matrix of winding numbers, we immediately conclude that for each grid state $\mathbf{x}$,

$$-\frac{3}{2} \leq A_i(\mathbf{x}) \leq \frac{1}{2}. \tag{11.10}$$

Moreover, for each $h \in \{(\frac{1}{2}, \frac{1}{2}), (-\frac{3}{2}, \frac{1}{2}), (\frac{1}{2}, -\frac{3}{2}), (-\frac{3}{2}, -\frac{3}{2})\}$, there is a unique grid state $\mathbf{x}$ with $\mathbb{A}(\mathbf{x}) = h$. For $h = (\frac{1}{2}, \frac{1}{2})$, $(-\frac{3}{2}, \frac{1}{2})$, $(\frac{1}{2}, -\frac{3}{2})$, and $(-\frac{3}{2}, -\frac{3}{2})$, the states are $(1, 2, 3, 4)$, $(4, 3, 2, 1)$, $(2, 1, 4, 3)$, and $(3, 4, 1, 2)$ respectively. It is now straightforward to compute

$$M((1,2,3,4)) = 1 \quad M((4,3,2,1)) = -1 = M((2,1,4,3)) \quad M((3,4,1,2)) = -3.$$

This argument computes $\widetilde{\mathbf{GH}}(\mathbb{G})$ in these Alexander multi-gradings. Combining with Proposition 11.1.9, it follows that $\widehat{\mathbf{GH}}(\mathbb{G})$ is 1-dimensional in Alexander multi-gradings $(\pm\frac{1}{2}, \pm\frac{1}{2})$, which, by Equation 11.10, determines $\widehat{\mathbf{GH}}(H_-)$.

To calculate $\mathbf{GH}^-(H_-)$, consider a model multi-graded chain complex C over $\mathbb{F}[V_3, V_2]$, with generators p, q, r, s (with multi-gradings given by the statement of the proposition), and differential given by

$$\partial s = V_3 \cdot r + V_2 \cdot q \qquad \partial p = \partial q = \partial r = 0.$$

Consider the graded map $\Phi\colon C \to \mathbf{GC}^-(\mathbb{G})$ of $\mathbb{F}[V_3, V_2]$-modules defined by

$$\begin{aligned}\Phi(p) &= (1,2,3,4)\\ \Phi(q) &= (4,2,3,1) + (1,3,2,4)\\ \Phi(r) &= (1,2,4,3) + (2,1,3,4)\\ \Phi(s) &= (3,2,4,1) + (3,1,2,4).\end{aligned}$$

It is straightforward to check that Φ is a chain map. Consider $\widehat{C} = \frac{C}{V_2=V_3=0}$. The induced map $\widehat{\Phi}\colon \widehat{C} \to \widehat{\mathbf{GC}}(\mathbb{G})$ is injective on homology. This follows from a direct analysis of the four generators $\Phi(p)$, $\Phi(q)$, $\Phi(r)$, and $\Phi(s)$: none of them appear as the boundary of a chain in $\widehat{\mathbf{GC}}(\mathbb{G})$. It follows from the above computation of $\widehat{\mathbf{GH}}(\mathbb{G})$ that $\widehat{\Phi}$ induces an isomorphism $H(\widehat{C}) \to \widehat{\mathbf{GH}}(\mathbb{G})$. Thus, Φ induces an isomorphism $H(C) \to \mathbf{GH}^-(\mathbb{G})$, as well; see for example Proposition A.3.5. Recall that on $\mathbf{GH}^-(\mathbb{G})$, U_1 and U_2 act as multiplication by V_3 and V_2 respectively. □

See Figure 11.2 for a picture of the grid homology polytope of H_-.

In a similar manner, for the positive Hopf link H_+, the multi-graded chain complex $\widehat{\mathbf{GC}}(H_+)$ is homotopy equivalent to the four-dimensional vector space spanned by four generators p, q, r, and s, with

$$\mathbb{A}(p) = (\frac{1}{2}, \frac{1}{2}) \qquad \mathbb{A}(q) = (-\frac{1}{2}, \frac{1}{2}) \qquad \mathbb{A}(r) = (\frac{1}{2}, -\frac{1}{2}) \qquad \mathbb{A}(s) = (-\frac{1}{2}, -\frac{1}{2})$$

and

$$M(p) = 0 \qquad M(q) = M(r) = -1 \qquad M(s) = -2.$$

These data compute $\widehat{\mathbf{GH}}(H_+)$.

The module $\mathbf{GH}^-(H_+)$ is generated by p, s, and t, where the gradings of p and s are as above, and

$$\mathbb{A}(t) = (-\frac{1}{2}, -\frac{1}{2}) \qquad M(t) = -3$$

modulo the relations $U_1 \cdot p = U_2 \cdot p = 0$.

EXERCISE 11.3.2. **(a)** Verify the above formulas for the grid homology of H_+. **(b)** Compute the grid homologies for the n-component unlink.

11.4. Symmetries of grid homology for links

Grid homology is an invariant of an oriented link. We study now the dependence of the homology on the orientation; compare also Section 7.1.

PROPOSITION 11.4.1. *The grid homology modules of the oriented link $\vec{L}$ (both $\widehat{\mathbf{GH}}$ and $\mathbf{GH}^-$) are isomorphic to the corresponding grid homology modules of the link $-\vec{L}$, in the sense that, for any $h \in H_1(S^3 \setminus L; \mathbb{Q})$,*

$$\widehat{\mathbf{GH}}(\vec{L}, h) \cong \widehat{\mathbf{GH}}(-\vec{L}, -h) \qquad \textit{and} \qquad \mathbf{GH}^-(\vec{L}, h) \cong \mathbf{GH}^-(-\vec{L}, -h).$$

Proof. Following the proof of Proposition 5.3.2, consider two grid diagrams $\mathbb{G}$ and $\mathbb{G}'$, for $\vec{L}$ and $-\vec{L}$ respectively, that differ by reflection across the diagonal. This reflection induces an identification of grid states $\phi\colon \mathbf{S}(\mathbb{G}) \to \mathbf{S}(\mathbb{G}')$ that also induces an isomorphism of grid complexes. The Maslov gradings and the components of the two Alexander vectors are identified under ϕ; $M_{\mathbb{O}}(\mathbf{x}) = M_{\mathbb{O}'}(\phi(\mathbf{x}))$ and $A_i^{\mathbb{G}}(\mathbf{x}) = A_i^{\mathbb{G}'}(\phi(\mathbf{x}))$ for each $i = 1, \dots, \ell$. The Alexander grading, thought of as an element of $H_1(S^3 \setminus L;\mathbb{Q})$, is defined by $\sum_{i=1}^{\ell} A_i \cdot \mu_i$, where μ_i is the ith meridian. Since the ith meridian μ_i for $\vec{L}$ is (-1) times the ith meridian for $-\vec{L}$, the claims now follow. □

The dependence of link homology on the orientation of the individual link components is more interesting, as follows. (Compare also Proposition 10.2.1.)

PROPOSITION 11.4.2. *Let $\vec{L}$ be an oriented ℓ-component link, and let $\vec{L}'$ be the oriented link obtained from $\vec{L}$ by reversing the orientation of its ith component L_i. Then, writing $h = \sum s_j \cdot \mu_j$, we have*

$$\widehat{\mathbf{GH}}_d(\vec{L},h) \cong \widehat{\mathbf{GH}}_{d-2s_i+\kappa_i}(\vec{L}',h),$$

where κ_i denotes the linking number of $\vec{L}_i$ with $\vec{L} \setminus \vec{L}_i$.

Proof. We wish to verify that for a given $(s_1,\dots,s_\ell) \in \mathbb{Q}^\ell$,

$$\text{(11.11)} \quad \widehat{\mathbf{GH}}_d(\vec{L},(s_1,\dots,s_\ell)) \cong \widehat{\mathbf{GH}}_{d-2s_i+\kappa_i}(\vec{L}',(s_1,\dots,s_{i-1},-s_i,s_{i+1},\dots,s_\ell)).$$

From a grid diagram $\mathbb{G}$ for $\vec{L}$ we can obtain a grid diagram $\mathbb{G}'$ for $\vec{L}'$ by switching the roles of those markings that correspond to the ith component of the link, $\mathbb{O}_i$ and $\mathbb{X}_i$. The complexes $\widetilde{\mathbf{GC}}$ for the two diagrams agree, but the Maslov and Alexander gradings change as follows. Let A_j be the jth component of the Alexander multi-grading and $M = M_{\mathbb{O}}$ be the Maslov grading calculated for the grid states using $\mathbb{G}$, and let A'_j and $M' = M_{\mathbb{O}'}$ be computed using $\mathbb{G}'$.

By a direct application of Equation (11.1) we see that

$$A'_j(\mathbf{x}) = \begin{cases} A_j(\mathbf{x}) & i \neq j \\ -A_i(\mathbf{x}) - n_i + 1 & i = j. \end{cases}$$

From Equation (11.6), we also have

$$M_{\mathbb{O}'}(\mathbf{x}) = M_{\mathbb{O}}(\mathbf{x}) - 2A_i(\mathbf{x}) + \ell k(L_i, L \setminus L_i) + 1 - n_i.$$

Putting these together, we get

$$\text{(11.12)} \quad \widetilde{\mathbf{GH}}_d(\mathbb{G},(s_1,\dots,s_\ell)) \cong \widetilde{\mathbf{GH}}_{d-2s_i+\kappa_i+1-n_i}(\mathbb{G}',(s_1,\dots,s_{i-1},-s_i+1-n_i,s_{i+1},\dots,s_\ell)).$$

To see that this induces the stated symmetry of $\widehat{\mathbf{GH}}$, we resort to Poincaré polynomials, as in the proof of Proposition 7.1.1, writing

$$\widetilde{\mathbf{P}}_{\mathbb{G}}(q,t_1,\dots,t_\ell) = \sum \dim\ \widetilde{\mathbf{GH}}_d(\mathbb{G},(s_1,\dots,s_\ell))q^d t_1^{s_1}\cdots t_\ell^{s_\ell}$$
$$\widehat{\mathbf{P}}_{\mathbb{G}}(q,t_1,\dots,t_\ell) = \sum \dim\ \widehat{\mathbf{GH}}_d(\mathbb{G},(s_1,\dots,s_\ell))q^d t_1^{s_1}\cdots t_\ell^{s_\ell}.$$

Equation (11.12) can be expressed as the relation:

$$q^{\kappa_i}\widetilde{\mathbf{P}}_{\mathbb{G}}(q,t_1,\dots,t_i,\dots,t_\ell) = (qt_i)^{1-n_i} \cdot \widetilde{\mathbf{P}}_{\mathbb{G}'}(q,t_1,\dots,(q^{-2}t_i^{-1}),\dots,t_\ell).$$

Proposition 11.1.9 (applied to $\mathbb{G}$ and $\mathbb{G}'$ respectively) gives

$$\widetilde{\mathbf{P}}_{\mathbb{G}}(q,t_1,\dots,t_\ell) = \widehat{\mathbf{P}}_{\mathbb{G}}(q,t_1,\dots,t_\ell)\cdot\prod_{j=1}^{\ell}(1+q^{-1}t_j^{-1})^{n_j-1}$$

$$\widetilde{\mathbf{P}}_{\mathbb{G}'}(q,t_1,\dots,t_\ell) = \widehat{\mathbf{P}}_{\mathbb{G}'}(q,t_1,\dots,t_\ell)\cdot\prod_{j=1}^{\ell}(1+q^{-1}t_j^{-1})^{n_j-1}.$$

These equations together give

$$q^{\kappa_i}\widehat{\mathbf{P}}_{\mathbb{G}}(q,t_1,\dots,t_\ell) = \widehat{\mathbf{P}}_{\mathbb{G}'}(q,t_1,\dots,t_{i-1},(q^{-2}t_i^{-1}),t_{i+1},\dots,t_\ell),$$

verifying Equation (11.11). To see that this verifies the proposition, note that the homology class in $H_1(S^3\setminus L;\mathbb{Q})$ corresponding to the element $(s_1,\dots,s_\ell)\in\mathbb{Q}^\ell$, for the orientation $\vec{L}$, agrees with the homology class corresponding to the element $(s_1,\dots,s_{i-1},-s_i,s_{i+1},\dots,s_\ell)$ for the orientation $\vec{L}'$. ☐

COROLLARY 11.4.3. *The simply blocked grid homology groups satisfy the symmetry* $\widehat{\mathbf{GH}}_d(\vec{L},h)\cong\widehat{\mathbf{GH}}_{d-2\sum s_i}(\vec{L},-h)$, *where* $h=\sum_{i=1}^{\ell}s_i\cdot\mu_i$. *In particular, the grid homology polytope is preserved by the map* $h\mapsto -h$.

Proof. Apply Proposition 11.4.2 successively to each component of $\vec{L}$, to get $\widehat{\mathbf{GH}}_d(\vec{L},h)\cong\widehat{\mathbf{GH}}_{d-2\sum s_i}(-\vec{L},h)$. Note that the linking number of $\vec{L}_i$ with $\vec{L}_j$ appears in both κ_i and κ_j, but with opposite signs; thus, the correction terms κ_i do not appear in the above expression. Next, apply Proposition 11.4.1 to see that $\widehat{\mathbf{GH}}_{d-2\sum s_i}(-\vec{L},h)=\widehat{\mathbf{GH}}_{d-2\sum s_i}(\vec{L},-h)$. ☐

Finally, Proposition 7.1.2 has the following straightforward generalization to links:

PROPOSITION 11.4.4. *If* $\vec{L}$ *is an oriented link and* $m(\vec{L})$ *is its mirror, then for all* $d\in\mathbb{Z}$ *and* $h\in\mathbf{H}(\vec{L})$, $\widehat{\mathbf{GH}}_d(\vec{L},h)\cong\widehat{\mathbf{GH}}_{1-d-\ell+2\sum s_i}(m(\vec{L}),h)$, *using the identification* $\mathbf{H}(\vec{L})\cong\mathbf{H}(m(\vec{L}))$ *that identifies an oriented meridian for* $\vec{L}$ *with an oriented meridian for* $m(\vec{L})$.

Proof. We modify the proof of Proposition 7.1.2. If $\mathbb{G}$ is a grid diagram for $\vec{L}$, let $\mathbb{G}^*$ be the grid diagram for $m(\vec{L})$ obtained by reflecting $\mathbb{G}$ through a horizontal axis, and let $\mathbf{x}\mapsto\mathbf{x}^*$ be the induced identification of grid states $\mathbf{S}(\mathbb{G})\cong\mathbf{S}(\mathbb{G}^*)$. Recall from Equation (7.2) that $M_{\mathbb{O}}(\mathbf{x})+M_{\mathbb{O}^*}(\mathbf{x}^*)=1-n$. A similar argument shows that $A_i^{\mathbb{G}}(\mathbf{x})+A_i^{\mathbb{G}^*}(\mathbf{x}^*)=1-n_i$. Since the reflection induces an isomorphism of complexes, it follows that

$$\widetilde{\mathbf{GH}}_d(\mathbb{G},(s_1,\dots,s_\ell))\cong\widetilde{\mathbf{GH}}_{1-n-d}(\mathbb{G}^*,(1-n_1-s_1,\dots,1-n_\ell-s_\ell)).$$

But now Proposition 11.4.2 (see especially Equation (11.12)) gives an isomorphism

$$\widetilde{\mathbf{GH}}_{1-n-d}(\mathbb{G}^*,(1-n_1-s_1,\dots,1-n_\ell-s_\ell))\cong \widetilde{\mathbf{GH}}_{1-n-d+\sum_{i=1}^{\ell}(2s_i+n_i-1)}(\mathbb{G}_2^*,(s_1,\dots,s_\ell)),$$

where $\mathbb{G}_2^*$ is obtained from $\mathbb{G}^*$ by switching $\mathbb{O}$ and $\mathbb{X}$. Since $1-n+\sum_{i=1}^{\ell}(n_i-1) = 1-\ell$, it follows that

$$\widetilde{\mathbf{GH}}_d(\mathbb{G},(s_1,\dots,s_\ell)) \cong \widetilde{\mathbf{GH}}_{1-d-\ell+2\sum s_i}(\mathbb{G}^*,(s_1,\dots,s_\ell)),$$

which by Propositions 11.1.9 and 11.4.1 implies the corresponding isomorphism for $\widehat{\mathbf{GH}}$. □

11.5. The multi-variable Alexander polynomial

In order to be able to identify the Euler characteristic of the multi-graded grid homology, we generalize the Alexander polynomial given earlier in Section 2.4 to the *multi-variable Alexander polynomial.*

Let $\vec{L}$ be an oriented link with ℓ components. Consider the link group $\pi_1(S^3\setminus\vec{L})$, and the first homology $H_1(S^3\setminus\vec{L};\mathbb{Z})\cong\mathbb{Z}^\ell$, equipped with generators given by the oriented meridians $\{\mu_i\}_{i=1}^{\ell}$ for the components of $\vec{L}$. The Hurewicz homomorphism $\pi_1(S^3\setminus\vec{L})\to H_1(S^3\setminus\vec{L};\mathbb{Z})\cong\mathbb{Z}^\ell$ induces an isomorphism from the abelianization of the fundamental group to the first homology. (See [**83**, Theorem 2A.1].)

Consider the *maximal abelian cover* of $S^3\setminus\vec{L}$; that is, the connected cover $p\colon X_\infty\to X=S^3\setminus\vec{L}$ with the property that the image of $\pi_1(X_\infty)$ in $\pi_1(S^3\setminus\vec{L})$ is the commutator subgroup of $\pi_1(S^3\setminus\vec{L})$. Since the commutator subgroup is normal, its quotient $\mathbb{Z}^\ell$ acts on the covering space by deck transformations. The induced action on the first homology of the the covering space gives a $\mathbb{Z}^\ell$-action on $H_1(X_\infty;\mathbb{Z})$; equivalently, $H_1(X_\infty;\mathbb{Z})$ is given the structure of a module over $\mathbb{Z}[\mathbb{Z}^\ell]$, which in turn can be thought of as the ring $\Lambda=\mathbb{Z}[t_1^{\pm1},\dots,t_\ell^{\pm1}]$ of Laurent polynomials. This module is called the ***Alexander module*** of the link $\vec{L}$.

The multi-variable Alexander polynomial can be extracted from the Alexander module via the following algebraic procedure. Observe that the Alexander module is finitely generated over the ring Λ. (See [**18**, Proposition 9.2]; see also Section 11.5.1 for an explicit, finite presentation.) Since Λ is a Noetherian ring, it follows that the module has a finite presentation; i.e. there are integers m and n and a map Q that fit into the exact sequence

$$\Lambda^m\stackrel{Q}{\to}\Lambda^n\to H_1(X_\infty;\mathbb{Z})\to 0.$$

If $m<n$, the Alexander polynomial is defined to be 0. Otherwise, let I denote the ideal in Λ generated by the determinants of the $n\times n$ minors of the presentation matrix Q. The greatest common divisor of the elements of this ideal is denoted $\Delta_{\vec{L}}(t_1,\dots,t_\ell)\in\Lambda$. By its definition, $\Delta_{\vec{L}}(t_1,\dots,t_\ell)$ is well-defined up to multiplication by units, i.e. by elements of the form $\pm t_1^{\pm a_1}\cdots t_\ell^{\pm a_\ell}$. With a little more work (see for example [**119**, Theorem 6.1]), one can see that $\Delta_{\vec{L}}$ does not depend on the choice of the presentation of $H_1(X_\infty;\mathbb{Z})$; it is called the *multi-variable Alexander polynomial* of $\vec{L}$. At present, this object is well-defined only up multiplication by units; we remove most of this indeterminacy below, cf. Definition 11.5.4.

REMARK 11.5.1. If M is any Λ-module with finite presentation

$$\Lambda^m\stackrel{Q}{\to}\Lambda^n\to M\to 0,$$

let I be the ideal defined as above. The gcd of the generators of I is the *order* $\Delta_0(M)$ of the module M; this notion generalizes the usual notion of the determinant of a presentation matrix: M is non-torsion if and only if $\Delta_0(M) = 0$. By considering the ith *elementary ideal* ϵ_i generated by the $(n-i) \times (n-i)$ minors of Q, further invariants $\Delta_i(M)$ can be defined.

The multivariable Alexander polynomial is related to the Alexander polynomial defined earlier by the following result. For a proof see [**119**]; for an alternative approach, one can use the grid matrix as in Section 11.5.2.

PROPOSITION 11.5.2. *If $\ell = 1$, the above definition gives back the Alexander polynomial of a knot from Section 2.4, up to a multiple of $\pm t^a$. If $\ell > 1$, specializing the multi-variable Alexander polynomial to $t = t_1 = \cdots = t_\ell$ gives a Laurent polynomial $\Delta_{\vec{L}}(t, \dots, t)$ in t, so that $(t-1) \cdot \Delta_{\vec{L}}(t, \dots, t)$ is the Alexander polynomial of the oriented link constructed in Section 2.4, again up to a multiple of $\pm t^a$.* □

The symmetry of the Alexander polynomial extends in the multi-variable case, which we state using the following shorthand. Given two Laurent polynomials $p_1(t_1, \dots, t_\ell)$ and $p_2(t_1, \dots, t_\ell)$, write $p_1 \doteq p_2$ if there is some $(a_1, \dots, a_\ell) \in \mathbb{Z}^\ell$ with $p_1(t_1, \dots, t_\ell) = \pm t_1^{a_1} \cdots t_\ell^{a_\ell} \cdot p_2(t_1, \dots, t_\ell)$.

THEOREM 11.5.3 (Torres, [**218**]). *For any oriented, ℓ-component link $\vec{L}$, we have $\Delta_{\vec{L}}(t_1, \dots, t_\ell) \doteq \Delta_{\vec{L}}(t_1^{-1}, \dots, t_\ell^{-1})$.* □

We do not spell out a proof of the above theorem; but it follows from some properties of grid homology we prove later (see Corollary 11.4.3 and Theorem 11.6.1).

DEFINITION 11.5.4. By Torres' theorem, we can multiply the earlier definition of the multi-variable Alexander polynomial by a monomial $t_1^{\frac{a_1}{2}} \cdots t_\ell^{\frac{a_\ell}{2}}$ chosen so that the resulting Laurent polynomial in $t_1^{\frac{1}{2}}, \dots, t_\ell^{\frac{1}{2}}$ is symmetric under the operation sending each $t_i^{\frac{1}{2}}$ to $t_i^{-\frac{1}{2}}$. The result, which is now defined up to multiplication by ± 1, is called the ***symmetrized multi-variable Alexander polynomial*** of the oriented link; it is also denoted $\Delta_{\vec{L}}(t_1, \dots, t_\ell)$.

The above construction defined the multi-variable Alexander polynomial up to multiplication by ± 1. This indeterminacy can be removed; see [**221**].

11.5.1. Fox calculus. Like in the single variable case, Fox calculus can be used to compute the multi-variable Alexander polynomial of an oriented link. For this purpose, consider a presentation

$$\pi_1(S^3 \setminus \vec{L}) = \langle x_1, \dots, x_n \mid r_1, \dots, r_m \rangle.$$

The presentation gives a surjective group homomorphism $F_n \to \pi_1(S^3 \setminus \vec{L})$, where F_n is the free group generated by the letters $x_1, \dots, x_n$. There is an induced map $\mathbb{Z}[F_n] \to \mathbb{Z}[\pi_1(S^3 \setminus \vec{L})]$ of group rings, which we compose with the map on group rings induced by the Hurewicz homomorphism $\pi_1(S^3 \setminus \vec{L}) \to H_1(S^3 \setminus \vec{L})$. Thinking of the group ring of $H_1(S^3 \setminus \vec{L}) \cong \mathbb{Z}^\ell$ as the ring of Laurent polynomials $\mathbb{Z}[t_1^{\pm 1}, \dots, t_\ell^{\pm 1}]$, we have a map $\phi \colon \mathbb{Z}[F_n] \to \mathbb{Z}[t_1^{\pm 1}, \dots, t_\ell^{\pm 1}]$.

Consider the $n \times m$ *Jacobi matrix* $J = (J_{i,j})$ over $\mathbb{Z}[t_1^{\pm 1}, \dots, t_\ell^{\pm 1}]$, where $J_{i,j}$ is obtained by applying ϕ to the free derivative $\frac{\partial r_i}{\partial x_j}$ (defined in Section 2.4). The relevance of this Jacobi matrix to the Alexander module is the following result, which we do not prove here. (See [**18**, Theorem 9.10].) Recall that $p \colon X_\infty \to S^3 \setminus \vec{L}$ is the connected infinite cover associated to the commutator subgroup of $\pi_1(S^3 \setminus \vec{L})$. Let $x_0 \in S^3 \setminus \vec{L}$ be a fixed basepoint, and let $Y_\infty \subset X_\infty$ denote $p^{-1}(x_0)$.

THEOREM 11.5.5. *The matrix J gives a presentation of the Alexander module $H_1(X_\infty, Y_\infty; \mathbb{Z})$, thought of as a module over Λ.* □

The long exact sequence of the pair (X_∞, Y_∞) shows that the rank of $H_1(X_\infty, Y_\infty; \mathbb{Z})$ as a Λ-module is at least one, and either $H_1(X_\infty; \mathbb{Z})$ is non-torsion (in which case its order, and hence the Alexander polynomial of $\vec{L}$ is zero), or it is isomorphic to the torsion part of $H_1(X_\infty, Y_\infty; \mathbb{Z})$. By [**86**, Theorem 3.4], for a Λ-module M of rank 1, and its torsion submodule $T(M)$ we have (using notation from Remark 11.5.1)

$$\Delta_0(T(M)) = \Delta_1(M).$$

This implies the following generalization of Theorem 2.4.16:

COROLLARY 11.5.6. *The multi-variable Alexander polynomial is zero if $m < n-1$; otherwise it is the greatest common divisor of the $(n-1) \times (n-1)$ minors of the Jacobi matrix J of $\vec{L}$.* □

The above definition of the Alexander polynomial involves computing the greatest common divisor of all the $(n-1) \times (n-1)$ minors in the $m \times n$ Jacobi matrix. When the link group is presented so that the number of generators is one bigger than the number of relations (as it is, for example, for a grid presentation as in Section 3.5), one can compute the Alexander polynomial using any one of the appropriate minors, according to the following:

LEMMA 11.5.7. *Let $\vec{L}$ be a link of $\ell > 1$ components, and fix a presentation for the link group of the form $\langle x_1, \dots, x_n \big| r_1, \dots, r_{n-1} \rangle$. Let D_i be the $(n-1) \times (n-1)$ minor of the Jacobi matrix obtained by deleting the ith column (corresponding to the generator x_i). Then, there is an i with $\phi(x_i) - 1 \neq 0$ and for such an i the Alexander polynomial can be computed by the formula*

$$D_i \doteq (1 - \phi(x_i)) \cdot \Delta_{\vec{L}}.$$

Proof. An inductive argument on the length of a word w in the generators $x_1, \dots, x_n$ shows that the Fox derivative satisfies

$$w - 1 = \sum_{j=1}^{n} \frac{\partial w}{\partial x_j} \cdot (x_j - 1)$$

(see also [**18**, Proposition 9.8]). In the case where $w = r_i$ is a relation, we can apply ϕ to the above formula to get that

$$0 = \sum_{j=1}^{n} \phi\left(\frac{\partial r_i}{\partial x_j}\right) \cdot (\phi(x_j) - 1). \tag{11.13}$$

Let $\{J_j\}_{j=1}^{n}$ denote the columns of the matrix J. Since Equation (11.13) holds for each relation, it follows that

$$0 = \sum_{j=1}^{n} J_j \cdot (\phi(x_j) - 1).$$

Thus, for any $1 \leq i < j \leq n$,

$$\begin{aligned}(\phi(x_i)-1)\cdot D_j &= (\phi(x_i)-1)\cdot \det(J_1,\dots,J_{j-1},J_{j+1},\dots,J_n)\\ &= \det(J_1,\dots,J_{i-1},(\phi(x_i)-1)\cdot J_i,J_{i+1},\dots,J_{j-1},J_{j+1},\dots,J_n)\\ &= \det(J_1,\dots,J_{i-1},-\sum_{k\neq i}(\phi(x_k)-1)\cdot J_k,J_{i+1},\dots,J_{j-1},J_{j+1}\dots,J_n)\\ &= \pm\det(J_1,\dots,J_{i-1},J_{i+1},\dots,J_{j-1},(1-\phi(x_j))\cdot J_j,J_{j+1},\dots,J_n)\\ &= \pm(\phi(x_j)-1)\cdot D_i.\end{aligned}$$

Since ϕ is surjective, there is some j for which $\phi(x_j) \neq 1$. It follows from the above formula that the rational function $p = \pm\frac{D_j}{\phi(x_j)-1}$ is independent of such a j. Similarly, we can choose i and j so that $\phi(x_i)$ and $\phi(x_j)$ are not part of the same cyclic subgroup, so that $\phi(x_i)-1$ and $\phi(x_j)-1$ are relatively prime. The above formula shows that $\phi(x_i)-1$ divides D_i; i.e. p is a Laurent polynomial, and in fact it is the greatest common divisor of $\{(\phi(x_i)-1)\cdot p\}_{i=1}^n$; this is also the multi-variable Alexander polynomial $\Delta_{\vec{L}}$ as defined earlier, up to multiplication by units. □

11.5.2. Grid diagrams and the multi-variable Alexander polynomial. When the grid diagram $\mathbb{G}$ represents an oriented link $\vec{L}$ with ℓ components, we refine the grid matrix from Definition 3.3.2 to compute the multi-variable Alexander polynomial. For $k = 1,\dots,\ell$, let $w_{i,j}^k$ denote the winding number of the projection of the kth component of the link $\vec{L}$ around the point (i,j). The refined grid matrix $\mathbb{M}(\mathbb{G})$ is the matrix whose $(n-j,i+1)^{th}$ entry is

$$\prod_{k=1}^{\ell} t_k^{-w_{i,j}^k}.$$

(Specializing $t = t_1 = \dots = t_\ell$ gives back the matrix $\mathbf{M}(\mathbb{G})$ of Definition 3.3.2.)

PROPOSITION 11.5.8 ([**136**]). *Let $\mathbb{G}$ be a grid diagram representing an oriented link $\vec{L}$ with $\ell > 1$, and let n_j denote the number of O-markings associated to the jth component of $\vec{L}$. Then, the determinant of the matrix $\mathbb{M}(\mathbb{G})$ of $\mathbb{G}$ is related to the multi-variable Alexander polynomial by*

$$\det \mathbb{M}(\mathbb{G}) \doteq \left(\prod_{j=1}^{\ell}(1-t_j)^{n_j}\right)\cdot \Delta_{\vec{L}}(t_1,\dots,t_\ell).$$

Proof. Consider the presentation of $\pi_1(S^3 \setminus \vec{L})$ given by Equation (3.6). Apply Fox calculus using this presentation to determine the Alexander polynomial of $\vec{L}$. Notice that in each relation each generator appears at most once, and with power $+1$. Map the resulting Jacobi matrix to the abelianization. The columns of the resulting $(n-1)\times n$ matrix correspond to the generators in the presentation, which in turn correspond to vertical segments in the link projection associated to the grid diagram. For any $i = 1,\dots,n$, let D_i be the $(n-1)\times(n-1)$ minor of this matrix, obtained by deleting the ith column. By Lemma 11.5.7,

$$D_i \doteq (1-t_{\pi(i)})\cdot \Delta_{\vec{L}}, \tag{11.14}$$

where $\pi\colon \{1,\dots,n\} \to \{1,\dots,\ell\}$ is the map which associates to a vertical segment its corresponding link component.

To relate this quantity to $\det(\mathbb{M}(\mathbb{G}))$, subtract each column from the next one in $\mathbb{M}(\mathbb{G})$. In the resulting matrix, the elements away from the first column are either zero (when the two neighbouring positions in the matrix are not separated by $\vec{L}$) or they are some unit times $t_{\pi(i-1)}^{\pm 1} - 1$ (when a component of $\vec{L}$ separates the entry in the $(i-1)^{st}$ and the ith positions). Therefore, we can factor $\prod_{i=1}^{n-1}(1-t_{\pi(i)})$ out from the determinant. In addition, the bottom row contains a single 1 (in its first column) and 0's afterwards. Consequently, up to multiplication by ± 1, we can delete the first column and the bottom row to compute the determinant. It is now straightforward to see that

$$D_n(t_1^{-1},\dots,t_\ell^{-1}) \cdot \prod_{i=1}^{n-1}(1-t_{\pi(i)}) \doteq \det(\mathbb{M}(\mathbb{G})).$$

The claimed equality now follows from Equation (11.14) and Theorem 11.5.3. □

EXERCISE 11.5.9. **(a)** Show that the multi-variable Alexander polynomial of the Hopf link $H_\pm$ is equal to ± 1.
(b) Prove that the multi-variable Alexander polynomial of the Borromean rings B (from the right of Figure 2.7) is given by

$$\Delta_B(t_1,t_2,t_3) = \pm(t_1^{\frac{1}{2}} - t_1^{-\frac{1}{2}})(t_2^{\frac{1}{2}} - t_2^{-\frac{1}{2}})(t_3^{\frac{1}{2}} - t_3^{-\frac{1}{2}}).$$

(c) Determine the multi-variable Alexander polynomial of the Whitehead link L_{Wh} (of Figure 2.7). Compare it to the single variable polynomial of the same link.

11.6. The Euler characteristic of multi-graded grid homology

The graded Euler characteristic of a finite dimensional $\mathbb{Z}\oplus\mathbf{H}(L)$-graded vector space X is

$$\chi(X) = \sum_{d\in\mathbb{Z},h=(s_1,\dots,s_\ell)\in\mathbf{H}(L)} (-1)^d t_1^{s_1}\cdots t_\ell^{s_\ell}\cdot \dim X_{d,h} \in \mathbb{Z}[t_1^{\pm\frac{1}{2}},\dots,t_\ell^{\pm\frac{1}{2}}].$$

If X is a finitely generated, bigraded module over $\mathbb{F}[V_1,\dots,V_n]$, then the graded Euler characteristic still makes sense, as a formal Laurent series.

The Euler characteristics of the various versions of multi-graded grid homologies can be expressed in terms of the multi-variable Alexander polynomial as follows:

THEOREM 11.6.1. *Let $\vec{L}$ be an ℓ-component oriented link with $\ell > 1$, let $\mathbb{G}$ be a grid diagram representing $\vec{L}$, and let n_i be the number of O-markings on its ith component, for $i = 1,\dots,\ell$. Then,*

$$\begin{aligned}
\chi(\widetilde{\mathbf{GH}}(\mathbb{G})) &= \pm\Delta_{\vec{L}}(t_1,\dots,t_\ell)\cdot\left(\prod_{i=1}^{\ell}(1-t_i^{-1})^{n_i}t_i^{\frac{1}{2}}\right),\\
\chi(\widehat{\mathbf{GH}}(\vec{L})) &= \pm\Delta_{\vec{L}}(t_1,\dots,t_\ell)\cdot\left(\prod_{i=1}^{\ell}(t_i^{\frac{1}{2}}-t_i^{-\frac{1}{2}})\right),\\
\chi(\mathbf{GH}^-(\vec{L})) &= \pm\Delta_{\vec{L}}(t_1,\dots,t_\ell)\cdot\left(\prod_{i=1}^{\ell}t_i^{\frac{1}{2}}\right).
\end{aligned} \tag{11.15}$$

Here, $\Delta_{\vec{L}}(t_1, \dots, t_\ell)$ denotes the multi-variable Alexander polynomial for $\vec{L}$ with its symmetric normalization $\Delta_{\vec{L}}(t_1, \dots, t_\ell) = \pm\Delta_{\vec{L}}(t_1^{-1}, \dots, t_\ell^{-1})$.

Proof. Using Lemma 11.2.5 and Proposition 11.5.8, we have that

$$\chi(\widetilde{\mathbf{GH}}(\mathbb{G})) = \chi(\widetilde{\mathbf{GC}}(\mathbb{G})) = \sum_{\mathbf{x}\in\mathbf{S}(\mathbb{G})} (-1)^{M(\mathbf{x})} t^{\mathbb{A}(\mathbf{x})} \doteq \sum_{\mathbf{x}\in\mathbf{S}(\mathbb{G})} (-1)^{M(\mathbf{x})} t^{\mathbb{A}'(\mathbf{x})}$$
$$\doteq \det(\mathbb{M}(\mathbb{G})) \doteq \Delta_{\vec{L}}(t_1, \dots, t_\ell) \left(\prod_{i=1}^{\ell} (1-t_i)^{n_i} \right).$$

This gives the first equation, up to multiplication by a monomial in the t_i.

The other two equations hinge on the relationship between the chain complex $\widetilde{\mathbf{GC}}(\vec{L})$ and the other complexes. For example, the relations on the Euler characteristics resulting from the short exact sequences

$$0 \to \frac{\mathbf{GC}^-(\mathbb{G})}{V_{j_1} = \dots = V_{j_{i-1}} = 0}[\![2, \mu_i]\!] \xrightarrow{V_{j_i}} \frac{\mathbf{GC}^-(\mathbb{G})}{V_{j_1} = \dots = V_{j_{i-1}} = 0} \to \frac{\mathbf{GC}^-(\mathbb{G})}{V_{j_1} = \dots = V_{j_i} = 0} \to 0$$

give the relations

$$\left(\prod_{i=1}^{\ell} (1-t_i^{-1}) \right) \cdot \chi(\mathbf{GC}^-(\mathbb{G})) = \chi(\widehat{\mathbf{GC}}(\mathbb{G})),$$
$$\left(\prod_{i=1}^{\ell} (1-t_i^{-1})^{n_i-1} \right) \cdot \chi(\widehat{\mathbf{GC}}(\mathbb{G})) = \chi(\widetilde{\mathbf{GC}}(\mathbb{G})).$$

(Compare Proposition 11.1.9.) Corollary 11.4.3 shows that $\chi(\widehat{\mathbf{GH}}(\vec{L}))$ is symmetric in $t_i \mapsto t_i^{-1}$, hence Equation (11.15) follows, and the other two equations follow from it. ☐

REMARK 11.6.2. In view of Theorem 11.6.1, Corollary 11.4.3 is a lift of Torres' symmetry in Theorem 11.5.3 to grid homology.

11.7. Seifert genus bounds from grid homology for links

Seifert genus bounds from grid homology are generalized to links, as follows:

PROPOSITION 11.7.1. *For any ℓ-component link $\vec{L}$, let F be any (not necessarily connected) oriented surface with $\partial F = \vec{L}$ and with no spherical components. We have*

$$\max\{h_1 + \dots + h_\ell \mid \widehat{\mathbf{GH}}(\vec{L}, (h_1, \dots, h_\ell)) \neq 0\} \leq \frac{\ell - \chi(F)}{2};$$

in particular,

$$\max\{h_1 + \dots + h_\ell \mid \widehat{\mathbf{GH}}(\vec{L}, (h_1, \dots, h_\ell)) \neq 0\} \leq g(\vec{L}) + \ell - 1, \tag{11.16}$$

where $g(\vec{L})$ is the Seifert genus of $\vec{L}$.

Proof. By Proposition 11.1.13, this is equivalent to

$$\max\{s \big| \widehat{GH}(\vec{L},s)\neq 0\} \leq \frac{\ell-\chi(F)}{2},$$

whose proof proceeds as the proof of Proposition 7.2.2. For a link, Proposition 11.2.6 shows that the Alexander grading is computed by

$$A(\mathbf{x}) = -\sum_{x\in\mathbf{x}} w_{\mathcal{D}}(x) + \frac{1}{8}\sum_{j=1}^{8n} w_{\mathcal{D}}(p_j) - \left(\frac{n-\ell}{2}\right)$$

(for the diagram $\mathcal{D}$ of $\vec{L}$ given by the grid); compare Equation (4.26) for the case of knots. As in the proof of Proposition 7.2.2, we pass to an equivalent, minimal-complexity non-negative matrix W, where the elements of minimal weight have weight equal to zero. In the case of links, the above formula shows that the actual Alexander grading and the weight of any generator differ by a correction term

$$\frac{1}{8}\sum_k w(p_k) - \frac{n-\ell}{2},$$

where here $w(p_k)$ denotes the $p_k = (i,j)$ entry of the matrix W (see the proof of Proposition 7.2.1), so the correction term computes the maximal Alexander gradings of grid states for our diagram. As in the proof of Proposition 3.4.9, this term is related to the Euler characteristic of the associated surface by

$$\frac{1}{8}\sum_k w(p_k) - \frac{n-\ell}{2} = \frac{\ell-\chi(F_H)}{2}.$$

The result follows from the proof of Proposition 3.4.11, which constructs, starting from F, a grid diagram $\mathbb{G}$ and another surface F_H with $\partial F_H = \vec{L}$ and no closed components, and $\chi(F_H)\geq\chi(F)$. □

REMARK 11.7.2. Although we do not prove this here, just like in the case for knots, the genus bounds from Proposition 11.7.1 are sharp, provided that the link has no unknotted, unlinked components; see [**159**]. A refinement of the above result is stated in Theorem 11.9.9.

11.8. Further examples

11.8.1. Alternating links. The following description of the grid homology of alternating links follows from a combination of Theorems 10.3.3 and 11.6.1 and Proposition 11.1.13:

THEOREM 11.8.1. *Let $\vec{L}$ be an oriented link with $\ell > 1$ components that has a connected, alternating projection. Write*

$$\left(\prod_{i=1}^{\ell}(t_i^{\frac{1}{2}} - t_i^{-\frac{1}{2}})\right)\Delta_{\vec{L}}(t_1,\dots,t_\ell) = \pm\sum_{s_1,\dots,s_\ell} a_{s_1,\dots,s_\ell} t^{s_1}\cdots t^{s_\ell},$$

and let $\sigma=\sigma(\vec{L})$ denote the signature of $\vec{L}$. The multi-graded grid homology groups of $\vec{L}$ are determined by

$$\widehat{\mathbf{GH}}_d(\vec{L},(s_1,\dots,s_\ell)) = \begin{cases} \mathbb{F}^{|a_{s_1,\dots,s_\ell}|} & \text{if } d = (\sum_{i=1}^{\ell} s_i) + \frac{\sigma-\ell+1}{2} \\ 0 & \text{otherwise.}\end{cases}$$

□

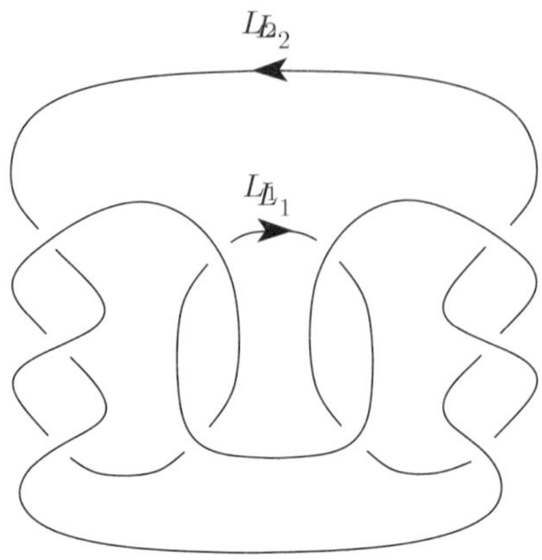

FIGURE 11.4. **The** $(3, -2, 2, -3)$ **pretzel link.**

11.8.2. The $(3, -2, 2, -3)$ **pretzel link.** As an example, we describe now some computations of grid homology for the $(3, -2, 2, -3)$ pretzel link, the two-component link pictured in Figure 11.4. (The orientation and labelling pictured here will be used to identify $H_1(S^3 \setminus \vec{L}; \mathbb{Z}) \cong \mathbb{Z}^2$.) This link naturally shows up when studying the Kinoshita-Terasaka knot appearing on the left of Figure 2.7: there is a skein triple connecting that knot, the unknot, and this pretzel link, with the orientation $\vec{L}$ specified in the figure.

We now study the grid homology of this link, to contrast it with the multi-variable Alexander polynomial $\Delta_{\vec{L}}(t_1, t_2)$, which is identically zero. A computer can calculate that the Poincaré polynomial of the simply blocked grid homology is

$$\widehat{\mathbf{P}}_{\vec{L}}(q, t_1, t_2) = \begin{array}{lllll} (1+q)t_1^{-1}t_2^2 & + & (q+q^2)t_2^2 & + & \\ 2(q^{-1}+1)t_1^{-1}t_2 & + & 2(1+q)t_2 & + & \\ (q^{-2}+q^{-1})t_1^{-1} & + & 3(q^{-1}+1) & + & (1+q)t_1 \\ & + & 2(q^{-2}+q^{-1})t_2^{-1} & + & 2(q^{-1}+1)t_1t_2^{-1} \\ & + & (q^{-3}+q^{-2})t_2^{-2} & + & (q^{-2}+q^{-1})t_1t_2^{-2}. \end{array}$$

(The dimensions of the Alexander bigraded groups, i.e. the coefficients of this Poincaré polynomial at $q = 1$, are also displayed on the right in Figure 11.5.) Computing grid homology by hand at all values of h is unwieldy, although it can be done with a computer. By contrast, the grid homology polytope can be computed by hand, as we do in the following:

PROPOSITION 11.8.2. *The grid homology polytope for the* $(3, -2, 2, -3)$ *pretzel link is the convex hull of the six points* $(-1, 2)$, $(0, 2)$, $(1, 0)$, $(1, -2)$, $(0, -2)$, $(-1, 0)$.

Before attacking the above, we start with some special cases.

PROPOSITION 11.8.3. *The groups* $\widehat{\mathbf{GH}}(\vec{L}, (-1, 2))$, $\widehat{\mathbf{GH}}(\vec{L}, (1, -2))$, $\widehat{\mathbf{GH}}(\vec{L}, (0, 2))$, *and* $\widehat{\mathbf{GH}}(\vec{L}, (0, -2))$ *all have dimension two. If* $\widehat{\mathbf{GH}}(\vec{L}, (s_1, s_2)) \neq 0$ *then* (s_1, s_2) *satisfies the following constraints:*

- $|s_1| \leq 1$
- $|s_1 + s_2| \leq 2$*; and equality holds iff* $(s_1, s_2) = (0, \pm 2)$
- $|s_1 - s_2| \leq 3$*; and equality holds iff* $(s_1, s_2) = \pm(1, -2)$.

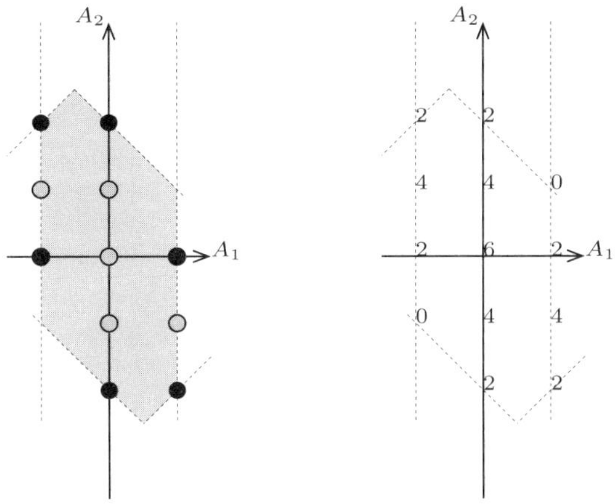

FIGURE 11.5. **Grid homology for the** $(3, -2, 2, -3)$ **pretzel link.** Points in $h \in \mathbf{H}(\vec{L})$ where the dimension $\widehat{\mathbf{GH}}(\vec{L}, h)$ can be non-zero are indicated by lightly shaded dots on the left. Points where the dimension is shown to be non-zero are indicated by solid dots. The dimensions of $\widehat{\mathbf{GH}}(\vec{L}, h)$ are displayed on the right.

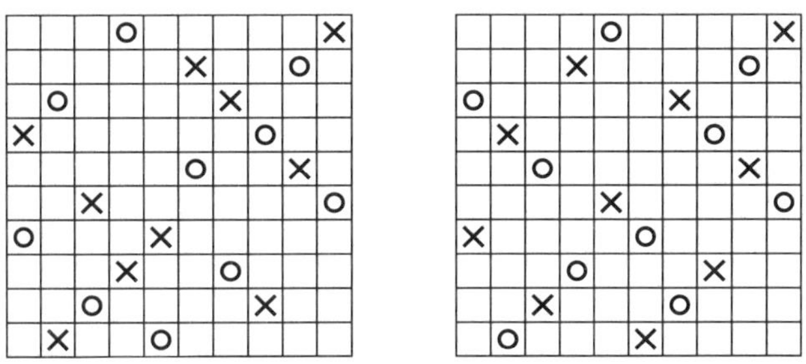

FIGURE 11.6. **Two grid diagrams for the** $(3, -2, 2, -3)$**.** The one on the left is $\mathbb{G}$ and the one on the right is $\mathbb{G}'$.

The upper bounds on the grid polytope from Proposition 11.8.3 are indicated by the shaded region on the left in Figure 11.5. The above proposition also justifies four out of the six solid dots displayed in that picture.

The proof of the above proposition is broken up into a sequence of lemmas, which use two different grid diagrams for the link. These diagrams are denoted $\mathbb{G}$, with

$$\mathbb{X} = (7, 1, 5, 3, 4, 9, 8, 2, 6, 10), \qquad \mathbb{O} = (4, 8, 2, 10, 1, 6, 3, 7, 9, 5)$$

appearing on the left in Figure 11.6; and $\mathbb{G}'$, with

$$\mathbb{X} = (4, 7, 2, 9, 5, 1, 8, 3, 6, 10), \qquad \mathbb{O} = (8, 1, 6, 3, 10, 4, 2, 7, 9, 5)$$

appearing on the right in Figure 11.6.

EXERCISE 11.8.4. Show that both $\mathbb{G}$ and $\mathbb{G}'$ above represent the $(3, -2, 2, -3)$ pretzel link from Figure 11.4.

LEMMA 11.8.5. *If* $\widehat{\mathbf{GH}}(\vec{L}, (s_1, s_2)) \neq 0$, *then* $|s_1| \leq 1$.

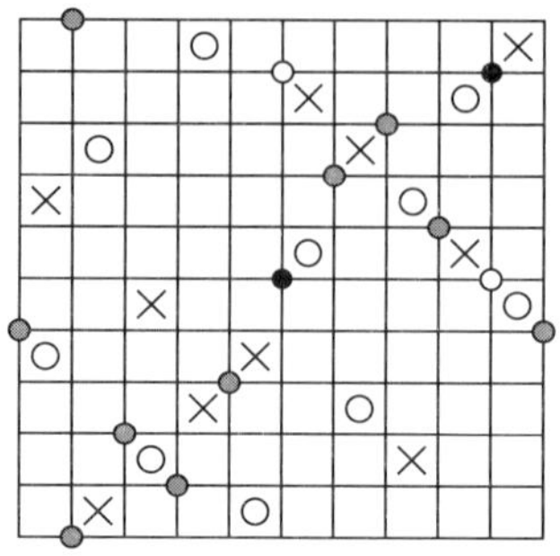

FIGURE 11.7. **The $(3, -2, 2, -3)$ pretzel link generators at $(-2, 2)$ in the grid $\mathbb{G}$.** These are the only two generators in the given Alexander multi-grading.

Proof. On the grid diagram $\mathbb{G}$, the unknot winds around a 3×3 square of lattice points; so for any $\mathbf{x} \in \mathbf{S}(\mathbb{G})$, the sum of the winding numbers of the unknotted component L_1 around the components of $\mathbf{x}$ lies between 0 and 3. Using Equation (11.9), it follows that if $\widetilde{\mathbf{GH}}(\vec{L}, (k_1, k_2)) \neq 0$, then $-2 \leq k_1 \leq 1$. The lemma now follows from Proposition 11.1.9. $\square$

LEMMA 11.8.6. *For the grid diagram $\mathbb{G}$, there are exactly two generators with $-A_1(\mathbf{x}) + A_2(\mathbf{x}) \leq 4$, and those two generators are*

$$(5, 1, 3, 2, 4, 6, 8, 9, 7, 10), \qquad (5, 1, 3, 2, 4, 10, 8, 9, 7, 6),$$

pictured in Figure 11.7, both with $A_1(\mathbf{x}) = -2$ and $A_2(\mathbf{x}) = 2$, and they cannot be connected by an empty rectangle, disjoint from $\mathbb{X}$ and $\mathbb{O}$.

Proof. Let $\mathcal{D}_1$ and $\mathcal{D}_2$ be the diagrams of the components L_1 and L_2 given by the grid $\mathbb{G}$. We consider the difference of winding numbers $-w_{\mathcal{D}_1} + w_{\mathcal{D}_2}$ (compare Section 3.4). The matrix of these winding number differences (given on the left) is equivalent (by adding and subtracting rows and columns of 1) to the non-negative matrix $\mathbf{W}$ on the right:

$$\begin{pmatrix} 0 & 0 & 0 & 0 & -1 & -1 & -1 & -1 & -1 & -1 \\ 0 & 0 & 0 & 0 & -1 & -1 & -2 & -2 & -2 & -1 \\ 0 & 0 & -1 & -1 & -2 & -2 & -3 & -2 & -2 & -1 \\ 0 & 1 & 0 & 0 & -1 & -1 & -2 & -1 & -2 & -1 \\ 0 & 1 & 0 & 0 & -1 & -1 & -1 & 0 & -1 & -1 \\ 0 & 1 & 0 & 1 & 0 & 0 & 0 & 1 & 0 & 0 \\ 0 & 0 & -1 & 0 & -1 & 0 & 0 & 1 & 0 & 0 \\ 0 & 0 & -1 & 0 & 0 & 1 & 1 & 1 & 0 & 0 \\ 0 & 0 & -1 & -1 & -1 & 0 & 0 & 0 & 0 & 0 \\ 0 & 0 & 0 & 0 & 0 & 0 & 0 & 0 & 0 & 0 \end{pmatrix} \sim \begin{pmatrix} 1 & 1 & 1 & 1 & 0 & 0 & 0 & 0 & 0 & 0 \\ 2 & 2 & 2 & 2 & 1 & 1 & 0 & 0 & 0 & 1 \\ 3 & 3 & 2 & 2 & 1 & 1 & 0 & 1 & 1 & 2 \\ 2 & 3 & 2 & 2 & 1 & 1 & 0 & 1 & 0 & 1 \\ 1 & 2 & 1 & 1 & 0 & 0 & 0 & 1 & 0 & 0 \\ 0 & 1 & 0 & 1 & 0 & 0 & 0 & 1 & 0 & 0 \\ 1 & 1 & 0 & 1 & 0 & 1 & 1 & 2 & 1 & 1 \\ 1 & 1 & 0 & 1 & 1 & 2 & 2 & 2 & 1 & 1 \\ 1 & 1 & 0 & 0 & 0 & 1 & 1 & 1 & 1 & 1 \\ 0 & 0 & 0 & 0 & 0 & 0 & 0 & 0 & 0 & 0 \end{pmatrix}$$

There are exactly two states with weight equal to zero (with respect to the second non-negative matrix). These generators can be found by visiting rows in the following order: 3, 4, 2, 8, 7, 9; and columns in the order 2, 1. At each stage, we choose the unique available position in $\mathbf{W}$ with a 0 entry to build up a partial state. There are two ways to complete this partial state to a grid state, giving the two generators from the lemma.

Obviously, there are no grid states with negative weight. By Proposition 4.7.2, this shows that for some c, the two states are the only states $\mathbf{x}$ with $A_1(\mathbf{x}) - A_2(\mathbf{x}) \geq c$. The lemma is proved by computing the Alexander grading of the two states. $\square$

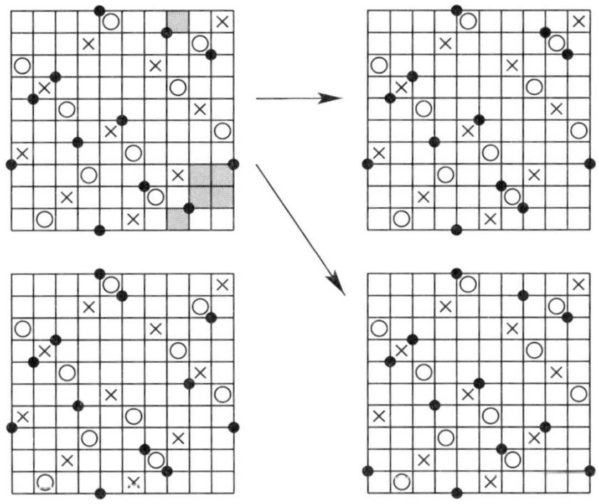

FIGURE 11.8. **The $(3, -2, 2, -3)$ pretzel link generators at $(0, 2)$ in the grid $\mathbb{G}'$.** Four generators are pictured. Two empty rectangles are shaded, inducing the maps are indicated by arrows.

LEMMA 11.8.7. *For the grid diagram $\mathbb{G}'$, there are four generators with $A(\mathbf{x}) = A_1(\mathbf{x}) + A_2(\mathbf{x}) \geq 2$, and those are*

$$\begin{array}{cc} (2,7,8,5,1,6,3,10,4,9), & (4,7,8,5,1,6,3,2,10,9) \\ (4,7,8,5,1,6,3,10,2,9), & (4,7,8,5,1,10,3,2,6,9) \end{array}$$

pictured in Figure 11.8, all with $A(\mathbf{x}) = 2$. These generators have $\mathbb{A}(\mathbf{x}) = (0, 2)$. Moreover, the homology of the corresponding complex is two-dimensional.

Proof. The matrix of $w_{\mathcal{D}_1} + w_{\mathcal{D}_2}$ and an equivalent non-negative matrix are:

$$\begin{pmatrix} 0 & 0 & 0 & 0 & 0 & -1 & -1 & -1 & -1 & -1 \\ 0 & 0 & 0 & 0 & 1 & 0 & 0 & 0 & 0 & -1 \\ 0 & -1 & -1 & -1 & 0 & -1 & -1 & 0 & 0 & -1 \\ 0 & -1 & 0 & 0 & 1 & 0 & 0 & 1 & 0 & -1 \\ 0 & -1 & 0 & -1 & 0 & -1 & -1 & 0 & -1 & -1 \\ 0 & -1 & 0 & -1 & 0 & 0 & 0 & 1 & 0 & 0 \\ 0 & 0 & 1 & 0 & 1 & 1 & 0 & 1 & 0 & 0 \\ 0 & 0 & 1 & 0 & 0 & 0 & -1 & 0 & 0 & 0 \\ 0 & 0 & 1 & 1 & 1 & 1 & 0 & 0 & 0 & 0 \\ 0 & 0 & 0 & 0 & 0 & 0 & 0 & 0 & 0 & 0 \end{pmatrix} \sim \begin{pmatrix} 1 & 1 & 1 & 1 & 1 & 0 & 0 & 0 & 0 & 0 \\ 1 & 1 & 1 & 1 & 2 & 1 & 1 & 1 & 1 & 0 \\ 1 & 0 & 0 & 0 & 1 & 0 & 0 & 1 & 1 & 0 \\ 1 & 0 & 1 & 1 & 2 & 1 & 1 & 2 & 1 & 0 \\ 1 & 0 & 1 & 0 & 1 & 0 & 0 & 1 & 0 & 0 \\ 1 & 0 & 1 & 0 & 1 & 1 & 1 & 2 & 1 & 1 \\ 0 & 0 & 1 & 0 & 1 & 1 & 0 & 1 & 0 & 0 \\ 1 & 1 & 2 & 1 & 1 & 1 & 0 & 1 & 1 & 1 \\ 0 & 0 & 1 & 1 & 1 & 1 & 0 & 0 & 0 & 0 \\ 0 & 0 & 0 & 0 & 0 & 0 & 0 & 0 & 0 & 0 \end{pmatrix}$$

Build a grid state with weight zero, visiting columns: 5, 3, and rows 8, 2, 4, 6. So far, each component is determined uniquely. The remaining components are found by a straightforward case analysis. The resulting four generators (given in the statement) obviously maximize $A = A_1 + A_2$; it is straightforward to check that they have $A = 2$. Looking at the picture, one finds two rectangles connecting these generators, as pictured in Figure 11.8. The non-trivial differentials are specified by

$$\widetilde{\partial}_{\mathbb{O},\mathbb{X}}(4,7,8,5,1,6,3,10,2,9) = (2,7,8,5,1,6,3,10,4,9) + (4,7,8,5,1,6,3,2,10,9)$$

Thus, the homology is two-dimensional, as claimed. □

Proof of Proposition 11.8.3. The implication $\widehat{\mathbf{GH}}(\vec{L}, (s_1, s_2)) \neq 0 \Rightarrow |s_1| \leq 1$ was verified in Lemma 11.8.5. The implication $\widehat{\mathbf{GH}}(\vec{L}, (s_1, s_2)) \neq 0 \Rightarrow s_1 + s_2 \leq 2$ with equality if and only if $(s_1, s_2) = (0, 2)$ follows immediately from Lemma 11.8.7

and the relationship between $\widetilde{\mathbf{GH}}$ and $\widehat{\mathbf{GH}}$ given in Proposition 11.1.9. By Lemma 11.8.6, $\widetilde{\mathbf{GH}}(\vec{L},(s_1,s_2)) \neq 0 \Rightarrow -s_1+s_2 \leq 4$ with equality if and only if $(s_1,s_2)=(-2,2)$. It now follows from Proposition 11.1.9 that $\widehat{\mathbf{GH}}(\vec{L},(s_1,s_2)) \neq 0 \Rightarrow -s_1+s_2 \leq 3$ with equality if and only if $(s_1,s_2)=(-1,2)$. Corollary 11.4.3 now completes the argument. □

EXERCISE 11.8.8. Show that the pretzel link $(3,-2,2,-3)$, oriented as in Figure 11.4, has Seifert genus one; while the same pretzel link, with the orientation of exactly one of its two components reversed, has Seifert genus two.

In Proposition 11.8.2 we computed $\widehat{\mathbf{GH}}(\vec{L},h)$ for some values of h. We can now complete the computation of the grid homology polytope:

Proof of Proposition 11.8.2. The proof follows from Proposition 11.8.3, and one additional computation, showing that $\widehat{\mathbf{GH}}(\vec{L},(\pm 1,0)) \neq 0$. To this end, consider the grid states for the grid diagram $\mathbb{G}'$ used to prove Lemma 11.8.7, pictured in Figure 11.9. Consider the states

$$\begin{aligned}
\mathbf{p}_1 &= (2,5,8,4,1,9,3,10,6,7) & \mathbf{q}_1 &= (2,5,9,4,1,8,3,10,6,7)\\
\mathbf{p}_2 &= (2,4,8,5,1,10,3,9,6,7) & \mathbf{q}_2 &= (2,5,8,4,1,10,3,9,6,7)\\
\mathbf{p}_3 &= (4,2,8,5,1,9,3,10,6,7) & \mathbf{q}_3 &= (4,2,8,5,1,10,3,9,6,7)\\
\mathbf{p}_4 &= (2,4,9,5,1,8,3,10,6,7) & \mathbf{q}_4 &= (4,2,9,5,1,8,3,10,6,7).
\end{aligned}$$

We claim that for $i=1,\dots,4$, $\tilde{\partial}_{\mathbb{O},\mathbb{X}}\mathbf{q}_i = 0$ and $\tilde{\partial}_{\mathbb{O},\mathbb{X}}\mathbf{p}_i = \mathbf{q}_i + \mathbf{q}_{i+1}$, where the subscripts are taken modulo 4; moreover, if $\mathbf{q}_i$ appears with non-zero multiplicity in $\tilde{\partial}_{\mathbb{O},\mathbb{X}}(\mathbf{x})$ for some grid state $\mathbf{x}$, then $\mathbf{x}=\mathbf{p}_j$ for some j. All of these claims are easily verified by glancing at Figure 11.9. We conclude that $\mathbf{q}_1$ is a homologically non-trivial cycle, which suffices to prove that $\widehat{\mathbf{GH}}(\vec{L},(1,0)) \neq 0$. □

11.9. Link polytopes and the Thurston norm

The grid homology polytope in $H_1(S^3 \setminus L;\mathbb{R})$ (Definition 11.1.12) is a link invariant. Another link invariant, which is a polytope in the same space, is determined by the Alexander polynomial, as follows. Let $\vec{L}$ be an oriented ℓ-component link, and label its components $\vec{L}_1,\dots,\vec{L}_\ell$. In the Alexander polynomial $\Delta_{\vec{L}}(t_1,\dots,t_\ell)$ the variable t_i corresponds to $\vec{L}_i$ through its oriented meridian μ_i. The monomial $t_1^{a_1}\cdots t_\ell^{a_\ell}$ corresponds to $\sum_{i=1}^{\ell} a_i\cdot\mu_i \in H_1(S^3\setminus\vec{L};\mathbb{R})$.

DEFINITION 11.9.1. The ***Alexander polytope*** B_Δ is the convex hull of all the points $h \in H_1(S^3\setminus\vec{L};\mathbb{R})$ whose corresponding monomial has non-zero coefficient in the multi-variable Alexander polynomial.

We describe now a third polytope: the Thurston polytope. Here we give a terse exposition of this construction; for more, see [**217**]. At the end of this section, we will spell out the relationships between these polytopes.

11.9.1. The Thurston norm.

Just as the Alexander polynomial and grid homology have multi-variable refinements, there is a further refinement of the knot genus to the case of links, called the *Thurston norm*. In this section we recall the construction of the Thurston norm and refer the reader to [**217**] for a detailed account of this beautiful story; see also [**19, 20, 21, 140**].

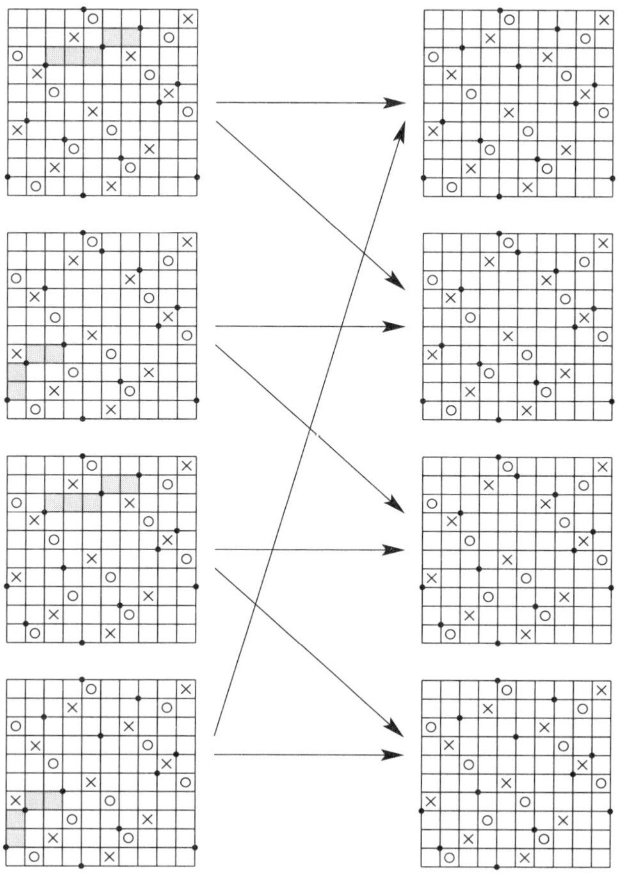

FIGURE 11.9. **Some generators with $h = (1, 0)$.** The generators in the left columns are, from top to bottom, denoted $\mathbf{p}_1, \ldots, \mathbf{p}_4$ in the proof of Proposition 11.8.2; the generators in the right column are labelled $\mathbf{q}_1, \ldots, \mathbf{q}_4$.

Let Y be a compact oriented three-manifold whose boundary is a union of tori. The relevant case for us is when Y is the complement of a tubular neighborhood of a link in S^3, but that specialization plays no role for our immediate purposes.

Let F be a compact, oriented surface-with-boundary, and let $i\colon (F, \partial F) \to (Y, \partial Y)$ be an embedding. The orientation on F gives rise to a fundamental class $[F, \partial F] \in H_2(F, \partial F; \mathbb{Z})$, whose image $a = i_*[F, \partial F]$ is an element of $H_2(Y, \partial Y; \mathbb{Z})$. In this case, we say that the embedded surface F *represents* the homology class $a \in H_2(Y, \partial Y; \mathbb{Z})$. (We will typically drop i from the notation, thinking of F as equipped with an embedding.) For example, if Y is the complement of a neighborhood of a knot $K \subset S^3$, a Seifert surface for K can be thought of as a representative of a generator of $H^2(Y, \partial Y) \cong H_2(S^3, K; \mathbb{Z}) \cong \mathbb{Z}$. The genus of a knot is the genus of a minimal genus representative of a generator of $H_2(S^3, K)$. Thurston generalizes this notion, using the following:

DEFINITION 11.9.2. If F is a connected, oriented surface-with-boundary, the ***complexity*** of F, denoted $\chi_-(F)$, is the quantity $\chi_-(F) = \max\{-\chi(F), 0\}$, where $\chi(F)$ denotes the Euler characteristic of the surface. If $F_1, \dots, F_k$ are the connected components of a compact surface F, then define $\chi_-(F) = \sum_{i=1}^k \chi_-(F_i)$.

Consider the map $\|\cdot\|_T\colon H_2(Y, \partial Y; \mathbb{Z}) \to \mathbb{Z}$ defined by

$$\|a\|_T = \inf\{\chi_-(F) \mid F \text{ is an oriented surface representing } a\}. \tag{11.17}$$

For non-negative integers n and m and $a, b \in H_2(Y, \partial Y; \mathbb{Z})$, it can be shown that

$$\|na\|_T = n\|a\|_T \qquad \text{and} \qquad \|na + mb\|_T \leq n\|a\|_T + m\|b\|_T.$$

(See [**19**, Lemma 5.8].) It follows quickly that $\|\cdot\|_T$ extends continuously to a convex function, called the ***Thurston semi-norm***, denoted $\|\cdot\|_T\colon H_2(Y, \partial Y; \mathbb{R}) \to \mathbb{R}$.

THEOREM 11.9.3. (Thurston, [**217**, Theorem 1]) *The function* $\|\cdot\|_T\colon H_2(Y, \partial Y; \mathbb{R}) \to \mathbb{R}$ *is a semi-norm that vanishes exactly on the subspace spanned by the embedded surfaces of non-negative Euler characteristic.* □

The seminorm $\|\cdot\|_T$ is uniquely specified by its unit ball B_T, defined by

$$B_T = \{a \in H_2(Y, \partial Y; \mathbb{R}) \mid \|a\|_T \leq 1\}.$$

Obviously B_T is a convex subset of $H_2(Y, \partial Y; \mathbb{R}) \cong \mathbb{R}^n$, and it is symmetric under the reflection $a \mapsto -a$. If $\|\cdot\|_T$ is a norm (i.e. when $\|a\|_T = 0$ implies that $a = 0$), then B_T is compact; otherwise, it is non-compact.

The next theorem places significant restrictions on the geometry of B_T, stated in terms of the naturally induced *dual Thurston norm* $\|\cdot\|_T^*$. In general, a norm on a vector space induces a norm on its dual space, defined by $\|\alpha\|_T^* = \sup_{b \in B_T} \alpha(b)$. In the case at hand, Poincaré duality identifies the dual space $H_2(Y, \partial Y; \mathbb{R})^* \cong H^2(Y, \partial Y; \mathbb{R}) \cong H_1(Y; \mathbb{R})$. The unit ball of the dual Thurston norm makes sense even when the Thurston semi-norm is not a norm; in this case, however, the dual norm can take ∞ as a value.

Recall that a (possibly non-compact) subset of $\mathbb{R}^n$ is called a *polytope* if it can be written as the intersection of finitely many half-spaces. A compact polytope is the convex hull of finitely many points, called the *vertices* of the polytope.

The following result is a consequence of integrality properties of the function $\|\cdot\|_T$:

THEOREM 11.9.4. (Thurston, [**217**, Theorem 2]) *The unit ball* B_T^* *of the dual Thurston norm is a compact polytope with integral lattice points* $\pm\beta_1, \dots, \pm\beta_k$ *as vertices. The unit ball* B_T *is a (possibly non-compact) polytope specified by* $B_T = \{a \in H_2(Y, \partial Y; \mathbb{R}) \mid |\beta_i(a)| \leq 1 \quad (1 \leq i \leq k)\}$. □

DEFINITION 11.9.5. The unit ball of the Thurston norm of the complement Y of a tubular neighborhood of the link $\vec{L}$ is called the ***Thurston polytope*** of $\vec{L}$ and it is denoted by B_T. The unit ball of its dual norm B_T^*, thought of as a norm on $H_2(Y, \partial Y; \mathbb{R})^* \cong H_1(Y; \mathbb{R})$, is called the ***dual Thurston polytope***.

It is more convenient to work with the unit ball of the dual Thurston norm, since that is always compact. The dual Thurston polytope determines the Thurston norm: for $a \in H_2(Y, \partial Y; \mathbb{R})$, it is easy to see that $\|a\|_T = \max_{h \in B_T^*} |\langle a, h\rangle|$.

11.9.2. Relationships between the polytopes. We now state relationships between the three polytopes we have encountered: the Alexander polytope, the grid polytope, and the dual Thurston polytope, all viewed as polytopes in $H_1(S^3\setminus L;\mathbb{R})$.

Given two convex sets A and B in a vector space, their Minkowski sum $A+B$ is the set of points which can be written as $a+b$ for some $a\in A$ and $b\in B$.

DEFINITION 11.9.6. Let $\vec{L}$ be an oriented link. The ***symmetric unit hypercube*** in $H_1(S^3\setminus\vec{L};\mathbb{R})$ is the convex hull of the 2^ℓ points $\pm\frac{1}{2}\mu_1+\cdots+\pm\frac{1}{2}\mu_\ell$.

PROPOSITION 11.9.7. *Let $\vec{L}$ be a link. The Minkowski sum of the Alexander polytope and the symmetric unit hypercube is contained inside the grid polytope.*

Proof. Clearly, if the Euler characteristic of $\widehat{\mathbf{GH}}(\vec{L},h)$ is non-zero, then the group $\widehat{\mathbf{GH}}(\vec{L},h)$ itself is non-trivial. Thus, the corollary follows from the relationship between $\widehat{\mathbf{GH}}(\vec{L})$ and the multi-variable Alexander polynomial stated in Theorem 11.6.1. The unit hypercube in the statement arises due to the multiplicative factor of $\left(\prod_{i=1}^{\ell}(t_i^{\frac{1}{2}}-t_i^{-\frac{1}{2}})\right)$ appearing in Equation (11.15). □

The following result of McMullen (which we have specialized here to links in S^3) generalizes the bound on the genus by the degree of the Alexander polynomial:

THEOREM 11.9.8 (McMullen, [**140**]). *If $\vec{L}$ is a link with $\ell>1$ components, then twice the Alexander polytope is contained in the dual Thurston polytope; equivalently, if $a\in H_2(S^3,\vec{L};\mathbb{R})$ is a non-trivial homology class and $\mathrm{PD}(a)\in H^1(S^3\setminus\vec{L};\mathbb{R})$ is its Poincaré dual, then $2\max_{h\in B_\Delta}\langle\mathrm{PD}(a),h\rangle\leq\|a\|_T$.* □

The relationship between the Thurston and the grid polytopes is expressed by the following result, which we state without proof here.

THEOREM 11.9.9 ([**182**]). *Let $\vec{L}$ be an oriented link which contains no components which are both unknotted and unlinked from the rest of $\vec{L}$. Then, the Minkowski sum of the dual Thurston polytope, scaled down by a factor of 2, and the symmetric unit hypercube equals the grid homology polytope; i.e., for each $\alpha\in H^1(S^3\setminus\vec{L};\mathbb{R})$, $\|\mathrm{PD}(\alpha)\|_T+\sum_{i=1}^{\ell}|\langle\alpha,\mu_i\rangle|=2\max_{\{h\in\mathbf{H}(L)\,|\,\widehat{\mathbf{GH}}(L,h)\neq 0\}}|\langle\alpha,h\rangle|$.* □

One consequence of the above theorem, which was conjectured by McMullen [**140**], can be viewed as a generalization of Murasugi's theorem about the genus of alternating knots (see Theorem 2.4.13):

COROLLARY 11.9.10 ([**182**]). *If $\vec{L}$ is an oriented link with more than one component, and it has a connected, alternating projection, then the convex hull of the non-zero terms in the multi-variable Alexander polynomial, scaled up by a factor of two, gives the dual Thurston polytope.*

Proof. Combine Theorem 11.8.1 with 11.9.9. Note that the unit hypercube factor $\left(\prod_{i=1}^{\ell}(t_i^{\frac{1}{2}}-t_i^{-\frac{1}{2}})\right)$ appearing in Theorem 11.8.1 cancels with the hypercube appearing in Theorem 11.9.9. □

Combining Proposition 11.8.2 with Theorem 11.9.9, we get the following:

PROPOSITION 11.9.11. *The dual Thurston polytope of the* $(3, -2, 2, -3)$ *pretzel link is the convex hull of the points* $(-1, 3)$, $(1, -1)$, $(1, -3)$, *and* $(-1, 1)$. □

See [**182**, Section 5.4] for a verification of Proposition 11.9.11 using the holomorphic theory, and [**118**] for a generalization.

EXERCISE 11.9.12. **(a)** Use Proposition 11.9.11 to express the Thurston polytope of the $(3, -2, 2, -3)$ pretzel link.
(b) Find minimal complexity surfaces representing the vertices of the Thurston polytope of the $(3, -2, 2, -3)$ pretzel link. You can appeal to Proposition 11.9.11 to show that your examples minimize complexity. (*Hint:* Use the grid diagrams $\mathbb{G}$ and $\mathbb{G}'$ from Subsection 11.8.2.)
(c)*Find the grid homology polytope for the pretzel link $(3, -2, -3, 2)$.

CHAPTER 12

Invariants of Legendrian and transverse knots

In this chapter we explore the connection between grid homology and contact geometry. Contact geometry offers a fortuitous blend of flexibility (e.g. every three-manifold admits a contact structure) and rigidity (often detected through the rich theory of pseudo-holomorphic curves). The power of contact geometry as a tool for studying the topology of three-manifolds has been magnified by its interaction with gauge theory. For example, Kronheimer and Mrowka proved that all knots have "Property P" [**109**] using contact structures as a stepping stone. Rudolph's "slice Bennequin inequality" from [**202**] (see Equation (12.9)) uses contact topology to estimate the slice genus of knots. Work of Eliashberg [**38**], Ding-Geiges [**31, 69**], Eliashberg-Thurston [**42**] and Gabai [**63**] provide further connections between contact and low-dimensional topology. Furthermore, as a tool within Floer theory, contact geometry also plays a fundamental role; see [**91, 115, 216**]. Finally, contact geometry is a rich subject in its own right; the tools of Floer homology and its combinatorial cousins can be brought to bear on purely contact geometric questions, as we do in the present chapter.

Informally, a *contact structure* on a three-manifold Y is a field of smoothly varying oriented two-planes in the tangent space of Y that are so twisted that there are no embedded surfaces in Y that are tangent to the two-planes. More formally, notice that if Y is an oriented three-manifold, then any smoothly varying oriented two-plane field ξ can be realized as the kernel of a nowhere vanishing one-form α. In terms of this presentation, the twisting condition is the condition that the three-form $\alpha \wedge d\alpha$ vanishes nowhere. For the structure to be compatible with the orientation of Y, we require that $\alpha \wedge d\alpha$ takes positive values on all positively oriented 3-frames. A *contact three-manifold* is an oriented three-manifold equipped with a contact structure ξ that can be specified as $\ker \alpha$ for α as above. Two contact three-manifolds (Y, ξ) and (Y', ξ') are said to be *contactomorphic*, and considered equivalent, if there is an orientation-preserving diffeomorphism $\phi\colon Y \to Y'$ which identifies the contact fields; equivalently, if $\xi = \ker \alpha$ and $\xi' = \ker \alpha'$, then there is a positive, smooth function f on Y so that $\phi^*(\alpha') = f \cdot \alpha$.

There are natural refinements of knot theory that arise in the presence of a contact structure. In a contact three-manifold it is natural to consider knots that are everywhere tangent to the plane field — the *Legendrian knots* — and those that are everywhere transverse to it — the *transverse knots*. We regard two Legendrian (or transverse) knots as equivalent if they are isotopic through Legendrian (or transverse) knots. Any knot type contains Legendrian and transverse representatives. Moreover, Legendrian and transverse knots have coarse numerical invariants (*Thurston-Bennequin* and *rotation numbers* for Legendrian knots, and *self-linking numbers* for transverse ones), which show that Legendrian and transverse knot theories are richer than traditional knot theory. The next challenge, to go beyond these classical invariants, provides an important impetus for exploring deeper methods in contact geometry; for example, the theory of pseudo-holomorphic disks [**23, 41, 81**] provided the first tools which showed that Legendrian knots have invariants beyond the classical ones. In the present chapter, we describe another non-classical invariant for Legendrian and transverse knots, defined in the framework of grid homology.

Grid diagrams can be used to study Legendrian and transverse knots, since a grid naturally determines a Legendrian and a transverse knot in the standard contact $\mathbb{R}^3$. In fact, toroidal grid diagrams, up to the equivalence generated by

- commutation moves and stabilizations of types *X:NW*, *X:SE*,
- commutation moves and stabilizations of types *X:NW*, *X:SE*, and *X:SW*,
- and commutation moves and stabilizations of types *X:NW*, *X:SE*, *X:SW*, and *X:NE*,

give models for Legendrian, transverse, and classical knot theory, respectively. (The first of these facts is verified in Proposition 12.2.6 below; the second combines Proposition 12.2.6 with Theorem 12.5.9 and Proposition 12.2.7; and the third is Cromwell's Theorem 3.1.9.)

At the same time, a grid diagram determines two distinguished grid states (which we first met in Section 6.4), which can be thought of as generators of the grid chain complex. Since these generators are cycles, they give rise to two distinguished elements in grid homology. The pair of homology classes turns out to be invariant under commutation moves and stabilizations of types *X:NW* and *X:SE*, so the pair provides a Legendrian invariant. One of the homology classes is also invariant under stabilizations of type *X:SW*, so it induces a transverse invariant.

The aim of this chapter is to introduce and study these grid invariants for Legendrian and transverse knots. We will start in Section 12.1 by briefly recalling the basics of Legendrian knot theory in $\mathbb{R}^3$ with its standard contact structure; we relate these notions with grid diagrams in Section 12.2. In Section 12.3 the grid invariants for Legendrian knots are defined. Section 12.4 gives applications of the Legendrian invariants, proving, among other things, Theorem 1.2.2 from the Introduction. In Section 12.5, we study transverse knots, and their relationship with Legendrian knots. This is used to define the transverse invariant. In Section 12.6, we give sample computations of transverse invariants, proving Theorem 1.2.3 from the Introduction. Section 12.7 discusses the fairly straightforward extension of these invariants to links. In Section 12.8 we sketch another way a grid diagram gives rise to a transverse knot (through braids). Section 12.9 discusses related constructions in gauge theory and Heegaard Floer homology.

The Legendrian and transverse invariants were first defined in [**185**]; the applications considered here are taken from [**25, 102, 157, 185**].

12.1. Legendrian knots in $\mathbb{R}^3$

We start by recalling some standard definitions from contact topology; for a more thorough discussion see [**45, 68**]. Consider the one-form $\alpha = dz - y\,dx$ on $\mathbb{R}^3$. The three-form $\alpha \wedge d\alpha$ is the standard volume form on $\mathbb{R}^3$, and so the two-plane field $\xi_{\mathrm{st}} = \ker \alpha$ gives a contact structure on $\mathbb{R}^3$. The choice of α orients the two-plane field: since $\alpha \wedge d\alpha$ vanishes nowhere, the restriction of $d\alpha$ gives a non-zero two-form, and hence an orientation, on $\xi_{\mathrm{st}} = \ker \alpha$. The structure ξ_{st} is called the *standard contact structure on* $\mathbb{R}^3$. The two-plane field ξ_{st} extends to the one-point compactification S^3 (cf. [**68**, Proposition 2.1.8]), inducing *the standard* *contact structure on* S^3. Just as knot theory in $\mathbb{R}^3$ is equivalent to knot theory in S^3, Legendrian knot theory in $(\mathbb{R}^3, \xi_{\mathrm{st}})$ is equivalent to the corresponding theory in the standard contact structure on S^3; we prefer to work in $(\mathbb{R}^3, \xi_{\mathrm{st}})$.

EXERCISE 12.1.1. Regard S^3 as the unit sphere in $\mathbb{C}^2$ and equip it with the (oriented) two-plane field defined by the complex lines in the tangent spaces. Show that the resulting oriented two-plane field is a contact structure, and the stereographic projection is a contactomorphism between this contact structure (restricted to $S^3 - \{\text{north pole}\}$) and the standard contact structure on $\mathbb{R}^3$.

DEFINITION 12.1.2. A ***Legendrian knot*** is a smooth knot $\mathcal{K} \subset \mathbb{R}^3$ whose tangent vectors are contained in the contact planes of ξ_{st}. Two Legendrian knots are ***Legendrian isotopic*** if they can be connected by a smooth one-parameter family of Legendrian knots.

There are two classical invariants of oriented Legendrian knots, the *Thurston–Bennequin invariant* and the *rotation number*, defined as follows. Let $\vec{\mathcal{K}}$ denote a given oriented Legendrian knot. The *contact framing* is the framing of $\vec{\mathcal{K}}$ that is contained in the two-plane field; i.e. it is specified by rotating the tangent vector $90°$ within the two-plane field. This condition specifies the contact framing uniquely up to isotopy. The contact framing naturally induces the following numerical invariant:

DEFINITION 12.1.3. Let $\vec{\mathcal{K}}'$ denote a push-off of $\vec{\mathcal{K}}$ with respect to the contact framing. The ***Thurston–Bennequin invariant*** of $\vec{\mathcal{K}}$, denoted $\mathrm{tb}(\vec{\mathcal{K}})$, is the linking number $\ell k(\vec{\mathcal{K}}, \vec{\mathcal{K}}')$. The Thurston-Bennequin number is independent of the choice of orientation on $\vec{\mathcal{K}}$.

The definition of the rotation number will use the following basic construction:

DEFINITION 12.1.4. Let E be an oriented two-plane bundle over a surface-with-boundary S, together with a trivialization along its boundary ∂S, thought of as specified by a nowhere vanishing section σ of $E|_{\partial S} \to \partial S$. The ***relative Euler number*** $e(E, \sigma)$ of $E \to S$ relative to σ is the signed count of zeros of any generic section s of $E \to S$ that extends σ.

DEFINITION 12.1.5. Let $\vec{\mathcal{K}}$ be an oriented, Legendrian knot. Fix a Seifert surface Σ for $\vec{\mathcal{K}}$. The restriction of ξ_{st} to Σ determines an oriented two-plane bundle over Σ, with a trivialization along the boundary induced by tangent vectors to the oriented knot. The ***rotation number*** $\mathrm{r}(\vec{\mathcal{K}})$ is the relative Euler number of the two-plane field ξ_{st} over Σ, relative to the trivialization over $\partial\Sigma$.

EXERCISE 12.1.6. **(a)** Verify that the rotation number of $\vec{\mathcal{K}}$ is independent of the choice of Seifert surface Σ used in its definition.
(b) Show that the sign of the rotation number depends on the choice of the orientation for $\vec{\mathcal{K}}$.

Legendrian knots can be conveniently studied via their *front projections*, determined by the projection map $(x, y, z) \mapsto (x, z)$. The front projection of a Legendrian embedding has no vertical tangencies, and in the generic case its only singularities are double points and cusps, locally modelled by the map $t \mapsto (t^2, t^3)$, cf. Section B.2.2 in the Appendix. For an example of a front projection (of the right-handed trefoil knot in Legendrian position), see Figure 12.1. A generic front

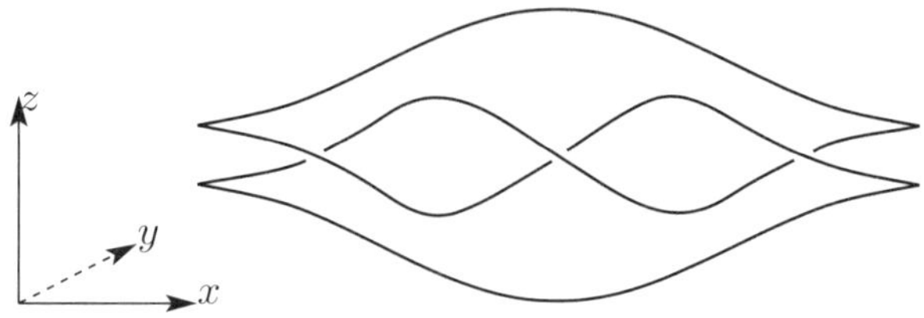

FIGURE 12.1. **Front projection of a Legendrian trefoil knot.**

projection determines the Legendrian embedding: i.e. if $(x(t), y(t), z(t))$ is a Legendrian knot, then the curves $x(t)$ and $z(t)$ determine $y(t)$ by the formula

$$y(t) = \begin{cases} \frac{z'(t)}{x'(t)} & \text{if } x'(t) \neq 0 \\ \frac{z''(t)}{x''(t)} & \text{if } x'(t) = 0. \end{cases}$$

In particular, at a crossing the strand with larger slope is under the strand with smaller slope. (To make sense of this convention, recall that the front projection is on the (x, z) plane, and note that the y axis points into this plane.) Since the projection determines the nature of the crossing, in Legendrian knot theory it is customary not to indicate the over- and under-crossings in front projections.

Any knot type can be represented by a Legendrian knot. This can be seen by working directly with the projection: replace vertical tangencies by cusps and apply a local move similar to the one from Figure 3.2 near crossings where the strand with larger slope is over. A variation on this argument can be used to show that $K \subset S^3$ can be approximated arbitrarily closely in the C^0 topology by a Legendrian representative of K; see [**45**, Theorem 2.5].

Legendrian knots can be studied through their Legendrian front projections in a way that is a natural enrichment of the way one studies ordinary knots via their generic projections. The basic moves on Legendrian front projections, called the *Legendrian Reidemeister moves*, are illustrated in Figure 12.2. A further move is *Legendrian planar isotopy*; such a move is an isotopy of the front projection that does not introduce vertical tangencies. These moves appear in the following Legendrian analogue of Reidemeister's Theorem 2.1.4. (For a proof see Section B.2.2.)

THEOREM 12.1.7 (Legendrian Reidemeister theorem, Światkowski, [**214**]). *Two front projections correspond to Legendrian isotopic Legendrian knots if and only if the projections can be connected by Legendrian planar isotopies and by Legendrian Reidemeister moves (shown in Figure* 12.2*).* □

The classical invariants of an oriented Legendrian knot $\vec{\mathcal{K}}$ can be easily read off from its oriented front projection $\mathcal{D}(\vec{\mathcal{K}})$ by the following formulas. Let $\mathrm{wr}(\mathcal{D}(\vec{\mathcal{K}}))$ denote the *writhe* of the projection, cf. Definition 2.1.14. Then,

$$\mathrm{tb}(\vec{\mathcal{K}}) = \mathrm{wr}(\mathcal{D}(\vec{\mathcal{K}})) - \frac{1}{2}\#\{\text{cusps in } \mathcal{D}(\vec{\mathcal{K}})\} \tag{12.1}$$

$$\mathrm{r}(\vec{\mathcal{K}}) = \frac{1}{2}\Big(\#\{\text{downward-oriented cusps}\} - \#\{\text{upward-oriented cusps}\}\Big). \tag{12.2}$$

The terms "downward-oriented" and "upward-oriented" are self-explanatory for a horizontal cusp; otherwise, isotop the cusp by less than 90° to make it horizontal. See [**68**, Section 3.5] or [**45**, Subsection 2.6.2] for the verification of these formulas.

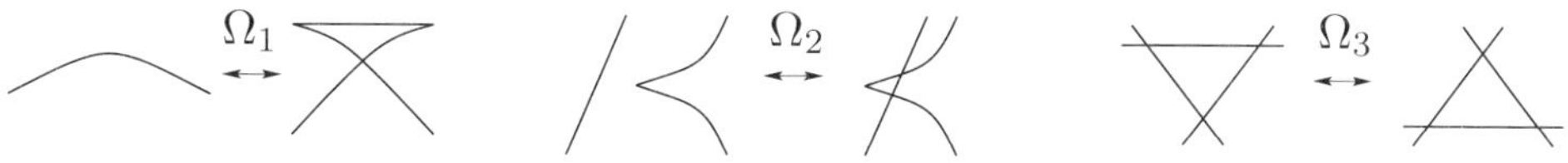

FIGURE 12.2. **Legendrian analogues of the Reidemeister moves.** Ω_1 has another variant obtained by rotating the picture by 180°. Ω_2 has three further variants, depending on whether the cusp is left or right, and whether the sign of the slope of the other strand is positive or negative. The local crossings are not indicated, as they are determined by the Legendrian condition.

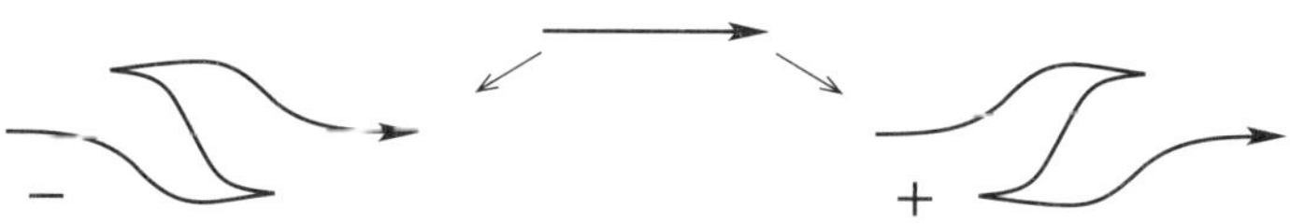

FIGURE 12.3. **Introducing cusps.** Given an arc in an oriented Legendrian knot, as in the middle, we can introduce two cusps locally, in two ways: left is the negative and right is the positive stabilization.

EXAMPLE 12.1.8. The Thurston-Bennequin number of the Legendrian right-handed trefoil knot illustrated in Figure 12.1 is 1, while the rotation number (with either orientation) is equal to 0.

EXERCISE 12.1.9. Find a Legendrian realization $\vec{\mathcal{K}}_{p,q}$ of the positive (p,q) torus knot with $\mathrm{r}(\vec{\mathcal{K}}_{p,q}) = 0$ and $\mathrm{tb}(\vec{\mathcal{K}}_{p,q}) = pq - p - q$.

Legendrian knots can be locally changed by introducing two new cusps, as shown in Figure 12.3. This operation is called *Legendrian stabilization*. Legendrian stabilization leaves the knot type unchanged, but it drops the Thurston-Bennequin number by one and changes the rotation number by $+1$ or -1, depending on how the cusps are oriented. If the stabilization increases rotation number by 1, it is called a *positive stabilization*, otherwise it is called a *negative stabilization*. See Figure 12.3. The inverse of a stabilization is called a *destabilization*. If $\vec{\mathcal{K}}$ is an oriented Legendrian knot, let $\vec{\mathcal{K}}^+$ resp. $\vec{\mathcal{K}}^-$ denote the knots obtained by a positive resp. negative stabilization. Stabilizations performed on different portions of the front projections produce Legendrian isotopic knots.

EXERCISE 12.1.10. **(a)** Using the Legendrian Reidemeister moves, show that the two Legendrian knots represented by the solid curves in Figure 12.4 are Legendrian isotopic.
(b) Verify that the Legendrian Reidemeister moves do not change the classical invariants of a Legendrian knot.
(c) Using Legendrian Reidemeister moves, show that a stabilization performed at different portions the front projections produce Legendrian isotopic knots, provided that the signs of the stabilizations are the same.

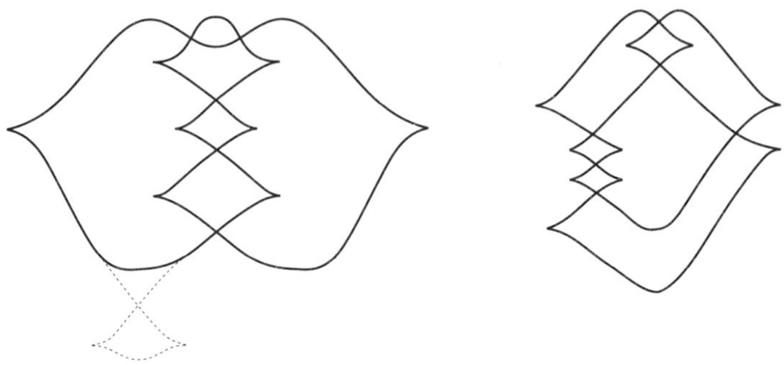

FIGURE 12.4. **Legendrian isotopic knots.** The two solid curves represent Legendrian isotopic knots. Performing the move Ω_1 indicated by the dashed curve, a sequence of Ω_2 and Ω_3 moves transforms one diagram into the other.

12.1.1. Legendrian simplicity. The classical invariants of a Legendrian knot (i.e. its knot type, Thurston-Bennequin and rotation numbers) are invariant under Legendrian isotopy. It is natural to wonder whether the classical invariants determine the Legendrian isotopy class of a knot.

DEFINITION 12.1.11. An oriented knot type $\vec{K}$ is ***Legendrian simple*** if for any two oriented Legendrian knots with equal Thurston-Bennequin and rotation numbers, both representing $\vec{K}$, there is a Legendrian isotopy connecting them. Otherwise the knot type $\vec{K}$ is called ***Legendrian non-simple***.

Clearly, $\vec{K}$ and $-\vec{K}$ are Legendrian simple at the same time.

Legendrian simplicity of various knot types (including the unknot [**39**], torus knots and the figure-eight knot [**44, 46**]) has been established using cut-and-paste methods in contact topology, relying on convex surface theory [**73**] and bypasses [**90**]. The first example of a Legendrian non-simple knot type was found by Chekanov [**23**]. His proof associates a suitable equivalence class of differential graded algebras to a Legendrian knot, presented by its Legendrian projection (i.e. its projection to the (x, y)-plane; see Section B.2.2). Although the definition was motivated by deep analytic ideas coming from contact homology [**40**], the presentations of the algebras are combinatorial. Using these algebras, Chekanov distinguished two Legendrian realizations of the mirror image of the 5_2 knot, shown by Figure 1.2. In [**43**], further extensions of this idea were used to verify Legendrian non-simplicity for an infinite family of twist knots.

12.2. Grid diagrams for Legendrian knots

A planar grid presentation G determines an oriented Legendrian front projection, via the following construction.

Let G be a planar grid diagram representing an oriented knot $\vec{K}$. The projection of $\vec{K}$ given by the grid G has corner points, which can be classified into four types: northwest (NW), southwest (SW), southeast (SE), and northeast (NE). Smooth all the northwest and southeast corners of the projection, and turn the southwest and northeast corners into cusps. To avoid vertical tangencies, tilt the diagram 45° clockwise. After this tilting, the cusps become horizontal, and the NE (respectively

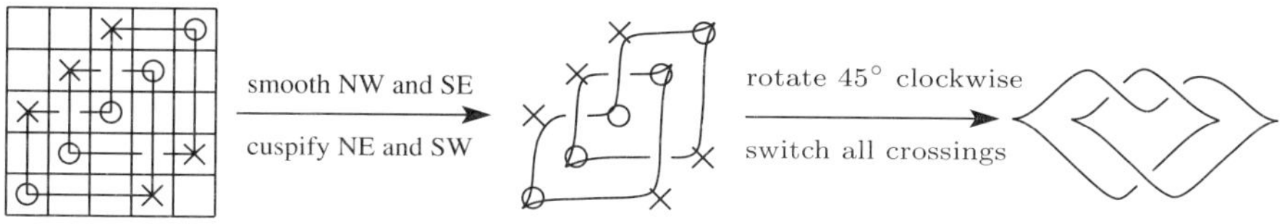

FIGURE 12.5. **The Legendrian knot determined by a grid diagram.** In this example we consider a grid diagram for the left-handed trefoil knot and get a Legendrian representative of the right-handed trefoil.

SW) corners become right (respectively left) cusps. At each crossing in the grid diagram, the vertical strand crosses over the horizontal one, but this is opposite to the crossing convention for a Legendrian front projection; so we reverse all the crossings. Thus, the above algorithm turns the planar grid G of the knot $\vec{K}$ into an oriented Legendrian front projection representing the mirror knot $m(\vec{K})$.

DEFINITION 12.2.1. Let G be a planar grid diagram. The oriented Legendrian knot $\vec{\mathcal{K}}_G$ represented by the Legendrian front projection constructed from G via the above procedure is called the ***Legendrian knot associated to the planar grid diagram*** G.

Figure 12.5 illustrates the above procedure, when G is a grid diagram for the left-handed trefoil, and so $\vec{\mathcal{K}}_G$ is a Legendrian representative of the right-handed trefoil.

REMARK 12.2.2. Alternatively, one could rotate the grid in the counterclockwise direction and smooth and cuspify the opposite corners to get a Legendrian projection of the knot $\vec{K}$, rather than its mirror, as in [**158**]. The present conventions are consistent with [**126, 185**]; see also [**108, 175**].

Although Definition 12.2.1 associates a Legendrian knot to a planar grid diagram, the Legendrian knot can be thought of as associated to a toroidal diagram:

PROPOSITION 12.2.3. *All Legendrian knots associated to the various planar realizations of a fixed toroidal grid diagram are Legendrian isotopic.*

The proof of this proposition relies on the following lemma.

LEMMA 12.2.4. *Commutations and stabilizations of types X:NW, X:SE, O:SE, and O:NW on a planar grid give isotopic Legendrian knots.*

Proof. Each elementary move of the grid diagram listed above can be realized as a sequence of Legendrian Reidemeister moves of the front projection. For example, Figure 12.6 demonstrates how the four versions of an *X:SE* stabilization become either a Legendrian planar isotopy or a Legendrian Reidemeister move Ω_1. (The stabilizations provide different diagrams based on the different positions of the O-markings in the row and column of the stabilized X.) Similarly, Figure 12.7 shows a commutation move inducing a Legendrian Reidemeister move Ω_2. (A commutation move might also involve a sequence of Legendrian Reidemeister moves Ω_3.) The verification of the statement for the further cases proceeds in a similar manner. □

FIGURE 12.6. **The four different ways an *X:SE* stabilization can appear after converting into a Legendrian front.** Three of them are Legendrian planar isotopies, while the fourth is a Legendrian Reidemeister move Ω_1 on the front projection.

FIGURE 12.7. **A commutation move giving a Legendrian Reidemeister move Ω_2.**

Proof of Proposition 12.2.3. Any two planar grids representing the same toroidal grid diagram differ by a sequence of vertical and horizontal cyclic permutations. In turn, by Lemma 3.2.4, a cyclic permutation can be decomposed into a sequence of *X:NW*, *X:SE*, *O:NW*, *O:SE* (de)stabilizations and commutations. The result now follows from Lemma 12.2.4. □

DEFINITION 12.2.5. Let $\mathbb{G}$ be a toroidal grid diagram for an oriented knot $\vec{K}$. The ***Legendrian knot associated to the toroidal grid diagram***, denoted $\vec{\mathcal{K}}_{\mathbb{G}}$ is the Legendrian representative of $m(\vec{K})$ associated to any planar realization of $\mathbb{G}$, in the sense of Definition 12.2.1. By Proposition 12.2.3, the oriented Legendrian knot type of $\vec{\mathcal{K}}_{\mathbb{G}}$ is well-defined.

The following converse to Lemma 12.2.4 is fundamental to the construction of Legendrian invariants from grid diagrams. (For more details, see Section B.4.1.)

PROPOSITION 12.2.6. *Any Legendrian knot type can be represented by some toroidal grid diagram. Two toroidal grid diagrams represent the same Legendrian knot type if and only if they can be connected by a sequence of commutations and (de)stabilizations of types X:NW and X:SE on the torus.*

Proof. First we claim that any Legendrian front projection can be turned into a grid diagram. Stretch the Legendrian front horizontally until no portion of the diagram is at an angle of more than 45° from the horizon. After this modification the curve can be approximated by a sequence of straight segments at an angle of ±45°, in such a way that we do not create new cusps. After rotating by 45° counter-clockwise and adjusting the segments to have consecutive integer coordinates, we have a grid diagram corresponding to the front projection. (A slightly different approach is explained in detail in Section B.4.1.)

Suppose next that the toroidal grids $\mathbb{G}_1$ and $\mathbb{G}_2$ can be connected by a sequence of moves as given in the statement. By Proposition 12.2.3 and Lemma 12.2.4, the diagrams represent the same Legendrian knot type.

For the converse direction we approximate Legendrian Reidemeister moves by grid moves. Approximations of Legendrian planar isotopies, and the Legendrian

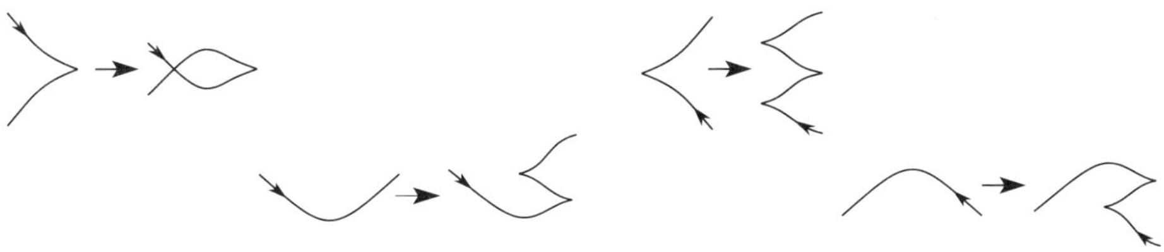

FIGURE 12.8. **Stabilizations of type** $X{:}SW$**, after conversion to a Legendrian front.** In each case, the Legendrian knot type is changed by a negative stabilization.

Reidemeister moves can all be realized by a sequence of commutations and stabilizations and destabilizations of the four possible types *X:NW*, *X:SE*, *O:NW*, and *O:SE*. The moves of type *O:NW* and *O:SE* can be eliminated via Lemma 3.2.2. For details on this approximation, see Theorem B.4.15. □

The stabilizations of a grid not appearing in the above proposition also have contact geometric interpretation:

PROPOSITION 12.2.7. *A stabilization of type X:SW (and similarly, of type O:NE) is a negative stabilization of the corresponding Legendrian knot. In a similar manner, stabilizatons of types X:NE and O:SW provide positive stabilizations on the Legendrian knots defined by the grid diagrams.*

Proof. Figure 12.8 shows the four possible cases of stabilizations of type *X:SW*; similar diagrams verify the statements for the other stabilizations. □

12.3. Legendrian grid invariants

Recall the definition of the two canonical grid states $\mathbf{x}^+ = \mathbf{x}^+(\mathbb{G})$ and $\mathbf{x}^- = \mathbf{x}^-(\mathbb{G})$ from Definition 6.4.1: the components of $\mathbf{x}^+(\mathbb{G})$ are the northeast corners of those squares which contain an X-marking, and the components of $\mathbf{x}^-(\mathbb{G})$ are the southwest corners of the same squares. In Section 6.4, we showed that these states are cycles in the grid complex, so they represent homology classes in grid homology. Our aim here is to show that these homology classes give invariants of the underlying Legendrian knot determined by the grid.

DEFINITION 12.3.1. Let $\mathbb{G}$ be a toroidal grid diagram for an oriented knot $\vec{K}$ and let $\vec{\mathcal{K}}_{\mathbb{G}}$ be the oriented Legendrian knot representing $m(\vec{K})$ associated to $\mathbb{G}$ (as in Definition 12.2.5). The homology classes in $GH^-(K)$ represented by the grid states $\mathbf{x}^+(\mathbb{G})$ and $\mathbf{x}^-(\mathbb{G})$ are denoted by $\lambda^+(\mathbb{G})$ and $\lambda^-(\mathbb{G})$, respectively; and these classes are called the ***Legendrian grid invariants*** of $\mathbb{G}$.

Let $\mathbb{G}$ be a toroidal grid diagram representing $\vec{K}$, and let $\vec{\mathcal{K}}_{\mathbb{G}}$ denote the induced oriented Legendrian representative of $m(\vec{K})$. The bigradings of the grid states $\mathbf{x}^{\pm}(\mathbb{G})$ are expressed in terms of the classical invariants of $\vec{\mathcal{K}}_{\mathbb{G}}$ as follows:

THEOREM 12.3.2. *The two cycles* $\mathbf{x}^+ = \mathbf{x}^+(\mathbb{G})$ *and* $\mathbf{x}^- = \mathbf{x}^-(\mathbb{G})$ *associated to* $\mathbb{G}$ *are supported in bigradings*

$$M(\mathbf{x}^+) = \mathrm{tb}(\vec{\mathcal{K}}) - \mathrm{r}(\vec{\mathcal{K}}) + 1, \qquad M(\mathbf{x}^-) = \mathrm{tb}(\vec{\mathcal{K}}) + \mathrm{r}(\vec{\mathcal{K}}) + 1,$$

$$A(\mathbf{x}^+) = \frac{\mathrm{tb}(\vec{\mathcal{K}}) - \mathrm{r}(\vec{\mathcal{K}}) + 1}{2}, \qquad A(\mathbf{x}^-) = \frac{\mathrm{tb}(\vec{\mathcal{K}}) + \mathrm{r}(\vec{\mathcal{K}}) + 1}{2},$$

where $\vec{\mathcal{K}} = \vec{\mathcal{K}}_{\mathbb{G}}$. *Furthermore, the homology elements* $\lambda^{\pm}(\mathbb{G}) \in GH^-(\mathbb{G})$ *represented by the cycles* $\mathbf{x}^{\pm}(\mathbb{G})$ *are non-torsion.*

We prove the above result after a little preparation. Consider a planar realization G of $\mathbb{G}$. The markings of G are classified into four types, according to whether they are northwest, southwest, southeast or northeast corners of the knot projection: a marking is, for example, of northwest type if one of the two neighbouring markings is to the south of it, and the other one is to the east. Let x_{NW} (and similarly, x_{SW}, x_{SE}, x_{NE}) denote the number of northwest (southwest, southest, northeast) X-markings, and define o_{NW}, o_{SW}, o_{SE}, o_{NE} similarly, now using O-markings instead of X-markings. There are obvious relations among these: for example, counting all upward pointing vertical segments in the projection first according to the starting X-marking, then according to the terminal O-marking gives:

$$x_{SW} + x_{SE} = o_{NW} + o_{NE}. \tag{12.3}$$

The *bridge index* $b(G)$ (from Definition 10.1.3) is equal to $x_{SE} + o_{SE}$ (which is the same as $x_{NW} + o_{NW}$). The number of upward-oriented cusps of $\vec{\mathcal{K}}_{\mathbb{G}}$ is equal to $o_{NE} + x_{SW}$, while the number of downward-oriented cusps is $x_{NE} + o_{SW}$. Thus, the rotation number of $\vec{\mathcal{K}}_{\mathbb{G}}$ can be written as

$$\mathrm{r}(\vec{\mathcal{K}}_{\mathbb{G}}) = \frac{1}{2}(x_{NE} + o_{SW} - x_{SW} - o_{NE}),$$

compare to Equation (12.2).

With the help of Equation (12.1), we see that

$$\mathrm{tb}(\vec{\mathcal{K}}_{\mathbb{G}}) + \mathrm{r}(\vec{\mathcal{K}}_{\mathbb{G}}) = \mathrm{wr}(\vec{\mathcal{K}}_{\mathbb{G}}) - o_{NE} - x_{SW}, \tag{12.4}$$

and

$$\mathrm{tb}(\vec{\mathcal{K}}_{\mathbb{G}}) - \mathrm{r}(\vec{\mathcal{K}}_{\mathbb{G}}) = \mathrm{wr}(\vec{\mathcal{K}}_{\mathbb{G}}) - x_{NE} - o_{SW}. \tag{12.5}$$

Note that the writhe $\mathrm{wr}(\vec{\mathcal{K}}_{\mathbb{G}})$ appearing here is the writhe of the front projection of the Legendrian knot $\vec{\mathcal{K}}_G$ determined by G, which is mirror to the knot projection associated to G in Definition 10.1.4; consequently $\mathrm{wr}(\vec{\mathcal{K}}_{\mathbb{G}}) = -\mathrm{wr}(G)$.

Now we are ready to determine the bigradings of the Legendrian grid invariants:

Proof of Theorem 12.3.2. Consider first $\mathbf{x}^-$. Fix a planar realization G for $\mathbb{G}$. Recall that the Maslov grading $M(\mathbf{x}^-)$ was computed by Equation (4.5) to be

$$M(\mathbf{x}^-) = \mathcal{J}(\mathbf{x}^-, \mathbf{x}^-) - 2\mathcal{J}(\mathbf{x}^-, \mathbb{O}) + \mathcal{J}(\mathbb{O}, \mathbb{O}) + 1 = \mathcal{J}(\mathbf{x}^- - \mathbb{O}, \mathbf{x}^- - \mathbb{O}) + 1.$$

According to Lemma 10.1.5, with the notation introduced above, we have that

$$\mathcal{J}(\mathbb{X} - \mathbb{O}, \mathbb{X} - \mathbb{O}) = -\mathrm{wr}(G) + b(\mathbb{G}) = -\mathrm{wr}(G) + x_{SE} + o_{SE}.$$

Since $\mathcal{J}(\mathbf{x}^-, \mathbf{x}^-) = \mathcal{J}(\mathbb{X}, \mathbb{X})$, it follows that

$$\mathcal{J}(\mathbf{x}^- - \mathbb{O}, \mathbf{x}^- - \mathbb{O}) = \mathcal{J}(\mathbb{X} - \mathbb{O}, \mathbb{X} - \mathbb{O}) + \mathcal{I}(\mathbb{X} - \mathbf{x}^-, \mathbb{O}) + \mathcal{I}(\mathbb{O}, \mathbb{X} - \mathbf{x}^-).$$

In computing $\mathcal{I}(\mathbb{X} - \mathbf{x}^-, \mathbb{O})$, notice that O-markings with type north or east contribute -1 (and the O-markings of northeast type contribute -2), so

$$\mathcal{I}(\mathbb{X}, \mathbb{O}) - \mathcal{I}(\mathbf{x}^-, \mathbb{O}) = -(o_{SE} + o_{NW} + 2o_{NE}).$$

In a similar fashion, we get that $\mathcal{I}(\mathbb{O}, \mathbb{X}) - \mathcal{I}(\mathbb{O}, \mathbf{x}^-) = 0$. Combining these observations, we find that

$$\mathcal{J}(\mathbf{x}^- - \mathbb{O}, \mathbf{x}^- - \mathbb{O}) = -\operatorname{wr}(G) + x_{SE} + o_{SE} - (o_{SE} + o_{NW} + 2o_{NE}). \tag{12.6}$$

By Equation (12.3), we have $x_{SE} - o_{NW} - 2o_{NE} = -x_{SW} - o_{NE}$, so Equation (12.6) shows

$$M(\mathbf{x}^-) = \mathcal{J}(\mathbf{x}^- - \mathbb{O}, \mathbf{x}^- - \mathbb{O}) + 1 = -\operatorname{wr}(G) - x_{SW} - o_{NE} + 1.$$

Combining this with Equation (12.4), the identity $\operatorname{wr}(\vec{\mathcal{K}}_{\mathbb{G}}) = -\operatorname{wr}(G)$ gives the claimed result. Proposition 6.4.8 gives

$$A(\mathbf{x}^-) = \frac{1}{2} M(\mathbf{x}^-),$$

concluding the computation of the bigrading of $\mathbf{x}^-$.

The argument for $\mathbf{x}^+$ can be reduced to the argument for $\mathbf{x}^-$, as follows. Let $\mathbb{G}'$ denote the grid diagram obtained by rotating $\mathbb{G}$ through $180°$. The same rotation induces a one-to-one correspondence between grid states, $\mathbf{x} \mapsto \mathbf{x}'$. Let M denote the Maslov grading on $\mathbf{S}(\mathbb{G})$ and M' the Maslov grading on $\mathbf{S}(\mathbb{G}')$. We claim that $M(\mathbf{x}) = M'(\mathbf{x}')$; in view of Proposition 4.3.1, it suffices to show that $M(\mathbf{x}^{SEO}) = 0$, which can be verified as in the proof of Lemma 4.3.5. Thus, $M(\mathbf{x}^+) = M'((\mathbf{x}^+)') = M'(\mathbf{x}^-)$; and by the result for $\mathbf{x}^-$, $M'(\mathbf{x}^-) = \operatorname{tb}(\vec{\mathcal{K}}') + r(\vec{\mathcal{K}}') + 1$, where $\vec{\mathcal{K}}'$ is the Legendrian knot associated to $\mathbb{G}'$; i.e. $\vec{\mathcal{K}}'$ has front projection obtained from the front for $\vec{\mathcal{K}}$ by rotation through $180°$. This $180°$ rotation preserves the writhe, and the total number of cusps, and hence by Equation (12.1), it preserves tb; but it switches the upward- and downward-oriented cusps, hence by Equation (12.2), it reverses the sign of r. This argument computes $M(\mathbf{x}^+)$ in terms of tb and r, and $A(\mathbf{x}^+) = \frac{1}{2}M(\mathbf{x}^+)$ again follows from Proposition 6.4.8.

Finally, the fact that the elements $\lambda^{\pm}(\mathbb{G})$ are non-torsion was proved in Proposition 6.4.8. □

We are now ready to state the invariance properties of $\lambda^{\pm}$. In fact, we already did most of the work in Section 6.4.

THEOREM 12.3.3. *Suppose that the two grid diagrams $\mathbb{G}$ and $\mathbb{G}'$ represent Legendrian isotopic knots. Then, there is a bigraded isomorphism $\phi\colon GH^-(\mathbb{G}) \to GH^-(\mathbb{G}')$ with $\phi(\lambda^+(\mathbb{G})) = \lambda^+(\mathbb{G}')$ and $\phi(\lambda^-(\mathbb{G})) = \lambda^-(\mathbb{G}')$.*

Proof. By Proposition 12.2.6, the assumption on $\mathbb{G}$ and $\mathbb{G}'$ implies that there is a sequence of commutations and (de)stabilizations of types *X:NW* and *X:SE* transforming the toroidal grid $\mathbb{G}$ to $\mathbb{G}'$. Lemma 6.4.4 shows that the grid invariants are taken to one another by the quasi-isomorphisms induced by commutations. Lemma 6.4.6 Case (S-1) identifies the homology classes for type *X:NW* and *X:SE* stabilizations. These results then conclude the proof of invariance. □

The above theorem justifies calling the homology classes $\lambda^{\pm}(\mathbb{G})$ *Legendrian grid invariants*, as we did in Definition 12.3.1. Sometimes, we write the Legendrian grid invariants as $\lambda^{\pm}(\vec{\mathcal{K}}) \in GH^{-}(K)$, where K is the mirror of the knot type of the Legendrian knot $\vec{\mathcal{K}}$.

Legendrian grid invariants behave in a controlled manner under stabilizations of the Legendrian knot.

THEOREM 12.3.4. *Let $\mathbb{G}$, $\mathbb{G}^+$, and $\mathbb{G}^-$ be grid diagrams whose associated Legendrian knots are $\vec{\mathcal{K}}$ and its stabilizations $\vec{\mathcal{K}}^+$ and $\vec{\mathcal{K}}^-$, respectively. Then, there are bigraded isomorphisms*

$$\phi^- : GH^-(\mathbb{G}) \longrightarrow GH^-(\mathbb{G}^-), \qquad \phi^+ : GH^-(\mathbb{G}) \longrightarrow GH^-(\mathbb{G}^+)$$

satisfying

$$\phi^-(\lambda^+(\vec{\mathcal{K}})) = \lambda^+(\vec{\mathcal{K}}^-), \qquad U \cdot \phi^+(\lambda^+(\vec{\mathcal{K}})) = \lambda^+(\vec{\mathcal{K}}^+),$$
$$U \cdot \phi^-(\lambda^-(\vec{\mathcal{K}})) = \lambda^-(\vec{\mathcal{K}}^-), \qquad \phi^+(\lambda^-(\vec{\mathcal{K}})) = \lambda^-(\vec{\mathcal{K}}^+).$$

Proof. By Theorem 12.3.3, it suffices to verify the statement for any choice of $\mathbb{G}^-$ and $\mathbb{G}^+$ representing $\vec{\mathcal{K}}^-$ and $\vec{\mathcal{K}}^+$, respectively. By Proposition 12.2.7, we can choose $\mathbb{G}^-$ and $\mathbb{G}^+$ to be obtained from $\mathbb{G}$ by stabilizations of type *X:SW* and *X:NE*, respectively. The required quasi-isomorphisms are supplied by Lemma 6.4.6: Case (S-2) verifies it for $\mathbb{G}^-$ and Case (S-3) for $\mathbb{G}^+$. □

The grid states $\mathbf{x}^{\pm}(\mathbb{G})$ can also be viewed as generators of the chain complex $\widehat{GC}(\mathbb{G})$ of the simply blocked theory. The corresponding homology classes are denoted $\widehat{\lambda}^{\pm}(\vec{\mathcal{K}}) \in \widehat{GH}(\mathbb{G})$; these, of course, can be thought of as the images of $\lambda^{\pm}(\mathbb{G})$ under the canonical homomorphism $GH^-(\mathbb{G}) \to \widehat{GH}(\mathbb{G})$.

THEOREM 12.3.5. *Suppose that $\mathbb{G}$ and $\mathbb{G}'$ are two grid diagrams whose associated Legendrian knots are Legendrian isotopic. Then, there is an isomorphism $\widehat{\phi}$ between $\widehat{GH}(\mathbb{G})$ and $\widehat{GH}(\mathbb{G}')$ that identifies the Legendrian invariants $\widehat{\lambda}^+(\mathbb{G})$ and $\widehat{\lambda}^-(\mathbb{G})$ with $\widehat{\lambda}^+(\mathbb{G}')$ and $\widehat{\lambda}^-(\mathbb{G}')$, respectively.*

Proof. Observe that the isomorphisms ϕ from Theorem 12.3.3 are induced by sequences of quasi-isomorphisms between the chain complexes GC^- (over $\mathbb{F}[V_1, \dots, V_n]$) and their inverses. Specializing these complexes and chain maps to $\widehat{GC}$ gives the isomorphism $\widehat{\phi}$ with the stated properties. □

EXERCISE 12.3.6. Suppose that the oriented Legendrian knot $\vec{\mathcal{K}}$ is the positive stabilization of another Legendrian knot. Show that $\widehat{\lambda}^+(\vec{\mathcal{K}}) = 0$.

The disadvantage of the simply blocked theory $\widehat{GH}$ is that it has less algebraic structure than GH^-, and hence it carries less information; its advantage is that it is easier to compute.

In practice, when trying to decide whether or not the homology elements $\widehat{\lambda}^{\pm}(\mathbb{G})$ vanish, it is easier to work with the fully blocked chain complex $\widetilde{GC}(\mathbb{G})$, which is a finite dimensional vector space, rather than with the simply blocked chain complex $\widehat{GC}(\mathbb{G})$, which is an infinite dimensional vector space. The following lemma justifies this simplification.

LEMMA 12.3.7. *The natural map* $\frac{GC^-(\mathbb{G})}{V_1=0} = \widehat{GC}(\mathbb{G}) \to \frac{GC^-(\mathbb{G})}{V_1=\cdots=V_n=0} = \widetilde{GC}(\mathbb{G})$ *induces an injective map on homology.*

Proof. We show that for all $i \geq 1$, the quotient map

$$q_i \colon \frac{GC^-(\mathbb{G})}{V_1 = \cdots = V_i = 0} \to \frac{GC^-(\mathbb{G})}{V_1 = \cdots = V_{i+1} = 0}$$

is an injection on homology. (Here the case $i = 1$ is interpreted as $\widehat{GC}(\mathbb{G}) \to \frac{GC^-(\mathbb{G})}{V_1=V_2=0}$.) This follows from the short exact sequence

$$0 \to \frac{GC^-(\mathbb{G})}{V_1=\cdots=V_i=0} \xrightarrow{V_{i+1}} \frac{GC^-(\mathbb{G})}{V_1=\cdots=V_i=0} \xrightarrow{q_i} \frac{GC^-(\mathbb{G})}{V_1=\cdots=V_{i+1}=0} \to 0. \tag{12.7}$$

Since V_{i+1} is homotopic to V_i (Lemma 4.6.9), which in turn acts trivially on $\frac{GC^-(\mathbb{G})}{V_1=\cdots=V_i=0}$, V_{i+1} acts trivially on $H(\frac{GC^-(\mathbb{G})}{V_1=\cdots=V_i=0})$. Thus, the long exact sequence associated to Equation (12.7) implies the injectivity of the map induced on homology by the quotient map q_i. The projection map $\widehat{GC}(\mathbb{G}) \to \widetilde{GC}(\mathbb{G})$ can be thought of as the composition $q_{n-1} \circ q_{n-2} \circ \cdots \circ q_1$, so the stated injectivity follows by the injectivity of the map on homology induced by each q_i. □

LEMMA 12.3.8. *The Legendrian invariant* $\widehat{\lambda}^+(\mathbb{G})$ *resp.* $\widehat{\lambda}^-(\mathbb{G})$ *vanishes if and only if the grid state* $\mathbf{x}^+(\mathbb{G})$ *resp.* $\mathbf{x}^-(\mathbb{G})$ *represents the trivial homology class in* $\widetilde{GH}(\mathbb{G})$.

Proof. The quotient map induces a chain map $q\colon \widehat{GC}(\mathbb{G}) \to \widetilde{GC}(\mathbb{G})$. The element $\mathbf{x}^+(\mathbb{G})$, thought of as a chain in $\widetilde{GC}(\mathbb{G})$, can be viewed as the image of $\mathbf{x}^+(\mathbb{G}) \in \widehat{GC}(\mathbb{G})$ under this map q. The result now follows from Lemma 12.3.7. □

Like the rotation number, the invariants $\lambda^+(\vec{\mathcal{K}})$ and $\lambda^-(\vec{\mathcal{K}})$ depend on the orientation of the Legendrian knot. Now, if $\vec{\mathcal{K}}$ is an oriented Legendrian knot, we can consider its reverse $-\vec{\mathcal{K}}$, which is the same Legendrian knot, equipped with the reversed orientation.

LEMMA 12.3.9. *Let G be a planar grid diagram representing $\vec{\mathcal{K}}$. Then, a grid diagram representing $-\vec{\mathcal{K}}$ can be obtained by reflecting G through the anti-diagonal.*

Proof. Recall that the Legendrian front projection defined by a grid diagram is given by rotating the projection 45° clockwise. Therefore, reflecting G through the antidiagonal corresponds to reflecting the Legendrian front projection through the z axis in the (x, z) plane and reversing the orientation of the knot; which in turn corresponds to rotating the Legendrian knot 180° around the z axis in $\mathbb{R}^3$ and reversing the orientation of the knot. To complete the lemma, we prove that rotation through 180° about the z axis takes any Legendrian knot to a Legendrian isotopic knot.

To construct this isotopy, use a radially symmetric model for the standard contact structure, with contact form (in cylindrical coordinates) given by $\alpha' = dz + r^2 d\theta$; i.e. $dz - ydx + xdy$. Consider the diffeomorphism $F\colon \mathbb{R}^3 \to \mathbb{R}^3$ given by $F(x, y, z) = (x, y, z + \frac{xy}{2})$. A straightforward computation verifies $F^*(\alpha') = \alpha$. Clearly, rotation R_θ by θ about the z axis preserves the contact form α'; and

$F^{-1} \circ R_\pi \circ F = R_\pi$. Thus, $t \mapsto F^{-1} \circ R_t \circ F$ gives an isotopy between the Legendrian knot specified by a front projection $\mathcal{D}$ and the one specified by $R_\pi(\mathcal{D})$. □

PROPOSITION 12.3.10. *If* $\mathbb{G}$ *is a grid diagram representing* $\vec{\mathcal{K}}$, *then there is a grid diagram* $\mathbb{G}'$ *representing* $-\vec{\mathcal{K}}$ *and an isomorphism* $\phi\colon GC^-(\mathbb{G}) \to GC^-(\mathbb{G}')$ *that sends the classes* $\lambda^+(\mathbb{G})$ *and* $\lambda^-(\mathbb{G})$ *to* $\lambda^-(\mathbb{G}')$ *and* $\lambda^+(\mathbb{G}')$, *respectively.*

Proof. Suppose that $\mathbb{G}$ represent $\vec{\mathcal{K}}$, and let $\mathbb{G}'$ be obtained from $\mathbb{G}$ by reflecting through the $\{x = -y\}$ axis. By Lemma 12.3.9, $\mathbb{G}'$ represents $-\vec{\mathcal{K}}$. The corresponding reflection induces a map on grid states, which extends to an isomorphism of chain complexes $\Phi\colon GC^-(\mathbb{G}) \to GC^-(\mathbb{G}')$. This isomorphism carries the states $\mathbf{x}^-(\mathbb{G})$ and $\mathbf{x}^+(\mathbb{G})$ to the states $\mathbf{x}^+(\mathbb{G}')$ and $\mathbf{x}^-(\mathbb{G}')$, respectively. □

The above proposition can be interpreted as saying that the pair of elements $\lambda^+(\mathbb{G})$ and $\lambda^-(\mathbb{G})$, thought of as an unordered pair of elements of $GH^-(\mathbb{G})$, is an invariant of the *unoriented* Legendrian isotopy class of the (oriented) Legendrian knot determined by $\mathbb{G}$.

12.4. Applications of the Legendrian invariants

We give two applications of the Legendrian invariants. In one, we give examples of Legendrian non-simple knots that are distinguished by their Legendrian invariants. In the second, we give a relationship between the classical invariants of a Legendrian knot and τ (and hence the slice genus) of the underlying smooth knot.

12.4.1. Legendrian non-simple knots. The first example of a pair of Legendrian knots that are not Legendrian isotopic but have the same classical invariants was found by Chekanov [**23**]. We reprove his result here, verifying Theorem 1.2.2 from the Introduction. The two candidate Legendrian knots $\vec{\mathcal{K}}_1$ and $\vec{\mathcal{K}}_2$ are shown in Figure 12.9, and their corresponding grid diagrams are shown in Figure 12.10. Note that $\vec{\mathcal{K}}_1$ is Legendrian isotopic to $-\vec{\mathcal{K}}_1$; and $\vec{\mathcal{K}}_2$ is Legendrian isotopic to $-\vec{\mathcal{K}}_2$. (We have already met these two diagrams in Section 3.1, cf. Figure 3.6.)

PROPOSITION 12.4.1 (Chekanov, [**23**]). *Consider the unoriented Legendrian knots* $\mathcal{K}_1$ *and* $\mathcal{K}_2$ *with front projections illustrated in Figure* 12.9, *both of topological type* $m(5_2)$ *(from Figure* 1.2*). Both Legendrian knots have* $\mathrm{tb} = 1$ *and* $\mathrm{r} = 0$*; but* $\mathcal{K}_1$ *and* $\mathcal{K}_2$ *are not Legendrian isotopic.*

Proof. The two Legendrian knots from Figure 12.9, each equipped with some orientation, are represented by the two grid diagrams $\mathbb{G}_1$ and $\mathbb{G}_2$ shown Figure 12.10. (The grid state $\mathbf{x}^+(\mathbb{G}_1)$ is also shown in the picture on the left.) It is easy to show that for all grid states $\mathbf{x}$, there are no rectangles in $\mathrm{Rect}^\circ(\mathbf{x}, \mathbf{x}^+(\mathbb{G}_1))$ with empty intersection with $\mathbb{X}$ and $\mathbb{O}$. By the symmetry of the diagram, the same is true for $\mathbf{x}^-(\mathbb{G}_1)$, immediately implying that $\lambda^+(\vec{\mathcal{K}}_1) \neq \lambda^-(\vec{\mathcal{K}}_1)$. Consider next the grid state $\mathbf{y}$ shown on $\mathbb{G}_2$ in Figure 12.10. Glancing at the figure, one can see that $\partial^-_{\mathbb{X}} \mathbf{y} = \mathbf{x}^+(\mathbb{G}_2) + \mathbf{x}^-(\mathbb{G}_2)$, which implies that $\lambda^+(\vec{\mathcal{K}}_2) = \lambda^-(\vec{\mathcal{K}}_2)$.

Thus, there is no isomorphism $GH^-(\mathbb{G}_1) \to GH^-(\mathbb{G}_2)$ taking the unordered pair of homology classes $\{\lambda^-(\vec{\mathcal{K}}_1), \lambda^+(\vec{\mathcal{K}}_1)\}$ (two classes which are distinct) to the

FIGURE 12.9. **Two different Legendrian representatives of the knot $m(5_2)$.**

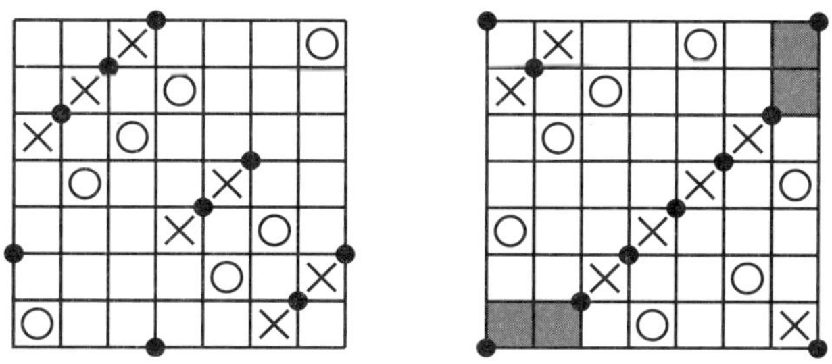

FIGURE 12.10. **Grid diagrams for the two different Legendrian representatives of the knot $m(5_2)$ of Figure 12.9.** The left diagram shows the grid state $\mathbf{x}^+(\mathbb{G}_1)$. The right diagram shows a grid state $\mathbf{y}$ and the two shaded rectangles which express $\partial^-_{\mathbb{X}}\mathbf{y} = \mathbf{x}^+(\mathbb{G}_2) + \mathbf{x}^-(\mathbb{G}_2)$.

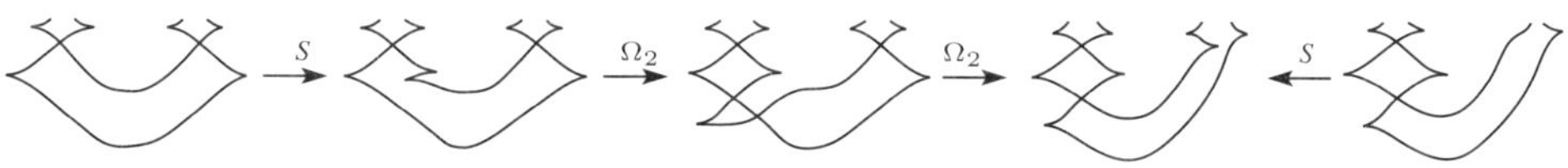

FIGURE 12.11. **Stabilizing Figure 12.9.** The sequence of diagrams shows that after one stabilization we can move one crossing from left to right; stabilization moves are labelled with an "S". Use Exercise 12.1.10 to see that the two knots in Figure 12.9 become Legendrian isotopic after one stabilization.

unordered pair of homology classes $\{\lambda^-(\vec{\mathcal{K}}_2), \lambda^+(\vec{\mathcal{K}}_2)\}$ (two classes which coincide). By Proposition 12.3.10, it follows that $\mathcal{K}_1$ and $\mathcal{K}_2$ are not Legendrian isotopic for any choice of orientations. □

The above argument distinguishes the Legendrian knots defined by the grid diagrams. Note that the Legendrian knots from Figure 12.9 become isotopic after one positive or one negative stabilization; see Figure 12.11 for a pictorial proof based on the Legendrian Reidemeister moves. The Legendrian knots in the knot type under examination are classified in [**48**]: in particular, there are exactly two distinct Legendrian knots in the knot type $m(5_2)$ with $tb = 1$ and $r = 0$. (Indeed, [**48**] provides a complete classification of Legendrian/transverse twist knots; see also [**163**].)

12.4.2. Bounds for $\tau(K)$ from the Legendrian invariants. Some properties of the Legendrian invariants can be used to get a bound on $\tau(K)$ (from Definition 6.1.5) in terms of classical invariants of a Legendrian realization of K or $m(K)$. These bounds, stated below, were first proved by Plamenevskaya [**187**] using the contact invariant in Heegaard Floer homology [**175**].

In the statement, we suppress orientations on our Legendrian knots: the Thurston-Bennequin invariant is insensitive to this data and, although the rotation number does depends on this choice, its absolute value does not.

PROPOSITION 12.4.2. *Let $\mathcal{K}$ be a Legendrian knot with knot type K. Then,*

$$\operatorname{tb}(\mathcal{K}) + |\operatorname{r}(\mathcal{K})| \leq 2\tau(K) - 1. \tag{12.8}$$

Proof. Let $\mathbb{G}$ be a grid diagram whose associated Legendrian knot is $\vec{\mathcal{K}}$ (whose existence is ensured by Proposition 12.2.6). According to Proposition 6.4.8 the elements $\lambda^-(\vec{\mathcal{K}}) = [\mathbf{x}^-(\mathbb{G})]$ and $\lambda^+(\vec{\mathcal{K}}) = [\mathbf{x}^+(\mathbb{G})]$ in $GH^-(m(K))$ are both non-torsion. The combination of this fact with the grading formula of Theorem 12.3.2 both for $\mathbf{x}^+(\mathbb{G})$ and for $\mathbf{x}^-(\mathbb{G})$ gives the inequality

$$\frac{1}{2}(\operatorname{tb}(\mathcal{K}) + |\operatorname{r}(\mathcal{K})| + 1) \leq -\tau(m(K)).$$

By Corollary 7.4.5, $-\tau(m(K)) = \tau(K)$, so we get the stated inequality. □

Proposition 12.4.2 is essentially a restatement of Corollary 6.4.9; except that now we have replaced the grid-diagrammatic quantities there with their Legendrian interpretations.

The combination of the above bound with the bound on the slice genus coming from τ (see Chapter 8) gives the following "slice Bennequin inequality" first proved using methods of gauge theory [**106, 202**] (generalizing an inequality discovered by Bennequin [**10**], stated for the Seifert genus in place of the slice genus):

COROLLARY 12.4.3 (Rudolph, [**202**]). *Let $\mathcal{K}$ be a Legendrian knot. Then,*

$$\operatorname{tb}(\mathcal{K}) + |\operatorname{r}(\mathcal{K})| \leq 2g_s(K) - 1. \tag{12.9}$$

Proof. This is an immediate combination of Proposition 12.4.2 with the bound $\tau(K) \leq g_s(K)$ from Corollary 8.1.2. □

As an illustration of Corollary 12.4.3, it is easy to find a Legendrian realization of the positive torus knot $T_{p,q}$ (cf. Exercise 12.1.9) for which the slice Bennequin inequality gives

$$(p-1)(q-1) \leq 2g_s(T_{p,q});$$

which, combined with the elementary inequalities

$$2g_s(T_{p,q}) \leq 2g(T_{p,q}) \leq (p-1)(q-1),$$

shows that

$$g_s(T_{p,q}) = g(T_{p,q}) = \frac{(p-1)(q-1)}{2}.$$

(This is essentially a recasting of the proof of Corollary 8.1.3.)

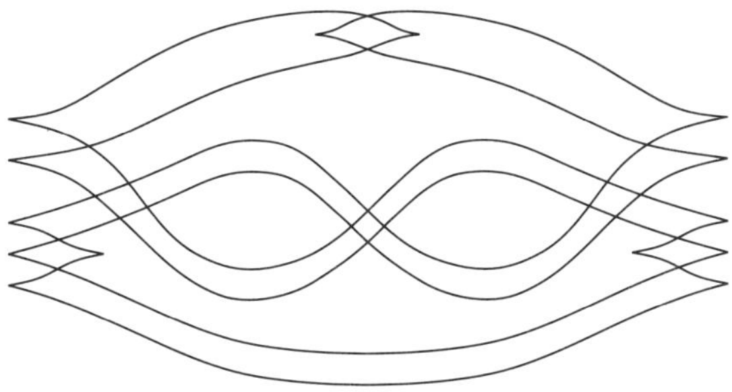

FIGURE 12.12. **A Legendrian representation of $W_0^+(T_{2,3})$.**

For another application in the same spirit, consider the 0-framed positive Whitehead double $K = W_0^+(T_{2,3})$ of the right-handed trefoil knot $T_{2,3}$. A Legendrian realization $\mathcal{K}$ of this knot with $\mathrm{tb}(\mathcal{K}) = 1$ and $\mathrm{r}(\mathcal{K}) = 0$ is shown in Figure 12.12. Thus, Equation (12.8) gives the bound $\tau(K) \geq 1$, implying that K is not smoothly slice. (This is essentially a reinterpretation of the computation appearing in the proof of Corollary 8.6.4.)

12.5. Transverse knots in $\mathbb{R}^3$

DEFINITION 12.5.1. A ***transverse knot*** is a smooth knot $\mathcal{T} \subset S^3$ whose tangent vectors are transverse to the contact planes of ξ_{st}. Two transverse knots are ***transverse isotopic*** if they are isotopic through transverse knots.

A transverse knot $\mathcal{T}$ naturally inherits an orientation from the (oriented) contact distribution: a tangent vector v to the transverse knot at $p \in \mathcal{T} \subset S^3$ is positive if the one-form α has positive value on it.

Recall that Legendrian knots have two classical numerical invariants; by contrast, transverse knots have only one:

DEFINITION 12.5.2. Let Σ be a Seifert surface for $\mathcal{T}$. The tangent space to Σ along $\mathcal{T}$ and the contact distribution ξ_{st} intersect in a line field. Choose a non-vanishing section σ of this line field along $\partial\Sigma$ that everywhere points away from Σ. Let $\mathcal{T}'$ be the push-off of $\mathcal{T}$ along this vector field. The ***self-linking number*** $\mathrm{sl}(\mathcal{T})$ of $\mathcal{T}$ is the linking number $\ell k(\mathcal{T}, \mathcal{T}')$.

EXERCISE 12.5.3. Let Σ be a Seifert surface for $\mathcal{T}$ and σ a trivialization of $\xi_{\mathrm{st}}|_{\partial\Sigma}$ as above. Show that $\mathrm{sl}(\mathcal{T})$ is the relative Euler number of the restriction of ξ_{st} to Σ, relative to the trivialization σ; cf. Definition 12.1.4.

12.5.1. Front diagrams of transverse knots. Like in the Legendrian case, transverse knots can be studied through their front projections, i.e., their image under the projection map $(x, y, z) \mapsto (x, z)$. We follow the discussion in [**43**], and refer the reader to [**43, 45**] for more information.

Unlike the Legendrian case, the front projection of a knot does not determine the transverse knot uniquely; rather, the transverse condition provides a bound on the $y(t)$ coordinate:

$$z'(t) - y(t) \cdot x'(t) > 0.$$

This inequality implies that the front projection of a transverse knot satisfies the following two constraints:

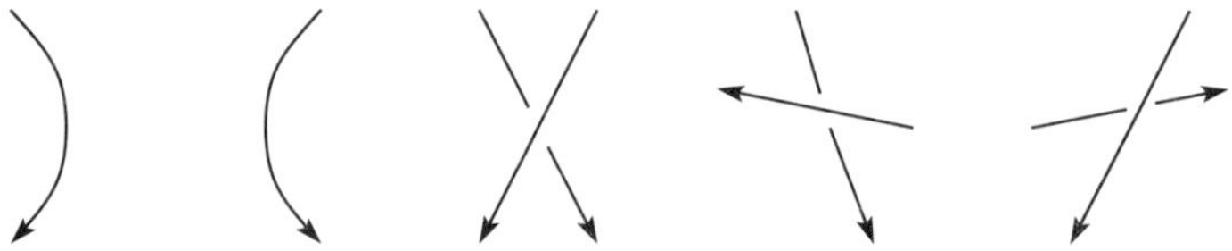

FIGURE 12.13. **Disallowed behaviour in a transverse knot projection.** A generic transverse front projection is not allowed to have vertical, downward tangencies (first two pictures); the next three crossings are also disallowed.

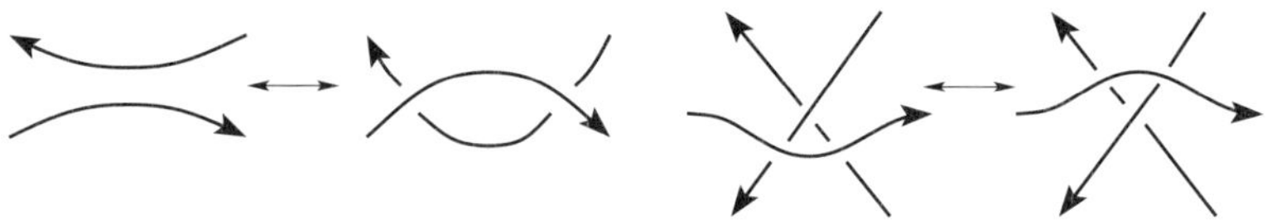

FIGURE 12.14. **Examples of transverse Reidemeister moves.**

- There are no vertical, downward-pointing tangencies. Indeed, at a vertical tangency we have $x'(t) = 0$, so the transverse condition gives
$$z'(t) - y(t) \cdot x'(t) = z'(t) > 0,$$
i.e. $z(t)$ points upwards.
- At each crossing where one strand $s_1 = (x_1(t), y_1(t), z_1(t))$ is oriented to the left (i.e. $x_1'(t) < 0$) while the other strand $s_2 = (x_2(t), y_2(t), z_2(t))$ is oriented to the right (i.e. $x_2'(t) > 0$), if the slope of s_1 is greater than the slope of s_2 (i.e. $\frac{z_1'(t)}{x_1'(t)} > \frac{z_2'(t)}{x_2'(t)}$), then s_2 crosses over s_1 (i.e. $y_2(t) < y_1(t)$). (Recall that the y axis points into the (x, z) plane, cf. Figure 12.1.)

Figure 12.13 shows the behaviour that is excluded by these constraints. A knot projection satisfying the above two properties specifies a transverse knot, unique up to transverse isotopy.

Similarly to Legendrian knots, transverse knots can be studied through their front projections. The *transverse Reidemeister moves* are simply those ordinary Reidemeister 2 and 3 moves that do not contain disallowed configurations as in Figure 12.13; see Figure 12.14 for some examples.

THEOREM 12.5.4. ([**43**], see also [**45**, Theorem 2.9], [**46**]) *Suppose that $\mathcal{D}_1$ and $\mathcal{D}_2$ are front projections of the transverse knots $\mathcal{T}_1$ and $\mathcal{T}_2$. The two transverse knots are transverse isotopic if and only if the diagrams $\mathcal{D}_1$ and $\mathcal{D}_2$ can be connected by a sequence of transverse Reidemeister moves and planar isotopies avoiding the configurations of Figure* 12.13. □

The proof of the above result is similar to the proof of Reidemeister's Theorem 2.1.4; see Section B.2.1.

Like the Thurston-Bennequin and rotation numbers of a Legendrian knot, the self-linking number of a transverse knot can be computed from a front projection: suppose that $\mathcal{T}$ is a transverse knot with front diagram $\mathcal{D}$. Then the self-linking number of $\mathcal{T}$ is given by

$$\mathrm{sl}(\mathcal{T}) = \mathrm{wr}(\mathcal{D}). \tag{12.10}$$

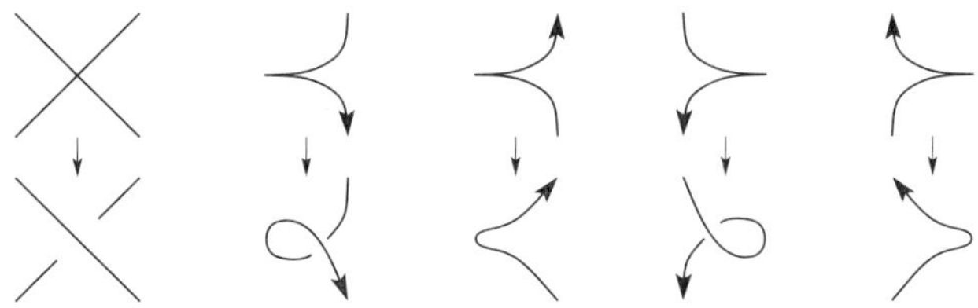

FIGURE 12.15. **Constructing the projection of a transverse push-off from an oriented Legendrian knot projection.** Given a Legendrian projection (top row), we construct a generic projection for the transverse push-off (second row), via the local modifications shown.

12.5.2. Approximation theorems. A Legendrian knot admits a transverse approximation, according to the following:

PROPOSITION 12.5.5. *Given any oriented Legendrian knot $\vec{\mathcal{K}}$, there are transverse knots $\mathcal{T} = \mathcal{T}(\vec{\mathcal{K}})$ that are arbitrarily close to $\vec{\mathcal{K}}$ in the C^1 topology. Furthermore, there is a neighborhood of $\vec{\mathcal{K}}$ in the C^1 topology with the property that any two transverse knots $\mathcal{T}$ and $\mathcal{T}'$ in this neighborhood are isotopic through transverse knots.*

Proof. Given a Legendrian knot projection, consider the transverse knot projection by smoothing the upward cusps and introducing Reidemeister 1 moves at the downward cusps; see Figure 12.15. Any generic front projection of a transverse knot which is sufficiently close to the Legendrian knot (in the C^1 topology) is of the form specified by this figure. This proves the uniqueness statement of the proposition.

Conversely, by making the Reidemeister 1 loop sufficiently small and choosing the function $y(t)$ so that $z'(t)-y(t)x'(t) > 0$ is sufficiently small, we can arrange for the transverse knot to approximate the initial Legendrian knot arbitrarily closely. This proves the existence statement. □

DEFINITION 12.5.6. The ***transverse push-off*** $\mathcal{T}(\vec{\mathcal{K}})$ of an oriented Legendrian knot $\vec{\mathcal{K}}$ is the transverse knot type which can be represented by transverse knot arbitrarily close (in the C^1 topology) to $\vec{\mathcal{K}}$.

REMARK 12.5.7. In [**45**], $\mathcal{T}(\vec{\mathcal{K}})$ is denoted $\vec{\mathcal{K}}_+$, and called the ***positive transverse push-off***, to distinguish it from the ***negative transverse push-off***, which, in our notation, would be denoted $\mathcal{T}(-\vec{\mathcal{K}})$. As a point set, the negative transverse push-off also has representatives arbitrarily close to $\vec{\mathcal{K}}$; but these representatives are oriented oppositely to $\vec{\mathcal{K}}$, and hence do not appear in any sufficiently small neighborhood of $\vec{\mathcal{K}}$ in the C^1 topology.

We can assume that at a crossing neither strand is vertical. We say that the strand s_1 with larger slope is *upward-pointing* if it points from left to right (that is, $x_1'(t) > 0$ at the crossing); and the strand s_{-1} with smaller slope is *upward-pointing* if it points from right to left (that is, $x_2'(t) < 0$ at the crossing). Otherwise the strands are *downward-pointing.* Notice that by a transverse isotopy avoding vertical tangencies the two strands at a crossing can be isotoped to strands with slopes $+1$

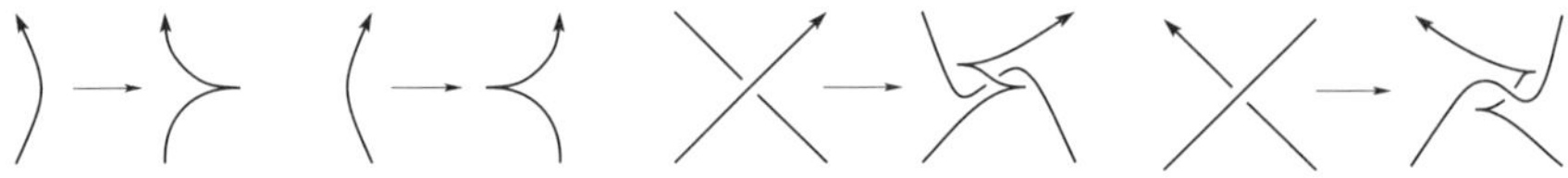

FIGURE 12.16. **Approximating a transverse projection with a Legendrian diagram.** At each Legendrian disallowed crossing, we stabilize an upward pointing strand; when both strands point upward, we choose one.

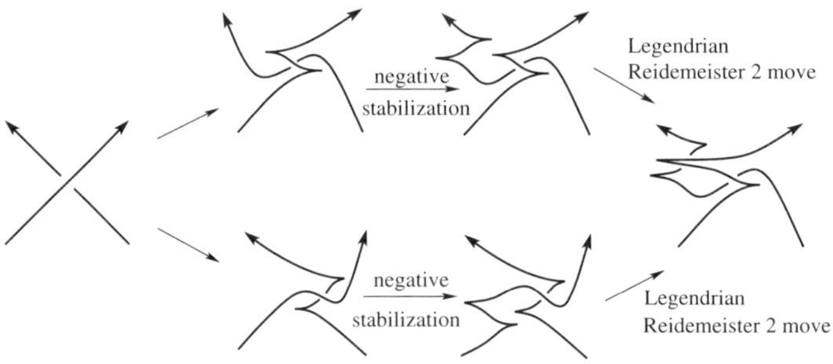

FIGURE 12.17. **The approximation is unique up to negative stabilization at a crossing of two upward pointing strands.**

and -1, and in this position the definition of up/down coincides with the natural up/down directions.

Any transverse knot $\mathcal{T}$ is transversely isotopic to the transverse push-off of some oriented Legendrian knot $\vec{\mathcal{K}}$, called a *Legendrian approximation* of $\mathcal{T}$. We construct a diagram for the Legendrian approximation by modifying the diagram $\mathcal{D}(\mathcal{T})$ of $\mathcal{T}$ as follows:

- at each isolated vertical tangency, introduce a cusp, as in the left two pictures in Figure 12.16,
- at each crossing not allowed in a Legendrian projection, stabilize one of the upward-pointing strands, as in the right two pictures in Figure 12.16.

There is a choice in the above procedure, at each Legendrian disallowed crossing where both strands are oriented upward: we chose one strand for stabilization. However, the two choices here become Legendrian isotopic once the other strand is also negatively stabilized, as shown in Figure 12.17. Thus, if we think of the algorithm as producing a Legendrian knot up to negative stabilizations, it becomes well-defined. It is also straightforward to see that the transverse push-off of the Legendrian approximation of a transverse knot $\mathcal{T}$ is transversely isotopic to $\mathcal{T}$.

REMARK 12.5.8. In keeping with the general spirit of the present work, we have defined Legendrian approximation in terms of diagrams. A different construction can be given using standard contact models for neighbourhoods of transverse knots and finding the Legendrian representatives in these local models.

Analyzing the Legendrian and transverse Reidemeister moves leads to the following result, which is of crucial importance in our definition of the transverse invariant. (See Section B.2.3 for a proof following [**43**]; see [**47**] for a more general statement.)

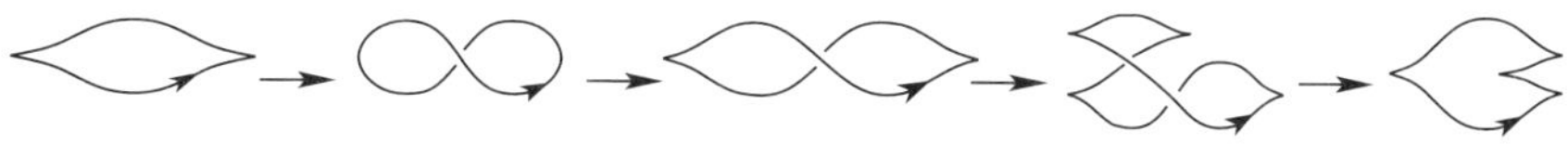

FIGURE 12.18. **Approximating the transverse push-off of the Legendrian unknot by a Legendrian knot.** The first arrow turns the Legendrian front into a transverse front, and the second arrow approximates the transverse knot with a Legendrian knot again. The last two arrows represent Legendrian Reidemeister moves Ω_1 and Ω_2 respectively.

FIGURE 12.19. **Stabilization of a transverse knot.**

THEOREM 12.5.9. (Epstein-Fuchs-Meyer, [**43**, Theorem 2.1]) *Every transverse knot type can be realized as the transverse push-off of some Legendrian knot type. Moreover, two oriented Legendrian knots $\vec{\mathcal{K}}_1$ and $\vec{\mathcal{K}}_2$ become Legendrian isotopic after some number of negative stabilizations if and only if their transverse push-offs are transversely isotopic.* □

The above theorem sets up a one-to-one correspondence between transverse knots and Legendrian knots, up to negative stabilizations. The correspondence associates to each transverse knot a Legendrian push-off, and to each Legendrian knot its transverse push-off. The Legendrian push-off of a transverse approximation is a negative stabilization of a Legendrian knot; see Figure 12.18 for an illustration.

LEMMA 12.5.10. *If $\mathcal{T}(\vec{\mathcal{K}})$ is the transverse push-off of the oriented Legendrian knot $\vec{\mathcal{K}}$, then*

$$\mathrm{sl}(\mathcal{T}(\vec{\mathcal{K}}))) = \mathrm{tb}(\vec{\mathcal{K}}) - \mathrm{r}(\vec{\mathcal{K}}). \tag{12.11}$$

Proof. Let $\mathcal{D}$ denote the front projection of $\mathcal{T}(\vec{\mathcal{K}})$ and let $\mathcal{D}(\vec{\mathcal{K}})$ denote the front projection of the Legendrian knot $\vec{\mathcal{K}}$. Recall that the transverse push-off introduces Reidemeister 1 moves at down cusps (see Figure 12.15); so

$$\mathrm{sl}(\mathcal{T}(\vec{\mathcal{K}})) = \mathrm{wr}(\mathcal{D}) = \mathrm{wr}(\mathcal{D}(\vec{\mathcal{K}})) - \#\{\text{down cusps in } \mathcal{D}(\vec{\mathcal{K}})\},$$

which by Equations (12.1) and (12.2) is equal to $\mathrm{tb}(\vec{\mathcal{K}}) - \mathrm{r}(\vec{\mathcal{K}})$. □

A transverse knot $\mathcal{T}$ also admits a stabilization $\mathcal{T}^+$, defined through the local modification of the front diagram of $\mathcal{T}$ as given by Figure 12.19.

EXERCISE 12.5.11. **(a)** Using the interpretation of the self-linking number of a transverse knot as the writhe of its front diagram, show that for the stabilization $\mathcal{T}^+$ of the transverse knot $\mathcal{T}$, we have $\mathrm{sl}(\mathcal{T}^+) = \mathrm{sl}(\mathcal{T}) - 2$.
(b) Show that the transverse push-off $\mathcal{T}(\vec{\mathcal{K}}^+)$ of the positive stabilization $\vec{\mathcal{K}}^+$ of a Legendrian knot $\vec{\mathcal{K}}$ is transverse isotopic to the transverse stabilization of the transverse push-off $\mathcal{T}(\vec{\mathcal{K}})$ of $\vec{\mathcal{K}}$.

Grid diagrams can be used to study transverse knots, by combining Theorem 12.5.9 with the material from Section 12.2. We summarize how:

- A grid diagram $\mathbb{G}$ uniquely specifies a transverse knot $\mathcal{T}_{\mathbb{G}}$ by taking the transverse push-off $\mathcal{T}(\vec{\mathcal{K}}_{\mathbb{G}})$ of the Legendrian knot $\vec{\mathcal{K}}_{\mathbb{G}}$ given by the grid.
- Any transverse knot $\mathcal{T}$ can be given by this way: approximate $\mathcal{T}$ by a Legendrian knot $\vec{\mathcal{K}}$ and take a grid representation of $\vec{\mathcal{K}}$ (as it is described in Proposition 12.2.6).
- If two grids represent transversely isotopic transverse knots, then (since the Legendrian knots given by the grids admit common negative stabilizations), the grids can be transformed into each other by a sequence of commutation moves and (de)stabilizations of types *X:NW*, *X:SE* and *X:SW*.

12.5.3. Transverse invariants in grid homology. The presentation of transverse knots through grid diagrams provides an invariant of transverse knot types in grid homology.

DEFINITION 12.5.12. Fix a grid diagram $\mathbb{G}$, let $\vec{\mathcal{K}}$ be its associated Legendrian knot, and $\mathcal{T}$ be its transverse push-off (as in Proposition 12.5.5). Define the ***transverse grid invariant*** $\theta(\mathcal{T})$ of the transverse knot $\mathcal{T}$ to be $\lambda^+(\mathbb{G}) \in GH^-(\mathbb{G})$.

The above transverse invariant is well-defined:

THEOREM 12.5.13. *The transverse grid invariant* $\theta(\mathcal{T}) \in GH^-_{\mathrm{sl}(\mathcal{T})+1}(\mathbb{G}, \frac{\mathrm{sl}(\mathcal{T})+1}{2})$ *is an invariant of the transverse isotopy class of the transverse knot* $\mathcal{T}$*; i.e., if* $\mathbb{G}$ *and* $\mathbb{G}'$ *are two grid diagrams whose associated Legendrian knots have transversely isotopic transverse push-offs, then there is an isomorphism*

$$\phi\colon GH^-(\mathbb{G}) \longrightarrow GH^-(\mathbb{G}')$$

with the property that $\phi(\theta(\mathcal{T})) = \theta(\mathcal{T}')$.

Proof. The gradings follow from Theorem 12.3.2 and Equation (12.11). The invariance statement is an immediate combination of Theorem 12.5.9 and Theorem 12.3.4, for negative stabilizations. □

Sometimes we write $\theta(\mathcal{T}) \in GH^-(m(K))$, where K is the knot type of $\mathcal{T}$. We can also consider the transverse invariant in the simply blocked theory, to get the invariant $\widehat{\theta}(\mathcal{T}) \in \widehat{GH}(m(K))$; compare Theorem 12.3.5.

12.6. Applications of the transverse invariant

As a concrete application, we use the transverse invariant to distinguish transversely non-isotopic knots, when the classical invariants fail to do so.

DEFINITION 12.6.1. A smooth oriented knot type K is ***transversely simple*** if for any two transverse representatives of K with equal self-linking numbers, there is a transverse isotopy connecting them. Otherwise the knot type K is called ***transversely non-simple***.

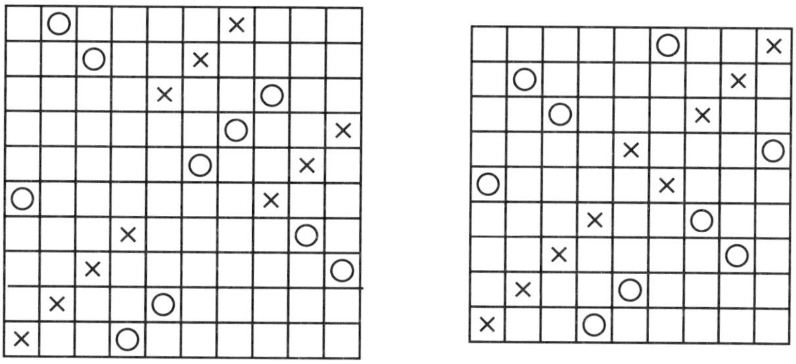

FIGURE 12.20. **The two grid diagrams for Proposition 12.6.2.** After appropriately stabilizing the diagram on the right, the resulting grid diagrams provide transversely non-isotopic transverse knots with identical classical invariants.

The first transversely non-simple knot type was found by Etnyre and Honda. In [**47**] they showed that the $(2,3)$-cable of the $(2,3)$ torus knot has a Legendrian representative which does not maximize Thurston-Bennequin number but nonetheless it is not the stabilization of another knot; see also [**157**, Proposition 5]. Transversely non-simple knot types were also found by Birman and Menasco [**13**] by using braid theory.

There are pairs of transverse knots with identical classical invariants, with the property that for one the transverse invariant $\widehat{\theta} \in \widehat{GH}$ vanishes while for the other it does not. By Theorem 12.5.13 it follows that the two knots are not transversely isotopic; and so examples of this kind verify Theorem 1.2.3 from the Introduction. We give two such pairs presently.

We start with an example taken from [**25**], for which the computation turns out to be very simple. The example gives two transverse representatives of the topological knot type $m(10_{161})$, with equal classical invariants but distinct transverse grid invariants. The examples are specified by the grid diagrams $\mathbb{G}_1$ and $\mathbb{G}_2$ shown in Figure 12.20.

PROPOSITION 12.6.2. *Let $\mathcal{T}_1$ be the transverse push-off of $\vec{\mathcal{K}}_{\mathbb{G}_1}$ and $\mathcal{T}_2'$ be the transverse push-off of the positive stabilization of $\vec{\mathcal{K}}_{\mathbb{G}_2}$. Both $\mathcal{T}_1$ and $\mathcal{T}_2'$ represent the topological knot type $m(10_{161})$ (with the same orientation) and have $\mathrm{sl}(\mathcal{T}_1) = \mathrm{sl}(\mathcal{T}_2') = 3$; but $\widehat{\theta}(\mathcal{T}_1) \neq 0$ and $\widehat{\theta}(\mathcal{T}_2') = 0$. Thus, the two transverse knots are not transversely isotopic, and the knot type $m(10_{161})$ is transversely non-simple.*

Proof. It is straightforward to compute that the Legendrian knots $\vec{\mathcal{K}}_1 = \vec{\mathcal{K}}_{\mathbb{G}_1}$ and $\vec{\mathcal{K}}_2 = \vec{\mathcal{K}}_{\mathbb{G}_2}$ have

$$\mathrm{tb}(\vec{\mathcal{K}}_1) - \mathrm{r}(\vec{\mathcal{K}}_1) = 3, \qquad \mathrm{tb}(\vec{\mathcal{K}}_2) - \mathrm{r}(\vec{\mathcal{K}}_2) = 5;$$

thus, the self-linking numbers of the transverse knots $\mathcal{T}_1 = \mathcal{T}_{\mathbb{G}_1}$ and $\mathcal{T}_2 = \mathcal{T}_{\mathbb{G}_2}$ are 3 and 5 respectively; and so, $\mathrm{sl}(\mathcal{T}_2') = 3$.

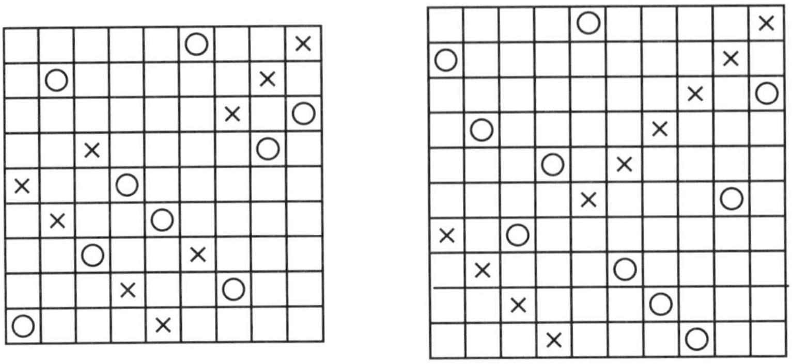

FIGURE 12.21. **Grid diagrams $\mathbb{G}_1$ (left) and $\mathbb{G}_2$ (right) representing distinct transverse knots with equal classical invariants.**

By Theorem 12.3.4, $\theta(\mathcal{T}_2^+) \in U \cdot GH^-(K)$ (where $K = 10_{161}$), so $\widehat{\theta}(\mathcal{T}_2^+) = 0$. Glancing at the grid diagram $\mathbb{G}_1$, it is clear that there is no $\mathbf{y} \in \mathbf{S}(\mathbb{G}_1)$ and $r \in \mathrm{Rect}^\circ(\mathbf{y}, \mathbf{x}^+(\mathbb{G}_1))$ with $r \cap \mathbb{O} = r \cap \mathbb{X} = \emptyset$; so $\widehat{\theta}(\mathcal{T}_1) \in \widehat{GH}(\mathbb{G}_1)$ is non-zero. □

The above computation is deceptively simple. For a more realistic example, we give a pair of transverse representatives for the smooth knot type of $m(10_{132})$, the mirror of the knot 10_{132} from [**198**]. These are the transverse push-offs of the Legendrian knots specified by the grid diagrams $\mathbb{G}_1$ and $\mathbb{G}_2$ from Figure 12.21. (This example appears in [**157**]; we use the grid presentation from [**102**].)

PROPOSITION 12.6.3. *Let $\mathcal{T}_1$ and $\mathcal{T}_2$ be the transverse push-offs of the Legendrian knots $\vec{\mathcal{K}}_1 = \vec{\mathcal{K}}_{\mathbb{G}_1}$ and $\vec{\mathcal{K}}_2 = \vec{\mathcal{K}}_{\mathbb{G}_2}$ from Figure 12.21. Both $\mathcal{T}_1$ and $\mathcal{T}_2$ represent the topological knot type $m(10_{132})$, both have self-linking numbers $\mathrm{sl}(\mathcal{T}_1) = \mathrm{sl}(\mathcal{T}_2) = -1$, but $\widehat{\theta}(\mathcal{T}_1) = 0$ and $\widehat{\theta}(\mathcal{T}_2) \neq 0$. Thus, $\mathcal{T}_1$ and $\mathcal{T}_2$ are not transversely isotopic, and hence the knot type $m(10_{132})$ is transversely non-simple.*

Before giving the details of the computation, we describe the overall strategy. Instead of working with the simply blocked grid homology $\widehat{GH}$, we work with the fully blocked groups $\widetilde{GH}$, as justified by Lemma 12.3.8. To determine whether or not the homology class $\widehat{\theta}(\mathbb{G})$ is trivial, we proceed as follows. If the grid state $\mathbf{x}$ appears with nonzero multiplicity in the boundary $\tilde{\partial}_{\mathbb{O},\mathbb{X}}(\mathbf{y})$ of the grid state $\mathbf{y}$, write $\mathbf{y} \to \mathbf{x}$. Define sets of grid states by the following inductive procedure. Let $\mathcal{A}_0 = \emptyset$ and $\mathfrak{B}_0 = \{\mathbf{x}^+(\mathbb{G})\}$. For $k = 1, 2, \ldots$, let

$$\mathcal{A}_k = \{\mathbf{y} \in \mathbf{S}(\mathbb{G}) \setminus \mathcal{A}_{k-1} \mid \mathbf{y} \to \mathbf{x} \quad \text{for some} \quad \mathbf{x} \in \mathcal{B}_{k-1}\},$$

$$\mathcal{B}_k = \{\mathbf{x} \in \mathbf{S}(\mathbb{G}) \setminus \mathcal{B}_{k-1} \mid \mathbf{y} \to \mathbf{x} \quad \text{for some} \quad \mathbf{y} \in \mathcal{A}_k\}.$$

Since there are finitely many grid states, this process terminates, giving the sets $\mathcal{A} = \cup_k \mathcal{A}_k$ and $\mathcal{B} = \cup_k \mathcal{B}_k$. Let $A(\mathbb{G})$ and $B(\mathbb{G})$ denote the linear subspaces of $\widetilde{GC}(\mathbb{G})$ spanned by the grid states of $\mathcal{A}$ and $\mathcal{B}$, respectively.

LEMMA 12.6.4. *Let $\mathbb{G}$ be a grid diagram. The homology class $[\mathbf{x}^+(\mathbb{G})] \in \widetilde{GH}(\mathbb{G})$ is non-zero if and only if the chain $\mathbf{x}^+(\mathbb{G}) \in B(\mathbb{G})$ is not in $\tilde{\partial}_{\mathbb{O},\mathbb{X}}(A(\mathbb{G}))$.*

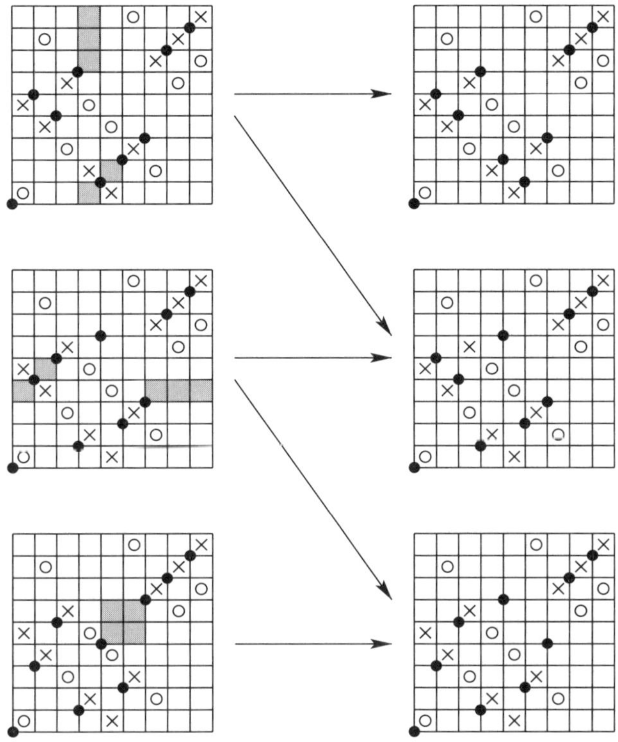

FIGURE 12.22. **Vanishing of the transverse invariant of $\mathcal{T}(\vec{\mathcal{K}}_1)$ from Proposition 12.6.3.** These grid states and rectangles demonstrate the vanishing of $[\mathbf{x}^+(\mathbb{G}_1)] \in GH^-(\mathbb{G}_1)$.

Proof. By the construction of these subsets, it follows that $\tilde{\partial}_{\mathbb{O},\mathbb{X}}(\widetilde{GC}(\mathbb{G})) \cap B(\mathbb{G}) = \tilde{\partial}_{\mathbb{O},\mathbb{X}}(A(\mathbb{G}))$, and $\mathbf{x}^+(\mathbb{G})$ is contained in $B(\mathbb{G})$. □

The above criterion is useful for grid diagrams where the sets $\mathcal{A}$ and $\mathcal{B}$ are not too large. As we will see, in the case of the diagram $\mathbb{G}_2$ of Figure 12.21 this computation is manageable: $\mathcal{A}$ has 12 elements, while $\mathcal{B}$ has 10. (As a point of comparison, the grid diagram itself has $10! = 3628800$ grid states.)

Proof of Proposition 12.6.3. It is straightforward to verify that both $\mathcal{T}_1$ and $\mathcal{T}_2$ represent the knot type $m(10_{132})$. It is easy to compute $\mathrm{sl}(\mathcal{T}_1) = \mathrm{sl}(\mathcal{T}_2) = -1$.

We claim that $[\mathbf{x}^+(\mathbb{G}_1)] = 0$ and $[\mathbf{x}^+(\mathbb{G}_2)] \neq 0$ as homology classes in the respective fully blocked grid homologies $\widetilde{GH}(\mathbb{G}_1)$ and $\widetilde{GH}(\mathbb{G}_2)$. In the light of Lemma 12.3.8, this property is sufficient to distinguish the transverse knots.

Applying the algorithm from Lemma 12.6.4, we can find three grid states $\mathbf{y}_1$, $\mathbf{y}_2$, and $\mathbf{y}_3$ with the property that

$$\mathbf{x}^+(\mathbb{G}_1) = \tilde{\partial}_{\mathbb{O},\mathbb{X}}(\mathbf{y}_1 + \mathbf{y}_2 + \mathbf{y}_3);$$

the relevant boundary maps are displayed in Figure 12.22.

For $\mathbf{x}^+(\mathbb{G}_2)$, the computation is slightly longer. First notice that the upper right corners of the five X-markings in the diagonal of $\mathbb{G}_2$ (but not in the upper right corner of the grid) are fixed for all elements of $\mathcal{A}$ and $\mathcal{B}$. This follows from the fact that no rectangle disjoint from the X-markings can have these coordinates as initial corners, and (because of the presence of the three O-markings in the upper right 6×6 subsquare) no rectangle disjoint from the O-markings can have these coordinates as terminal corners. This property of the diagram significantly decreases the number of potential elements in the corresponding sets $\mathcal{A}$ and $\mathcal{B}$.

We specify grid states by their corresponding permutations. For instance,

$$\mathbf{x}^+(\mathbb{G}_2) = (1,5,4,3,2,6,7,8,9,10).$$

Since the last 5 coordinates for all the permutations we consider are the same, we will drop those coordinates from the descriptions of all other grid states. With this convention, let

$$\begin{array}{lll} \mathbf{y}_1 = (1,4,5,3,2) & \mathbf{y}_2 = (1,5,3,4,2) & \mathbf{y}_3 = (1,5,4,2,3) \\ \mathbf{y}_4 = (4,1,5,2,3) & \mathbf{y}_5 = (3,5,1,2,4) & \mathbf{y}_6 = (2,4,5,1,3) \\ \mathbf{y}_7 = (2,5,3,1,4) & \mathbf{y}_8 = (1,4,2,5,3) & \mathbf{y}_9 = (4,1,2,3,5) \\ \mathbf{y}_{10} = (5,1,2,4,3) & \mathbf{y}_{11} = (2,4,1,3,5) & \mathbf{y}_{12} = (5,2,1,3,4) \end{array}$$

and

$$\begin{array}{lll} \mathbf{x}_1 = (4,1,5,3,2) & \mathbf{x}_2 = (3,5,1,4,2) & \mathbf{x}_3 = (2,5,4,1,3) \\ \mathbf{x}_4 = (4,2,5,1,3) & \mathbf{x}_5 = (3,5,2,1,4) & \mathbf{x}_6 = (4,1,2,5,3) \\ \mathbf{x}_7 = (2,4,1,5,3) & \mathbf{x}_8 = (4,2,1,3,5) & \mathbf{x}_9 = (5,2,1,4,3). \end{array}$$

It is straightforward to verify that

$$\begin{array}{lll} \mathcal{A}_1 = \{\mathbf{y}_1, \mathbf{y}_2, \mathbf{y}_3\} & \mathcal{A}_2 = \{\mathbf{y}_4, \mathbf{y}_5, \mathbf{y}_6, \mathbf{y}_7\} & \mathcal{A}_3 = \{\mathbf{y}_8, \mathbf{y}_9, \mathbf{y}_{10}, \mathbf{y}_{11}\} \\ \mathcal{A}_4 = \{\mathbf{y}_{12}\} & \mathcal{A}_5 = \emptyset & \\ \mathcal{B}_0 = \{\mathbf{x}^+(\mathbb{G}_2)\} & \mathcal{B}_1 = \{\mathbf{x}_1, \mathbf{x}_2, \mathbf{x}_3\} & \mathcal{B}_2 = \{\mathbf{x}_4, \mathbf{x}_5, \mathbf{x}_6, \mathbf{x}_7\} \\ \mathcal{B}_3 = \{\mathbf{x}_8, \mathbf{x}_9\} & \mathcal{B}_4 = \emptyset. & \end{array}$$

and in fact

$$\begin{array}{lll} \tilde{\partial}_{\mathbb{O},\mathbb{X}}(\mathbf{y}_1) = \mathbf{x}^+(\mathbb{G}_2) + \mathbf{x}_1 & \tilde{\partial}_{\mathbb{O},\mathbb{X}}(\mathbf{y}_2) = \mathbf{x}^+(\mathbb{G}_2) + \mathbf{x}_2 & \tilde{\partial}_{\mathbb{O},\mathbb{X}}(\mathbf{y}_3) = \mathbf{x}^+(\mathbb{G}_2) + \mathbf{x}_3 \\ \tilde{\partial}_{\mathbb{O},\mathbb{X}}(\mathbf{y}_4) = \mathbf{x}_1 + \mathbf{x}_4 + \mathbf{x}_6 & \tilde{\partial}_{\mathbb{O},\mathbb{X}}(\mathbf{y}_5) = \mathbf{x}_2 + \mathbf{x}_5 & \tilde{\partial}_{\mathbb{O},\mathbb{X}}(\mathbf{y}_6) = \mathbf{x}_3 + \mathbf{x}_4 + \mathbf{x}_7 \\ \tilde{\partial}_{\mathbb{O},\mathbb{X}}(\mathbf{y}_7) = \mathbf{x}_3 + \mathbf{x}_5 & \tilde{\partial}_{\mathbb{O},\mathbb{X}}(\mathbf{y}_8) = \mathbf{x}_6 + \mathbf{x}_7 & \tilde{\partial}_{\mathbb{O},\mathbb{X}}(\mathbf{y}_9) = \mathbf{x}_6 + \mathbf{x}_8 \\ \tilde{\partial}_{\mathbb{O},\mathbb{X}}(\mathbf{y}_{10}) = \mathbf{x}_6 + \mathbf{x}_9 & \tilde{\partial}_{\mathbb{O},\mathbb{X}}(\mathbf{y}_{11}) = \mathbf{x}_7 + \mathbf{x}_8 & \tilde{\partial}_{\mathbb{O},\mathbb{X}}(\mathbf{y}_{12}) = \mathbf{x}_8 + \mathbf{x}_9. \end{array}$$

Clearly, $\mathbf{x}^+(\mathbb{G}_2) \notin \tilde{\partial}_{\mathbb{O},\mathbb{X}}(A(\mathbb{G}))$; so $[\mathbf{x}^+(\mathbb{G}_2)] \neq 0$ by Lemma 12.6.4. □

Ng and Khandhawit [**102**] have fit the above knot into an infinite family of transversely non-simple knot types that can be detected by the transverse grid invariant, via a computation with the same complexity as the one explained above. Different infinite families of transversely non-simple knot types were found by Vértesi [**223**], using a formula that determines the behaviour of the transverse grid invariant under connected sum.

We will study another transversely non-simple knot type in Section 14.4; there the verification of transverse non-simplicity will use additional structure on the transverse invariant, defined in Section 14.3.

12.7. Invariants of Legendrian and transverse links

The previous constructions and results admit natural extensions to Legendrian and transverse links. This section describes the mostly straightforward changes needed for this generalization.

The discussion from Section 12.2 readily extends to links, associating Legendrian links to grid diagrams.

The Thurston-Bennequin number $\mathrm{tb}(\vec{\mathcal{L}})$ and rotation number $\mathrm{r}(\vec{\mathcal{L}})$ for and oriented Legendrian link $\vec{\mathcal{L}}$ can be defined exactly as in the case of knots (Definitions 12.1.3 and 12.1.5): $\mathrm{tb}(\vec{\mathcal{L}})$ is defined as the linking number with the contact framing, and $\mathrm{r}(\vec{\mathcal{L}})$ as the relative Euler number of ξ_{st} over a Seifert surface (with the trivialization on the possibly disconnected boundary given by Definition 12.1.5). The formulas for these numbers in terms of Legendrian fronts (Equations (12.1) and (12.2)) remain valid for links. These quantities admit suitable vector-valued extensions, as well, which we describe presently.

DEFINITION 12.7.1. Let $\vec{\mathcal{L}}$ be an oriented Legendrian link with ℓ components. Let $\vec{L} = (L_1, \dots, L_\ell)$ denote the underlying smooth (oriented) link, and let $\vec{\mathcal{L}}'$ be the push-off of $\vec{\mathcal{L}}$ along its contact framing. Define $\mathrm{tb}_i = \mathrm{tb}_i(\vec{\mathcal{L}})$ to be the linking number of L_i with the Legendrian push-off $\vec{\mathcal{L}}'$. The resulting homology class

$$\sum_{i=1}^{\ell} \mathrm{tb}_i \cdot \mu_i \in H_1(S^3 \setminus \vec{L}; \mathbb{Z})$$

is called the ***Thurston-Bennequin invariant*** of $\vec{\mathcal{L}}$.

As for knots, the Thurston-Bennequin numbers tb_i for links can be computed in terms of a generic front projection $\mathcal{D}(\vec{\mathcal{L}}) = \bigcup_{i=1}^{\ell} \mathcal{D}(\vec{\mathcal{L}})_i$ of $\vec{\mathcal{L}}$:

$$\mathrm{tb}_i(\vec{\mathcal{L}}) = \mathrm{wr}(\mathcal{D}(\vec{\mathcal{L}})_i) + \ell k(\mathcal{D}(\vec{\mathcal{L}})_i, \mathcal{D}(\vec{\mathcal{L}}) \setminus \mathcal{D}(\vec{\mathcal{L}})_i) - \frac{1}{2}\#\{\text{cusps in } \mathcal{D}(\vec{\mathcal{L}})_i\} \tag{12.12}$$

(Recall that $\ell k(\mathcal{D}(\vec{\mathcal{L}})_i, \mathcal{D}(\vec{\mathcal{L}}) \setminus \mathcal{D}(\vec{\mathcal{L}})_i)$ is computed from the signed crossings of $\mathcal{D}(\vec{\mathcal{L}})_i$ with $\mathcal{D}(\vec{\mathcal{L}}) \setminus \mathcal{D}(\vec{\mathcal{L}})_i$.) We can also use the front projection to define rotation numbers

$$\mathrm{r}_i(\vec{\mathcal{L}}) = \frac{1}{2}\Big(\#\{\text{downward cusps in } \mathcal{D}(\vec{\mathcal{L}})_i\} - \#\{\text{upward cusps in } \mathcal{D}(\vec{\mathcal{L}})_i\}\Big). \tag{12.13}$$

REMARK 12.7.2. These numbers can be thought of as the components of a homology class, $\sum \mathrm{r}_i \cdot \mu_i$, which can be thought of as Poincaré dual to a class $H^2(S^3, \vec{L}; \mathbb{Z})$ which is the relative Euler class of the two plane field ξ, relative to the trivialization given by the tangent space of $\vec{L}$.

The vector of Thurston–Bennequin invariants of the individual components of $\vec{\mathcal{L}}$, together with the matrix of linking numbers, determines the Thurston–Bennequin invariant $\mathrm{tb}(\vec{\mathcal{L}})$. The present packaging, though, is more convenient in the upcoming formulas.

EXERCISE 12.7.3. **(a)** Check that $\mathrm{r}_i(\vec{\mathcal{L}})$ is invariant under Legendrian isotopies. **(b)** Show that for any Legendrian link $\vec{\mathcal{L}}$, $\mathrm{tb}(\vec{\mathcal{L}}) = \sum_i \mathrm{tb}_i(\vec{\mathcal{L}})$ and $\mathrm{r}(\vec{\mathcal{L}}) = \sum_i \mathrm{r}_i(\vec{\mathcal{L}})$. **(c)** Compute the Thurston-Bennequin invariants of the Legendrian links (representing the torus link $T_{2,4}$ and the Borromean rings B) of Figure 12.23.

REMARK 12.7.4. In the Legendrian realization of the Borromean rings from Figure 12.23, the sum of the Thurston-Bennequin invariants of the components is -4. While each component is individually unknotted, it was shown in [**145**] that there is no Legendrian representation for the link where the Thurston-Bennequin number of each component is -1.

(a) (b)

FIGURE 12.23. **A Legendrian representations of (a) the $T_{2,4}$ torus link and (b) the Borromean rings B.**

Let $\mathbb{G}$ be a grid diagram and let $\vec{\mathcal{L}}$ be its associated oriented Legendrian link. Definition 6.4.1 extends to the case of links, giving the pair of elements $\mathbf{x}^+(\mathbb{G}), \mathbf{x}^-(\mathbb{G}) \in \mathbf{GC}^-(\mathbb{G})$, both of which are cycles. (See the proof of Lemma 6.4.2.)

THEOREM 12.7.5. *Suppose that $\vec{\mathcal{L}}$ is a given oriented Legendrian link of ℓ components and $\mathbb{G}$ is a grid diagram representing it. Then, the two cycles $\mathbf{x}^+ = \mathbf{x}^+(\mathbb{G})$ and $\mathbf{x}^- = \mathbf{x}^-(\mathbb{G})$, representing the homology elements $\lambda^+(\mathbb{G}), \lambda^-(\mathbb{G}) \in \mathbf{GH}^-(\mathbb{G})$, are supported in gradings*

$$(12.14) \qquad A_i(\mathbf{x}^+) = \frac{\mathrm{tb}_i(\vec{\mathcal{L}}) - \mathrm{r}_i(\vec{\mathcal{L}}) + 1}{2} \qquad A_i(\mathbf{x}^-) = \frac{\mathrm{tb}_i(\vec{\mathcal{L}}) + \mathrm{r}_i(\vec{\mathcal{L}}) + 1}{2}$$

$$(12.15) \qquad M(\mathbf{x}^+) = \mathrm{tb}(\vec{\mathcal{L}}) - \mathrm{r}(\vec{\mathcal{L}}) + 1 \qquad M(\mathbf{x}^-) = \mathrm{tb}(\vec{\mathcal{L}}) + \mathrm{r}(\vec{\mathcal{L}}) + 1.$$

Moreover, if $\mathbb{G}$ and $\mathbb{G}'$ are two different grid diagrams which represent Legendrian isotopic oriented links, then there is an isomorphism $\phi\colon \mathbf{GH}^-(\mathbb{G}) \longrightarrow \mathbf{GH}^-(\mathbb{G}')$ of $\mathbb{Z} \oplus \mathbf{H}(L)$-graded modules over $\mathbb{F}[U_1, \dots, U_\ell]$ with $\phi(\lambda^+(\mathbb{G})) = \lambda^+(\mathbb{G}')$ and $\phi(\lambda^-(\mathbb{G})) = \lambda^-(\mathbb{G}')$.

Proof. This is a mostly straightforward generalization of Theorems 12.3.2 and 12.3.3. For example, Equation (12.15) is an easy adaptation of the argument from Theorem 12.3.2. Computing the Alexander multi-gradings requires a little extra care, as follows.

Fix a planar realization G of a toroidal grid diagram $\mathbb{G}$ whose associated Legendrian link is $\vec{\mathcal{L}}$, and let $\mathcal{D}(\vec{\mathcal{L}})$ be the Legendrian front projection associated to G. Refine the numbers x_{SW}, x_{NE}, etc. to x^i_{SW}, x^i_{NE} (and so on), where, for example, x^i_{SW} is the number of those X-markings which are of type southwest and lie on the ith component of the link. Let $\mathbb{X}_i$, resp. $\mathbb{O}_i$, denote the X-markings, resp. O-markings on the ith component of the link. Let $\mathcal{D}(\vec{\mathcal{L}})_i$ be the projection of the ith component in the Legendrian front projection $\mathcal{D}(\vec{\mathcal{L}})$. Let $\mathbb{O}'_i = \mathbb{O} \setminus \mathbb{O}_i$, $\mathbb{X}'_i = \mathbb{X} \setminus \mathbb{X}_i$, and $\mathcal{D}(\vec{\mathcal{L}})'_i = \mathcal{D}(\vec{\mathcal{L}}) \setminus \mathcal{D}(\vec{\mathcal{L}})_i$. From the definition of A_i and bilinearity of $\mathcal{J}$, it follows that

$$\begin{aligned} 2A_i(\mathbf{x}^-) - 1 = 2\mathcal{J}(\mathbf{x}^- - \mathbb{X}, \mathbb{X}_i) - 2\mathcal{J}(\mathbf{x}^- - \mathbb{X}, \mathbb{O}_i) \\ + \mathcal{J}(\mathbb{X}_i - \mathbb{O}_i, \mathbb{X}_i - \mathbb{O}_i) + \mathcal{J}(\mathbb{X}'_i - \mathbb{O}'_i, \mathbb{X}_i - \mathbb{O}_i) - n_i. \end{aligned}$$

Arguing as in the proof of Theorem 12.3.2, we find that

$$2\mathcal{J}(\mathbf{x}^- - \mathbb{X}, \mathbb{O}_i) = o^i_{SE} + o^i_{NW} + 2o^i_{NE}.$$

An easy computation shows that

$$2\mathcal{J}(\mathbf{x}^- - \mathbb{X}, \mathbb{X}_i) = n_i.$$

The proof of Lemma 10.1.5 can be adapted to the planar subgrid G_i given by the markings on the ith component, to give

$$\mathrm{wr}(\mathcal{D}(\vec{\mathcal{L}})_i) = -\mathrm{wr}(G_i) = \mathcal{J}(\mathbb{X}_i - \mathbb{O}_i, \mathbb{X}_i - \mathbb{O}_i) - b(G_i) = \mathcal{J}(\mathbb{X}_i - \mathbb{O}_i, \mathbb{X}_i - \mathbb{O}_i) - x^i_{SE} - o^i_{SE}.$$

Using Lemma 10.1.5 for G and the diagram G' obtained by reversing the orientation of $\vec{L}_i$ gives

$$\mathcal{J}(\mathbb{X}'_i - \mathbb{O}'_i, \mathbb{X}_i - \mathbb{O}_i) = \ell k(\mathcal{D}(\vec{\mathcal{L}})_i, \mathcal{D}(\vec{\mathcal{L}})'_i).$$

Combining all of this, we find that

$$\begin{aligned} 2A_i(\mathbf{x}^-) - 1 &= -(o^i_{SE} + o^i_{NW} + 2o^i_{NE}) + x^i_{SE} + o^i_{SE} \\ &\quad + \mathrm{wr}(\mathcal{D}(\vec{\mathcal{L}})_i) + \ell k(\mathcal{D}(\vec{\mathcal{L}})_i, \mathcal{D}(\vec{\mathcal{L}})'_i) \\ &= \mathrm{wr}(\mathcal{D}(\vec{\mathcal{L}})_i) + \ell k(\mathcal{D}(\vec{\mathcal{L}})_i, \mathcal{D}(\vec{\mathcal{L}})'_i) - x^i_{SW} - o^i_{NE} \\ &= \mathrm{tb}(\vec{L}_i) + \mathrm{r}(\vec{L}_i), \end{aligned}$$

verifying Equation (12.14) for $\mathbf{x}^-$. The computation for $\mathbf{x}^+$ follows by rotating the grid diagram 180°, exactly as in the proof of Theorem 12.3.2.

Legendrian invariance of $\lambda^\pm$ follows as in the proof of Theorem 12.3.3. □

Let $\lambda^+(\vec{\mathcal{L}})$ and $\lambda^-(\vec{\mathcal{L}})$ denote the homology classes of $\mathbf{x}^+(\mathbb{G})$ and $\mathbf{x}^-(\mathbb{G})$ in the $\mathbb{F}[U_1, \dots, U_\ell]$-module $\mathbf{GH}^-(\vec{L})$. Theorem 12.3.4 has the following straightforward generalization:

THEOREM 12.7.6. *Let $\vec{\mathcal{L}}$ be an oriented Legendrian link, and let $\vec{\mathcal{L}}^-$ (respectively $\vec{\mathcal{L}}^+$) denote the oriented Legendrian links obtained as a single negative (respectively positive) stabilization of $\vec{\mathcal{L}}$ on the ith component. Then, there are isomorphisms*

$$\phi^- : \mathbf{GH}^-(\mathbb{G}) \longrightarrow \mathbf{GH}^-(\mathbb{G}^-), \qquad \phi^+ : \mathbf{GH}^-(\mathbb{G}) \longrightarrow \mathbf{GH}^-(\mathbb{G}^+)$$

with the properties

$$\begin{aligned} \phi^-(\lambda^+(\vec{\mathcal{L}})) &= \lambda^+(\vec{\mathcal{L}}^-) & \qquad U_i \cdot \phi^+(\lambda^+(\vec{\mathcal{L}})) &= \lambda^+(\vec{\mathcal{L}}^+) \\ U_i \cdot \phi^-(\lambda^-(\vec{\mathcal{L}})) &= \lambda^-(\vec{\mathcal{L}}^-) & \phi^+(\lambda^-(\vec{\mathcal{L}})) &= \lambda^-(\vec{\mathcal{L}}^+). \end{aligned}$$ □

As an application, we give two Legendrian non-isotopic links that are isotopic as smooth links, have identical classical invariants, and have Legendrian isotopic components (as both components of both links are isotopic to the Legendrian unknot).

PROPOSITION 12.7.7. *The oriented Legendrian links $\vec{\mathcal{L}}_1$ and $\vec{\mathcal{L}}_2$ with front projections given by Figure 12.24 both of topological type $m(6^2_3)$ and having $\mathrm{tb}_i = 1$ and $\mathrm{r}_i = 0$ for $i = 1, 2$, are not Legendrian isotopic.*

Proof. Consider the grid diagrams $\mathbb{G}_1$ and $\mathbb{G}_2$ of Figure 12.25 for the two oriented links in Figure 12.24. Looking at the diagrams of Figure 12.25, we see that there is no grid state $\mathbf{y}$ with the property that there is some $r \in \mathrm{Rect}^\circ(\mathbf{y}, \mathbf{x}^+(\mathbb{G}_1))$ with $r \cap \mathbb{O} = r \cap \mathbb{X} = \emptyset$, so $\lambda^+(\mathbb{G}_1) \neq \lambda^-(\mathbb{G}_1)$. On the other hand, the grid state $\mathbf{y}$ shown on $\mathbb{G}_2$ in Figure 12.24 has the property that $\partial^-_{\mathbb{X}}\mathbf{y} = \mathbf{x}^+(\mathbb{G}_2) + \mathbf{x}^-(\mathbb{G}_2)$, hence $\lambda^+(\mathbb{G}_2) = \lambda^-(\mathbb{G}_2)$. It follows at once that the two oriented Legendrian links are not Legendrian isotopic. □

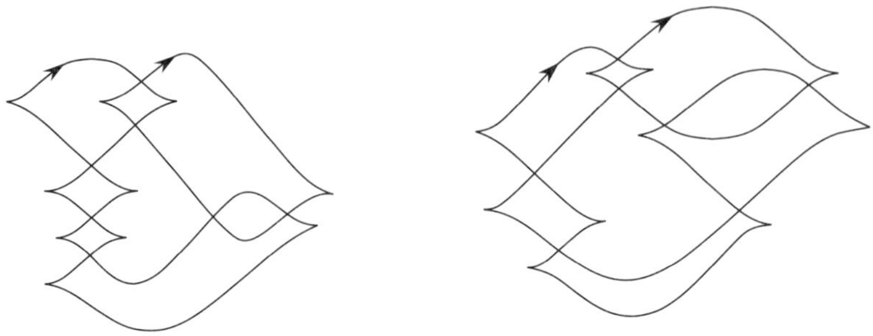

FIGURE 12.24. **Two different Legendrian representatives $\vec{\mathcal{L}}_1$ and $\vec{\mathcal{L}}_2$ of the link $m(6_3^2)$ (the mirror of 6_3^2 of [198]).**

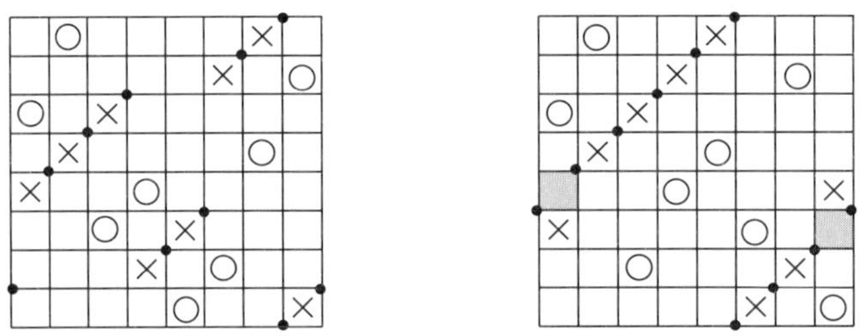

FIGURE 12.25. **Grid diagrams $\mathbb{G}_1$ and $\mathbb{G}_2$ for $\vec{\mathcal{L}}_1$ and $\vec{\mathcal{L}}_2$.** The left diagram shows $\mathbf{x}^+(\mathbb{G}_1)$. The right diagram shows a grid state $\mathbf{y}$ with $\partial_{\mathbb{X}}^- \mathbf{y} = \mathbf{x}^+(\mathbb{G}_2) + \mathbf{x}^-(\mathbb{G}_2)$.

With more work, grid homology can be used to detect links that are transversely non-simple; see [**4**]. Chongchitmate and Ng's *Legendrian knot atlas* [**25**] gives a conjectural Legendrian classification of knots and links with small crossing number.

12.8. Transverse knots, grid diagrams, and braids

We sketch another construction of transverse knots from grid diagrams; see [**102, 158**]. A planar grid diagram naturally defines a closed braid: connect the O- and X-markings horizontally as before, but when connecting them vertically, always go up from the X-marking to the O-marking. When this procedure runs out of the square in the fundamental domain (that is, when the X is above the O-marking), connect the markings by going up and around a large circular path, as in Figure 12.26. The picture can be slightly perturbed to eliminate horizontal segments without introducing new local maxima or local minima. It is easy to see that the resulting knot is isotopic to the knot associated to the grid diagram in the sense of Chapter 3. This description presents our knot as a *closed braid.* A closed braid, on the other hand, can be made into a transverse knot, by drawing it in a sufficiently small neighborhood of the standard transverse unknot.

It can be shown that the transverse knot associated to the braid via the above procedure is transversely isotopic to the mirror of the one given by the construction $\mathbb{G} \mapsto \mathcal{T}(\vec{\mathcal{K}}_{\mathbb{G}})$. See [**102**, Section 2.4]; see also [**158**] for further discussion.

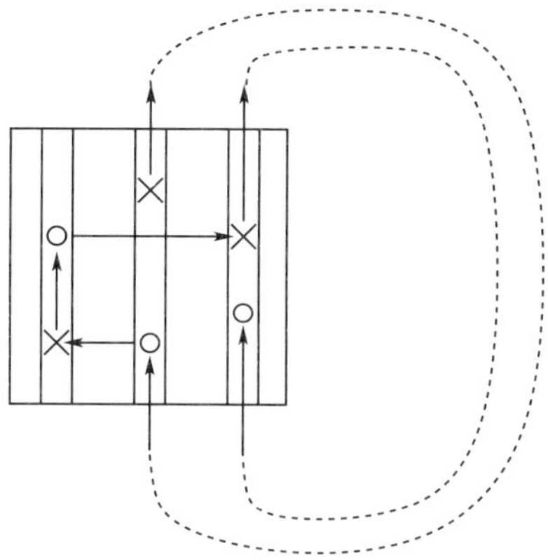

FIGURE 12.26. **From grid diagrams to braids.**

12.9. Further remarks

This chapter offers a glimpse into the deep interaction between contact geometry and Floer-type invariants in low dimensional topology. The relationship between gauge theory and contact geometry was discovered in work of Lisca and Matić [**125**] and Kronheimer and Mrowka [**108**], using Seiberg-Witten theory.

Kronheimer and Mrowka define a Floer homology class that is a contact invariant with values in monopole Floer homology [**110**]. Building on work of Giroux [**74**], an analogous construction in Heegaard Floer homology was defined in [**175**], giving an invariant of contact structures with values in the Heegaard Floer homology. The Heegaard Floer contact invariant is useful for classifying contact structures on specific three-manifolds; see for example [**72**].

The Legendrian and transverse invariants can be thought of as a variation on this construction. Legendrian and transverse invariants in an arbitrary oriented, contact three-manifold were defined in [**126**]; and they were shown to generalize the grid invariants studied here, in work of Baldwin, Vela-Vick, and Vértesi [**7**].

Lenny Ng [**156**] has constructed another useful combinatorial invariant for transverse knots, inspired by methods from symplectic geometry.

CHAPTER 13

The filtered grid complex

The aim of the present chapter is to endow grid homology with more structure. We define a chain complex similar to the grid complex from Chapter 4, except that now the Alexander function on grid states induces a filtration, rather than a grading. While the knot invariant studied in earlier chapters is a bigraded module over $\mathbb{F}[U]$, we now get a knot invariant that is a more complicated algebraic object: it is the filtered quasi-isomorphism type of a $\mathbb{Z}$-filtered, $\mathbb{Z}$-graded chain complex over $\mathbb{F}[U]$.

In Section 13.1, we formulate the relevant notions of filtered complexes and their quasi-isomorphism types, and discuss some related algebraic operations. (For more background on homological algebra, see Appendix A.) In Section 13.2, we define the filtered grid complex, generalizing the construction from Chapter 4. In Section 13.3, we explain how to adapt the invariance proof from Chapter 5 to our present framework. This requires some more work, especially in verifying stabilization invariance (see Lemma 13.3.13 below).

The enrichment presented here is, indeed, stronger than the grid homology from Chapter 4. We will describe some of its applications in Chapter 14.

13.1. Some algebraic background

As mentioned above, the grid complex can be used to get more than just a bigraded homology group; it can be used to give an equivalence class of filtered chain complexes (over $\mathbb{F}[U]$). The aim of the present section is to explain the meaning of this algebraic object. The impatient reader can skip to Section 13.2, and refer back here as needed.

For the purposes of this background section, we fix a base ring $\mathbb{K}$. (In the rest of the chapter, we will consider $\mathbb{K} = \mathbb{Z}/2\mathbb{Z}$.) Objects equipped by filtrations will be denoted by calligraphic letters, to distinguish them from their graded or bigraded analogues.

DEFINITION 13.1.1. A ***$\mathbb{Z}$-filtered, $\mathbb{Z}$-graded chain complex over*** $\mathbb{K}$ is a $\mathbb{K}$-module $\mathcal{C}$ with the following additional structure:

- a $\mathbb{K}$-module endomorphism $\partial\colon \mathcal{C} \to \mathcal{C}$ satisfying $\partial \circ \partial = 0$;
- a $\mathbb{Z}$-grading, which is a splitting of $\mathcal{C}$ as a $\mathbb{K}$-module $\mathcal{C} = \bigoplus_{d\in\mathbb{Z}} \mathcal{C}_d$, that is compatible with the differential in the sense that $\partial(\mathcal{C}_d) \subseteq \mathcal{C}_{d-1}$;
- a sequence of $\mathbb{K}$-submodules $\mathcal{F}_s\mathcal{C} \subset \mathcal{C}$ with $\mathcal{F}_s\mathcal{C} \subset \mathcal{F}_{s+1}\mathcal{C}$, called the ***$\mathbb{Z}$-filtration***, that exhausts $\mathcal{C}$, in the sense that $\bigcup_{s\in\mathbb{Z}} \mathcal{F}_s\mathcal{C} = \mathcal{C}$;
- the filtration is compatible with the $\mathbb{Z}$-grading, in the following sense: letting $\mathcal{F}_s\mathcal{C}_d = (\mathcal{F}_s\mathcal{C}) \cap \mathcal{C}_d$, then $\mathcal{F}_s\mathcal{C} = \bigoplus_{d\in\mathbb{Z}} \mathcal{F}_s\mathcal{C}_d$;
- the filtration is compatible with the differential, in the sense that $\partial(\mathcal{F}_s\mathcal{C}) \subseteq \mathcal{F}_s\mathcal{C}$ (i.e. the $\mathcal{F}_s\mathcal{C}$ are subcomplexes);

- the filtration is bounded below in the sense that for any given $d \in \mathbb{Z}$, there is an n_d so that $\mathcal{F}_s\mathcal{C}_d = 0$ for $s \leq n_d$.

We will typically consider $\mathbb{Z}$-filtered, $\mathbb{Z}$-graded chain complexes with the following additional structure:

DEFINITION 13.1.2. Fix some integer $n \geq 0$. A ***$\mathbb{Z}$-filtered, $\mathbb{Z}$-graded chain complex over*** $\mathbb{K}[V_1, \ldots, V_n]$ is a $\mathbb{Z}$-filtered, $\mathbb{Z}$-graded chain complex $\mathcal{C}$ equipped with $\mathbb{K}$-module endomorphisms $V_i \colon \mathcal{C} \to \mathcal{C}$ for $i = 1, \ldots, n$ that are compatible with the above structures in the following ways:

- for all $1 \leq i, j \leq n$, V_i and V_j commute with each other;
- each V_i is compatible with the differential, in the sense that $\partial \circ V_i = V_i \circ \partial$;
- each V_i is compatible with the gradings, in the sense that $V_i(\mathcal{C}_d) \subseteq \mathcal{C}_{d-2}$;
- each V_i is compatible with the filtration, in the sense that $V_i(\mathcal{F}_s\mathcal{C}) \subseteq \mathcal{F}_{s-1}\mathcal{C}$.

The endomorphisms V_i for $i = 1, \ldots, n$ equip $\mathcal{C}$ with the structure of a chain complex over $\mathbb{K}[V_1, \ldots, V_n]$. Obviously, the case where $n = 0$ in Definition 13.1.2 gives back Definition 13.1.1.

The only non-standard aspect of Definition 13.1.2 is the interaction of the grading and the filtration with the algebra action: for us, multiplication by V_i is required to change both the grading and the filtration. This is the choice that turns out to be natural in grid homology.

In a complex $\mathcal{C}$ as above, the *filtration level* of a non-zero element $x \in \mathcal{C}$ is the minimal s for which $x \in \mathcal{F}_s\mathcal{C} \subset \mathcal{C}$.

Typically, our complexes will be finitely generated free modules over $\mathbb{K}[V_1, \ldots, V_n]$. The Alexander filtration will be specified by a function, the "Alexander function", defined on a preferred generating set $\{x_i\}_{i=1}^m$ over $\mathbb{K}[V_1, \ldots, V_n]$. There is a corresponding basis for $\mathcal{C}$ as a $\mathbb{K}$-module, consisting of elements of the form $V_1^{k_1} \cdots V_n^{k_n} \cdot x$, where x lies in the generating set and $k_1, \ldots, k_n$ are non-negative integers. The Alexander function is extended to these $\mathbb{K}$-module generators by the formula

$$A(V_1^{k_1} \cdots V_n^{k_n} \cdot x) = A(x) - \sum_{j=1}^{n} k_j.$$

The subcomplex $\mathcal{F}_s\mathcal{C}$ is the span (over $\mathbb{K}$) of those generators whose Alexander function is less than or equal to s. To check that this definition gives a subcomplex, is equivalent to verify that for each preferred generator x with $A(x) = a$, ∂x is in the span of $\mathbb{K}$-module generators whose Alexander function evaluates $\leq a$. Clearly, such finitely generated complexes are automatically bounded below; they are also bounded above in the sense that $\mathcal{F}_s\mathcal{C} = \mathcal{C}$ for all sufficiently large s.

DEFINITION 13.1.3. Given a $\mathbb{Z}$-filtered, $\mathbb{Z}$-graded chain complex $(\mathcal{C}, \partial)$, the ***associated graded object*** is the chain complex

$$\mathrm{gr}(\mathcal{C}) = \bigoplus_{d,s \in \mathbb{Z}} (\mathcal{F}_s\mathcal{C}_d / \mathcal{F}_{s-1}\mathcal{C}_d),$$

equipped with the bigrading $\mathrm{gr}(\mathcal{C})_{d,s} = \mathcal{F}_s\mathcal{C}_d / \mathcal{F}_{s-1}\mathcal{C}_d$ and differential $\mathrm{gr}(\partial) \colon \mathrm{gr}(\mathcal{C}) \to \mathrm{gr}(\mathcal{C})$ induced by $\partial \colon \mathcal{C} \to \mathcal{C}$; i.e. $\mathrm{gr}(\partial) = \sum_{d,s} \mathrm{gr}(\partial)_{d,s}$ is a sum of components

$$\mathrm{gr}(\partial)_{d,s} \colon \mathcal{F}_s\mathcal{C}_d / \mathcal{F}_{s-1}\mathcal{C}_d \to \mathcal{F}_s\mathcal{C}_{d-1} / \mathcal{F}_{s-1}\mathcal{C}_{d-1}.$$

The result $(\mathrm{gr}(\mathcal{C}), \mathrm{gr}(\partial))$ is a bigraded chain complex over $\mathbb{K}[V_1, \dots, V_n]$, in the sense of Definition 4.5.1.

To formulate our knot invariant, we need a suitable equivalence relation on filtered complexes. Before defining this, it is useful to formulate a notion of morphisms.

DEFINITION 13.1.4. Fix two $\mathbb{Z}$-filtered, $\mathbb{Z}$-graded chain complexes $\mathcal{C}$ and $\mathcal{C}'$ over $\mathbb{K}[V_1, \dots, V_n]$. A ***chain map*** $f\colon \mathcal{C} \to \mathcal{C}'$ is a $\mathbb{K}[V_1, \dots, V_n]$-module homomorphism with $\partial' \circ f = f \circ \partial$. A ***$\mathbb{Z}$-graded, $\mathbb{Z}$-filtered chain map*** is one that satisfies the following two additional properties:

- $f(\mathcal{C}_d) \subseteq \mathcal{C}'_d$; i.e. f is a graded map.
- $f(\mathcal{F}_i\mathcal{C}) \subset \mathcal{F}_i\mathcal{C}'$; i.e. f is a filtered map.

More generally, for fixed integers (m, t), a chain map $f.\ \mathcal{C} \to \mathcal{C}'$ is said to be ***homogeneous of degree*** (m, t) if

- $f(\mathcal{C}_d) \subseteq \mathcal{C}'_{d+m}$;
- $f(\mathcal{F}_s\mathcal{C}) \subset \mathcal{F}_{s+t}\mathcal{C}'$.

DEFINITION 13.1.5. Fix two $\mathbb{Z}$-filtered, $\mathbb{Z}$-graded chain complexes $\mathcal{C}$ and $\mathcal{C}'$ over $\mathbb{K}[V_1, \dots, V_n]$, and fix two $\mathbb{Z}$-filtered, $\mathbb{Z}$-graded chain maps $f\colon \mathcal{C} \to \mathcal{C}'$ and $g\colon \mathcal{C} \to \mathcal{C}'$. A ***filtered chain homotopy*** from g to f is a map $h\colon \mathcal{C} \to \mathcal{C}'$ that satisfies the following properties:

(h-1) h is compatible with gradings and filtrations, in the sense that h maps $\mathcal{F}_s\mathcal{C}_d$ into $\mathcal{F}_s\mathcal{C}'_{d+1}$;
(h-2) h satisfies the homotopy relation $\partial' \circ h + h \circ \partial = f - g$;
(h-3) h is a homomorphism of $\mathbb{K}[V_1, \dots, V_n]$-modules.

Two maps $f\colon \mathcal{C} \to \mathcal{C}'$ and $g\colon \mathcal{C} \to \mathcal{C}'$ are said to be ***filtered chain homotopic*** if there is a filtered chain homotopy between them. More generally, if $f\colon \mathcal{C} \to \mathcal{C}'$ and $g\colon \mathcal{C} \to \mathcal{C}'$ are two chain maps that are homogeneous of degree (m, t), a ***homogeneous chain homotopy*** from f to g is a map as above, except that Condition (h-1) is replaced by the following:

(h-1′) h maps $\mathcal{F}_s\mathcal{C}_d$ into $\mathcal{F}_{s+t}\mathcal{C}'_{d+m+1}$.

DEFINITION 13.1.6. Two chain complexes $\mathcal{C}$ and $\mathcal{C}'$ are said to be ***filtered chain homotopy equivalent*** if there are maps $f\colon \mathcal{C} \to \mathcal{C}'$ and $g\colon \mathcal{C}' \to \mathcal{C}$ with the property that the maps $f \circ g$ and $g \circ f$ are filtered chain homotopic to the respective identity maps. In this case, the map f is called a ***filtered chain homotopy equivalence***, and the maps f and g are said to be filtered chain homotopy inverses of one another.

A filtered chain map f naturally induces a chain map $\mathrm{gr}(f)\colon \mathrm{gr}(\mathcal{C}) \to \mathrm{gr}(\mathcal{C}')$ between the associated graded objects. It is easy to see that if $f\colon \mathcal{C} \to \mathcal{C}'$ is a filtered chain homotopy equivalence, then $\mathrm{gr}(f)\colon \mathrm{gr}(\mathcal{C}) \to \mathrm{gr}(\mathcal{C}')$ is a chain homotopy equivalence between the associated graded objects.

It will be convenient to formulate another equivalence relation between chain complexes. To formulate this notion, we find it helpful to go back first to the bigraded context considered previously. Recall (Definition 5.2.9) that a quasi-isomorphism is a bigraded chain map $f\colon C \to C'$ between bigraded chain complexes that induces an isomorphism in homology.

DEFINITION 13.1.7. Two chain complexes (C, ∂) and (C', ∂') are said to be ***quasi-isomorphic*** if there is a chain complex (C'', ∂'') and two quasi-isomorphisms $f: C'' \to C$ and $g: C'' \to C'$. In this case we write $(C, \partial) \simeq (C', \partial')$. In cases where (C, ∂) and (C', ∂') are bigraded complexes over $\mathbb{K}[V_1, \ldots, V_n]$, we require our quasi-isomorphisms to be bigraded chain maps over $\mathbb{K}[V_1, \ldots, V_n]$.

Clearly, if there is a quasi-isomorphism $f: C' \to C$, then C is quasi-isomorphic to C': let $C'' = C'$, and g be the identity map.

EXAMPLE 13.1.8. Let the chain complex C be freely generated over $\mathbb{F}[U]$ by two generators x and y with bigradings $(-1, -1)$ and $(0, 0)$ respectively. Equip C with the differential $\partial x = U \cdot y$. Let C' be the chain complex with a single generator z in bigrading $(0, 0)$, so that $U \cdot z = \partial z = 0$. Consider the $\mathbb{F}[U]$-module map $f: C \to C'$ determined by $f(x) = 0$ and $f(y) = z$. Then f is a quasi-isomorphism, but there is no non-trivial $\mathbb{F}[U]$-module map from C' to C.

Quasi-isomorphism is an equivalence relation. (See Proposition A.3.11.) Thus, when considering a bigraded chain complex (C, ∂) such as $GC^-(\mathbb{G})$, one can consider the quasi-isomorphism type of the complex.

Generalizing to the filtered case, we have the following:

DEFINITION 13.1.9. Let $\mathcal{C}$ and $\mathcal{C}'$ be two $\mathbb{Z}$-filtered, $\mathbb{Z}$-graded chain complexes. A ***filtered quasi-isomorphism*** $f: \mathcal{C} \to \mathcal{C}'$ is a $\mathbb{Z}$-filtered, $\mathbb{Z}$-graded chain map whose associated graded map $\mathrm{gr}(f)$ induces an isomorphism on homology. Two complexes $\mathcal{C}$ and $\mathcal{C}'$ are said to be ***filtered quasi-isomorphic*** if there is a third complex $\mathcal{C}''$ and filtered quasi-isomorphisms from $\mathcal{C}''$ to $\mathcal{C}$ and from $\mathcal{C}''$ to $\mathcal{C}'$. In this case, we write $\mathcal{C} \simeq \mathcal{C}'$.

When $\mathcal{C}$ and $\mathcal{C}'$ are $\mathbb{Z}$-filtered, $\mathbb{Z}$-graded chain complexes over $\mathbb{K}[V_1, \ldots V_n]$, the quasi-isomorphisms are required to be $\mathbb{Z}$-filtered, $\mathbb{Z}$-graded chain maps of chain complexes over $\mathbb{K}[V_1, \ldots, V_n]$.

If $\mathcal{C}$ is a $\mathbb{Z}$-filtered, $\mathbb{Z}$-graded chain complex, the ***filtered quasi-isomorphism type of*** $\mathcal{C}$ is the equivalence class of $\mathcal{C}$, under the equivalence relation of filtered quasi-isomorphism.

As in the bigraded case, the third complex $\mathcal{C}''$ in Definition 13.1.9 is introduced to make the notion of "filtered quasi-isomorphic" into an equivalence relation.

Let $f: \mathcal{C} \to \mathcal{C}'$ be a filtered chain map. There is an associated mapping cone in this context, $\mathcal{C}one(f: \mathcal{C} \to \mathcal{C}')$, whose underlying chain complex is defined exactly as it was in Definition 5.2.10: as a $\mathbb{K}[V_1, \ldots, V_n]$-module, the cone is specified as $\mathcal{C}one(f: \mathcal{C} \to \mathcal{C}') = \mathcal{C} \oplus \mathcal{C}'$, and it is endowed with the differential from Equation (5.17). In the present case, though, we give $\mathcal{C}one(f: \mathcal{C} \to \mathcal{C}')$ the structure of a $\mathbb{Z}$-filtered, $\mathbb{Z}$-graded chain complex over $\mathbb{K}[V_1, \ldots, V_n]$, with

$$\mathcal{F}_s(\mathcal{C} \oplus \mathcal{C}')_d = \mathcal{F}_s\mathcal{C}_{d-1} \oplus \mathcal{F}_s\mathcal{C}'_d.$$

More generally, if $f: \mathcal{C} \to \mathcal{C}'$ is homogeneous of degree (m, t), let

$$\mathcal{F}_s(\mathcal{C} \oplus \mathcal{C}')_d = \mathcal{F}_{s-t}\mathcal{C}_{d-m-1} \oplus \mathcal{F}_s\mathcal{C}'_d.$$

(Compare Equation (5.18).)

EXERCISE 13.1.10. An *isomorphism* $\phi: \mathcal{C} \to \mathcal{C}'$ between two $\mathbb{Z}$-filtered, $\mathbb{Z}$-graded chain complexes over $\mathbb{K}[V_1, \ldots, V_n]$ is an invertible, $\mathbb{Z}$-filtered, $\mathbb{Z}$-graded chain map whose inverse is also $\mathbb{Z}$-filtered and $\mathbb{Z}$-graded. If there is an isomorphism from $\mathcal{C}$ to $\mathcal{C}'$, we say that $\mathcal{C}$ and $\mathcal{C}'$ are *isomorphic*.

(a) Show that isomorphism in the above sense is an equivalence relation.
(b) Show that isomorphic $\mathbb{Z}$-filtered, $\mathbb{Z}$-graded chain complexes are filtered chain homotopy equivalent.
(c) Show that if $\mathcal{C}$ and $\mathcal{C}'$ are filtered quasi-isomorphic, then their homology groups $H(\mathcal{C},\partial)$ and $H(\mathcal{C}',\partial')$ are isomorphic.
(d) Show that the notion of "filtered quasi-isomorphic" is an equivalence relation.

A $\mathbb{Z}$-filtered, $\mathbb{Z}$-graded chain complex $\mathcal{C}$ over $\mathbb{K}[V_1,\dots,V_n]$ can be viewed as a $\mathbb{Z}$-filtered, $\mathbb{Z}$-graded chain complex over $\mathbb{K}[U]$, where $U=V_i$ for some $i\in\{1,\dots,n\}$. We denote the resulting complex over $\mathbb{K}[U]$ by $(\mathcal{C},V_i)$. The relevance of the following result in the grid context will become clear in Lemma 13.2.8.

PROPOSITION 13.1.11. *Let $\mathcal{C}$ be a $\mathbb{Z}$-filtered, $\mathbb{Z}$-graded chain complex over $\mathbb{K}[V_1,V_2]$, and suppose that V_1 and V_2 are filtered chain homotopic, as maps of degree $(-2,-1)$. Then, the two $\mathbb{Z}$-filtered, $\mathbb{Z}$-graded chain complexes $(\mathcal{C},V_1)$ and $(\mathcal{C},V_2)$ over $\mathbb{K}[U]$ are filtered quasi-isomorphic.*

We warm up with the bigraded analogue:

LEMMA 13.1.12. *Let (C,∂) be a bigraded chain complex over $\mathbb{K}[V_1,V_2]$, and suppose that multiplication by V_1 is chain homotopic to V_2, as homogeneous chain maps of degree $(-2,-1)$. Then the two bigraded chain complexes (C,V_1) and (C,V_2) over $\mathbb{K}[U]$ are quasi-isomorphic to each other.*

Proof. Promote (C,∂) to a module over $\mathbb{K}[V_0,V_1,V_2]$, denoted $C[V_0]$ (as in Definition 5.2.15); and for $i=1,2$, abbreviate $\mathrm{Cone}(V_0-V_i\colon C[V_0]\to C[V_0])$ to $\mathrm{Cone}(V_0-V_i)$. Now,

$$\begin{aligned}(C,V_1)\cong\left(\frac{C[V_0]}{V_0-V_1},V_1\right)\cong\left(\frac{C[V_0]}{V_0-V_1},V_0\right)\simeq(\mathrm{Cone}(V_0-V_1),V_0)\\ \cong(\mathrm{Cone}(V_0-V_2),V_0)\simeq(\frac{C[V_0]}{V_0-V_2},V_2)\cong(C,V_2)\end{aligned}$$

The first two of these isomorphisms are tautologies; the third quasi-isomorphism follows from the fact that V_0-V_1 is an injective endomorphism of $C[V_0]$ (cf. Lemma 5.2.13); the fourth follows from the fact that V_0-V_1 is chain homotopic to V_0-V_2 (and chain homotopic maps have isomorphic mapping cones; see Lemma A.3.7); and the remaining two (quasi-)isomorphisms are as before. □

There are filtered analogues of the constructions going into the above proof. A filtered complex $\mathcal{C}$ over $\mathcal{R}=\mathbb{K}[V_2,\dots,V_n]$ can be promoted to a filtered complex over $\mathcal{R}[V_1]=\mathbb{K}[V_1,\dots,V_n]$, denoted $\mathcal{C}[V_1]$. The construction is as in the bigraded case (Definition 5.2.15), with the understanding that $(m\cdot V_1^k)\in\mathcal{F}_s\mathcal{C}[V_1]_d$ precisely when $m\in\mathcal{F}_{s+k}\mathcal{C}_{d+2k}$. Obviously, $\mathrm{gr}(\mathcal{C}[V_1])\cong\mathrm{gr}(\mathcal{C})[V_1]$. Also, if $f\colon\mathcal{C}\to\mathcal{C}'$ is a filtered chain map, we can form the quotient $\frac{\mathcal{C}'}{f(\mathcal{C})}$, which naturally inherits the structure of a filtered chain complex.

LEMMA 13.1.13. *If $f\colon\mathcal{C}\to\mathcal{C}'$ is a filtered chain map that is homogeneous of degree (m,t), and which induces an injective map on the associated graded object, then there is a quasi-isomorphism from $\mathrm{Cone}(f\colon\mathcal{C}\to\mathcal{C}')$ to $\frac{\mathcal{C}'}{f(\mathcal{C})}$.*

Proof. Let $C = \mathrm{gr}(\mathcal{C})$ and $C' = \mathrm{gr}(\mathcal{C}')$. It is straightforward to see that $\mathrm{gr}(\mathcal{C}one(f\colon \mathcal{C} \to \mathcal{C}')) \cong \mathrm{Cone}(\mathrm{gr}(f)\colon C \to C')$. Moreover, the quotient map $\mathcal{C}one(f\colon \mathcal{C} \to \mathcal{C}') \to \frac{\mathcal{C}'}{f(\mathcal{C})}$ induces the map $\mathrm{Cone}(\mathrm{gr}(f)\colon C \to C') \to \frac{C'}{\mathrm{gr}(f)(C)}$, which is a quasi-isomorphism by Lemma 5.2.13. □

PROPOSITION 13.1.14. *Let $\mathcal{C}$ and $\mathcal{C}'$ be two $\mathbb{Z}$-filtered, $\mathbb{Z}$-graded chain complexes, and $f, g\colon \mathcal{C} \to \mathcal{C}'$ be two homogeneous chain maps of degree (m,t) that are chain homotopic. Then $\mathcal{C}one(f)$ is filtered quasi-isomorphic to $\mathcal{C}one(g)$.*

Proof. As in the bigraded case (see Lemma A.3.7), a filtered chain homotopy h from f to g induces a map $\Phi_h\colon \mathcal{C}one(f) \to \mathcal{C}one(g)$, defined by $\Phi_h(x,x') = (x, x' + h(x))$. This map is easily seen to be a filtered chain map, with inverse Φ_{-h}; so $\mathcal{C}one(f)$ is in fact filtered isomorphic to $\mathcal{C}one(g)$. □

Proof of Proposition 13.1.11. The two key tools in the proof of Lemma 13.1.12 are Lemmas 5.2.13 and A.3.7. Having generalized these to the filtered setting in Lemma 13.1.13 and Proposition 13.1.14, the proof of Lemma 13.1.12 applies. □

13.2. Defining the invariant

The new variant of the grid complex counts more empty rectangles in the definition of the differential. For notational simplicity, we focus on the case of knots, returning to the case of links in Section 14.5.

DEFINITION 13.2.1. Let $\mathbb{G}$ be a toroidal grid diagram with grid number n representing some knot K. The ***filtered grid complex*** $\mathcal{GC}^-(\mathbb{G})$ is generated over $\mathbb{F}[V_1,\dots,V_n]$ by the same grid states $\mathbf{S}(\mathbb{G})$ as $GC^-(\mathbb{G})$, and it is endowed with the differential

$$\partial^- \mathbf{x} = \sum_{\mathbf{y}\in\mathbf{S}(\mathbb{G})} \sum_{r\in\mathrm{Rect}^\circ(\mathbf{x},\mathbf{y})} V_1^{O_1(r)} \cdots V_n^{O_n(r)} \cdot \mathbf{y}. \tag{13.1}$$

Comparing this definition with the differential for the unblocked grid complex $GC^-(\mathbb{G})$ (as defined in Equation (4.10)), observe that in the present case we drop the requirement that the counted rectangles r satisfy $r \cap \mathbb{X} = \emptyset$. Lemma 4.6.7 has the following generalization:

LEMMA 13.2.2. *The operator $\partial^-\colon \mathcal{GC}^-(\mathbb{G}) \to \mathcal{GC}^-(\mathbb{G})$ satisfies $\partial^- \circ \partial^- = 0$.*

Proof. We follow the proof of Lemma 4.6.7. The key difference with that proof is that now we can no longer exclude Case (R-3): the presence of some X-marking in a thin annulus does not serve to eliminate it from the counts. The proof of Lemma 4.6.7 shows that $\partial^- \circ \partial^-(\mathbf{x})$ counts (with suitable V_i-powers) all the thin annuli from Case (R-3). Indeed, for each generator $\mathbf{x}$, each row of squares in the torus determines a thin annulus starting at $\mathbf{x}$, as does each column. An annulus is counted (in the $\mathbf{x}$ component of $\partial^- \circ \partial^-(\mathbf{x})$) with coefficient V_i, where O_i is the O-marking in the given annulus. Thus, the contribution of the row through O_i cancels the contribution of the column through O_i, implying $\partial^- \circ \partial^-(\mathbf{x}) = 0$. □

The Maslov grading on the grid complex (defined in Equations (4.5) and (4.11)) induces a grading on $\mathcal{GC}^-(\mathbb{G})$. The Alexander function A (defined in Equation (4.3) and (4.12)) defines a function on the $\mathbb{F}$-generators of $\mathcal{GC}^-(\mathbb{G})$. According to the following lemma, this function induces a filtration:

LEMMA 13.2.3. *The differential ∂^- drops Maslov grading by one, and it is compatible with the Alexander filtration.*

Proof. The proof of Lemma 4.6.8 applies, with minor modifications. If $V_1^{k_1} \cdots V_n^{k_n} \cdot \mathbf{y}$ appears in $\partial^- \mathbf{x}$, then there is a rectangle $r \in \operatorname{Rect}^\circ(\mathbf{x}, \mathbf{y})$ with $O_i(r) = k_i$ for $i = 1, \dots, n$. (We no longer require that $r \cap \mathbb{X} = \emptyset$.) The argument from the proof of Lemma 4.6.8 applies, showing that ∂^- drops Maslov grading by one. For the Alexander filtration, Equation (4.14) shows that

$$A(V_1^{k_1} \cdots V_n^{k_n} \cdot \mathbf{y}) - A(\mathbf{y}) - \#(r \cap \mathbb{O}) = A(\mathbf{x}) - \#(r \cap \mathbb{X}) \leq A(\mathbf{x}). \tag{13.2}$$

This is what is needed to see that ∂^- is compatible with the Alexander filtration: the terms appearing in $\partial^- \mathbf{x}$ have Alexander filtration level at most $A(\mathbf{x})$. □

We have the following analogue of Theorem 4.6.3:

THEOREM 13.2.4. *The object $(\mathcal{GC}^-(\mathbb{G}), \partial^-)$ is a $\mathbb{Z}$-filtered, $\mathbb{Z}$-graded chain complex over $\mathbb{F}[V_1, \dots, V_n]$, in the sense of Definition* 13.1.2.

Proof. According to Lemma 13.2.2, $(\mathcal{GC}^-(\mathbb{G}), \partial^-)$ is a complex, and according to Lemma 13.2.3 the differential drops Maslov grading by one and is compatible with the Alexander filtration. Equations (4.11) and (4.12) ensure that multiplication by V_i changes Maslov gradings and Alexander filtrations as required by Definition 13.1.1. The filtration exhausts $\mathcal{GC}^-$ by definition. In fact, the filtration is bounded, since the complex is freely generated by the finite set $\mathbf{S}(\mathbb{G})$. □

Lemma 4.6.9 has the following generalization:

LEMMA 13.2.5. *For any pair of integers $i, j \in \{1, \dots, n\}$, multiplication by V_i is filtered chain homotopic to multiplication by V_j.*

Proof. The proof of Lemma 4.6.9 adapts readily. We need only to extend the homotopy operator $\mathcal{H}_i$ counting rectangles that cross X_i (defined in Equation (4.16)) to the filtered case, by relaxing the requirement that $r \cap \mathbb{X} = X_i$ to simply $X_i \in r$. It is easy to see that these maps give the needed filtered chain homotopies. □

The relationship between $\mathcal{GC}^-$ and GC^- is summarized in the following:

PROPOSITION 13.2.6. *The chain complex $(GC^-(\mathbb{G}), \partial_{\mathbb{X}}^-)$ is the graded complex associated to the Alexander filtration on $(\mathcal{GC}^-(\mathbb{G}), \partial^-)$.*

Proof. The differentials in both $\partial_{\mathbb{X}}^-$ and ∂^- are suitable counts of rectangles. Indeed, the rectangles that contribute in $\partial_{\mathbb{X}}^-$ are those that contribute to ∂^-, and that satisfy the hypothesis $\#(\mathbb{X} \cap r) = 0$. Looking at Equation (4.14) (cf. also Equation (13.2)), these are precisely those terms in $\partial^-(\mathbf{x})$ whose Alexander level is the same as that of $\mathbf{x}$, verifying the claim in the lemma. □

DEFINITION 13.2.7. Fix some $i \in \{1, \dots, n\}$. The ***simply blocked filtered grid complex*** is the quotient complex $\widehat{\mathcal{GC}}(\mathbb{G}) = \frac{\mathcal{GC}^-(\mathbb{G})}{V_i}$.

LEMMA 13.2.8. *Choose any $i \in \{1, \dots, n\}$. The filtered quasi-isomorphism type of the simply blocked filtered grid complex $\frac{\mathcal{GC}^-(\mathbb{G})}{V_i}$, thought of as $\mathbb{Z}$-filtered, $\mathbb{Z}$-graded chain complex over $\mathbb{F}$, is independent of the choice of i. Similarly, the filtered quasi-isomorphism type of the unblocked grid complex $\mathcal{GC}^-(\mathbb{G})$, thought of as a $\mathbb{Z}$-filtered, $\mathbb{Z}$-graded chain complex over $\mathbb{F}[U]$, where U acts as multiplication by V_i, is independent of the choice of i.*

Proof. Both statements follow from Lemma 13.2.5 and general algebraic considerations. Specifically, Lemma 13.1.13 and Proposition 13.1.14 give filtered quasi-isomorphisms for any $1 \le i, j \le n$:

$$\frac{\mathcal{GC}^-(\mathbb{G})}{V_i} \simeq \mathcal{C}one(V_i) \simeq \mathcal{C}one(V_j) \simeq \frac{\mathcal{GC}^-(\mathbb{G})}{V_j},$$

while Proposition 13.1.11 shows that the quasi-isomorphism type of $\mathcal{GC}^-(\mathbb{G})$ is independent of the choice of i. □

Now we are ready to state the invariance result for the filtered grid complexes. The proof of this result will be given in the next section.

THEOREM 13.2.9. *It $\mathbb{G}$ is a grid diagram representing a knot K, the filtered quasi-isomorphism types of $\widehat{\mathcal{GC}}(\mathbb{G})$ and of $\mathcal{GC}^-(\mathbb{G})$ (in the sense of Definition 13.1.9), thought of as a $\mathbb{Z}$-filtered, $\mathbb{Z}$-graded chain complexes over $\mathbb{F}$ and $\mathbb{F}[U]$ respectively, depend on the grid $\mathbb{G}$ only through its underlying unoriented knot K.*

13.3. Topological invariance of the filtered quasi-isomorphism type

The invariance of the filtered quasi-isomorphism type of $\mathcal{GC}^-(\mathbb{G})$ is verified by following the scheme from Chapter 5. Again, we appeal to Cromwell's theorem and adapt the commutation maps and stabilization maps to the filtered setting. It turns out that the commutation maps extend in a rather straightforward way, while the destabilization maps require some further work.

13.3.1. Commutation invariance. Suppose that $\mathbb{G}$ and $\mathbb{G}'$ are two grid diagrams that differ by a commutation move, as in Figure 5.1. Extend the pentagon counting map from Section 5.1 to a map $\mathcal{P}\colon \mathcal{GC}^-(\mathbb{G}) \to \mathcal{GC}^-(\mathbb{G}')$, defined by

$$\mathcal{P}(\mathbf{x}) = \sum_{\mathbf{y}' \in \mathbf{S}(\mathbb{G}')} \sum_{p \in \mathrm{Pent}^\circ(\mathbf{x}, \mathbf{y}')} V_1^{O_1(p)} \cdots V_n^{O_n(p)} \cdot \mathbf{y}'. \tag{13.3}$$

(This is the same formula as in Equation (5.2), except now we have dropped the requirement that $p \cap \mathbb{X} = \emptyset$.)

LEMMA 13.3.1. *The map $\mathcal{P}$ is compatible with the Maslov grading and the Alexander filtration.*

Proof. Adapt the argument from Lemma 5.1.3. The Maslov grading is preserved exactly as before. To verify that $\mathcal{P}$ is compatible with the Alexander filtration, we argue as follows. Suppose that $V_1^{k_1}\cdots V_n^{k_n}\mathbf{y}'$ appears in $\mathcal{P}(\mathbf{x})$, so that there is some $p\in\mathrm{Pent}^\circ(\mathbf{x},\mathbf{y}')$ with $\#(p\cap\mathbb{O})=k_1+\cdots+k_n$. According to Equation (5.5),

$$A(V_1^{k_1}\cdots V_n^{k_n}\mathbf{y}')=A(\mathbf{y}')-\#(p\cap\mathbb{O})=A(\mathbf{x})-\#(p\cap\mathbb{X})\leq A(\mathbf{x}),$$

verifying the claim. □

A straightforward adaptation of the proof of Lemma 5.1.4 now gives:

LEMMA 13.3.2. *$\mathcal{P}$ defined as in Equation* (13.3) *is a chain map of $\mathbb{F}[V_1,\dots,V_n]$-modules.* □

LEMMA 13.3.3. *$\mathcal{P}$ induces the map P from Equation* (5.2) *on the associated graded object.*

Proof. According to the proof of Lemma 13.3.1, the terms in $\mathcal{P}$ preserving Alexander filtration count exactly those pentagons for which $\#(p\cap\mathbb{X})=0$; i.e. $\mathcal{P}$ induces P on the associated graded object. □

PROPOSITION 13.3.4. *If $\mathbb{G}$ and $\mathbb{G}'$ are two grid diagrams that differ by a commutation or a switch, then their associated filtered complexes $\mathcal{GC}^-(\mathbb{G})$ and $\mathcal{GC}^-(\mathbb{G}')$ are filtered quasi-isomorphic, as filtered complexes over $\mathbb{F}[U]$.*

Proof. According to Lemmas 13.3.1 and 13.3.2, $\mathcal{P}$ is a filtered chain map. Since $\mathcal{P}$ induces the map P on the associated graded level (Lemma 13.3.3), which is a quasi-isomorphism by Lemma 5.1.6, it follows that $\mathcal{P}$ is a filtered quasi-isomorphism between the filtered chain complexes $\mathcal{GC}^-(\mathbb{G})$ and $\mathcal{GC}^-(\mathbb{G}')$. These are quasi-isomorphisms over all of $\mathbb{F}[V_1,\dots,V_n]$, and hence over $\mathbb{F}[U]$ for any $U=V_i$. □

The above argument shows that $\mathcal{P}$ and the natural extension $\mathcal{P}'$ of P' (defined by dropping the requirement that $p\cap\mathbb{X}=\emptyset$ in Equation (5.7)) are both quasi-isomorphisms; but it does not show, for example, that they induce inverses to one another in homology. This is true, though, according to the following exercise. (Compare Section 13.4; see also Section A.8.)

EXERCISE 13.3.5. Verify that $\mathcal{P}$ and $\mathcal{P}'$ are filtered homotopy inverses to one another.

13.3.2. Stabilization invariance. Our next aim is to show that if the grid $\mathbb{G}'$ is the stabilization of $\mathbb{G}$, then the filtered chain complexes $\mathcal{GC}^-(\mathbb{G}')$ and $\mathcal{GC}^-(\mathbb{G})$ are filtered quasi-isomorphic. Before going into the details, first we make a few remarks about the associated graded case; that is, the case of the grid complex considered in Section 5.2.

In Proposition 5.2.1, stabilization invariance for grid homology was stated as an identification of the grid homology modules $GH^-(\mathbb{G}')\cong GH^-(\mathbb{G})$. As we now recall, its proof in fact constructed quasi-isomorphisms associated to stabilization moves, which we wish to promote to the filtered case.

For concreteness, let $\mathbb{G}'$ bet obtained from $\mathbb{G}$ by a stabilization of type *X:SW*; so there is a distinguished 2×2 square in $\mathbb{G}'$ marked by $\begin{array}{c|c} X_1 & O_1 \\ \hline & X_2 \end{array}$. As before, let O_2 be the marking in the same row as X_2, and let c be the crossing in the square.

Consider the chain map

$$D\colon GC^-(\mathbb{G}') \to \mathrm{Cone}(V_1 - V_2\colon GC^-(\mathbb{G})[V_1] \to GC^-(\mathbb{G})[V_1]) \tag{13.4}$$

defined by

$$D(\mathbf{x}) = (e(\mathbf{x}), e \circ \mathcal{H}^{\mathbf{I}}_{X_2}(\mathbf{x})),$$

in the notation of Proposition 5.2.17 (cf. Equation (5.22)). This map is called the *graded destabilization map*. Although it was not stated this way, the proof of Proposition 5.2.17 verifies the following:

PROPOSITION 13.3.6. *The graded destabilization map D is a bigraded quasi-isomorphism of modules over $\mathbb{F}[V_1,\dots,V_n]$.* □

To obtain a quasi-isomorphism from $GC^-(\mathbb{G}')$ to $GC^-(\mathbb{G})$ over $\mathbb{F}[V_2,\dots,V_n]$, compose D with the quotient map

$$\mathrm{Cone}(V_1 - V_2\colon GC^-(\mathbb{G})[V_1] \to GC^-(\mathbb{G})[V_1]) \to \frac{GC^-(\mathbb{G})[V_1]}{V_1 - V_2},$$

(which is a quasi-isomorphism by Lemma A.3.9), followed by the isomorphism of chain complexes over $\mathbb{F}[V_2,\dots,V_n]$, $\frac{GC^-(\mathbb{G})[V_1]}{V_1-V_2} \cong GC^-(\mathbb{G})$.

When constructing a filtered quasi-isomorphism for a stabilization of type *X:SW*, regard the mapping cone $\mathit{Cone}(V_1 - V_2\colon \mathcal{GC}^-(\mathbb{G})[V_1] \to \mathcal{GC}^-(\mathbb{G})[V_1])$, as a $\mathbb{Z}$-filtered, $\mathbb{Z}$-graded complex; we abbreviate it by $\mathit{Cone}(V_1 - V_2)$. (Note that this is the cone of $V_1 - V_2$ on all of $\mathcal{GC}^-(\mathbb{G})[V_1]$, whereas in Proposition 13.3.6 we were considering the cone of the induced map on the associated graded object $GC^-(\mathbb{G})[V_1]$.) We generalize the map of Equation (13.4) to a filtered quasi-isomorphism $\mathcal{D}\colon \mathcal{GC}^-(\mathbb{G}') \to \mathit{Cone}(V_1 - V_2)$, after setting up some notation.

Following Section 5.2, partition grid states for $\mathbb{G}'$ into $\mathbf{S}(\mathbb{G}') = \mathbf{I}(\mathbb{G}') \cup \mathbf{N}(\mathbb{G}')$, according to whether or not the grid state $\mathbf{x}$ contains the distinguished point c. There is a corresponding $\mathbb{F}[V_1,\dots,V_n]$-module splitting of the complex $\mathcal{GC}^-(\mathbb{G}') \cong \mathcal{I} \oplus \mathcal{N}$. Unlike in Section 5.2, it is no longer true that $\mathcal{N}$ is a subcomplex. There is still a one-to-one correspondence between generators in $\mathbf{I}(\mathbb{G}')$ and generators for $\mathbb{G}$, which we denote $e\colon \mathbf{I}(\mathbb{G}') \to \mathbf{S}(\mathbb{G})$; but unlike the case of Lemma 5.2.18, the map e does not lift to a chain map.

The map $\mathcal{D}$ counts certain domains that terminate at states in $\mathbf{I}(\mathbb{G}')$. We separate two types of domains: those whose oriented boundary approaches c from the left and those whose oriented boundary approaches c from the right – *type iL* and *type iR*, respectively. (To make the definitions uniform, think of the trivial domain as having type *iL*.) Those two types of domains are organized into a map mapping into the two different bigraded $\mathbb{F}[V_1,\dots,V_n]$-module summands of

$$\mathit{Cone}(V_1 - V_2) = \mathcal{GC}^-(\mathbb{G})[V_1][\![1,1]\!] \oplus \mathcal{GC}^-(\mathbb{G})[V_1].$$

The domains are defined as follows.

DEFINITION 13.3.7. Fix $\mathbf{x} \in \mathbf{S}(\mathbb{G}')$ and $\mathbf{y} \in \mathbf{I}(\mathbb{G}')$. A domain $p \in \pi(\mathbf{x},\mathbf{y})$ is said to be ***into** L* resp. ***into** R* or, more succinctly, of type *iL* or type *iR*, if it is trivial, in which case it is of type *iL*, or if it satisfies the following conditions:

(d-1) All the local multiplicities of p are non-negative.
(d-2) At each corner point in $\mathbf{x} \cup \mathbf{y} \setminus \{c\}$, at least three of the four adjoining squares have vanishing local multiplicities.

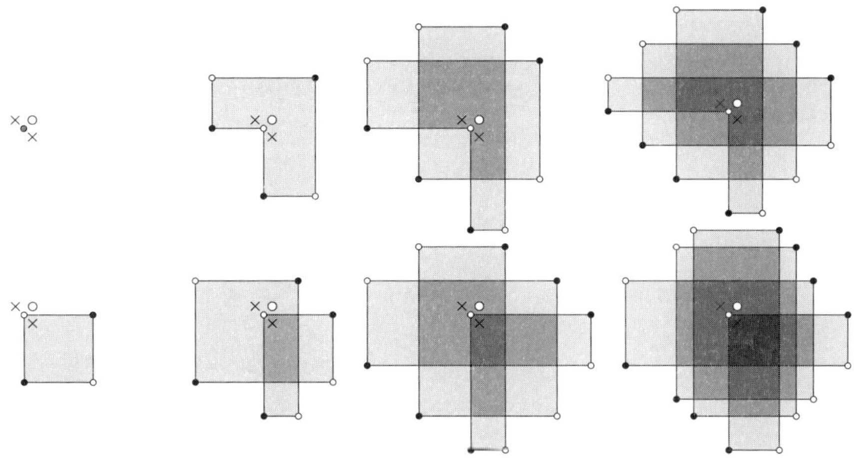

FIGURE 13.1. **Types of domains.** We have listed here domains in the stabilized diagram, labeling the initial points by dark circles, and terminal points by empty circles. The top row lists domains of type iL, while the bottom row lists some of type iR.

(d-3) The domain has the same local multiplicity k at three of the four squares that share the corner c. When p is of type iL, the local multiplicity of p at the southwest square meeting c is $k-1$, while for p of type iR, the local multiplicity of p at the southeast square containing c is $k+1$.

(d-4) If the domain is of type iL, then $\mathbf{y}$ has $2k+1$ components that are not in $\mathbf{x}$; if the domain is of type iR, then $\mathbf{y}$ has $2k+2$ components that are not in $\mathbf{x}$.

The set of domains of type iL resp. iR from $\mathbf{x}$ to $\mathbf{y}$ is denoted $\pi^{iL}(\mathbf{x}, \mathbf{y})$ resp. $\pi^{iR}(\mathbf{x}, \mathbf{y})$. The domains of type iL or iR are called ***destabilization domains***, and the set of destabilization domains is written $\pi^{\mathcal{D}} = \pi^{iL} \cup \pi^{iR}$.

Some destabilization domains are shown in Figure 13.1.

DEFINITION 13.3.8. We define the ***complexity*** of a destabilization domain to be one if it is the trivial domain; otherwise, the complexity counts the number of horizontal segments in its boundary. For example, the rectangle from Figure 13.1 is a destabilization domain of complexity two.

Fix a non-trivial destabilization domain p from $\mathbf{x}$ to $\mathbf{y}$. Let β_i be the vertical circle through c, and let $x_1 = \mathbf{x} \cap \beta_i$. (Note that c is a component of $\mathbf{y}$, but it is not a component of $\mathbf{x}$.) The length of the horizontal segment in ∂p containing x_1 is called the ***innermost width*** of p. Similarly, let α_i be the horizontal circle through c, and $x_k = \mathbf{x} \cap \alpha_i$. Then, the length of the vertical segment in ∂p containing x_k is called the ***innermost height*** of p.

EXERCISE 13.3.9. Draw destabilization domains with complexity 9 and 10.

Note that the domains of type iL resp. iR are the destabilization domains with odd resp. even complexity. The destabilization maps are defined by counting domains of type iL and iR, without factors of V_1, in the following sense.

DEFINITION 13.3.10. Let $\mathbb{G}'$ be a grid diagram obtained from $\mathbb{G}$ by a stabilization of type *X:SW*. For $\mathbf{y} \in \mathbf{I}(\mathbb{G}')$, let $e(\mathbf{y}) \in \mathbf{S}(\mathbb{G})$ be the corresponding state, i.e. $e(\mathbf{y}) = \mathbf{y} \setminus \{c\}$. Consider the $\mathbb{F}[V_1, \ldots, V_n]$-module maps

$$\mathcal{D}^{iL}\colon \mathcal{GC}^-(\mathbb{G}') \to \mathcal{GC}^-(\mathbb{G})[V_1][\![1,1]\!] \qquad \text{resp.} \qquad \mathcal{D}^{iR}\colon \mathcal{GC}^-(\mathbb{G}') \to \mathcal{GC}^-(\mathbb{G})[V_1],$$

with

$$\mathcal{D}^{iL}(\mathbf{x}) = \sum_{\mathbf{y}\in\mathbf{I}(\mathbb{G}')} \sum_{p\in\pi^{iL}(\mathbf{x},\mathbf{y})} V_2^{O_2(p)} \cdots V_n^{O_n(p)} \cdot e(\mathbf{y})$$

$$\mathcal{D}^{iR}(\mathbf{x}) = \sum_{\mathbf{y}\in\mathbf{I}(\mathbb{G}')} \sum_{p\in\pi^{iR}(\mathbf{x},\mathbf{y})} V_2^{O_2(p)} \cdots V_n^{O_n(p)} \cdot e(\mathbf{y}),$$

for any $\mathbf{x} \in \mathbf{S}(\mathbb{G}')$. Put these together to define the ***filtered destabilization map***

$$\mathcal{D}\colon \mathcal{GC}^-(\mathbb{G}') \to \mathit{Cone}(V_1 - V_2) \tag{13.5}$$

$$\mathcal{D}(\mathbf{x}) = (\mathcal{D}^{iL}(\mathbf{x}), \mathcal{D}^{iR}(\mathbf{x})) \in \mathcal{GC}^-(\mathbb{G})[V_1][\![1,1]\!] \oplus \mathcal{GC}^-(\mathbb{G})[V_1] \cong \mathit{Cone}(V_1 - V_2),$$

where the latter is an isomorphism of bigraded modules (not of chain complexes).

Note that for $\mathbf{x} \in \mathbf{I}(\mathbb{G}')$, $\mathcal{D}^{iL}(\mathbf{x}) = e(\mathbf{x})$, and $\mathcal{D}^{iR}(\mathbf{x}) = 0$.

We have the following schematic picture for the map $\mathcal{D}$, where the top row represents $\mathcal{GC}^-(\mathbb{G}')$ (with its decomposition as $\mathcal{I} \oplus \mathcal{N}$), the bottom row represents $\mathit{Cone}(V_1 - V_2)$, the bottom horizontal arrow represents a portion of the boundary operator in $\mathit{Cone}(V_1 - V_2)$, the top horizontal arrow indicates that there are differentials in $\mathcal{GC}^-(\mathbb{G}')$ between $\mathcal{I}$ and $\mathcal{N}$ (both ways), and the arrows connecting the two rows together represent $\mathcal{D}$:

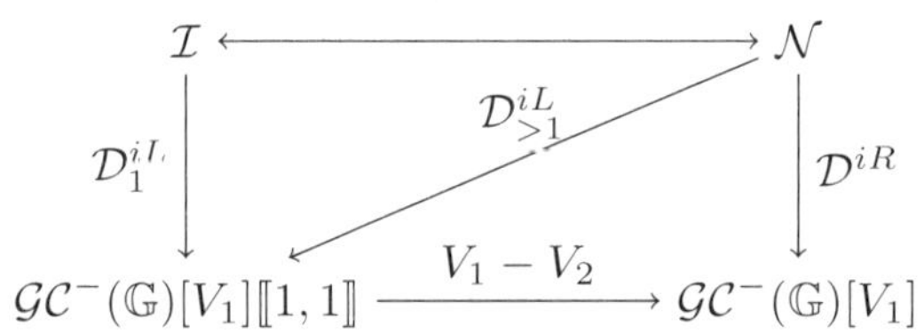

(13.6)

We have written $\mathcal{D}^{iL} = \mathcal{D}^{iL}_1 + \mathcal{D}^{iL}_{>1}$, where the subscript indicates the restriction on the complexity of the domains.

LEMMA 13.3.11. *Let $p \in \pi(\mathbf{x}, \mathbf{y})$ be a destabilization domain of complexity k. Then there is a sequence of states $\{\mathbf{x}_i\}_{i=1}^k$ with $\mathbf{x}_1 = \mathbf{x}$ and $\mathbf{x}_k = \mathbf{y}$ and empty rectangles $\{r_i\}_{i=1}^{k-1}$ with $r_i \in \mathrm{Rect}^\circ(\mathbf{x}_i, \mathbf{x}_{i+1})$, so that $p \in \pi(\mathbf{x}, \mathbf{y})$ is the juxtaposition $r_1 * \cdots * r_{k-1}$. Among such sequences of rectangles, there is a unique one with the property that each rectangle r_i has an edge on the distinguished vertical circle β_i through c.*

Proof. Consider the case where p is of type iL, so its complexity is odd. We prove the existence result by induction on the complexity. The case where the complexity is 1 is trivially true.

Assume the complexity k is greater than 1. Consider the two k-element subsets

$$\{x_1, \ldots, x_k\} \subset \mathbf{x} \setminus (\mathbf{x} \cap \mathbf{y}) \qquad \text{and} \qquad \{y_1, \ldots, y_k\} \subset \mathbf{y} \setminus (\mathbf{x} \cap \mathbf{y}).$$

Number $\{x_j\}_{j=1}^k$ and $\{y_j\}_{j=1}^k$ successively by the following rule: x_1 is the component of $\mathbf{x}$ on β_i (the β-circle containing c), x_j and y_j share the same horizontal circle, and

y_j and x_{j+1} share the same vertical circle. The rest of the components $\{x_j\}_{j=k+1}^n = \{y_j\}_{j=k+1}^n$ of $\mathbf{x}$ (and of $\mathbf{y}$) are ordered arbitrarily. Let s be the intersection of β_i with the horizontal circle through x_2 and t be the intersection of β_i with the horizontal circle through x_3. Let $\mathbf{x}_2 = \{y_1, s, x_3, \dots, x_n\}$ and $\mathbf{x}_3 = \{y_1, y_2, t, x_4, \dots, x_n\}$.

It is easy to see that there is a unique rectangle r_1 with the following three properties:

- the initial state of r_1 is $\mathbf{x}$,
- r_1 is contained in the support of p,
- r_1 has an edge on β_i.

That rectangle r_1 is in $\mathrm{Rect}^\circ(\mathbf{x}_1, \mathbf{x}_2)$; and there is a positive domain q from $\mathbf{x}_2$ to $\mathbf{y}$ with $p = r_1 * q$. Similarly, there is a unique rectangle r_2 whose initial state is $\mathbf{x}_2$, whose support is contained in the support of q, and which has an edge on β_i; and that is a rectangle $r_2 \in \mathrm{Rect}^\circ(\mathbf{x}_2, \mathbf{x}_3)$. We can now decompose $p = r_1 * r_2 * p'$. Observe that p' is a domain whose complexity is $k-2$, so the induction hypothesis applies, verifying both existence and uniqueness of the stated decomposition.

For domains of type iR, the argument is similar. □

LEMMA 13.3.12. *The filtered destabilization map from Definition* 13.3.10 *respects the Maslov grading and the Alexander filtrations.*

Proof. For any destabilization domain $p \in \pi(\mathbf{x}, \mathbf{y})$,

$$A(\mathbf{x}) - A(V_2^{O_2(p)} \cdots V_n^{O_n(p)} \cdot \mathbf{y}) = \#\mathbb{X}(p) - O_1(p),$$

where $\#\mathbb{X}(p)$ denotes the total multiplicity of p at all the points in $\mathbb{X}$. This follows readily from Lemma 13.3.11 and the manner in which the Alexander grading changes under rectangles (Equation (4.14)). Note that the correction from $O_1(p)$ appears because there is no compensating V_1-power in the definition of $\mathcal{D}$. For each destabilization domain p, it is straightforward to see that

$$X_2(p) - O_1(p) = \begin{cases} 0 & \text{if } p \in \pi^{iL} \\ 1 & \text{if } p \in \pi^{iR}. \end{cases}$$

Taking into account the shift in the Alexander filtrations on the mapping cone and the comparison of the Alexander functions on $\mathbb{G}$ and $\mathbb{G}'$ (from Lemma 5.2.4), it follows that if $p \in \pi(\mathbf{x}, \mathbf{y})$ is a destabilization domain, then the Alexander filtration of $\mathbf{x}$ minus the Alexander filtration of the contribution of p is the sum of the local multiplicities of p at $X_1, X_3, \dots, X_n$. It follows that $\mathcal{D}$ is compatible with the Alexander filtrations.

Maslov gradings are preserved by a similar argument. According to Lemma 13.3.11, a destabilization domain p with complexity m can be written as a composition of $m-1$ empty rectangles. It follows from the manner in which Maslov gradings change under rectangles (Equation (4.2)) that

$$M(\mathbf{x}) - M(V_2^{O_2(p)} \cdots V_n^{O_n(p)} \cdot \mathbf{y}) = m - 1 - 2\#(p \cap O_1).$$

When p is of type iR, the complexity m is even, and p contains O_1 with multiplicity $\frac{m}{2} - 1$, so we conclude that the Maslov grading of $\mathbf{x}$ agrees with that of its image under $\mathcal{D}$, bearing in mind the Maslov grading shift in the identification between $\mathbf{I}(\mathbb{G}')$ and $\mathbf{S}(\mathbb{G})$ from Lemma 5.2.4. The case of domains of type iL are handled similarly. □

Most of the work in this section goes into verifying the following result, using an argument that can be thought of a complicated variant of familiar proofs (such as those of Lemmas 4.6.7, 5.1.4, 13.2.2, and 13.3.2). The difficulty, of course, arises since the regions counted in the destabilization maps are more complex than the ones considered until now.

LEMMA 13.3.13. *The map $\mathcal{D}$ from Definition* 13.3.10 *is a chain map.*

Proof. For clarity of exposition, we distinguish the notation for differentials on the various complexes considered in the lemma:

- ∂' denotes the differential on the grid complex $\mathcal{GC}^-(\mathbb{G}')$,
- ∂ denotes the differential on the grid complex $\mathcal{GC}^-(\mathbb{G})$,
- ∂_{cone} denotes the differential on $\mathcal{C}one(V_1 - V_2 \colon \mathcal{GC}^-(\mathbb{G})[V_1] \to \mathcal{GC}^-(\mathbb{G})[V_1])$,

so our goal is to prove

$$\partial_{cone} \circ \mathcal{D} = \mathcal{D} \circ \partial'. \tag{13.7}$$

To this end, we identify the terms appearing on both sides of this equation as a weighted count of regions in the diagram $\mathbb{G}'$. Recall that there is a one-to-one correspondence e between grid states for $\mathbb{G}$ and those in $\mathbf{I}(\mathbb{G}')$, and that rectangles connecting grid states in $\mathbf{S}(\mathbb{G})$ correspond to rectangles in $\mathbb{G}'$ connecting the corresponding grid states in $\mathbf{I}(\mathbb{G}')$. So far, this is similar to the earlier stabilization invariance proof from Chapter 5. However, there is one new feature: in the present case, we must also consider empty rectangles in $\mathbb{G}$ that cross over the stabilization region, and the corresponding rectangles in $\mathbb{G}'$ contain the point c. In the proof, we will find it convenient to label five types of rectangles in $\mathbb{G}'$:

(A) rectangles that are disjoint from c and empty,
(B) rectangles that contain c as a corner and are empty,
(C) rectangles that contain c in the interior of their boundary and are empty,
(D) rectangles r with initial state $\mathbf{x} \in \mathbf{I}(\mathbb{G}')$ satisfying $\mathrm{Int}(r) \cap \mathbf{x} = \{c\}$ (in particular, r is non-empty in $\mathbb{G}'$),
(E) rectangles that are empty, but $c \in \mathrm{Int}(r)$.

Thus, rectangles of Types (A) and (D) connecting generators of $\mathbf{I}(\mathbb{G}')$ give a model for $\mathcal{GC}^-(\mathbb{G})$. The differential of the chain complex $\mathcal{GC}^-(\mathbb{G}')$ counts rectangles of Types (A), (B), (C), and (E). (Recall that rectangles of Type (D) are non-empty.)

Equation (13.7) is verified by analyzing how these rectangles can interact with the domains enumerated in the definition of $\mathcal{D}$. The generators for $\mathcal{C}one(V_1 - V_2)$ are two copies of the grid states $\mathbf{S}(\mathbb{G}) \cong \mathbf{I}(\mathbb{G}')$. That is, we think of generators of $\mathcal{C}one(V_1 - V_2)$ as grid states in $\mathbf{I}(\mathbb{G}')$, further labelled by their types, which can be either $\mathcal{L}$ or $\mathcal{R}$. Those labelled with $\mathcal{L}$ generate the summand $\mathcal{GC}^-(\mathbb{G})[V_1][\![1,1]\!] \subset \mathcal{C}one(V_1 - V_2)$, while those labelled with $\mathcal{R}$ generate the summand $\mathcal{GC}^-(\mathbb{G})[V_1] \subset \mathcal{C}one(V_1 - V_2)$. The differentials connecting two generators both labelled $\mathcal{L}$ or both labelled $\mathcal{R}$ (i.e. the differential on $\mathcal{GC}^-(\mathbb{G})[V_1][\![1,1]\!]$ or on $\mathcal{GC}^-(\mathbb{G})[V_1]$) count rectangles of Type (A) or (D) in $\mathbb{G}'$. There is one more type of term in the differential ∂_{cone}, and that is the differential connecting generators labelled $\mathcal{L}$ to those labelled $\mathcal{R}$. We think of this term as a count of a particularly degenerate kind of domain: one that is empty, but labelled with the algebra element $V_1 - V_2$. We call these domains *invisible*.

Next we classify the juxtapositions of domains (consisting of a rectangle and a destabilization domain) that contribute to $\partial_{cone} \circ \mathcal{D} + \mathcal{D} \circ \partial'$. The classification records the type of rectangle and the number of shared moving corners between the rectangle and the destabilization domain, in the following sense. If $p \in \pi(\mathbf{x}, \mathbf{y})$ is a domain, we say that the *moving corners of* p is the set $\mathbf{x} \cup \mathbf{y} \setminus (\mathbf{x} \cap \mathbf{y})$. If $p \in \pi(\mathbf{x}, \mathbf{y})$ and $q \in \pi(\mathbf{y}, \mathbf{w})$ are two domains, ***a shared moving corner*** is an element in the intersection of the set of moving corners of p with the set of moving corners of q. The juxtapositions of domains with non-zero contribution to $\partial_{cone} \circ \mathcal{D} + \mathcal{D} \circ \partial'$ are of the following eleven possible types:

(A-0) A composition $p * r$ or $r * p$ of a rectangle r of Type (A) and a domain of p in $\pi^{\mathcal{D}}$ with complexity ≥ 1 with no shared moving corners.

(A-1) A composition $p * r$ or $r * p$ of a rectangle r of Type (A) and a domain $p \in \pi^{\mathcal{D}}$, with complexity > 1, with one shared moving corner.

(A-2) A composition $p * r$ or $r * p$ where r is a rectangle of Type (A), and p is a domain in $\pi^{\mathcal{D}}$, with complexity ≥ 4, where r and p share two moving corners. (For examples of pairs of domains (A-0), (A-1) and (A-2), see the upper row of Figure 13.2.)

(A-3) A composition $p * r$ or $r * p$ where r is a rectangle of Type (A), and p is a domain in $\pi^{\mathcal{D}}$ with complexity ≥ 3, where r and p share three moving corners. When this happens, the composite domain wraps around the torus, the rectangle has height or width equal to one, and the domain p has innermost height or width equal to one. This can happen in three ways:
- the domain goes vertically around the torus, passing through O_1;
- the domain goes horizontally around the torus, containing the row through O_1; in this case, the complexity of p is odd,
- the composite domain goes horizontally around the torus, containing the row through O_2; in this case, the complexity of p is even.

Domains of this third type are called *exceptional*, or of Type $(\text{A-3})^e$. The first two types are *ordinary*, or of Type $(\text{A-3})^o$. (See the lower row of Figure 13.2.)

(B-0) a composition $r * p$, where r is a rectangle of Type (B) and p is a domain in $\pi^{\mathcal{D}}$ with complexity 1,

(B-3) a composition $r * p$ where r is a rectangle of Type (B) and p is a domain with complexity 3, and r and p have three shared moving corners,

(B-4) A composition $r * p$ where r is a rectangle of Type (B), and p is a domain in $\pi^{\mathcal{D}}$ with complexity 2, and r and p share four moving corners. When this happens, the composite domain is either a thin vertical annulus through O_1, or a thin horizontal annulus through O_2. See Figure 13.3.

(C-1) A composition $r * p$, where r is of Type (C) and $p \in \pi^{\mathcal{D}}$ has complexity ≥ 2. There are three subtypes:
- $r \cap \{O_1, X_1, X_2\} = \{X_1, O_1\}$ and p has even complexity,
- $r \cap \{O_1, X_1, X_2\} = \{X_1\}$,
- $r \cap \{O_1, X_1, X_2\} = \{X_2\}$ and p has odd complexity.

Observe that the case where $O_1 \in r$ is the case where the composite $r * p$ itself looks like a destabilization domain with odd complexity. In the first case, we call the domain of Type (C-1) *exceptional*, or of Type $(\text{C-1})^e$;

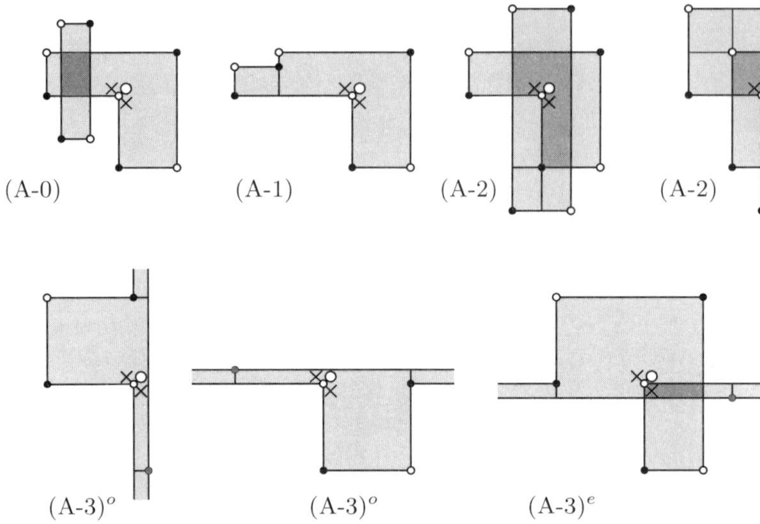

FIGURE 13.2. **Types of juxtapositions using rectangles of Type (A).** The black components are in the initial generator, the white components are in the terminal generator, and the gray ones are in both; darker shadings indicate regions with multiplicity two.

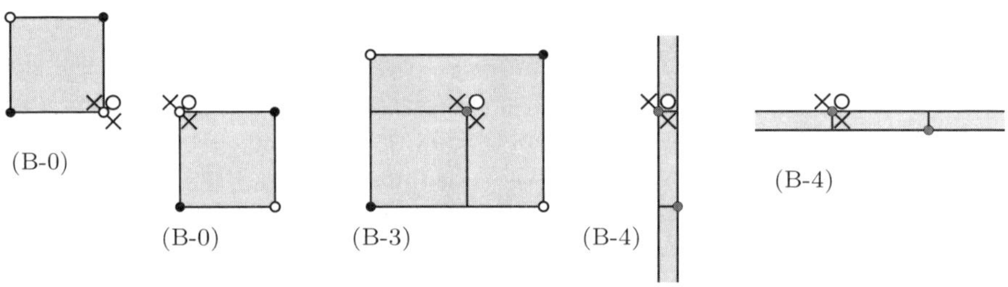

FIGURE 13.3. **Types of juxtapositions using rectangles of Type (B).** We use the same conventions as in Figure 13.2.

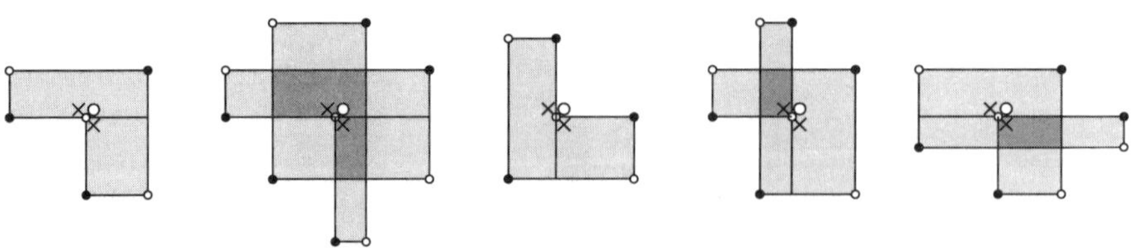

FIGURE 13.4. **Types of juxtapositions using rectangles of Type (C)**. The first two are exceptional; all the others are ordinary.

otherwise, we call it *ordinary*, or of Type (C-1)^o. See Figure 13.4. When $r \cap \{O_1, X_1, X_2\} = \{X_1\}$, we let δ denote the innermost width of p; otherwise, it is the innermost height.

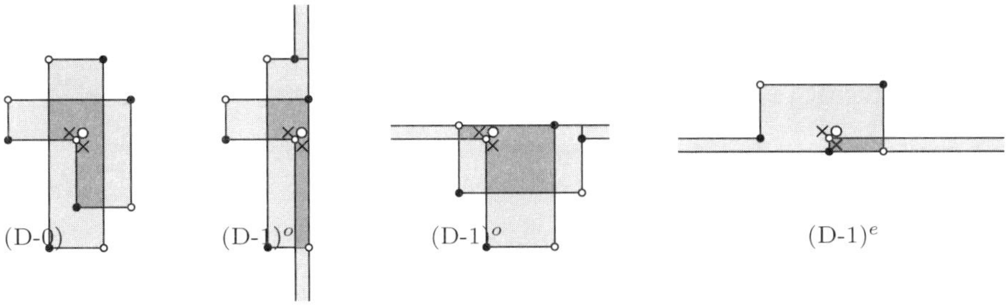

FIGURE 13.5. **Compositions involving a rectangle of Type (D).**

(D-0) A composition $p*r$, where r is of Type (D), and p is in $\pi^{\mathcal{D}}$ with complexity ≥ 1, and p and r have no shared moving corners.

(D-1) A composition $p*r$, where r is of Type (D), p is a destabilization domain with complexity ≥ 2, and p and r share one moving corner. In this case, the composite domain must wrap around the torus, and it contains a thin (horizontal or vertical) annulus that goes through O_1; or it contains a thin horizontal annulus that goes through O_2. When the annulus goes through O_1, we call the domain *ordinary*, or of Type (D-1)^o; when it goes through O_2, we call it *exceptional*, or of Type (D-1)^e. See Figure 13.5.

(I) A domain p of type iL, multiplied by the weight $V_1 - V_2$. This can be thought of as a decomposition of the domain as $p*i$, where i is an invisible domain counted in the differential of the mapping cone.

All of the above types of juxtapositions contain a destabilization domain p. We define the *complexity* of a juxtaposition to be complexity of p. Note that configurations of Type (I) always have odd complexity, and they always land in $\mathcal{R}$. For all the other types, the target is in $\mathcal{L}$ if and only if the complexity is odd.

It is easy to see that we have enumerated all possible pairs of domains with non-trivial contribution to $\partial_{cone} \circ \mathcal{D} + \mathcal{D} \circ \partial'$. In particular, we claim that rectangles of Type (E), which count in the differential for $\mathcal{GC}^-(\mathbb{G}')$, cannot be followed by a domain p from $\pi^{\mathcal{D}}$: the geometry forces some component of the initial state to lie in the interior of r or some component of the terminal state to lie in p.

The verification of Equation (13.7) amounts to seeing how the contributions of juxtaposed domains in the above classification can be paired off. We will verify that contributions of Type (A-0) and those of Type (A-1) cancel in pairs. The remaining cancellations are summarized in the following table; contributions from each row cancel with each other:

(D-0)$_{m>1}$	(A-2)$_{m+2}$	
(D-0)$_{m=1}$	(B-3)$_{m=3}$	
(B-4)$_{m=2}$	(I)$_{m=1}$	
(B-0)$_{m=1}$	(A-3)$^o_{m=3}$	
${}^{\delta=1}$(C-1)$^o_{m>1}$	(D-1)^o_m	
${}^{\delta>1}$(C-1)$^o_{m>1}$	(A-3)$^o_{m+2}$	
${}^{\delta=1}$(C-1)$^e_{m=2k}$	(D-1)$^e_{m=2k}$	(I)$_{m=2k+1}$
${}^{\delta>1}$(C-1)$^e_{m=2k}$	(A-3)$^e_{m=2k+2}$	

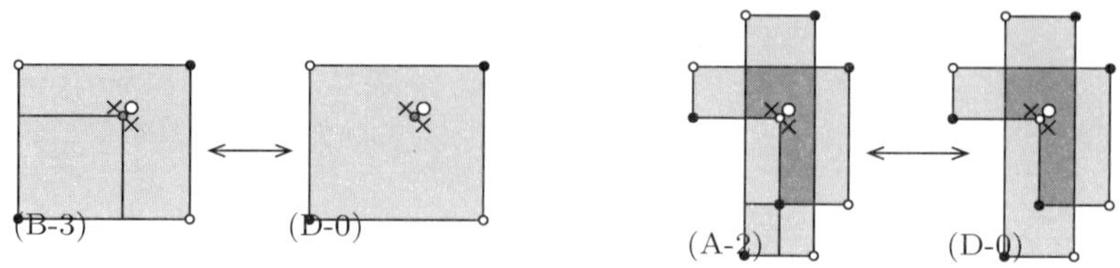

FIGURE 13.6. **Contributions of (A-2) and (B-3) cancel contributions of (D-0).** The degenerate case of Type (D-0) where the destabilization domain p has complexity 1 is shown in the left; the generic case is in the right.

In the above table, the subscript represents the complexity of the composite domain.

We verify the above cancellation scheme as follows.

Contributions of Type (A-0) cancel in pairs: for each such decomposition $r * p$, there is an alternative decomposition $p' * r'$ of the same type, where the order of the rectangle and the destabilization domain are reversed. (This is reminiscent of how the contributions of disjoint rectangles cancel in pairs in Case (R-1) of the proof of Lemma 4.6.7.)

Contributions of Type (A-1) also cancel in pairs. For this, we observe that the composite domain has a single corner (other than possibly c) with a 270° angle. Cutting this domain in two different ways gives two different decompositions of the same domain as a juxtaposition of domains of Type (A-1). (This is reminiscent of how the contributions of rectangles sharing one corner cancel in pairs in Case (R-2) of the proof of Lemma 4.6.7.)

Compositions $r * p$ or $p * r$ of Type (A-2), where p has complexity $m \geq 4$, cancel with compositions $p' * r'$ of Type (D-0), where now p' has complexity $m - 2$. The remaining domains $p' * r'$ of Type (D-0) with complexity $m = 1$ cancel with the decomposition of Type (B-3). See Figure 13.6 for both cancellations.

For the remaining cases, we explain how to group juxtapositions so that their contributions cancel. As we shall see, these pairings do not occur simply as pairs of alternate decompositions of the same domain. Nonetheless, the different composite domains differ by thin annuli. In verifying that the contributions cancel, the reader should bear in mind that destabilization domains do not contribute powers of V_1 (even if the domain crosses O_1); and domains of Type (I) contribute $V_1 - V_2$ (even though the domain contains neither O_1 nor O_2).

With the preceding understood, the contributions of Type (I) with complexity $= 1$ are clearly cancelled by the contributions of the two domains of Type (B-4).

Contributions of Type (B-0) cancel the complexity 3 contributions of Type (A-3). When c is the lower right resp. upper left corner of the rectangle in the Type (B-0) decomposition, the corresponding destabilization domain in the Type (A-3) juxtaposition has innermost width resp. height equal to one. The order of the rectangle and the destabilization domain in the Type (A-3) decomposition depends on the placement of the components of the initial generator $\mathbf{x}$ in the column or row through O_1. See Figure 13.7 for an illustration.

It remains to verify that contributions $r * p$ of Type (C-1) cancel with contributions of Type (D-1) or (A-3), with possible contributions of Type (I).

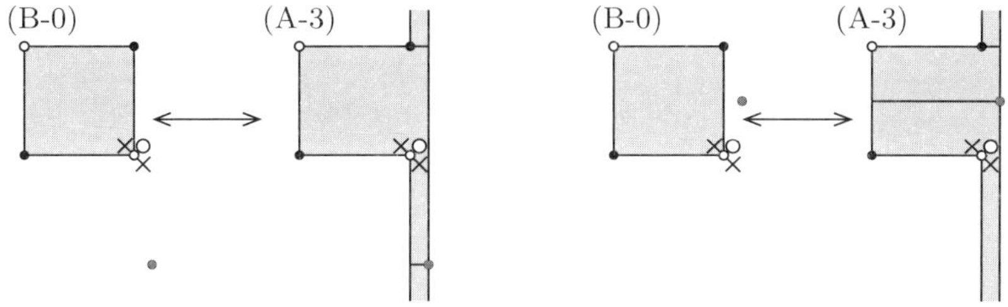

FIGURE 13.7. **Contributions of (B-0) cancel contributions of (A-3) with complexity** 3. We have drawn cases where the rectangle in the Type (B-0) decomposition has lower right corner at c. The placement of the initial generator in the column immediately to the right of O_1 determines the order of the polygons, as indicated. Similar pictures can be drawn when the rectangle has upper left corner at c; in these cases, the domain of Type (A-3) contains a horizontal annulus through O_1.

We start with the subcase where $r * p$ is of Type (C-1)^o and $r \cap \{O_1, X_1, X_2\} = \{X_2\}$. In this case, the domain underlying $r * p$ has a unique decomposition that contributes to $\partial_{cone} \circ \mathcal{D} + \mathcal{D} \circ \partial'$. Form a new domain by adding the thin annulus of height one through O_1. We claim that this new domain has a unique decomposition that contributes to $\partial_{cone} \circ \mathcal{D} + \mathcal{D} \circ \partial'$, and that cancels the contribution of $r * p$. In cases where the innermost height of p is 1 (i.e. $\delta = 1$), the decomposition is of Type (D-1); for more general p, the decomposition is of Type (A-3), and the order (of the rectangle and the domain) is determined by the positions of the generator in the newly added row. If m denotes the complexity of the decomposition of Type (C-1), then the complexity of a corresponding decomposition of Type (D-1) is also m; while a corresponding decomposition of Type (A-3) is $m+2$. See Figure 13.8 for an illustration.

When $r \cap \{O_1, X_1, X_2\} = \{X_1\}$, the same argument works, after adding a vertical (rather than a horizontal) thin annulus through O_1 to the original domain $r * p$. This completes the case where $r * p$ is of Type (C-1)^o.

When $r * p$ is of Type (C-1)^e, the domain underlying $r * p$ itself looks like a destabilization domain of type iL; i.e. it has another contribution to $\partial_{cone} \circ \mathcal{D} + \mathcal{D} \circ \partial'$, when viewed as a decomposition of Type (I). (Note that this contribution of Type (I) has complexity ≥ 3.) We can also construct an alternate domain by adjoining the horizontal, height one annulus through O_2. When the innermost height δ of p is one, the new domain has a unique decomposition of Type (D-1)^e; when $\delta > 1$, the new domain has a unique decomposition of Type (A-3)^e. Now the three different contributions (of Types (C-1)^e, (D-1)^e or (A-3), and (I)) cancel. See Figure 13.9. This argument concludes the case analysis and proves that the destabilization map $\mathcal{D}$ of Definition 13.3.10 is a chain map, as claimed. □

EXERCISE 13.3.14. **(a)** Draw a domain of Type (C-1)^e with complexity 4. Find the cancelling contribution.
(b) Draw a complexity 5 domain. Find all the rectangles that can be added to it to give a composite domain of Type (A-2). Identify the cancelling domains.

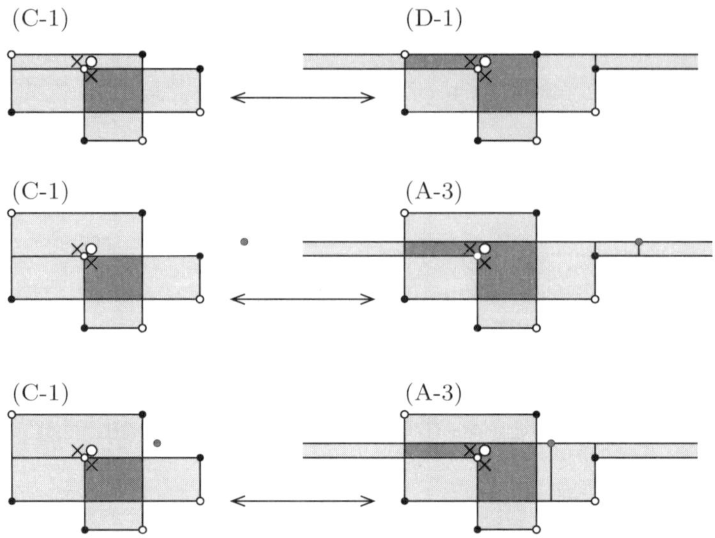

FIGURE 13.8. **Contributions of (C-1)^o cancel contributions of (D-1)^o or (A-3)^o.** In the pictures, for $r * p$ of Type (C-1), we have drawn the case where $r \cap \{O_1, X_1, X_2\} = \{X_2\}$ and the complexity of p is 3.

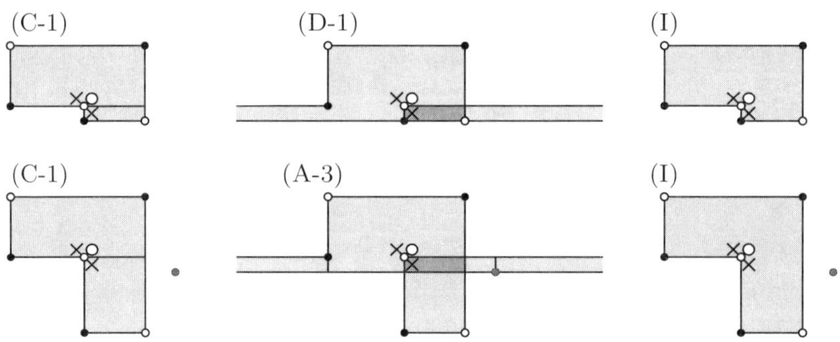

FIGURE 13.9. **Cancellation of (C-1)^e with contributions of (D-1)^e, (A-3)^e, and (I) with complexity > 1.**

(c) Draw a complexity 6 composite domain of Type (A-3)^e. Find the cancelling composite domains.

LEMMA 13.3.15. *The filtered destabilization map $\mathcal{D}$ from Definition* 13.3.10 *induces the destabilization map D from Proposition* 13.3.6 *on the associated graded level.*

Proof. By Lemma 13.3.12, the map on the associated graded level induced by $\mathcal{D}$ counts destabilization domains that do not go through $X_1, X_3, \ldots, X_n$. It is easy to see that these destabilization domains have complexity ≤ 2; and hence, that these are the domains counted in the destabilization map D (cf. Equation (13.4)). □

We now have the following generalization of Proposition 5.2.1:

PROPOSITION 13.3.16. *Label the O-markings in $\mathbb{G}$ by $\{2,\dots,n\}$, and fix some $i\in\{2,\dots,n\}$. Suppose that $\mathbb{G}'$ is obtained form $\mathbb{G}$ by stabilization. Then, there is a filtered quasi-isomorphism from $\mathcal{GC}^-(\mathbb{G}')$ to $\mathcal{GC}^-(\mathbb{G})$, viewing both as chain complexes over $\mathbb{F}[V_i]$.*

Proof. We start with the case of stabilizations of type *X:SW*. The filtered destabilization map $\mathcal{D}\colon \mathcal{GC}^-(\mathbb{G}')\to \mathcal{C}one(V_1-V_2)$ of Definition 13.3.10 is a graded and filtered chain map according to Lemmas 13.3.12 and 13.3.13, and it is a quasi-isomorphism over $\mathbb{F}[V_2,\dots,V_n]$ by Lemma 13.3.15 and Proposition 13.3.6.

Apply Lemma 13.1.13 with $f=V_1-V_2$, to get a a filtered quasi-isomorphism

$$\mathcal{C}one(V_1-V_2)\to \frac{\mathcal{GC}^-(\mathbb{G})[V_1]}{V_1-V_2}. \tag{13.8}$$

Finally, there is an isomorphism of filtered chain complexes

$$\frac{\mathcal{GC}^-(\mathbb{G})[V_1]}{V_1-V_2}\cong \mathcal{GC}^-(\mathbb{G}) \tag{13.9}$$

over $\mathbb{F}[V_2,\dots,V_n]$. The desired filtered quasi-isomorphism is obtained by composing the filtered quasi-isomorphisms induced by $\mathcal{D}$ with the quasi-isomorphisms from Equations (13.8) and (13.9), and then restricting to the subring $\mathbb{F}[V_i]$.

As usual, we can use switches to reduce all stabilizations to the above case. (See Corollary 3.2.3 and Proposition 13.3.4.) □

13.3.3. The invariance statement. We can synthesize the above results to get the invariance of the quasi-isomorphism type of the grid complex:

Proof of Theorem 13.2.9. Both the construction of $\widehat{\mathcal{GC}}(\mathbb{G})$ and the restriction of $\mathcal{GC}^-(\mathbb{G})$ to a module over $\mathbb{F}[U]$ involve choosing some $i\in\{1,\dots,n\}$. Lemma 13.2.8 proves that the filtered quasi-isomorphism types of the complexes are independent of this choice; i.e. they depend only on the underlying toroidal grid diagram $\mathbb{G}$.

Combine Cromwell's Theorem 3.1.9, commutation invariance (Proposition 13.3.4), and stabilization invariance (Proposition 13.3.16) to see that the filtered quasi-isomorphism type $\mathcal{GC}^-(\mathbb{G})$ depends on the underlying oriented knot only. By the five lemma, if $\mathcal{GC}^-(\mathbb{G})$ is quasi-isomorphic to $\mathcal{GC}^-(\mathbb{G}')$, then $\frac{\mathcal{GC}^-(\mathbb{G})}{U}$ is quasi-isomorphic to $\frac{\mathcal{GC}^-(\mathbb{G}')}{U}$. (Compare Proposition A.3.5.) To see the independence of the choice of orientation, reflect across the diagonal, as in the proof of Proposition 5.3.2. □

DEFINITION 13.3.17. For a knot K let $\mathbb{G}$ be a grid diagram representing K. The filtered quasi-isomorphism type of the $\mathbb{Z}$-filtered, $\mathbb{Z}$-graded $\mathbb{F}[U]$-module $\mathcal{GC}^-(\mathbb{G})$ is called the ***filtered grid invariant*** of K and it is denoted by $\mathcal{GC}^-(K)$. Similarly, the filtered quasi-isomorphism type of the $\mathbb{Z}$-filtered, $\mathbb{Z}$-graded chain complex (over $\mathbb{F}$) $\widehat{\mathcal{GC}}(\mathbb{G})$ is the ***simply blocked filtered grid invariant*** of K, and it will be denoted by $\widehat{\mathcal{GC}}(K)$.

13.4. Filtered homotopy equivalences

The filtered grid invariant as defined above is the filtered quasi-isomorphism type of $\mathcal{GC}^-(\mathbb{G})$ for any grid diagram for K, where multiplication by U is defined to be multiplication by V_i for any $i = 1, \dots, n$. In the above topological invariance proof, we gave filtered quasi-isomorphisms between the grid complexes for various choices: varying the choice of V_i, performing commutation moves on $\mathbb{G}$, and performing stabilization moves.

Our aim here is to promote these filtered quasi-isomorphisms to explicit filtered chain homotopy equivalences. Although this is not strictly needed to set up the theory (see Section A.8), the explicit forms of the maps can be useful for computations. In Section 13.4.1, we show that different choices of V_i give filtered chain homotopic grid complexes. Filtered homotopy equivalences for commutation moves are supplied by the pentagon counting maps from Section 13.3.1; to see they are homotopy equivalences, extend the hexagon counting homotopies from Section 5.1 in a straightforward manner. The more interesting stabilization maps are sketched in Section 13.4.2.

13.4.1. Varying the choice of V_i. We construct homotopy equivalences between the grid complexes, thought of as chain complexes over the polynomial algebra with variables V_i, for various choices of i.

Let $\mathcal{H}_{O_i} : \mathcal{GC}^-(\mathbb{G}) \to \mathcal{GC}^-(\mathbb{G})$ be the map

$$\mathcal{H}_{O_i}(\mathbf{x}) = \sum_{\mathbf{y}\in\mathbf{S}(\mathbb{G})} \sum_{\{r\in\mathrm{Rect}^\circ(\mathbf{x},\mathbf{y})\,\big|\,O_i\in r\}} V_1^{O_1(r)} \cdots\cdot V_{i-1}^{O_{i-1}(r)} \cdot V_{i+1}^{O_{i+1}(r)} \cdots V_n^{O_n(r)} \cdot \mathbf{y},$$

There is a similar operator $\mathcal{H}_{X_i}$, counting rectangles that contain the X-marking X_i (which is in the same row as O_i). If O_j is in the same column as X_i (i.e. V_i and V_j are consecutive, as in the proof of Lemma 4.6.9), consider the map $f\colon (\mathcal{GC}^-(\mathbb{G}), V_i) \to (\mathcal{GC}^-(\mathbb{G}), V_j)$ determined on grid states by the formula

$$\phi(\mathbf{x}) = \mathbf{x} - \mathcal{H}_{X_i} \circ \mathcal{H}_{O_i}(\mathbf{x}) \tag{13.10}$$

and the relations $\phi(V_k \cdot \xi) = V_k \cdot \phi(\xi)$ for $k \neq i$ and $\phi(V_i \cdot \xi) = V_j \cdot \phi(\xi)$. This map was introduced and studied by Sarkar [**203**]. We prove here the following:

PROPOSITION 13.4.1. *Let $(\mathcal{GC}^-(\mathbb{G}), V_i)$ be the grid complex of a knot, thought of as a module over $\mathbb{F}[U]$, where U acts as V_i. If V_i and V_j are consecutive, the map f defined above gives a filtered homotopy equivalence from $(\mathcal{GC}^-(\mathbb{G}), V_i)$ to $(\mathcal{GC}^-(\mathbb{G}), V_j)$. Thus, the filtered chain homotopy type of $(\mathcal{GC}^-(\mathbb{G}), V_i)$ is independent of the choice of i.*

Proof. Let $(\mathcal{C}, \partial) = (\mathcal{GC}^-(\mathbb{G}), \partial^-)$. Number the variables so that V_1 and V_2 are consecutive. Any element of $c \in \mathcal{C}$ can be uniquely written as a sum $\sum c_k V_1^k$, where c_k are in the $\mathbb{F}[V_2, \dots, V_n]$-submodule $\mathcal{C}^1$ of $\mathcal{C}$ generated by grid states. Define a filtered $\mathbb{F}[V_2, \dots, V_n]$-module map $h\colon \mathcal{C} \to \mathcal{C}$ by

$$h(\sum c_k V_1^k) = \sum_{p+q=k-1} V_1^p V_2^q c_k,$$

where $c_k \in \mathcal{C}^1$. It is elementary to verify that

$$h \circ V_1 - V_2 \circ h = \mathrm{Id}\,.$$

Let $\mathcal{H} = \mathcal{H}_{X_1}$ be the operator from Lemma 13.2.5 satisfying

$$\partial \circ \mathcal{H} + \mathcal{H} \circ \partial = V_1 - V_2.$$

Consider the map $\phi\colon \mathcal{C} \to \mathcal{C}$ defined by

$$\phi = \mathrm{Id}_{\mathcal{C}} - \partial \circ \mathcal{H} \circ h - \mathcal{H} \circ h \circ \partial. \tag{13.11}$$

Obviously, ϕ is a chain map; and

$$\begin{aligned} \phi \circ V_1 - V_2 \circ \phi &= (V_1 - V_2) - \partial \circ \mathcal{H} \circ (h \circ V_1 - V_2 \circ h) - \mathcal{H} \circ (h \circ V_1 - V_2 \circ h) \circ \partial \\ &= (V_1 - V_2) - \partial \circ \mathcal{H} - \mathcal{H} \circ \partial = 0; \end{aligned}$$

i.e. ϕ can be viewed as a filtered chain map $\phi\colon (\mathcal{C}, V_1) \to (\mathcal{C}, V_2)$ over $\mathbb{F}[U]$.

We show that a homotopy inverse to ϕ is given by $\psi\colon \mathcal{C} \to \mathcal{C}$ defined by

$$\psi = \mathrm{Id}_{\mathcal{C}} + \partial \circ \mathcal{H} \circ h' + \mathcal{H} \circ h' \circ \partial,$$

where h' is defined like h, with the roles of V_1 and V_2 reversed, so that

$$h' \circ V_2 - V_1 \circ h' = \mathrm{Id}.$$

The same reasoning as before shows that $\psi \circ V_2 - V_1 \circ \psi = 0$.

Letting

$$\mathcal{K} = -\mathcal{H} \circ h + \mathcal{H} \circ h' - \mathcal{H} \circ h' \circ \partial \circ \mathcal{H} \circ h - \mathcal{H} \circ h' \circ \mathcal{H} \circ h \circ \partial,$$

it is straightforward to verify that

$$\partial \circ \mathcal{K} + \mathcal{K} \circ \partial = \psi \circ \phi - \mathrm{Id};$$

i.e. $\psi \circ \phi$ is chain homotopic to the identity. We want to check that $\psi \circ \phi$ is homotopic to the identity as a chain map over $\mathbb{F}[V_1]$. For any endomorphism f, let $[f, V_1] = f \circ V_1 - V_1 \circ f$. Observe that

$$\begin{aligned} [\mathcal{H} \circ h, V_1] &= (V_2 - V_1) \circ \mathcal{H} \circ h + \mathcal{H} \\ [\mathcal{H} \circ h', V_1] &= 0 \\ [\mathcal{H} \circ h' \circ \partial \circ \mathcal{H} \circ h, V_1] &= \mathcal{H} \circ \partial \circ \mathcal{H} \circ h + \mathcal{H} \circ h' \circ \partial \circ \mathcal{H} \\ &= -\mathcal{H} \circ \mathcal{H} \circ \partial \circ h + (V_1 - V_2) \circ \mathcal{H} \circ h + \mathcal{H} \circ h' \circ \partial \circ \mathcal{H} \\ [\mathcal{H} \circ h' \circ \mathcal{H} \circ h \circ \partial, V_1] &= \mathcal{H} \circ \mathcal{H} \circ h \circ \partial + \mathcal{H} \circ h' \circ \mathcal{H} \circ \partial \end{aligned}$$

and hence

$$\begin{aligned} -[\mathcal{K}, V_1] &= \mathcal{H} + \mathcal{H} \circ \mathcal{H} \circ (h \circ \partial - \partial \circ h) + \mathcal{H} \circ h' \circ (V_1 - V_2) \\ &= \mathcal{H} \circ \mathcal{H} \circ (h \circ \partial - \partial \circ h). \end{aligned}$$

The same considerations as in the proof of Lemma 13.2.2 show that $\mathcal{H} \circ \mathcal{H} = 0$: that operator counts decompositions of regions that cover the distinguished X marking twice, and those decompositions come in pairs. Thus, $\mathcal{K}$ commutes with V_1, as needed. Also, $\phi \circ \psi$ is homotopic to the identity map by the same reasoning.

Note that for each grid state $\mathbf{x}$, $h(\mathbf{x}) = 0$ and $h(\partial \mathbf{x}) = \mathcal{H}_{O_1}(\mathbf{x})$, so Equation (13.10) follows from Equation (13.11). □

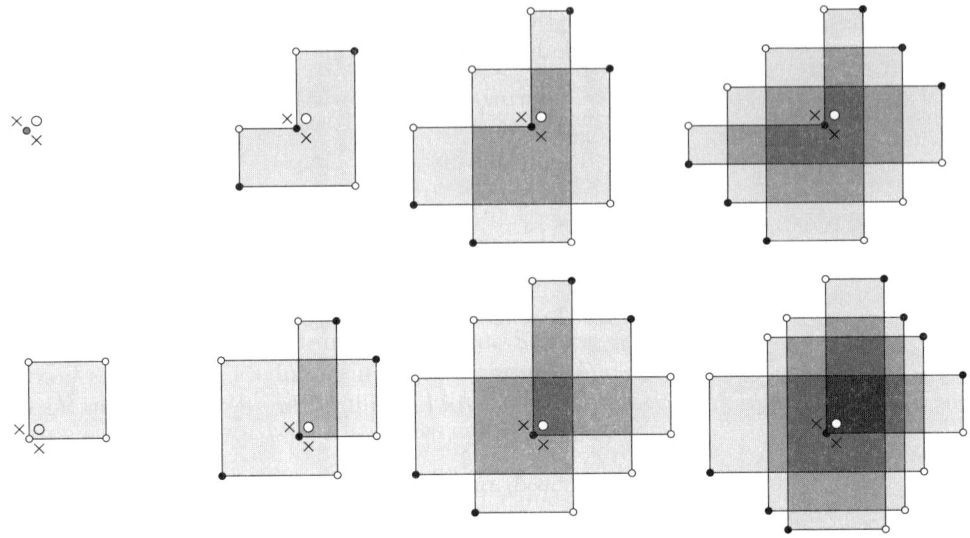

FIGURE 13.10. **Examples of stabilization domains.** The upper row shows examples of elements of type oL, the lower row examples of type oR.

13.4.2. Filtered stabilization maps. In the proof of stabilization invariance, we constructed filtered destabilization maps for stabilizations of type $X{:}SW$, and proved that they are filtered quasi-isomorphisms. Here we invite the interested reader to construct explicit homotopy inverses for these destabilization maps, giving hints along the way. This discussion is a filtered generalization of Section 5.7; other stabilizations will be considered in Section 14.3.

The stabilization maps for $X{:}SW$ stabilizations are defined by counting domains which go out of L or out of R, in the following sense.

DEFINITION 13.4.2. Fix $\mathbf{x} \in \mathbf{I}(\mathbb{G}')$ and $\mathbf{y} \in \mathbf{S}(\mathbb{G}')$. A domain $p \in \pi(\mathbf{x}, \mathbf{y})$ is said to be ***out of*** L resp. ***out of*** R or, more succinctly, of type oL or oR, if it is trivial, in which case it is of type oL, or it satisfies the following conditions:

(s-1) All the local multiplicities of p are non-negative.
(s-2) At each corner in $\mathbf{x} \cup \mathbf{y} \setminus \{c\}$, at least three of the four adjoining squares have vanishing local multiplicities.
(s-3) In a neighborhood of c, the local multiplicities in three of the adjoining squares are the same number k. When p has type oL, the domain has local multiplicity $k - 1$ at the northwest square meeting c; when p is of type oR, it has multiplicity $k + 1$ at the northeast square meeting c.
(s-4) If the domain has type oL, then $\mathbf{y}$ has $2k + 1$ coordinates not in $\mathbf{x}$; if the domain has oR, then $\mathbf{y}$ has $2k + 2$ coordinates not in $\mathbf{x}$.

The set of domains of type oL resp. oR from $\mathbf{x}$ to $\mathbf{y}$ is denoted $\pi^{oL}(\mathbf{x}, \mathbf{y})$ resp. $\pi^{oR}(\mathbf{x}, \mathbf{y})$. Domains of type oL or oR are called ***stabilization*** domains.

Stabilization domains look like destablization domains as in Definition 13.3.7, reflected through a horizontal axis. See Figure 13.10.

DEFINITION 13.4.3. Consider the maps

$$\mathcal{S}^{oL}\colon \mathcal{GC}^-(\mathbb{G})[V_1][\![1,1]\!] \to \mathcal{GC}^-(\mathbb{G}') \qquad \text{and} \qquad \mathcal{S}^{oR}\colon \mathcal{GC}^-(\mathbb{G})[V_1] \to \mathcal{GC}^-(\mathbb{G}'),$$

defined by

$$\mathcal{S}^{oL}(\mathbf{x}) = \sum_{\mathbf{y}\in\mathbf{S}(\mathbb{G}')} \sum_{p\in\pi^{oL}(e'(\mathbf{x}),\mathbf{y})} V_2^{O_2(p)} \cdots V_n^{O_n(p)} \cdot \mathbf{y},$$

$$\mathcal{S}^{oR}(\mathbf{x}) = \sum_{\mathbf{y}\in\mathbf{S}(\mathbb{G}')} \sum_{p\in\pi^{oR}(e'(\mathbf{x}),\mathbf{y})} V_2^{O_2(p)} \cdots V_n^{O_n(p)} \cdot \mathbf{y},$$

where $e'\colon \mathbf{S}(\mathbb{G}) \xrightarrow{\cong} \mathbf{I}(\mathbb{G}')$ is the identification of grid states (i.e. it is the inverse of the map e considered earlier).

These maps can be assembled to form a map $\mathcal{C}one(V_1 - V_2) \to \mathcal{GC}^-(\mathbb{G}')$, specified by the following diagram:

$$\begin{array}{ccc}
\mathcal{GC}^-(\mathbb{G})[V_1][\![1,1]\!] & \xrightarrow{\;V_1 - V_2\;} & \mathcal{GC}^-(\mathbb{G})[V_1] \\
\Big\downarrow \mathcal{S}_1^{oL} & \searrow^{\mathcal{S}_{>1}^{oL}} & \Big\downarrow \mathcal{S}^{oR} \\
\mathcal{I} & \xleftrightarrow{\;\partial^-\;} & \mathcal{N}
\end{array}$$

(where the top row specifies $\mathcal{C}one(V_1 - V_2)$ and the bottom specifies $\mathcal{GC}^-(\mathbb{G}')$).

EXERCISE 13.4.4. **(a)** Show that the stabilization map is a chain map that respects the Maslov gradings and Alexander filtrations.
(b)*If $\mathbb{G}'$ is obtained from $\mathbb{G}$ by a stabilization of type *X:SW*, find a map

$$\mathcal{K}\colon \mathcal{GC}^-(\mathbb{G}') \to \mathcal{GC}^-(\mathbb{G}')$$

so that

$$\mathcal{D} \circ \mathcal{S} = \mathrm{Id}$$
$$\mathcal{S} \circ \mathcal{D} + \partial \circ \mathcal{K} + \mathcal{K} \circ \partial = \mathrm{Id},$$

verifying that the stabilization and destabilization maps are homotopy inverses of one another.

Stabilization maps for other types of stabilizations will be discussed in Section 14.3.

CHAPTER 14

More on the filtered chain complex

The aim of this chapter is twofold: to make the filtered invariant introduced in the last chapter more concrete, and to develop some of the applications of this new structure. In Section 14.1, we explain how to extract numerical invariants from the filtered quasi-isomorphism type, some of which go beyond the information present in the grid homology groups. One of the results from this section (Proposition 14.1.1) presents certain constraints on the filtered quasi-isomorphism type of $\mathcal{GC}^-$. In Section 14.2 these constraints (combined with computations of grid homology groups) are used to explicitly determine the quasi-isomorphism type of $\mathcal{GC}^-$ for certain specific knots. In Section 14.3 we use the filtration to give more structure to the Legendrian and transverse invariants defined in Chapter 12. In Section 14.4, we show how this extra structure can be used in practice to distinguish Legendrian and transverse knots. In Section 14.5, we make a few remarks on the generalization of the filtered invariant to the case of links. Section 14.6 contains a brief account of the relationship between the constructions described here and the pseudo-holomorphic theory.

14.1. Information in the filtered grid complex

We have so far defined the filtered quasi-isomorphism type of the grid complexes $\widehat{\mathcal{GC}}(\mathbb{G})$ and $\mathcal{GC}^-(\mathbb{G})$, which are both topological invariants of knots. One might wonder how to make these refined invariants more concrete: i.e. what is a reasonable invariant of a filtered quasi-isomorphism type of a complex?

We will construct numerical invariants from the filtered quasi-isomorphism type by looking at the interplay of two extremes. In one extreme, we can take the homology of the associated graded object, but that simply recaptures the grid homology groups from Chapter 4. In the other extreme, we can take the homology of the total object. This is a knot invariant, since the filtered quasi-isomorphism type of $\mathcal{GC}^-(\mathbb{G})$ is a knot invariant (Theorem 13.2.9), and the homology groups of filtered quasi-isomorphic chain complexes are isomorphic (by Proposition A.6.1). That group loses the (Alexander) filtration, but it still retains a (Maslov) grading. Since the homology ignores the X-markings entirely, it is independent of the knot, and we have the following:

PROPOSITION 14.1.1. *For any knot K and any grid diagram $\mathbb{G}$ representing K, $H(\widehat{\mathcal{GC}}(\mathbb{G})) \cong \mathbb{F}$, supported in Maslov grading 0; and similarly $H(\mathcal{GC}^-(\mathbb{G})) \cong \mathbb{F}[U]$, where the generator corresponding to $1 \in \mathbb{F}[U]$ has Maslov grading equal to 0.*

Proof. If we forget the filtration on $\mathcal{GC}^-(\mathbb{G})$, the resulting chain complex is independent of the placement of $\mathbb{X}$. By moving around the X-markings, we can find a new grid diagram with the same grid number as $\mathbb{G}$, and indeed with the same placement of the O-markings, but representing the unknot. Such a diagram $\mathbb{G}'$ can be found by placing X-markings immediately south of each O-marking. It follows that $H(\mathcal{GC}^-(\mathbb{G})) \cong H(\mathcal{GC}^-(\mathbb{G}'))$ and $H(\widehat{\mathcal{GC}}(\mathbb{G})) \cong H(\widehat{\mathcal{GC}}(\mathbb{G}'))$. By the invariance of $H(\mathcal{GC}^-(\mathbb{G}))$, the result now follows from a calculation in a 2×2 grid diagram for the unknot. □

The above proposition is a homological lift of the fact that $\Delta_K(1) = 1$: we have a homology group $\widehat{GH}(K)$ endowed with an Alexander grading and, after taking homology with respect to differentials which collapse the Alexander grading, we obtain a complex $\widehat{\mathcal{GC}}(\mathbb{G})$ whose homology is one-dimensional. (The above proposition is in a certain sense dual to Proposition 6.1.4; see Lemma 14.1.9.)

Proposition 14.1.1 suggests the following numerical invariant of knots K:

$$t(K) = \min\{s \mid \text{the image of } H(\mathcal{F}_s\widehat{\mathcal{GC}}) \text{ in } H(\widehat{\mathcal{GC}}) \cong \mathbb{F} \text{ is non-zero}\}. \tag{14.1}$$

By Theorem 13.2.9 (and the fact that the right-hand-side is canonically associated to the filtered quasi-isomorphism type of $\widehat{\mathcal{GC}}$; see Corollary A.6.5), this is a knot invariant. It is, however, one we have seen already:

PROPOSITION 14.1.2. *The invariants $t(K)$ and $\tau(K)$ are the same.*

We will return to the proof of the above proposition at the end of this section.

So far, we have shown how the filtered object can be used to reformulate objects which we have already met; but the filtered theory contains new information, as well. For instance, there is the following naturally defined infinite sequence of invariants $\{\tau_i\}_{i\in\mathbb{Z}^{\geq 0}}$ defined in the spirit of τ (as reformulated in Equation (14.1)).

DEFINITION 14.1.3. Let $\{\tau_i\}_{i=0}^{\infty}$ be the knot invariants defined as follows:

$$\tau_i(K) = \min\{s \mid \text{the image of } H(\mathcal{F}_s\mathcal{GC}^-) \text{ in } H(\mathcal{GC}^-) \cong \mathbb{F}[U] \text{ contains } U^i\}. \tag{14.2}$$

It is convenient to repackage the information contained in the above τ_i, defining knot invariants $\{h_k\}_{k\in\mathbb{Z}}$ by

$$h_k(K) = \dim_{\mathbb{F}} \operatorname{Coker}(H(\mathcal{F}_k(\mathcal{GC}^-)) \to H(\mathcal{GC}^-)). \tag{14.3}$$

EXAMPLE 14.1.4. For the unknot, $h_k(\mathcal{O}) = \max(-k, 0)$.

The h_k can be expressed in terms of the τ_i, according to the following:

EXERCISE 14.1.5. **(a)** Show that $h_{k+1} \leq h_k \leq h_{k+1} + 1$.
(b) Show that $h_k(K) = \#\{i \mid \tau_i > k\}$.

The invariants h_k are the "local h-invariants" considered by Rasmussen in [**191, 192**] (compare also [**61**]). The invariant τ_0 is called ν^+ in [**89**]. Work of Rasmussen shows that τ_0 gives slice genus bounds, generalizing the bounds from τ. Work of Hom and Wu [**89**] give examples of knots where the τ_0-bounds are stronger than the τ-bounds.

EXERCISE 14.1.6. **(a)** Show that for all $i \in \mathbb{Z}$, $\operatorname{rk}_{\mathbb{F}[U]} H(\mathcal{F}_i\mathcal{GC}^-) = 1$.
(b) Define h_i' to be the maximal Maslov grading of any homogeneous, non-torsion generator in $H(\mathcal{F}_i\mathcal{GC}^-)$. Show that $-h_i' = 2h_i$.
(c) Show that if K_+ and K_- are two knots that differ by a crossing change, then $0 \leq h_i(K_+) - h_i(K_-) \leq 1$.

The filtered quasi-isomorphism type provides further information beyond what was discussed above. A useful packaging of this information is given in the form of the *spectral sequence of a filtration*, as follows. Given a filtered complex such as $\mathcal{GC}^-(\mathbb{G})$, form its associated graded object which, in this case, is $GC^-(\mathbb{G})$. That in turn is equipped with a map $\partial_1^- : GC_d^-(\mathbb{G}, s) \to GC_{d-1}^-(\mathbb{G}, s-1)$ induced by the part of the differential of $\mathcal{GC}^-(\mathbb{G})$ that drops filtration level by exactly one. In the present case, this map counts rectangles that cross exactly one $X \in \mathbb{X}$. This map induces a map δ_1^- on homology, with the property that $\delta_1^- \circ \delta_1^- = 0$. Take the homology of $GH^-(\mathbb{G})$ with respect to δ_1^-, to get a new graded $\mathbb{F}[U]$-module, denoted E^2. In fact, this procedure can be iterated to obtain a sequence of chain complexes E^i, equipped with differentials δ_i^-, whose homology is the module underlying the next complex E^{i+1}. A sequence of complexes with this property is called a *spectral sequence* and the terms E^i are called its *pages*. A filtered complex naturally gives rise to such a spectral sequence, which is an invariant of the filtered quasi-isomorphism type of the initial complex. In view of this general construction, combined with the invariance of the filtered quasi-isomorphism type of $\mathcal{GC}^-(\mathbb{G})$ (Theorem 13.2.9), each page in the spectral sequence for the filtered grid complex is a knot invariant.

In the interest of simplicity we will avoid the explicit use of spectral sequences throughout the rest of this book, although the techniques from Section 14.4 can be interpreted in this language. For more on spectral sequences, see [**139**].

14.1.1. The identification of $t(K)$ with $\tau(K)$. We establish a symmetry of the grid complex generalizing Proposition 7.1.1, and use it to prove Proposition 14.1.2.

To formulate the symmetry, think of a $\mathbb{Z}$-filtered complex $\mathcal{C}$ over $\mathbb{F}[U]$ as equipped with two filtrations. One of these is given by the original $\mathbb{Z}$-filtration, which we now call the *initial filtration*. The second, the *algebraic filtration*, is induced by powers of U; i.e. it corresponds to the sequence of subcomplexes

$$\dots \subset U^m\mathcal{C} \subset U^{m-1}\mathcal{C} \subset \dots \mathcal{C}.$$

To describe a filtration that increases in the parameter, we say that $U^m\mathcal{C}$ is the part of $\mathcal{C}$ with algebraic filtration level $-m$. We make the further assumption that $\mathcal{C}$ is freely and finitely generated over $\mathbb{F}[U]$. Consider the complex $\mathcal{C}\otimes_{\mathbb{F}[U]}\mathbb{F}[U,U^{-1}]$, which inherits both the algebraic and the initial filtrations. The original complex $\mathcal{C}$ can be thought of as the subcomplex of $\mathcal{C}\otimes_{\mathbb{F}[U]}\mathbb{F}[U,U^{-1}]$ generated by elements whose algebraic filtration level is ≤ 0.

DEFINITION 14.1.7. Let $\mathcal{C}$ be a $\mathbb{Z}$-filtered, $\mathbb{Z}$-graded complex that is a free, finitely-generated chain complex over $\mathbb{F}[U]$. The ***conjugate complex*** $\overline{\mathcal{C}}$ is the subcomplex of $\mathcal{C}\otimes_{\mathbb{F}[U]}\mathbb{F}[U,U^{-1}]$ with initial filtration ≤ 0, equipped with the $\mathbb{Z}$-filtration induced by the algebraic filtration of $\mathcal{C}$. The conjugate complex $\overline{\mathcal{C}}$ inherits its $\mathbb{Z}$-grading from $\mathcal{C}\otimes_{\mathbb{F}[U]}\mathbb{F}[U,U^{-1}]$.

Note that $\overline{\mathcal{C}}$ is also finitely generated over $\mathbb{F}[U]$.

EXERCISE 14.1.8. Consider the free bigraded $\mathbb{F}[U]$ module generated by three elements $\mathbf{x}_{-1}$, $\mathbf{x}_0$, and $\mathbf{x}_1$ with bigradings specified by $M(\mathbf{x}_i) = i-1$ and $A(\mathbf{x}_i) = i$ for $i = -1, 0, 1$. We will consider three different differentials on this module, ∂_1, ∂_2, and ∂_3. The three differentials will be distinguished by $\partial_i\mathbf{x}_0$. For $i = 1, 2$, and

3, $\partial_i(\mathbf{x}_1) = \partial_i(\mathbf{x}_{-1}) = 0$ and

$$\partial_1(\mathbf{x}_0) = U \cdot \mathbf{x}_1 \qquad \partial_2(\mathbf{x}_0) = \mathbf{x}_{-1} \qquad \partial_3(\mathbf{x}_0) = U \cdot \mathbf{x}_1 + \mathbf{x}_{-1}.$$

(a) Show that (C, ∂_1) is filtered quasi-isomorphic to the conjugate of (C, ∂_2).
(b) Show that (C, ∂_1) is not filtered quasi-isomorphic to (C, ∂_2).
(c) Show that (C, ∂_3) is filtered quasi-isomorphic to its own conjugate.

The following symmetry of the filtered grid invariant is stated in terms of the specialization $\frac{\mathcal{GC}^-(\mathbb{G})}{V_1=\dots=V_n}$ of $\mathcal{GC}^-(\mathbb{G})$. In the statement, we adapt the shift operator to the $\mathbb{Z}$-graded, $\mathbb{Z}$-filtered setting, as follows. If $\mathcal{C}$ is a $\mathbb{Z}$-graded, $\mathbb{Z}$-filtered complex, then $\mathcal{C}[\![a,b]\!] = \mathcal{C}'$ is the filtered complex with $\mathcal{F}_s\mathcal{C}'_d = \mathcal{F}_{s+b}\mathcal{C}_{d+a}$.

LEMMA 14.1.9. *If $\mathbb{G}$ is a grid diagram with grid number n representing a knot, then the conjugate of $\frac{\mathcal{GC}^-(\mathbb{G})}{V_1=\dots=V_n}$ is filtered quasi-isomorphic to $\frac{\mathcal{GC}^-(\mathbb{G})}{V_1=\dots=V_n}[\![1-n,1-n]\!]$.*

Proof. Let $\mathbb{G}_1 = \mathbb{G}$. Explicitly, $\mathcal{C}^1 = \frac{\mathcal{GC}^-(\mathbb{G}_1)}{V_1=\dots=V_n}$ is the chain complex generated freely over $\mathbb{F}[U]$ by $\mathbf{S}(\mathbb{G}_1)$ and equipped with the differential

$$\partial_1\mathbf{x} = \sum_{\mathbf{y}\in\mathbf{S}(\mathbb{G}_1)} \sum_{r\in\mathrm{Rect}^\circ(\mathbf{x},\mathbf{y})} U^{\#(\mathbb{O}\cap r)} \cdot \mathbf{y},$$

where U is equal to $V_1 = \dots = V_n$.

Switching the roles of $\mathbb{X}$ and $\mathbb{O}$, we obtain a new grid diagram $\mathbb{G}_2$ and a new complex $\mathcal{C}^2 = \frac{\mathcal{GC}^-(\mathbb{G}_2)}{V_1=\dots=V_n}$ with the same generators $\mathbf{S}(\mathbb{G}_2) \cong \mathbf{S}(\mathbb{G}_1)$. For $i = 1, 2$, let $A_i\colon \mathbf{S}(\mathbb{G}_1) \cong \mathbf{S}(\mathbb{G}_2) \to \mathbb{Z}$ be the Alexander function induced by $\mathbb{G}_i$. As we saw in the proof of Proposition 7.1.1,

$$M_1(\mathbf{x}) - M_2(\mathbf{x}) = 2A_1(\mathbf{x}) + n - 1 \qquad \text{and} \qquad A_1(\mathbf{x}) + A_2(\mathbf{x}) = 1 - n. \tag{14.4}$$

The differential on $\mathcal{C}^2$ can be interpreted as

$$\partial_2\mathbf{x} = \sum_{\mathbf{y}\in\mathbf{S}(\mathbb{G}_2)} \sum_{r\in\mathrm{Rect}^\circ(\mathbf{x},\mathbf{y})} U^{\#(\mathbb{X}\cap r)} \cdot \mathbf{y},$$

where the $\mathbb{X}$ are as in $\mathbb{G}_1$.

Consider the homomorphism of $\mathbb{F}[U]$-modules $\phi\colon \mathcal{C}^2 \to \mathcal{C}^1 \otimes_{\mathbb{F}[U]} \mathbb{F}[U,U^{-1}]$ that, for each $\mathbf{x} \in \mathbf{S}(\mathbb{G}_2)$, is given by $\phi(\mathbf{x}) = U^{A_1(\mathbf{x})} \cdot \mathbf{x}$. Let $\mathcal{E} \subset \mathcal{C}^1 \otimes_{\mathbb{F}[U]} \mathbb{F}[U,U^{-1}]$ be the subcomplex generated by elements of Alexander filtration level ≤ 0; i.e. this is the conjugate complex for $\mathcal{C}^1$. For each $\mathbf{x} \in \mathbf{S}(\mathbb{G}_1)$, the A_1-Alexander filtration level of $U^{A_1(\mathbf{x})} \cdot \mathbf{x}$ is equal to zero, so ϕ maps $\mathcal{C}^2$ into $\mathcal{E}$. The elements of the form $U^{A_1(\mathbf{x})} \cdot \mathbf{x}$ form a basis (over $\mathbb{F}$) of all elements in $\mathcal{E}$ with vanishing Alexander filtration, so ϕ is an isomorphism of $\mathbb{F}[U]$-modules onto $\mathcal{E}$. To see that ϕ is a chain map, observe that

$$\begin{aligned}\partial_1(U^{A_1(\mathbf{x})}\mathbf{x}) &= \sum_{\mathbf{y}\in\mathbf{S}(\mathbb{G}_1)} \sum_{r\in\mathrm{Rect}^\circ(\mathbf{x},\mathbf{y})} U^{\#(\mathbb{O}\cap r) - A_1(\mathbf{y}) + A_1(\mathbf{x})}(U^{A_1(\mathbf{y})}\mathbf{y}) \\ &= \sum_{\mathbf{y}\in\mathbf{S}(\mathbb{G}_1)} \sum_{r\in\mathrm{Rect}^\circ(\mathbf{x},\mathbf{y})} U^{\#(\mathbb{X}\cap r)}(U^{A_1(\mathbf{y})}\mathbf{y}),\end{aligned}$$

(where the last equality follows by Equation (4.4)); so ϕ induces an isomorphism of chain complexes $\mathcal{C}^2 \cong \mathcal{E}$.

Compare gradings using Equation (14.4). Since

$$M_1(\phi(\mathbf{x})) = M_1(U^{A_1(\mathbf{x})}(\mathbf{x})) = M_1(\mathbf{x}) - 2A_1(\mathbf{x}) = M_2(\mathbf{x}) + n - 1,$$

and the algebraic filtration level $\overline{A}$ of $U^{A_1(\mathbf{x})} \cdot \mathbf{x}$ is given by

$$\overline{A}(U^{A_1(\mathbf{x})} \cdot \mathbf{x}) = -A_1(\mathbf{x}) = A_2(\mathbf{x}) + n - 1,$$

ϕ induces a $\mathbb{Z}$-graded, $\mathbb{Z}$-filtered isomorphism $\Phi\colon \mathcal{C}^2[\![1-n, 1-n]\!] \to \overline{\mathcal{C}}^1$. Since $\mathbb{G}_2$ represents K with the reversed orientation, Theorem 13.2.9 shows that $\mathcal{C}^1 = \frac{\mathcal{GC}^-(\mathbb{G}_1)}{V_1=\ldots=V_n}$ is filtered quasi-isomorphic to $\mathcal{C}^2 = \frac{\mathcal{GC}^-(\mathbb{G}_2)}{V_1=\ldots=V_n}$. □

REMARK 14.1.10. A version of Lemma 14.1.9 holds without the $V_1 = \cdots = V_n$ specialization; i.e. $\mathcal{GC}^-(\mathbb{G})$ is filtered quasi-isomorphic to its own conjugate. But the above version is sufficient for our present purposes.

The symmetry was stated in terms of the specialization $\frac{\mathcal{GC}^-(\mathbb{G})}{V_1=\cdots=V_n}$. To draw conclusions about the unspecialized theory, we establish a relationship between the specialization and $\mathcal{GC}^-(\mathbb{G})$, generalizing Proposition 4.6.15. To state this relationship, consider the two-dimensional vector space $\mathcal{W}$ with one generator in grading zero and filtration level 0, and another one in grading -1 and filtration level -1. We will tensor the $\mathbb{Z}$-filtered, $\mathbb{Z}$-graded complex $\mathcal{C}$ over $\mathbb{F}[U]$ with $\mathcal{W}$, to get a new graded and filtered complex $\mathcal{C} \otimes \mathcal{W}$, with

$$\mathcal{F}_i(\mathcal{C} \otimes \mathcal{W})_d = \mathcal{F}_i\mathcal{C}_d \oplus \mathcal{F}_{i+1}\mathcal{C}_{d+1}.$$

LEMMA 14.1.11. *Let $\mathbb{G}$ be a grid diagram representing some knot K. Then, there are filtered quasi-isomorphisms*

$$\frac{\mathcal{GC}^-(\mathbb{G})}{V_1 = \cdots = V_n} \simeq \mathcal{GC}^-(\mathbb{G}) \otimes \mathcal{W}^{\otimes(n-1)} \tag{14.5}$$

$$\frac{\mathcal{GC}^-(\mathbb{G})}{V_1 = \cdots = V_n = 0} \simeq \widehat{\mathcal{GC}}(\mathbb{G}) \otimes \mathcal{W}^{\otimes(n-1)}, \tag{14.6}$$

where $\mathcal{W}$ is the two-dimensional graded and filtered vector space described above.

Proof. This is a straightforward adaptation of Lemma 7.4.2 to the filtered context, taking into account Lemma 13.2.5 (the filtered analogue of Lemma 4.6.9). For example, setting $V_1 = V_2$ corresponds up to quasi-isomorphism to taking the mapping cone of multiplication by $V_1 - V_2$ (see Equation (13.8)). Now, Lemma 13.2.5 states that $V_1 - V_2$ is filtered homotopic to the zero map, so the mapping cone in turn is filtered quasi-isomorphic to $\mathcal{GC}^-(\mathbb{G}) \otimes \mathcal{W}$; i.e. we have a filtered quasi-isomorphism

$$\frac{\mathcal{GC}^-(\mathbb{G})}{V_1 = V_2} \simeq \mathcal{GC}^-(\mathbb{G}) \otimes \mathcal{W}.$$

Iterating this argument for the other relations $V_2 = \cdots = V_n$, the quasi-isomorphism from Equation (14.5) follows. Further setting $V_1 = 0$ gives the quasi-isomorphism from Equation (14.6). □

We will also use the following formalities. Suppose that $\mathcal{C}$ is a $\mathbb{Z}$-filtered chain complex over $\mathbb{F}[U]$. There is an induced $\mathbb{Z}$-filtration on $\widehat{\mathcal{C}} = \mathcal{C}/U$, which we denote by $\widehat{\mathcal{F}}_s \subset \widehat{\mathcal{C}}$. When $\widehat{\mathcal{C}}$ has non-zero homology, let $t(\mathcal{C})$ be the minimal s so that the map on homology induced by $\widehat{\mathcal{F}}_s \subset \widehat{\mathcal{C}}$ is non-trivial. Similarly, consider $\mathrm{gr}(\mathcal{C})$. When $H(\mathrm{gr}(\mathcal{C}))$ contains non-torsion elements, let $\tau(\mathcal{C})$ be -1 times the maximal grading (induced by the filtration) of any homogeneous, non-torsion element in $H(\mathrm{gr}(\mathcal{C}))$.

These notions are natural generalization of the knot invariants we saw earlier; i.e. $\tau(K) = \tau(\mathcal{GC}^-(\mathbb{G}))$ and $t(K) = t(\mathcal{GC}^-(\mathbb{G}))$.

LEMMA 14.1.12. *Let $\mathcal{C}$ be a $\mathbb{Z}$-filtered chain complex over $\mathbb{F}[U]$ that is free, and finitely generated as an $\mathbb{F}[U]$-module, and let $\overline{\mathcal{C}}$ be the conjugate complex. Then, $H(\mathcal{C}/U) \cong H(\frac{\mathrm{gr}(\overline{\mathcal{C}})}{U-1})$, and if that vector space is non-zero, then $t(\mathcal{C}) = \tau(\overline{\mathcal{C}})$.*

Proof. Let $\mathcal{C}' = \mathcal{C} \otimes_{\mathbb{F}[U]} \mathbb{F}[U, U^{-1}]$ with its algebraic and initial filtrations induced from $\mathcal{C}$. Given $x, y \in \mathbb{Z}$, let $\mathcal{F}_{x,*}(\mathcal{C}) \subset \mathcal{C}'$ be the subcomplex with algebraic filtration level $\leq x$; $\mathcal{F}_{*,y}(\mathcal{C}) \subset \mathcal{C}'$ be the subcomplex with initial filtration level $\leq y$; $\mathcal{F}_{x,y}(\mathcal{C}) = \mathcal{F}_{x,*}(\mathcal{C}) \cap \mathcal{F}_{*,y}(\mathcal{C})$.

In this notation,

$$\frac{\mathcal{C}}{U} \cong \frac{\mathcal{F}_{0,*}(\mathcal{C})}{\mathcal{F}_{-1,*}(\mathcal{C})}. \tag{14.7}$$

$\mathcal{F}_y(\frac{\mathcal{C}}{U}) = \frac{\mathcal{F}_{0,y}(\mathcal{C})}{\mathcal{F}_{-1,y}(\mathcal{C})}$, and the inclusion $\mathcal{F}_y(\frac{\mathcal{C}}{U}) \subset \frac{\mathcal{C}}{U}$ is the map $\frac{\mathcal{F}_{0,y}}{\mathcal{F}_{-1,y}} \to \frac{\mathcal{F}_{0,*}}{\mathcal{F}_{-1,*}}$. Thus,

$$t(\mathcal{C}) = \min\left\{ y \middle| H\left(\frac{\mathcal{F}_{0,y}}{\mathcal{F}_{-1,y}}\right) \to H\left(\frac{\mathcal{F}_{0,*}}{\mathcal{F}_{-1,*}}\right) \text{ is non-trivial}\right\}. \tag{14.8}$$

Similarly, $\mathrm{gr}(\mathcal{C}) = \bigoplus_y \frac{\mathcal{F}_{0,y}}{\mathcal{F}_{0,y-1}}$,

$$\frac{\mathrm{gr}(\mathcal{C})}{U-1} \cong \frac{\mathcal{F}_{*,0}(\mathcal{C})}{\mathcal{F}_{*,-1}(\mathcal{C})}, \tag{14.9}$$

and multiplication by U^y induces an isomorphism $\frac{\mathcal{F}_{0,y}}{\mathcal{F}_{0,y-1}} \cong \frac{\mathcal{F}_{-y,0}}{\mathcal{F}_{-y,-1}}$ which, when composed with the inclusion $\frac{\mathcal{F}_{-y,0}}{\mathcal{F}_{-y,-1}} \subset \frac{\mathcal{F}_{*,0}}{\mathcal{F}_{*,-1}}$, gives the map $\mathrm{gr}(\mathcal{C}) \to \frac{\mathrm{gr}(\mathcal{C})}{U-1}$. By the universal coefficient theorem, a homology class in $\mathrm{gr}(\mathcal{C})$ is non-torsion if and only if its image in $H(\frac{\mathrm{gr}(\mathcal{C})}{U-1})$ is non-zero, so

$$\tau(\mathcal{C}) = -\max\left\{ y \middle| H\left(\frac{\mathcal{F}_{-y,0}}{\mathcal{F}_{-y,-1}}\right) \to H\left(\frac{\mathcal{F}_{*,0}}{\mathcal{F}_{*,-1}}\right) \text{ is non-trivial}\right\}. \tag{14.10}$$

Since $\mathcal{F}_{x,y}(\mathcal{C}) = \mathcal{F}_{y,x}(\overline{\mathcal{C}})$, the first statement follows from Equations (14.7) and (14.9), and the second follows from Equations (14.8) and (14.10). □

Proof of Proposition 14.1.2. Let $\mathcal{C} = \frac{\mathcal{GC}^-(\mathbb{G})}{V_1 = \cdots = V_n}$. Apply, in order Equation (14.5), Lemmas 14.1.12, 14.1.9, and Equation (14.6) to get:

$$\tau(K) = \tau(\mathcal{C}) = t(\overline{\mathcal{C}}) = t(\mathcal{C}) + n - 1 = t(\mathcal{GC}^-(\mathbb{G})) = t(K).$$ □

14.2. Examples of filtered grid complexes

A tidy description of the quasi-isomorphism type of $\mathcal{GC}^-(\mathbb{G})$ is supplied by the following lemma:

LEMMA 14.2.1. *$\mathcal{GC}^-(\mathbb{G})$ and $\widehat{\mathcal{GC}}(\mathbb{G})$ are filtered quasi-isomorphic to filtered complexes that are free modules (over $\mathbb{F}[U]$ and $\mathbb{F}$ respectively) with rank $\dim_{\mathbb{F}} \widehat{GH}(K)$.*

Proof. Clearly, $\mathcal{GC}^-(\mathbb{G})$ is a finitely generated, free module over $\mathbb{F}[V_1,\dots,V_n]$. Thanks to Lemma 13.2.5, we can apply Proposition A.7.2 to get the result for $\mathcal{GC}^-(\mathbb{G})$. The result for $\widehat{\mathcal{GC}}(\mathbb{G})$ now follows formally. □

With the help of the above lemma, in some cases the knowledge of grid homology, combined with Proposition 14.1.1, suffices to determine the filtered quasi-isomorphism type of the grid complex.

14.2.1. The trefoil. Consider the right-handed trefoil knot $T_{2,3}$. By Equation (4.31) and Lemma 14.2.1, there is a representative for $\mathcal{GC}^-(T_{2,3})$ generated by three generators, denoted x_{-1}, x_0, and x_1, so that $M(x_i) = i - 1$ and $A(x_i) = i$ for for $i = -1, 0, 1$. We claim that $\mathcal{GC}^-/U$ (representing $\widehat{\mathcal{GC}}(T_{2,3})$) is the filtered complex with these three generators, and with differential given by

$$\widehat{\partial}(x_i) = \begin{cases} x_{-1} & \text{if } i = 0 \\ 0 & \text{otherwise.} \end{cases}$$

To see this, we argue as follows. There are three possible differentials that are compatible with the Maslov gradings and with $(\widehat{\partial})^2 = 0$:

$$\widehat{\partial}_1(x_i) \equiv 0;$$

$$\widehat{\partial}_2(x_i) = \begin{cases} x_0 & \text{if } i = 1 \\ 0 & \text{otherwise;} \end{cases}$$

$$\widehat{\partial}_3(x_i) = \begin{cases} x_{-1} & \text{if } i = 0 \\ 0 & \text{otherwise;} \end{cases}$$

The first of these has three-dimensional homology; the second has one-dimensional homology supported in Maslov grading -2. By Proposition 14.1.1, neither can represent $\widehat{\mathcal{GC}}(T_{2,3})$. Thus, the only remaining candidate is the third, which was the stated answer.

We can extend the above argument to give a computation of $\mathcal{GC}^-(T_{2,3})$. Specifically, there are only two extensions of the above complex $(\widehat{\mathcal{GC}}(T_{2,3}), \widehat{\partial})$ that are compatible with the Maslov grading: either $\partial^- x_0 = U \cdot x_1 + x_{-1}$, or $\partial^- x_0 = x_{-1}$. The second of these answers is incompatible with the computation of $GH^-(T_{2,3})$ displayed in Equation (4.32).

To summarize, we have shown that $\mathcal{GC}^-(T_{2,3})$ has three generators x_1, x_0, and x_{-1}, with $M(x_i) = i - 1$ and $A(x_i) = i$ for $i = -1, 0, 1$; equipped with the differential

$$\partial^- x_i = \begin{cases} U \cdot x_1 + x_{-1} & \text{if } i = 0 \\ 0 & \text{otherwise.} \end{cases} \tag{14.11}$$

This complex and the next two are illustrated in Figure 14.2.2.

EXERCISE 14.2.2. Show that

$$h_i(T_{2,3}) = \begin{cases} 1 & \text{if } i = 0 \\ 0 & \text{if } i > 0 \\ -i & \text{if } i < 0. \end{cases}$$

For the left-handed trefoil knot $T_{-2,3}$, the complex $\mathcal{GC}^-(T_{-2,3})$ is generated by three generators y_{-1}, y_0, and y_1 with $A(y_i) = i$ and $M(y_i) = i + 1$, and differential

$$\partial^- y_1 = y_0 \qquad \partial^- y_{-1} = U \cdot y_0 \qquad \partial^- y_0 = 0.$$

EXERCISE 14.2.3. Show that for the left-handed trefoil knot, $h_i(T_{-2,3}) = \max(0,-i)$.

14.2.2. The figure-eight knot. For the figure-eight knot W_2, the complex $\mathcal{GC}^-(W_2)$ has five generators, which we label a, b, c, d, and e, with gradings

$$M(b) = 1 \qquad M(a) = M(c) = M(e) = 0 \qquad M(d) = -1,$$

filtration specified by $A(x) = M(x)$ for $x \in \{a, \ldots, e\}$; and differential given as

$$\partial^- a = U \cdot b + d \qquad \partial^- b = c \qquad \partial^- d = U \cdot c \qquad \partial^- c = \partial^- e = 0.$$

Like for the unknot, $h_i(K) = \max(0, -i)$.

EXERCISE 14.2.4. Verify that $\mathcal{GC}^-(W_2)$ has the stated form.

We can illustrate the filtered quasi-isomorphism type $\mathcal{GC}^-(K)$ of a knot K in the plane as follows. Consider a representative of $\mathcal{GC}^-(K)$ provided by Lemma 14.2.1: start with a homogeneous generating set for $\widehat{GH}(K)$, and form all the translates by U^i ($i \in \mathbb{Z}$) to get a homogeneous generating set for $\mathcal{GC}^-(K) \otimes_{\mathbb{F}[U]} \mathbb{F}[U, U^{-1}]$ over $\mathbb{F}$. Place these generators on lattice points in the plane, so that the horizontal coordinate records the algebraic filtration level of the generator (i.e. -1 times the U-power), and the vertical coordinate records the Alexander filtration. The U-action is represented by a translation of the generators in the direction $(-1, -1)$. The non-trivial components of $\partial^-(U^i \cdot x)$ for a homogeneous generator $U^i \cdot x$ (with $x \in \widehat{GH}(K)$ homogeneous generator) are indicated by arrows pointing from the lattice point of $U^i \cdot x$.

In this picture the complex $\mathcal{GC}^-(K)$ corresponds to the left half plane. To get the associated graded object $GC^-(K)$, we keep only those differentials in $\mathcal{GC}^-(K)$ that correspond to horizontal arrows; to get $\widehat{\mathcal{GC}}(K)$, we keep only those lattice points that lie on the vertical axis, together with the vertical arrows that connect them. Note that, in general, the Maslov grading is missing from this picture. However, for each of the above three examples, there is a constant c with the property that if a lattice point is supported at coordinates (x, y), then its Maslov grading is given by $x + y + c$. (For the three examples, the constants are -1, $+1$, and 0 respectively.)

14.2.3. Grid complexes for alternating knots. By building on Theorem 10.3.1, models for the filtered quasi-isomorphism type of any alternating knot K can be extracted explicitly from its signature and Alexander polynomial.

EXERCISE 14.2.5. **(a)** Let $n \geq 1$ be an integer. Describe a model complex, as in Lemma 14.2.1, for $\mathcal{GC}^-(T_{2,2n+1})$. .
(b) For integers $n \geq 1$, describe a model complex for $\mathcal{GC}^-(T_{-2,2n+1})$.
(c)*Let K be an alternating knot. Describe the model complex of $\mathcal{GC}^-(K)$ in terms of $\sigma(K)$ and $\Delta_K(t)$.

Torus knots are another general class of knots for which the filtered quasi-isomorphism type can be explicitly described; but this computation uses the holomorphic perspective (see Theorem 16.2.6).

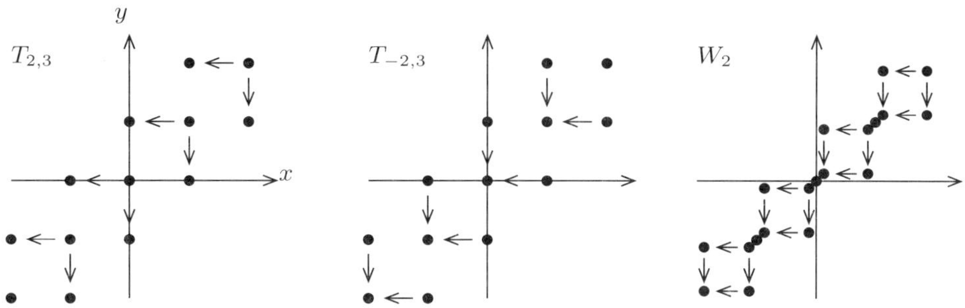

FIGURE 14.1. **Filtered chain complexes for the right- and left-handed trefoil knots, and the figure-eight knot.** We have drawn here a finite portion of models for $\mathcal{GC}^- \otimes \mathbb{F}[U, U^{-1}]$. (The pictures are periodic with diagonal translation.) The x axis represents the algebraic filtration (negative of the U power), the y axis represents the Alexander filtration.

14.3. Refining the Legendrian and transverse invariants: definitions

The aim of the present section is to show how the filtered quasi-isomorphism type can be used to place extra structure on the Legendrian and transverse invariants from Chapter 12. Applications of this strengthening will be given in the next section. To formulate this extra structure, we give the following:

DEFINITION 14.3.1. Fix chain complexes C and C'. An isomorphism $\phi\colon H(C) \to H(C')$ is said to be ***covered by a quasi-isomorphism*** if there is a third chain complex C'', and a pair of quasi-isomorphisms $f\colon C'' \to C$ and $g\colon C'' \to C'$, so that ϕ satisfies $\phi = H(g) \circ H(f)^{-1}$. More generally, if $\mathcal{C}$ and $\mathcal{C}'$ are $\mathbb{Z}$-filtered chain complexes, an isomorphisms $\phi\colon H(\mathrm{gr}(\mathcal{C})) \to H(\mathrm{gr}(\mathcal{C}'))$ is said to be ***covered by a filtered quasi-isomorphism*** if there is a $\mathbb{Z}$-filtered complex $\mathcal{C}''$ and a pair of filtered quasi-isomorphisms $F\colon \mathcal{C}'' \to \mathcal{C}$ and $G\colon \mathcal{C}'' \to \mathcal{C}'$ so that $\phi = H(\mathrm{gr}(G)) \circ H(\mathrm{gr}(F))^{-1}$.

Our aim in this section is to give the following refinements of Theorems 12.3.3 and 12.3.4, which are proved in Subsection 14.3.2:

THEOREM 14.3.2. *If $\mathbb{G}$ and $\mathbb{G}'$ represent Legendrian isotopic knots, then there is an isomorphism $\phi\colon GH^-(\mathbb{G}) \to GH^-(\mathbb{G}')$ of bigraded $\mathbb{F}[U]$-modules that is covered by a filtered quasi-isomorphism, with $\phi(\lambda^+(\mathbb{G})) = \lambda^+(\mathbb{G}')$ and $\phi(\lambda^-(\mathbb{G})) = \lambda^-(\mathbb{G}')$.*

THEOREM 14.3.3. *Let $\mathbb{G}$, $\mathbb{G}^+$, and $\mathbb{G}^-$ be grid diagrams whose associated Legendrian knots are $\vec{\mathcal{K}}$ and its stabilizations $\vec{\mathcal{K}}^+$, and $\vec{\mathcal{K}}^-$ respectively. Then, there are isomorphisms*

$$\phi^-\colon GH^-(\mathbb{G}) \longrightarrow GH^-(\mathbb{G}^-), \qquad \phi^+\colon GH^-(\mathbb{G}) \longrightarrow GH^-(\mathbb{G}^+)$$

covered by filtered quasi-isomorphisms, that satisfy

$$\phi^-(\lambda^+(\vec{\mathcal{K}})) = \lambda^+(\vec{\mathcal{K}}^-), \qquad U \cdot \phi^+(\lambda^+(\vec{\mathcal{K}})) = \lambda^+(\vec{\mathcal{K}}^+),$$
$$U \cdot \phi^-(\lambda^-(\vec{\mathcal{K}})) = \lambda^-(\vec{\mathcal{K}}^-), \qquad \phi^+(\lambda^-(\vec{\mathcal{K}})) = \lambda^-(\vec{\mathcal{K}}^+).$$

14.3.1. Filtered stabilization maps. In the proofs of Theorems 12.3.3 and 12.3.4 we investigated how the canonical cycles transformed under maps induced by the four types of stabilization of type X. For their generalizations, Theorems 14.3.2 and 14.3.3, we need to describe four filtered quasi-isomorphisms that lift the isomorphisms on homology (studied in Proposition 5.4.1).

We start by stating simpler versions for the grid complex GC^-. Following notation from Section 5.4, we have the following slightly stronger form of Proposition 5.4.1, which was actually proved in the original proof of that proposition:

PROPOSITION 14.3.4. *Let $\mathbb{G}'$ be obtained from $\mathbb{G}$ by stabilization.*

- *Suppose the stabilization is of type X:SW or X:NE, so that* $GC^-(\mathbb{G}') = \mathrm{Cone}(\partial_{\mathbf{I}}^{\mathbf{N}}\colon \mathbf{I}\to\mathbf{N})$. *Then, the commutative square of chain maps*

$$\begin{array}{ccc} \mathbf{I} & \xrightarrow{\quad\partial_{\mathbf{I}}^{\mathbf{N}}\quad} & \mathbf{N} \\ {\scriptstyle e}\Big\downarrow & & \Big\downarrow{\scriptstyle e\circ\mathcal{H}^{\mathbf{I}}_{X_2}} \\ GC^-(\mathbb{G})[V_1][\![1,1]\!] & \xrightarrow{\quad V_1-V_2\quad} & GC^-(\mathbb{G})[V_1] \end{array}$$

  induces a quasi-isomorphism $D\colon GC^-(\mathbb{G}')\to \mathrm{Cone}(V_1-V_2)$.
- *Suppose that the stabilization is of type X:SE or X:NW, so that* $GC^-(\mathbb{G}') = \mathrm{Cone}(\partial_{\mathbf{N}}^{\mathbf{I}}\colon \mathbf{N}\to\mathbf{I})$. *Then, the commutative square*

$$\begin{array}{ccc} \mathbf{N} & \xrightarrow{\quad\partial_{\mathbf{N}}^{\mathbf{I}}\quad} & \mathbf{I} \\ {\scriptstyle \mathcal{H}^{\mathbf{N}}_{X_2}\circ e'}\Big\uparrow & & \Big\uparrow{\scriptstyle e'} \\ GC^-(\mathbb{G})[V_1][\![1,1]\!] & \xrightarrow{\quad V_1-V_2\quad} & GC^-(\mathbb{G})[V_1] \end{array}$$

  induces a quasi-isomorphism $S\colon \mathrm{Cone}(V_1-V_2)\to GC^-(\mathbb{G}')$. □

Compose the quasi-isomorphism from Lemma A.3.9 with an isomorphism of chain complexes over $\mathbb{F}[V_2,\dots,V_n]$ to obtain the quasi-isomorphism over $\mathbb{F}[V_2,\dots,V_n]$

$$\mathrm{Cone}(V_1-V_2)\to \frac{GC^-(\mathbb{G})[V_1]}{V_1-V_2}\cong GC^-(\mathbb{G}).$$

The map on homology induced by this quasi-isomorphism can be composed with the map on homology induced by D or the inverse of the map on homology induced by S from Proposition 5.4.1 to give an isomorphism $GH^-(\mathbb{G}')\cong GH^-(\mathbb{G})$ covered by a quasi-isomorphism.

We now turn to the filtered chain complexes $\mathcal{GC}^-$. Filtered quasi-isomorphisms for destabilizations of type *X:SW* were constructed in Section 13.3. The stabilization maps (counting domains specified by Definition 13.4.2) were described in Section 13.4.2. For stabilizations of type *X:SE*, we describe stabilization maps, defined by counting domains which go out of L or out of R, defined as follows:

DEFINITION 14.3.5. Suppose that $\mathbb{G}'$ is obtained from $\mathbb{G}$ by a stabilization of type *X:SE*. Fix $\mathbf{x}\in\mathbf{I}(\mathbb{G}')$ and $\mathbf{y}\in\mathbf{S}(\mathbb{G}')$. A domain $p\in\pi(\mathbf{x},\mathbf{y})$ is said to be of type oL or oR, if it satisfies the conditions specified in Definition 13.4.2, except that Property (s-3) is replaced with

(s-3)' The domain has the same local multiplicity k at three of the four squares that share the corner c. When p has type oL, the domain has multiplicity $k-1$ at the the southeast square containing c, while for p of type oR, the domain has local multiplicity $k+1$ at the southwest square containing c.

The set of domains of type oL resp. oR from $\mathbf{x}$ to $\mathbf{y}$ is denoted $\pi^{oL}(\mathbf{x},\mathbf{y})$ resp. $\pi^{oR}(\mathbf{x},\mathbf{y})$.

The domains encountered by the above definition are obtained by reflecting the domains in Figure 13.1 through a vertical axis; specifically, the reflection carries domains of type iL and iR to domains of type oL and oR respectively.

DEFINITION 14.3.6. Consider the maps

$$\mathcal{S}^{oL}\colon \mathcal{GC}^-(\mathbb{G}) \to \mathcal{GC}^-(\mathbb{G}') \qquad \text{and} \qquad \mathcal{S}^{oR}\colon \mathcal{GC}^-(\mathbb{G})[\![1,1]\!] \to \mathcal{GC}^-(\mathbb{G}'),$$

defined by

$$\mathcal{S}^{oL}(\mathbf{x}) = \sum_{\mathbf{y}\in\mathbf{S}(\mathbb{G}')} \sum_{p\in\pi^{oL}(e'(\mathbf{x}),\mathbf{y})} V_2^{O_2(p)}\cdots V_n^{O_n(p)}\cdot\mathbf{y},$$
$$\mathcal{S}^{oR}(\mathbf{x}) = \sum_{\mathbf{y}\in\mathbf{S}(\mathbb{G}')} \sum_{p\in\pi^{oR}(e'(\mathbf{x}),\mathbf{y})} V_2^{O_2(p)}\cdots V_n^{O_n(p)}\cdot\mathbf{y},$$

where $e'\colon \mathbf{S}(\mathbb{G}) \xrightarrow{\cong} \mathbf{I}(\mathbb{G}')$ is the identification of grid states.

The ***filtered stabilization map*** is the map $\mathcal{S}\colon \mathit{Cone}(V_1 - V_2) \to \mathcal{GC}^-(\mathbb{G}')$ given in terms of the splitting of filtered $\mathbb{F}[V_1,\dots,V_n]$-modules

$$\mathit{Cone}(V_1 - V_2) \cong \mathcal{GC}^-(\mathbb{G})[V_1][\![1,1]\!] \oplus \mathcal{GC}^-(\mathbb{G})[V_1],$$

by $\mathcal{S} = (\mathcal{S}^{oR},\mathcal{S}^{oL})$.

More generally:

DEFINITION 14.3.7. Suppose that $\mathbb{G}'$ is obtained form $\mathbb{G}$ by a stabilization. We define maps between $\mathcal{GC}^-(\mathbb{G}')$ and $\mathit{Cone}(V_1 - V_2)$ in the following cases.

- If $\mathbb{G}'$ is obtained from $\mathbb{G}$ by a stabilization of type *X:SW*, then define $\mathcal{D}\colon \mathcal{GC}^-(\mathbb{G}') \to \mathit{Cone}(V_1 - V_2)$ as in Definition 13.3.10.
- If the stabilization is of type *X:NE*, define $\mathcal{D}\colon \mathcal{GC}^-(\mathbb{G}') \to \mathit{Cone}(V_1 - V_2)$ as in Definition 13.3.10, only now modifying the definitions of the domains as follows: define domains to be of type iR as in Definition 13.3.7 except now it is the northwest square with corner c whose local multiplicity is one bigger than the other three, and define domains to be of type iL as in Definition 13.3.7 except now the northeast square meeting c has one smaller local multiplicity than the other three.
- If $\mathbb{G}'$ is obtained from $\mathbb{G}$ by a stabilization of type *X:SE*, then define $\mathcal{S}\colon \mathit{Cone}(V_1 - V_2) \to \mathcal{GC}^-(\mathbb{G}')$ as in Definition 14.3.6.
- If $\mathbb{G}'$ is obtained from $\mathbb{G}$ by a stabilization of type *X:NW*, then define $\mathcal{S}\colon \mathit{Cone}(V_1 - V_2) \to \mathcal{GC}^-(\mathbb{G}')$ as in Definition 13.4.3, only now modifying the definitions of the domains as follows: define domains to be of type oR as in Definition 14.3.5, except now it is the northeast square with corner c whose local multiplicity is $k+1$, and define domains to be of type oL

as in Definition 14.3.5, except now it is the northwest square with corner c that has multiplicity $k-1$.

More succinctly, the domains for *X:NE* are obtained by rotating the domains for *X:SW* by 180°, while the domains for *X:NW* are obtained by rotating the domains for *X:SE* by 180°.

LEMMA 14.3.8. *The maps $\mathcal{D}$ and $\mathcal{S}$ from Definition* 14.3.7 *are filtered chain maps.*

Proof. For stabilizations of type *X:SW*, the result is Lemma 13.3.13; the other cases are proved in the same way, noting the relationship between the domains. □

Lemma 13.3.15 has the following straightforward generalization:

LEMMA 14.3.9. *Let D and S be the functions from Proposition* 14.3.4. *On the associated graded level, the map $\mathcal{D}$ induces D; and $\mathcal{S}$ induces S.* □

EXERCISE 14.3.10. In Section 13.4, we described the homotopy inverse to the destabilization map of type *X:SW*. Find homotopy inverses to destabilization maps of type *X:NE* and the stabilization maps of type *X:NW* and *X:SE*.

14.3.2. Refined Legendrian and transverse invariants. We can now prove the refined Legendrian and transverse invariance theorems:

Proof of Theorem 14.3.2. The isomorphisms on homology induced by commutations, which preserve the Legendrian invariants according to Lemma 6.4.4, are covered by filtered quasi-isomorphisms according to the proof of Proposition 13.3.4.

The isomorphisms on homology induced by Legendrian stabilizations (i.e. of type *X:SE* and *X:NW*), which preserve the Legendrian invariants by Lemma 6.4.6, are covered by filtered quasi-isomorphisms by Lemma 14.3.9.

The result now follows from Proposition 12.2.6. □

Proof of Theorem 14.3.3. Consider the isomorphisms $\phi^{\pm}$ on homology induced by the stabilization identifications (as in Proposition 5.4.1). According to Theorem 12.3.4, these transform the transverse invariants as stated; by Lemma 14.3.9 they are covered by filtered quasi-isomorphisms. □

As usual, it is often easier to work with $\widehat{GH}$. The analogues of Theorems 14.3.2 and 14.3.3 hold in this context, as well:

COROLLARY 14.3.11. *If $\mathbb{G}$ and $\mathbb{G}'$ represent Legendrian isotopic knots, then there is an isomorphism $\widehat{\phi}\colon \widehat{GH}(\mathbb{G}) \to \widehat{GH}(\mathbb{G}')$ that is covered by a filtered quasi-isomorphism, with $\widehat{\phi}(\widehat{\lambda}^+(\mathbb{G})) = \widehat{\lambda}^+(\mathbb{G}')$ and $\widehat{\phi}(\widehat{\lambda}^-(\mathbb{G})) = \widehat{\lambda}^-(\mathbb{G}')$.*

Proof. By the five lemma, a quasi-isomorphism $\mathcal{C} \to \mathcal{C}'$ between free chain complexes over $\mathbb{F}[U]$ induces a quasi-isomorphism $\frac{\mathcal{C}}{U} \to \frac{\mathcal{C}'}{U}$. Thus, the result follows immediately from Theorem 14.3.2. □

COROLLARY 14.3.12. *If $\mathbb{G}$, $\mathbb{G}^+$, and $\mathbb{G}^-$ are grid diagrams whose associated Legendrian knots are $\vec{\mathcal{K}}$ and its stabilizations $\vec{\mathcal{K}}^+$, and $\vec{\mathcal{K}}^-$ respectively, then there are isomorphisms*

$$\widehat{\phi}^-\colon \widehat{GH}(\mathbb{G}) \longrightarrow \widehat{GH}(\mathbb{G}^-), \qquad \widehat{\phi}^+\colon \widehat{GH}(\mathbb{G}) \longrightarrow \widehat{GH}(\mathbb{G}^+)$$

covered by filtered quasi-isomorphisms, that satisfy

$$\widehat{\phi}^-(\widehat{\lambda}^+(\vec{\mathcal{K}})) = \widehat{\lambda}^+(\vec{\mathcal{K}}^-), \qquad 0 = \widehat{\lambda}^+(\vec{\mathcal{K}}^+),$$
$$0 = \widehat{\lambda}^-(\vec{\mathcal{K}}^-), \qquad \widehat{\phi}^+(\widehat{\lambda}^-(\vec{\mathcal{K}})) = \widehat{\lambda}^-(\vec{\mathcal{K}}^+).$$

Proof. As in the proof of Corollary 14.3.11, we specialize quasi-isomorphisms over $\mathbb{F}[U]$ (in this case, induced by Theorem 14.3.3) to $U = 0$. The corollary follows, with the observation that $\lambda^\pm(\vec{\mathcal{K}}^\pm)$ being U-times another homology class is equivalent to the condition that the induced invariant in $\widehat{GH}$ is trivial. □

14.4. Applications of the refined Legendrian and transverse invariants

According to Theorem 12.3.3, if $\mathcal{K}_1$ and $\mathcal{K}_2$ are two Legendrian isotopic knots, then their Legendrian invariants are identified via an isomorphism between their associated grid homology groups. Theorem 14.3.2 lifts the isomorphism from Theorem 12.3.3 to a filtered quasi-isomorphism of the filtered complexes; Theorem 14.3.3 lifts the corresponding isomorphism for the transverse invariant. This extra constraint allows us to use further algebraic properties of the Legendrian invariants to give stronger obstructions to Legendrian and transverse isotopy. We start with some generalities.

DEFINITION 14.4.1. Let $\mathcal{C}$ be a $\mathbb{Z}$-filtered, $\mathbb{Z}$-graded chain complex, and fix a positive integer j. Let $\xi \in H(\mathcal{F}_i/\mathcal{F}_{i-1})$ be an element in the homology of the associated graded object. We say that ξ represents a ***cycle to order*** j (for $j \geq 1$ and integer) if there is some $x \in \mathcal{F}_i\mathcal{C}$ with the following two properties:

- $\partial x \in \mathcal{F}_{i-j}$, and
- the induced element $[x] \in \mathcal{F}_i/\mathcal{F}_{i-1}$, which is a cycle in the associated graded object by the previous condition, represents the homology class ξ.

Any homology class ξ represents a cycle to order 1. The condition that a class is a cycle to order 2 is invariant under quasi-isomorphisms:

PROPOSITION 14.4.2. *Let $\mathcal{C}$ and $\mathcal{C}'$ be two $\mathbb{Z}$-filtered, $\mathbb{Z}$-graded chain complexes and $\phi\colon H(\mathrm{gr}(\mathcal{C})) \to H(\mathrm{gr}(\mathcal{C}'))$ be an isomorphism between the homologies of their associated graded objects that is covered by a quasi-isomorphism. Then, $\xi \in H(\mathrm{gr}(\mathcal{C}))$ represents a cycle to order 2 if and only if $\phi(\xi)$ does.*

The above proposition will follow quickly from the next lemma, which is phrased in terms of the connecting homomorphism $\delta\colon H(\mathcal{F}_i/\mathcal{F}_{i-1}) \to H(\mathcal{F}_{i-1}/\mathcal{F}_{i-2})$ associated to the short exact sequence

$$0 \longrightarrow \mathcal{F}_{i-1}/\mathcal{F}_{i-2} \longrightarrow \mathcal{F}_i/\mathcal{F}_{i-2} \longrightarrow \mathcal{F}_i/\mathcal{F}_{i-1} \longrightarrow 0.$$

LEMMA 14.4.3. *A homology class $\xi \in H(\mathcal{F}_i/\mathcal{F}_{i-1})$ represents a cycle to order 2 if and only if ξ is contained in the kernel of δ.*

Proof. This follows quickly from the definition of the connecting homomorphism (which in turn is recalled in the proof of Lemma A.2.1). □

Proof of Proposition 14.4.2. If Φ is a quasi-isomorphism from $\mathcal{C}$ to $\mathcal{C}'$ then the naturality of the connecting homomorphism (cf. Lemma A.2.2) gives a commutative square

$$\begin{array}{ccc} H(\frac{\mathcal{F}_i\mathcal{C}}{\mathcal{F}_{i-1}\mathcal{C}}) & \xrightarrow{\delta} & H(\frac{\mathcal{F}_{i-1}\mathcal{C}}{\mathcal{F}_{i-2}\mathcal{C}}) \\ \phi\downarrow & & \downarrow\phi \\ H(\frac{\mathcal{F}_i\mathcal{C}'}{\mathcal{F}_{i-1}\mathcal{C}'}) & \xrightarrow{\delta'} & H(\frac{\mathcal{F}_{i-1}\mathcal{C}'}{\mathcal{F}_{i-2}\mathcal{C}'}), \end{array}$$

where the vertical maps are the isomorphisms induced by the quasi-isomorphism. In particular, the kernel of δ and the kernel of δ' are identified. The result now follows from Lemma 14.4.3. □

With the help of Proposition 14.4.2, the results of Chapter 12 obstructing Legendrian (and transverse) isotopies can be strengthened to:

COROLLARY 14.4.4. *If $\vec{\mathcal{K}}_1$ and $\vec{\mathcal{K}}_2$ are Legendrian isotopic knots, then*

- *$\lambda^+(\vec{\mathcal{K}}_1)$ is a cycle to order 2 if and only if $\lambda^+(\vec{\mathcal{K}}_2)$ is a cycle to order 2;*
- *$\widehat{\lambda}^+(\vec{\mathcal{K}}_1)$ is a cycle to order 2 if and only if $\widehat{\lambda}^+(\vec{\mathcal{K}}_2)$ is a cycle to order 2.*

Similar remarks hold for λ^- in place of λ^+. Also, if $\mathcal{T}_1$ and $\mathcal{T}_2$ are transverse isotopic knots, then

- *$\theta(\mathcal{T}_1)$ is a cycle to order 2 if and only if $\theta(\mathcal{T}_2)$ is a cycle to order 2;*
- *$\widehat{\theta}(\mathcal{T}_1)$ is a cycle to order 2 if and only if $\widehat{\theta}(\mathcal{T}_2)$ is a cycle to order 2.*

Proof. For the Legendrian invariants λ^+ and λ^- combine Proposition 14.4.2 with Theorem 14.3.2. For the transverse case, in addition, appeal to Theorem 14.3.3 (cf. Theorem 12.5.13 regarding the definition of θ). For $\widehat{\lambda}^+$, appeal to Corollary 14.3.11 in the Legendrian and Corollary 14.3.12 in the transverse case. □

For computations in $\widehat{GC}$, it is in practice preferable to work with $\widetilde{GC}$; compare Lemma 12.3.8. This is justified by the following:

LEMMA 14.4.5. *An element $\zeta \in \widehat{GC}(\mathbb{G})$ is a cycle to order 2 if and only if its image in $\widetilde{GC}(\mathbb{G})$ is a cycle to order 2.*

Proof. Lemma 14.1.11 gives a filtered quasi-isomorphism

$$\frac{\mathcal{GC}^-(\mathbb{G})}{V_1 = \cdots = V_n = 0} \simeq \widehat{\mathcal{GC}}(\mathbb{G}) \otimes \mathcal{W}^{\otimes(n-1)}.$$

Following the proof of that lemma, we see that the filtered quasi-isomorphism carries the projection $\widehat{\mathcal{GC}}(\mathbb{G}) = \frac{\mathcal{GC}^-(\mathbb{G})}{V_1=0} \to \frac{\mathcal{GC}^-(\mathbb{G})}{V_1=\cdots=V_n=0}$ to the inclusion of $\widehat{\mathcal{GC}}(\mathbb{G})$ into a direct summand in $\widehat{\mathcal{GC}}(\mathbb{G}) \otimes \mathcal{W}^{\otimes(n-1)}$. □

We illustrate Corollary 14.4.4 with a specific example from [**157**]: we give two transverse pretzel knots that are smoothly isotopic, have the same self-linking numbers, and have non-vanishing transverse invariants; but the knots are not transverse isotopic. Let $\mathcal{K}_1$ and $\mathcal{K}_2$ be the Legendrian $P(-4,-3,3)$ pretzel knots shown in Figure 14.2. Both of these knots represent the knot type $m(10_{140})$, have

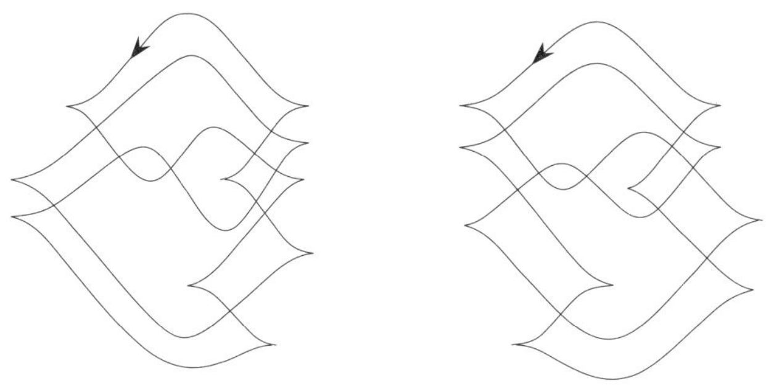

FIGURE 14.2. **Two Legendrian realizations ($\mathcal{K}_1$ on the left, and $\mathcal{K}_2$ on the right) of the pretzel knot $P(-4,-3,3)$.**

$\mathrm{tb} = -1$ and $r = 0$, and thus their positive transverse push-offs (for either orientation) have self-linking number -1. We can represent both of these Legendrian knots by grid diagrams with grid number 9, with $\mathbb{X}_1 = (9,8,1,4,6,5,7,2,3)$ and $\mathbb{O}_1 = (4,2,5,7,9,8,3,6,1)$ for $\mathcal{K}_1$ and $\mathbb{X}_2 = (9,8,2,4,6,5,3,7,1)$ and $\mathbb{O}_2 = (4,3,5,7,9,8,1,2,6)$ for $\mathcal{K}_2$.

PROPOSITION 14.4.6. *The transverse push-offs $\mathcal{T}(\vec{\mathcal{K}}_1)$ and $\mathcal{T}(\vec{\mathcal{K}}_2)$ of the Legendrian knots $\vec{\mathcal{K}}_1$ and $\vec{\mathcal{K}}_2$ are not transversely isotopic.*

Proof. Applying the algorithm from Lemma 12.6.4 (and Lemma 12.3.8, to compute in $\widetilde{GH}$), we find that the transverse invariants $\widehat{\theta}(\vec{\mathcal{K}}_1)$ and $\widehat{\theta}(\vec{\mathcal{K}}_2)$ are both non-zero. The invariants are distinguished by

$$\widehat{\delta}_1(\widehat{\theta}(\mathcal{T}(\vec{\mathcal{K}}_1)) = \widehat{\delta}_1(\widehat{\lambda}^+(\mathcal{K}_1)) = 0 \qquad \text{and} \qquad \widehat{\delta}_1(\widehat{\theta}(\mathcal{T}(\vec{\mathcal{K}}_2))) = \widehat{\delta}_1(\widehat{\lambda}^+(\mathcal{K}_2)) \neq 0.$$

In fact, to do these computations in practice, we use $\widetilde{GH}$ (cf. Lemma 14.4.5), and the induced connecting homomorphism $\widetilde{\delta}_1 \colon \widetilde{GH}(\mathbb{G},s) \to \widetilde{GH}(\mathbb{G},s-1)$, which is induced by the chain map

$$\widetilde{\partial}_1(\mathbf{x}) = \sum_{\mathbf{y}\in\mathbf{S}(\mathbb{G})} \#\{r \in \mathrm{Rect}^\circ(\mathbf{x},\mathbf{y}) \big| r\cap\mathbb{O} = \emptyset, \#(r\cap\mathbb{X}) = 1\}\cdot\mathbf{y}.$$

To establish non-triviality of $\widetilde{\delta}_1$, we slightly modify the algorithm from Lemma 12.6.4 to start with $\mathcal{B}_0$ consisting of the set of grid states appearing in $\widetilde{\partial}_1(\theta(\mathcal{T}(\vec{\mathcal{K}}_i))$. (This application of Lemma 12.6.4 involves hundreds of grid states. Although one would not wish to perform these by hand, a computer can handle them easily.) □

14.5. Filtrations in the case of links

It is routine to adapt the grid homology for links to the filtered context. We will sketch the construction here, and leave the details to the interested reader.

The primary challenge is the formulation: we must speak of complexes which are $\mathbf{H}$-filtered, $\mathbb{Z}$-graded complexes over $\mathbb{F}[U_1,\dots,U_\ell]$, where $\mathbf{H} \subset \mathbb{Q}^\ell$ is an affine space for $\mathbb{Z}^\ell$, and it plays the role of $\mathbf{H}(\vec{L})$ (the Alexander grading set, from Definition 11.1.3). We give the following straightforward generalization of Definition 13.1.1:

DEFINITION 14.5.1. Let $\mathbf{H} \subset \mathbb{Q}^\ell$ be an affine space for $\mathbb{Z}^\ell$, and write $\{\mathbf{e}_i\}_{i=1}^\ell$ for the standard basis of $\mathbb{Z}^\ell$. We can think of $\mathbf{H}$ as partially ordered, where $h_1 \geq h_2$ if there are non-negative integers $a_1, \ldots, a_\ell$ so that

$$h_1 = h_2 + \sum_{i=1}^{\ell} a_i \cdot \mathbf{e}_i.$$

An ***$\mathbf{H}$-filtered, $\mathbb{Z}$-graded chain complex*** $(\mathcal{C}, \partial)$ over $\mathbb{F}[V_1, \ldots, V_n]$ is a vector space, equipped with the following additional structures:

- an endomorphism $\partial\colon \mathcal{C} \to \mathcal{C}$, called the ***differential***, satisfying the relation $\partial \circ \partial = 0$;
- a ***$\mathbb{Z}$-grading***, which is a splitting of $\mathcal{C}$ as a vector space $\mathcal{C} = \bigoplus_{d\in\mathbb{Z}} \mathcal{C}_d$ which is compatible with the differential, in the sense that $\partial \mathcal{C}_d \subseteq \mathcal{C}_{d-1}$;
- a collection of vector subspaces $\{\mathcal{F}_h\mathcal{C} \subset \mathcal{C}\}_{h\in\mathbf{H}}$, called the $\mathbf{H}$-filtration, which satisfy the following further properties:
  - they are ascending; i.e. if $h \leq h'$, then $\mathcal{F}_h\mathcal{C} \subseteq \mathcal{F}_{h'}\mathcal{C}$,
  - they are graded; i.e. if $\mathcal{F}_h\mathcal{C}_d = (\mathcal{F}_h\mathcal{C}) \cap \mathcal{C}_d$, then $\mathcal{F}_h\mathcal{C} = \bigoplus_{d\in\mathbb{Z}} \mathcal{F}_h\mathcal{C}_d$,
  - they exhaust $\mathcal{C}$; i.e. $\bigcup_{h\in\mathbf{H}} \mathcal{F}_h\mathcal{C} = \mathcal{C}$,
  - they are bounded below in the sense that for any given $d \in \mathbb{Z}$, there is an h_d so that $\mathcal{F}_h\mathcal{C}_d = 0$ for $h \leq h_d$,
  - they are compatible with the differential; i.e. $\partial\mathcal{F}_h\mathcal{C} \subset \mathcal{F}_h\mathcal{C}$;
- a map $\pi\colon \{1, \ldots, n\} \to \{1, \ldots, \ell\}$
- endomorphisms V_i for $i = 1, \ldots, n$ that commute with one another, are compatible with the differential in the sense that $\partial \circ V_i = V_i \circ \partial$ for $i = 1, \ldots, n$, and that respect the grading and filtration, in the sense that $V_i \cdot \mathcal{C}_d \subset \mathcal{C}_{d-2}$ and $V_i \cdot \mathcal{F}_h\mathcal{C} \subset \mathcal{F}_{h-\mathbf{e}_{\pi(i)}}\mathcal{C}$.

DEFINITION 14.5.2. Suppose that $\vec{L}$ is an oriented link with Alexander grading set $\mathbf{H}(\vec{L})$ given by Definition 11.1.3, and fix a grid diagram $\mathbb{G}$ representing $\vec{L}$, with grid number n. The ***multi-filtered grid complex*** $\mathcal{GC}^-(\mathbb{G})$ of $\mathbb{G}$ is the $\mathbb{F}[V_1, \ldots, V_n]$-module freely generated by grid states $\mathbf{S}(\mathbb{G})$, equipped with its usual Maslov $\mathbb{Z}$-grading induced by M, an Alexander multi-filtration with values in $\mathbf{H}(\vec{L})$ induced by the Alexander vector (Definition 11.1.5), and the differential ∂^- defined in Equation (13.1).

Theorem 13.2.4 has the following adaptation to the case of links:

THEOREM 14.5.3. *If $\vec{L}$ is an oriented link with ℓ components and $\mathbb{G}$ is a grid diagram with grid number n representing $\vec{L}$, then $\mathcal{GC}^-(\mathbb{G})$ is an $\mathbf{H}(\vec{L})$-filtered, $\mathbb{Z}$-graded chain complex over $\mathbb{F}[V_1, \ldots, V_n]$, in the sense of Definition* 14.5.1. □

For an invariance statement, we generalize the notion of "filtered quasi-isomorphism" (from Definition 13.1.9). An $\mathbf{H}$-filtered, $\mathbb{Z}$-graded chain complex $\mathcal{C}$ has an associated graded object, which inherits a grading by $\mathbb{Z} \oplus \mathbf{H}$:

$$\mathrm{gr}(\mathcal{C}) = \bigoplus_{d\in\mathbb{Z}, h\in\mathbf{H}} \frac{\mathcal{F}_h\mathcal{C}_d}{\bigcup_{h'<h} \mathcal{F}_{h'}\mathcal{C}_d}.$$

DEFINITION 14.5.4. An $\mathbf{H}$-filtered, $\mathbb{Z}$-graded ***quasi-isomorphism*** $f\colon \mathcal{C} \to \mathcal{C}'$ is an $\mathbf{H}$-filtered, $\mathbb{Z}$-graded chain map whose associated graded map $\mathrm{gr}(f)$ induces an isomorphism on the homology of the associated graded object. Two complexes

$\mathcal{C}$ and $\mathcal{C}'$ are said to be ***filtered quasi-isomorphic*** if there is a third complex $\mathcal{C}''$ and filtered quasi-isomorphisms from $\mathcal{C}''$ to $\mathcal{C}$ and from $\mathcal{C}''$ to $\mathcal{C}'$.

Theorem 13.2.9 has the following straightforward generalization to the case of links:

THEOREM 14.5.5. *Fix an oriented link $\vec{L}$. For any grid diagram $\mathbb{G}$ representing $\vec{L}$, choose a sequence $V_{i_1},\dots V_{i_\ell}$, one on each component. Then, the filtered quasi-isomorphism type of $\mathcal{GC}^-(\mathbb{G})$ (in the sense of Definition 14.5.4), over $\mathbb{F}[U_1,\dots,U_\ell]$ where U_j acts as multiplication by V_{i_j}, is independent of the choice of $\mathbb{G}$; and the homology of its associated graded object is the multi-graded, unblocked grid homology $\mathbf{GH}^-(\vec{L})$ from Chapter 11.* □

EXAMPLE 14.5.6 (The two-component unlink). The example of the two-component unlink L from Section 11.3 can be enriched as follows. A model for $\mathcal{GC}^-(L)$ is generated by two generators p and q with $\mathbb{A}(p) = \mathbb{A}(q) = (0,0)$ and $M(p) = 0$, $M(q) = -1$. The differential is determined by

$$\partial^- q = U_1 \cdot p + U_2 \cdot p \qquad \partial^- p = 0.$$

EXAMPLE 14.5.7 (The Hopf link H_-). The example of the negative Hopf link H_- from Section 11.3 can be enriched as follows. $\mathcal{GC}^-(H_-)$ is generated by four generators p, q, r, and s, with

$$\mathbb{A}(p) = (\frac{1}{2},\frac{1}{2}) \qquad \mathbb{A}(q) = (-\frac{1}{2},\frac{1}{2}) \qquad \mathbb{A}(r) = (\frac{1}{2},-\frac{1}{2}) \qquad \mathbb{A}(s) = (-\frac{1}{2},-\frac{1}{2})$$

$$M(p) = 1 \qquad M(q) = M(r) = 0 \qquad M(s) = -1;$$

and differential determined by

$$\partial^- p = q + r \qquad \partial^- q = 0 = \partial^- r \qquad \partial^- s = U_1 \cdot r + U_2 \cdot q.$$

EXAMPLE 14.5.8 (The Hopf link H_+). The example of the positive Hopf link H_+ from Section 11.3 can be enriched as follows. $\mathcal{GC}^-(H_+)$ is generated by four generators p, q, r, and s, with

$$\mathbb{A}(p) = (\frac{1}{2},\frac{1}{2}) \qquad \mathbb{A}(q) = (-\frac{1}{2},\frac{1}{2}) \qquad \mathbb{A}(r) = (\frac{1}{2},-\frac{1}{2}) \qquad \mathbb{A}(s) = (-\frac{1}{2},-\frac{1}{2})$$

$$M(p) = 0 \qquad M(q) = M(r) = -1 \qquad M(s) = -2;$$

and differential determined by

$$\partial^- p = 0 = \partial^- s \qquad \partial^- q = U_1 \cdot p + s \qquad \partial^- r = U_2 \cdot p + s$$

EXERCISE 14.5.9. Verify the above descriptions of $\mathcal{GC}^-$ for the two-component unlink and the two Hopf links.

14.6. Remarks on three-manifold invariants

We have seen in this chapter certain aspects of grid homology which tie in with more general constructions in Heegaard Floer homology.

Proposition 14.1.1 states that the total homology of the filtered chain complex is independent of the knot. This generalizes neatly to null-homologous knots in more general three-manifolds: in that case, the knot filtration is a filtration on the Heegaard Floer complex of the ambient three-manifold; i.e. its total homology is the Heegaard Floer homology of the ambient three-manifold. From this perspective,

Proposition 14.1.1 illustrates the fact that the Heegaard Floer homology of S^3 is particularly simple.

As a more subtle point, the homology groups of the filtered pieces $\mathcal{F}_s\mathcal{GC}^-$ also have interpretations in terms of Heegaard Floer homology groups for closed three-manifolds. Indeed, the homology of $\mathcal{F}_s\mathcal{GC}^-$ is a summand of the Heegaard Floer homology of the three-manifold obtained by any sufficiently large surgery on $K \subset S^3$. The invariants h_s from Equation (14.3), measure the absolute grading of a canonically defined homology class in this Heegaard Floer homology group. The inclusion map on homology $H(\mathcal{F}_s\mathcal{GC}^-) \to H(\mathcal{GC}^-)$ (which is also measured by the h_s) is a four-manifold invariant, associated to the four-dimensional cobordism defined by attaching a two-handle to $S^3 \times [0, 1]$ along the knot.

It follows from work of Rasmussen [**191**, Corollary 7.4] that

$$\min\{s \big| h_s = 0\} \leq g_s(K).$$

This inequality generalizes the bound given by $\tau(K)$; see [**89**, Proposition 2.3].

CHAPTER 15

Grid homology over the integers

In the previous chapters we defined and worked with versions of grid homologies and filtered complexes over $\mathbb{F}$ or $\mathbb{F}[U]$ (or even $\mathbb{F}[U_1, \dots, U_\ell]$), where $\mathbb{F} \cong \mathbb{Z}/2\mathbb{Z}$ denotes the field with two elements. The advantages of working mod 2 are that computations are simpler and the definition of the complexes are more straightforward. In fact, this apparently simpler theory is sufficient for the topological applications stated in Section 1.2.

The aim of the present chapter is to explain how to lift the grid complexes to complexes with integer coefficients. In this construction, we associate a sign ± 1 to each rectangle in a compatible way, via a *sign assignment*, and define the differential to be the signed count of rectangles. Although the sign assignment represents a further choice in the construction of the grid complex, the associated homology groups turn out to be independent of this choice, giving knot invariants.

In Section 15.1 we formalize the properties of a sign assignment and then use sign assignments to lift the grid complex to the integers. In Section 15.2 we verify the relevant existence and uniqueness results for sign assignments for grid diagrams, except for a technical fact about the "spin group" that is verified in a later section. In Section 15.3 we prove the invariance properties of the grid homology groups over $\mathbb{Z}$. In Chapter 13, we defined a refinement of grid homology to the filtered quasi-isomorphism type of a filtered complex. In Section 15.4, we describe how to lift that construction to $\mathbb{Z}$, and establish invariance properties of this sign-refined filtered complex. In Section 15.5, we sketch integral lifts of some of the other constructions from this book. In Section 15.6 we use the theory over $\mathbb{Z}$ to define further numerical invariants of knots, that are variants of $\tau(K)$ over $\mathbb{Z}/p\mathbb{Z}$ for any prime p. In Section 15.7 we complete the existence statement from Section 15.2, using properties of quaternions. Finally in Section 15.8 we close with some further remarks on grid homology over $\mathbb{Z}$.

The sign refinement for grid homology was first constructed in [**136**]; our discussion here follows ideas of Gallais [**66**], see also [**166**]. Knot Floer homology has an analogous lift over $\mathbb{Z}$, defined by consistently orienting the appropriate moduli spaces. In all known cases (including alternating knots, torus knots, and knots with small crossing number), the grid homology is a free $\mathbb{Z}$-module, and hence it is determined by its $\mathbb{Z}/2\mathbb{Z}$ specialization. This is unlikely to hold in general: the sign refinement is expected to contain further interesting information about knots. (See also Section 15.6 and Problem 17.2.8.)

15.1. Signs assignments and grid homology over $\mathbb{Z}$

The grid complex is lifted to $\mathbb{Z}$ by consistently associating signs to the rectangles counted by the boundary map, so that the associated signed count of rectangles still gives a differential. Our immediate aim is to formalize such sign assignments.

Fix a grid diagram $\mathbb{G}$. Although we will need to associate signs only to empty rectangles in $\mathbb{G}$, it will be convenient to associate them to all rectangles. Following Section 4.4, if $\mathbf{x}$ and $\mathbf{y}$ are any two grid states, the set of rectangles $\mathrm{Rect}(\mathbf{x}, \mathbf{y})$ from $\mathbf{x}$ to $\mathbf{y}$ is either empty, or it contains exactly two elements, in which case $\mathrm{Rect}(\mathbf{y}, \mathbf{x})$ also contains exactly two elements. Let $\mathrm{Rect}(\mathbb{G}) = \bigcup_{\mathbf{x},\mathbf{y}\in\mathbf{S}(\mathbb{G})} \mathrm{Rect}(\mathbf{x}, \mathbf{y})$.

DEFINITION 15.1.1. Fix rectangles $r_1 \in \mathrm{Rect}(\mathbf{x}, \mathbf{y})$ and $r_2 \in \mathrm{Rect}(\mathbf{y}, \mathbf{z})$ and $r_1' \in \mathrm{Rect}(\mathbf{x}, \mathbf{w})$ and $r_2' \in \mathrm{Rect}(\mathbf{w}, \mathbf{z})$ with $\mathbf{y} \neq \mathbf{w}$, and suppose that the domain associated to the composition $r_1 * r_2$ (in the sense of Definition 4.6.4) is the same as the domain associated to $r_1' * r_2'$. We call the two pairs of rectangles (r_1, r_2) and (r_1', r_2') ***alternative pairs*** if one of the two following conditions holds: either the eight corners of r_1 and r_2 are distinct; or the intersection of r_1 and r_2 is an interval. (These are the pairs of rectangles we have to take into account when proving $\partial^2 = 0$, cf. Cases (R-1) and (R-2) in the proof of Lemma 4.6.7.)

Consider next the case where there is a rectangle $r_1 \in \mathrm{Rect}(\mathbf{x}, \mathbf{y})$ and another rectangle $r_2 \in \mathrm{Rect}(\mathbf{y}, \mathbf{x})$. In this case the support of $r_1 * r_2$ is an annulus. We distinguish two subcases, according to whether the annulus is horizontal or vertical.

DEFINITION 15.1.2. A ***sign assignment*** for the grid diagram $\mathbb{G}$ is a function $S\colon \mathrm{Rect}(\mathbb{G}) \to \{\pm 1\}$ satisfying the following properties:

(1) if (r_1, r_2) and (r_1', r_2') are alternative pairs of rectangles, then
$$S(r_1) \cdot S(r_2) = -S(r_1') \cdot S(r_2');$$
(2) if (r_1, r_2) is a pair of rectangles forming a horizontal annulus, then
$$S(r_1) \cdot S(r_2) = 1;$$
(3) if (r_1, r_2) is a pair of rectangles forming a vertical annulus, then
$$S(r_1) \cdot S(r_2) = -1.$$

REMARK 15.1.3. Notice that the set $\mathrm{Rect}(\mathbb{G})$, and any sign assignment on it is insensitive to the placement of the O- and X-markings. Indeed, the only relevant information about the grid in the definition of a sign assignment is its size.

Given a sign assignment $S\colon \mathrm{Rect}(\mathbb{G}) \to \{\pm 1\}$ and a function $g\colon \mathbf{S}(\mathbb{G}) \to \{\pm 1\}$, we define a new function $S^g\colon \mathrm{Rect}(\mathbb{G}) \to \{\pm 1\}$ by
$$S^g(r) = g(\mathbf{x}) \cdot S(r) \cdot g(\mathbf{y}),$$
for $r \in \mathrm{Rect}(\mathbf{x}, \mathbf{y})$. It is easy to see that S^g is also a sign assignment.

DEFINITION 15.1.4. A function $g\colon \mathbf{S}(\mathbb{G}) \to \{\pm 1\}$ on the set $\mathbf{S}(\mathbb{G})$ of grid states is called a ***gauge transformation***. Multiplication in $\{\pm 1\} \cong \mathbb{Z}/2\mathbb{Z}$ naturally gives the set of gauge transformations a group structure, identifying it with the direct sum of $n!$ many copies of $\mathbb{Z}/2\mathbb{Z}$. The map $(g, S) \mapsto S^g$ induces an action of the group of gauge transformations on the set of sign assignments. For a gauge transformation g the signs assignments S and S^g are called ***gauge equivalent***.

The construction of grid homology to $\mathbb{Z}$ hinges on the following:

THEOREM 15.1.5. *For any grid diagram $\mathbb{G}$, there exits a sign assignment, and it is unique up to gauge equivalence.*

EXERCISE 15.1.6. **(a)** For a 2×2 grid diagram, there are two grid states and four rectangles. Show that a sign assignment is uniquely determined by its value on one of the four rectangles. Show that the two sign assignments are gauge equivalent.
(b) Consider a 3×3 grid diagram. Construct a sign assignment for $\mathbb{G}$. Show that any two sign assignments on a 3×3 grid are gauge equivalent.
(c) Show that any sign assignment is uniquely determined by its values on empty rectangles.

Theorem 15.1.5 is proved in Section 15.2. We explain now how a sign assignment S can be used to lift the definition of the grid complex to a chain complex with coefficients in $\mathbb{Z}[V_1, \dots, V_n]$, starting with the lift of the chain complex $GC^-(\mathbb{G})$ from Chapter 4, when $\mathbb{G}$ represents a knot. The modifications for lifting grid homology for links (from Chapter 11) and the filtered chain complex (from Chapter 13) will be given in Sections 15.5 and 15.4, respectively.

Let $\mathbb{G}$ be a toroidal grid diagram with grid number n representing a knot K, and let $\mathcal{R}_{\mathbb{Z}} = \mathbb{Z}[V_1, \dots, V_n]$ be the polynomial ring in n variables, now with integer coefficients. Let $GC^-(\mathbb{G};\mathbb{Z})$ denote the $\mathcal{R}_{\mathbb{Z}}$-module freely generated by the grid states of $\mathbb{G}$. Given a sign assignment S on $\mathbb{G}$, we define the endomorphism

$$\partial^-_{\mathbb{X},S}\mathbf{x} = \sum_{\mathbf{y}\in\mathbf{S}(\mathbb{G})} \sum_{\{r\in \mathrm{Rect}^\circ(\mathbf{x},\mathbf{y}) | r\cap\mathbb{X}=\emptyset\}} S(r)\cdot V_1^{O_1(r)}\cdots V_n^{O_n(r)}\cdot \mathbf{y}. \tag{15.1}$$

The mod 2 reduction of the above formula gives the differential $\partial^-_{\mathbb{X}}$ on the grid complex from Equation (4.10).

LEMMA 15.1.7. *The operator $\partial^-_{\mathbb{X},S}$ satisfies $\partial^-_{\mathbb{X},S}\circ\partial^-_{\mathbb{X},S}=0$; i.e. $(GC^-(\mathbb{G};\mathbb{Z}),\partial^-_{\mathbb{X},S})$ is a chain complex.*

Proof. The proof of this result follows the proof of Lemma 4.6.7, taking signs into account. The contributions to $\partial^-_{\mathbb{X},S}\circ\partial^-_{\mathbb{X},S}$ from Cases (R-1) and (R-2) (in the notation of the proof of Lemma 4.6.7) cancel in pairs, since for the alternative pairs (r_1, r_2) and (r_1', r_2'), we have $S(r_1)\cdot S(r_2) + S(r_1')\cdot S(r_2') = 0$. Since the boundary operator counts rectangles disjoint from $\mathbb{X}$, Case (R-3) does not occur. □

DEFINITION 15.1.8. The bigraded chain complex

$$GC^-_S(\mathbb{G};\mathbb{Z}) = (GC^-(\mathbb{G};\mathbb{Z}),\partial^-_{\mathbb{X},S}),$$

is called the ***sign-refined grid complex*** of the grid diagram $\mathbb{G}$.

Just as in the case with coefficients in $\mathbb{F}$, the Maslov and Alexander functions induce a bigrading on the $\mathcal{R}_{\mathbb{Z}}$-module $GC^-(\mathbb{G};\mathbb{Z})$. As before (see Lemma 4.6.8), the differential $\partial^-_{\mathbb{X},S}$ respects this bigrading.

Actions by the various V_i variables are homotopic to one another as bigraded module maps, as shown by the following straightforward adaptation of Lemma 4.6.9.

LEMMA 15.1.9. *Suppose that the grid diagram $\mathbb{G}$ gives rise to a knot K, and S is a fixed sign assignment on $\mathbb{G}$. Then, on the chain complex $GC^-_S(\mathbb{G};\mathbb{Z})$, as a module over $\mathbb{Z}[V_1, \dots, V_n]$, multiplication by V_i is chain homotopic to multiplication by V_j.*

Proof. As in the proof of Lemma 4.6.9, it suffices to consider the case where V_i and V_j correspond to consecutive O-markings O_i and O_j, for which there is some X-marking X_i in the same row as O_i and the same column as O_j. The homotopy operator of Equation (4.16) lifts to the sign refined version

$$\mathcal{H}_{i,S}(\mathbf{x}) = \sum_{\mathbf{y}\in\mathbf{S}(\mathbb{G})} \sum_{\{r\in \text{ Rect}^\circ(\mathbf{x},\mathbf{y}) \big| r\cap\mathbb{X}=X_i\}} S(r)\cdot V_1^{O_1(r)}\cdots V_n^{O_n(r)}\cdot \mathbf{y}. \tag{15.2}$$

As in the proof of Lemma 4.6.9, terms in the sum $\partial^-_{\mathbb{X},S}\circ\mathcal{H}_{i,S} + \mathcal{H}_{i,S}\circ\partial^-_{\mathbb{X},S}$ count alternative pairs of rectangles (which cancel pairwise in the sum) and the two annuli with width one and height one through X_i. The horizontal thin annulus through X_i contributes $V_i\cdot\mathbf{x}$ while the vertical one through X_i contributes $-V_j\cdot\mathbf{x}$; so

$$\mathcal{H}_{i,S}\circ\partial^-_{\mathbb{X},S} + \partial^-_{\mathbb{X},S}\circ\mathcal{H}_{i,S} = V_i - V_j, \tag{15.3}$$

as needed. □

To define the chain complex $GC^-_S(\mathbb{G};\mathbb{Z})$, we must choose a sign assignment S. As the next proposition shows, the isomorphism type of the chain complex is independent of this choice.

PROPOSITION 15.1.10. *Given any two sign assignments S_1 and S_2 for the grid diagram $\mathbb{G}$, there is an isomorphism $(GC^-(\mathbb{G};\mathbb{Z}),\partial^-_{\mathbb{X},S_1}) \cong (GC^-(\mathbb{G};\mathbb{Z}),\partial^-_{\mathbb{X},S_2})$ of bigraded chain complexes over $\mathbb{Z}[V_1,\dots,V_n]$.*

Proof. By the uniqueness statement in Theorem 15.1.5, there is a gauge transformation $g\colon \mathbf{S}(\mathbb{G})\to\{\pm1\}$ so that $S_2 = S_1^g$. Since $S_2(r) = g(\mathbf{x})\cdot S_1(r)\cdot g(\mathbf{y})$, the $\mathcal{R}_\mathbb{Z}$-module homomorphism $f\colon (GC^-(\mathbb{G};\mathbb{Z}),\partial^-_{\mathbb{X},S_1}) \to (GC^-(\mathbb{G};\mathbb{Z}),\partial^-_{\mathbb{X},S_2})$, whose value on a grid state $\mathbf{x}\in\mathbf{S}(\mathbb{G})$ is $f(\mathbf{x}) = g(\mathbf{x})\cdot\mathbf{x}$, is a chain map, giving the stated isomorphism. □

The isomorphisms on grid homology over $\mathbb{F}$ associated to grid moves in Chapter 5 can be lifted to integer coefficients, proving the following invariance of the homology $H(GC^-(\mathbb{G};\mathbb{Z}),\partial^-_{\mathbb{X},S})$, which extends Theorem 4.6.19:

THEOREM 15.1.11. *Let K be a knot, choose a grid diagram $\mathbb{G}$ with grid number n representing K, a sign assignment S for $\mathbb{G}$, and some integer $i\in\{1,\dots,n\}$. The simply blocked grid homology $\widehat{GH}(\mathbb{G};\mathbb{Z}) = H(\frac{GC^-_S(\mathbb{G};\mathbb{Z})}{V_i})$, thought of as a bigraded abelian group, and the unblocked grid homology $GH^-(\mathbb{G};\mathbb{Z}) = H(GC^-_S(\mathbb{G};\mathbb{Z}))$, thought of as a bigraded $\mathbb{Z}[U]$-module, where the U-action is given by multiplication by V_i, are both invariants of the unoriented knot K.*

DEFINITION 15.1.12. Let K be a knot, and choose $\mathbb{G}$, S, and i as in Theorem 15.1.11. The homology of $GC^-_S(\mathbb{G};\mathbb{Z})$, thought of as a bigraded module over $\mathbb{Z}[U]$, where the U-action is induced by multiplication by V_i, is called the ***sign-refined (unblocked) grid homology of*** K; and it is denoted $GH^-(K;\mathbb{Z})$. The homology of $GC^-_S(\mathbb{G};\mathbb{Z})/V_i$, thought of as a bigraded abelian group, is called the ***sign-refined simply blocked grid homology of*** K; and it is denoted $\widehat{GH}(K;\mathbb{Z})$.

Theorem 15.1.11 is proved in Section 15.3, after we construct the sign assignments for rectangles used to define the grid complex over $\mathbb{Z}$.

EXERCISE 15.1.13. Show that for the unknot $\mathcal{O}$, $\widehat{GH}(\mathcal{O};\mathbb{Z}) \cong \mathbb{Z}_{(0,0)}$, and $GH^-(\mathcal{O};\mathbb{Z}) \cong \mathbb{Z}[U]_{(0,0)}$, adapting notation from Section 7.3.

15.2. Existence and uniqueness of sign assignments

The aim of this section is to prove Theorem 15.1.5. Our treatment (based on Gallais' approach in [**66**]) uses a little bit of group theory for the symmetric group. We start with the following standard terminology of group theory.

DEFINITION 15.2.1. Let K be an abelian group and G, H be groups. The group H is a ***central extension of** G **by** K* if the groups fit into a short exact sequence

$$1 \to K \xrightarrow{i} H \xrightarrow{p} G \to 1$$

with the property that K is a subgroup of the center $Z(H)$ of H.

We recall the relevance of the symmetric group $\mathfrak{S}_n$ to the study of grid states and the rectangles that connect them. Fix a toroidal grid diagram $\mathbb{G}$ with grid number n. Fix a labeling of the horizontal circles by $\boldsymbol{\alpha} = \{\alpha_1, \dots, \alpha_n\}$ and the vertical circles by $\boldsymbol{\beta} = \{\beta_1, \dots, \beta_n\}$, both of which are numbered compatibly with the cyclic ordering on the torus. As explained in Section 4.1, this data specifies a one-to-one correspondence $\mathbf{x} \mapsto \sigma_{\mathbf{x}}$ between grid states and elements of $\mathfrak{S}_n$: the permutation $\sigma_{\mathbf{x}}$ maps i to j if the component of $\mathbf{x}$ on the ith vertical circle β_i is on the jth horizontal circle α_j. For two grid states $\mathbf{x}, \mathbf{y}$ the set $\mathrm{Rect}(\mathbf{x}, \mathbf{y})$ is non-empty if and only if $\sigma_{\mathbf{x}}^{-1}\sigma_{\mathbf{y}}$ is a transposition, in which case $\mathrm{Rect}(\mathbf{x}, \mathbf{y})$ contains exactly two elements (and $\mathrm{Rect}(\mathbf{y}, \mathbf{x})$ also contains two elements). Thus, the set of rectangles starting at $\mathbf{x}$ is in two-to-one correspondence with the set of transpositions in $\mathfrak{S}_n$.

We will study a group $\widetilde{\mathfrak{S}}_n$, called the *spin extension of the symmetric group*, that is twice as big as $\mathfrak{S}_n$, constructed presently. To avoid degenerate cases, we always assume that $n \geq 2$. Consider the n vectors $(v_1, \dots, v_n)$ in $\mathbb{R}^n$ defined by

$$v_i = ne_i - \sum_{k=1}^{n} e_k, \tag{15.4}$$

where e_i are the standard basis vectors for $\mathbb{R}^n$. Embed $\mathfrak{S}_n$ into the group of rigid rotations $\mathrm{SO}(n)$ of $\mathbb{R}^n$ as the subgroup that permutes the vectors $v_1, \dots, v_n$. Recall [**29**] that there is a unique connected topological group, the *spin group*, that fits into a short exact sequence

$$\{1\} \longrightarrow \mathbb{Z}/2\mathbb{Z} \longrightarrow \mathrm{Spin}(n) \xrightarrow{\ p\ } \mathrm{SO}(n) \longrightarrow \{1\}, \tag{15.5}$$

where $\mathbb{Z}/2\mathbb{Z}$ is contained in the center of $\mathrm{Spin}(n)$. For $n = 2$, $\mathrm{SO}(2) = S^1$ (the group of unit complex numbers) and $\mathrm{Spin}(2) = S^1$, where $p\colon S^1 \to S^1$ is the map $z \mapsto z^2$. For $n \geq 3$, an element in $\mathrm{Spin}(n)$ can be viewed as a homotopy class of paths (with fixed endpoints) in $\mathrm{SO}(n)$, starting at the identity element. (Recall that $\pi_1(\mathrm{SO}(n)) = \mathbb{Z}/2\mathbb{Z}$ once $n \geq 3$.) Now, $\widetilde{\mathfrak{S}}_n$ is the preimage of $\mathfrak{S}_n \subset \mathrm{SO}(n)$ under p. By its construction, $\widetilde{\mathfrak{S}}_n$ is a central extension of $\mathfrak{S}_n$ by $\mathbb{Z}/2\mathbb{Z}$, so $|\widetilde{\mathfrak{S}}_n| = 2n!$.

We will give a set of generators for $\widetilde{\mathfrak{S}}_n$ that are in one-to-one correspondence with rectangles starting from $\mathbf{x}$. Before doing this, we recall some familiar preliminaries about the symmetric group $\mathfrak{S}_n$.

Let $\tau_{i,j} \in \mathfrak{S}_n$ with $1 \leq i < j \leq n$ denote the transposition that switches i and j. The following result is standard.

PROPOSITION 15.2.2. *The symmetric group $\mathfrak{S}_n$ on n letters admits a presentation with generators $\{\tau_{i,j} \mid 1 \leq i < j \leq n\}$ (where $\tau_{i,j}$ is the transposition of the ith and the jth letters), and relations*

(1) $\tau_{i,j}^2 = 1$ *for* $1 \leq i < j \leq n$,
(2) $\tau_{i,j} \cdot \tau_{k,l} = \tau_{k,l} \cdot \tau_{i,j}$ *for any* $1 \leq i, j, k, l \leq n$ *satisfying* $\{i,j\} \cap \{k,l\} = \emptyset$,
(3) $\tau_{i,j} \cdot \tau_{j,k} \cdot \tau_{i,j} = \tau_{j,k} \cdot \tau_{i,j} \cdot \tau_{j,k} = \tau_{i,k}$ *for any* $1 \leq i < j < k \leq n$ *distinct indices.*

Proof. The fact that $\tau_{i,j}$ generate $\mathfrak{S}_n$ is well known. Also, simple computation shows that transpositions satisfy items (1)-(3) of the proposition. It follows that there is a surjective homomorphism from Γ_n, the group with generators and relations stated in the proposition, to the symmetric group $\mathfrak{S}_n$. By induction on n we will prove that this surjection is an isomorphism.

The base case $n = 2$ is obvious. For the inductive step, suppose that $\Gamma_n \cong \mathfrak{S}_n$, and consider the subgroup $H \subset \Gamma_{n+1}$ generated by $\tau_{i,j}$ with $1 \leq i < j \leq n$. By the inductive hypothesis, H is a quotient of $\mathfrak{S}_n$, so that H is a finite group with

$$|H| \leq n! \tag{15.6}$$

Label $n+1$ of the left cosets of H in Γ_{n+1} as

$$L_{n+1} = H, \quad L_n = \tau_{n,n+1} \cdot H, \quad \dots \quad L_i = \tau_{i,n+1} \cdot H, \quad \dots \quad L_1 = \tau_{1,n+1} \cdot H.$$

(We have not yet verified that these are distinct or exhaustive; but it will follow from the proof.) It follows immediately from the relations that left translation by $\tau_{i,j}$ permutes the cosets $L_1, \dots, L_{n+1}$. (In fact, $\tau_{i,j}$ permutes L_i and L_j and fixes all other L_k for $k \neq i, j$.) Since by definition the $\tau_{i,j}$ generate Γ_{n+1}, we conclude that left translation by any element of Γ_{n+1} permutes the $L_1, \dots, L_{n+1}$; in particular, the index of H in Γ_{n+1} is bounded:

$$[\Gamma_{n+1} : H] \leq n + 1. \tag{15.7}$$

Equations (15.6) and (15.7) show that Γ_{n+1} is a finite group with

$$|\Gamma_{n+1}| \leq (n+1)! = |\mathfrak{S}_{n+1}|,$$

forcing the surjection from Γ_{n+1} to $\mathfrak{S}_{n+1}$ to be an isomorphism, as claimed. $\square$

We would like to give explicit lifts of the $\tau_{i,j}$ in $\widetilde{\mathfrak{S}}_n$. We do this with the help of the following geometric construction involving the groups $\mathrm{SO}(n)$ and $\mathrm{Spin}(n)$.

DEFINITION 15.2.3. Given a 2-plane P in $\mathbb{R}^n$, there is a well-defined 180° rotation R_P of that plane, fixing its orthogonal complement. Fixing an orientation on P gives an associated ***spin rotation***, $\widetilde{R}_P \in \mathrm{Spin}(n)$, which is the element associated to the path $\{r_t\}_{t \in [0,\pi]}$ in $\mathrm{SO}(n)$, where r_t rotates P (with respect to the fixed orientation) through an angle of t.

We will use the following lemma, whose proof is relegated to Section 15.7:

LEMMA 15.2.4. *Let $z \in \mathrm{Spin}(n)$ be the generator of* $\operatorname{Ker} p$ *of the short exact sequence of Equation* (15.5). *For any oriented two-dimensional subspace P, the spin rotation $\widetilde{R}_P$ satisfies*

$$\widetilde{R}_P^2 = z. \tag{15.8}$$

When u_1, u_2, w are three unit orthogonal vectors in $\mathbb{R}^n$, then

$$\widetilde{R}_{\langle u_1,w\rangle} \cdot \widetilde{R}_{\langle u_2,w\rangle} = z \cdot \widetilde{R}_{\langle u_2,w\rangle} \cdot \widetilde{R}_{\langle u_1,w\rangle}, \tag{15.9}$$

where the plane $\langle u_i, w\rangle$ is oriented so that u_i and w form a positive basis.

When u_1, u_2, w are three unit vectors so that u_1 and u_2 meet at an angle of $\frac{2\pi}{3}$ and w is orthogonal to both u_1 and u_2, then

$$\widetilde{R}_{\langle u_1,w\rangle} \cdot \widetilde{R}_{\langle u_2,w\rangle} \cdot \widetilde{R}_{\langle u_1,w\rangle} = \widetilde{R}_{\langle u_1+u_2,w\rangle}. \tag{15.10}$$

For fixed $i, j \in \{1, \ldots, n\}$, consider the plane P spanned by $v_i - v_j = n(e_i - e_j)$ and $\sum_{k=1}^n e_k$, where v_i are as in Equation (15.4). The 180° rotation around P clearly fixes v_k for $k \neq i, j$, and it switches v_i and v_j. Thus, when $i < j$, this rotation realizes the element $\tau_{i,j}$, thought of as an element of $\mathrm{SO}(n)$. Orient P by taking the ordered basis $v_i - v_j$ and $\sum_{k=1}^n e_k$, and let $\widetilde{\tau}_{i,j}$ be the corresponding element of $\mathrm{Spin}(n)$. The element $\widetilde{\tau}_{j,i}$ corresponds to the same plane, now with the orientation given by $v_j - v_i$ and $\sum_{k=1}^n e_k$. The elements $\widetilde{\tau}_{i,j} \in \widetilde{\mathfrak{S}}_n \subset \mathrm{Spin}(n)$ are called *generalized transpositions.*

PROPOSITION 15.2.5. *The spin extension $\widetilde{\mathfrak{S}}_n$ of $\mathfrak{S}_n$ has by the following group presentation: the generators are $\widetilde{\tau}_{i,j}$ $(1 \leq i \neq j \leq n)$ and z, which are subject to the following relations:*

- *$z^2 = 1$ and z is central, that is, $z \cdot \widetilde{\tau}_{i,j} = \widetilde{\tau}_{i,j} \cdot z$ for all $i \neq j$,*
- *$\widetilde{\tau}_{i,j}^2 = z$ and $\widetilde{\tau}_{i,j} = z \cdot \widetilde{\tau}_{j,i}$ for $1 \leq i \neq j \leq n$,*
- *for $1 \leq i, j, k, l \leq n$ satisfying $\{i, j\} \cap \{k, l\} = \emptyset$,*

$$\widetilde{\tau}_{i,j} \cdot \widetilde{\tau}_{k,l} = z \cdot \widetilde{\tau}_{k,l} \cdot \widetilde{\tau}_{i,j}, \tag{15.11}$$

- *For any three distinct integers $1 \leq i, j, k \leq n$,*

$$\widetilde{\tau}_{i,j} \cdot \widetilde{\tau}_{j,k} \cdot \widetilde{\tau}_{i,j} = \widetilde{\tau}_{j,k} \cdot \widetilde{\tau}_{i,j} \cdot \widetilde{\tau}_{j,k} = \widetilde{\tau}_{i,k}. \tag{15.12}$$

REMARK 15.2.6. Notice that transpositions $\tau_{i,j} \in \mathfrak{S}_n$ are indexed by pairs of integers with $i < j$, while generalized transpositions $\widetilde{\tau}_{i,j}$ are indexed by pairs of integers with $i \neq j$.

Proof. We verify first that the stated relations hold for $\widetilde{\tau}_{i,j}$ and z as defined above. By construction, z is central and it satisfies $z^2 = 1$. The other relations follow from Lemma 15.2.4: specifically, $\widetilde{\tau}_{i,j}^2 = z$ follows from Equation (15.8), the third relation follows from Equation (15.9) (with $u_1 = v_i - v_j$ and $u_2 = v_k - v_l$, $w = \sum_{m=1}^n e_m$), and the fourth relation follows from Equation (15.10) (with $u_1 = v_i - v_j$, $u_2 = v_j - v_k$ and $w = \sum_{m=1}^n e_m$). Since the generators $\widetilde{\tau}_{i,j}$ project to a generating set of $\mathfrak{S}_n$; it follows that $\widetilde{\tau}_{i,j}$ and z generate $\widetilde{\mathfrak{S}}_n$.

Let $\widetilde{\Gamma}_m$ be the group defined with the generators z and $\widetilde{\tau}_{i,j}$ and relations given in the proposition. Since the corresponding generators of $\widetilde{\mathfrak{S}}_n$ satisfy the stated relations, there is a surjection $\widetilde{\Gamma}_n \to \widetilde{\mathfrak{S}}_n$. We will apply the logic of the proof of Proposition 15.2.2 to show that

$$\widetilde{\Gamma}_n \cong \widetilde{\mathfrak{S}}_n. \tag{15.13}$$

The proof is by induction on n. The base case $n = 2$ is easy: $\widetilde{\Gamma}_2 \cong \mathbb{Z}/4\mathbb{Z} \cong \widetilde{\mathfrak{S}}_2$. Next, suppose that $\widetilde{\Gamma}_n \cong \widetilde{\mathfrak{S}}_n$, and consider the group $\widetilde{\Gamma}_{n+1}$. Let $\widetilde{H}$ be the subgroup of $\widetilde{\mathfrak{S}}_{n+1}$ generated by $\widetilde{\tau}_{i,j}$ with $1 \leq i, j \leq n$. Since the generators of $\widetilde{H}$ satisfy the relations from $\widetilde{\mathfrak{S}}_n$, it follows that $\widetilde{H}$ is a quotient of $\widetilde{\mathfrak{S}}_n$; so it is a finite group with $|\widetilde{H}| \leq 2n!$. As in the proof of Proposition 15.2.2, by considering the action of the $\widetilde{\tau}_{i,j}$ on the coset space, we find that $[\widetilde{\Gamma}_{n+1} : \widetilde{H}] \leq n+1$. It follows that $\widetilde{\Gamma}_{n+1}$ is a finite group with at most $2(n+1)!$ elements. Since $\widetilde{\Gamma}_{n+1}$ surjects onto $\widetilde{\mathfrak{S}}_{n+1}$, a group with $2(n+1)!$ elements, we conclude that the surjection is an isomorphism, establishing the inductive step needed to prove Equation (15.13). □

Maps in the central extension

$$1 \to \mathbb{Z}/2\mathbb{Z} \xrightarrow{i} \widetilde{\mathfrak{S}}_n \xrightarrow{p} \mathfrak{S}_n \to 1, \tag{15.14}$$

can be written in terms of the above generating sets: i takes the generator of $\mathbb{Z}/2\mathbb{Z}$ to z, and p is the homomorphism with

$$p(\widetilde{\tau}_{i,j}) = \begin{cases} \tau_{i,j} & \text{if } i < j, \\ \tau_{j,i} & \text{if } j < i. \end{cases}$$

A function $\gamma\colon \mathfrak{S}_n \to \widetilde{\mathfrak{S}}_n$ is called a *section* of $p\colon \widetilde{\mathfrak{S}}_n \to \mathfrak{S}_n$ if $p \circ \gamma = \mathrm{id}_{\mathfrak{S}_n}$. Define

$$\mathcal{S}ec_n = \{\gamma\colon \mathfrak{S}_n \to \widetilde{\mathfrak{S}}_n \mid p \circ \gamma = \mathrm{id}_{\mathfrak{S}_n}\}.$$

REMARK 15.2.7. In the special case where a central extension admits a section that is also a group homomorphism, the extension is called a *split* extension. Already in the case where $n = 2$, $\widetilde{\mathfrak{S}}_n \cong \mathbb{Z}/4\mathbb{Z}$, so the spin central extension is not split.

The proof of Theorem 15.1.5 proceeds by giving a correspondence between sections (whose existence and uniqueness properties are obvious; see Lemma 15.2.8) and sign assignments (see Propositions 15.2.12 and 15.2.13).

LEMMA 15.2.8. *The group of maps $g\colon \mathfrak{S}_n \to \mathbb{Z}/2\mathbb{Z}$ acts transitively and freely on $\mathcal{S}ec_n$.*

Proof. Think of a map $g\colon \mathfrak{S}_n \to \mathbb{Z}/2\mathbb{Z}$ as a map into $\{1, z\}$, the center of $\widetilde{\mathfrak{S}}_n$. With this understanding, a section $\gamma \in \mathcal{S}ec_n$ can be multiplied (pointwise) with the map g to give a new section. This operation gives the required transitive, free group action of the abelian group of maps $\mathfrak{S}_n \to \mathbb{Z}/2\mathbb{Z}$ on the set of sections. □

For any two grid states $\mathbf{x}$ and $\mathbf{y}$ for which the set $\mathrm{Rect}(\mathbf{x},\mathbf{y})$ is non-empty, there is a unique transposition $\tau_{i,j} = \sigma_{\mathbf{x}}^{-1} \cdot \sigma_{\mathbf{y}}$ connecting the states, but there are two rectangles from $\mathbf{x}$ to $\mathbf{y}$. We define now a one-to-one correspondence between the two rectangles and the two lifts $\widetilde{\tau}_{i,j}$ and $\widetilde{\tau}_{j,i} = z \cdot \widetilde{\tau}_{i,j}$ of $\tau_{i,j} \in \mathfrak{S}_n$ in $\widetilde{\mathfrak{S}}_n$:

DEFINITION 15.2.9. For $r \in \mathrm{Rect}(\mathbf{x},\mathbf{y})$, the ***generalized transposition corresponding to*** r, denoted by $\widetilde{\tau}(r)$, is $\widetilde{\tau}_{i,j}$, where the southwest corner of r is on β_i and the northeast corner is on β_j.

Let $\widetilde{T}$ denote the set $\{\widetilde{\tau}_{i,j} | 1 \leq i \neq j \leq n\}$ of generalized transpositions in $\widetilde{\mathfrak{S}}_n$. For a fixed grid state $\mathbf{x} \in \mathbf{S}(\mathbb{G})$, the assignment $r \mapsto \widetilde{\tau}(r)$ induces a bijection

$$\widetilde{\tau}\colon \bigcup_{\mathbf{y}} \mathrm{Rect}(\mathbf{x},\mathbf{y}) \to \widetilde{T}. \tag{15.15}$$

Moreover, if $\mathrm{Rect}(\mathbf{x},\mathbf{y})$ is non-empty, then for the two rectangles $r, r' \in \mathrm{Rect}(\mathbf{x},\mathbf{y})$, $\widetilde{\tau}(r) = z \cdot \widetilde{\tau}(r')$.

LEMMA 15.2.10. *The correspondence* $r \mapsto \widetilde{\tau}(r)$ *from Definition* 15.2.9 *satisfies the following identities:*

- *if* (r_1, r_2) *and* (r_1', r_2') *are alternative pairs of rectangles, then*

$$\widetilde{\tau}(r_1) \cdot \widetilde{\tau}(r_2) = z \cdot \widetilde{\tau}(r_1') \cdot \widetilde{\tau}(r_2'), \tag{15.16}$$

- *if* (r_1, r_2) *is a pair of rectangles forming a horizontal annulus then*

$$\widetilde{\tau}(r_1) \cdot \widetilde{\tau}(r_2) = 1, \tag{15.17}$$

- *if* (r_1, r_2) *is a pair of rectangles forming a vertical annulus then*

$$\widetilde{\tau}(r_1) \cdot \widetilde{\tau}(r_2) = z. \tag{15.18}$$

Proof. Suppose first that for the alternative pairs (r_1, r_2) and (r_1', r_2'), the rectangles r_1 and r_2 do not share sides. Let $\widetilde{\tau}(r_1) = \widetilde{\tau}_{i,j}$ and $\widetilde{\tau}(r_2) = \widetilde{\tau}_{k,\ell}$. Since r_1 and r_2 have distinct sides, $\{i, j\} \cap \{k, l\} = \emptyset$. Since the support of r_1' is the same as the support of r_2, it follows that $\widetilde{\tau}(r_1') = \widetilde{\tau}_{k,\ell}$ and $\widetilde{\tau}(r_2') = \widetilde{\tau}_{i,j}$, as well. In this case, Equation (15.16) is an immediate consequence of the relation in $\widetilde{\mathfrak{S}}_n$ from Equation (15.11).

Next consider the alternative pairs (r_1, r_2) and (r_1', r_2'), where r_1 and r_2 (and hence r_1' and r_2') share a side. Let r_1 correspond to the element $\widetilde{\tau}_{i,j}$. For the further three elements in the alternative pairs there are eight combinatorially distinct possibilities, corresponding to the following choices: either terminal corner of r_1 can be an initial corner for r_2, the rectangles r_1 and r_2 can meet along a horizontal or vertical segment, and this segment can be properly contained in an edge of exactly one of r_1 or r_2. After possibly switching the roles of (r_1, r_2) and (r_1', r_2'), we are left with the four cases illustrated in Figure 15.1. Equation (15.16) in these four cases reduce to the identities:

$$\text{(1)}\quad \widetilde{\tau}_{i,j} \cdot \widetilde{\tau}_{j,k} = z \cdot \widetilde{\tau}_{i,k} \cdot \widetilde{\tau}_{i,j}, \qquad \text{(2)}\quad \widetilde{\tau}_{i,j} \cdot \widetilde{\tau}_{j,k} = z \cdot \widetilde{\tau}_{j,k} \cdot \widetilde{\tau}_{i,k},$$
$$\text{(3)}\quad \widetilde{\tau}_{i,j} \cdot \widetilde{\tau}_{k,j} = z \cdot \widetilde{\tau}_{k,j} \cdot \widetilde{\tau}_{i,k}, \qquad \text{(4)}\quad \widetilde{\tau}_{i,j} \cdot \widetilde{\tau}_{i,k} = z \cdot \widetilde{\tau}_{j,k} \cdot \widetilde{\tau}_{i,j};$$

which in turn follow from the relations $z^2 = 1$, $\widetilde{\tau}_{i,j}^2 = z$, and Equation (15.12).

Suppose now that the two rectangles (r_1, r_2) form a horizontal annulus. Let $r_1 \in \mathrm{Rect}(\mathbf{x}, \mathbf{y})$ correspond to $\widetilde{\tau}_{i,j}$. Then the rectangle $r_2 \in \mathrm{Rect}(\mathbf{y}, \mathbf{x})$ corresponds to the element $\widetilde{\tau}_{j,i} = \widetilde{\tau}_{i,j}^{-1}$, and hence Equation (15.17) follows.

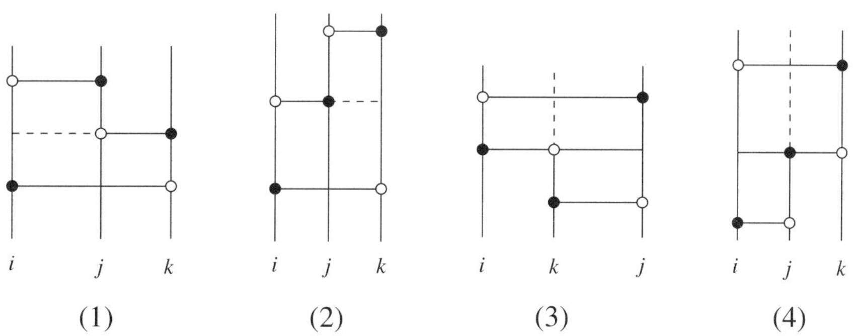

FIGURE 15.1. **The four basic combinatorial possibilities for domains given by composing two rectangles which share a side.**

Suppose finally that the rectangles (r_1, r_2) form a vertical annulus. As before, suppose that $r_1 \in \mathrm{Rect}(\mathbf{x}, \mathbf{y})$ corresponds to $\widetilde{\tau}_{i,j}$. Then the rectangle $r_2 \in \mathrm{Rect}(\mathbf{y}, \mathbf{x})$ also corresponds to $\widetilde{\tau}_{i,j}$. Equation (15.18) follows from the relation $\widetilde{\tau}_{i,j}^2 = z$. □

DEFINITION 15.2.11. A section $\gamma \in \mathcal{S}ec_n$ and a sign assignment S are ***compatible*** if for any pair of grid states $\mathbf{x}, \mathbf{y}$ and any rectangle $r \in \mathrm{Rect}(\mathbf{x}, \mathbf{y})$,

$$\gamma(\sigma_{\mathbf{y}}) = \begin{cases} \gamma(\sigma_{\mathbf{x}}) \cdot \widetilde{\tau}(r) & \text{if } S(r) = +1 \\ z \cdot \gamma(\sigma_{\mathbf{x}}) \cdot \widetilde{\tau}(r) & \text{if } S(r) = -1. \end{cases}$$

Equivalently, thinking of $\{\pm 1\}$ as the subgroup of $\widetilde{\mathfrak{S}}_n$ generated by the element z,

$$\gamma(\sigma_{\mathbf{y}}) = \gamma(\sigma_{\mathbf{x}}) \cdot \widetilde{\tau}(r) \cdot S_\gamma(r). \tag{15.19}$$

PROPOSITION 15.2.12. *Given any section γ, there is a unique sign assignment S compatible with γ. Moreover, if γ and γ' differ by an overall translation by z, then their compatible sign assignments coincide.*

Proof. For a fixed section γ, Equation (15.19) clearly specifies the function $S_\gamma \colon \mathrm{Rect}(\mathbb{G}) \to \{\pm 1\}$ uniquely.

The fact that S_γ is a sign assignment follows from Lemma 15.2.10. For example, suppose that (r_1, r_2) and (r_1', r_2') are alternative pairs of rectangles; let $\mathbf{x}$ be the initial state of r_1 (and r_1') and $\mathbf{z}$ be the terminal state of r_2 (and r_2'). Applying Equation (15.19) four times, we see that

$$\gamma(\sigma_{\mathbf{x}}) \cdot \widetilde{\tau}(r_1) \cdot \widetilde{\tau}(r_2) \cdot S_\gamma(r_1) \cdot S_\gamma(r_2) = \gamma(\sigma_{\mathbf{z}}) = \gamma(\sigma_{\mathbf{x}}) \cdot \widetilde{\tau}(r_1') \cdot \widetilde{\tau}(r_2') \cdot S_\gamma(r_1') \cdot S_\gamma(r_2').$$

The desired relation $S_\gamma(r_1) \cdot S_\gamma(r_2) = -S_\gamma(r_1') \cdot S_\gamma(r_2')$ now follows from Equation (15.16). The remaining two relations for a sign assignment follow similarly from the other two cases of Lemma 15.2.10 (given in Equations (15.17) and (15.18)).

The sign assignment S_γ is compatible with the section γ by construction. It is also clear from the construction that $S_{z \cdot \gamma} = S_\gamma$. □

PROPOSITION 15.2.13. *If S is a sign assignment, then there are exactly two sections that are compatible with S.*

Proof. We will show first that the value of γ at the permutation $\sigma_{\mathbf{x}}$ (for a fixed grid state $\mathbf{x}$), and compatibility with S, determine the value of γ uniquely at all permutations of the form $\sigma_{\mathbf{y}} = \sigma_{\mathbf{x}} \cdot \tau$, where τ is some transposition.

Let r and r' denote the two rectangles in $\mathrm{Rect}(\mathbf{x}, \mathbf{y})$. We claim that for either choice of $\gamma(\sigma_{\mathbf{x}}) \in p^{-1}(\sigma_{\mathbf{x}})$

$$\gamma(\sigma_{\mathbf{x}}) \cdot \widetilde{\tau}(r) \cdot S(r) = \gamma(\sigma_{\mathbf{x}}) \cdot \widetilde{\tau}(r') \cdot S(r'). \tag{15.20}$$

This is true because $\widetilde{\tau}(r) = z \cdot \widetilde{\tau}(r')$, and also $S(r) = -S(r')$, since S is a sign assignment and there is some other rectangle $r'' \in \mathrm{Rect}(\mathbf{y}, \mathbf{x})$ with the property that r and r'' form a vertical annulus and r'' and r' form a horizontal one. Thus, given $\gamma(\sigma_{\mathbf{x}})$, compatibility forces us to define

$$\gamma(\sigma_{\mathbf{y}}) = \gamma(\sigma_{\mathbf{x}}) \cdot \widetilde{\tau}(r) \cdot S(r) = \gamma(\sigma_{\mathbf{x}}) \cdot \widetilde{\tau}(r') \cdot S(r').$$

Since transpositions generate the symmetric group, it follows readily (by moving around the role of $\mathbf{x}$) that the value of γ at $\sigma_{\mathbf{x}}$ (and compatibility with S) determines the value of the section γ at all elements of $\mathfrak{S}_n$. Concretely, given any

$\sigma_{\mathbf{y}} \in \mathfrak{S}_n$, factor $\sigma_{\mathbf{x}}^{-1} \cdot \sigma_{\mathbf{y}}$ into transpositions $\tau_{i_1,j_1} \cdots \tau_{i_m,j_m}$. There are grid states $\mathbf{x} = \mathbf{x}_1, \mathbf{x}_2, \dots, \mathbf{x}_{m+1} = \mathbf{y}$ uniquely determined by the equations $\sigma_{\mathbf{x}_{k+1}} = \sigma_{\mathbf{x}_k} \cdot \tau_{i_k,j_k}$. Choose a sequence of rectangles $r_k \in \mathrm{Rect}(\mathbf{x}_k, \mathbf{x}_{k+1})$ with $k = 1, \dots, m$ and define

$$\gamma(\sigma_{\mathbf{y}}) = \gamma(\sigma_{\mathbf{x}}) \cdot \left(\prod_{k=1}^{m} \widetilde{\tau}(r_i) \right) \cdot \left(\prod_{k=1}^{m} S(r_i) \right). \tag{15.21}$$

Equation (15.20) shows that the right-hand-side is independent of the 2^m choices of rectangles $r_k \in \mathrm{Rect}(\mathbf{x}_k, \mathbf{x}_{k+1})$ for $k = 1, \dots, m$. However, the right-hand-side *a priori* depends on the factorization $\sigma_{\mathbf{x}}^{-1} \cdot \sigma_{\mathbf{y}} = \tau_{i_1,j_1} \cdots \tau_{i_m,j_m}$. So to see that Equation (15.21) is consistent, we must show that the right-hand-side is also independent of the factorization.

Now, according to Proposition 15.2.2, any two factorizations of a fixed element of the symmetric group can be connected by a sequence of the following three moves (and their inverses) (1) insert $\tau_{i,j}^2$ in the sequence, (2) replace consecutive letters in the sequence $\tau_{i,j} \cdot \tau_{k,\ell}$ by $\tau_{k,\ell} \cdot \tau_{i,j}$ (provided $\{i,j\} \cap \{k,\ell\} = \emptyset$), (3) replace $\tau_{i,j} \cdot \tau_{j,k}$ by $\tau_{i,k} \cdot \tau_{i,j}$ (with $1 \le i,j,k \le n$ distinct), or $\tau_{j,k} \cdot \tau_{i,j}$ by $\tau_{i,k} \cdot \tau_{j,k}$ (again, with $1 \le i,j,k \le n$ distinct).

In Case (1), suppose we have two sequences, and one is obtained from the other by inserting $\tau_{i,j}^2$. Lift $\tau_{i,j}^2 = \tau_{i,j} \cdot \tau_{i,j}$ to two rectangles r_1 and r_2, which form a horizontal annulus. It follows at once that $\widetilde{\tau}(r_1) \cdot \widetilde{\tau}(r_2) = 1$ and $S(r_1) \cdot S(r_2) = 1$, and hence (choosing all other rectangles the same) the two possible values of $\gamma(\sigma_{\mathbf{y}})$ specified by Equation (15.21) using the two sequences coincide.

In the other cases, we can find alternative pairs of rectangles (r_1, r_2) and (r_1', r_2'), that correspond to exchanging the products of transpositions. In all these cases,

$$\widetilde{\tau}(r_1) \cdot \widetilde{\tau}(r_2) \cdot S(r_1) \cdot S(r_2) = \widetilde{\tau}(r_1') \cdot \widetilde{\tau}(r_2') \cdot S(r_1') \cdot S(r_2')$$

in view of Lemma 15.2.10. It follows that the two possible values of $\gamma(\sigma_{\mathbf{y}})$ computed on the right-hand-side of Equation (15.21) coincide under this replacement, as well.

Thus, having fixed S and the value $\gamma(\sigma_{\mathbf{x}})$ for some fixed grid state $\mathbf{x}$, Equation (15.21) specifies uniquely a function $\gamma\colon \mathfrak{S}_n \to \widetilde{\mathfrak{S}}_n$. It follows immediately from its definition that γ is a section and it is compatible with S. The two choices for γ correspond to the two possible values of $\gamma(\sigma_{\mathbf{x}})$; and these two sections are easily seen to differ by an overall translation by z. □

Proof of Theorem 15.1.5. For the existence statement, consider a section $\gamma \in \mathcal{S}ec_n$ (which obviously exists) and take the corresponding sign assignment S_γ provided by Proposition 15.2.12.

For the uniqueness statement (up to gauge equivalence), consider two sign assignments S_1 and S_2. Proposition 15.2.13 allows us to choose sections γ_1 and γ_2 so that γ_i is compatible with S_i for $i = 1, 2$. Lemma 15.2.8 gives a function $v\colon \mathfrak{S}_n \to \{\pm 1\}$ with the property that $\gamma_1 = v \cdot \gamma_2$. Under the identification $\mathbf{S}(\mathbb{G}) \cong \mathfrak{S}_n$, v corresponds to a gauge transformation g. The uniqueness statement in Proposition 15.2.12 shows that the sign assignment S_1 compatible with γ_1 coincides with the sign assignment corresponding to $v \cdot \gamma_2$. Since the sign assignment corresponding to $v \cdot \gamma_2$ is equal to S_2^g, it follows that $S_1 = S_2^g$, concluding the proof of uniqueness. □

15.3. The invariance of grid homology over $\mathbb{Z}$

Our aim now is to show that the sign-refined grid homology, thought of as a bigraded module over $\mathbb{Z}[U]$, is a knot invariant. This result was stated in Theorem 15.1.11.

Since we already verified that these groups are independent of the choice of sign assignments (Proposition 15.1.10), by Cromwell's theorem, it remains to verify that grid moves induce isomorphisms on grid homology with integer coefficients. We will start with the invariance under commutation moves, and consider the effect of stabilizations afterwards.

15.3.1. Commutation invariance. Suppose that $\mathbb{G}, \mathbb{G}'$ are two grid diagrams differing by a commutation or a switch, and S is a fixed sign assignment on $\mathbb{G}$, which can be naturally viewed also as a sign assignment on $\mathbb{G}'$.

PROPOSITION 15.3.1. *If $\mathbb{G}$ and $\mathbb{G}'$ are two grid diagrams which differ by a commutation or a switch, and S is a sign assignment on $\mathbb{G}$ and on $\mathbb{G}'$, then there is an isomorphism of bigraded $\mathbb{Z}[U]$-modules*

$$H(GC^-(\mathbb{G};\mathbb{Z}), \partial^-_{\mathbb{X},S}) \cong H(GC^-(\mathbb{G}';\mathbb{Z}), \partial^-_{\mathbb{X}',S}).$$

In the proof we will use a sign-refined version of the pentagon counting map of Section 5.1, so we need to explain how to associate a sign to a pentagon. We focus on the case of column commutations, and return to the modifications needed for row commutations at the end of this subsection.

For fixed grid states $\mathbf{x} \in \mathbf{S}(\mathbb{G})$ and $\mathbf{y}' \in \mathbf{S}(\mathbb{G}')$, and a pentagon $p \in \text{Pent}(\mathbf{x}, \mathbf{y}')$, there is a naturally associated rectangle $R(p) \in \text{Rect}(\mathbf{x}, \mathbf{y})$, where $\mathbf{y} = I(\mathbf{y}')$ (as in Equation (5.3)). The rectangle $R(p)$ is determined by the property that all of its local multiplicities away from the two bigons (between β_i and γ_i) agree with those of p. Informally, this map straightens the pentagon p into the rectangle $R(p)$, cf. Figure 15.2. Given $p' \in \text{Pent}(\mathbf{y}', \mathbf{x})$, there is a similarly defined $R'(p') \in \text{Rect}(\mathbf{y}', \mathbf{x}')$, where $I(\mathbf{x}') = \mathbf{x}$.

Note that for any pentagon $p \in \text{Pent}(\mathbf{x}, \mathbf{y}')$, the associated rectangle $R(p)$ contains a segment of β_i on its boundary, so it lies either to the right or to the left of β_i. We call such pentagons *right pentagons* or *left pentagons*, respectively.

DEFINITION 15.3.2. Fix $p \in \text{Pent}(\mathbf{x}, \mathbf{y}')$. The ***sign of*** p, denoted $S_{pent}(p)$, is $(-1)^{M(\mathbf{x})+1} S(R(p))$ if p is a left pentagon and $(-1)^{M(\mathbf{x})} S(R(p))$ if p is right.

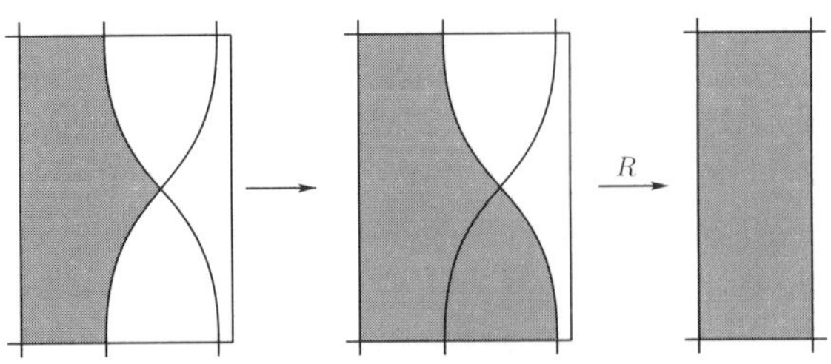

FIGURE 15.2. **Straightening a pentagon to a rectangle.** By this operation, we can apply the sign assignments defined for rectangles and (after suitable modification) we can equip pentagons with signs.

The $\mathcal{R}_{\mathbb{Z}}$-module map $P_S\colon GC_S^-(\mathbb{G};\mathbb{Z}) \to GC_S^-(\mathbb{G}';\mathbb{Z})$ is specified by its value on any grid state $\mathbf{x} \in \mathbf{S}(\mathbb{G})$ by the formula

$$P_S(\mathbf{x}) = \sum_{\mathbf{y}'\in\mathbf{S}(\mathbb{G}')} \sum_{\{p\in\mathrm{Pent}^\circ(\mathbf{x},\mathbf{y}')|p\cap\mathbb{X}=\emptyset\}} S_{pent}(p)\cdot V_1^{O_1(p)}\cdots V_n^{O_n(p)}\cdot\mathbf{y}'. \tag{15.22}$$

LEMMA 15.3.3. *The map $P_S\colon GC_S^-(\mathbb{G};\mathbb{Z}) \to GC_S^-(\mathbb{G}';\mathbb{Z})$ is a chain map.*

Proof. Let $\partial^-_{\mathbb{X},S}$ and $\partial^-_{\mathbb{X}',S}$ be the differentials for $GC_S^-(\mathbb{G};\mathbb{Z})$ and $GC_S^-(\mathbb{G}';\mathbb{Z})$, respectively. In the verification of the claimed identity $P_S\circ\partial^-_{\mathbb{X},S} = \partial^-_{\mathbb{X}',S}\circ P_S$, we need to analyze the interaction of pentagons and rectangles — now with signs. As in the proof of Lemma 5.1.4, we distinguish three cases:

The analogue of (P-1) of Lemma 5.1.4 works with minor changes, and provides the desired result. Indeed, the domain of the composition $r * p$ admits a unique alternate decomposition as $p' * r'$, and the corresponding pairs $(r, R(p))$ and $(R'(p'), r')$ provide alternative pairs of rectangles, hence we get

$$S(r)\cdot S(R(p)) + S(R'(p'))\cdot S(r') = 0.$$

Since the initial grid states of p and p' have Maslov gradings of different parity, and the pentagons are either both right pentagons or both left pentagons, it follows that the contributions cancel in $P_S\circ\partial^-_{\mathbb{X},S} - \partial^-_{\mathbb{X}',S}\circ P_S$, proving the desired equality.

The analogue of (P-2) of Lemma 5.1.4 gives two different decompositions of the same domain as a pentagon and a rectangle (possibly in either order). After straightening the pentagons in the decomposition, either we get alternative pairs of rectangles or we get the same two pairs of rectangles. When straightening the pentagons gives alternative pairs of rectangles, both pentagons are on the same side, and hence their contributions cancel in $P_S\circ\partial^-_{\mathbb{X},S} - \partial^-_{\mathbb{X}',S}\circ P_S$. When straightening the pentagons gives the same two rectangles, the decompositions have the form $r*p$ and $p'*r'$, and the two pentagons p and p' are on opposite sides of the commutation (i.e. one is a left pentagon and the other is a right pentagon; see Figure 15.3 for an example); so their contributions to $P_S\circ\partial^-_{\mathbb{X},S} - \partial^-_{\mathbb{X}',S}\circ P_S$ cancel.

In Case (P-3)(h), the domain has two decompositions, of the form $r * p$ and $p' * r'$, whose straightenings give the same decomposition of a horizontal annulus.

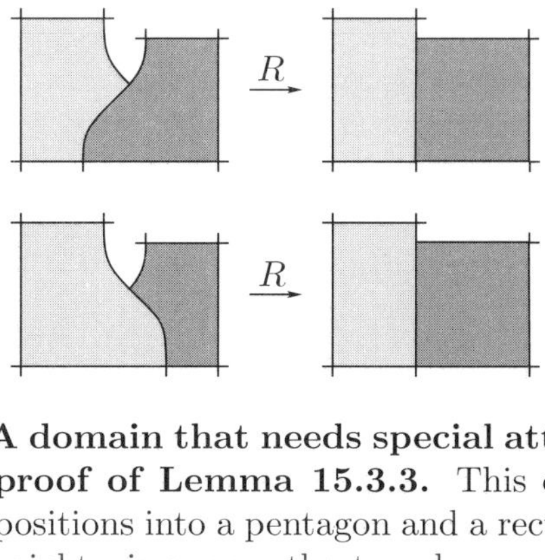

FIGURE 15.3. **A domain that needs special attention in case (P-2) in the proof of Lemma 15.3.3.** This domain has two different decompositions into a pentagon and a rectangle, but after applying the straightening map, the two decompositions become the same. In this case the two pentagons are necessarily on different sides of the commutation.

Since the pentagons are on opposite sides of the commutation, the contributions to $P_S \circ \partial^-_{\mathbb{X},S} - \partial^-_{\mathbb{X}',S} \circ P_S$ cancel.

Finally, in the analogue of (P-3)(v) of Lemma 5.1.4, the proof of that lemma shows that if the domain $p * r$ (or $r * p$) admits a unique decomposition as a juxtaposition of a rectangle and a pentagon, then there is a unique similar domain with a unique decomposition $p' * r'$ or $r' * p'$, now on the other side of the commutation. Observe that, for the two juxtapositions, the straightening of the pentagon and the rectangle gives decompositions of two vertical annuli, and hence the product of the signs of their straightenings are the same. It follows now that the contributions to $P_S \circ \partial^-_{\mathbb{X},S} - \partial^-_{\mathbb{X}',S} \circ P_S$, once again, cancel. □

The same basic idea provides a map from $GC^-_S(\mathbb{G}';\mathbb{Z})$ to $GC^-_S(\mathbb{G};\mathbb{Z})$ as follows. Given a pentagon $p \in \mathrm{Pent}(\mathbf{x}',\mathbf{y})$, its induced sign $S_{pent}(p)$ is defined as in Definition 15.3.2; and define $P'_S\colon GC^-_S(\mathbb{G}';\mathbb{Z}) \to GC^-_S(\mathbb{G};\mathbb{Z})$ as in Equation (15.22). The argument from Lemma 15.3.3 shows that P'_S is a chain map. The proof of Lemma 5.1.3 shows that the maps P_S and P'_S both preserve the Maslov and Alexander gradings.

In order to complete the proof of Proposition 15.3.1, we show that P_S and $-P'_S$ are homotopy inverses of one another. As in Section 5.1, the homotopies count hexagons, but now they do so as follows. To associate signs to hexagons, observe that to each hexagon $h \in \mathrm{Hex}(\mathbf{x},\mathbf{y})$ for $\mathbf{x},\mathbf{y} \in \mathbf{S}(\mathbb{G})$ (in the sense of Definition 5.1.5), there is a naturally associated rectangle $R(h) \in \mathrm{Rect}(\mathbf{x},\mathbf{y})$, whose local multiplicities away from the two bigons agree with those of h. This operation induces a one-to-one correspondence between hexagons and rectangles that have a boundary arc along β_i, which crosses the marked points b and a in that order.

With this understood, we define $S_{hex}(h) = S(R(h))$, and consider

$$H_S(\mathbf{x}) = \sum_{\mathbf{y}\in\mathbf{S}(\mathbb{G})} \sum_{\{h\in\mathrm{Hex}^\circ(\mathbf{x},\mathbf{y}) | h\cap\mathbb{X}=\emptyset\}} S_{hex}(h) \cdot V_1^{O_1(h)} \cdots V_n^{O_n(h)} \cdot \mathbf{y}. \tag{15.23}$$

LEMMA 15.3.4. *The maps*

$$P_S\colon GC^-_S(\mathbb{G};\mathbb{Z}) \to GC^-_S(\mathbb{G}';\mathbb{Z})$$

and

$$-P'_S\colon GC^-_S(\mathbb{G}';\mathbb{Z}) \to GC^-_S(\mathbb{G};\mathbb{Z})$$

are homotopy inverses of one another.

Proof. We show that the map of Equation (15.23) provides a chain homotopy between $-P'_S \circ P_S$ and $\mathrm{Id}_{GC^-_S(\mathbb{G};\mathbb{Z})}$; that is,

$$\mathrm{Id}_{GC^-_S(\mathbb{G};\mathbb{Z})} + P'_S \circ P_S + H_S \circ \partial^-_{\mathbb{X},S} + \partial^-_{\mathbb{X},S} \circ H_S = 0. \tag{15.24}$$

The terms appearing in $P'_S \circ P_S + H_S \circ \partial^-_{\mathbb{X},S} + \partial^-_{\mathbb{X},S} \circ H_S$ are in one-to-one correspondence with decompositions of domains consisting of two rectangles, where the composite domain contains an arc in β_i along its boundary that crosses both b and a in that order. (This correspondence is set up using the straightening maps for pentagons and hexagons; for example, for a domain $p' * p$ counted in $P'_S \circ P_S$, the corresponding domain is $R(p') * R(p)$.)

Grouping juxtapositions according to their underlying domains when $N(\psi) = 2$ (as in the proof of Lemma 5.1.6), there are two cases: the two decompositions give alternative pairs of rectangles after straightening or the two decompositions give the same pair of rectangles. In the first case there are either 0 or 2 pentagons, and they appear on the same side. In the second case, each decomposition contains one pentagon, and the two pentagons are on opposite sides. It is easy to see that in both cases, the signs cancel.

The remaining term with $N(\psi) \neq 0$ corresponds to the thin vertical annulus, which has a unique decomposition, and its contribution is cancelled by the identity map appearing in Equation (15.24).

The fact that $-P_S \circ P'_S$ is homotopic to the identity is verified analogously, now counting hexagons that connect $\mathbf{x}', \mathbf{y}' \in \mathbf{S}(\mathbb{G}')$. □

Proof of Proposition 15.3.1. Suppose that $\mathbb{G}$ and $\mathbb{G}'$ differ by a commutation of two columns. According to Lemma 15.3.3 the maps P_S and $-P'_S$ are chain maps; and by Lemma 15.3.4, they are chain homotopy inverses of each other. Thus, they both induce isomorphisms on homology.

In the case of row commutations, sign assignments for pentagons (Definition 15.3.2) use the notions of "up pentagons" and "down pentagons" in place of "left pentagons" and "right pentagons"; i.e. the sign of a pentagon is the sign of its straightened rectangle times $(-1)^{M(\mathbf{x})}$ for an up and $(-1)^{M(\mathbf{x})+1}$ for a down pentagon. These sign assignments are used to define maps $P_S \colon GC^-_S(\mathbb{G};\mathbb{Z}) \to GC^-_S(\mathbb{G}';\mathbb{Z})$ and $P'_S \colon GC^-_S(\mathbb{G}';\mathbb{Z}) \to GC^-_S(\mathbb{G};\mathbb{Z})$ for row commutations. Adapting the proof of Lemma 15.3.4, we verify that these maps are homotopy inverses of one another. The signs in the analogue of Equation (15.24) are modified, as follows:

$$\mathrm{Id}_{GC^-_S(\mathbb{G};\mathbb{Z})} - P'_S \circ P_S + H_S \circ \partial^-_{\mathbb{X},S} + \partial^-_{\mathbb{X},S} \circ H_S = 0.$$

With these remarks in place, it follows that the sign refined grid homology is invariant under row commutations.

Since the argument works for switches equally well, the proof of the proposition is complete. □

15.3.2. Stabilization invariance. Suppose that $\mathbb{G}$ is a given grid diagram and $\mathbb{G}'$ is a grid diagram obtained by a stabilization on $\mathbb{G}$. Let $S_{\mathbb{G}'}$ be a sign assignment on $\mathbb{G}'$. This sign assignment can be restricted to give a sign assignment on $\mathbb{G}$ as follows. Recall from Section 5.2 that there is a one-to-one correspondence between generators $\mathbf{x} \in \mathbf{S}(\mathbb{G})$ and generators $\mathbf{x}' \in \mathbf{I}(\mathbb{G}') \subset \mathbf{S}(\mathbb{G}')$ with $c \in \mathbf{x}'$. Similarly, rectangles $r \in \mathrm{Rect}(\mathbf{x},\mathbf{y})$ for $\mathbb{G}$ are in natural one-to-one correspondence with rectangles $r' \in \mathrm{Rect}(\mathbf{x}',\mathbf{y}')$ connecting $\mathbf{x}', \mathbf{y}' \in \mathbf{I}(\mathbb{G}')$ in $\mathbb{G}'$. Define $S_{\mathbb{G}}(r)$ to be $S_{\mathbb{G}'}(r')$. It is easy to see that the function $S_{\mathbb{G}} \colon \mathrm{Rect}(\mathbb{G}) \to \mathbb{Z}$ defined in this way is a sign assignment. For simplicity, we will denote both $S_{\mathbb{G}'}$ and $S_{\mathbb{G}}$ by S.

PROPOSITION 15.3.5. *Suppose that $\mathbb{G}'$ is given by a stabilization of the grid diagram $\mathbb{G}$. Let S be a fixed sign assignment on $\mathbb{G}'$, inducing a sign assignment (also denoted by S) on $\mathbb{G}$. Then, the homology groups of $(GC^-(\mathbb{G};\mathbb{Z}), \partial^-_S)$ and $(GC^-(\mathbb{G}';\mathbb{Z}), \partial^-_S)$ are isomorphic as bigraded $\mathbb{Z}[U]$-modules.*

Proof. The proof of Proposition 5.2.1 carries over with minor modifications. For instance, in the case of stabilizations of type *X:SW*, there is a quotient complex $\mathbf{I} \subset GC^-_S(\mathbb{G}';\mathbb{Z})$ generated by the grid states in $\mathbf{I}(\mathbb{G}')$, with induced differential

$\partial^{\mathbf{I}}_{\mathbf{I},S}$. We can replace Lemma 5.2.18 by the statement that the natural one-to-one correspondence between $\mathbf{I}(\mathbb{G}')$ and $\mathbf{S}(\mathbb{G})$ induces an isomorphism

$$e_{\mathbb{Z}}\colon (\mathbf{I},\partial^{\mathbf{I}}_{\mathbf{I},S}) \to GC^-_S(\mathbb{G};\mathbb{Z})[V_1][\![1,1]\!]$$

of bigraded chain complexes. This statement in turn follows immediately from our definition of the induced sign assignment on $\mathbb{G}$.

Note that the stabilization invariance proof follows from the homotopy formula connecting multiplication by V_1 with multiplication by V_2. For instance, in the proof of Proposition 5.2.17 (again, in the case of $X{:}SW$ stabilizations), there was a commutative square (Equation (5.22)), which is now replaced by the following diagram

$$\begin{array}{ccc} (\mathbf{I},\partial^{\mathbf{I}}_{\mathbf{I},S}) & \xrightarrow{\partial^{\mathbf{N}}_{\mathbf{I},S}} & (\mathbf{N},\partial^{\mathbf{N}}_{\mathbf{N},S}) \\ \downarrow e_{\mathbb{Z}} & & \downarrow e_{\mathbb{Z}}\circ\mathcal{H}^{\mathbf{I}}_{X_2,S} \\ (GC^-_S(\mathbb{G};\mathbb{Z})[V_1][\![1,1]\!],-\partial^-_{\mathbb{X}}) & \xrightarrow{V_1-V_2} & (GC^-_S(\mathbb{G};\mathbb{Z})[V_1],\partial^-_{\mathbb{X}}) \end{array} \tag{15.25}$$

where the operator $\mathcal{H}^{\mathbf{I}}_{X_2,S}$ is the natural modification of $\mathcal{H}^{\mathbf{I}}_{X_2}$, incorporating the sign assignment in the usual manner:

$$\mathcal{H}^{\mathbf{I}}_{X_2,S}(\mathbf{x}) = \sum_{\mathbf{y}\in\mathbf{I}(\mathbb{G}')}\ \sum_{\{r\in\mathrm{Rect}^\circ(\mathbf{x},\mathbf{y})\,\big|\,r\cap\mathbb{X}=X_2\}} S(r)\cdot V_1^{O_1(r)}\cdots V_n^{O_n(r)}\cdot\mathbf{y}. \tag{15.26}$$

This square is not a commutative square: rather it is a chain complex, in the following sense. Each of the edge maps anti-commutes with the differential, and the two compositions of the edge maps add up to zero (rather than being equal). This latter fact follows from Equation (15.3).

Anti-commutativity of the square can be interpreted as the statement that the map

$$D_S\colon GC^-_S(\mathbb{G}') \to \mathrm{Cone}(V_1-V_2).$$

defined by

$$D_S(\mathbf{x}) = (-1)^{M(\mathbf{x})}(e_{\mathbb{Z}}(\mathbf{x}), e_{\mathbb{Z}}\circ\mathcal{H}^{\mathbf{I}}_{X_2,S}(\mathbf{x})) \tag{15.27}$$

is a chain map.

The first vertical arrow in Equation (15.25) induces a quasi-isomorphism; the second vertical map is also a quasi-isomorphism by Equation (15.3); so the map D_S is a quasi-isomorphism by Lemma 5.2.12.

Any stabilization can be reduced to one of type $X{:}SW$ by commutations and switches, so the general case is reduced to the above case via Proposition 15.3.1. □

DEFINITION 15.3.6. The map defined in Equation (15.27), giving a quasi-isomorphism from $GC^-_S(\mathbb{G}')$ to the sign-refined mapping cone of the chain map

$$V_1-V_2\colon GC^-_S(\mathbb{G})[V_1] \to GC^-_S(\mathbb{G})[V_1],$$

is called the ***sign-refined destabilization map***.

The sign-refined destabilization map is the chain map induced from the diagram in Equation (15.25). The proof of the above proposition in fact shows that the sign-refined destabilization map is a quasi-isomorphism of $\mathbb{Z}[V_2,\dots,V_n]$-modules, and in particular it induces an isomorphism of $\mathbb{Z}[U]$-modules in homology.

15.3.3. Completion of the invariance proof. As usual, the invariance proof is given in two steps: first, proving that grid homology is an oriented knot invariant, and then observing that it is independent of the orientation.

PROPOSITION 15.3.7. *The grid homology module $GH^-(\mathbb{G};\mathbb{Z})$, as a bigraded module over $\mathbb{Z}[U]$, depends on the grid $\mathbb{G}$ only through its underlying oriented knot $\vec{K}$.*

Proof. Recall that the bigraded isomorphism type of $GC^-_S(\mathbb{G};\mathbb{Z})$ is independent of the choice of sign assignment, by Proposition 15.1.10 (so we suppress S from the notation). Lemma 15.1.9 shows that the quasi-isomorphism type of $GC^-(\mathbb{G};\mathbb{Z})$, as a module over $\mathbb{Z}[U]$, is independent of the choice of V_i. Thus, invariance of the bigraded grid homology module $GH^-(\mathbb{G};\mathbb{Z})$ follows from Cromwell's theorem (Theorem 3.1.9), together with the fact that the module is invariant under commutations (by Propositions 15.3.1), and stabilizations (by Proposition 15.3.5). □

Independence of orientation is verified in the following:

PROPOSITION 15.3.8. *The grid homology $GH^-(K;\mathbb{Z})$ is independent of the choice of orientation on K.*

Proof. Consider a grid diagram $\mathbb{G}$ representing the oriented knot $\vec{K}$. As in Proposition 5.3.2, the grid $\mathbb{G}'$ obtained by reflecting $\mathbb{G}$ across the diagonal provides a diagram for $-\vec{K}$. The map ϕ on grid states induced by this reflection induces a module isomorphism $GC^-(\mathbb{G};\mathbb{Z}) \to GC^-(\mathbb{G}';\mathbb{Z})$. The map ϕ can be used to identify $\text{Rect}^\circ(\mathbf{x},\mathbf{y})$ with $\text{Rect}^\circ(\phi(\mathbf{x}),\phi(\mathbf{y}))$ for any pair of grid states $\mathbf{x},\mathbf{y} \in \mathbf{S}(\mathbb{G})$. We want to define a sign assignment for $\mathbb{G}'$ by pulling back the sign assignment S of $\mathbb{G}$ by this identification. The pull-back, however, is not quite a sign assignment: it does satisfy the first property of Definition 15.1.2, but since reflection switches horizontal and vertical, for rectangles forming an annulus, the product of values of the pull-back are the opposite to what they would be for a sign assignment. To remedy this, consider instead the function S' on the rectangles of $\mathbb{G}'$

$$S'(\phi(r)) = (-1)^{\mathbb{X}(r)+\#(\text{Int}(r)\cap\mathbf{x})} \cdot S(r),$$

for all $r \in \text{Rect}(\mathbf{x},\mathbf{y})$. It is easy to see that S' is a sign assignment on $\mathbb{G}'$, and hence that the function $\mathbf{x} \mapsto \phi(\mathbf{x})$ defines an isomorphism between the chain complexes $GC^-_S(\mathbb{G};\mathbb{Z})$ and $GC^-_{S'}(\mathbb{G}';\mathbb{Z})$, concluding the proof. □

Proof of Theorem 15.1.11. Invariance of $GH^-(\mathbb{G};\mathbb{Z})$ is an immediate consequences of Propositions 15.3.7 and 15.3.8.

Since invariance of GH^- is proved by a sequence of quasi-isomorphisms, invariance of $\widehat{GH}$ follows formally: a quasi-isomorphism between two free chain complexes over $\mathbb{Z}[U]$ induces a quasi-isomorphism between their $U=0$ specializations, by the five lemma. (See Proposition A.3.5 for more details.) □

EXERCISE 15.3.9. **(a)** Compute $\widehat{GH}(T_{2,3};\mathbb{Z})$ and $GH^-(T_{2,3};\mathbb{Z})$.
(b)* Let K be an alternating knot. Express the modules $\widehat{GH}(K;\mathbb{Z})$ and $GH^-(K;\mathbb{Z})$ in terms of the Alexander polynomial and the signature of K.

15.4. Invariance in the filtered theory

We construct the sign-refined version of the filtered grid invariant from Chapter 13.

Fix a grid diagram $\mathbb{G}$ representing an oriented knot $\vec{K}$ and a sign assignment S on it. Let $\mathcal{GC}^-_S(\mathbb{G})$ be the free $\mathbb{Z}[V_1,\dots,V_n]$-module generated by grid states, equipped with the Maslov grading and the Alexander filtration of Chapter 13, and consider the $\mathbb{Z}[V_1,\dots,V_n]$-module endomorphism specified on grid states $\mathbf{x}\in\mathbf{S}(\mathbb{G})$ by

$$\partial^-_S \mathbf{x} = \sum_{\mathbf{y}\in\mathbf{S}(\mathbb{G})}\ \sum_{r\in\mathrm{Rect}^\circ(\mathbf{x},\mathbf{y})} S(r)\cdot V_1^{O_1(r)}\cdots V_n^{O_n(r)}\cdot\mathbf{y} \tag{15.28}$$

(the lift of Equation (13.1)).

LEMMA 15.4.1. *The operator ∂^-_S satisfies $\partial^-_S\circ\partial^-_S=0$; i.e. the pair $\mathcal{GC}^-_S(\mathbb{G};\mathbb{Z})=(GC^-(\mathbb{G};\mathbb{Z}),\partial^-_S)$ is a chain complex.*

Proof. The proof of the lemma follows the same line of reasoning as the proof of Lemma 15.1.7. Indeed, for Cases (R-1) and (R-2) of Lemma 4.6.7 the same argument applies.

Now, however, we must consider also Case (R-3) (in the notation of Lemma 13.2.2). In the proof of Lemma 13.2.2, the contributions to $(\partial^-)^2$ cancel mod 2, since the contribution from the vertical thin annulus through an $\mathbb{O}$-marking O_i cancels with the contribution of the horizontal thin annulus through the same marking. This cancellation also holds over the integers, by the axioms of a sign assignment. □

Actions by the various V_i variables corresponding to different O-markings are filtered chain homotopic to one another, by the following straigthforward generalization of Lemma 15.1.9, verified using the homotopy operator:

$$\mathcal{H}_{i,S}(\mathbf{x}) = \sum_{\mathbf{y}\in\mathbf{S}(\mathbb{G})}\ \sum_{\{r\in\ \mathrm{Rect}^\circ(\mathbf{x},\mathbf{y})\,\big|\,X_i\in r\cap\mathbb{X}\}} S(r)\cdot V_1^{O_1(r)}\cdots V_n^{O_n(r)}\cdot\mathbf{y}. \tag{15.29}$$

View the filtered chain complex $\mathcal{GC}^-_S(\mathbb{G};\mathbb{Z})$, as a module over $\mathbb{Z}[U]$, where multiplication by U is defined to be multiplication by some fixed V_i. In the rest of the section we will show that the filtered quasi-isomorphism type of this chain complex depends only on $\vec{K}$, and is independent of the chosen sign assignment and the grid diagram representing $\vec{K}$.

THEOREM 15.4.2. *Let $\vec{K}$ be an oriented knot and $\mathbb{G}$ be a grid diagram that represents $\vec{K}$. The filtered quasi-isomorphism type of $\mathcal{GC}^-_S(\mathbb{G};\mathbb{Z})$, thought of as a $\mathbb{Z}$-filtered, $\mathbb{Z}$-graded chain complex over $\mathbb{Z}[U]$, is an invariant of the oriented knot $\vec{K}$. Similarly, the filtered quasi-isomorphism type of $\widehat{\mathcal{GC}}_S(\mathbb{G};\mathbb{Z})$, thought of as a $\mathbb{Z}$-filtered, $\mathbb{Z}$-graded chain complex, is an invariant of the oriented knot $\vec{K}$.*

The filtered chain complex over the integers is independent of the choice of the sign assignment: for two sign assignments S_1 and S_2 the map f of Proposition 15.1.10 induces an isomorphism between $\mathcal{GC}^-_{S_1}(\mathbb{G};\mathbb{Z})$ and $\mathcal{GC}^-_{S_2}(\mathbb{G};\mathbb{Z})$.

Independence from the choice of the grid diagram follows from an adaptation of our earlier arguments with coefficients in $\mathbb{F}=\mathbb{Z}/2\mathbb{Z}$: we verify commutation invariance first and then the rather more involved stabilization invariance.

PROPOSITION 15.4.3. *If $\mathbb{G}, \mathbb{G}'$ are two grid diagrams differing by a commutation or a switch, and S is a fixed sign assignment on $\mathbb{G}$ (and hence on $\mathbb{G}'$), then $\mathcal{GC}^-_S(\mathbb{G};\mathbb{Z})$ and $\mathcal{GC}^-_S(\mathbb{G}';\mathbb{Z})$ are filtered quasi-isomorphic chain complexes.*

Using the extension of a sign assignment to pentagons (as described in Definition 15.3.2) we get the map

$$\mathcal{P}_S\colon \mathcal{GC}^-_S(\mathbb{G};\mathbb{Z}) \to \mathcal{GC}^-_S(\mathbb{G}';\mathbb{Z})$$

specified on the grid state $\mathbf{x} \in \mathbf{S}(\mathbb{G})$ by the formula

$$\mathcal{P}_S(\mathbf{x}) = \sum_{\mathbf{y}'\in\mathbf{S}(\mathbb{G}')} \sum_{p\in\mathrm{Pent}^\circ(\mathbf{x},\mathbf{y}')} S_{pent}(p)\cdot V_1^{O_1(p)}\cdots V_n^{O_n(p)}\cdot \mathbf{y}'.$$

Proof of Proposition 15.4.3. A simple adaptation of Lemma 15.3.3 shows that $\mathcal{P}_S$ is a chain map. The proof of Lemma 13.3.1 shows that $\mathcal{P}_S$ respects the grading and the filtration; and the argument from Lemma 13.3.3 shows that $\mathcal{P}_S$ induces the map P_S on the associated graded object, which in turn is a quasi-isomorphism, according to Lemma 15.3.4. □

Next we turn to the invariance under stabilizations. Suppose that the grid diagram $\mathbb{G}'$ is given as a stabilization of type *X:SW* on $\mathbb{G}$ and S is a sign assignment on $\mathbb{G}'$ inducing a sign assignment (also denoted by S) on $\mathbb{G}$.

LEMMA 15.4.4. *Suppose that the grids $\mathbb{G}, \mathbb{G}'$ and the sign assignment S are as above. Then, the filtered chain complex $\mathcal{GC}^-_S(\mathbb{G}';\mathbb{Z})$ is quasi-isomorphic to $\mathcal{GC}^-_S(\mathbb{G};\mathbb{Z})$.*

For the proof of this result, we define a sign-refined destabilization map

$$\mathcal{D}_S\colon \mathcal{GC}^-_S(\mathbb{G}';\mathbb{Z}) \to (\mathit{Cone}_S(V_1 - V_2);\mathbb{Z}),$$

and we check that it is a filtered quasi-isomorphism. The key point here is to associate signs to the destabilization domains $p \in \pi^{iL}(\mathbf{x},\mathbf{y}) \cup \pi^{iR}(\mathbf{x},\mathbf{y})$ from Definition 13.3.7.

DEFINITION 15.4.5. Let p be a destabilization domain. The ***canonical decomposition*** of a domain p is a factorization $p = r_1 * \cdots * r_{k-1}$ into empty rectangles with the property that each rectangle r_j meets the distinguished (destabilized) circle β_i in the side of the rectangle.

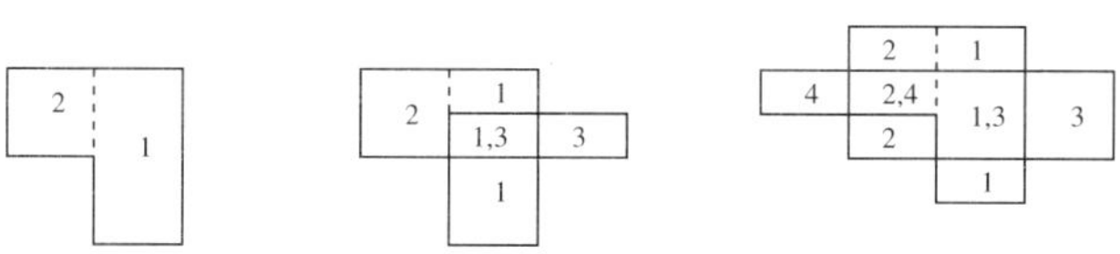

FIGURE 15.4. **Canonical decompositions.** We have illustrated here canonical decompositions of destabilization domains with complexity 3, 4, and 5. The rectangles in the decomposition are labelled by their order, and each region is marked by a sequence of integers representing the rectangles that contain the region; thus, the local multiplicity of p at a region is the number of labels it has.

The existence and uniqueness of canonical decompositions was verified in Lemma 13.3.11. If p is a destabilization domain, define the sign of p as follows: $S(p) = 1$ if p has complexity 1, and if p has complexity $k > 1$,

$$S(p) = \prod_{i=1}^{k-1} S(r_i), \tag{15.30}$$

where $p = r_1 * \cdots * r_{k-1}$ is its canonical decomposition.

DEFINITION 15.4.6. Fix a sign assignment S for $\mathbb{G}'$ and extend the sign assignment to destabilization domains as above. The maps

$$\mathcal{D}_S^{iL}(\mathbf{x}) = \sum_{\mathbf{y}\in\mathbf{I}(\mathbb{G}')} \sum_{p\in\pi^{iL}(\mathbf{x},\mathbf{y})} (-1)^{M(\mathbf{x})} \cdot S(p) \cdot V_2^{O_2(p)} \cdots V_n^{O_n(p)} \cdot e(\mathbf{y})$$

$$\mathcal{D}_S^{iR}(\mathbf{x}) = \sum_{\mathbf{y}\in\mathbf{I}(\mathbb{G}')} \sum_{p\in\pi^{iR}(\mathbf{x},\mathbf{y})} (-1)^{M(\mathbf{x})} \cdot S(p) \cdot V_2^{O_2(p)} \cdots V_n^{O_n(p)} \cdot e(\mathbf{y}),$$

are components of the ***sign-refined destabilization map***

$$\mathcal{D}_S\colon \mathcal{GC}_S^-(\mathbb{G}';\mathbb{Z}) \to \mathcal{C}one_S(V_1 - V_2)$$

$$\mathcal{D}_S(\mathbf{x}) = (\mathcal{D}_S^{iL}(\mathbf{x}), \mathcal{D}_S^{iR}(\mathbf{x})) \in \mathcal{GC}_S^-(\mathbb{G};\mathbb{Z})[V_1][\![1,1]\!] \oplus \mathcal{GC}_S^-(\mathbb{G};\mathbb{Z})[V_1] \cong \mathcal{C}one_S(V_1 - V_2),$$

where the latter is an isomorphism of filtered modules.

We would like to verify that $\mathcal{D}_S$ is a chain map. To this end, it is useful to have the following:

DEFINITION 15.4.7. Let r be a rectangle, and suppose that r can be realized as the juxtaposition of three rectangles $r = r_1 * r_2 * r_3$; in this case, one of the edges of r_2 is the union of an edge e of r_1 and an edge of r_3. If the edge e is horizontal resp. vertical, we call the factorization $r = r_1 * r_2 * r_3$ of ***horizontal type*** resp. of ***vertical type***. For examples of such factorizations, see Figure 15.5.

LEMMA 15.4.8. *Suppose that r factors into three rectangles $r = r_1 * r_2 * r_3$. If the factorization is of horizontal type, then $S(r) = -S(r_1) \cdot S(r_2) \cdot S(r_3)$, and if the factorization is of vertical type, then $S(r) = S(r_1) \cdot S(r_2) \cdot S(r_3)$.*

Proof. Suppose that e is horizontal. Choose r_4 so that r and r_4 form a horizontal annulus, so that $S(r) \cdot S(r_4) = 1$, and consider the juxtaposition of rectangles $r_1 * r_2 * r_3 * r_4$. Replace (r_3, r_4) by the alternative pair (r_3', r_4'), then notice that (r_2, r_3') and (r_1, r_4') form horizontal annuli; thus $S(r_1) \cdot S(r_2) \cdot S(r_3) \cdot S(r_4) = -1$, verifying that $S(r) = -S(r_1) \cdot S(r_2) \cdot S(r_3)$, as claimed.

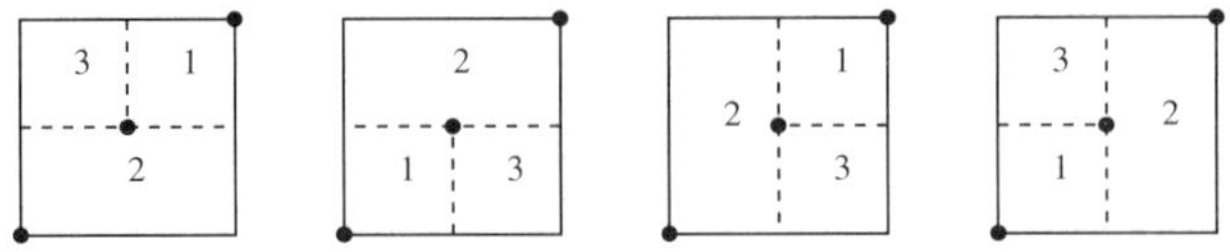

FIGURE 15.5. **Factoring rectangles.** The left two pictures are horizontal type factorizations of the larger rectangle; the right two pictures are vertical type factorizations of the larger rectangle.

When e is vertical, we can replace (r_1, r_2) with an alternative pair (r_1', r_2'), so that $r = r_1' * r_2' * r_3$ and r_2' meets r_1' and r_3 in a horizontal edge, so we have reduced to the previous case. □

LEMMA 15.4.9. *The sign-refined destabilization map*

$$\mathcal{D}_S \colon \mathcal{GC}_S^-(\mathbb{G}';\mathbb{Z}) \to \mathcal{C}one_S(V_1 - V_2)$$

of Definition 15.4.6 *is a chain map.*

Proof. The proof follows the logic of Lemma 13.3.13. The key point is to show that the cancellation of terms modulo two verified in that proof still hold with signs. In that proof, we investigated various juxtapositions involving one destabilization domain and a rectangle. For each such juxtaposition, we found either one other juxtaposition with the same contribution, or two other juxtapositions (where one of the juxtapositions is of Type (I)) so that the contributions of all three together cancel. In the present proof, we will show that within these groupings, the contributions also cancel with sign.

To help with the verification, we set some notation. Recall the natural splitting of the $\mathbb{Z}[V_1, \dots, V_n]$-modules:

$$\mathcal{C}one_S(V_1 - V_2 \colon \mathcal{GC}_S^-(\mathbb{G};\mathbb{Z})[V_1] \to \mathcal{GC}_S^-(\mathbb{G};\mathbb{Z})[V_1]) \cong \mathcal{GC}_S^-(\mathbb{G};\mathbb{Z})[V][\![1,1]\!] \oplus \mathcal{GC}_S^-(\mathbb{G};\mathbb{Z}).$$

There is also a $\mathbb{Z}[V_1, \dots, V_n]$-module splitting of $\mathcal{GC}_S^-(\mathbb{G}';\mathbb{Z})$ as $\mathcal{I} \oplus \mathcal{N}$. In terms of the above two splittings, the destabilization map and the differentials on $\mathcal{GC}_S^-(\mathbb{G}';\mathbb{Z})$ and on $\mathcal{C}one_S(V_1 - V_2 \colon \mathcal{GC}_S^-(\mathbb{G};\mathbb{Z})[V_1] \to \mathcal{GC}_S^-(\mathbb{G};\mathbb{Z})[V_1])$ can be represented by the 2×2 matrices:

$$\begin{pmatrix} \mathcal{D}^{iL}_{1,S} & \mathcal{D}^{iL}_{>1,S} \\ 0 & \mathcal{D}^{iR}_{S} \end{pmatrix} \qquad \begin{pmatrix} \partial^{\mathbf{I}}_{\mathbf{I},S} & \partial^{\mathbf{I}}_{\mathbf{N},S} \\ \partial^{\mathbf{N}}_{\mathbf{I},S} & \partial^{\mathbf{N}}_{\mathbf{N},S} \end{pmatrix} \qquad \begin{pmatrix} -\partial^-_S & 0 \\ V_1 - V_2 & \partial^-_S \end{pmatrix}.$$

(For the destablization map, we are using notation from Equation (13.6).) Thus, the lemma is equivalent to the vanishing of the operator

$$\mathcal{GC}_S^-(\mathbb{G}';\mathbb{Z}) \to \mathcal{C}one_S(V_1 - V_2 \colon \mathcal{GC}_S^-(\mathbb{G};\mathbb{Z})[V_1] \to \mathcal{GC}_S^-(\mathbb{G};\mathbb{Z})[V_1])$$

specified by the 2×2 matrix (suppressing the sign assignment S from the notation):

$$\begin{pmatrix} -\partial^- \circ \mathcal{D}^{iL}_1 - \mathcal{D}^{iL}_1 \circ \partial^{\mathbf{I}}_{\mathbf{I}} - \mathcal{D}^{iL}_{>1} \circ \partial^{\mathbf{N}}_{\mathbf{I}} & -\partial^- \circ \mathcal{D}^{iL}_{>1} - \mathcal{D}^{iL}_1 \circ \partial^{\mathbf{I}}_{\mathbf{N}} - \mathcal{D}^{iL}_{>1,} \circ \partial^{\mathbf{N}}_{\mathbf{N}} \\ (V_1 - V_2) \cdot \mathcal{D}^{iL}_1 - \mathcal{D}^{iR} \circ \partial^{\mathbf{N}}_{\mathbf{I}} & (V_1 - V_2) \cdot \mathcal{D}^{iL}_{>1} + \partial^- \circ \mathcal{D}^{iR} - \mathcal{D}^{iR} \circ \partial^{\mathbf{N}}_{\mathbf{N}} \end{pmatrix}. \tag{15.31}$$

Juxtapositions of domains give contributions to the above matrix. These contributions are grouped together to cancel. Most of the groupings involve pairs of juxtapositions of a rectangle and a destabilization domain. In those pairings, the order of the rectangle can be different (i.e. the two juxtapositions have the form $r * p$ and $p' * r'$); or the same (i.e. the two juxtapositions have the form $r * p$ and $r' * p'$; or $p * r$ and $p' * r'$). To show that the contributions of the pairs cancel, we will prove the following: if the juxtapositions are of the form $r * p$ and $r' * p'$, then

$$S(r) \cdot S(p) = -S(r') \cdot S(p'); \tag{15.32}$$

if they are of the form $p * r$ and $p' * r'$, then

$$S(p) \cdot S(r) = -S(p') \cdot S(r'); \tag{15.33}$$

and if they are of the form $r * p$ and $p' * r'$, then

$$S(r) \cdot S(p) = \begin{cases} S(p') \cdot S(r') & \text{if } p \text{ and } p' \text{ are of type } iL \\ -S(p') \cdot S(r') & \text{if } p \text{ and } p' \text{ are of type } iR. \end{cases}$$

The above equation can be more succinctly reformulated as follows: if p has complexity m, then

$$S(r) \cdot S(p) = (-1)^{m-1} S(p') \cdot S(r'). \tag{15.34}$$

Equations (15.32), (15.33), and (15.34) will be proved using the following method. Consider the two sequences of rectangles obtained by taking the rectangle and the canonical decomposition of the destabilization domain, taken in the specified order. (Recall that this canonical decomposition was used in the definition of the sign associated to a destabilization domain, cf. Equation (15.30).) We will exhibit a sequence of moves transforming one sequence into the other, where each move is one of the following three types:

- *Rectangle swaps.* Replace some pair of consecutive rectangles in a sequence by their alternative pair of rectangles, in the sense of Definition 15.1.1.
- *Rectangle merges.* Replace some sequence of three consecutive rectangles r_i, r_{i+1}, r_{i+2} in a sequence by a single rectangle r', in the case where $r' = r_i * r_{i+1} * r_{i+2}$. As in Lemma 15.4.8, there are two different kinds of such moves, which can be *horizontal type rectangle merges* and *vertical type rectangle merges.*
- *Annulus moves.* Between two rectangles in the sequence, insert a factorization of a thin horizontal or vertical annulus (as a juxtaposition of two rectangles).

The signs attached to the two sequences differ by the parity of the number of rectangle swaps, horizontal type rectangle merges, and vertical annulus moves needed in the sequence of moves. The equations are verified by keeping track of this parity. To this end, we introduce some notational shorthand for the above moves. Write

$$\mathbf{s}_i \colon r_1 * \cdots * r_n \to r_1' * \cdots * r_n'$$

if the sequence of rectangles $(r_1', \ldots, r_n')$ is obtained from the sequence $(r_1, \ldots, r_n)$ by swapping the ith pair of rectangles. Similarly, write

$$\mathbf{m}_i^h \colon r_1 * \cdots * r_n \to r_1 * \ldots r_{i-1} * r' * r_{i+3} * \cdots * r_n$$

if $r' = r_i * r_{i+1} * r_{i+2}$ is a horizontal type rectangle merge; or replace the symbol with $\mathbf{m}_i^v$ if it is a vertical type rectangle merge. Finally, write

$$\mathbf{a}_i^h \colon r_1 * \cdots * r_n \to r_1 * \cdots * r_{i-1} * q_1 * q_2 * r_i * \cdots * r_n$$

if $q_1 * q_2$ is a factorization of a thin horizontal annulus, whose initial state is the initial state of r_i; or replace the symbol with $\mathbf{a}_i^v$ if it is a vertical one. When moves of the above types are in succession, we write words representing operations separated by asterisks, in the order they are performed from left to right.

We now verify the above signed cancellations, going through the various cases covered in the proof of Lemma 13.3.13.

Domains of Type (A-0) cancel in pairs. Observe that in this case, one of these domains is of the form $r * p$ and the other is of the form $p' * r'$, where r and r' have the same underlying two-chain, and p and p' have the same underlying two-chain. Suppose that p has complexity m and let $p = r_1 * \cdots * r_{m-1}$ be its canonical decomposition. Consider the sequence of rectangles $(r, r_1, \dots, r_{m-1})$. We can successively swap each pair of rectangles, starting with the first pair, and moving along until each pair has been swapped exactly once. The resulting sequence of rectangles is $(r'_1, \dots, r'_{m-1}, r')$, obtained from the canonical decomposition of $p' = r'_1 * \cdots * r'_{m-1}$; or, more succinctly

$$\mathbf{s}_1 * \cdots * \mathbf{s}_{m-1} \colon r * p \to p' * r'$$

Since there were $m-1$ rectangle swaps, Equation (15.34) follows in this case.

Domains of Type (A-1) cancel in pairs. A juxtaposition of Type (A-1) can be either of the form $r * p$ or $p * r$, and two such pairings with either form can be paired off with one another. When one of the pairings is of the form $r * p$ and the other is of the form $p' * r'$, the verification of the cancellation proceeds exactly as in the case of Type (A-0); i.e., we can write

$$\mathbf{s}_1 * \cdots * \mathbf{s}_{m-1} \colon r * p \to p' * r';$$

see the top row of Figure 15.6.

When $r * p$ is paired with $r' * p'$, let $p = r_1 * \cdots * r_{m-1}$ be the canonical decomposition, and consider the first i with the property that r and r_i share an edge. Connect

$$\mathbf{s}_1 * \cdots * \mathbf{s}_{i-1} * \mathbf{s}_i * \mathbf{s}_{i-1} * \cdots * \mathbf{s}_1 \colon r * p \to r' * p';$$

see the bottom row of Figure 15.6 for an example. Note that there are $2i-1$ swaps in the above sequence, verifying Equation (15.32). When $p * r$ is paired with $p' * r'$, let $p = r_1 * \cdots * r_{m-1}$ be the canonical decomposition. Consider the first i with the property that r and r_i share an edge. Equation (15.33) follows from the sequence

$$\mathbf{s}_{m-1} * \cdots * \mathbf{s}_{i+1} * \mathbf{s}_i * \mathbf{s}_{i+1} \cdots * \mathbf{s}_{m-1} \colon p * r \to p' * r'.$$

Juxtapositions of Type (A-2)$_{m\geq 4}$ cancel with those of Type (D-0)$_{m-2}$. To verify that they cancel with sign, note that juxtapositions of Type (D-0) are always of the form $p' * r'$, while there are two types of juxtapositions of Type (A-2)$_{m+2}$: $p * r$ and $r * p$. We analyze the two subcases separately.

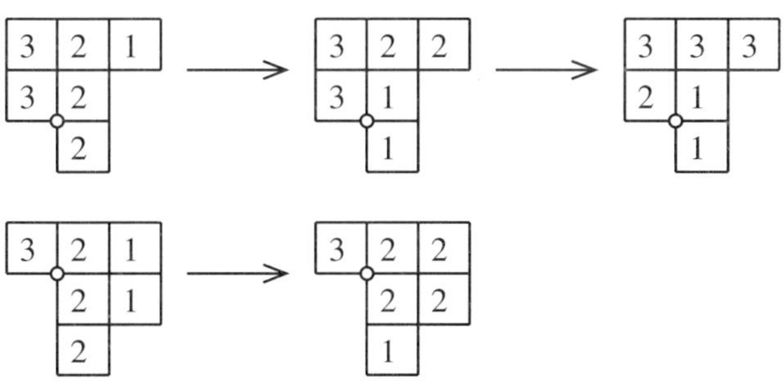

FIGURE 15.6. **Cancellation of terms of Type (A-1).** In the top row we have sequences connecting $r * p$ and $p' * r'$, both of Type (A-1); in the bottom row we have a sequence connecting $r * p$ with $r' * p'$. The union of squares decorated with i give the support of the rectangle r_i.

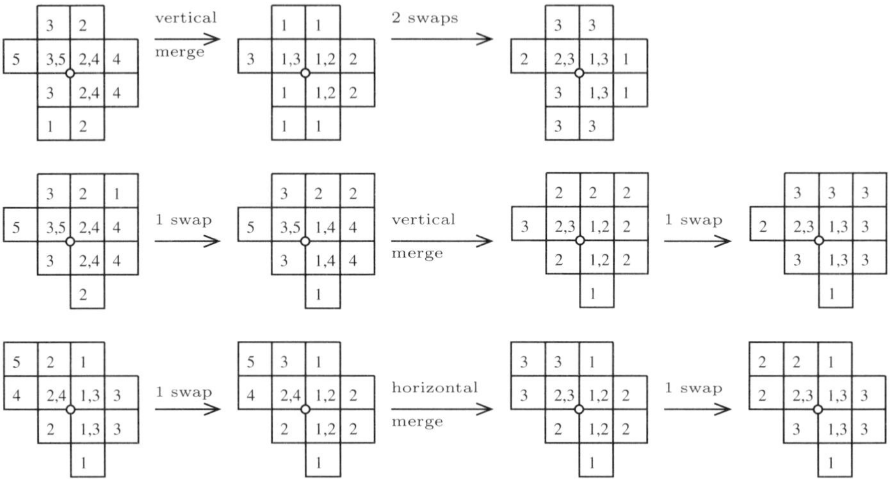

FIGURE 15.7. **Cancellation of terms of Type (A-2).** The first of these is a special subcase of connecting $r * p$ to $p' * r'$; the second two are generic subcases.

If the juxtaposition of Type (A-2) has the form $r * p$, consider the canonical decomposition $p = r_1 * \cdots * r_{m-1}$. When both r_1 and r_2 meet r, the three rectangles r, r_1, and r_2 form the vertical type decomposition of another rectangle. In this case, we can connect

$$\mathbf{m}_1^v * \mathbf{s}_1 * \cdots * \mathbf{s}_{m-3} \colon r * p \to p' * r';$$

see the first line of Figure 15.7. When r does not meet both r_1 and r_2, consider the first r_i in the above decomposition which meets r, and connect

$$\mathbf{s}_1 * \cdots * \mathbf{s}_i * \mathbf{m}_{i+1}^v * \mathbf{s}_{i+1} * \cdots * \mathbf{s}_{m-3} \colon r * p \to p' * r';$$

see the second line of Figure 15.7. In this process, we have done a total of $m-3$ swaps and a vertical type rectangle merge, verifying Equation (15.34).

If the juxtaposition of Type (A-2) has the form $p * r$, let r_i be the first rectangle in the canonical decomposition of p that meets r along an edge. We now connect the two sequences:

$$\mathbf{s}_{m-1} * \cdots * \mathbf{s}_{i+3} * \mathbf{s}_i * \mathbf{m}_{i+1}^h * \mathbf{s}_i * \cdots * \mathbf{s}_{m-3} \colon p * r \to p' * r';$$

see the third line of Figure 15.7. The total number of swap moves connecting these sequences is $2m - 2i - 4$, and there is another horizontal rectangle merge, hence Equation (15.33) follows. This completes the verification that terms of Type (A-2) with $m \geq 4$ cancel terms of Type (D-0) with complexity $m-2$.

Terms of Type (B-4)$_{m=2}$ cancel terms of Type (I)$_{m=1}$. This is straightforward: terms of Type (B-4)$_{m=2}$ are obtained as $(-1)^{M(\mathbf{x})-1}$ times the sum of the contribution of the vertical annulus, which is $-V_1$, with the sum of the horizontal one, which is V_2; while the contribution of Type (I)$_{m=1}$ is $(-1)^{M(\mathbf{x})}$ times $(V_1 - V_2) \cdot \mathbf{x}$. (Note that this was a grouping of three terms, rather than the usual two.)

Terms of Type (B-0)$_{m=1}$ cancel terms of Type (A-3)$^o_{m=3}$. Let $r * p$ be the term of Type (B-0)$_{m=1}$ (so that p is the trivial domain). When the term of

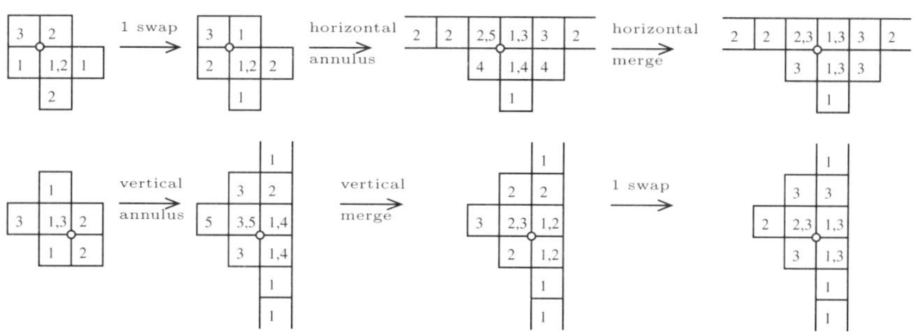

FIGURE 15.8. **Cancellation of terms of Types $^{\delta=1}$(C-1)$^o_{m\geq 2}$ with (D-1)^o_m.**

Type (A-3)$^o_{m=3}$ is of the form $r' * p'$, connect

$$\mathbf{a}_1^v : r * p \to r' * p' \qquad \text{or} \qquad \mathbf{a}_1^h * \mathbf{s}_2 : r * p \to r' * p',$$

depending on whether r goes through X_1 or X_2; here the annulus goes through O_1. When the term of Type (A-3)$^o_{m=3}$ is of the form $p' * r'$, connect

$$\mathbf{a}_1^v * \mathbf{s}_2 : r * p \to p' * r' \qquad \text{or} \qquad \mathbf{a}_2^h * \mathbf{s}_1 * \mathbf{s}_2 : r * p \to p' * r',$$

again depending on whether r goes through X_1 or X_2; the annulus is also as above. These sequences verify the claimed cancellations.

Terms of Type (D-0)$_{m=1}$ cancel terms of Type (B-3)$_{m=3}$. Interpret the decomposition of Type (B-3) as the vertical type decomposition of the rectangle appearing in the Type (D-0)$_{m=1}$ decomposition, and apply Lemma 15.4.8.

Terms of Type $^{\delta=1}$(C-1)$^o_{m\geq 2}$ cancel terms of Type (D-1)^o_m. The first is of the form $r * p$ and the second is of the form $p' * r'$. When r and p meet along a horizontal edge, connect

$$\mathbf{s}_1 * \cdots * \mathbf{s}_{m-2} * \mathbf{a}^h_{m-1} * \mathbf{m}^h_m : r * p \to p' * r',$$

where $\mathbf{a}^h_{m-1}$ refers to a move involving the horizontal annulus through X_1; see the top line of Figure 15.8 for an illustration. During these moves, we make $m-1$ sign changes, verifying Equation (15.34).

When r and p meet along a vertical edge, connect

$$\mathbf{a}_1^v * \mathbf{m}_2^v * \mathbf{s}_2 * \cdots * \mathbf{s}_{m-1} : r * p \to p' * r',$$

where the annulus goes through X_2; see the second line of Figure 15.8. Equation (15.34) follows once again.

Terms of Type $^{\delta>1}$(C-1)$^o_{m\geq 2}$ cancel those of Type (A-3)$^o_{m+2}$. The first term is of the form $r * p$, while the second can be either $r' * p'$ or $p' * r'$. When the second term has the form $r' * p'$, and r and p meet along a horizontal edge, connect the sequences

$$\mathbf{s}_1 * \cdots * \mathbf{s}_{m-1} * \mathbf{a}^h_{m+1} * \mathbf{s}_m * \cdots * \mathbf{s}_1 : r * p \to r' * p',$$

where $\mathbf{a}^h_{m+1}$ inserts the factorization of the horizontal annulus through X_1; see the first line of Figure 15.9. When the second term has the form $r' * p'$ and r and p meet along a vertical edge, connect

$$\mathbf{a}_1^v : r * p \to r' * p';$$

see the second line of Figure 15.9. In both cases, Equation (15.32) follows.

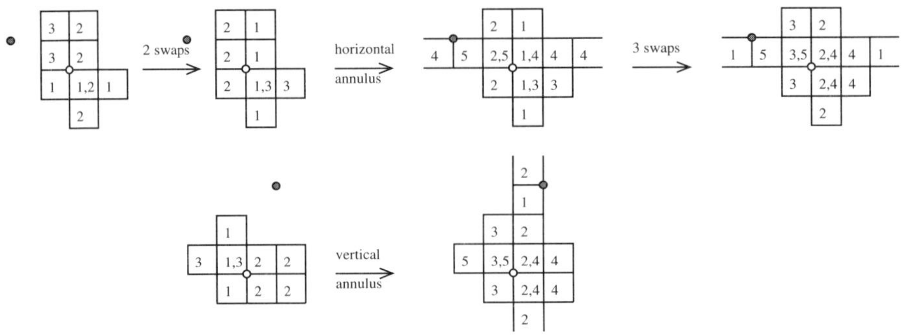

FIGURE 15.9. **Cancellation of terms of Type $^{\delta>1}$(C-1)$^o_{m\geq 2}$ and (A-3)$^o_{m+2}$ of the form $r' * p'$.**

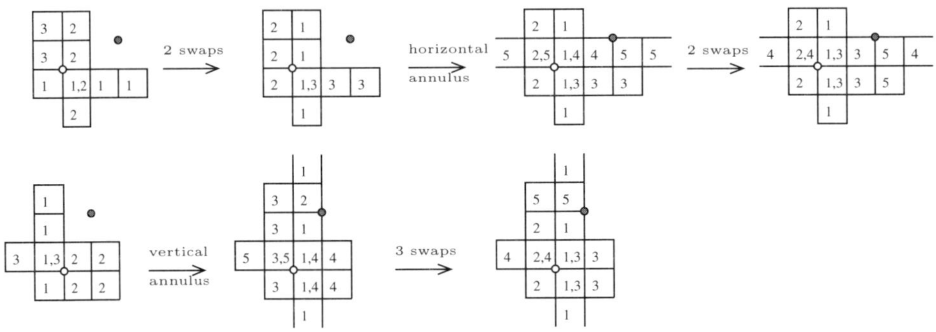

FIGURE 15.10. **Cancellation of terms of Type $^{\delta>1}$(C-1)$^o_{m\geq 2}$ and (A-3)$^o_{m+2}$ of the form $p' * r'$.**

When the second term has the form $p' * r'$, and r and p meet along a horizontal edge, connect

$$\mathbf{s}_1 * \cdots * \mathbf{s}_{m-1} * \mathbf{a}^h_{m+1} * \mathbf{s}_m * \mathbf{s}_{m+1} \colon r * p \to p' * r',$$

where $\mathbf{a}^h_{m+1}$ again involves the horizontal annulus through X_1; as illustrated in the first line of Figure 15.10. When the second term has the form $p' * r'$ and r and p meet along a vertical edge, connect

$$\mathbf{a}^v_1 * \mathbf{s}_2 * \cdots * \mathbf{s}_{m+1} \colon r * p \to p' * r',$$

where $\mathbf{a}^v_1$ involves the vertical annulus through X_2; see the second line of Figure 15.10.

Some terms of Type (I)$_{m=2k+1}$ cancel terms of Type $^{\delta=1}$(C-1)$^e_{m=2k}$ and those of Type (D-1)$^e_{m=2k}$. Write the three terms as p, $r' * p'$, and $p'' * r''$ respectively. Connect

$$\mathbf{s}_{2k-1} * \cdots * \mathbf{s}_1 \colon p \to r' * p',$$

as illustrated in Figure 15.11. This shows the cancellation of the contribution of the part of the Type (I) involving V_1 with the contribution of Type (C-1). Next, connect

$$\mathbf{a}^h_{2k-1} * \mathbf{m}^v_{2k} \colon p \to p' * r',$$

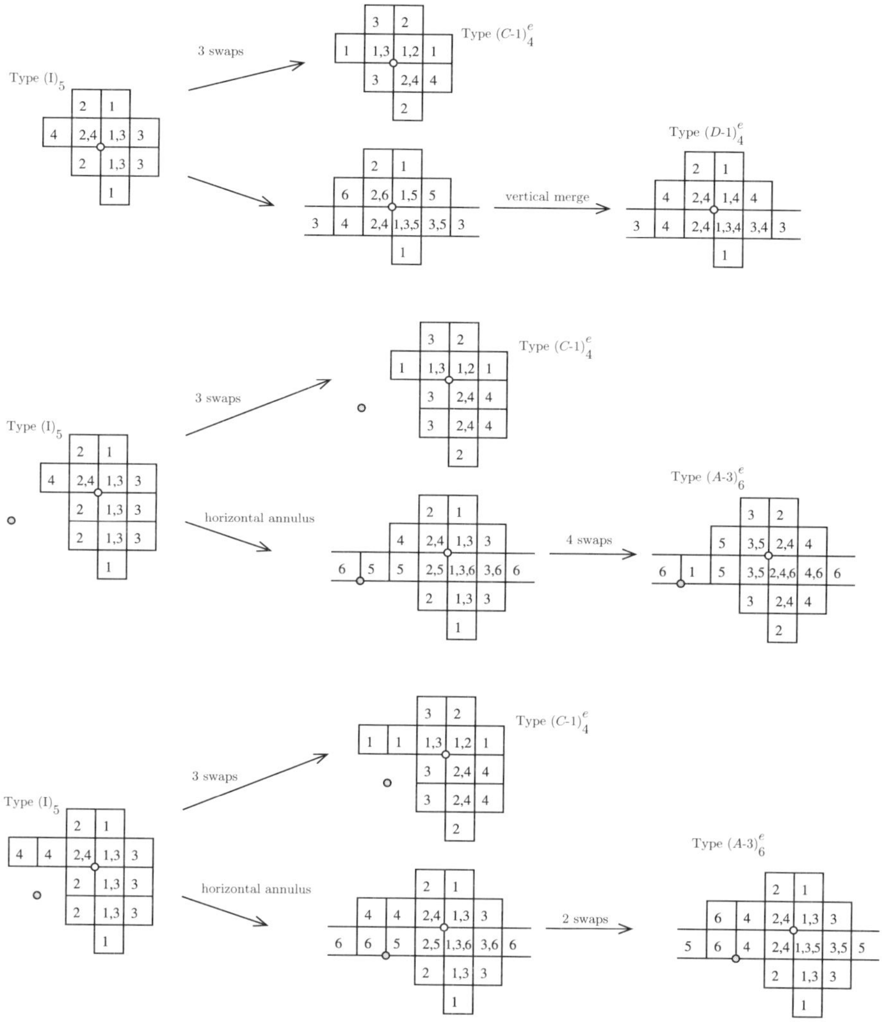

FIGURE 15.11. **Cancellation of terms of Type $(\mathbf{I})_{m=2k+1}$ with one of Type $(\mathbf{C\text{-}1})^e_{m=2k}$ and one of Type $(\mathbf{D\text{-}1})^e_{m=2k}$ or $(\mathbf{A\text{-}3})^e_{m=2k+2}$.**

where $\mathbf{a}^h_{2k-2}$ corresponds to the horizontal annulus through X_2; see Figure 15.11. There is no sign change in this path, giving the remaining cancellation. (See the top row of Figure 15.11.)

The remaining terms of Type (I) cancel with terms of Type $^{\delta>1}(\mathbf{C\text{-}1})^e_{m=2k}$ and those of Type $(\mathbf{A\text{-}3})^e_{m=2k+2}$. Write the terms as p, $r' * p'$, and $p'' * r''$ or $r'' * p''$ respectively. Connect

$$\mathbf{s}_{2k-1} * \cdots * \mathbf{s}_1 : p \to r' * p',$$

as illustrated in Figure 15.11. This readily verifies that the multiple of V_1 in Type (I) cancels with the corresponding contribution of Type $^{\delta>1}(\text{C-1})^e$. When the Type (A-3) juxtaposition is of the form $r'' * p''$, connect

$$\mathbf{a}^h_{2k+1} * \mathbf{s}_{2k} * \cdots * \mathbf{s}_1 \colon p \to r'' * p'',$$

where now $\mathbf{a}^h_{2k+1}$ corresponds to the horizontal annulus through X_2; see the second row of Figure 15.11. If the Type (A-3) juxtaposition is of the form $p'' * r''$, connect

$$\mathbf{a}^h_{2k+1} * \mathbf{s}_{2k} * \mathbf{s}_{2k+1} \colon p \to p'' * r'',$$

where again $\mathbf{a}^h_{2k+1}$ corresponds to the horizontal annulus through X_2; see the third row of Figure 15.11. This argument concludes the proof of Lemma 15.4.9. □

Proof of Lemma 15.4.4. Since the chain map $\mathcal{D}_S$ induces the sign-refined destabilization map D_S (from Definition 15.3.6) on the associated graded level, and the latter map is a quasi-isomorphism, it follows that for type *X:SW* stabilization the filtered chain complexes $\mathcal{GC}^-_S(\mathbb{G};\mathbb{Z})$ and $\mathcal{GC}^-_S(\mathbb{G}';\mathbb{Z})$ are quasi-isomorphic. □

Proof of Theorem 15.4.2. The proof follows the usual strategy. Lemma 15.1.9 implies the independence of the choice of V_i in defining the $\mathbb{Z}[U]$-module structure. The proof of Proposition 15.1.10 shows that the complex is independent of the choice of signs assignments. Invariance under commutations and switches was proved in Proposition 15.4.3. Invariance under *X:SW* stabilization is verified in Lemma 15.3.5, and the invariance under other stabilizations are reduced to this case by the invariance under commutations and switches. This verifies that grid homology gives an oriented knot invariant. □

Orientation dependence of the filtered chain complex is expressed as follows:

PROPOSITION 15.4.10. *Let $\mathcal{GC}^-(\vec{K})'$ be the filtered chain complex over $\mathbb{Z}[U]$, where the action of U is induced by $-V_i$. Then $\mathcal{GC}^-(\vec{K})'$ is quasi-isomorphic to $\mathcal{GC}^-(-\vec{K})$.*

Proof. As in Proposition 15.3.8, if $\mathbb{G}$ is a grid diagram representing $\vec{K}$, then its reflection $\mathbb{G}'$ represents $-\vec{K}$. If S is a sign assignment for $\mathbb{G}$ used to define $\mathcal{GC}^-(\vec{K})$, then for $\mathrm{Rect}^\circ(\mathbf{x},\mathbf{y})$

$$S''(\phi(r)) = (-1)^{\mathbb{O}(r)+\#(\mathrm{Int}(r)\cap\mathbf{x})} \cdot S(r)$$

defines a sign assignment for $\mathbb{G}'$ where $\phi\colon \mathbb{G} \to \mathbb{G}'$ is the reflection map. (Note that this differs from the one used in the proof of Proposition 15.3.8.) The map $\mathbf{x} \mapsto \phi(\mathbf{x})$ extends uniquely to a linear map $\Phi\colon \mathcal{GC}^-_S(\mathbb{G}) \to \mathcal{GC}^-_{S''}(\mathbb{G}')$ satisfying

$$\Phi(V_i \cdot \xi) = -V_i \cdot \Phi(\xi), \tag{15.35}$$

for all $i = 1,\dots,n$ and $\xi \in \mathcal{GC}^-_S(\mathbb{G})$. Clearly, Φ is an isomorphism of filtered chain complexes over $\mathbb{Z}$, so Equation (15.35) gives the needed isomorphism $\mathcal{GC}^-_S(\mathbb{G}) \cong \mathcal{GC}^-_{S''}(\mathbb{G}')'$ of filtered chain complexes over $\mathbb{Z}[U]$. □

EXERCISE 15.4.11. Show that for any knot K, $H(\widehat{\mathcal{GC}}(K;\mathbb{Z})) \cong \mathbb{Z}$ is supported in (Maslov) grading 0; and $H(\mathcal{GC}^-(K;\mathbb{Z})) \cong \mathbb{Z}[U]$, whose generator has grading 0.

15.5. Other grid homology constructions over $\mathbb{Z}$

Many of the earlier constructions given in this book have integral lifts, some of which can be done with fairly little effort.

For instance, for an oriented link $\vec{L}$ we can form the collapsed grid complex

$$cGC^-_S(\vec{L};\mathbb{Z}) = \frac{GC^-_S(\vec{L};\mathbb{Z})}{V_{i_1} = \cdots = V_{i_\ell}}.$$

The homology of this chain complex, the *sign-refined, collapsed grid homology* $cGH^-(\vec{L};\mathbb{Z})$, thought of as a bigraded module over $\mathbb{Z}[U]$, is a link invariant. The skein sequence from Chapter 9 has a straightforward integral lift:

THEOREM 15.5.1. *Let $(\vec{L}_+, \vec{L}_-, \vec{L}_0)$ be an oriented skein triple as in Figure 9.1. If the two strands meeting at the distinguished crossing in $\vec{L}_+$ belong to the same component, then there is an exact triangle of $\mathbb{Z}[U]$-modules*

$$\begin{array}{ccc} cGH^-(\vec{L}_+;\mathbb{Z}) & \xrightarrow{f^-} & cGH^-(\vec{L}_-;\mathbb{Z}) \\ & {\scriptstyle h^-}\nwarrow \quad \swarrow{\scriptstyle g^-} & \\ & cGH^-(\vec{L}_0;\mathbb{Z}) & \end{array}$$

If the strands at $\vec{L}_+$ belong to different components, there is a triangle

$$\begin{array}{ccc} cGH^-(\vec{L}_+;\mathbb{Z}) & \xrightarrow{f^-} & cGH^-(\vec{L}_-;\mathbb{Z}) \\ & {\scriptstyle h^-}\nwarrow \quad \swarrow{\scriptstyle g^-} & \\ & cGH^-(\vec{L}_0;\mathbb{Z}) \otimes J & \end{array}$$

where J is the free abelian group of rank four with one generator in bigrading $(0,1)$ one in bigrading $(-2,-1)$, and two generators in bigrading $(-1,0)$. In both diagrams, f^- and h^- are bigraded homomorphisms of $\mathbb{Z}[U]$-modules, and g^- is a homomorphism of $\mathbb{Z}[U]$-modules that is homogeneous of degree $(-1,0)$.

Proof. We follow the proof of Theorem 9.1.1. In particular, the methods from Chapter 9 (more specifically, Lemma 9.2.4, whose statement remains true over $\mathbb{Z}$) show that the following diagram of chain complexes

$$\begin{array}{ccc} \mathbf{I}' & \xrightarrow{\partial_{\mathbf{I}'}^{\mathbf{N}'}} & \mathbf{N}' \\ {\scriptstyle \partial_{\mathbf{I}}^{\mathbf{N}} \circ T}\downarrow & & \downarrow{\scriptstyle -T\circ\partial_{\mathbf{N}'}^{\mathbf{I}'}} \\ \mathbf{N} & \xrightarrow{\partial_{\mathbf{N}}^{\mathbf{I}}} & \mathbf{I} \end{array} \tag{15.36}$$

with the understanding that the complexes and maps are the $\mathbb{Z}$-lifts of the complexes and maps from Chapter 9. The left column of Equation (15.36) is isomorphic to $cGC^-(\mathbb{G}_+;\mathbb{Z})$, the right is isomorphic to $cGC^-(\mathbb{G}_-;\mathbb{Z})$, the top row is isomorphic to $cGC^-(\mathbb{G}_0';\mathbb{Z})$ and the bottom is isomorphic to $cGC^-(\mathbb{G}_0;\mathbb{Z})$. Moreover, the above 2×2 complex is isomorphic to the mapping cone of the chain map

$$(-1)^M(\partial_{\mathbf{I}'}^{\mathbf{N}'} \circ T - T \circ \partial_{\mathbf{N}'}^{\mathbf{I}'})\colon GC^-(\mathbb{G}_0') \to GC^-(\mathbb{G}_0),$$

where M denotes the Maslov grading of the output in $GC^-(\mathbb{G}_0)$.

The proof of Lemma 9.2.7 verifies the following sign-refined analogue of Equation (9.6):

$$(-1)^M P \circ \left(\partial_{\mathbf{I}}^{\mathbf{N}} \circ T - T \circ \partial_{\mathbf{N}'}^{\mathbf{I}'}\right) = h_Y \circ h_{X_2} + h_{X_2} \circ h_Y.$$

In the proof of Lemma 9.2.7, we gave a correspondence between juxtapositions contributing to the left and those contributing to the right; see Figure 9.4. In the present case, each rectangle and pentagon comes with a sign, and under the correspondence, the signs work out. For instance, in the terms for the composite map $-(-1)^M P \circ T \circ \partial_{\mathbf{N}'}^{\mathbf{I}'}$, all the pentagons are left pentagons; see the second column of the figure. The sign of a left pentagon is $(-1)^{M+1}$ times the sign of its straightening, and their straightenings give exactly the terms in $h_{X_2} \circ h_Y$; see the first column of Figure 9.4. When verifying $(-1)^M P \circ \partial_{\mathbf{N}}^{\mathbf{I}} \circ T = h_Y \circ h_{X_2}$ we compare terms in the fourth and the third column of Figure 9.4. In the first row of the figure, after straightening the pentagon, we get an alternative pair of rectangles, which is compensated with the left pentagon, carrying an extra sign. In the other three cases, the pentagon is on the right, and their contributions are evidently the same.

Since Lemma 9.2.8 holds with $\mathbb{Z}$ coefficients, the theorem follows. □

In another direction, the multi-graded grid homology group $\mathbf{GH}^-(\vec{L})$ of an ℓ-component oriented link $\vec{L}$ (defined in Chapter 11 over the ring $\mathbb{F}[U_1, \dots, U_\ell]$) can be lifted to the ring $\mathbb{Z}[U_1, \dots, U_\ell]$. Suppose that the grid diagram $\mathbb{G}$ represents $\vec{L}$, and fix a sign assignment S. The formula of Equation (15.1) gives a boundary map in the case of links, as well. This differential is compatible with the Alexander multi-grading defined in Definition 11.1.5. The proof of Lemma 15.1.9 now gives homotopies between the actions of variables V_i and V_j provided that they correspond to O-markings O_i and O_j on the same component of $\vec{L}$. Thus, the homology groups of $GC_S^-(\mathbb{G};\mathbb{Z}) = (GC^-(\mathbb{G};\mathbb{Z}), \partial_{\mathbb{X},S}^-)$ naturally inherit the structure of a multi-graded module over $\mathbb{Z}[U_1, \dots, U_\ell]$, where the Alexander multi-grading takes values in the Alexander grading set of Definition 11.1.3. We denote the complex with this extra structure by $\mathbf{GC}_S^-(\mathbb{G};\mathbb{Z})$. Adapting the arguments from Propositions 15.1.10, 15.3.1 and 15.3.5 to the multi-graded context gives the following:

THEOREM 15.5.2. *Suppose that the grid diagram $\mathbb{G}$ represents the oriented link $\vec{L}$ of ℓ components. Let S be a chosen sign assignment on $\mathbb{G}$. Then the homology $\mathbf{GH}^-(\mathbb{G};\mathbb{Z})$ of $\mathbf{GC}_S^-(\mathbb{G};\mathbb{Z})$, thought of as a multi-graded $\mathbb{Z}[U_1, \dots, U_\ell]$- module, is an invariant of the oriented link $\vec{L}$.* □

In a similar manner, the multi-filtered grid complex of a link from Section 14.5 can be lifted to an $\mathbf{H}(L)$-filtered, $\mathbb{Z}$-graded chain complex $\mathcal{GC}^-(\mathbb{G};\mathbb{Z})$ over $\mathbb{Z}[U_1, \dots, U_\ell]$, by using the differential from Equation (15.28). Simple adaptations of our previous arguments show that the filtered chain homotopy type of this graded, $\mathbf{H}(L)$-filtered chain complex over $\mathbb{Z}[U_1, \dots, U_\ell]$ is an invariant of the oriented link $\vec{L}$.

EXERCISE 15.5.3. * Use sign assignments to define a natural $\mathbb{Z}$ lift of the double-point enhanced grid complex from Section 5.5, and prove its homology is a knot invariant.

15.6. On the τ-invariant

Let p be a prime number, and consider the ring $\mathbb{F}_p = \mathbb{Z}/p\mathbb{Z}$. It follows quickly from Theorem 15.1.11 that the homology of the chain complex $GC^-_S(\mathbb{G};\mathbb{Z}) \otimes_{\mathbb{Z}} \mathbb{F}_p$ is a knot invariant. We denote this homology by $GH^-(K;\mathbb{F}_p)$; in the case where $p=2$, this is exactly the grid homology from Chapter 4. An analogous construction can be done with $\mathbb{Q}$ in place of $\mathbb{F}_p$, giving the invariant $GH^-(K;\mathbb{Q})$.

Since $\mathbb{F}_p[U]$ is a principal ideal domain, $GH^-(\mathbb{G};\mathbb{F}_p)$ is a finite sum of cyclic modules. The argument from Proposition 6.1.4 shows that if $\mathbb{G}$ represents a knot, there is a single free summand in $GH^-(\mathbb{G};\mathbb{F}_p)$, and its generator has bigrading $(-2t,-t)$ for some integer t. Define $\tau(K;\mathbb{F}_p)$ to be this integer t. Note that $\tau(K;\mathbb{F}_2)$ is $\tau(K)$ as defined in Definition 6.1.5. A similar construction can be given using the rationals $\mathbb{Q}$ in place of $\mathbb{F}_p$, giving $\tau(K;\mathbb{Q})$. By repeating the discussion from Chapter 8, using $GH^-(K;\mathbb{F}_p)$ or $GH^-(K;\mathbb{Q})$ in place of $GH^-(K)$ we get:

THEOREM 15.6.1. *For each prime p, $\tau(K;\mathbb{F}_p)$ gives a bound on the slice genus $g_s(K)$ of K: $|\tau(K;\mathbb{F}_p)| \leq g_s(K)$. Similarly, $|\tau(K;\mathbb{Q})| \leq g_s(K)$.* □

One might wonder if this construction provides new genus bounds beyond the one we have from $\tau(K)$. At the writing of this book, for all the computed examples, the invariants $\widehat{GH}(K;\mathbb{Z})$ and $GH^-(K;\mathbb{Z})$, thought of as bigraded abelian groups, are free. For these knots K, $\widehat{GH}(K;\mathbb{F})$ and $GH^-(K;\mathbb{F})$ immediately determine the abelian groups $\widehat{GH}(K;\mathbb{Z})$ and $GH^-(K;\mathbb{Z})$, and hence, by the universal coefficient theorem, all the further specializations. Even for these knots, though, the $\mathbb{F}[U]$-module structure on $GH^-(K;\mathbb{F})$ does not determine the $\mathbb{Z}[U]$-module structure on $GH^-(K;\mathbb{Z})$ (or the $\mathbb{F}_p[U]$-module structure for other p).

We give an example that illustrates this algebraic subtlety. In this formal setting, for a bigraded chain complex C over $\mathbb{Z}[U]$, let $\tau(C;\mathbb{F}_p)$ (resp. $\tau(C;\mathbb{Q})$) be minus one times the maximal Alexander grading of any homogeneous non-torsion element in the homology group $H(C \otimes_{\mathbb{Z}} \mathbb{F}_p)$ (resp. $H(C \otimes_{\mathbb{Z}} \mathbb{Q})$), thought of as a module over $\mathbb{F}_p[U]$ (resp. $\mathbb{Q}[U]$).

EXAMPLE 15.6.2. Consider the graded chain complex C over $\mathbb{Z}[U]$ with three generators, x_1, x_2, and x_3, with $A(x_i) = i$ for $i = 1, \dots, 3$, and differential specified by $\partial x_1 = 2U \cdot x_2 + U^2 \cdot x_3$, $\partial x_2 = 0$, and $\partial x_3 = 0$. A straightforward computation shows that the homology $H(C)$, thought of as a module over $\mathbb{Z}[U]$, has two generators x_2 and x_3 satisfying the relation $2Ux_2 + U^2 x_3 = 0$; in particular, it is free as a $\mathbb{Z}$-module. Similarly, the homology $H(\frac{C}{U}) \cong \mathbb{Z}^3$ is a free $\mathbb{Z}$-module. However, $\tau(C;\mathbb{Q}) = -3 = \tau(C,\mathbb{F}_p)$ for all $p \neq 2$, and $\tau(C,\mathbb{F}_2) = -2$.

At the moment, very little is known about $\tau(K;\mathbb{F}_p)$ for $p \neq 2$, see Problems 17.2.10 and 17.2.11. In particular, there is no known knot K with $\tau(K;\mathbb{F}_p) \neq \tau(K;\mathbb{F}_q)$ for two primes $p \neq q$.

15.7. Relations in the spin group

Our aim in this section is to verify Lemma 15.2.4. We will use an explicit model for the group Spin(3), based on the quaternion algebra $\mathbb{H}$.

Recall that the quaternion algebra $\mathbb{H}$ is the four-dimensional non-commutative algebra over $\mathbb{R}$ with generators 1, $\mathbf{i}$, $\mathbf{j}$, and $\mathbf{k}$, subject to the relations:

$$\mathbf{i}^2 = \mathbf{j}^2 = \mathbf{k}^2 = -1; \quad \mathbf{i}\cdot\mathbf{j} = -\mathbf{j}\cdot\mathbf{i} = \mathbf{k}; \quad \mathbf{j}\cdot\mathbf{k} = -\mathbf{k}\cdot\mathbf{j} = \mathbf{i}; \quad \mathbf{k}\cdot\mathbf{i} = -\mathbf{i}\cdot\mathbf{k} = \mathbf{j}.$$

A typical quaternion q has the form $q = a + b\mathbf{i} + c\mathbf{j} + d\mathbf{k}$, where a, b, c, d are real numbers. The *real part* $\mathrm{Re}(q)$ of the quaternion q is the component a. When q has $\mathrm{Re}(q) = 0$, the quaternion q is called *purely imaginary.*

There is a conjugation action on the quaternions, defined by

$$\overline{a + b\mathbf{i} + c\mathbf{j} + d\mathbf{k}} = a - b\mathbf{i} - c\mathbf{j} - d\mathbf{k}.$$

The *norm* of a quaternion is the real number $|q| = q \cdot \overline{q}$. More generally, the standard Euclidean metric on $\mathbb{H} \cong \mathbb{R}^4$ can be described as the map sending $p, q \in \mathbb{H}$ to $\mathrm{Re}(p \cdot \overline{q})$. Quaternions with $|q| = 1$ are called *unit quaternions.* The space of unit quaternions is naturally identified with the three-sphere S^3.

If q is a unit quaternion and h is a purely imaginary one, it is easy to check that $q \cdot h \cdot \overline{q}$ is purely imaginary as well, with the same norm as h. Using the identification between purely imaginary quaternions and $\mathbb{R}^3$, the map $(q, h) \mapsto q \cdot h \cdot \overline{q}$ specifies the double cover from the space of unit quaternions to $\mathrm{SO}(3)$.

Given a unit, purely imaginary quaternion q, let $q^\perp$ be its orthogonal complement in $\mathbb{R}^3$. This two-plane inherits a natural orientation, and hence it determines the spin rotation $\widetilde{R}_{q^\perp}$ as in Definition 15.2.3.

LEMMA 15.7.1. *Let q be a unit, purely imaginary quaternion. Then,*

$$\widetilde{R}_{q^\perp} = q. \tag{15.37}$$

Moreover, if q_1 and q_2 are two unit quaternions, then

$$\widetilde{R}_{q_1^\perp} \cdot \widetilde{R}_{q_2^\perp} = \widetilde{R}_{(q_1 \cdot q_2)^\perp}. \tag{15.38}$$

Proof. Let $q = \mathbf{i}$. Consider the path $t \mapsto e^{t\mathbf{i}}$ in the unit quaternions with $t \in [0, \frac{\pi}{2}]$. This path projects to the path $v \mapsto e^{t\mathbf{i}} \cdot v \cdot e^{-t\mathbf{i}}$ in $\mathrm{SO}(3)$, which fixes $\mathbf{i}$ and sends $\mathbf{j}$ to $e^{2t\mathbf{i}} \cdot \mathbf{j}$. Thus, the path in $\mathrm{SO}(3)$ rotates by an angle of $2t$ in the $\mathbf{j}$-$\mathbf{k}$ plane. It follows that the endpoint $\mathbf{i}$ of the path gives a model for $\widetilde{R}_{\mathbf{i}^\perp}$. The above computation remains valid when $\mathbf{i}$ and $\mathbf{j}$ are replaced by any two orthogonal quaternions q and q'.

For the second statement, observe that the curve $t \mapsto e^{tq_1}$ in the unit quaternions, parameterized by $t \in [0, \frac{\pi}{2}]$, connects 1 to q_1; and $t \mapsto e^{tq_2}$ connects 1 to q_2; so the curve $t \mapsto e^{tq_1}e^{tq_2}$, whose endpoint specifies $\widetilde{R}_{q_1^\perp} \cdot \widetilde{R}_{q_2^\perp}$, connects 1 to $q_1 \cdot q_2$. □

Proof of Lemma 15.2.4. To see Equation (15.8), observe that the oriented plane P lies in some $\mathbb{R}^3 \subset \mathbb{R}^n$. View $\mathbb{R}^3$ as the purely imaginary quaternions, and realize $\widetilde{R}_P$ as $\widetilde{R}_{q^\perp}$ for some purely imaginary unit quaternion q. By Equation (15.37), $\widetilde{R}_{q^\perp}$ corresponds to q, and by Equation (15.38), $\widetilde{R}_P^2$ is represented by q^2, which in turn equals -1, since q is a unit, purely imaginary quaternion, verifying Equation (15.8).

To verify Equation (15.9), note that the vectors u_1, u_2, w span a three-dimensional subspace in $\mathbb{R}^n$. Thus, it suffices to show that Equation (15.9) holds in $\mathrm{Spin}(3)$. After rotating, we can assume that $u_1 = \mathbf{i}$, $u_2 = \mathbf{j}$, and $w = \mathbf{k}$. By Equation (15.37), $\widetilde{R}_{\langle u_1, w\rangle}$ corresponds to $-\mathbf{j}$ and $\widetilde{R}_{\langle u_2, w\rangle}$ corresponds to $\mathbf{i}$, so by Equation (15.38), Equation (15.9) is equivalent to the relation $\mathbf{i} \cdot \mathbf{j} = -\mathbf{j} \cdot \mathbf{i}$.

For Equation (15.10), again we can consider the span of u_1, u_2, w, identified with the purely imaginary quaternions. After rotating, we can assume that $u_1 = \mathbf{i}$, $u_2 = \mathbf{i} \cdot e^{\frac{2\pi}{3}\mathbf{j}}$, and $w = \mathbf{j}$. Using Equation (15.37), $\widetilde{R}_{\langle u_1, w\rangle}$ corresponds to the quaternion $\mathbf{k}$; $\widetilde{R}_{\langle u_2,w\rangle}$ corresponds to $\mathbf{k}\cdot e^{\frac{2\pi}{3}\mathbf{j}}$; and $\widetilde{R}_{\langle -u_1-u_2,w\rangle}$ corresponds to $\mathbf{k}\cdot e^{-\frac{2\pi}{3}\mathbf{j}}$. The relation $\mathbf{k}\cdot\mathbf{k}\cdot e^{\frac{2\pi}{3}\mathbf{j}}\cdot\mathbf{k} = -\mathbf{k}\cdot e^{\frac{-2\pi}{3}\mathbf{j}}$ in $\mathbb{H}$ gives

$$\widetilde{R}_{\langle u_1,w\rangle}\cdot\widetilde{R}_{\langle u_2,w\rangle}\cdot\widetilde{R}_{\langle u_1,w\rangle} = z\cdot\widetilde{R}_{\langle -u_1-u_2,w\rangle} = \widetilde{R}_{\langle u_1+u_2,w\rangle}. \qquad \square$$

15.8. Further remarks

Theorem 15.1.5 was proved in [**136**], using a slightly different construction for sign assignments. For that construction, consider a fundamental domain for the grid, and first consider a rectangle $r\in \mathrm{Rect}(\mathbf{x},\mathbf{y})$ that has connected support in the fundamental domain, with corners given by the coordinates $(i,a),(i+1,h)$ (as part of the initial grid state $\mathbf{x}$) and $(i+1,a),(i,h)$ (as part of the terminal grid state $\mathbf{y}$). In particular, r is of width one. Define the value of the function S on $r\in\mathrm{Rect}(\mathbf{x},\mathbf{y})$ by the formula $(-1)^{I(r)}$, where $I(r)$ is given by

$$I(r) = \mathcal{I}(\mathbf{x},\{(x_1,x_2)\in\mathbf{x}\mid x_2\leq h\}). \tag{15.39}$$

Using the defining properties of a sign assignment, this function extends uniquely to a sign assignment for empty rectangles; compare also [**166**].

The interpretation of sign assignments in terms of $\widetilde{\mathfrak{S}}_n$ is due to Gallais [**66**]. He also gives the following slightly different construction of the chain complex. Let $\widetilde{\mathbf{S}}(\mathbb{G})$ be the set of pairs $\widetilde{\mathbf{x}} = (\mathbf{x},\widetilde{\sigma})\in\mathbf{S}(\mathbb{G})\times\widetilde{\mathfrak{S}}_n$ with $\sigma_{\mathbf{x}} = p(\widetilde{\sigma})$. There is a natural quotient $q\colon\widetilde{\mathbf{S}}(\mathbb{G})\to\mathbf{S}(\mathbb{G})$ sending $(\mathbf{x},\widetilde{\sigma})$ to $\mathbf{x}$, which is a $2:1$ map. The set $\widetilde{\mathbf{S}}(\mathbb{G})$ admits an action by $\mathbb{Z}/2\mathbb{Z}$, translating by z on the second factor. Also, $\mathbb{Z}/2\mathbb{Z}$ acts on the polynomial ring $\mathbb{Z}[V_1,\dots,V_n]$, so that the non-trivial element multiplies by -1. Thus, we can form the $\mathbb{Z}[V_1,\dots,V_n]$-module

$$\mathbb{Z}[V_1,\dots,V_n]\times_{\mathbb{Z}/2\mathbb{Z}}\widetilde{\mathbf{S}}(\mathbb{G}) = \frac{\mathbb{Z}[V_1,\dots,V_n]\times\widetilde{\mathbf{S}}(\mathbb{G})}{\mathbb{Z}/2\mathbb{Z}},$$

where the action by $\mathbb{Z}/2\mathbb{Z}$ is diagonal on both factors. This module inherits an action by $\widetilde{\mathfrak{S}}_n$. It can be equipped with an endomorphism

$$\partial\widetilde{\mathbf{x}} = \sum_{\widetilde{\mathbf{y}}\in\widetilde{\mathbf{S}}(\mathbb{G})}\ \sum_{r\in\mathrm{Rect}^\circ(q(\widetilde{\mathbf{x}}),q(\widetilde{\mathbf{y}}))} V_1^{O_1(r)}\cdots V_n^{O_n(r)}\cdot\widetilde{\mathbf{y}}\cdot\widetilde{\tau}(r),$$

where $\widetilde{\tau}$ is the map from Equation (15.15). By Lemma 15.2.10, combined with the considerations in the proof of Lemma 13.2.2, we get that $\partial^2=0$. By the usual arguments showing independence under grid moves, the homology, thought of as a bigraded $\mathbb{Z}[U]$-module, can be shown to be an invariant.

In this set-up a section $\gamma\in\mathcal{S}ec_n$ induces an isomorphism of free $\mathbb{Z}[V_1,\dots,V_n]$-modules $\mathbb{Z}[V_1,\dots,V_n]\otimes\mathbf{S}(\mathbb{G})\to\mathbb{Z}[V_1,\dots,V_n]\otimes_{\mathbb{Z}/2\mathbb{Z}}\widetilde{\mathbf{S}}(\mathbb{G})$, that identifies this construction with the chain complex $\mathcal{GC}^-_S(K;\mathbb{Z})$ defined using the sign assignment S compatible with γ.

CHAPTER 16

The holomorphic theory

Grid homology is a special case of a holomorphic construction of *knot Floer homology*. The purpose of this chapter is to highlight some of the key features of this more general construction. Before giving the definition in Section 16.2, we start in Section 16.1 with a combinatorial representation of three-manifolds (and knots) via *Heegaard diagrams*, closely related to grid diagrams. In Section 16.3, we describe the generalization to links. Finally, in Section 16.4 we explain how the holomorphic construction generalizes grid homology.

16.1. Heegaard diagrams

An *oriented handlebody* is an oriented three-manifold-with-boundary obtained by attaching three-dimensional one-handles to a three-ball. It is a classical result in three-manifold topology that any closed, connected, oriented three-manifold can be decomposed along a separating surface Σ as a union of two handlebodies [**195, 212**]; see also [**207**]. Such a decomposition is called a *Heegaard decomposition.*

A Heegaard decomposition can be specified by a combinatorial object, called a *Heegaard diagram*, which we describe presently. Let Σ be a closed, oriented surface of genus g, and fix a g-tuple of homologically linearly independent, mutually disjoint curves $\gamma = \{\gamma_1, \dots, \gamma_g\}$; equivalently, assume that $\Sigma \backslash (\cup_{i=1}^{g} \gamma_i)$ is a connected, planar surface. We call such a collection of curves a *system of attaching circles*. A system of attaching circles specifies a handlebody with boundary Σ: the circles γ_i bound mutually disjoint, embedded disks D_i in the handlebody.

A *Heegaard diagram* is an oriented surface Σ, called the *Heegaard surface*, equipped with two systems of attaching circles $\boldsymbol{\alpha} = \{\alpha_1, \dots, \alpha_g\}$ and $\boldsymbol{\beta} = \{\beta_1, \dots, \beta_g\}$. We sometimes abbreviate the data by $\mathcal{H} = (\Sigma, \boldsymbol{\alpha}, \boldsymbol{\beta})$. The associated three-manifold Y and its Heegaard splitting consists of the handlebodies U_α and $-U_\beta$, specified by the g-tuples $\boldsymbol{\alpha}$ and $\boldsymbol{\beta}$, glued to Σ.

Any closed, connected, oriented three-manifold can be specified by a Heegaard diagram, and two Heegaard diagrams for the same three-manifold can be connected by a sequence of standard moves [**195, 207, 212**].

EXAMPLE 16.1.1. Consider the Heegaard diagram $(\Sigma, \{\alpha\}, \{\beta\})$ where Σ is the two-dimensional torus, and the simple closed curves α and β intersect (transversely) in a single point. This Heegaard diagram specifies the three-sphere S^3.

To specify a knot K in Y, fix two basepoints w and z in Σ, that are disjoint from all the α_i and the β_j. These two points specify a knot as follows: connect w to z by a smoothly embedded, unknotted arc inside U_α. Here the term "unknotted" should be interpreted as follows: the complement of the attaching disks for the α_i in U_α is a three-ball, w and z specify two points on the boundary of that three-ball, and our path is an unknotted arc in that three-ball. Next, connect z and w by an

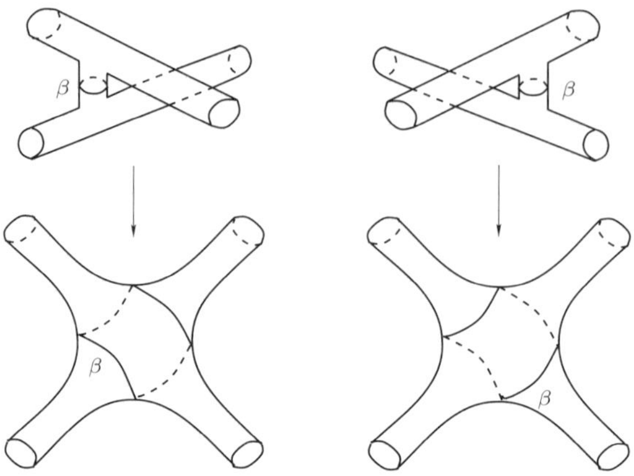

FIGURE 16.1. **The choice of the β-curve at a crossing of the projection.** The two possible choices of β reflect the two possible crossings (under- or over-crossing).

analogous curve in U_β. The union of these two unknotted arcs specifies a knot in Y. In some sense (made precise in Example 16.3.2) the w marking is analogous to the O-marked square in a grid diagram, and the z marking is analogous to the X-marked square in a grid diagram.

A *doubly-pointed Heegaard diagram* for a knot $K \subset Y$ is a Heegaard diagram $(\Sigma, \boldsymbol{\alpha}, \boldsymbol{\beta}, w, z)$ with the property that $(\Sigma, \boldsymbol{\alpha}, \boldsymbol{\beta})$ specifies Y, and the two basepoints w and z together specify the knot $K \subset Y$.

Every pair (Y, K), where Y is a closed, connected, oriented three-manifold, and $K \subset Y$ is a knot can be represented by a doubly-pointed Heegaard diagram. There is a set of standard moves that can be used to connect any two doubly-pointed Heegaard diagrams for the fixed pair (Y, K); see [**172**].

EXAMPLE 16.1.2. Let $K \subset \mathbb{R}^3 \subset S^3$ be a given knot and fix a projection of K to a generic plane $P \subset \mathbb{R}^3$. We can associate a Heegaard diagram for (S^3, K) to this projection as follows. Consider a small neighborhood of the projection in $\mathbb{R}^3$. The closure of this neighborhood will be the β-handlebody U_β; its complement in S^3 is also a handlebody, denoted by U_α. The Heegaard surface Σ is the oriented boundary of U_α. If the projection has n crossings, the genus g of Σ is $n+1$. Observe that Σ intersects the plane containing the projection in $g+1$ simple closed curves. Choose g of these curves, and call them $\alpha_1, \dots, \alpha_g$. For each crossing choose a curve β_i supported in a neighborhood of this crossing as indicated in Figure 16.1. Finally, take β_g to be a meridian of the knot that meets only one of the α_j. Pick z, w on two sides of β_g. It is straightforward to see that the resulting Heegaard diagram represents the knot K in S^3. See Figure 16.2 for a picture of this diagram, in the case where K is the left-handed trefoil knot.

EXAMPLE 16.1.3. Let Σ be the torus, and choose two embedded curves α and β whose signed intersection number is 1. Place two basepoints w and z in their complement. The resulting diagram $(\Sigma, \{\alpha\}, \{\beta\}, w, z)$ represents a knot in S^3. The knots that can be represented by such diagrams are called $(1, 1)$ *knots*.

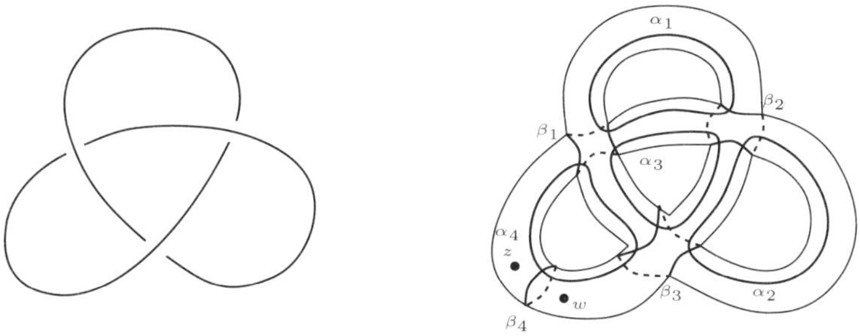

FIGURE 16.2. **Doubly-pointed Heegaard diagram for the left-handed trefoil.**

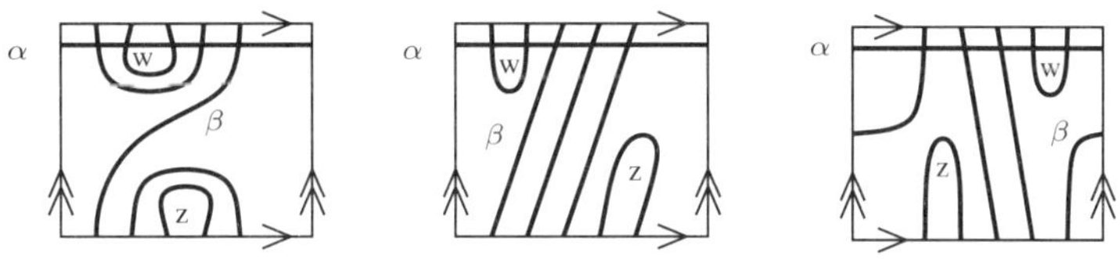

FIGURE 16.3. **Some** $(1,1)$ **diagrams.** We have here pictures for the figure-eight knot4_1, and the torus knots $T_{2,5}$ and $T_{3,4}$, respectively.

See Figure 16.3 for some pictures. The class of $(1,1)$ knots includes for example all 2-bridge knots and all torus knots. See [**32**] for the generalization of the notion of $(1,1)$ knots; and see [**24**] for a parametrization of $(1,1)$ knots. For more on doubly-pointed Heegaard diagrams, see [**169, 172**].

EXERCISE 16.1.4. **(a)** Show that any torus knot has a $(1,1)$ diagram.
(b) Draw a $(1,1)$ diagram for the $(4,5)$ torus knot.
(c) Show that all twist knots are $(1,1)$ knots.

16.2. From Heegaard diagrams to holomorphic curves

One can associate a chain complex to a Heegaard diagram, with the help of *Lagrangian Floer homology*, a construction from symplectic geometry discovered by Andreas Floer [**50**]; see also [**62, 210**]. We sketch this construction in the special case relevant to us.

If Σ is a two-manifold, equipped with a complex structure (a *Riemann surface*), one can consider its *d-fold symmetric product* $\mathrm{Sym}^d(\Sigma)$, which is the space of unordered d-tuples of points on Σ counted with multiplicity. This symmetric product can be thought of as the quotient of the d-fold Cartesian product $\Sigma^{\times d}$ of Σ, divided out by the action of the symmetric group $\mathfrak{S}_d$ on d letters. Although the action is not free, this quotient nonetheless inherits from Σ the structure of a smooth, d-dimensional complex manifold (and hence a $2d$-dimensional real manifold). Locally, the complex structure on $\mathrm{Sym}^d(\Sigma)$ is supplied by the fundamental theorem of algebra, which identifies $\mathrm{Sym}^d(\mathbb{C})$ with $\mathbb{C}^d$ (i.e. a monic polynomial of degree d is uniquely determined either by its unordered d-tuple of roots or by its ordered d-tuple of coefficients). See [**128**] for more on the symmetric product of a Riemann surface.

Starting from a genus g Heegaard diagram $\mathcal{H} = (\Sigma, \boldsymbol{\alpha}, \boldsymbol{\beta})$ for Y, consider the g-fold symmetric product of Σ, equipped with the pair of submanifolds

$$\mathbb{T}_\alpha = \alpha_1 \times \cdots \times \alpha_g \qquad \text{and} \qquad \mathbb{T}_\beta = \beta_1 \times \cdots \times \beta_g,$$

which are two smoothly embedded, g-dimensional tori in $\mathrm{Sym}^g(\Sigma)$. We will always assume that the $\boldsymbol{\alpha}$- and the $\boldsymbol{\beta}$-curves intersect transversely in Σ, and hence that the two tori intersect transversely in $\mathrm{Sym}^g(\Sigma)$. In particular, the intersection $\mathbb{T}_\alpha \cap \mathbb{T}_\beta$ consists of finitely many points.

Let $\mathbf{S}(\mathcal{H})$ denote the set of intersection points $\mathbb{T}_\alpha \cap \mathbb{T}_\beta$; equivalently,

$$\mathbb{T}_\alpha \cap \mathbb{T}_\beta = \bigcup_{\sigma \in \mathfrak{S}_g} \prod_{i=1}^{g} \alpha_{\sigma(i)} \cap \beta_i.$$

There is a chain complex whose generators are the intersection points of these two tori and whose differential uses a complex structure J on $\mathrm{Sym}^g(\Sigma)$ (which in turn is induced from the choice of a complex structure on Σ). The differential counts holomorphic strips, in the following sense:

DEFINITION 16.2.1. A ***holomorphic strip*** connecting $\mathbf{x}, \mathbf{y} \in \mathbf{S}(\mathcal{H})$ is a continuous map $u\colon [0,1] \times \mathbb{R} \to \mathrm{Sym}^g(\Sigma)$ satisfying the following conditions:

(HS-1) u maps $\{0\} \times \mathbb{R}$ into $\mathbb{T}_\beta$,
(HS-2) u maps $\{1\} \times \mathbb{R}$ into $\mathbb{T}_\alpha$,
(HS-3) The one-parameter family of paths $u_t = u|_{[0,1]\times\{t\}}$ converges uniformly to the constant path at $\mathbf{x}$ resp. $\mathbf{y}$ as $t \to -\infty$ resp. $+\infty$.
(HS-4) u is holomorphic; i.e. u satisfies the Cauchy-Riemann equations

$$\frac{\partial u}{\partial s} + J_{u(s+it)} \frac{\partial u}{\partial t} = 0.$$

Note that the strip $[0,1] \times \mathbb{R}$ is conformally equivalent to the unit disk. This equivalence identifies holomorphic strips with holomorphic disks. We use the two notions interchangeably.

A *Whitney disk* is a map that satisfies Properties (HS-1)-(HS-3) of the conditions in Definition 16.2.1. Whitney disks can be collected into homotopy classes: two Whitney disks from $\mathbf{x}$ to $\mathbf{y}$ are *homotopic* if there is a continuous, one parameter family of Whitney disks from $\mathbf{x}$ to $\mathbf{y}$ that connects the given two Whitney disks. We let $\pi(\mathbf{x}, \mathbf{y})$ denote the space of homotopy classes of Whitney disks from $\mathbf{x}$ to $\mathbf{y}$. There is a natural juxtaposition map $\pi(\mathbf{x}_1, \mathbf{x}_2) \times \pi(\mathbf{x}_2, \mathbf{x}_3) \to \pi(\mathbf{x}_1, \mathbf{x}_3)$; the juxtaposition of ϕ_1 and ϕ_2 is written $\phi_1 * \phi_2$.

The algebraic topology of Whitney disks can be concretely understood in terms of the Heegaard diagram. To explain this, note that the $\boldsymbol{\alpha}$- and the $\boldsymbol{\beta}$-curves divide Σ into path-connected components, which we label $\mathcal{D}_1, \dots, \mathcal{D}_m$, i.e.

$$\Sigma \setminus (\boldsymbol{\alpha} \cup \boldsymbol{\beta}) = \mathcal{D}_1 \coprod \cdots \coprod \mathcal{D}_m.$$

DEFINITION 16.2.2. A ***domain*** ψ from $\mathbf{x}$ to $\mathbf{y}$ is a formal linear combination of the $\mathcal{D}_1, \dots, \mathcal{D}_m$, thought of as a two-chain $\psi = \sum_{i=1}^m a_i \cdot \overline{\mathcal{D}}_i$ with $a_i \in \mathbb{Z}$, satisfying the property that the portion of $\partial\psi$ inside $\alpha_1 \cup \cdots \cup \alpha_g$, which we write as $\partial_\alpha \psi$, is a one-chain with

$$\partial(\partial_\alpha \psi) = \mathbf{y} - \mathbf{x} = \sum_{y \in \mathbf{y}} y - \sum_{x \in \mathbf{x}} x.$$

Defining $\partial_\beta \psi$ analogously, it follows that $\partial(\partial_\beta \psi) = \mathbf{x} - \mathbf{y}$.

Domains can be composed: if ϕ is a domain from $\mathbf{x}$ to $\mathbf{y}$, and ψ is a domain from $\mathbf{y}$ to $\mathbf{z}$, then their sum is a domain from $\mathbf{x}$ to $\mathbf{z}$.

A Whitney disk u induces a domain in the above sense, according to the following construction. For any point $p \in \Sigma \backslash (\boldsymbol{\alpha} \cup \boldsymbol{\beta})$, let $n_p(u)$ be the algebraic intersection number of u with the submanifold $\{p\} \times \mathrm{Sym}^{g-1}(\Sigma) \subset \mathrm{Sym}^g(\Sigma)$. For any $\mathcal{D}_i$, let $n_i(u)$ denote $n_q(u)$ for any $q \in \mathcal{D}_i$. Observe that $n_i(u)$ is independent of the choice of $q \in \mathcal{D}_i$; and it is called the *local multiplicity of u at $\mathcal{D}_i$*. We can put these local multiplicities together to give a domain associated to u: $u \mapsto \sum_i n_i(u) \cdot \overline{\mathcal{D}}_i$. Homotopic Whitney disks induce the same domain; and if $\mathcal{D}(\phi)$ denotes the domain associated to the homotopy class ϕ, then $\mathcal{D}(\phi * \psi) = \mathcal{D}(\phi) + \mathcal{D}(\psi)$. In fact, when $g > 2$, the map $\phi \mapsto \mathcal{D}(\phi)$ gives a one-to-one correspondence between homotopy classes of Whitney disks and domains from $\mathbf{x}$ to $\mathbf{y}$. A key property we will use is the following positivity:

LEMMA 16.2.3. ([**174**, Lemma 3.2]) *The local multiplicities of a holomorphic strip are all non-negative.* □

Holomorphic strips can be naturally collected into moduli spaces, indexed by $\phi \in \pi(\mathbf{x}, \mathbf{y})$. These moduli spaces are denoted by $\mathcal{M}(\phi)$. Vertical translation of the strip induces an action of $\mathbb{R}$ on $\mathcal{M}(\phi)$. There is a function $\mu\colon \pi(\mathbf{x}, \mathbf{y}) \to \mathbb{Z}$, the *Maslov index*, that measures the expected dimension of the moduli space $\mathcal{M}(\phi)$; see for instance [**51, 197**]. The Maslov index satisfies the following two key properties:

(1) $\mu(\phi_1 * \phi_2) = \mu(\phi_1) + \mu(\phi_2)$, and
(2) if $\mathcal{M}(\phi)$ is a smooth manifold in a neighborhood of some $u \in \mathcal{M}(\phi)$ (in a suitable technical sense), then its dimension is computed by $\mu(\phi)$.

With these preliminaries, we can sketch the construction of the knot Floer complex. Fix a Heegaard diagram $\mathcal{H} = (\Sigma, \boldsymbol{\alpha}, \boldsymbol{\beta}, w, z)$ for (Y, K). Although the construction of knot Floer homology can be set up for any closed, oriented, connected three-manifold Y, the case of non-trivial first homology $H_1(Y;\mathbb{Z})$ requires special attention. For simplicity of exposition, we assume that Y is an integral homology sphere, meaning that $H_1(Y;\mathbb{Z}) = 0$. In fact, the case of primary interest in this book is when $Y \cong S^3$.

Let $\widehat{\mathrm{CFK}}(\mathcal{H})$ be the finite dimensional $\mathbb{F}$-vector space generated by $\mathbf{S}(\mathcal{H})$. There is a differential $\widehat{\partial}_K$ on $\widehat{\mathrm{CFK}}(\mathcal{H})$ with the property that the coefficient of $\mathbf{y} \in \mathbf{S}(\mathcal{H})$ in $\widehat{\partial}_K \mathbf{x}$ counts points (mod 2) in the zero-dimensional moduli spaces $\mathcal{M}(\phi)/\mathbb{R}$ (i.e. only for those ϕ for with $\mu(\phi) = 1$) that satisfy the further constraints that $n_z(\phi) = n_w(\phi) = 0$. Explicitly:

$$\widehat{\partial}_K \mathbf{x} = \sum_{\mathbf{y} \in \mathbf{S}(\mathcal{H})} \sum_{\{\phi \in \pi(\mathbf{x},\mathbf{y}) \,\big|\, \mu(\phi)=1, n_w(\phi)=n_z(\phi)=0\}} \# \left(\frac{\mathcal{M}(\phi)}{\mathbb{R}}\right) \mathbf{y}. \tag{16.1}$$

In the above formula, the point counts in the moduli spaces are to be taken only mod 2, so that these homology groups are defined over $\mathbb{F} = \mathbb{Z}/2\mathbb{Z}$. Lifting to a chain complex over $\mathbb{Z}$ can be defined with more work, using orientation systems on the moduli spaces [**174**], see also [**62, 210**].

In the above discussion, we have pretended that the moduli spaces of holomorphic strips $\mathcal{M}(\phi)$ with $\mu(\phi) = 1$ are always 1-dimensional manifolds. This is not true in general: to ensure the necessary transversality properties, we relax the notion of holomorphicity. This is a standard practice in symplectic geometry [**81**] (and

Lagrangian Floer homology [**50, 55, 161**]) where one fixes an open set of almost-complex structures that are suitably compatible with a given symplectic structure, and then considers perturbed-holomorphic curves satisfying the equation

$$\frac{\partial u}{\partial s} + J(s)_{u(s+it)}\frac{\partial u}{\partial t} = 0,$$

where now $\{J(s)\}_{s\in[0,1]}$ is a generic path of such almost complex structures. Such maps are called *pseudo-holomorphic* strips.

We wish to apply Gromov's compactness theorem to show that the point counts appearing in the right-hand-side of Equation (16.1) are finite. Gromov's compactness applies when there are energy bounds on the curves. When fixing a homotopy class of pseudo-holomorphic strips with Lagrangian boundary condition, these energy bounds come for free [**81**]. In [**174**], the energy bounds came from a related mechanism, after lifting to the cartesian product; another approach, choosing a suitable symplectic form on the symmetric product, was found by T. Perutz [**186**]. Moreover, the condition that $b_1(Y) = 0$ provides a universal energy bound on all homotopy classes with Maslov index one, ensuring that the sum appearing on the right-hand-side in Equation (16.1) has finitely many terms.

It is a surprisingly subtle analytic fact (which is fundamental to Lagrangian Floer homology) that the resulting map $\widehat{\partial}_K$ is, indeed, a differential; i.e. that $(\widehat{\partial}_K)^2 = 0$ holds. The proof is achieved by considering, for each pair of generators $\mathbf{x}$ and $\mathbf{z}$ and for each non-negative domain $\phi \in \pi(\mathbf{x},\mathbf{z})$ with $\mu(\phi) = 2$, and $n_w(\phi) = n_z(\phi) = 0$ the moduli space $\mathcal{M}(\phi)/\mathbb{R}$. There is a structure theorem describing the compactification of $\mathcal{M}(\phi)/\mathbb{R}$; see [**52, 55, 62, 174**]. Specifically, it is proved that there is a one-dimensional compact manifold-with-boundary whose interior is $\mathcal{M}(\phi)/\mathbb{R}$, and whose boundary is identified with

$$\bigcup_{\mathbf{y}} \bigcup_{\left\{\begin{smallmatrix}\phi_1\in\pi(\mathbf{x},\mathbf{y})\\ \phi_2\in\pi(\mathbf{y},\mathbf{z})\end{smallmatrix}\,\middle|\,\begin{smallmatrix}\mu(\phi_1)=\mu(\phi_2)=1,\ \phi_1*\phi_2=\phi,\\ n_z(\phi_1)=n_z(\phi_2)=n_w(\phi_1)=n_w(\phi_2)=0\end{smallmatrix}\right\}} \left(\frac{\mathcal{M}(\phi_1)}{\mathbb{R}}\right)\times\left(\frac{\mathcal{M}(\phi_2)}{\mathbb{R}}\right).$$

The structure theorem consists of several steps: a transversality argument, showing that $\mathcal{M}(\phi)/\mathbb{R}$ is a smooth manifold; a Gromov compactness argument, that compactifies the moduli space; and a gluing argument that identifies its boundary as $\left(\frac{\mathcal{M}(\phi_1)}{\mathbb{R}}\right)\times\left(\frac{\mathcal{M}(\phi_2)}{\mathbb{R}}\right)$ with $\phi_1 * \phi_2 = \phi$. Further restrictions can be placed on ϕ_1 and ϕ_2 when $\mathcal{M}(\phi_1)$ and $\mathcal{M}(\phi_2)$ are non-empty: Additivity of the Maslov index under juxtaposition ensures that $\mu(\phi_1) + \mu(\phi_2) = 2$, and transversality again ensures $\mu(\phi_i) \geq 1$, forcing $\mu(\phi_1) = 1 = \mu(\phi_2)$. Similarly, additivity of n_w gives $n_w(\phi_1) + n_w(\phi_2) = 0$, while Lemma 16.2.3 gives $n_w(\phi_i) \geq 0$, so we can conclude $n_w(\phi_1) = 0 = n_w(\phi_2)$. Similarly, $n_z(\phi_1) = n_z(\phi_2) = 0$. Since a compact one-manifold has an even number of boundary points, we can conclude that

$$\sum_{\mathbf{y}} \sum_{\left\{\begin{smallmatrix}\phi_1\in\pi(\mathbf{x},\mathbf{y})\\ \phi_2\in\pi(\mathbf{y},\mathbf{z})\end{smallmatrix}\,\middle|\,\begin{smallmatrix}\mu(\phi_1)=\mu(\phi_2)=1,\ \phi_1*\phi_2=\phi,\\ n_z(\phi_1)=n_z(\phi_2)=n_w(\phi_1)=n_w(\phi_2)=0\end{smallmatrix}\right\}} \#\left(\frac{\mathcal{M}(\phi_1)}{\mathbb{R}}\right)\cdot\#\left(\frac{\mathcal{M}(\phi_2)}{\mathbb{R}}\right) = 0.$$

Adding up the left-hand-side over all $\phi \in \pi(\mathbf{x},\mathbf{z})$ with $\mu(\phi) = 2$ computes the $\mathbf{z}$ component of $\widehat{\partial}_K^2(\mathbf{x})$, so $\widehat{\partial}_K \circ \widehat{\partial}_K = 0$.

It turns out that the homology of this chain complex is independent of the analytic choices made in its construction. This is again familiar in Lagrangian

Floer homology. More surprising is the fact that the homology groups turn out to be independent of the choice of the Heegaard diagram going into their construction, and the homology depends only on the underlying pair (Y, K).

The chain complex is bigraded, with bigrading determined by functions M and A (the *Maslov* and *Alexander* gradings) on the generators satisfying the property that for fixed $\mathbf{x}, \mathbf{y} \in \mathbf{S}(\mathcal{H})$ and $\phi \in \pi(\mathbf{x}, \mathbf{y})$, the following relations hold:

$$M(\mathbf{x}) - M(\mathbf{y}) = \mu(\phi) - 2n_w(\phi) \tag{16.2}$$

$$A(\mathbf{x}) - A(\mathbf{y}) = n_z(\phi) - n_w(\phi). \tag{16.3}$$

These equations determine M and A up to overall translations. When $Y \cong S^3$, this indeterminacy can be removed as follows.

We start with the indeterminacy in M. Observe that the differential in Equation (16.1) can be modified, to allow disks to cross z; i.e. we can consider

$$\widehat{\partial}\mathbf{x} = \sum_{\mathbf{y} \in \mathbf{S}(\mathcal{H})} \sum_{\{\phi \in \pi(\mathbf{x}, \mathbf{y}) \big| \mu(\phi)=1, n_w(\phi)=0\}} \# \left(\frac{\mathcal{M}(\phi)}{\mathbb{R}} \right) \mathbf{y}. \tag{16.4}$$

(Compare the simply-blocked grid complex $\widehat{\mathcal{GC}}(\mathbb{G})$ from Definition 13.2.7.) This complex no longer depends on the knot, as it refers to only one of the two basepoints. In general, the homology of this complex is the Heegaard Floer homology $\widehat{\mathrm{HF}}$ of the ambient closed three-manifold Y. In the case at hand, where $Y \cong S^3$, the homology turns out to be one-dimensional. The complex retains its relative grading from M; and so, we can pin down the additive indeterminacy of M by requiring that the one-dimensional homology is supported in Maslov grading 0.

Next, we study A. It turns out that the graded Euler characteristic

$$\sum_{d,s} (-1)^d \dim_{\mathbb{F}} \widehat{\mathrm{CFK}}_d(K, s) t^s = \sum_{\mathbf{x} \in \mathbf{S}(\mathcal{H})} (-1)^{M(\mathbf{x})} t^{A(\mathbf{x})} \tag{16.5}$$

agrees, up to an overall multiple of some power of t, with the (symmetrized) Alexander polynomial of $K \subset S^3$. The additive indeterminacy of A is eliminated by the requirement that this Euler characteristic is symmetric in t; or, equivalently, by the condition that, for each $s \in \mathbb{Z}$,

$$\#\{\mathbf{x} \in \mathbf{S}(\mathcal{H}) \big| A(\mathbf{x}) = s\} \equiv \#\{\mathbf{x} \in \mathbf{S}(\mathcal{H}) \big| A(\mathbf{x}) = -s\} \pmod 2.$$

With these grading conventions, we have the following:

THEOREM 16.2.4. *The homology of the chain complex* $(\widehat{\mathrm{CFK}}(\mathcal{H}), \widehat{\partial}_K)$ *is a bigraded* $\mathbb{F}$*-vector space, which is an invariant of the knot* K. □

The bigraded homology group $H(\widehat{\mathrm{CFK}}(\mathcal{H}), \widehat{\partial}_K)$ considered in the above theorem is called the *knot Floer homology* of K, and it is denoted $\widehat{\mathrm{HFK}}(K)$.

There are several variations on this theme, as should be clear by now from the analogy with grid homology. For example, one could consider the chain complex $\mathrm{CFK}^-(\mathcal{H})$ over $\mathbb{F}[U]$ freely generated by $\mathbf{S}(\mathcal{H})$, and equipped with the differential

$$\partial_K^- \mathbf{x} = \sum_{\mathbf{y} \in \mathbf{S}(\mathcal{H})} \sum_{\{\phi \in \pi(\mathbf{x}, \mathbf{y}) \big| \mu(\phi)=1, n_z(\phi)=0\}} \# \left(\frac{\mathcal{M}(\phi)}{\mathbb{R}} \right) U^{n_w(\phi)} \mathbf{y}. \tag{16.6}$$

Specializing $U = 0$ recaptures $\widehat{\mathrm{CFK}}(\mathcal{H})$. Taking homology, we obtain the knot invariant $\mathrm{HFK}^-(K)$, which is a module over $\mathbb{F}[U]$.

In a similar vein, consider the $\mathbb{F}[U]$-module $\mathrm{CF}^-(\mathcal{H})$ generated freely by $\mathbf{S}(\mathcal{H})$ and equip it with a differential ∂^- counting even more disks, defined by

$$\partial^- \mathbf{x} = \sum_{\mathbf{y}\in\mathbf{S}(\mathcal{H})} \sum_{\{\phi\in\pi(\mathbf{x},\mathbf{y})\big|\mu(\phi)=1\}} \#\left(\frac{\mathcal{M}(\phi)}{\mathbb{R}}\right) U^{n_w(\phi)}\mathbf{y}. \tag{16.7}$$

The homology of the resulting chain complex does not depend on the chosen knot K (since its definition does not use the second basepoint z at all); it is the Heegaard Floer homology HF^- of S^3, which in turn is isomorphic to $\mathbb{F}[U]$. (Compare the filtered grid complex $\mathcal{GC}^-(\mathbb{G})$ from Definition 13.2.1.) The Alexander grading A defined above induces a $\mathbb{Z}$-filtration on $\mathrm{CF}^-(\mathcal{H})$. The filtered chain homotopy type of the resulting filtered chain complex is a knot invariant; the knot Floer homology group $\mathrm{HFK}^-(K)$ is the homology of the associated graded object.

It is, in general, difficult to compute the knot Floer homology directly from its definition. In some cases, computations can be done with a fortuitous choice of Heegaard diagram. For instance, the Heegaard diagram from Example 16.1.2 can be used to give an easy computation of the knot Floer homology for alternating knots in terms of their Alexander polynomials and signatures, cf. Theorem 10.3.1; see [**169**].

EXERCISE 16.2.5. **(a)** Find the generators for the chain complex associated to the Heegaard diagram from Figure 16.2. Compute their Alexander gradings.
(b)* Show that for an alternating knot K, $\dim \widehat{\mathrm{HFK}}(K) = \det(K)$.

Torus knots form another class of knots for which knot Floer homology groups can be computed easily. Let (p, q) be a pair of relatively prime, positive integers, and let $T_{p,q}$ denote the (p, q) torus knot. Recall that the (symmetrized) Alexander polynomial $\Delta_{T_{p,q}}(t)$ of $T_{p,q}$ is given by $t^{\frac{pq-p-q+1}{2}}\Delta_{T_{p,q}}(t) = \frac{(t^{pq}-1)(t-1)}{(t^p-1)(t^q-1)}$.

THEOREM 16.2.6. *Let $T_{p,q} \subset S^3$ be the positive (p, q) torus knot, and write its (symmetrized) Alexander polynomial as $\Delta_{T_{p,q}}(t) = (-1)^k \sum_{i=-k}^{k}(-1)^i t^{n_i}$ for some increasing sequence of integers $\{n_i\}_{i=-k}^{k}$. Consider the associated sequence of integers $\{\delta_i\}_{i=-k}^{k}$ inductively defined by*

$$\delta_i = \begin{cases} 0 & \text{if } i = k, \\ \delta_{i+1} - 2(n_{i+1} - n_i) + 1 & \text{if } k - i \text{ is odd}, \\ \delta_{i+1} - 1 & \text{if } k - i > 0 \text{ is even}. \end{cases}$$

Then, $\widehat{\mathrm{HFK}}(T_{p,q}) \cong \bigoplus_{i=-k}^{k} \mathbb{F}_{(\delta_i, n_i)}$. □

While the proof of the above theorem from [**177**] uses the relationship between knot Floer homology and Heegaard Floer homology, a more direct way to study knot Floer homology for torus knots uses their $(1, 1)$ diagrams, where the differential on the chain complex is combinatorial, as we shall describe presently.

Holomorphic disk counting in a $(1, 1)$ diagram for a knot can be done as follows (cf. [**172**, Section 6.2]). Start from a $(1, 1)$ diagram for a knot, and fix a homotopy class of disks $\phi \in \pi(x, y)$ in the torus. In a sufficiently small neighborhood of each corner point x and y, the curves α and β divide the neighborhood into four quadrants. Suppose that at both corner points x and y the local multiplicity of ϕ is zero at three of the four quadrants (and hence 1 at the remaining one). Suppose moreover that the lift of ϕ to the universal cover $\mathbb{R}^2$ is an embedded disk. Then,

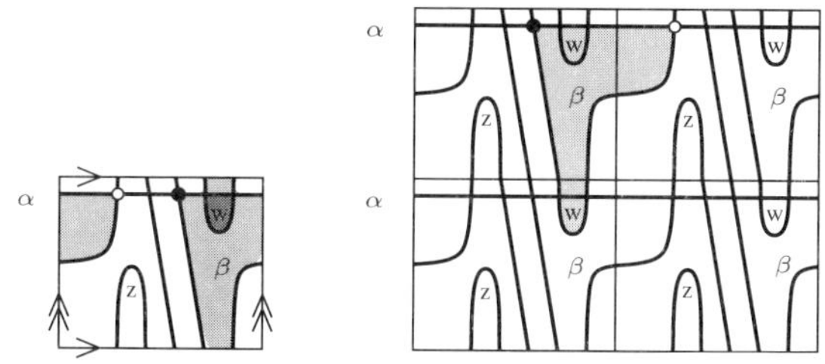

FIGURE 16.4. **Lifting disks.** The homotopy class indicated on the left (where the darker shaded region has multiplicity 2 and the lighter one multiplicity 1) can be lifted to the universal cover to give the embedded disk on the right.

$\mu(\phi) = 1$ and, by the Riemann mapping theorem, ϕ has a unique holomorphic representative up to translation. See Figure 16.4.

EXERCISE 16.2.7. **(a)** Verify Theorem 16.2.6 for $T_{3,4}$ using its $(1,1)$ diagram given in Figure 16.3.
(b) More generally, verify Theorem 16.2.6 for $T_{3,3n+1}$ using its $(1,1)$ diagram.
(c)*Prove Theorem 16.2.6.

16.3. Multiple basepoints

To bridge the above construction with grid diagrams, and even to consider links with more than one component, we must broaden the types of Heegaard diagrams we use. In the notion of doubly-pointed Heegaard diagrams, we considered knots that meet the Heegaard surface in exactly two points, so that the surface divides the knot into two unknotted arcs in the handlebodies. This notion can be generalized to allow for diagrams representing a knot that is divided into $2n$ unknotted arcs by the Heegaard surface.

DEFINITION 16.3.1. A ***$2n$-pointed Heegaard diagram*** is a surface Σ of genus g equipped with two $(g+n-1)$-tuples of pairwise disjoint, embedded, closed curves $\{\alpha_1, \dots, \alpha_{g+n-1}\}$ and $\{\beta_1, \dots, \beta_{g+n-1}\}$, and two n-tuples of basepoints $\mathbf{w} = w_1, \dots, w_n$ and $\mathbf{z} = z_1, \dots, z_n$, satisfying the following conditions:

- $\boldsymbol{\alpha} = \{\alpha_1, \dots, \alpha_{g+n-1}\}$ spans a g-dimensional subspace in $H_1(\Sigma)$; equivalently, $\Sigma \setminus \boldsymbol{\alpha}$ decomposes as a disjoint union of n connected sets $A_1, \dots, A_n$ and each A_i is a planar surface.
- The analogous property holds for $\boldsymbol{\beta} = \{\beta_1, \dots, \beta_{g+n-1}\}$, so that $\Sigma \setminus \boldsymbol{\beta}$ decomposes as a disjoint union of planar surfaces $B_1, \dots, B_n$.
- Each component A_i contains exactly one of the basepoints in $\mathbf{w}$, which we label w_i.
- Each component B_j contains exactly one of the basepoints in $\mathbf{w}$.
- Each component A_i contains exactly one of the basepoints in $\mathbf{z}$.
- Each component B_j contains exactly one of the basepoints in $\mathbf{z}$.

A $2n$-pointed Heegaard diagram gives rise to a three-manifold Y, together with an oriented link $\vec{L}$ in Y as follows. Start from $[-1,1] \times \Sigma$, and attach three-dimensional 2-handles along copies of the $\boldsymbol{\alpha}$-curves in $\{-1\} \times \Sigma$, and then attach three-dimensional 2-handles along the copies of the $\boldsymbol{\beta}$-curves in $\{1\} \times \Sigma$. The result has $2n$ S^2-boundaries, which we can close off with $2n$ three-dimensional 3-handles to get the closed three-manifold Y. Connecting the basepoints that share components A_i (or B_j) in the handlebody given by the $\boldsymbol{\alpha}$-curves ($\boldsymbol{\beta}$-curves, resp.) with unknotted, unlinked arcs, we get an embedded closed one-manifold L in Y. Next, orient the link L so that the portions of the link in the α-handlebody point away from the w-basepoints and into the z-basepoints. Depending on the combinatorics of the curves and basepoints, the resulting oriented one-manifold $\vec{L}$ can have as many as n components. Indeed, any oriented link in any three-manifold can be presented by some $2n$-pointed Heegaard diagram.

EXAMPLE 16.3.2. A basic example of a $2n$-pointed Heegaard diagram is given by a toroidal grid diagram. The grid torus provides the Heegaard surface, the horizontal and vertical circles are the $\boldsymbol{\alpha}$- and $\boldsymbol{\beta}$-curves, the O-markings correspond to the basepoints $\mathbf{w}$ and the X-markings correspond to the basepoints $\mathbf{z}$.

When working with multi-pointed Heegaard diagrams, we consider the chain complex $\mathrm{CFK}^-(\mathcal{H})$ freely generated by the elements of $\mathbb{T}_\alpha \cap \mathbb{T}_\beta$ over the ring $\mathbb{F}[V_1, \dots, V_n]$. The variable V_i corresponds to the basepoint $w_i \in \mathbf{w}$.

In the differential we count disks in $\mathrm{Sym}^{g+n-1}(\Sigma)$, and consider

(16.8)
$$\partial^-_{\vec{L}}\mathbf{x} = \sum_{\mathbf{y}\in\mathbf{S}(\mathcal{H})} \sum_{\{\phi\in\pi(\mathbf{x},\mathbf{y})\,\big|\,\mu(\phi)=1, n_{z_1}(\phi)=\dots=n_{z_n}(\phi)=0\}} \#\left(\frac{\mathcal{M}(\phi)}{\mathbb{R}}\right)\cdot V_1^{n_{w_1}(\phi)}\cdots V_n^{n_{w_n}(\phi)}\cdot\mathbf{y}.$$

In the case where the diagram represents a knot, the homology of this complex is isomorphic to $\mathrm{HFK}^-(K)$ of Theorem 16.4.1, where the action by U is defined to be multiplication by any V_i.

The multi-pointed setting allows us to consider also links with more than one components. To ensure finite sums in Equation (16.8), we need the diagram to have the additional *admissibility requirement*: if $P = \sum a_i A_i + b_i B_i$ is a non-zero two-chain with $n_{w_i}(P) = n_{z_i}(P) = 0$ for all $i = 1, \dots, n$, then there are points p and q where the local multiplicity of P is positive and negative respectively. (A Heegaard diagram derived from a grid automatically satisfies this property.)

With this constraint, Equation (16.8) specifies a differential. If an ℓ-component, oriented link $\vec{L} = (L_1, \dots, L_\ell)$ is represented by a $2n$-pointed diagram, then the chain complex is naturally an $\mathbb{F}[V_1, \dots, V_n]$-module, while the homology inherits the structure of a module over the ring $\mathbb{F}[U_1, \dots, U_\ell]$, where U_i acts by multiplication by some V_{j_i} corresponding to the ith component $\vec{L}_i$ of the link $\vec{L}$. The homology of the resulting chain complex, thought of as a module over $\mathbb{F}[U_1, \dots, U_\ell]$, is an invariant of the underlying oriented link $\vec{L} \subset S^3$, and it is called the *link Floer homology* $\mathrm{HFL}^-(\vec{L})$ of $\vec{L}$. We can further specialize the complex, setting $V_{j_1} = \dots = V_{j_\ell} = 0$, to get a simpler version of link Floer homology, $\widehat{\mathrm{HFL}}(\vec{L})$. For a link $\vec{L}$ the Alexander grading naturally takes values in the Alexander grading set $\mathbf{H}(\vec{L})$ from Definition 11.1.3. For more on this construction, see [**180**].

Through an appropriate Euler characteristic, link Floer homology computes the multi-variable Alexander polynomial. To state this result, for a given $h \in \mathbf{H}(\vec{L})$, we let $\mathbf{T}^h$ denote the monomial in $t_1, \dots, t_\ell$, specified so that the exponent of t_i is the coefficient of μ_i in h.

THEOREM 16.3.3. ([**180**, Theorem 1.3]) *Let $\vec{L}$ be an oriented ℓ-component link in S^3 with $\ell > 1$, and consider the identification $H_1(S^3 - \vec{L};\mathbb{Z}) \cong \mathbb{Z}^\ell$ as in Equation* 2.1. *Then,*

$$\sum_{h\in\mathbf{H}} \chi(\widehat{\mathrm{HFL}}_*(\vec{L},h)) \cdot \mathbf{T}^h = \pm \prod_{i=1}^{\ell} (t_i^{\frac{1}{2}} - t_i^{-\frac{1}{2}}) \cdot \Delta_{\vec{L}}(t_1,\dots,t_\ell),$$

where $\Delta_{\vec{L}}$ denotes the (symmetrized) multi-variable Alexander polynomial of $\vec{L}$. □

By considering the boundary map

$$\partial^- \mathbf{x} = \sum_{\mathbf{y}\in\mathbf{S}(\mathcal{H})} \sum_{\{\phi\in\pi(\mathbf{x},\mathbf{y})\big|\mu(\phi)=1\}} \#\left(\frac{\mathcal{M}(\phi)}{\mathbb{R}}\right) \cdot V_1^{n_{w_1}(\phi)} \cdots V_n^{n_{w_n}(\phi)} \cdot \mathbf{y}$$

we get a chain complex associated to the multi-pointed Heegaard diagram whose homology is independent of the choice of the link $\vec{L}$. The multi-valued Alexander function equips this chain complex with a multi-filtration. The multi-filtered chain homotopy type of the resulting multi-filtered, $\mathbb{Z}$-graded chain complex is a link invariant. This invariant, denoted $\mathrm{CFK}^{-,*}(\vec{L})$ in [**180**], corresponds to the filtered link invariant $\mathcal{GC}^-(\vec{L})$ from Section 14.5. The link Floer homology group $\mathrm{HFL}^-(\vec{L})$ is the homology of the associated graded object of the multi-filtered chain complex $\mathrm{CFK}^{-,*}(\vec{L})$.

16.4. Equivalence of knot Floer homology with grid homology

In the case where the $2n$-pointed Heegaard diagram $\mathcal{H}$ for a knot K is a grid diagram $\mathbb{G}$ (cf. Example 16.3.2), the differential on the knot Floer complex, which appears to depend on analytic choices, turns out to agree with the differential on the grid complex, according to the following:

THEOREM 16.4.1 ([**135**]). *If $\mathcal{H}$ is a Heegaard diagram induced from a grid diagram $\mathbb{G}$, then the chain complex $(\mathrm{CFK}^-(\mathcal{H}),\partial_K^-)$ is isomorphic to $(GC^-(\mathbb{G}),\partial_{\mathbb{X}}^-)$. In a similar vein, the filtered complex $(\mathrm{CFK}^{-,*}(\mathcal{H}),\partial^-)$ is isomorphic to $(\mathcal{GC}^-(\mathbb{G}),\partial^-)$.*

The key step in the proof of this result is the following explicit determination of the corresponding moduli space.

LEMMA 16.4.2. *Fix a Heegaard diagram $\mathcal{H}$ derived from a grid diagram $\mathbb{G}$ as in Example* 16.3.2. *If $\phi \in \pi(\mathbf{x},\mathbf{y})$ is a homotopy class whose associated domain is an empty rectangle, then $\mu(\phi) = 1$ and $\#(\frac{\mathcal{M}(\phi)}{\mathbb{R}}) = 1$.*

Proof. The lemma is verified using the following interpretation of holomorphic disks in the symmetric product. A holomorphic disk in $\mathrm{Sym}^d(\Sigma)$ can be viewed as the following collection (F, P, f) of data: (1) a surface-with-boundary F, equipped with a complex structure, (2) a degree-d holomorphic map $P\colon F \to D$, where D is the standard disk in $\mathbb{C}$, and (3) a holomorphic map $f\colon F \to \Sigma$. See [**174**, Lemma 3.6]; see also [**120**] for a development of the entire theory from this perspective. In this correspondence, the domain $\mathcal{D}$ of the holomorphic disk can be thought of as the two-chain induced by the map f from F to Σ.

Consider the case where $R \subset \Sigma$ is an empty rectangle, and $d = 2$. By the Riemann mapping theorem, the rectangle R is conformally equivalent to the (closed) upper half plane $\overline{H}$ (with its point at infinity), via an equivalence that carries its four corners to the four real numbers $y_1 < x_1 < y_2 < x_2$ (in this order). Since by our assumption R appears with multiplicity 1 in the domain associated to the holomorphic strip, the map f is a conformal equivalence, so we can choose $\overline{H}$ to be F.

Consider the map $Q\colon \overline{H} \to \overline{H}$ given by the formula

$$Q(z) = \frac{(z - x_1)(z - x_2)}{(z - y_1)(z - y_2)}. \tag{16.9}$$

Post-composing Q with a conformal equivalence $K\colon \overline{H} \to D$, sending $0 \in \overline{H}$ to $-i$ and $\infty \in \overline{H}$ to i gives $P = K \circ Q\colon F = \overline{H} \to D$. The conformal equivalence from F to R supplies the needed map f. The map Q above is the unique branched double cover $\overline{H} \to \overline{H}$ that vanishes on x_1 and x_2 and sends y_1 and y_2 to infinity. The map P is not unique, since K is characterized only up to an $\mathbb{R}$ action. This shows that $\frac{\mathcal{M}(\phi)}{\mathbb{R}}$ consists of one point.

More generally, when $R \subset \Sigma$ is an empty rectangle, and $d > 2$, we take F to be the disjoint union of $d-1$ disks, one of which we think of as $\overline{H}$, and the others are $d-2$ copies of D. On $\overline{H}$ we use the map as before; the other components D are mapped by f to constant points in Σ, and they are mapped by P by the identity map to D.

The lemma is verified after one shows that the single holomorphic disk exhibited above is transversally cut out in its moduli space; see for example [**174**, Proposition 3.9]. □

A domain is called *positive* if its local multiplicities are all non-negative, and at least one of them is positive. We will deduce Theorem 16.4.1 from Lemma 16.4.2 and the following factorization result for positive domains:

PROPOSITION 16.4.3. *Let $\mathbb{G}$ be a grid diagram, fix $\mathbf{x}, \mathbf{y} \in \mathbf{S}(\mathbb{G})$, and let $\psi \in \pi(\mathbf{x}, \mathbf{y})$ be a positive domain. Then, there is a sequence of grid states $\{\mathbf{x}_i\}_{i=1}^{k}$ with $\mathbf{x}_1 = \mathbf{x}$ and $\mathbf{x}_k = \mathbf{y}$, and empty rectangles $\{r_i\}_{i=1}^{k-1}$, with $r_i \in \mathrm{Rect}^\circ(\mathbf{x}_i, \mathbf{x}_{i+1})$, so that $\psi = r_1 * \cdots * r_{k-1}$.*

Proof. We prove this statement by induction on the total multiplicity of ψ. Given a positive domain ψ, we will give an algorithm for finding a (not necessarily empty) rectangle r and another non-negative domain ψ' with $\psi = r * \psi'$ or $\psi' * r$. Since a non-empty rectangle can be expressed as a juxtaposition of empty rectangles (see Figure 15.1), the result will follow.

Choose a vertical annulus where some local multiplicity is non-zero. If the local multiplicities of ψ are all positive, we can decompose the annulus as a juxtaposition of two rectangles $r_1 * r_2$ from $\mathbf{x}$ to $\mathbf{x}$, and then write $\psi = r_1 * r_2 * \psi'$.

After applying the same logic to the rows instead of the columns, we can assume that ψ has the property that each row and each column has some square where the local multiplicity is zero.

For $i = 1, \ldots, n$, consider $a_i = (\partial\psi) \cap \alpha_i$. For a general domain, a_i is an immersed interval in α_i. Since each row has a square where the local multiplicity of ψ is zero, it follows that a_i is either trivial or it is an embedded, oriented interval. We call the pair of α-circles (α_i, α_j) an *admissible pair* if the following holds:

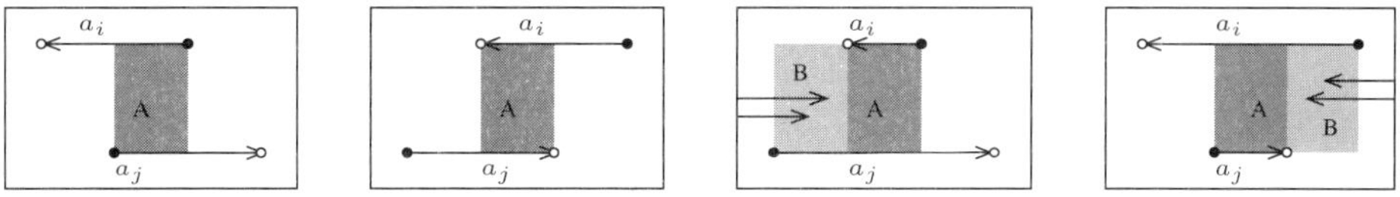

FIGURE 16.5. **The four types of admissible pairs.** When (α_i, α_j) has minimal height; the intervals can overlap in one of the four ways pictured. In all four cases, (by minimality) the region labelled A cannot meet any other a_k. In the first two cases, that region forms a rectangle that can be factored off (on the left or the right, respectively). In the third and fourth cases, the region B can meet other a_k; but (by minimality) all the arcs that meet B are oriented in the same direction as a_j and a_i, respectively. Thus, ψ has positive local multiplicities in $A \cup B$, and hence that region, thought of as a rectangle, can be factored off, leaving a non-negative domain.

- both a_i and a_j are intervals, a_i is oriented from east to west and a_j is oriented from west to east, and the horizontal projections $\mathrm{pr}_h(a_i)$ and $\mathrm{pr}_h(a_j)$ have non-trivial overlap.

If a given vertical annulus has local multiplicity zero somewhere and non-zero somewhere else, then there are two oppositely oriented intervals a_i and a_j that meet the given vertical annulus. In particular, the set of admissible pairs is non-empty.

Let (α_i, α_j) be an admissible pair. Draw them in a fundamental domain so that α_i is above α_j. Then the *height* of (α_i, α_j) is the height of α_i minus the height of α_j. Let (α_i, α_j) be an admissible pair with minimal height, which exists by the above considerations. Let a_i be an arc from x_i to y_i and a_j be an arc from x_j to y_j. There are four cases, as illustrated in Figure 16.5; and in each case, we can factor off a rectangle. Induction on the total multiplicity then concludes the proof. □

Proof of Theorem 16.4.1. First observe that the grid states generating $GC^-(\mathbb{G})$ correspond to intersection points between $\mathbb{T}_\alpha$ and $\mathbb{T}_\beta$. The desired isomorphism $\mathrm{CFK}^-(\mathcal{H}) \cong GC^-(\mathbb{G})$ of chain complexes identifies the boundary maps as follows.

By Lemma 16.2.3, the domain ϕ associated to a holomorphic strip has non-negative local multiplicities everywhere. Domains with zero local multiplicities everywhere correspond to constant maps, which have $\mu(\phi) = 0$, and so they are not counted in the differential. Any other such domain ϕ can be factored as $\phi = \phi_1 * \cdots * \phi_m$, where ϕ_i are all empty rectangles. Since μ is additive under composition, and by Lemma 16.4.2 we have $\mu(\phi_i) = 1$, we conclude that $m = \mu(\phi)$. Since the holomorphic strips counted in the differential have Maslov index 1, we conclude that only empty rectangles are counted in the differential. Lemma 16.4.2 ensures that each such rectangle is counted in the differential, completing the identification of the differential in $\mathrm{CFK}^-(\mathcal{H})$ with the differential in $GC^-(\mathbb{G})$.

Since the Maslov and Alexander gradings transform the same in both theories, when the homotopy class ϕ corresponds to an empty rectangle, it follows that the identification of complexes respects relative bigradings. In fact, the identification respects absolute bigradings, since in both theories, the Alexander grading is normalized to be symmetric, and the Maslov grading is normalized using the total homology of the complexes. □

The analogue of Theorem 16.4.1 holds for links, as well, with minor changes.

16.5. Further remarks

A key property of knot Floer homology is that it is related to the Heegaard Floer homology groups of three-manifolds obtained as surgeries on knots. More precisely, in [**181, 184**], the Heegaard Floer homology groups of surgeries on K are described in terms of the filtration induced by the knot (denoted $\mathrm{CFK}^{-,*}$ in the context of Heegaard Floer homology, generalizing the filtered complex $\mathcal{GC}^-$ from Chapter 13). See also [**172, 191**].

This interplay can be used to study both questions within knot theory and surgery problems. For example, Heegaard Floer homology can be used to study unknotting numbers of knots [**176**]. It can also be used to give obstructions on knots admitting lens space surgeries [**80, 177, 192**]; see also [**114**].

CHAPTER 17

Open problems

In this chapter we collect open problems that are naturally related to grid diagrams and grid homologies. We have divided these problems into two sections: in Section 17.1, we collected problems about grid diagrams and grid homology, and in Section 17.2, we discuss problems in knot Floer homology. Some of the problems in Section 17.1 have already been solved using the holomorphic theory; in that case, we are asking for a proof within the framework of grid homology (i.e. without appealing to the equivalence with the holomorphic theory).

17.1. Open problems in grid homology

Unknot detection. Knot Floer homology is known to detect the unknot. (See Theorem 1.3.1.) From the equivalence between grid homology and knot Floer homology, it follows that grid homology detects the unknot.

PROBLEM 17.1.1. Use grid diagrams directly to show that grid homology detects the unknot; that is, show that a knot $K \subset S^3$ with $\widehat{GH}(K) = \mathbb{F}$ is the unknot.

Seifert genus. In fact, knot Floer homology (and therefore grid homology) detects the Seifert genus of a knot. (See Theorem 1.3.2.) Once again, the proof of this result relies on the holomorphic version of the theory.

PROBLEM 17.1.2. Without appealing to the equivalence with the holomorphic theory, show that grid homology detects the Seifert genus of a knot; that is, for any knot $K \subset S^3$,

$$g(K) = \max\{s \mid \widehat{GH}_*(K, s) \neq 0\}.$$

Note that Dynnikov [**37**] has an algorithm for detecting the unknot using grid diagrams. This result prompts the following question:

PROBLEM 17.1.3. Is there a direct algorithm for detecting knot genus using grid diagrams, in the spirit of Dynnikov's unknot detection algorithm?

An optimistic version of the above is the following question: if $\mathbb{G}$ is a grid diagram for a knot K whose associated genus is minimal among all grid diagrams $\mathbb{G}'$ that differ from $\mathbb{G}$ by sequences of commutation moves, does it follow that either (1) $\mathbb{G}$ can be destabilized after a sequence of commutation moves or (2) the associated genus of $\mathbb{G}$ agrees with the Seifert genus of K?

Fiberedness. In a similar vein, Theorem 1.3.3 shows that knot Floer homology detects whether or not a knot is fibered. The proof relies on the holomorphic definition of knot Floer homology.

PROBLEM 17.1.4. Without appealing to the equivalence with the holomorphic theory, show that grid homology detects fiberedness of a knot; that is, any knot K of genus $g(K)$ is fibered if and only if $\dim \widehat{GH}(K, g(K)) = 1$.

A few easier, related problems along these lines are the following.

PROBLEM 17.1.5. Suppose that K is a fibered knot with genus $g(K)$. Is there a grid diagram $\mathbb{G}$ with the property that there is a unique grid state in Alexander grading $s = g(K)$ and no grid states in any larger grading?

For torus knots, a grid diagram with the above property was given in Lemma 4.8.4.

PROBLEM 17.1.6. Show directly that if $\mathbb{G}$ is a grid diagram with a unique grid state $\mathbf{x}$ in some grading s and no grid states in any larger grading, then K is fibered.

A theorem of Stallings [**213**] states that K is fibered if and only if the commutator subgroup of $\pi_1(S^3 \setminus K)$ is finitely generated. Perhaps the presentation of $\pi_1(S^3 \setminus K)$ described in Lemma 3.5.1 is useful in considering these questions.

Computations. For knots with sufficiently small grid number, grid homology can be explicitly computed, especially with the help of a computer. Computations of the grid homology groups of infinite families of knots is typically harder. Grid homology groups of certain infinite families of knots were computed in Chapters 9 and 10.

An important infinite family one might wonder about is the case of torus knots. According to Theorem 16.2.6, the knot Floer homology for a positive torus knot can be computed directly from the Alexander polynomial of the knot. That formula can be proved either by working with a suitable genus-1 Heegaard diagram, or by appealing to more abstract principles [**177**].

PROBLEM 17.1.7. Compute the grid homology of the torus knot $T_{p,q}$ purely within the framework of grid homology.

Naturality. We have shown that two grid diagrams representing isotopic knots have isomorphic grid homology groups.

PROBLEM 17.1.8. Does an isotopy between two knots induce a well-defined isomorphism between the corresponding unblocked grid homology groups?

There are analogous questions for the simply blocked theory, which involves choosing a particular point p on the knot (corresponding to the special O_i marking in the diagram). In this case, one would expect pointed isotopies to induce maps between the simply blocked invariants.

To put this into context, for $i = 1, 2$, let (S^3, K_i, p_i) be a knot equipped with a basepoint $p_i \in K_i$. In [**98**], it is shown that a diffeomorphism from S^3 to itself carrying K_1 to K_2 and p_1 to p_2 induces a well-defined isomorphism between the corresponding knot Floer homology groups $\widehat{\mathrm{HFK}}$. Sarkar [**203**] has defined and computed the action of moving the basepoint around the knot.

Maps associated to knot cobordisms.

PROBLEM 17.1.9. Does an oriented knot cobordism from K_1 to K_2 induce a map between the corresponding grid homology groups?

As noted earlier, it is natural to expect that the surfaces appearing above also should have some additional structure.

Candidate maps associated to one-handles appear in Chapters 8 and 9; compare also [**95**].

As a special case, a slice disk should induce an element of knot Floer homology.

PROBLEM 17.1.10. Can knot Floer homology be used to distinguish pairwise non-isotopic slice disks for a given knot?

In a different direction:

PROBLEM 17.1.11. Does an unoriented knot cobordism from K_1 to K_2 induce a map between the corresponding simply blocked grid homology groups?

Candidate maps associated to one-handles, in a sufficiently stabilized setting, appear in the unoriented skein exact sequence from Chapter 10.

Spectrum-valued refinement.

PROBLEM 17.1.12. Is there a space $X_{\vec{L}}$ that can be associated to an oriented link $\vec{L}$, that is functorial under oriented saddle moves, and whose singular homology coincides with $\widehat{GH}(\vec{L})$?

Since the Maslov grading can take negative values, we need to have a variant of spaces that have homology in negative dimension. Such a generalized version of a space exists in algebraic topology: it is called a *spectrum*, see for example [**227**].

In [**205**], Sarkar constructed spaces that correspond to certain quotient complexes of $\widetilde{GC}(\mathbb{G})$. Sarkar conjectures that these could be fit together in a natural way to construct the spectrum asked for in Problem 17.1.12. More generally, one might hope to find a spectrum X_K^- with an S^1-action, whose S^1-equivariant cohomology is $GH^-(K)$. A further challenge would be to find a filtration on a spectrum, generalizing the filtered quasi-isomorphism type from Chapter 13.

Note that for Seiberg-Witten theory, and Y a rational homology three-sphere, Manolescu [**130**] constructed an S^1-spectrum whose S^1-equivariant cohomology is monopole Floer homology. This construction uses analysis of the Seiberg-Witten monopole equations; see also [**129**].

In a different direction, Lipshitz and Sarkar [**122**] constructed a spectrum associated to Khovanov homology.

17.2. Open problems in knot Floer homology

Knot Floer homology and the fundamental group. It would be very interesting to find a concrete relationship between the fundamental group of the complement of a knot and its knot Floer homology. One possible relationship is provided by a conjecture of Kronheimer and Mrowka [**111**], stating that the dimension of knot Floer homology (with coefficients in a field of characteristic zero) is equal to the dimension of instanton knot Floer homology [**53**]. Note that Floer's instanton homology is related to certain $SO(3)$ representations of the fundamental group of the knot complement. For other connections between Heegaard Floer homology and the fundamental group, see [**15**].

The Fox-Milnor condition. Many of the properties of knot Floer homology are lifts or generalizations of various familiar properties of the Alexander polynomial. Conspicuously missing from this list is the Fox-Milnor condition: if K is a slice knot, then there is a polynomial f in t with the property that $\Delta_K(t) = f(t)\cdot f(t^{-1})$. One might think that this generalizes to the statement that if K is a slice knot, then $\widehat{\mathrm{CFK}}(K) \cong C \otimes C^*$ for some chain complex C, where C^* denotes the dual complex of C. This would, in turn, imply that the total rank of the knot Floer homology of a slice knot is a perfect square. In fact, this is not the case. For example, the Kinoshita-Terasaka knot of Figure 2.7 is slice, but its total homology, which can be computed using grid diagrams, has rank 33 (see Equation (4.33)). This leaves open a vague question:

PROBLEM 17.2.1. What can be said about the structure of knot Floer homology for smoothly slice knots?

One might also hope to derive clues about potentially differentiating slice and ribbon knots (cf. Remark 2.6.3). This leads to the following (similarly vague) problem:

PROBLEM 17.2.2. What can be said about the structure of knot Floer homology for ribbon knots?

In a slightly different direction, a knot K is called *doubly slice* if there is an unknotted embedding of S^2 in S^4 whose intersection with an equatorial S^3 is K.

PROBLEM 17.2.3. What can be said about the structure of knot Floer homology for a doubly slice knot?

Counting more holomorphic curves. Knot Floer homology is defined as a version of Lagrangian Floer homology in the g-fold symmetric product. As such, it counts holomorphic disks in this symplectic manifold.

PROBLEM 17.2.4. Can moduli spaces of curves with genus $g > 0$ (and boundaries in $\mathbb{T}_\alpha$ and $\mathbb{T}_\beta$) be used to construct stronger knot invariants than knot Floer homology?

In [**120**], Lipshitz reformulates Heegaard Floer homology, so that the holomorphic curves counted in the differential correspond to embedded curves in $[0,1] \times \mathbb{R} \times \Sigma$. Lipshitz also formulates a version that counts curves with double-points, and includes a power series variable that records the number of double-points.

For grid diagrams, Lipshitz rephrases this in concrete terms, as described in Section 5.5. It remains an open problem to see if the double-point enhancement gives more information:

PROBLEM 17.2.5. For every knot K, is the double-point enhanced grid homology isomorphic to $GH^-(K)[v]$ (in the notation of Definition 5.2.15)?

For more on this proposed homology theory, see [**120, 121**].

Mutations. First, recall the operation of *(Conway) mutation*: suppose that K is a knot with a projection with a distinguished disk whose boundary circle meets the projection in four points, that we think of as equally spaced around the boundary circle. Let K' be the new knot obtained by cutting out the disk, rotating it 180° in the plane, and then regluing it. It is known that the Alexander polynomial is mutation invariant, that is, if K' is a mutant of K then $\Delta_{K'}(t) = \Delta_K(t)$.

Knot Floer homology is not mutation invariant: the Conway and the Kinoshita-Terasaka knots (shown in Figure 2.7) are mutants, but the knot Floer homologies of these two knots are different. (See Exercises 4.8.6 and 4.8.7.) More conceptually, the genera of the knots are different, so by Theorem 1.3.2 their knot Floer homologies cannot be isomorphic as bigraded groups. The total dimensions of the knot Floer homologies, however, are the same. In fact, if we collapse the Maslov grading M and Alexander grading A on knot Floer homology to a single grading $\delta = M - A$, the δ-graded grid homology groups of the Kinoshita-Terasaka and the Conway knots are the same. More generally, Baldwin and Levine [**6**] conjecture an affirmative answer to the following question:

PROBLEM 17.2.6. Is the δ-graded knot Floer homology invariant under mutation?

Related questions can be asked for Khovanov homology; see [**14, 225**]. An analogous problem can be considered for genus 2 mutations; see [**148**].

Linking and link Floer homology. The linking number places restrictions on link Floer homology. For example, if L is a link with two components, and $\vec{L}$ is any orientation on L, the δ-graded link Floer homology of $\vec{L}$, and the linking number of the two components, determine the link Floer homology $\widehat{\mathrm{HFL}}(L)$ of L, endowed with any of its four possible orientations. (See for example Proposition 10.2.1.)

There are higher order obstructions to linking, due to Milnor [**141**], which can be reexpressed in terms of Massey products [**138, 188**].

For example, let $\vec{L} = \vec{L}_1 \cup \vec{L}_2 \cup \vec{L}_3$ be an oriented link with three components, and suppose that the linking numbers of any two components of $\vec{L}$ vanishes. (An example to keep in mind here is the Borromean rings.) Then, there are Seifert surfaces F_i for $\vec{L}_i$ with $F_i \cap \vec{L}_j = \emptyset$ for $i \neq j$. The triple Milnor invariant is obtained as a signed number of triple points in $F_1 \cap F_2 \cap F_3$; see [**26**].

PROBLEM 17.2.7. Do the Milnor invariants place algebraic restrictions on the structure of link Floer homology?

Torsion in knot Floer homology. Consider knot Floer homology with integer coefficients.

PROBLEM 17.2.8. Is there a knot K with the property that the abelian group $\widehat{\mathrm{HFK}}(K;\mathbb{Z})$ has torsion?

Concordance invariants. The invariant $\tau(K)$ can be computed once one calculates $GH^-(\mathbb{G})$. It is natural to wonder if $\tau(K)$ is easier to compute than knot Floer homology. For example:

PROBLEM 17.2.9. Is there a direct way to compute the parity of $\tau(K)$ for a knot?

Of course, such a computation would lead to a computation of τ, just as one can compute the signature of a knot K from the Alexander polynomials of all the knots in an unknotting sequence; see Remark 2.3.12.

Using integer coefficients, we defined $\tau(K,\mathbb{Q})$ and, for each prime p, an invariant $\tau(K;\mathbb{Z}/p\mathbb{Z})$ in Section 15.6.

PROBLEM 17.2.10. Exhibit a knot K and two primes p and q, for which $\tau(K;\mathbb{Z}/p\mathbb{Z}) \neq \tau(K;\mathbb{Z}/q\mathbb{Z})$; or a knot K and a prime p for which $\tau(K;\mathbb{Z}/p\mathbb{Z}) \neq \tau(K;\mathbb{Q})$.

Note that Problem 17.2.8 is independent of Problem 17.2.10. (See Example 15.6.2.)

As a point of comparison, Khovanov homology can also be used to construct an invariant $s(K)$ similar to $\tau(K)$. Just as τ has a collection of variations, indexed by prime numbers p, there is also a corresponding collection of s invariants. The fact that the $\mathbb{Q}$-version and the $\mathbb{Z}/2\mathbb{Z}$-version of these s invariants can be different has been verified by C. Seed, using his program Knotkit. (For the 14-crossing knot $K = 14n19265$, s with coefficients in $\mathbb{Z}/2\mathbb{Z}$ is different from s with coefficients in $\mathbb{Q}$.) For more questions along these lines for Khovanov homology, see Section 6 of [**123**].

We formulate an optimistic variant of Problem 17.2.10 in terms of the *smooth concordance group of knots*, the group $\mathcal{C}$ of equivalence classes of knots, where $K_1 \sim K_2$ if $K_1 \# m(-K_2)$ is a slice knot. (Addition in this group is defined by taking connected sum.) It follows from a Künneth principle that $\tau(K, \mathbb{Z}/p\mathbb{Z})$ is additive under connected sums; since it vanishes for slice knots (Theorem 15.6.1), it follows that for each prime p, the map $K \mapsto \tau(K; \mathbb{Z}/p\mathbb{Z})$ induces a homomorphism $\tau_{\mathbb{Z}/p\mathbb{Z}} \colon \mathcal{C} \to \mathbb{Z}$ from the smooth concordance group to the integers. Similarly, $K \mapsto \tau(K; \mathbb{Q})$ induces a homomorphism $\tau_{\mathbb{Q}}$ from the smooth concordance group to the integers.

PROBLEM 17.2.11. Is the infinite collection of homomorphisms $\tau_{\mathbb{Z}/p\mathbb{Z}}$, indexed by primes p, together with $\tau_{\mathbb{Q}}$ linearly independent, as homomorphisms from the smooth concordance group to the integers?

Knot Floer homology in fact can be used to construct infinitely many linearly independent homomorphisms from the concordance group $\mathcal{C}$ to $\mathbb{Z}$. The first such construction is due to Hom [**88**]. We will describe a different method from [**164**] which rests on a simple modification of the construction of τ.

As a preliminary step, take a rational number $t \in [0,2]$, and consider the module $GC^t(\mathbb{G})$ over the algebra $\mathbb{F}[v^t, v^{2-t}]$ generated freely by grid states. Equip the module $GC^t(\mathbb{G})$ with a grading induced by $\mathrm{gr}_t(v^{tm}\mathbf{x}) = M(\mathbf{x}) - tA(\mathbf{x}) - tm$. For a rectangle $r \in \mathrm{Rect}(\mathbf{x},\mathbf{y})$, let $\mathbb{X}(r)$ denote the number of X-markings in r, and let $\mathbb{O}(r)$ denote the number of O-markings in r. Consider the $\mathbb{F}[v^t]$-module endomorphism specified by

$$\partial_t \mathbf{x} = \sum_{\mathbf{y}\in\mathbf{S}(\mathbb{G})} \sum_{\{r\in\mathrm{Rect}^\circ(\mathbf{x},\mathbf{y})\}} \# \left(\frac{\mathcal{M}(\phi)}{\mathbb{R}}\right) v^{t\mathbb{X}(r)+(2-t)\mathbb{O}(r)}\mathbf{y}. \tag{17.1}$$

Obviously, multiplication by v drops gr_t by one; it is also fairly easy to see that the endomorphism ∂_t is a differential that drops the grading gr_t by 1.

Although the homology of $GC^t(\mathbb{G})$ is not a knot invariant (because of stabilizations; i.e. like $\widetilde{GH}(\mathbb{G})$, there is an extra factor of a two-dimensional vector space, taken to the $(n-1)^{st}$ tensor power, where n is the grid number of the diagram $\mathbb{G}$), we can define $\Upsilon_K(t)$ to be the maximal gr_t of any gr_t-homogeneous, non-torsion class in $H(GC^t(\mathbb{G}))$. According to [**164**], for each rational number $t \in [0,2]$, that quantity is a knot invariant. The function $\Upsilon_K(t)$ can be naturally extended to a piecewise linear, continuous function on $[0,2]$; and indeed, Υ gives a homomorphism from the smooth concordance group of knots to the vector space of real-valued, piecewise linear, continuous functions on $[0,2]$. Thus, Υ gives plenty of room to detect infinitely many linearly independent knots.

Using sign assignments as in Chapter 15, the construction of Υ can be adapted to coefficients in $\mathbb{Z}$, and hence specialized once again to $\mathbb{Z}/p\mathbb{Z}$ and $\mathbb{Q}$. There are natural analogues of Problem 17.2.10, and more optimistically, Problem 17.2.11 for the resulting functions on $[0,2]$, where $\Upsilon_K(t;\mathbb{Z}/p\mathbb{Z})$ and $\Upsilon_K(t;\mathbb{Q})$ play the roles of the integers $\tau(K;\mathbb{Z}/p\mathbb{Z})$ and $\tau(K;\mathbb{Q})$.

Transverse invariants. The transverse invariant of a link gives an invariant in grid homology. As in Section 14.3, viewing grid homology as the associated graded object for the filtered knot invariant, the transverse invariant inherits extra structures.

Recall the language of Definition 14.4.1: the transverse invariant is said to be a cycle to order n if there is a chain $x \in \mathcal{GC}^-(\mathbb{G})$ with the following properties:

- if $a = \frac{\mathrm{sl}(\mathcal{T})+1}{2}$, then $x \in \mathcal{F}_a\mathcal{GC}^-(\mathbb{G})$;
- the projection of x to $GC^-(\mathbb{G}, \frac{\mathrm{sl}(\mathcal{T})+1}{2})$ is a cycle, and it represents $\theta(\mathcal{T}) \in GH^-(\mathbb{G}, \frac{\mathrm{sl}(\mathcal{T})+1}{2})$;
- $\partial x \in \mathcal{F}_{a-n}\mathcal{GC}^-(\mathbb{G})$.

PROBLEM 17.2.12. Given $n \geq 1$, is there a transverse knot $\mathcal{T}$ whose invariant $\theta(\mathcal{T})$ can be represented by a cycle to order n but not $n+1$?

An example with $n = 1$ is given in Proposition 14.4.6.

Given k, knot types with k distinct transverse representatives with the same self-linking number are found in [**48**]. Different examples would be supplied by an affirmation of the following:

PROBLEM 17.2.13. Given any $n > 2$, is there an n-tuple of transverse knots $\mathcal{T}_1, \dots, \mathcal{T}_n$ that are smoothly isotopic, and with the same self-linking number, so that $\theta(\mathcal{T}_i)$ can be represented by a cycle to order i but not $i+1$?

Module realization in knot Floer homology.

PROBLEM 17.2.14. Characterize the graded $\mathbb{F}[U]$-modules that arise as knot Floer homology groups of knots.

PROBLEM 17.2.15. Which graded $\mathbb{F}[U]$-modules arise as knot Floer homology groups of more than one knot?

Note that the unknot is the only knot that has $\widehat{\mathrm{HFK}}$ of rank one [**171**]. A theorem of Ghiggini [**71**] (see also Theorem 1.3.3) implies that the trefoil knots and the figure-eight knot are uniquely characterized by their knot homologies. On the other hand, infinitely many knots with the same knot Floer homology modules were described in Section 9.5; see also [**84**].

The obvious generalization of these problems is the following:

PROBLEM 17.2.16. Characterize the multi-graded $\mathbb{F}[U_1, \dots, U_\ell]$-modules that arise as link Floer homology groups of links.

A simpler question can be asked: what polytopes arise as grid homology polytopes? This is equivalent to the question of characterizing Thurston polytopes of links in $\mathbb{R}^3$.

Axiomatic characterizations of Floer homology. Let W be the two-dimensional bigraded vector space with one generator in bigrading $(0,0)$ and another in bigrading $(-1,-1)$, and J be the four-dimensional bigraded vector space with one generator in bigrading $(0, 1)$, one in $(-2, -1)$, and two generators in bigrading $(-1, 0)$.

DEFINITION 17.2.17. Let $\mathcal{H}(\vec{L})$ be an oriented link invariant, which has the form of a bigraded module over $\mathbb{F}[U]$. We say that $\mathcal{H}$ satisfies the ***oriented skein exact sequence*** if for each oriented skein triple $(\vec{L}_+, \vec{L}_-, \vec{L}_0)$, there are exact triangles of bigraded $\mathbb{F}[U]$-modules (with grading shifts indicated on the arrows):

$$\begin{array}{ccc} \mathcal{H}(\vec{L}_+) & \longrightarrow & \mathcal{H}(\vec{L}_-) \\ & \nwarrow \quad \swarrow_{(-1,0)} & \\ & \mathcal{H}(\vec{L}_0) & \end{array}$$

if the two strands at the distinguished crossing of $\vec{L}_+$ belong to the same component; and

$$\begin{array}{ccc} \mathcal{H}(\vec{L}_+) & \longrightarrow & \mathcal{H}(\vec{L}_-) \\ & \nwarrow \quad \swarrow_{(-1,0)} & \\ & \mathcal{H}(\vec{L}_0) \otimes J & \end{array}$$

if the two strands at the distinguished crossing of $\vec{L}_+$ belong to different components.

PROBLEM 17.2.18. Are there any bigraded link invariants $\mathcal{H}$, other than collapsed grid homology, that satisfy the following two properties:

- with $\mathcal{U}_n$ denoting the n-component unlink, $\mathcal{H}(\mathcal{U}_n) \cong \mathbb{F}[U] \otimes W^{\otimes n-1}$, and
- $\mathcal{H}$ satisfies the oriented skein sequence?

Analogous questions can be asked for the simply blocked grid homology, and coefficients in $\mathbb{Z}$ in place of $\mathbb{F}$.

Note that Khovanov and Khovanov-Rozansky have constructed other homology theories for knots [**103, 104, 105**] that satisfy similar skein exact sequences; compare also [**131, 211**]. There are various conjectures relating these invariants to knot Floer homology. There is a conjectured spectral sequence from reduced Khovanov homology to $\widehat{\mathrm{HFK}}$, see [**193**]; and from reduced HOMFLY homology to $\widehat{\mathrm{HFK}}$, see [**36**].

APPENDIX A

Homological algebra

For the sake of completeness, in this appendix we recall some basic notions and constructions of homological algebra. In fact, our discussion is slightly non-standard: the gradings natural in grid homology are somewhat different from the gradings that come up naturally in algebraic topology. We start with basics of chain complexes and their homologies in Sections A.1. In Section A.2 we describe exact triangles, and in Section A.3 we discuss mapping cones. In Section A.4 we describe the structure of the homology groups over the ring $\mathbb{F}[U]$. In Section A.5 we describe the relationship between a complex and its dual. In Section A.7 we discuss minimal models of filtered chain complexes; this result gives economical chain complexes which represent a fixed filtered quasi-isomorphism type. Finally, in Section A.8 we discuss the relation between chain homotopies and quasi-isomorphisms.

This appendix is intentionally brief; for a more detailed discussion of the topics the reader is advised to turn to [**70, 83, 200**].

A.1. Chain complexes and their homology

Let $\mathbb{K}$ denote either the finite field $\mathbb{Z}/p\mathbb{Z}$ for some prime $p \in \mathbb{N}$, or $\mathbb{Q}$ or the ring $\mathbb{Z}$. In the following, $\mathcal{R}$ will denote the polynomial ring $\mathbb{K}[V_1, \dots, V_n]$ of n variables. We include the $n = 0$ case with the understanding that in this case $\mathcal{R} = \mathbb{K}$. In particular, $\mathcal{R}$ is a field if $n = 0$ and $\mathbb{K}$ is not $\mathbb{Z}$; and $\mathcal{R}$ is a principal ideal domain (PID) if either $n = 0$, or $n = 1$ and $\mathbb{K}$ is not $\mathbb{Z}$.

DEFINITION A.1.1. A ***chain complex*** is an $\mathcal{R}$-module C, equipped with an $\mathcal{R}$-module homomorphism $\partial\colon C \to C$ with the property that $\partial \circ \partial = 0$. The map ∂ is called the ***boundary map*** or the ***differential*** for C. A ***cycle*** is an element $z \in C$ with $\partial z = 0$ and a ***boundary*** is an element b of the form $b = \partial a$ for some $a \in C$; i.e. the cycles are the elements in the kernel $\mathrm{Ker}\partial$, and the boundaries are the elements in the image $\mathrm{Im}\partial$.

An $\mathcal{R}$-submodule $C' \subset C$ of a chain complex (C, ∂) is a *subcomplex* if $\partial(C') \subset C'$. In this case, the pair $(C', \partial|_{C'})$ is a chain complex. Similarly, if $(C', \partial|_{C'})$ is a subcomplex, the quotient module C/C' inherits a boundary operator $\partial_{C/C'}$, induced from ∂. The pair $(C/C', \partial_{C/C'})$ is the *quotient complex* of C by C'.

The condition $\partial^2 = 0$ says $\mathrm{Im}\partial \subset \mathrm{Ker}\partial$, so we can make the following definition:

DEFINITION A.1.2. The ***homology*** $H(C, \partial)$ of the chain complex (C, ∂) is the quotient $\mathcal{R}$-module $\mathrm{Ker}\partial/\mathrm{Im}\partial$.

In standard homological algebra, it is customary to consider $\mathbb{Z}$-graded chain complexes, where the differential ∂ drops grading by 1 and where the action of the ring $\mathcal{R}$ preserves gradings. Such a complex admits a direct sum splitting $C = \bigoplus_{d \in \mathbb{Z}} C_d$, so that C_d is an $\mathcal{R}$-submodule for all d. In this case, the homology $H(C)$

inherits a grading, where $H_d(C)$ is the cycles modulo the boundaries in C_d. In the present volume, we consider a mild variation of this, where the action of the ring $\mathcal{R} = \mathbb{K}[V_1, \ldots, V_n]$ changes the grading on the chain complex. In fact, quite often our complexes come naturally with two different gradings (see Theorem 4.6.3, for example). We formalize this in the following:

DEFINITION A.1.3. Throughout this book, a ***bigraded chain complex over*** $\mathbb{K}[V_1, \ldots, V_n]$ is a chain complex (C, ∂) over $\mathbb{K}$, equipped with endomorphisms $V_i \colon C \to C$ for $i = 1, \ldots n$, called the ***algebra actions***, and a splitting $C = \bigoplus_{(d,s) \in \mathbb{Z} \oplus \mathbb{Z}} C_{d,s}$, satisfying the following compatibility conditions:

- for $i = 1, \ldots, n$, $\partial \circ V_i = V_i \circ \partial$;
- for all $i, j \in \{1, \ldots, n\}$, $V_i \circ V_j = V_j \circ V_i$;
- ∂ maps $C_{d,s}$ to $C_{d-1,s}$;
- V_i maps $C_{d,s}$ to $C_{d-2,s-1}$.

The first grading is called the ***Maslov grading***, and the second the ***Alexander grading***. The actions can be viewed as endowing C with the structure of a module over $\mathbb{K}[V_1, \ldots, V_n]$. The condition that ∂ commutes with V_i is equivalent to the condition that ∂ is a $\mathbb{K}[V_1, \ldots, V_n]$-module homomorphism.

REMARK A.1.4. The above definition of bigraded complexes fits naturally into a more general framework of bigraded complexes over a bigraded ring. For this purpose, consider the polynomial ring $\mathbb{K}[V_1, \ldots, V_n]$ to be bigraded, so that V_i has bigrading $(-2, -1)$. Our bigraded complexes, then, are bigraded complexes over this bigraded ring, equipped with a differential which has bigrading $(-1, 0)$.

When $n = 0$, Definition A.1.3 specializes to a bigraded complex over $\mathbb{K}$. Of particular relevance to us is the case when $n = 1$, in which case we often denote the single variable by U. According to Theorem 4.6.3, the grid complex is a bigraded complex as above with $\mathbb{K}$ chosen to be $\mathbb{F} = \mathbb{Z}/2\mathbb{Z}$. In Chapter 11 (see especially Theorem 11.1.7), the Alexander grading set is also enlarged. In Chapter 13, the Alexander grading is relaxed to an Alexander filtration. In Chapter 15, $\mathbb{K}$ is no longer the field $\mathbb{F}$, but the ring $\mathbb{Z}$ of integers.

For a bigraded chain complex, the homology $H(C)$ inherits both the structure of a $\mathbb{K}[V_1, \ldots, V_n]$-module, and a bigrading compatible with that action. The portion of $H(C)$ in bigrading (d, s) is given by the quotient of the cycles in C by the boundaries in bigrading (d, s), and multiplication by V_i on C induces a corresponding action on $H(C)$.

Let (C, ∂) and (C', ∂') be two chain complexes over $\mathcal{R}$. An $\mathcal{R}$-module map $f \colon C \to C'$ is a *chain map* if it commutes with the boundary operators, that is, for every $c \in C$, $f(\partial c) = \partial' f(c)$. If C and C' are bigraded complexes over $\mathcal{R}$ as in Definition A.1.3, then a *bigraded chain map* is a chain map f that maps $C_{d,s}$ into $C'_{d,s} \subset C'$. More generally, if C and C' are as above, $f \colon C \to C'$ is a chain map, and there is a pair of integers (m, t) so that f maps $C_{d,s}$ into $C'_{d+m,s+t}$, then we say that f is a *homogeneous map with bidegree* (m, t).

Suppose that C is a bigraded chain complex over $\mathcal{R}$, and fix integers m, t. The (m, t)*-shift of* C is the bigraded chain complex $C[\![m, t]\!]$ defined by $C[\![m, t]\!]_{d,s} = C_{d+m,s+t}$. A chain map from C to C' that is homogeneous of degree (m, t) is the same as a bigraded chain map from C to $C'[\![m, t]\!]$.

A chain map $f \colon C \to C'$ maps cycles to cycles and boundaries to boundaries, and so it induces an $\mathcal{R}$-module map $H(f) \colon H(C, \partial) \to H(C', \partial')$ on the homology modules. When C and C' are bigraded complexes over $\mathcal{R}$ and $f\colon C \to C'$ is a bigraded chain map, the induced map $H(f)$ respects the induced bigrading on the homology modules.

Given two bigraded chain maps $f\colon (C,\partial) \to (C',\partial')$ and $g\colon (C',\partial') \to (C'',\partial'')$, their composite $g\circ f$ is another bigraded chain map, whose induced map on homology satisfies

$$H(g\circ f) = H(g)\circ H(f). \tag{A.1}$$

The identity map Id_C is a chain map that induces the identity map on $H(C,\partial)$. Consequently, an isomorphism between two chain complexes obviously induces an isomorphism between the corresponding homology modules. More generally:

DEFINITION A.1.5. Let (C,∂) and (C',∂') be two chain complexes. A chain map $f\colon C\to C'$ is a ***quasi-isomorphism*** if it induces an isomorphism on homology. Two chain complexes (C,∂) and (C',∂') are said to be ***quasi-isomorphic*** if there is a chain complex (C'',∂'') and two quasi-isomorphisms $f\colon C''\to C$ and $g\colon C''\to C'$. In cases where (C,∂) and (C',∂') are bigraded complexes over $\mathcal{R}$, we require our quasi-isomorphisms to be bigraded maps over $\mathcal{R}$.

Special kinds of quasi-isomorphisms are supplied by the following:

DEFINITION A.1.6. Suppose that $f,g\colon (C,\partial)\to(C',\partial')$ are two chain maps between two chain complexes over $\mathcal{R}$. The maps f and g are said to be ***chain homotopic*** if there is an $\mathcal{R}$-module map $h\colon C\to C'$ satisfying

$$f-g=\partial'\circ h+h\circ\partial.$$

A map $f\colon (C,\partial)\to(C',\partial')$ is a ***chain homotopy equivalence*** if there is a map $g\colon (C',\partial')\to(C,\partial)$ so that $f\circ g$ and $g\circ f$ are both chain homotopic to the respective identity maps $\mathrm{Id}_{C'}$ and Id_C.

LEMMA A.1.7. *Let (C,∂) and (C',∂') be chain complexes. Then chain homotopic maps from C to C' induce the same map in homology. Consequently, a chain homotopy equivalence is a quasi-isomorphism.*

Proof. Suppose that $z\in C$ is a cycle. By definition, $H(f)([z])=[f(z)]$, hence

$$H(f)([z])=[g(z)+\partial'(h(z))+h(\partial(z))]=[g(z)]=H(g)([z]),$$

since $\partial(z)=0$ and $g(z)$ and $g(z)+\partial'(h(z))$ are homologous. Thus, a chain homotopy equivalence is a quasi-isomorphism in view of Equation (A.1). □

Note that not every quasi-isomorphism is a chain homotopy equivalence; compare Example 13.1.8, and see Proposition A.8.1. The definition of chain homotopy equivalence extends to the bigraded setting as follows.

DEFINITION A.1.8. Suppose that (C,∂) and (C',∂') are two bigraded chain complexes over $\mathcal{R}$. Suppose that $f,g\colon C\to C'$ are two chain maps of degree (m,t). We say that f and g are ***chain homotopic as homogeneous maps of degree*** (m,t) if there is an $\mathcal{R}$-module map $h\colon C\to C'$ that sends $C_{d,s}$ to $C'_{d+m+1,s+t}$ with

$$\partial'\circ h+h\circ\partial=f-g. \tag{A.2}$$

A.2. Exact sequences

A sequence $\{C^i\}_{i\in\mathbb{Z}}$ of $\mathcal{R}$-modules equipped with $\mathcal{R}$-module maps $f^i\colon C^i \to C^{i+1}$ is called an *exact sequence of* $\mathcal{R}$*-modules* if $\mathrm{Im} f^i = \mathrm{Ker} f^{i+1}$. A special case is when $\mathcal{C}$ is a *short exact sequence*, that is, $C^i = 0$ unless $i = 1, 2, 3$. In this case the maps $f^1\colon C^1 \to C^2$ and $f^2\colon C^2 \to C^3$ satisfy that

- f^1 is injective,
- f^2 is surjective, and
- $\mathrm{Im} f^1 = \mathrm{Ker} f^2$.

An *exact triangle* is a 3-periodic exact sequence; i.e. in which there are three $\mathcal{R}$-modules C^1, C^2, and C^3, and maps $f^1\colon C^1 \to C^2$, $f^2\colon C^2 \to C^3$ and $f^3\colon C^3 \to C^1$, with $\mathrm{Ker}(f^i) = \mathrm{Im}(f^{i-1})$, where $C^i = C^{i+3}$ and $f^i = f^{i+3}$.

An *exact sequence of chain complexes* (C^i, ∂^i) *over* $\mathcal{R}$ is an exact sequence of $\mathcal{R}$-modules, where the maps $f^i\colon C^i \to C^{i+1}$ are also chain maps.

A short exact sequence of chain complexes induces a long exact sequence in homology, according to the following standard result. (See also [**83**, Theorem 2.16].)

LEMMA A.2.1. *To each short exact sequence of chain complexes of* $\mathcal{R}$*-modules*

$$0 \to (C, \partial) \xrightarrow{f} (C', \partial') \xrightarrow{g} (C'', \partial'') \to 0,$$

there is an associated $\mathcal{R}$*-module homomorphism* $\delta\colon H(C'', \partial'') \to H(C, \partial)$*, called the connecting homomorphism, such that*

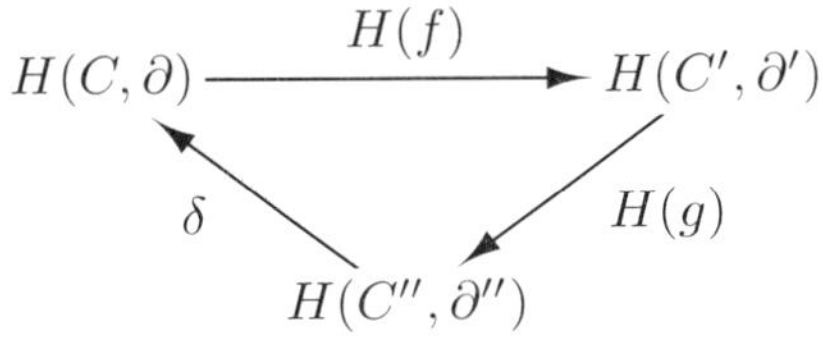

is an exact triangle. Moreover, if the three complexes are bigraded complexes of $\mathcal{R}$*-modules, and* f *and* g *are homogeneous* $\mathcal{R}$*-module homomorphisms of degree* (m_1, t_1) *and* (m_2, t_2) *respectively, then* δ *is a homogeneous map with degree* $(-m_1 - m_2 - 1, -t_1 - t_2)$.

Proof. Given a homology class $x \in H(C'', \partial'')$, define $\delta(x)$ as follows. Pick a cycle $c'' \in C''$ representing x, and find $c \in C$ and $c' \in C'$ so that

$$g(c') = c'' \qquad \text{and} \qquad f(c) = \partial' c'. \tag{A.3}$$

The element c' can be found since g is surjective; the element c can be found since

$$g(\partial' c') = \partial''(g(c')) = \partial''(c'') = 0;$$

i.e. $\partial c' \in \mathrm{Ker}(g) = \mathrm{Im}(f)$. Moreover, the element c is a cycle since

$$f(\partial c) = \partial'(f(c)) = \partial'(\partial' c') = 0,$$

and f is injective. We then define $\delta(x)$ to be the homology class represented by c. We made three choices above: the cycle c'' representing x, the choice of c' with $g(c') = c''$, and the choice of c with $f(c) = \partial c'$. It is straightforward to verify that different choices result in homologous cycles $c \in C$; i.e. δ is a well defined map in homology. Furthermore, the map δ is an $\mathcal{R}$-module map. For example, for fixed cycle c'', if the elements $c \in C$ and $c' \in C'$ solve Equations (A.3), then for the cycle $r \cdot c''$, the elements $r \cdot c$ and $r \cdot c'$ solve the corresponding version of Equation (A.3).

In the bigraded case, if c'' is supported in bigrading (d, s), then we can find c' in bigrading $(d - m_2, s - t_2)$, and hence c in bigrading $(d - m_1 - m_2 - 1, s - t_1 - t_2)$.

Next, we verify exactness of the triangle at $H(C')$. The exactness of the short exact sequence ensures that $g \circ f = 0$, and hence $H(g) \circ H(f) = 0$; i.e.

$$\mathrm{Im} H(f) \subseteq \mathrm{Ker} H(g). \tag{A.4}$$

We must check that this inclusion is an equality. An element in $\mathrm{Ker}H(g)$ is represented by an element $c' \in C'$ with $\partial' c' = 0$ and $g(c') = \partial'' c''$ for some $c'' \in C''$. Since g is surjective, there is some c'_2 with $g(c'_2) = c''$. Thus,

$$c' - \partial' c'_2 \in \mathrm{Ker}(g) = \mathrm{Im}(f)$$

so we can find $c \in C$ with $f(c) = c' - \partial' c'_2$. Since f is injective, it follows that c is a cycle; and so $[c'] = H(f)([c])$. Since $[c'] \in \mathrm{Ker}H(g)$ is arbitrary, we have verified that $\mathrm{Ker}H(g) \subseteq \mathrm{Im}H(f)$ which, together with Equation (A.4), implies that $\mathrm{Ker}H(g) = \mathrm{Im}H(f)$. Exactness at the other two terms can be verified by a similar diagram chase. □

A further elaboration on the proof of Lemma A.2.1 is the following:

LEMMA A.2.2. *The connecting homomorphism is natural, in the sense that for a map of short exact sequences of chain complexes, i.e. for a commutative diagram*

$$\begin{array}{ccccccccc} 0 & \longrightarrow & C & \xrightarrow{f} & C' & \xrightarrow{g} & C'' & \longrightarrow & 0 \\ & & \downarrow{\phi} & & \downarrow{\phi'} & & \downarrow{\phi''} & & \\ 0 & \longrightarrow & B & \xrightarrow{f'} & B' & \xrightarrow{g'} & B'' & \longrightarrow & 0 \end{array}$$

(where the rows are exact sequences and the squares commute), the following diagram commutes

$$\begin{array}{ccc} H(C'') & \xrightarrow{\delta} & H(C) \\ \downarrow{H(\phi'')} & & \downarrow{H(\phi)} \\ H(B'') & \xrightarrow{\delta'} & H(B) \end{array}$$

where the maps δ and δ' are the connecting homomorphisms for the two short exact sequences.

Proof. The proof is a straightforward application of the definition of the connecting homomorphism (Equation (A.3)). □

The above naturality gives a useful method for establishing isomorphisms between modules (see for example Lemma A.3.9 below), when combined with the following *five lemma*:

LEMMA A.2.3 (Five lemma). *Suppose that the diagram*

$$\begin{array}{ccccccccc} C^1 & \xrightarrow{f_1} & C^2 & \xrightarrow{f_2} & C^3 & \xrightarrow{f_3} & C^4 & \xrightarrow{f_4} & C^5 \\ \downarrow{\alpha_1} & & \downarrow{\alpha_2} & & \downarrow{\alpha_3} & & \downarrow{\alpha_4} & & \downarrow{\alpha_5} \\ B^1 & \xrightarrow{g_1} & B^2 & \xrightarrow{g_2} & B^3 & \xrightarrow{g_3} & B^4 & \xrightarrow{g_4} & B^5 \end{array}$$

is commutative, the two rows are exact and $\alpha_1, \alpha_2, \alpha_4$ and α_5 are isomorphisms. Then, the homomorphism α_3 is an isomorphism. □

Proof. A standard diagram chase shows that (a) if α_2 and α_4 are surjective and α_5 is injective then α_3 is surjective, and (b) if α_2 and α_4 are injective and α_1 is surjective then α_3 is injective. The lemma then easily follows. For more details, see [**83**, page 129]. □

A.3. Mapping cones

We recall now the *mapping cone* construction from homological algebra. For more on this construction, see [**226**, Chapter 1.5].

DEFINITION A.3.1. Let (C, ∂) and (C', ∂') be two chain complexes over $\mathcal{R}$ and let $f\colon C \to C'$ be a chain map. The ***mapping cone*** $\mathrm{Cone}(f\colon C \to C') = \mathrm{Cone}(f)$ is the chain complex whose underlying module is $C \oplus C'$, and whose differential is

$$D(c, c') = (-\partial(c), \partial(c') + f(c)). \tag{A.5}$$

There are variants of this construction in the presence of gradings. For instance, if the complexes C and C' are graded, so that $f\colon C \to C'$ preserves gradings, then $\mathrm{Cone}(f)$ is graded, by

$$\mathrm{Cone}(f)_d = C_{d-1} \oplus C'_d.$$

With this convention, the differential

$$D_d(c, c') = (-\partial_{d-1}c, \partial'_d(c') + f_{d-1}(c))$$

drops grading by one. More generally, if (C, ∂) and (C', ∂') are two bigraded complexes over $\mathcal{R}$, and $f\colon C \to C'$ is a chain map that is homogeneous of bidegree (m, t), then $\mathrm{Cone}(f)$ is a bigraded chain complex, with the differential of Equation (A.5) and bigrading given by

$$\mathrm{Cone}(f)_{d,s} = C_{d-m-1,s-t} \oplus C'_{d,s}.$$

LEMMA A.3.2. *Let C and C' be bigraded chain complexes, and $f\colon C \to C'$ be a bigraded chain map. Then there is a long exact sequence*

$$\dots \longrightarrow H_{d,s}(C) \overset{H(f)}{\longrightarrow} H_{d,s}(C') \longrightarrow H_{d,s}(\mathrm{Cone}(f)) \longrightarrow H_{d-1,s}(C) \longrightarrow \dots$$

If f is a homogeneous map of bidegree (m, t), then we have the long exact sequence with the following degree shifts

$$\dots \to H_{d,s}(C) \overset{H(f)}{\longrightarrow} H_{d+m,s+t}(C') \longrightarrow H_{d+m,s+t}(\mathrm{Cone}(f)) \longrightarrow H_{d-1,s}(C) \to \dots$$

Proof. For any chain map $f\colon C \to C'$ there is a short exact sequence of chain complexes

$$0 \longrightarrow C' \overset{i}{\longrightarrow} \mathrm{Cone}(f) \overset{p}{\longrightarrow} C \longrightarrow 0, \tag{A.6}$$

where the maps are defined by $i(c') = (0, c')$ and $p(c, c') = (-1)^d c$, when $c \in C_{d,s}$. It follows immediately from Equation (A.5) (and the choice of the sign in the definition of p) that i and p are chain maps, and it is easy to verify that they fit into the above short exact sequence. The grading shifts on the mapping cone are set up so that i preserves bigrading and p shifts it by $(-1, 0)$.

The long exact sequence appearing in the statement of the lemma follows from the long exact sequence associated to the above short exact sequence, once we verify that the connecting homomorphism in this associated long exact sequence is the map induced by f on homology, times $(-1)^d$.

To this end, recall the definition of the connecting homomorphism δ from Lemma A.2.1: given a cycle $c_1 \in C_{d,s}$, find $c_2 \in \mathrm{Cone}_{d+1,s}(f)$ and $c_3 \in C'_{d,s}$ with:

$$q(c_2) = c_1 \qquad \text{and} \qquad i(c_3) = \partial c_2.$$

It follows that c_3 is a cycle; and in fact $\delta([c_1])$ is represented by this cycle. By choosing $c_2 = ((-1)^d c_1, 0)$ and $c_3 = (-1)^d f(c_1)$, the connecting homomorphism is immediately seen to be equal to the map $(-1)^d H(f)$ induced by f on homology. We stated the exact sequence in the lemma without the factor of $(-1)^d$. This is justified, since the kernel (and the image) of $(-1)^d H(f)$ coincides with the kernel (and the image) of $H(f)$, hence the exactness in the lemma follows.

If f is of bidegree (m, t), the same argument provides the long exact sequence with the indicated degree shifts. □

This lemma has the following immediate corollary:

Corollary A.3.3. *A map f is a quasi-isomorphism if and only if* $\mathrm{Cone}(f)$ *has trivial homology.* □

As an application, we have the following result, which we state after a definition.

Definition A.3.4. A bigraded chain complex C of $\mathcal{R}$-modules is ***bounded above*** if for all sufficiently large d, $C_{d,s} = 0$.

For example, if the bigraded chain complex C is finitely generated as an $\mathcal{R}$-module, then C is bounded above. If C is a bigraded chain complex that is bounded above, then $H(C)$ is bounded above, as well.

If C is a bigraded chain complex over $\mathcal{R} = \mathbb{K}[V_1, \dots, V_n]$ with $n > 0$, we can construct the quotient complex $\frac{C}{V_1 \cdot C}$, a bigraded chain complex over $\mathcal{R}$, which we abbreviate $\frac{C}{V_1}$.

Proposition A.3.5. *Let C and C' be two bigraded chain complexes over $\mathcal{R} = \mathbb{K}[V_1, \dots, V_n]$ with $n \geq 1$, and suppose that C and C' are both free modules that are bounded above. A bigraded chain map $f\colon C \to C'$ is a quasi-isomorphism if and only if it induces a quasi-isomorphism $\overline{f}\colon \frac{C}{V_1} \to \frac{C'}{V_1}$ over $\mathcal{R}$.*

Before proving this result, we establish the following:

Lemma A.3.6. *Let C be a free, bigraded chain complex over $\mathcal{R} = \mathbb{K}[V_1, \dots, V_n]$ with $n \geq 1$; and suppose that C is bounded above. Then $H(C) \neq 0$ iff $H(\frac{C}{V_1}) \neq 0$.*

Proof. Since C is free, there is a short exact sequence

$$0 \longrightarrow C \xrightarrow{V_1} C \longrightarrow \frac{C}{V_1} \longrightarrow 0.$$

From the associated long exact sequence, it follows that $H(C) = 0$ implies that $H(\frac{C}{V_1}) = 0$. In the other direction, we use the fact that $H(C)$ is bounded above; so if $H(C) \neq 0$, there must be some homogeneous, non-zero element ξ with maximal Maslov grading. This element cannot be in the image of V_1, and hence it must inject into $H(\frac{C}{V_1})$. □

Proof of Proposition A.3.5. Observe that

$$\frac{\mathrm{Cone}(f\colon C \to C')}{V_1} \cong \mathrm{Cone}(\overline{f}\colon \frac{C}{V_1} \to \frac{C'}{V_1}).$$

The map f is a quasi-isomorphism $\iff$ $H(\mathrm{Cone}(f)) = 0$ (by Corollary A.3.3) $\iff$ $H(\frac{\mathrm{Cone}(f)}{V_1}) = 0$ (by Lemma A.3.6) $\iff$ $H(\mathrm{Cone}\overline{f}\colon \frac{C}{V_1} \to \frac{C'}{V_1}) = 0$ (by the above isomorphism) $\iff$ $\overline{f}$ is a quasi-isomorphism (by Corollary A.3.3). □

The long exact sequence of a mapping cone is natural in the following sense:

LEMMA A.3.7. *If two bigraded chain maps $f, g\colon C \to C'$ between bigraded complexes of $\mathcal{R}$-modules are chain homotopic, then their mapping cones are isomorphic.*

Proof. If $\partial' \circ h + h \circ \partial = f - g$, define a map $\Phi_h\colon \mathrm{Cone}(f) \to \mathrm{Cone}(g)$ by $\Phi_h(x, x') = (x, h(x) + x')$, and define $\Phi_{-h}\colon \mathrm{Cone}(g) \to \mathrm{Cone}(f)$ analogously, by $\Phi_{-h}(y, y') = (y, -h(y) + y')$. It is straightforward to check that Φ_h and Φ_{-h} are chain maps, with $\Phi_{-h} \circ \Phi_h = \mathrm{Id}_{\mathrm{Cone}(f)}$ and $\Phi_h \circ \Phi_{-h} = \mathrm{Id}_{\mathrm{Cone}(g)}$. □

More generally, maps between mapping cones can be induced as follows:

LEMMA A.3.8. *Let C, C', E, E' be four bigraded chain complexes, and suppose that there are chain maps fitting into the square*

$$\begin{array}{ccc} C & \xrightarrow{f} & C' \\ {\scriptstyle\phi}\downarrow & & \downarrow{\scriptstyle\phi'} \\ E & \xrightarrow{g} & E', \end{array}$$

that commutes up to homotopy; i.e. the map $\phi' \circ f$ is chain homotopic to $g \circ \phi$. Suppose moreover that ϕ and ϕ' are bigraded maps, f and g are homogeneous of bidegree (m, t), and the homotopies are compatible with these gradings. Then, there is an induced bigraded chain map $\Phi\colon \mathrm{Cone}(f) \to \mathrm{Cone}(g)$ that fits into the following commutative diagram of short exact sequences

$$\begin{array}{ccccccccc} 0 & \longrightarrow & C' & \xrightarrow{i} & \mathrm{Cone}(f) & \xrightarrow{q} & C & \longrightarrow & 0 \\ & & {\scriptstyle\phi'}\downarrow & & {\scriptstyle\Phi}\downarrow & & \downarrow{\scriptstyle\phi} & & \\ 0 & \longrightarrow & E' & \xrightarrow{j} & \mathrm{Cone}(g) & \xrightarrow{p} & E & \longrightarrow & 0. \end{array} \tag{A.7}$$

If ϕ and ϕ' are quasi-isomorphisms, then so is Φ.

Proof. By hypothesis, there is a map $h\colon C \to E'$ with

$$\partial_{E'} \circ h + h \circ \partial_C = \phi' \circ f - g \circ \phi. \tag{A.8}$$

We can now define a bigraded map $\Phi(c, c') = (\phi(c), h(c) + \psi'(c'))$. The verification that this is a chain map easily follows from Equation (A.8). Commutativity of Equation (A.7) is straightforward.

Suppose that ϕ and ϕ' are quasi-isomorphisms and that $(m,t)=(0,0)$. Consider the diagram

$$\begin{array}{ccccccccccc}
\cdots & \longrightarrow & H_{d,s}(C) & \xrightarrow{\delta} & H_{d,s}(C') & \longrightarrow & H_{d,s}(\mathrm{Cone}(f)) & \longrightarrow & H_{d-1,s}(C) & \longrightarrow & \cdots \\
 & & \Big\downarrow{\scriptstyle H(\phi)} & & \Big\downarrow{\scriptstyle H(\phi')} & & \Big\downarrow{\scriptstyle H(\Phi)} & & \Big\downarrow{\scriptstyle H(\phi)} & & \\
\cdots & \longrightarrow & H_{d,s}(E) & \xrightarrow{\delta'} & H_{d,s}(E') & \longrightarrow & H_{d,s}(\mathrm{Cone}(g)) & \longrightarrow & H_{d-1,s}(E) & \longrightarrow & \cdots
\end{array}$$

whose rows are the long exact squences from Lemma A.2.1. The squares involving δ commute by Lemma A.2.2; the other squares obviously commute; so $H(\Phi)$ is an isomorphism by the five lemma (Lemma A.2.3). The case of general (m,t) follows with minor notational changes. □

The mapping cone of $f\colon C\to C'$ can be thought of as a type of quotient of C' by C, according to the following:

LEMMA A.3.9. *If $f\colon C\to C'$ is an injective chain map, then there is a quasi-isomorphism $\phi\colon \mathrm{Cone}(f)\to \frac{C'}{f(C)}$. This is compatible with gradings: when C and C' are bigraded chain complexes over $\mathcal{R}$ and f is a bigraded chain map, then ϕ is a quasi-isomorphism of bigraded chain complexes over $\mathcal{R}$.*

Proof. For notational convenience, we write out the case where C and C' are $\mathbb{Z}$-graded, and f is a $\mathbb{Z}$-graded chain map. Consider $C''=\frac{C'}{f(C)}$ and form the short exact sequence

$$0\longrightarrow C\xrightarrow{f} C'\xrightarrow{q} C''\longrightarrow 0, \tag{A.9}$$

with $q\colon C'\to C''=\frac{C'}{f(C)}$ being the projection. Define a chain map $\phi\colon \mathrm{Cone}(f)\to C''$ by the formula $\phi(c,c')=q(c')$. To verify that it is a quasi-isomorphism, we fit together the long exact sequences associated to the short exact sequences in Equations (A.6) and in (A.9), as in the following diagram:
(A.10)

$$\begin{array}{ccccccccc}
\longrightarrow H_{d+1}(C') & \xrightarrow{H(i)} & H_{d+1}(\mathrm{Cone}(f)) & \xrightarrow{H(p)} & H_d(C) & \xrightarrow{(-1)^d\delta} & H_d(C') \longrightarrow \\
\Big\downarrow{\scriptstyle \mathrm{Id}} & & \Big\downarrow{\scriptstyle H(\phi)} & & \Big\downarrow{\scriptstyle \mathrm{Id}} & & \Big\downarrow{\scriptstyle \mathrm{Id}} \\
\longrightarrow H_{d+1}(C') & \xrightarrow{H(q)} & H_{d+1}(C'') & \xrightarrow{(-1)^{d+1}\delta'} & H_d(C) & \xrightarrow{H(f)} & H_d(C') \longrightarrow
\end{array}$$

(Note that we multiplied the coboundary map with a $(-1)^d$ to make the rightmost square commute; this does not affect exactness.) Here, $\delta'\colon H_{d+1}(C'')\to H_d(C)$ is the connecting homomorphism for the short exact sequence from Equation (A.9). Once we verify that the squares appearing above commute, the verification that ϕ induces an isomorphism in homology follows from the five lemma (Lemma A.2.3). Commutativity of the left-most square is straightforward: indeed, even on the chain level, it is true that $\phi\circ i=p$. Commutativity of the rightmost square was verified in the proofs of Lemma A.3.2.

Commutativity of the middle square follows once again from a careful look at the definition of the connecting homomorphism. A cycle in $\mathrm{Cone}_{d+1}(f)$ is a pair (c, c') with $\partial c = 0$ and $\partial c' = -f(c)$. Now $\phi(c, c') = q(c')$ and $\delta'[q(c')]$ is defined by finding some cycle c_1 representing $[q(c')]$, and next finding $c_2 \in C_{d+1}$ and $c_3 \in C_d$ so that

$$q(c_2) = c_1 \qquad \text{and } f(c_3) = \partial c_2.$$

Then, c_3 represents $[\delta' q(c')]$. Let $c_1 = q(c')$, $c_2 = c'$, and $c_3 = -c$. With these choices it immediately follows that $(-1)^{d+1}\delta'[q(c')] = (-1)^d c = p(c, c')$, as needed. The bigraded case follows similarly. □

The next lemma establishes an exact triangle that contains the homologies of the mapping cones of two maps and their composite. (We will use this result only in the graded setting, so we formulate it in this generality.)

LEMMA A.3.10. *Suppose that C, C', and C'' are three $\mathbb{Z}$-graded chain complexes, and $f\colon C' \to C''$ and $g\colon C \to C'$ are chain maps that are homogeneous of degrees a and b respectively. Then, there is a chain map $\Phi\colon \mathrm{Cone}(f) \to \mathrm{Cone}(g)$ which is homogeneous of degree $-a-1$ and whose induced map on homology fits into an exact triangle*

$$\begin{array}{ccc} H(\mathrm{Cone}(f)) & \xrightarrow{\;-a-1\;} & H(\mathrm{Cone}(g)) \\ & \nwarrow \qquad \swarrow_{a} & \\ & H(\mathrm{Cone}(f\circ g)) & \end{array}$$

(A.11)

where the integers indicate shifts on degree.

Proof. Define $\Phi\colon \mathrm{Cone}(f)_d \to \mathrm{Cone}(g)_{d-a-1}$ by $\Phi(c', c'') = (-1)^d(0, c') \in \mathrm{Cone}(g)_{d-a-1}$.

Obviously, Φ is a chain map which is homogeneous of degree $-a-1$. According to Lemma A.3.2, we have now an exact triangle:

$$\begin{array}{ccc} H(\mathrm{Cone}(f)) & \xrightarrow{\;-a-1\;} & H(\mathrm{Cone}(g)) \\ & {}^{a}\nwarrow \qquad \swarrow & \\ & H(\mathrm{Cone}(\Phi)) & \end{array}$$

(A.12)

We will denote elements of $\mathrm{Cone}(\Phi)$ by quadruples $((c_1', c''), (c, c_2'))$, where $(c_1', c'') \in \mathrm{Cone}(f)$ and $(c, c_2') \in \mathrm{Cone}(g)$. Thus, the differential is given by

$$D((c_1', c''), (c, c_2')) = ((\partial' c_1', -f(c_1') - \partial''(c'')), (-\partial c, (-1)^d c_1' + g(c) + \partial' c_2')).$$

Consider the map $\alpha\colon \mathrm{Cone}(f \circ g) \to \mathrm{Cone}(\Phi)$ defined by the formula

$$\alpha(c, c'') = (((-1)^{d-a+1} g(c), (-1)^{d-a} c''), (c, 0)),$$

when c'' is homogeneous of degree d; and the map $\beta\colon \mathrm{Cone}(\Phi) \to \mathrm{Cone}(f \circ g)$ defined by

$$\beta((c_1', c''), (c, c_2')) = (c, (-1)^d c'' - f(c_2'))$$

when c_2' is homogeneous of degree d. Define $h\colon \mathrm{Cone}(\Phi) \to \mathrm{Cone}(\Phi)$ by

$$h((c_1', c''), (c, c_2')) = ((-1)^{d+1} c_2', 0), (0, 0)),$$

when c_2' is homogeneous of degree d. The map α is homogeneous of degree $-a$, β is homogeneous of degree a, and h is homogeneous of degree $a+1$. It is straightforward to verify the identities:

$$0 = \mathrm{Id}_{\mathrm{Cone}(f\circ g)} - \beta \circ \alpha,$$
$$D \circ h + h \circ D = \mathrm{Id}_{\mathrm{Cone}(\Phi)} - \alpha \circ \beta.$$

Thus, we have a homotopy equivalence

$$\mathrm{Cone}(\Phi)[\![a]\!] \simeq \mathrm{Cone}(f \circ g).$$

The exact triangle of Diagram (A.11) is obtained from Diagram (A.12) by applying the above substitution on homology and the appropriate grading shifts. □

We can use mapping cones to show that the notion of "quasi-isomorphic chain complexes" (in the sense of Definition A.1.5) is an equivalence relation.

PROPOSITION A.3.11. *If A and B are quasi-isomorphic chain complexes and B and C are quasi-isomorphic chain complexes, then A and C are quasi-isomorphic.*

Proof. By hypothesis, there are chain complexes A' and B', and quasi-isomorphisms $f\colon A' \to A$, $f'\colon A' \to B$, $g\colon B' \to B$, and $g'\colon B' \to C$. Consider the map $F\colon A' \oplus B' \to B$ defined by $F(a', b') = f'(a') - g(b')$. It remains to check that the map $h\colon \mathrm{Cone}(F) \to A[\![-1]\!]$ defined by $h((a', b'), b) = (-1)^d \cdot f(a')$, where d denotes the degree of a' and b', is a quasi-isomorphism; as is the map $h'\colon \mathrm{Cone}(F) \to C[\![-1]\!]$ defined by $h((a', b'), b) = (-1)^d \cdot g'(b')$. To see this, consider the mapping cone of h. There is a natural short exact sequence of chain complexes

$$0 \longrightarrow \mathrm{Cone}(g) \to \mathrm{Cone}(h) \to \mathrm{Cone}(f) \longrightarrow 0,$$

where, by Corollary A.3.3 we have $H(\mathrm{Cone}(g)) = H(\mathrm{Cone}(f)) = 0$. By Lemma A.3.2 this implies $H(\mathrm{Cone}(h)) = 0$ so, again by Corollary A.3.3, h is a quasi-isomorphism. The other map h' is a quasi-isomorphism by a similar argument. □

A.4. On the structure of homology

PROPOSITION A.4.1. *If C is a finitely generated chain complex over the ring $\mathcal{R} = \mathbb{K}[V_1, \dots, V_n]$, then its homology $H(C)$ is also a finitely generated $\mathcal{R}$-module.*

Proof. It is a basic result in commutative algebra (Hilbert's Basis theorem; see for example [**3**, Theorem 7.5]) that $\mathcal{R} = \mathbb{K}[V_1, \dots, V_n]$ is a *Noetherian ring.* This means that every ideal is finitely generated. It follows that every submodule of a finitely generated module is also finitely generated (see for instance [**3**, Theorem 6.4]). In particular, the submodule of cycles in C is finitely generated. Since the quotient of a finitely generated module is finitely generated, $H(C)$ is finitely generated, too. □

Proposition A.4.1 applies to show that grid homology with coefficients in $\mathcal{R} = \mathbb{K}[V_1, \dots, V_n]$ is finitely generated. For the rest of the present section it is crucial to work with the special case where $\mathbb{K}$ is a field (either $\mathbb{Z}/p\mathbb{Z}$ or $\mathbb{Q}$), and where $\mathcal{R} = \mathbb{K}[U]$ is a polynomial algebra over the field $\mathbb{K}$ in a single variable.

We now state a structure theorem for bigraded modules which is relevant to the structure of grid homology. This result is a variant of the classification theorem for modules over a principal ideal domain, paying special attention to gradings. The statement will use the following notational shorthand. Let $\mathbb{K}[U]/U^n_{(d,s)}$ denote the bigraded cyclic $\mathbb{K}[U]$-module whose generator g has bigrading (d,s). (Note that the non-zero, homogeneous elements of this module have bigradings $(d-2i, s-i)$ for $i=0,\dots,n-1$.) Furthermore, let $\mathbb{K}[U]_{(d,s)}$ denote the bigraded free module of rank one, whose generator has Maslov grading d and Alexander grading s. The proof will use the following notion:

DEFINITION A.4.2. Let X be a bigraded $\mathbb{K}[U]$-module. An element $\xi \in X$ is called ***U-torsion***, or simply ***torsion***, if for all sufficiently large integers n, $U^n \cdot \xi = 0$. Let $\mathrm{Tors}(X) \subset X$ denote the subset of torsion elements.

PROPOSITION A.4.3. *Suppose that $\mathbb{K}$ is a field, and X is a finitely generated, bigraded $\mathbb{K}[U]$-module in the sense of Definition* A.1.3. *Then, there are collections of triples of integers $\{(d_i,s_i,n_i)\}_{i=1}^k$ and pairs of integers $\{(\delta_j,\sigma_j)\}_{j=1}^N$ so that*

$$X \cong \left(\bigoplus_{i=1}^k \mathbb{K}[U]/U^{n_i}_{(d_i,s_i)}\right) \oplus \left(\bigoplus_{j=1}^N \mathbb{K}[U]_{(\delta_j,\sigma_j)}\right). \tag{A.13}$$

Proof. Observe first that X/UX is a finite dimensional vector space over $\mathbb{K}$, since X is finitely generated. We prove Equation (A.13) by induction on the dimension of the vector space X/UX.

If the module X is non-zero, then any homogeneous element $x \in X$ with maximal Alexander grading induces a non-zero element of X/UX. It follows that if $X/UX=0$, then $X=0$.

If X is non-zero, choose some homogeneous element x in X with maximal Alexander grading a. Consider the cyclic module $\mathbb{K}[U]\cdot x$ generated by x. This submodule is either isomorphic to $\mathbb{K}[U]$, or it is identified with $\mathbb{K}[U]/p(U)$, where $p(U)$ is the polynomial of minimal degree for which $p(U)\cdot x = 0$. In the latter case, we claim that $p(U) = U^i$ for some i: for otherwise we could write $p(U) = U^i + U^{i+1}q(U)$ with $q \neq 0$. Since $U^i x$ has Alexander grading $a-i$ while $U^{i+1}q(U)$ is a sum of terms of Alexander grading $< a-i$, it follows that $U^i \cdot x = 0$, contradicting minimality of the degree of $p(U)$.

Next, let $X' = X/(\mathbb{K}[U]\cdot x)$. The natural projection from X/U to X'/UX' is surjective, and its kernel is generated by x; so $\dim(X'/UX') = \dim(X/U) - 1$, and hence by induction Equation (A.13) applies to X'. The generators of the cyclic summands in X' can be lifted to homogeneous elements $\{y_i\}_{i=1}^m$ of X. Depending on whether or not a generator in X' is torsion, for the lifts y_i we have either $U^{n_i}y_i \in \mathbb{K}[U]\cdot x$ (for some positive integer n_i) or $(\mathbb{K}[U]\cdot y_i)\cap(\mathbb{K}[U]\cdot x) = \{0\}$. In the first case, either $U^{n_i}y_i = 0$ or $U^{n_i}y_i = U^{k_i}x$. Let I be the set of those indices $i \in \{1,\dots,m\}$ for which $U^{n_i}y_i = U^{k_i}x \neq 0$. Observe that for all $i \in I$, $k_i \geq n_i$, following from the maximality of the Alexander grading of x. Next define elements $\{x_i\}_{i=1}^{m+1}$ in X by:

$$x_i = \begin{cases} y_i & \text{if } i \leq m \text{ and } i \notin I \\ y_i - U^{k_i-n_i}x & \text{if } i \in I \\ x & \text{if } i = m+1. \end{cases}$$

It is straightforward to verify that these elements generate X, and each generates a distinct cyclic summand of the stated form. □

The isomorphism from Equation (A.13) is not canonical. A more canonical formulation can be given, in terms of the torsion modules from Definition A.4.2, as follows. Note that the set $\mathrm{Tors}(X) \subset X$ is a $\mathbb{K}[U]$-submodule, and forms the first direct summand in Equation (A.13).

COROLLARY A.4.4. *Let X be a finitely generated bigraded module over $\mathbb{K}[U]$, and $\mathrm{Tors}(X) \subset X$ its torsion submodule. Then, the quotient $X/\mathrm{Tors}(X)$ is isomorphic to $\mathbb{K}[U]^N$ for some N.* □

DEFINITION A.4.5. The number of free summands in a finitely generated module X (i.e. the quantity N appearing in Equation (A.13) and in Corollary A.4.4), is called the ***rank*** of the module.

A.5. Dual complexes

Let X be a module over $\mathcal{R} = \mathbb{K}[V_1, \dots V_n]$. We can consider the *dual module* $\mathrm{Hom}_{\mathcal{R}}(X, \mathcal{R})$ of $\mathcal{R}$-module morphisms. This notion can be extended to chain complexes as follows. Let (C, ∂) be a chain complex over the ring $\mathcal{R}$, then its dual module $\mathrm{Hom}_{\mathcal{R}}(C, \mathcal{R})$ inherits a boundary map $\check{\partial}\colon \mathrm{Hom}_{\mathcal{R}}(C, \mathcal{R}) \to \mathrm{Hom}_{\mathcal{R}}(C, \mathcal{R})$, defined as follows. Given $\phi\colon C \to \mathcal{R}$, the homomorphism $\check{\partial}(\phi)$ is the homomorphism whose value on $c \in C$ is the value of ϕ on ∂c; i.e. $\check{\partial}(\phi)(c) = \phi(\partial c)$. The relation $\partial^2 = 0$ dualizes readily to give $\check{\partial}^2 = 0$. The chain complex $(\mathrm{Hom}_{\mathcal{R}}(C, \mathcal{R}), \check{\partial})$ is called the *dual complex* of (C, ∂).

EXAMPLE A.5.1. If (C, ∂) is a finitely generated, free $\mathcal{R}$-module, given with a basis $\{\mathbf{x}_i\}_{i=1}^N$, then the differential ∂ is specified by the matrix $A = (a_{i,j})$ so that

$$\partial \mathbf{x}_i = \sum_{i,j=1}^{N} a_{i,j} \mathbf{x}_j.$$

In this case, the dual complex is generated by $\{\mathbf{x}_i^*\}_{i=1}^N$, with $x_i^*(x_j) = \delta_{ij}$; and the differential $\check{\partial}$ is specified by the transpose of A:

$$\check{\partial} \mathbf{x}_i^* = \sum_{i,j=1}^{N} a_{j,i} \mathbf{x}_j^*.$$

Suppose now that (C, ∂) is a bigraded chain complex. The dual complex $\mathrm{Hom}_{\mathcal{R}}(C, \mathcal{R})$ inherits a bigrading from the complex C. We explain this in the case where $\mathcal{R} = \mathbb{K}[U]$, which we think of now as a bigraded $\mathcal{R}$-module, where the element U^m has bigrading $(-2m, -m)$. This induces a bigrading on the dual module: an element $\phi \in \mathrm{Hom}_{\mathcal{R}}(C, \mathcal{R})$ has bigrading (m, t) if ϕ maps $C_{d,s}$ to $\mathcal{R}_{d+m,s+t}$. With this convention, the dual of a bigraded chain complex C over $\mathbb{K}[U]$ is also a bigraded chain complex over $\mathbb{K}[U]$, and the induced differential on $\mathrm{Hom}_{\mathcal{R}}(C, \mathcal{R})$ drops grading by one, just as the differential for C did.

Note that the dual complex is equivalent to the usual notion of cohomology, with the understanding that the grading conventions differ by an overall multiplication by -1. We choose our grading conventions exactly so that the dual of a bigraded complex is also bigraded in the same sense.

The *Universal Coefficient Theorem* (for cohomology) provides the link between the homology of a chain complex and the homology of its dual. To state it, consider the pairing $\mathrm{Hom}_{\mathcal{R}}(C,\mathcal{R})\otimes_{\mathcal{R}} C \to \mathcal{R}$ given by the formula $(f,c)\mapsto f(c)$. This descends to homology, giving the *Kronecker pairing* $H(\mathrm{Hom}_{\mathcal{R}}(C,\mathcal{R}),\check{\partial})\otimes_{\mathcal{R}} H(C,\partial)\to \mathcal{R}$; and hence a duality map

$$H(\mathrm{Hom}_{\mathcal{R}}(C,\mathcal{R}),\check{\partial}) \to \mathrm{Hom}_{\mathcal{R}}(H(C,\partial),\mathcal{R}). \tag{A.14}$$

The Universal Coefficient Theorem can be conveniently formulated when $\mathcal{R}$ is either a field of a principal ideal domain. When $\mathcal{R}$ is a field (i.e, $n=0$ and $\mathbb{K}$ is $\mathbb{Z}/p\mathbb{Z}$ or $\mathbb{Q}$), the duality map is an isomorphism, according to the following.

THEOREM A.5.2. *If $\mathcal{R}$ is a field, then for a chain complex (C,∂),*

$$H(\mathrm{Hom}_{\mathcal{R}}(C,\partial),\check{\partial}) \cong \mathrm{Hom}_{\mathcal{R}}(H(C,\partial),\mathcal{R}).$$ □

If $\mathcal{R}$ is a PID, any module X over $\mathcal{R}$ can be fit into a short exact sequence (called a *free resolution*)

$$0 \longrightarrow F^2 \xrightarrow{\ r\ } F^1 \xrightarrow{\ p\ } X \longrightarrow 0, \tag{A.15}$$

where F^1 and F^2 are free $\mathcal{R}$-modules. Dualizing r gives a map $r^*\colon \mathrm{Hom}_{\mathcal{R}}(F^1,\mathcal{R})\to \mathrm{Hom}_{\mathcal{R}}(F^2,\mathcal{R})$, whose kernel is identified with $\mathrm{Hom}_{\mathcal{R}}(X,\mathcal{R})$, and whose cokernel, denoted $\mathrm{Ext}(X,\mathcal{R})$, turns out to be independent of the choice of the resolution. (This module is also denoted $\mathrm{Ext}^1(X;\mathcal{R})$ in some texts.)

EXAMPLE A.5.3. The following are the only Ext groups which will be of relevance in this text:

1. If X is a free module, then we can take $F^1=X$ and $F^2=0$. Thus, $\mathrm{Ext}(X,\mathcal{R})=0$.
2. If $\mathcal{R}$ is a field, then any $\mathcal{R}$-module is free, so $\mathrm{Ext}(X,\mathcal{R})=0$.
3. $\mathrm{Ext}(X\oplus Y,\mathcal{R})=\mathrm{Ext}(X,\mathcal{R})\oplus\mathrm{Ext}(Y,\mathcal{R})$.
4. Let $\mathcal{R}=\mathbb{K}[U]$ and $X=\mathbb{K}[U]/U^n$ for some n. Then, we can take $F^1=F^2=\mathbb{K}[U]$, and r to be the map which is multiplication by U^n. In this case, $\mathrm{Ext}(\mathbb{K}[U]/U^n,\mathbb{K}[U])\cong\mathbb{K}[U]/U^n$.

In fact, we will use a refinement of the Universal Coefficient Theorem, in the case where C is a bigraded chain complex over $\mathbb{K}[U]$, in the sense of Definition A.1.3.

As a first step, observe that if X is a bigraded module over $\mathbb{K}[U]$, then the Ext modules also inherit the structure of a bigraded module over $\mathbb{K}[U]$. To define the bigrading, consider first a free resolution (Equation (A.15)) so that F^1 ane F^2 are bigraded, and the maps p and r preserve bigradings. Such a resolution can be formed by the following tautological construction: F^1 is the free module over $\mathbb{F}[U]$ whose generating set is the set of homogeneous, non-zero elements of X. This comes equipped with a canonical bigraded map $p\colon F^1\to X$. Its kernel is a free module since $\mathbb{K}[U]$ is a principal ideal domain (hence any submodule of a free module is free), and it is bigraded, since p is bigraded.

Armed with these bigradings, the cokernel of

$$r^*\colon \mathrm{Hom}_{\mathcal{R}}(F^1,\mathcal{R})\to\mathrm{Hom}_{\mathcal{R}}(F^2,\mathcal{R}),$$

which is $\mathrm{Ext}(X,\mathcal{R})$, naturally inherits a bigrading.

EXAMPLE A.5.4. Consider the bigraded $\mathbb{K}[U]$-module $X=\mathbb{K}[U]/U^n_{(d,s)}$. Then, there is an isomorphism of bigraded modules over $\mathbb{K}[U]$

$$\mathrm{Ext}(X,\mathbb{K}[U])\cong\mathbb{K}[U]/U^n_{(2n-d,n-s)}.$$

EXAMPLE A.5.5. Consider the bigraded $\mathbb{K}[U]$-module $X = \mathbb{K}[U]_{(d,s)}$, whose generator has bigrading (d, s). Then, there is an isomorphism of bigraded $\mathbb{K}[U]$-modules

$$\mathrm{Hom}_{\mathcal{R}}(X, \mathbb{K}[U]) = \mathrm{Hom}(X, \mathbb{K}[U]) \cong \mathbb{K}[U]_{(-d,-s)}.$$

THEOREM A.5.6. *If C is a bigraded complex of free modules over $\mathbb{K}[U]$, then there is an isomorphism of bigraded modules*

$$H(\mathrm{Hom}(C, \mathbb{K}[U])) \cong \mathrm{Hom}(H(C), \mathbb{K}[U]) \oplus \mathrm{Ext}(H(C), \mathbb{K}[U])[\![1,0]\!]. \tag{A.16}$$

More explicitly, writing

$$H(C) \cong \left(\bigoplus_{i=1}^{k} \mathbb{K}[U]/U^{n_i}_{(d_i,s_i)}\right) \oplus \left(\bigoplus_{j=1}^{N} \mathbb{K}[U]_{(\delta_j,\sigma_j)}\right),$$

we have that

$$H(\mathrm{Hom}(C, \mathbb{K}[U])) \cong \left(\bigoplus_{i=1}^{k} \mathbb{K}[U]/U^{n_i}_{(2n_i-d_i-1,n_i-s_i)}\right) \oplus \left(\bigoplus_{j=1}^{N} \mathbb{K}[U]_{(-\delta_j,-\sigma_j)}\right). \tag{A.17}$$

Proof. The usual $\mathbb{Z}$-graded Universal Coefficient Theorem (see for example [**83**, Chapter 3.1]) adapts to prove Equation (A.16), as follows. Consider the subcomplex $Z \subset C$ of cycles and the subcomplex $B \subset C$ of boundaries. Since C is free, so are B and Z; so the short exact sequence

$$0 \longrightarrow B \xrightarrow{\ i\ } Z \longrightarrow H(C) \longrightarrow 0 \tag{A.18}$$

is a free resolution of $H(C)$. There is a short exact sequence of bigraded chain complexes

$$0 \longrightarrow Z \longrightarrow C \longrightarrow B[\![-1,0]\!] \longrightarrow 0. \tag{A.19}$$

Since B is free, the above short exact sequence dualizes to a short exact sequence of chain complexes

$$0 \longleftarrow \mathrm{Hom}(Z, \mathbb{K}[U]) \longleftarrow \mathrm{Hom}(C, \mathbb{K}[U]) \longleftarrow \mathrm{Hom}(B[\![-1,0]\!], \mathbb{K}[U]) \longleftarrow 0.$$

Here the boundary maps on $\mathrm{Hom}(Z, \mathbb{K}[U])$ and $\mathrm{Hom}(B[\![1,0]\!], \mathbb{K}[U])$ are identically zero. The long exact sequence in homology gives the short exact sequence

$$0 \leftarrow \mathrm{Hom}(H(C), \mathbb{K}[U]) \leftarrow H(\mathrm{Hom}(C, \mathbb{K}[U])) \leftarrow \mathrm{Ext}(H(C), \mathbb{K}[U][\![1,0]\!] \longleftarrow 0, \tag{A.20}$$

since dualizing Equation (A.18) we get that $\mathrm{Hom}(H(C), \mathbb{K}[U]) \cong \mathrm{Ker}(i^*)$, and

$$\frac{\mathrm{Hom}(B[\![-1,0]\!], \mathbb{K}[U])}{\mathrm{Hom}(Z[\![-1,0]\!], \mathbb{K}[U])} \cong \frac{\mathrm{Hom}(B, \mathbb{K}[U])}{\mathrm{Hom}(Z, \mathbb{K}[U])}[\![1,0]\!] \cong \mathrm{Ext}(H(C), \mathbb{K}[U])[\![1,0]\!],$$

in view of our grading conventions on the dual complex (which are opposite to the usual one on cohomology). As usual, a splitting of Equation (A.19) provides a chain map $C \to Z$ which, when composed with the quotient map to $H(C)$, dualizes to a map $\mathrm{Hom}(H(C), \mathbb{K}[U]) \to H(\mathrm{Hom}(C), \mathbb{K}[U])$ that provides a splitting of Equation (A.20).

The graded isomorphism of Equation (A.17) comes from our computation of the bigraded Ext for the relevant modules (see Examples A.5.3, A.5.4 and A.5.5). □

REMARK A.5.7. Note that if the ground ring is not a PID, much of the homological algebra discussed above is slightly more complicated: finitely generated modules might not decompose as direct sums of cyclic modules; and the universal coefficient theorem is much more involved [**226**]. This is already relevant for rings appearing in grid homology, such as $\mathbb{Z}[U]$ (as in Chapter 15) or $\mathbb{K}[U_1, \dots, U_n]$, $n \geq 2$ (as in Chapter 11).

A.6. On filtered complexes

In Chapter 13, grid diagrams are used to go beyond a bigraded homology group; they are used to define a quasi-isomorphism class of $\mathbb{Z}$-filtered, $\mathbb{Z}$-graded chain complexes, in the sense of Definition 13.1.1. We refer the reader back to Section 13.1 for the necessary definitions and algebraic constructions. The filtered complexes we consider in this section will be defined over $\mathbb{K}[V_1, \dots, V_n]$; $\mathbb{K}$ can be any base ring.

Let $\mathcal{C}$ and $\mathcal{C}'$ be two $\mathbb{Z}$-filtered, $\mathbb{Z}$-graded chain complexes over $\mathbb{K}[V_1, \dots, V_n]$, and let $\phi\colon \mathcal{C} \to \mathcal{C}'$ be a filtered chain map, in the sense of Definition 13.1.4. Such a map induces a bigraded chain map $\mathrm{gr}(\phi)$ on the associated graded chain complex; and ϕ is called a filtered quasi-isomorphism when $\mathrm{gr}(\phi)$ is a quasi-isomorphism.

PROPOSITION A.6.1. *Let $\phi\colon \mathcal{C} \to \mathcal{C}'$ be a filtered quasi-isomorphism. Then, ϕ induces isomorphisms $H(\mathcal{F}_i\mathcal{C}) \cong H(\mathcal{F}_i\mathcal{C}')$ and $H(\mathcal{C}) \cong H(\mathcal{C}')$.*

The proof will follow from the following special case:

LEMMA A.6.2. *If $\mathcal{C}$ is a filtered complex with $H(\mathrm{gr}(\mathcal{C})) = 0$, then $H(\mathcal{F}_i\mathcal{C}) = 0$ and $H(\mathcal{C}) = 0$.*

Proof. Fix an integer d. We show that $H_d(\mathcal{F}_i\mathcal{C}) = 0$ by increasing induction on i. The case where i is small is always true, because our filtered complexes are assumed to be bounded below; i.e. for sufficiently small i, $\mathcal{F}_i\mathcal{C}_d = 0$. For the inductive step, assume that $H_d(\mathcal{F}_i\mathcal{C}) = 0$, and note that $H_d(\frac{\mathcal{F}_{i+1}\mathcal{C}}{\mathcal{F}_i\mathcal{C}}) = 0$, by the hypothesis that $H(\mathrm{gr}(\mathcal{C})) = 0$. Thus, the vanishing of $H_d(\mathcal{F}_{i+1}\mathcal{C})$ follows from the long exact sequence associated to the short exact sequence

$$0 \longrightarrow \mathcal{F}_i\mathcal{C} \longrightarrow \mathcal{F}_{i+1}\mathcal{C} \longrightarrow \frac{\mathcal{F}_{i+1}\mathcal{C}}{\mathcal{F}_i\mathcal{C}} \longrightarrow 0.$$ □

Proof of Proposition A.6.1. The *filtered mapping cone* of ϕ, $\mathit{Cone}(\phi)$, is the mapping cone of ϕ endowed with the filtration where

$$\mathcal{F}_i\mathit{Cone}(\phi) \cong \mathrm{Cone}(\phi|_{\mathcal{F}_i\mathcal{C}}). \tag{A.21}$$

Observe that there is an isomorphism of chain complexes $\mathrm{gr}(\mathit{Cone}(\phi)) \cong \mathrm{Cone}(\mathrm{gr}(\phi))$. By Corollary A.3.3, $H(\mathrm{Cone}(\mathrm{gr}(\phi))) = 0$, and hence by Lemma A.6.2,

$$H(\mathcal{F}_i\mathit{Cone}(\phi)) = 0 = H(\mathit{Cone}(\phi)).$$

In view of Equation (A.21), Corollary A.3.3 ensures that ϕ induces isomorphisms $H(\mathcal{F}_i\mathcal{C}) \cong H(\mathcal{F}_i\mathcal{C}')$. Similarly, $H(\mathit{Cone}(\phi)) = 0$ implies that $H(\mathcal{C}) \cong H(\mathcal{C}')$. □

REMARK A.6.3. Recall that our filtered complexes are required to be bounded below, by definition. Without this hypothesis, the conclusion of Proposition A.6.1 does not hold. For example, consider the filtered complex $\mathcal{C}$ generated over $\mathbb{K}$ by two sequences of elements $\{x_i, y_i\}_{i\leq 0}$, with $x_i \in \mathcal{F}_i\mathcal{C}_1 \setminus \mathcal{F}_{i-1}\mathcal{C}_1$, $y_i \in \mathcal{F}_i\mathcal{C}_0 \setminus \mathcal{F}_{i-1}\mathcal{C}_{-1}$, and with a differential $\partial x_i = y_i + y_{i-1}$. Clearly, $H(\mathrm{gr}(\mathcal{C})) = 0$, but $H(\mathcal{C}) \cong \mathbb{K}$.

As a corollary to the above discussion, there are numerical invariants associated to the filtered quasi-isomorphism types of filtered complexes. Suppose that $\mathcal{C}$ is a $\mathbb{Z}$-filtered, $\mathbb{Z}$-graded chain complex over $\mathbb{F}$, and suppose that $H(\mathcal{C})$ is finite dimensional. Then, we claim that for sufficiently small i, the image of $H(\mathcal{C}_i)$ in $H(\mathcal{C})$ is trivial: $H(\mathcal{C})$ is finite dimensional, so since the filtration is bounded below, we can choose i small enough that for each d for which $H_d(\mathcal{C}) \neq 0$, $\mathcal{F}_i\mathcal{C}_d = 0$. Similarly, for i sufficiently large, $H(\mathcal{C}_i) \to H(\mathcal{C})$ is non-trivial; if $z \in \mathcal{C}$ is a non-trivial cycle representing some homology class, then since the filtration exhausts $\mathcal{C}$, we know that $z \in \mathcal{F}_i\mathcal{C}$ for sufficiently large i; i.e. $[z] \subset H(\mathcal{C})$ is the image of a homology class in $H(\mathcal{F}_i\mathcal{C})$. Thus, we can make the following;

DEFINITION A.6.4. Let $t(\mathcal{C})$ be the minimal i so that $H(\mathcal{F}_i\mathcal{C}) \to H(\mathcal{C})$ is non-zero.

COROLLARY A.6.5. *Suppose that $\mathcal{C}$ and $\mathcal{C}'$ are quasi-isomorphic complexes, and that $H(\mathcal{C}) \cong H(\mathcal{C}') \neq 0$. Then, $t(\mathcal{C}) = t(\mathcal{C}')$.*

Proof. It suffices to show that if $\phi\colon \mathcal{C} \to \mathcal{C}'$ is a filtered quasi-isomorphism, then $t(\mathcal{C}) = t(\mathcal{C}')$. To this end, consider the commutative square

$$\begin{array}{ccc} H(\mathcal{F}_s\mathcal{C}) & \xrightarrow{H(i_s)} & H(\mathcal{C}) \\ {\scriptstyle H(\phi|_{\mathcal{F}_s\mathcal{C}})}\downarrow & & \downarrow{\scriptstyle H(\phi)} \\ H(\mathcal{F}_s\mathcal{C}') & \xrightarrow{H(i'_s)} & H(\mathcal{C}'), \end{array}$$

where the horizontal maps are induced by inclusions. Both vertical maps are isomorphisms by Proposition A.6.1, so $H(i_s)$ is non-trivial exactly when $H(i'_s)$ is. □

EXERCISE A.6.6. Let $\mathcal{C}$ be a $\mathbb{Z}$-filtered, $\mathbb{Z}$-graded chain complex over $\mathbb{F}[U]$, and suppose that $H(\mathcal{C}) \cong \mathbb{F}[U]$. For any integer $i \geq 0$, let

$$\tau_i(\mathcal{C}) = \min\{s \big| \text{ the image of } H(\mathcal{F}_s\mathcal{GC}^-) \text{ in } H(\mathcal{GC}^-) \cong \mathbb{F}[U] \text{ contains } U^i\}.$$

Show that if $\mathcal{C}'$ is a $\mathbb{Z}$-filtered, $\mathbb{Z}$-graded chain complex over $\mathbb{F}[U]$ that is filtered quasi-isomorphic to $\mathcal{C}$, then for all i, $\tau_i(\mathcal{C}) = \tau_i(\mathcal{C}')$.

A.7. Small models for filtered grid complexes

There are infinitely many different chain complexes in a fixed quasi-isomorphism class of a $\mathbb{Z}$-filtered, $\mathbb{Z}$-graded chain complex; and the chain complexes coming from the filtered grid complex tend to have many generators. Thus, it is useful to get an economical complex representing a fixed filtered quasi-isomorphism class. Such model complexes were used in Section 14.2 to describe examples.

In Section 14.2 we represented $\mathcal{GC}^-(\vec{K})$ (the filtered quasi-isomorphism class of the filtered chain complexes $\mathcal{GC}^-(\mathbb{G})$ for grid diagrams representing $\vec{K}$) by a free chain complex over $\mathbb{F}[U]$ whose rank coincides with the dimension of the $\mathbb{F}$-vector space $\widehat{GH}(\vec{K})$. The existence of such a representative was stated in Lemma 14.2.1. Our aim here is to prove that lemma, after restating the result in more abstract terms. For this purpose, throughout this section we fix a field $\mathbb{K}$. We start with a special case.

LEMMA A.7.1. *Fix a field* $\mathbb{K}$, *and let* $\mathcal{C}$ *be a* $\mathbb{Z}$*-filtered,* $\mathbb{Z}$*-graded chain complex over* $\mathbb{K}[U]$ *that is a finitely generated, free module over* $\mathbb{K}[U]$. *Then,* $\mathcal{C}$ *is filtered homotopy equivalent to a free, finitely generated chain complex* $\mathcal{C}'$ *over* $\mathbb{K}[U]$ *whose induced differential on* $\mathrm{gr}(\frac{\mathcal{C}'}{U})$ *vanishes; i.e.* $\dim_{\mathbb{K}} H(\mathrm{gr}(\frac{\mathcal{C}'}{U})) = \dim_{\mathbb{K}} \mathrm{gr}(\frac{\mathcal{C}'}{U})$.

Proof. Suppose that the differential on $\mathrm{gr}(\frac{\mathcal{C}}{U})$ does not vanish. This means that there is a basis $\{[x_i]_{i=1}^m\}$ for $\mathrm{gr}(\frac{\mathcal{C}}{U})$ so that $\partial[x_1] = [x_2]$ with $[x_2] \neq 0$. Find lifts x_i of the $[x_i]$; i.e. if $[x_i] \in \mathcal{F}_{a_i}/\mathcal{F}_{a_i-1}$, choose $x_i \in \mathcal{F}_{a_i}$ whose projection is $[x_i]$. Clearly, $\{x_i\}_{i=1}^m$ generate $\mathcal{C}$ as a $\mathbb{K}[U]$-module. Thus, there is a $\mathbb{K}[U]$-module map $T\colon \mathcal{C} \to \mathcal{C}$ defined by

$$T(x_i) = \begin{cases} x_1 & \text{if } i = 2 \\ 0 & \text{otherwise.} \end{cases}$$

The map T is a filtered map; and so $\phi = \mathrm{Id} - \partial T - T\partial$ is a filtered chain map. Clearly $T^2 = 0$, and so it follows that $\phi \circ T = T \circ \phi$; hence the subcomplex $\phi(\mathcal{C})$ of $\mathcal{C}$ is chain homotopy equivalent to $\mathcal{C}$. Note that $\phi(x_1) = 0$, and so the dimension of $\mathrm{gr}(\phi(\frac{\mathcal{C}}{U}))$ is smaller than the the dimension of $\mathrm{gr}(\frac{\mathcal{C}}{U})$. This procedure must terminate at some point, giving a chain homotopy equivalent subcomplex $\mathcal{C}'$ whose rank agrees with $H(\mathrm{gr}(\frac{\mathcal{C}'}{U}))$. □

PROPOSITION A.7.2. *Fix a field* $\mathbb{K}$ *and let* $\mathcal{C}$ *be a* $\mathbb{Z}$*-filtered,* $\mathbb{Z}$*-graded chain complex over* $\mathbb{K}[U]$, *that is a free module over* $\mathbb{K}[U]$, *and suppose that* $H(\mathrm{gr}(\frac{\mathcal{C}}{U}))$ *is a finite dimensional* $\mathbb{K}$*-vector space. Then,* $\mathcal{C}$ *is filtered homotopy equivalent to a chain complex* $\mathcal{C}'$ *that is a finitely generated, free module over* $\mathbb{K}[U]$ *whose rank coincides with* $\dim_{\mathbb{K}} H(\mathrm{gr}(\frac{\mathcal{C}}{U}))$.

Proof. If $\mathcal{C}$ is infinitely generated, choose a generating set $\{[x_i]_{i=1}^\infty\}$ of $\mathrm{gr}(\frac{\mathcal{C}}{U})$, and fix some integer $d \geq 0$. By boundedness, there are only finitely many x_i with grading $\geq d$. Using the method from Lemma A.7.1, we can find a sequence of filtered subcomplexes $\mathcal{C}^{(i)}$ of $\mathcal{C}$, with the following properties:

- $\mathcal{C}^{(i)} \subseteq \mathcal{C}^{(i+1)}$
- $\mathcal{C}^{(i)} = \mathcal{C}$ for sufficiently large i,
- $\mathrm{gr}(\frac{\mathcal{C}^{(i)}}{U})$ has vanishing differentials in all gradings $\geq i$.
- There are maps $T^{(i)}\colon \mathcal{C}^{(i)} \to \mathcal{C}^{(i)}$, with $T^{(i)} \circ \phi^{(i)} = \phi^{(i)} \circ T^{(i)}$, so that $\mathcal{C}^{(i-1)}$ is the image of $\phi^{(i)}\colon \mathcal{C}^{(i)} \to \mathcal{C}^{(i)}$ by the map $\phi^{(i)} = \mathrm{Id} - \partial \circ T^{(i)} - T^{(i)} \circ \partial$.
- For any fixed d, $i \leq d-1$, $T^{(i)}$ vanishes on $\mathcal{C}^{(i)}_d$.

It follows from the above properties that the inclusion of $\mathcal{C}^{(i-1)}$ in $\mathcal{C}^{(i)}$ is a filtered chain homotopy equivalence. For each d, $T^{(i)}$ vanishes on $\mathcal{C}^{(i)}_d$ for all but finitely many i; and so, $\phi^{(i)}|_{\mathcal{C}^{(i)}}$ is the identity map for all but finitely many i. Thus, for a fixed d, there is an i_d so that $\mathcal{C}^{(i_d)}_d = \mathcal{C}^{(i_d-1)}_d = \dots$, so can form the infinite composite $\phi = \dots \circ \phi^{(i)} \circ \phi^{(i+1)} \circ \dots$. The image of ϕ is our desired subcomplex. □

A.8. Filtered quasi-isomorphism versus filtered homotopy type

In Section 13.1, we discussed two equivalence relations on filtered chain complexes: filtered quasi-isomorphism, and filtered chain homotopy equivalence. Two filtered homotopy equivalent complexes are necessarily filtered quasi-isomorphic (Exercise 13.1.10), but filtered quasi-isomorphic complexes need not be filtered chain homotopy equivalent (Example 13.1.8). Since our grid complexes are free modules over $\mathbb{K}[V_1,\dots,V_n]$, this distinction disappears, according to the present results.

Throughout this section, we fix once again $\mathbb{K}$ to be a field. We show the following:

PROPOSITION A.8.1. *Let $\mathcal{C}$ and $\mathcal{C}'$ be two $\mathbb{Z}$-filtered, $\mathbb{Z}$-graded chain complexes over $\mathbb{K}[U]$. Then, these complexes are filtered chain homotopy equivalent over $\mathbb{K}[U]$ if and only if they are filtered quasi-isomorphic over $\mathbb{K}[U]$.*

We return to the proof after reducing to a special case (where $\mathcal{C}'=0$).

DEFINITION A.8.2. A $\mathbb{Z}$-graded, $\mathbb{Z}$-filtered chain complex $\mathcal{C}$ over a ring $\mathcal{R}$ is ***chain contractible*** if there is an $\mathcal{R}$-module homomorphism $H\colon \mathcal{C}\to\mathcal{C}$, the ***contraction map***, that raises grading by 1 and satisfies the equation $\partial\circ H+h\circ\partial=\mathrm{Id}_{\mathcal{C}}$, so that $\mathcal{C}$ is filtered chain homotopy equivalent to the trivial chain complex 0.

LEMMA A.8.3. *Let $f\colon \mathcal{C}\to\mathcal{C}'$ be a $\mathbb{Z}$-filtered, $\mathbb{Z}$-graded chain map between two $\mathbb{Z}$-filtered, $\mathbb{Z}$-graded chain complexes over $\mathbb{K}[V_1,\dots,V_n]$. Then, $Cone(f)$ is contractible if and only if f is a filtered chain homotopy equivalence.*

Proof. Recall that $Cone(f)$ is the direct sum $\mathcal{C}\oplus\mathcal{C}'$, equipped with the differential

$$D=\begin{pmatrix} -\partial & 0\\ f & \partial'\end{pmatrix}.$$

A null-homotopy $H\colon Cone(f)\to Cone(f)$ of the identity map satisfies the equation

$$D\circ H+H\circ D-\mathrm{Id}_{\mathcal{C}\oplus\mathcal{C}'}=0. \tag{A.22}$$

Writing H in its components, we get $H=\begin{pmatrix} h & g\\ i & h'\end{pmatrix}$, where $g\colon \mathcal{C}'\to\mathcal{C}$, $h\colon\mathcal{C}\to\mathcal{C}$, $h'\colon\mathcal{C}'\to\mathcal{C}'$. Equation (A.22) can be interpreted as the vanishing of a 2×2 matrix; i.e. it gives four relations; and three of these are:

$$\partial\circ g-g\circ\partial'=0,\qquad -\partial\circ h-h\circ\partial=\mathrm{Id}_{\mathcal{C}}-g\circ f,\qquad \partial'\circ h'+h'\circ\partial'=\mathrm{Id}_{\mathcal{C}'}-f\circ g.$$

In words, g is a chain map, and f and g are homotopy inverses to one another. In particular, f is a homotopy equivalence.

Conversely, given f, g, h, and h' satisfying the above, it follows that if we let

$$H=\begin{pmatrix} h & g\\ 0 & h'\end{pmatrix},$$

then $\Phi=D\circ H+H\circ D$ is a lower triangular chain map, and hence invertible, and so $\Phi^{-1}\circ H$ is the desired contraction. □

LEMMA A.8.4. *Let $\mathcal{C}$ be a $\mathbb{Z}$-filtered, $\mathbb{Z}$-graded chain complex that is free over $\mathbb{K}[U]$, where $\mathbb{K}$ is a field. If $H(\mathrm{gr}(\mathcal{C}))=0$, then $\mathcal{C}$ is filtered homotopy equivalent over $\mathbb{K}[U]$ to the 0 complex.*

Proof. This is an immediate consequence of Proposition A.7.2: under our present hypotheses, the subcomplex is the 0 complex. □

Proof of Proposition A.8.1 One direction is easy: a filtered homotopy equivalence f induces a homotopy equivalence on the associated graded level, which is therefore a quasi-isomorphism, and so f is a filtered quasi-isomorphism.

For the converse direction, recall that for any filtered chain map

$$H(\mathrm{gr}(\mathcal{C}one(f))) \cong H(\mathrm{Cone}(\mathrm{gr}(f))),$$

so if f a quasi-isomorphism, then $H(\mathrm{Cone}(\mathrm{gr}(f)) = 0$ by the long exact sequence for the mapping cone of $\mathrm{gr}(f)$. Thus, by Lemma A.8.4, we conclude that $\mathcal{C}one(f)$ is chain contractible over $\mathbb{K}[V_i]$. The result now follows from Lemma A.8.3. □

APPENDIX B

Basic theorems in knot theory

This appendix covers some of the foundational material on knot theory used in this book. We start, in Section B.1, with a classical theorem of Reidemeister that characterizes, via local moves, those link diagrams that represent isotopic links. In Section B.2, we discuss the analogue of this theorem in the contact context, for transverse and Legendrian knots and links. In Section B.3 we prove the theorem of Reidemeister-Singer that relates different Seifert surfaces for the same link. Next, we prove Cromwell's theorem in Section B.4 that characterizes, via local moves, those grid diagrams that represent isotopic links. The Legendrian version of this theorem is also proved in this section. Finally, in Section B.5 we define the normal form of a cobordism between knots, and show that any cobordism can be isotoped to such a form.

B.1. The Reidemeister Theorem

The Reidemeister Theorem 2.1.4 allows us to study knots and links in $\mathbb{R}^3$ in terms of their diagrams. We restate this theorem as follows:

THEOREM B.1.1 (Reidemeister, [**196**]). *Two link diagrams represent equivalent links if and only if these diagrams can be transformed into each other by a finite sequence of Reidemeister moves (shown on Figure 2.2) and planar isotopies.*

One direction of the equivalence in the theorem is straightforward: Reidemeister moves and planar isotopies clearly preserve the isotopy type of the link.

There are a number of proofs of this fundamental result; below we will describe a proof using singularity theory from [**199**]. (For a proof of the piecewise linear version of the theorem, see [**18**].) Throughout the discussion, we will assume familiarity with standard transversality results in differential topology, as presented, for example in [**87, 151**]. We will concentrate on the case of knots; the general case of links can be proved along the same lines.

Suppose that two knots K_0 and K_1 are smoothly isotopic. An isotopy gives rise to a one-parameter family of knot diagrams. Keeping track of the isotopy parameter, this gives a map from the two-manifold $[0,1] \times S^1$ to the three-manifold $[0,1] \times \mathbb{R}^2$. A key local result we will use is the following theorem of Whitney:

THEOREM B.1.2 (Whitney, [**228**]). *Let W be a smooth two-manifold and Y a smooth three-manifold. Any smooth map $g_0\colon W \to Y$ can be approximated arbitrarily closely (in the C^2 topology) by a smooth map $g\colon W \to Y$ with the following property. Around each point $p \in W$, there are local coordinates (x, y) so that p corresponds to $(0,0)$, and there are local coordinates (u, v, z) around $g(p) \in Y$, so that* $(0,0,0)$

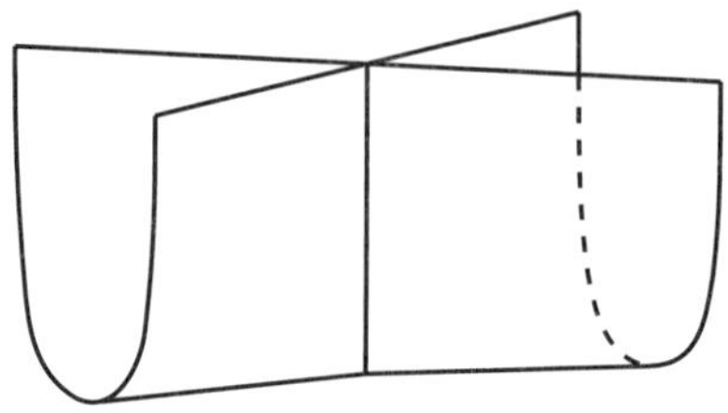

FIGURE B.1. **The Whitney umbrella.** It is the image of the singularity $(x, y) \mapsto (x^2, xy, y)$.

corresponds to $g(p)$, *with respect to which the function* g *has the form*

- $(x, y) \mapsto (x, y, 0)$ *or*
- $(x, y) \mapsto (x^2, xy, y)$.

More intrinsically, consider the Jacobian $J_g(p)\colon T_pW \to T_{g(p)}Y$ of g at $p \in W$. Points of the first kind are those for which the Jacobian is injective; at those points g is an *immersion*. Points of the second kind are the *singular points* of g, where the Jacobian has 1-dimensional kernel. Note that for a generic choice of g (as above), the rank of $J_g(p)$ is non-zero for all $p \in W$. Thus, Whitney's theorem gives a canonical form for the neighborhood of the singular points of g; such a singularity is called a *Whitney umbrella.* Its image is the locus of points $(u, v, z) \in \mathbb{R}^3$ with $v^2 - uz^2 = 0$ (and $u \geq 0$); see Figure B.1. A proof of Theorem B.1.2 is given in Subsection B.1.2.

Suppose that $\mathcal{D}_0$ and $\mathcal{D}_1$ are diagrams of the smoothly embedded knots K_0 and K_1, and assume that the two knots are isotopic. Regard the isotopy as a smoothly embedded surface-with-boundary $f\colon [0,1] \times S^1 \to [0,1] \times \mathbb{R}^3$ with the following properties:

- $\mathrm{Im}(f) \cap (\{i\} \times \mathbb{R}^3) = \{i\} \times K_i$ for $i = 0, 1$, and
- the intersection $f([0,1] \times S^1) \cap (\{t\} \times \mathbb{R}^3)$ is transverse for all $t \in [0,1]$.

Let $\mathrm{pr}_1\colon [0,1] \times \mathbb{R}^3 \to [0,1]$ denote the projection onto the first factor. The second condition above is equivalent to requiring that $\mathrm{pr}_1 \circ f\colon [0,1] \times S^1 \to [0,1]$ has no critical points; it is clearly an open condition.

Extend the projection map $\pi\colon \mathbb{R}^3 \to \mathbb{R}^2$ defining the knot diagram to the map $P = \mathrm{Id} \times \pi\colon [0,1] \times \mathbb{R}^3 \to [0,1] \times \mathbb{R}^2$. Compose the embedding f of the annulus $[0,1] \times S^1$ with P to get a map $\phi = P \circ f\colon [0,1] \times S^1 \to [0,1] \times \mathbb{R}^2$. Theorem B.1.1 is proved by applying Theorem B.1.2 to this map ϕ, as follows.

Proof of Theorem B.1.1. Consider two isotopic knots K_0, K_1 with the isotopy $f\colon [0,1] \times S^1 \to [0,1] \times \mathbb{R}^3$ between them, and with knot diagrams $\mathcal{D}_0$ and $\mathcal{D}_1$, and let $\phi\colon [0,1] \times S^1 \to [0,1] \times \mathbb{R}^2$ be the map defined as above. Since P is a submersion, ϕ can be put into general position by slightly perturbing the map f, so that it remains an isotopy between K_0 and K_1.

Theorem B.1.2 shows that there are finitely many points $B \subset [0,1] \times S^1$ with a Whitney umbrella singularity, away from which the map is an immersion. By a further general position argument, we can assume that every point in $[0,1] \times \mathbb{R}^2$ has at most three preimages. In fact, we can arrange that ϕ has only finitely many triple points T (i.e. points in $[0,1] \times \mathbb{R}^2$ with three preimages), and a union of one-dimensional submanifolds $D \subset [0,1] \times \mathbb{R}^2$ of double points. The closure of the set

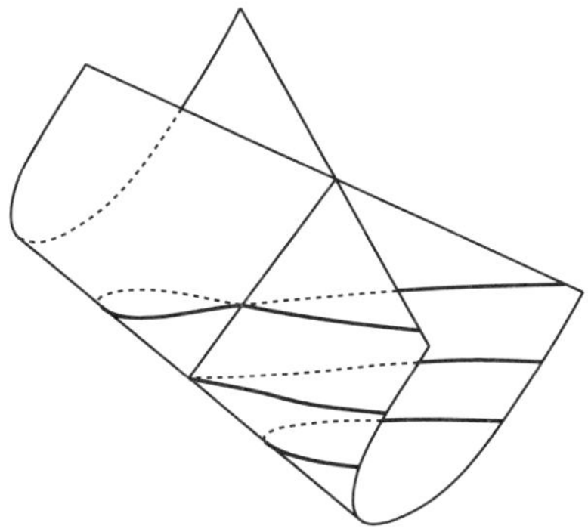

FIGURE B.2. **The Whitney umbrella and the first Reidemeister move.** We can assume that the singularity is transverse to the projection.

of double points includes the set of triple points and the set of Whitney umbrella singularities; its boundary also includes the double points $D \cap (\{i\} \times \mathbb{R}^2)$ $(i = 0, 1)$ of the two original knot diagrams $\mathcal{D}_0$ and $\mathcal{D}_1$.

Regard the intersections $\phi([0,1] \times S^1) \cap (\{t\} \times \mathbb{R}^2)$ as a one-parameter family of knots diagrams. Again, by general position, there are finitely many special $t \in [0,1]$ where these diagrams are not generic, and where exactly one of the following occurs:

- $\phi^{-1}(\{t\} \times \mathbb{R}^2)$ contains a Whitney umbrella singularity;
- $\phi^{-1}(\{t\} \times \mathbb{R}^2)$ contains a triple point;
- $\{t\} \times \mathbb{R}^2$ is tangent to D.

Consider $0 \leq t_1 < t_2 \leq 1$, and suppose that there are no special values of $t \in [t_1, t_2]$. In this case, the knot projections at t_1 and t_2 are planar isotopic.

Assume next that the interval $[t_1, t_2]$ contains a single special value, which corresponds to a Whitney umbrella singularity. Furthermore, assume that at that value t the slice $\{t\} \times \mathbb{R}^2$ is transverse to the one-dimensional image of the Jacobian. Then the diagrams $\{t_1\} \times \mathbb{R}^2$ and $\{t_2\} \times \mathbb{R}^2$ differ by a single Reidemeister 1 move; see Figure B.2. For example, intersecting the surface $v^2 - uz^2 = 0$ with the one-parameter family of planes $\{u = z + a\}$ with $|a| < \epsilon < 1$, gives the one-parameter family of plane curves given by the parametric equation $v^2 - z^2(z + a) = 0$ with $z + a \geq 0$, and these plane curves at $a = -\epsilon$ and $a = \epsilon$ differ by a local Reidemeister 1 move.

As the $\{t\} \times \mathbb{R}^2$ slice passes through a point of tangency with D, the knot projection undergoes a Reidemeister 2 move; see Figure B.3.

A triple point is locally modelled on three intersecting planes; intersecting the surface with transverse level sets immediately before and immediately after the triple point gives two diagrams that differ by a Reidemeister 3 move. This completes the proof of Reidemeister's theorem. □

B.1.1. Morse functions. The above proof involves a careful analysis of the possible local forms of generic functions, as described in Theorem B.1.2. A similar analysis appears in *Morse theory*, where one considers real-valued functions on n-dimensional manifolds. Although we do not use Morse theory for the proof of Whitney's Theorem B.1.2, since we will use facts on Morse functions later, we

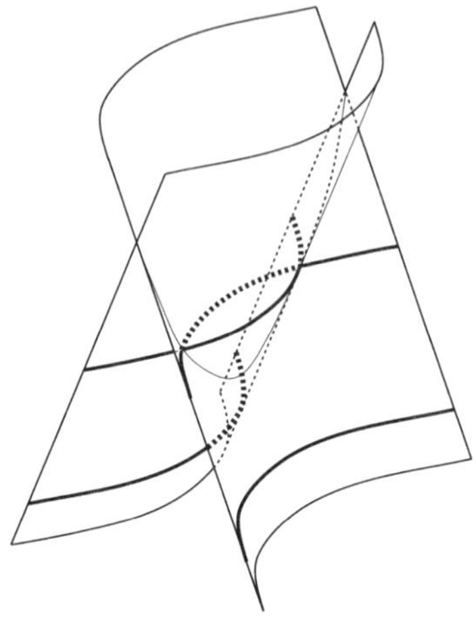

FIGURE B.3. **The double point set of the projection.** When crossing the double point set with a plane, we get crossings in the knot projection. When the plane $\{t\} \times \mathbb{R}^2$ is tangent to the double point set, we get a Reidemeister 2 move.

recall here some of the basic notions of this theory; for a thorough treatment of this beautiful subject, see [**142, 143**], cf. also [**77**].

Let $f\colon M \to \mathbb{R}$ be a smooth function on a compact, n-dimensional manifold. A *critical point* of f is a point $p \in M$ where the Jacobi matrix, written in a local coordinate system $(x_1, \ldots, x_n)$ as $J_f(p) = (\frac{\partial f}{\partial x_1}(p), \ldots, \frac{\partial f}{\partial x_n}(p))$, is zero. A critical point p corresponding to $(0, \ldots, 0)$ in the coordinate system $(x_1, \ldots, x_n)$ is called *non-degenerate* if the *Hessian matrix*

$$H(f)(p) = \left(\frac{\partial^2 f}{\partial x_i \partial x_j}(p)\right)$$

has non-zero determinant. (This property is independent of the choice of local coordinate system around p.) A function is called a *Morse function* if every critical point is non-degenerate. Morse functions form an open and dense set among smooth functions [**143**, Theorem 2.7].

The *Morse Lemma* [**142**, Lemma 2.2] states that around each non-degenerate critical point p, there are local coordinates $(x_1, \ldots, x_n)$ around p, so that $(0, \ldots, 0)$ corresponds to p and

$$f(x_1, \ldots, x_n) = f(p) - \sum_{i=1}^{\lambda_p} x_i^2 + \sum_{i=\lambda_p+1}^{n} x_i^2,$$

for some integer λ_p. The integer λ_p is the *index* of f at a critical point p, and it is independent of the chosen local coordinate system.

B.1.2. The proof of Theorem B.1.2. In our presentation we will follow Whitney's original argument from [**228**]; for a more modern treatment see [**75**, Theorem VI.4.6].

Since the theorem concerns the local behaviour of a generic map between a two- and a three-dimensional manifold, we will concentrate on maps $g\colon \mathbb{R}^2 \to \mathbb{R}^3$ with

$g(0,0) = (0,0,0)$, which we abbreviate $g\colon (\mathbb{R}^2, 0) \to (\mathbb{R}^3, 0)$. For a generic choice of g the Jacobian is not equal to the zero matrix (which is of codimension six among all 2×3 matrices) in any point p, hence rk $J_g(p)$ is at least 1 for all $p \in \mathbb{R}^2$.

The Jacobian $J_g(0)$ of g at the origin is a 2×3 matrix, and if its rank is 2, then the implicit function theorem gives coordinate systems where $g(x,y) = (x,y,0)$. In the following we will examine the case when rk $J_g(0) = 1$.

Choose a coordinate system (x,y) so that $\frac{\partial g}{\partial x}(0) = (0,0,0)$. Consider the 3×3 matrix

$$Q_g(0) = \begin{pmatrix} \frac{\partial g_1}{\partial y}(0) & \frac{\partial^2 g_1}{\partial x^2}(0) & \frac{\partial^2 g_1}{\partial x \partial y}(0) \\ \frac{\partial g_2}{\partial y}(0) & \frac{\partial^2 g_2}{\partial x^2}(0) & \frac{\partial^2 g_2}{\partial x \partial y}(0) \\ \frac{\partial g_3}{\partial y}(0) & \frac{\partial^2 g_3}{\partial x^2}(0) & \frac{\partial^2 g_3}{\partial x \partial y}(0) \end{pmatrix}. \tag{B.1}$$

LEMMA B.1.3. *For $g\colon (\mathbb{R}^2, 0) \to (\mathbb{R}^3, 0)$ with* rk $J_q(0) = 1$ *there are functions h arbitrarily close to g (in the C^2 topology) with $J_g(0) = J_h(0)$ and* $\det Q_h(0) \neq 0$. *If* $\det Q_g(0) \neq 0$ *then there is an open neighbourhood U of $0 \in \mathbb{R}^2$ with the property that* rk $J_g(p) = 2$ *for $p \in U \setminus \{0\}$.*

Proof. Let g_i $(i = 1,2,3)$ denote the coordinate functions of g. Choose a coordinate system (x,y) around 0 so that $\frac{\partial}{\partial x}$ spans the kernel of $J_g(0)$; i.e. for $i=1, \ldots, 3$,

$$\frac{\partial g_i}{\partial x}(0,0) = 0. \tag{B.2}$$

Since rk $J_g(0) = 1$, there is a linear transformation of the target so that

$$\left(\frac{\partial g_1}{\partial y}(0,0), \frac{\partial g_2}{\partial y}(0,0), \frac{\partial g_3}{\partial y}(0,0)\right) = (0,0,1). \tag{B.3}$$

Now either adding $\epsilon \cdot x^2$ to g_1 or $\delta \cdot xy$ to g_2 (or both) for sufficiently small ϵ and δ, we get a function h close to g with $J_h(0) = J_g(0)$ and $\det Q_h(0) \neq 0$.

For the second statement, let V be a neighborhood of the origin so that for $p \in V$, $\det Q_g(p) \neq 0$. Let L_p denote the line spanned by the first column of $Q_g(p)$ and let T_p be the plane spanned by the second and third columns, all viewed as vectors in $\mathbb{R}^3$. The condition on $Q_g(p)$ ensures that for $p \in V$, the line L_p and the plane T_p span $\mathbb{R}^3$; in fact, after passing to a smaller neighborhood $V' \subset V$ of 0 if needed, we can assume that for $p, p' \in V'$ the line L_p and the plane $T_{p'}$ intersect transversely. For such a sufficiently small neighborhood (by the mean value theorem) the vector $(\frac{\partial g_i}{\partial x}(p') - \frac{\partial g_i}{\partial x}(0))_{i=1}^3$ is in a plane close to $T_{p'}$, and hence is not in $L_{p'}$. Since $\frac{\partial g_i}{\partial x}(0) = 0$ for $i = 1,2,3$, this shows that $(\frac{\partial g_i}{\partial x}(p'))_{i=1}^3$ and $(\frac{\partial g_i}{\partial y}(p'))_{i=1}^3$ are linearly independent, hence rk $J_g(p') = 2$, concluding the proof. □

Proof of Theorem B.1.2. Suppose that $g\colon W^2 \to Y^3$ is a generic map, hence by our discussion above we can assume that rk $J_g(w) > 0$ for every point w of W. If this rank is equal to 2, then the implicit function theorem concludes the proof and shows that locally g is of the form $(x,y) \mapsto (x,y,0)$. If the rank is equal to 1 at a point $w \in W$, then again by genericity (and by Lemma B.1.3) we can assume that the determinant of Equation (B.1) is non-zero and w has a neighbourhood U with rk $J_g(p) = 2$ for all $p \in U \setminus \{w\}$ (that is, w is isolated).

Since we are working on coordinate charts, from now on we will work with g as a function $(\mathbb{R}^2, 0) \to (\mathbb{R}^3, 0)$. As in the proof of Lemma B.1.3, we can choose our local coordinates (x, y) on $\mathbb{R}^2$ so that Equation (B.2) and (B.3) holds. By the implicit function theorem, we can choose our coordinate system on $\mathbb{R}^3$ so that $g_3(x, y) = y$. The non-degeneracy of $Q_g(0)$ then implies that one of $\frac{\partial^2 g_i}{\partial x^2}(0)$ for $i = 1, 2$ is non-zero. Changing coordinates on $\mathbb{R}^3$ if necessary, we can arrange that

$$\frac{\partial^2 g_1}{\partial x^2}(0, 0) \neq 0. \tag{B.4}$$

In the next step we apply a coordinate transformation so that $g_1(x, y) = x^2$. To achieve this, let $F(x, y) = \frac{\partial g_1}{\partial x}(x, y)$, and consider the function $x = \phi(y)$ solving

$$F(\phi(y), y) = 0 \tag{B.5}$$

in a neighborhood of the origin. This solution can be found using the implicit function theorem, thanks to Equation (B.4). Consider the coordinate change on $\mathbb{R}^2$ specified by

$$x' = x - \phi(y), \quad y' = y.$$

In this new coordinate system, by the chain rule, Equation (B.3), and Equation (B.5),

$$\frac{\partial g_1}{\partial x'}(0, y') = \frac{\partial g_1}{\partial x}(\phi(y'), y') = 0.$$

Thus, when expanding g_1 in x' we get $g_1(x', y') = \psi_0(y') + (x')^2\psi_2(x', y')$. By Equation (B.4), it follows that $\psi_2(0, 0) \neq 0$. With the coordinate change

$$x'' = |\psi_2(x', y)|^{\frac{1}{2}} \cdot x', \quad y'' = y'$$

on $\mathbb{R}^2$, and the coordinate change

$$g_1' = g_1 - \psi_0(g_3), \qquad g_2' = g_2 \qquad g_3' = g_3,$$

on $\mathbb{R}^3$, after dropping primes, we get new coordinate systems where

$$g_1(x, y) = x^2, \qquad g_3(x, y) = y.$$

It remains to bring g_2 to the required form.

Since $\det Q_g(0) \neq 0$, the third column of the matrix from Equation (B.1) must be non-zero; but the top and the bottom entry clearly vanish, so it follows that $\frac{\partial^2 g_2}{\partial x \partial y}(0) \neq 0$; while $\frac{\partial g_2}{\partial x}(0, 0)$ and $\frac{\partial g_2}{\partial y}(0, 0) = 0$ by Equations (B.2) and (B.3). It follows that $g_2(x, y)$ can be written as

$$g_2(x, y) = xy + R(x, y),$$

where $\frac{\partial R}{\partial x}(0, 0) = \frac{\partial R}{\partial y}(0, 0) = \frac{\partial^2 R}{\partial x \partial y}(0, 0) = 0$. Let $g_2' = g_2 - R(0, g_3)$. Observe that $R(x, y) - R(0, y)$ is divisible by x, so dropping primes, we get $g_2(x, y) = x(y + R_1(x, y))$. Let

$$R_2(x, y) = \frac{R_1(x, y) + R_1(-x, y)}{2} \quad \text{and} \quad R_3(x, y) = \frac{R_1(x, y) - R_1(-x, y)}{2},$$

so

$$R_1(x, y) = R_2(x, y) + R_3(x, y),$$

with $R_2(-x, y) = R_2(x, y)$ and $R_3(x, y) = -R_3(-x, y)$.

Solve the implicit equation

$$y + R_2(x, y) = 0$$

for $y = y(x)$. Since $\frac{\partial g_2}{\partial y}(0,0) = 0$, it follows that the above implicit equation has a solution. By the symmetry $x \mapsto -x$ and local uniqueness, this solution can be written as $y = \eta(x^2)$, for a smooth function η. Define

$$\zeta(x^2) = xR_3(x, \eta(x^2)),$$

and consider the new coordinates

$$x' = x, \quad y' = y - \eta(x^2), \quad g_1' = g_1, \quad g_2' = g_2 - \zeta(g_1), \quad g_3' = g_3 - \eta(g_1).$$

Once again, this is a valid coordinate transformation, which keeps the shape of g_1 and g_3, and rewrites g_2 (again, after dropping primes) as $g_2(x, y) = x(y + R_4(x, y))$ with $R_4(x, 0) = 0$, hence

$$g_2(x, y) = xy(1 + R_5(x, y)).$$

Decomposing $R_5(x, y) = P(x^2, y) + xQ(x^2, y)$ into its even and odd parts in x (as we did with R_1), we find

$$g_2(x, y) = xy(1 + P(x^2, y)) + x^2yQ(x^2, y). \tag{B.6}$$

The coordinate transformation $g_2 = g_2 - g_1g_3Q(g_1, g_3)$ can be used to eliminate second term in Equation (B.6), hence we are left with $g_2(x, y) = xy(1 + P(x^2, y))$.

Let $G(x, y) = \frac{P(x,y)}{1+P(x,y)}$ and apply the following coordinate transformation:

$$g_1' = g_1, \quad g_2' = g_2 - G(g_1, g_3)g_2, \quad g_3' = g_3.$$

Since $G(0, 0) = 0$, this last transformation still gives coordinates on $\mathbb{R}^3$, and now

$$g_2'(x, y) = g_2(x, y) \cdot (1 - G(x^2, y)) = g_2(x, y) \cdot \frac{1}{1 + P(x^2, y)} = xy,$$

hence (after dropping primes) we get the desired from of $g = (g_1, g_2, g_3)$ around the origin. □

B.2. Reidemeister moves in contact knot theory

Recall that Legendrian and transverse knots can also be dealt with via their (front) projections, and there is a similar set of Reidemeister moves for these diagrams. In this section we adapt the argument from the smooth case to prove these analogues of the Reidemeister theorem in the contact case.

B.2.1. Transverse Reidemeister moves. Since the transverse condition is open, we can assume that the transverse isotopy between two transverse knots has only generic singularities. This leads quickly to the verification of the *transverse Reidemeister theorem*:

Proof of Theorem 12.5.4. Suppose that $\mathcal{T}_1$ and $\mathcal{T}_2$ are two transversely isotopic transverse knots, with front diagrams $\mathcal{D}_1$ and $\mathcal{D}_2$. Let $f\colon [0,1] \times S^1 \to [0,1] \times \mathbb{R}^3$ be the surface given by the transverse isotopy. Since transversality is an open condition, if we take a small perturbation of f, we can assume that the composition of f with the projection map P from Section B.1 is a generic map $\phi = P \circ f\colon [0,1] \times S^1 \to [0,1] \times \mathbb{R}^2$.

Recall that the self-linking number of a transverse knot is equal to the writhe of its front projection, so Reidemeister 1 moves (which change the writhe) cannot occur in the diagram-change along a transverse isotopy. This shows that the projection of a transverse isotopy cannot have Whitney umbrella singularities, and so for transverse knots we need only two types of Reidemeister moves: the adaptations of the second and of the third Reidemeister moves of Figure 2.2, as given in Figure 12.14. (Configurations from Figure 12.13 are disallowed by the transverse condition.) □

While transverse isotopies do not contain Reidemeister 1 moves, the converse is not true: an isotopy between two transverse knots giving a map without any Whitney umbrella singularity is not necessarily a transverse isotopy. In fact, using results from [**219**], two transverse representatives of the same knot type with the same self-linking number always can be connected by a smooth isotopy whose projection has no Whitney umbrella singularities, so their diagrams can be connected by a sequence of Reidemeister 2 and 3 moves and planar isotopies. As the transversely non-simple knot types demonstrate, such an isotopy may not be a transverse isotopy, and so the planar isotopies or the Reidemeister moves involve disallowed configurations (shown by Figure 12.13).

B.2.2. Legendrian Reidemeister moves. Since the Legendrian condition is not open, the appropriate adaptation of Reidemeister's theorem requires additional care. Our treatment here follows the discussion of Światkowski [**214**]. We start by restating the theorem:

THEOREM B.2.1 (Światkowski, [**214**]). *Two front projections correspond to Legendrian isotopic Legendrian links if and only if the projections can be connected by Legendrian planar isotopies and by Legendrian Reidemeister moves (shown in Figure* 12.2*).*

This subsection is devoted to a proof of this result; although the theorem holds for links, for simplicity in the proof we will deal with the case of knots only.

First we make more precise what we mean by generic Legendrian knot.

DEFINITION B.2.2. A Legendrian knot $\mathcal{K}$ is ***front generic*** if its front projection (projection to the (x, z) plane) has the following properties:

(fg-1) The only singularities of the projection are cusps, which can be given as $t \mapsto (t^2, t^3)$ in appropriate local coordinates;

(fg-2) the vertices of the cusps are distinct and are not on any other strand of the projection;

(fg-3) the strands of the projection meet transversely, without triple intersections.

As in Section 12.1, we denote the coordinate functions of a parametrization of the given Legendrian knot $\vec{\mathcal{K}}$ by $x(t), y(t), z(t)$. The Legendrian condition can be expressed as

$$z'(t) = y(t) \cdot x'(t). \tag{B.7}$$

LEMMA B.2.3. *Suppose that $\vec{\mathcal{K}}$ is an oriented Legendrian knot. The front projection of $\vec{\mathcal{K}}$ has singularities only as described in Property* (fg-1) *if and only if $x(t)$ is a Morse function.*

Proof. Notice first that the projection has a singularity at t if $x'(t) = 0$, since by Equation (B.7), this implies $z'(t) = 0$. If all singularities have the local form described in Property (fg-1), then $x''(t) \neq 0$ at the critical points of x, and hence $x(t)$ is obviously Morse.

Conversely, suppose that $x(t)$ is Morse, so at a critical point in local coordinates it is equal to t^2. (In these coordinates, the critical point corresponds to $t = 0$.) From Equation (B.7) we have

$$z''(t) = y'(t) \cdot x'(t) + y(t) \cdot x''(t),$$

which at the singular point $t = 0$ (that is, at a critical point of $x(t)$) gives $z''(0) = y(0) \cdot x''(0)$. If we choose the line parallel to the x axis, passing through the singularity as the first coordinate axis and the line with slope $y(0) = \frac{z''(0)}{x''(0)}$ (also passing through the singularity) as the second coordinate axis around the singularity, we get a local coordinate system in the (x, z) plane such that the projection is given by the map $t \mapsto (t^2, f(t))$, with $f(0) = f'(0) = f''(0) = 0$. Since $x'(0) = z'(0) = 0$ implies $y'(0) \neq 0$ (since $\vec{\mathcal{K}}$ is given by a smooth embedding), we have that $f'''(t) \neq 0$, so $f(t)$ can be written as $t^3(1 + h(t))$ with $h(0) = 0$. Decompose $h(t)$ into even and odd parts, as in the proof of Theorem B.1.2 so that $h(t) = g_1(t^2) + tg_2(t^2)$. Then, $f(t) = t^3(1 + g_1(t^2)) + t^4 g_2(t^2)$. The second term is a function of the first coordinate, hence can be easily eliminated, leaving us with $f(t) = t^3(1 + g_1(t^2))$. With the function $F(t) = \frac{g_1(t)}{1+g_1(t)}$ the new local coordinate

$$f_{new}(t) = f(t) - F(t) \cdot f(t)$$

shows that the singularity can be written in the form $t \mapsto (t^2, t^3)$. □

Front generic Legendrian knots form an open and dense set among Legendrian knots. In verifying this statement, we will use another projection of a Legendrian knot (usually called the *Legendrian projection*): the projection to the (x, y)-plane. Since all contact planes embed under the differential of this projection, it can be shown that (unlike the front projection, containing cusps) the Legendrian projection of a smooth Legendrian knot is a smooth immersion. Since the form $dz - ydx$ vanishes on the tangents of a Legendrian knot, the third coordinate function $z(t)$ can be recovered from the Legendrian projection by the formula

$$z(t) = z_0 + \int_0^t y(\theta) x'(\theta) d\theta.$$

Such a lift is not uniquely determined by the Legendrian projection (since it depends on the choice of z_0); also, not every smooth, immersed closed curve in the (x, y) plane gives rise to a Legendrian knot. Indeed, by choosing z_0 over a point (x_0, y_0) of the immersed closed curve, and parametrizing the diagram in the (x, y)-plane by $[0, 2\pi]$ (with the understanding that 0 and 2π both map to (x_0, y_0)), the value $z(2\pi)$ is not necessarily equal to z_0. If the integral from 0 to 2π vanishes, the diagram lifts to a closed, immersed curve, which furthermore is embedded if the two z-coordinates over any double points are different; see also [**45**]. This construction will be used in the proof of the next lemma, which justifies Definition B.2.2.

LEMMA B.2.4. ([**214**, Theorem A]) *The set of front generic Legendrian knots is open and dense in the set of all Legendrian knots.*

Proof. First we would like to arrange that the front projection has only cusp singularties. Let $\lambda \colon S^1 \to \mathbb{R}$ be a Morse function close to $x(t)$. Using the curve $C =$

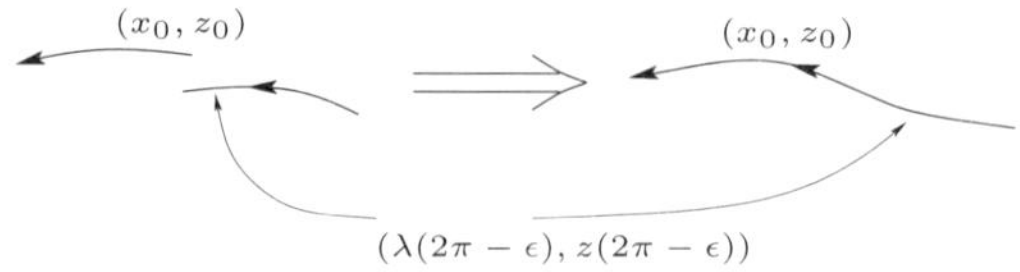

FIGURE B.4. **Closing up the front projection after replacing $x(t)$ by a nearby Morse function $\lambda(t)$.**

$(\lambda(t), y(t))$ in the (x, y)-plane as a Legendrian projection, we recover a Legendrian arc, with z-coordinate function (for $t \in [0, 2\pi]$) given by

$$z(t) = z_0 + \int_0^t y(\theta)\lambda'(\theta)d\theta. \tag{B.8}$$

Suppose that the function λ is sufficiently close to $x(t)$ and choose the point (x_0, y_0, z_0) on the Legendrian knot so that at this point the t-derivative of the z-coordinate is non-zero. Then the point (x_0, y_0, z_0) maps to a smooth point in the front projection, and we can close up the front diagram with a local modification near this image, as shown in Figure B.4. In this procedure we take the curve given by the integral for the values $t \in [0, 2\pi - \epsilon]$ and close up the resulting front projection. This then defines a Legendrian knot which is close to the given one and satisfies Property (fg-1) of Definition B.2.2.

The resulting front diagram now has the required cusp singularities. The smooth segments can be easily isotoped in the (x, z) plane using standard transversality arguments for maps of intervals to the plane.

Since Morse functions form an open and dense subset of all C^∞ functions, and the further properties (not having triple intersections and cusps on other branches) are also open and dense, the claims of the statement follow. □

Notice that the above procedure provides a Legendrian isotopy from any Legendrian knot to one which is front generic: just connect the function $x(t)$ to the chosen Morse function $\lambda(t)$ with a family of functions $\lambda^s(t)$ (with $x(t) = \lambda^0(t)$ and $\lambda(t) = \lambda^1(t)$) such that $\lambda^s(t)$ is Morse for $s > 0$, and do the final deformation (of closing the Legendrian arcs) parametrically.

In the next step we clarify what we mean by a generic isotopy in the Legendrian context. Suppose that H is a Legendrian isotopy between two Legendrian knots. The first step in achieving a generic front was to modify the coordinate function $x(t)$ to be generic (simply meaning Morse). For an isotopy, the similar step requires the genericity of a map $F\colon \mathbb{R}^2 \to \mathbb{R}^2$ derived from the first coordinate function of the isotopy, cf. Definition B.2.7. The following classification result of Whitney (similar in spirit to Theorem B.1.2) will be of crucial importance.

THEOREM B.2.5 (Whitney, [**229**]). *Let W and X be two smooth two-dimensional manifolds. A map $g_0\colon W \to X$ can be approximated arbitrarily close (in the C^2 topology) with a smooth map $g\colon W \to X$ with the following property: around each point $p \in W$ there are local coordinates (x, y) (so that p corresponds to $(0, 0)$) and local coordinates (u, v) around $g(p)$ in such a way that g has the form*

- $(x, y) \mapsto (x, y)$ *so g is a local diffeomorphism near p or*
- $(x, y) \mapsto (x^2, y)$ *so p is a fold point of g or*
- $(x, y) \mapsto (xy + x^3, y)$ *so p is a simple cusp.*

□

REMARK B.2.6. The proof of the above theorem can be given by adapting the methods of the proof of Theorem B.1.2. Indeed, for a generic map $g\colon \mathbb{R}^2 \to \mathbb{R}^2$ we can assume that the rank of the Jacobian J_g is non-zero at every point. At points when $\operatorname{rk} J_g = 2$, we have a local diffeomorphism, providing the first case in Theorem B.2.5. If $\operatorname{rk} J_g(p) = 1$ at a point p (say, $p = 0$), we can choose coordinates (x, y) so that x spans the kernel of $J_g(0)$, and the coordinate functions can be chosen so that $g_2(x, y) = y$. Assuming $\frac{\partial^2 g_1}{\partial x^2}(0) \neq 0$, the argument from the proof of Theorem B.1.2 normalizing the first coordinate function there (see text around Equation (B.5)) shows that, after appropriate coordinate changes, we get $(x, y) \mapsto (x^2, y)$, the second option of the theorem; see [**229**, Section 15]. In the final case, when $\frac{\partial^2 g_1}{\partial x^2}(p) = 0$, by genericity we can assume that $\frac{\partial^3 g_1}{\partial x^3}(p) \neq 0$, and a slightly longer argument (again, in the spirit of the proof of Theorem B.1.2, detailed in [**229**, Section 16]) concludes the proof of Theorem B.2.5. (For a more modern treatment of this result see [**75**, Chapter VI, Section 2].)

DEFINITION B.2.7. Suppose that $H = H(s,t)$ is a Legendrian isotopy between the two Legendrian knots $\vec{\mathcal{K}}_1$ and $\vec{\mathcal{K}}_2$ and write $H = (H_x, H_y, H_z)$, where the components denote the coordinate functions (now from $S^1 \times [0,1]$ to $\mathbb{R}$). The isotopy H is a ***front isotopy*** if it satisfies the following conditions:

- The singularities of the map $F\colon S^1 \times [0,1] \to \mathbb{R}^2$ given by $F(s,t) = (H_x(s,t), t)$ are folds and simple cusps,
- the double point set D of the composition $\phi = P \circ H$ of H with $P = \operatorname{Id} \times \pi$ (where $\pi\colon \mathbb{R}^3 \to \mathbb{R}^2$ is the projection to the (x, z) plane) is a one-dimensional submanifold and the restriction of the projection pr_1 to $[0,1]$ is a Morse function,
- the triple points of the map ϕ are isolated, and
- the values of the projection pr_1 at the cusp singularities, at the critical points of pr_1 on D and on the triple points are all distinct.

PROPOSITION B.2.8. ([**214**, Lemma 3.3]) *If $\vec{\mathcal{K}}_1$ and $\vec{\mathcal{K}}_2$ are Legendrian isotopic Legendrian knots with generic fronts, then there is a front isotopy between $\vec{\mathcal{K}}_1$ and $\vec{\mathcal{K}}_2$.*

Proof. The proof is similar in spirit to the proof of Lemma B.2.4. Let H be a Legendrian isotopy between $\vec{\mathcal{K}}_1$ and $\vec{\mathcal{K}}_2$ and suppose that $H = (H_x, H_y, H_z)$ are the coordinate functions of the isotopy. Consider the function $F(s,t) = (H_x(s,t), t)$, mapping from $S^1 \times [0,1]$ to $\mathbb{R}^2$. Consider a generic perturbation F_1 of this map, which (by an appropriate coordinate transformation) can be assumed to be of the form $F_1(s,t) = (G_x(s,t), t)$.

The pair (G_x, H_y) now provides a family of closed curves (parametrized by $t \in [0, 2\pi]$) on the (x, y) plane, hence we can use the integral formula of Equation (B.8) to lift this planar isotopy to an isotopy of knots in the three-space. The same problem as in the proof of Lemma B.2.4 arises: the resulting Legendrian curves might not close up. By performing the same closing operation (possibly in different places), we get an isotopy which now satisfies the first constraint given in Definition B.2.7. The usual genericity arguments conclude the proof. □

We can now prove the Legendrian Reidemeister theorem.

Proof of Theorem 12.1.7. As always, one direction of the theorem is easy: if two front projections differ by Legendrian Reidemeister moves and Legendrian planar isotopies, then the corresponding two knots are Legendrian isotopic.

Suppose now that two Legendrian knots are isotopic via a Legendrian isotopy H. By Lemma B.2.4 we can assume that the knots are front generic, and by Proposition B.2.8 we can assume that the isotopy is a front isotopy.

Suppose that $(s_0, t_0) \in S^1 \times [0, 1]$ is a point where the function $F(s, t) = (H_x(s, t), t)$ has a simple cusp singularity (here $s \in S^1$ and $t \in [0, 1]$). By Theorem B.2.5 the local model of the map F at such a point is given by $(st + s^3, t)$. Since $\frac{\partial H_y(s,t)}{\partial s}$ is non-zero at such a cusp point (as the Legendrian projection to the (x, y) plane is still an immersion), we get that the ratio $\frac{\partial H_z(s,t)}{\partial s} / \frac{\partial H_x(s,t)}{\partial s}$ is monotone there, providing a local model which shows that near the point (s_0, t_0) the diagrams given by the slices $\{t\} \times \mathbb{R}^2$ undergo a Legendrian Reidemeister 1 move. Indeed, the slice with $\{t\} \times \mathbb{R}^2$ gives the x-coordinate function $s \mapsto st + s^3$, which has no critical point for $t > 0$ and two critical points (giving rise to two cusps in the front projection) for $t < 0$.

As in the smooth case, we need to examine the points where $\{t\} \times \mathbb{R}^2$ is tangent to the double point set D. It is easy to see that in these points only a smooth point and a cusp can meet: by our genericity assumption two cusps cannot project to the same point, and two smooth branches of the projection cannot be tangent to each other, since then the Legendrian knot in the isotopy has a self-intersection. This means that the tangencies of the double point set give Legendrian Reidemeister 2 moves, as shown by Figure 12.2.

We can handle the case when the plane $\{t\} \times \mathbb{R}^2$ crosses a triple point exactly as in the smooth case, providing the Legendrian Reidemeister 3 moves. □

B.2.3. Approximations. In Section 12.5.2 we described how to approximate Legendrian knots by transverse knots, and conversely, transverse knots by Legendrian ones. We constructed the transverse push-off of a Legendrian knot, giving a transverse knot unique up to transverse isotopy associated to the Legendrian knot. (See Proposition 12.5.5.) A concrete description of this operation in terms of front projections is illustrated in Figure 12.15. Similarly, we explained how to construct a Legendrian approximation of a given transverse knot. The construction was defined in terms of front projections, using an algorithm which transformed a front projection of the transverse knot to the front projection of a Legendrian knot. This algorithm constructs a diagram that is defined uniquely up to negative stabilization.

According to Theorem 12.5.9, these constructions are inverses to one another, giving a one-to-one correspondence between transverse isotopy classes of transverse knots and equivalence classes of Legendrian isotopy classes of Legendrian knots modulo negative stabilization. Our present goal is to prove Theorem 12.5.9. The arguments below are based on [**43**, Section 2]. (Note that our convention differs from the one of [**43**]: we work with the contact structure given by the one-form $\alpha = dz - y dx$, while in the reference the contact structure is given by $y dx - dz = -\alpha$, giving different orientations in certain statements.)

PROPOSITION B.2.9. *Suppose that the two transverse knots $\mathcal{T}_1$ and $\mathcal{T}_2$ are transverse isotopic. Then their Legendrian approximations become isotopic after suitable negative stabilizations.*

Proof. Consider front projections $\mathcal{D}(\mathcal{T}_1)$ and $\mathcal{D}(\mathcal{T}_2)$ of $\mathcal{T}_1$ and $\mathcal{T}_2$. The algorithm described in Section 12.5.2 provides diagrams $\mathcal{D}(\vec{\mathcal{K}}_1)$ and $\mathcal{D}(\vec{\mathcal{K}}_2)$ of the Legendrian approximations, both determined up to negative stabilizations and Legendrian planar isotopy.

Since the two transverse knots are transverse isotopic, by the transverse Reidemeister theorem there is a sequence of transverse Reidemeister moves and transverse planar isotopies (so avoiding the disallowed configurations of Figure 12.13) transforming one diagram into the other. We will show that the Legendrian approximations of the diagrams are Legendrian isotopic, possibly after negative stabilizations.

A transverse planar isotopy either directly translates to a Legendrian planar isotopy, or it introduces a vertical tangency (pointing necessarily up), or it contains an isolated moment where a non-vertical segment crosses a vertical one, see Figure B.5. The figure also shows how these moves translate to Legendrian isotopies and negative stabilizations. In the first move (pictured on the left part of Figure B.5) there is one further case to analyze (when the strand points up and to the left), while in the second move (shown on the right part of Figure B.5) there are eight possibilities (depending on the orientation of the horizontal segment, the nature of the crossing, and direction of the vertical segment; i.e. whether the extremal point is on the left or the right). In the figure we only show one case, the remaining ones can be handled by similar means. The diagrams show how the Legendrian approximations of the results of the transverse planar isotopies can be realized by Legendrian Reidemeister moves and negative stabilizations.

In a similar manner, a transverse Reidemeister 2 move translates to a Legendrian Reidemeister 2 move — after possibly applying negative stabilizations. There are eight cases here, depending on the orientations of the strands and on the choice of the over-passing strand, but two of them contains a (transversally) disallowed

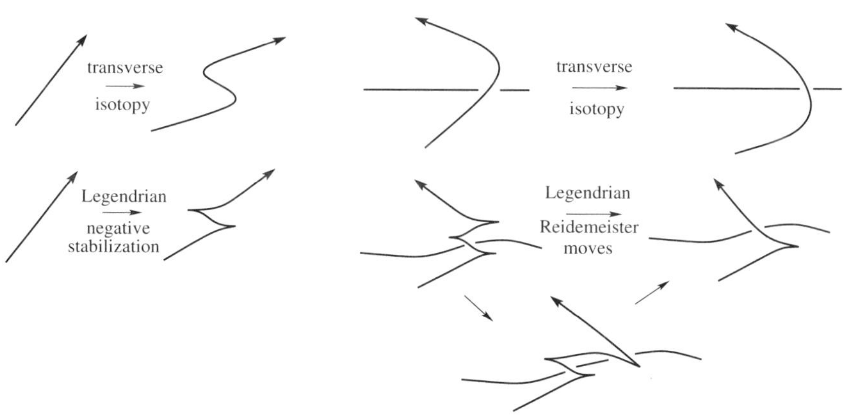

FIGURE B.5. **Transverse planar isotopies and Legendrian approximations.** On the left a transverse isotopy introducing (upward pointing) vertical tangencies is translated into negative stabilizations of the Legendrian approximation. On the right (one case of) an exchange of a vertical tangency and an intersection point is translated to the Legendrian approximation. The seven further cases are handled similarly.

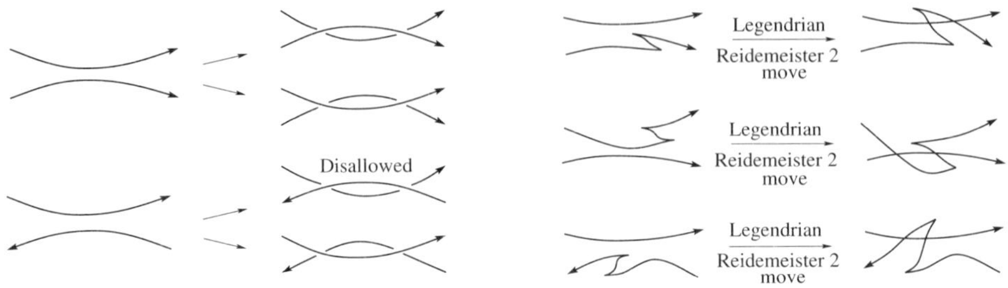

FIGURE B.6. **Reidemeister 2 moves and Legendrian approximations.** In the left we show four cases of the eight possible transverse Reidemeister 2 moves (one of which is disallowed), and on the right their Legendrian approximations, which require the application of a negative stabilization and a Legendrian Reidemeister 2 move.

FIGURE B.7. **The effect of a transverse Reidemeister 3 move on the Legendrian approximation in case $p = 1$.** The left diagram shows the transverse projection, the middle one depicts the front projection of the Legendrian approximation and (after a Legendrian Reidemeister 2 move) we get a new diagram in which a Legendrian Reidemeister 3 move can be performed (indicated by the gray triangle); just as in the $p = 0$ case.

portion. Figure B.6 verifies the statement for four cases; simple modifications provide the result for the remaining configurations.

In a transverse Reidemeister 3 move we have three crossings; let p denote the number of crossings that are disallowed in a Legendrian front projection (so in the Legendrian approximation, a negative stabilization is needed; see Figure 12.16). We will group the various configurations according to the value of p.

If $p = 0$ then the transverse Reidemeister 3 move translates directly into a Legendrian Reidemeister 3 move on the Legendrian approximation.

When $p = 1$, then when translating the transverse diagram to a Legendrian diagram, we need to apply the modification of Figure 12.16 at one of the crossings. This case can be reduced to the $p = 0$ case by a Legendrian Reidemeister 2 move. This is illustrated in Figure B.7 in one case; the further $p = 1$ possibilities can be handled similarly.

When $p = 2$, let s denote the strand passing through the two crossings disallowed in the Legendrian front diagram. In turning the transverse diagram into the Legendrian front, we need to stabilize twice. Depending on the local configuration there are two cases. When s points from left to right, we stabilize it twice, while if it points from right to left, we stabilize on the other two strands. In both cases a Legendrian Reidemeister 2 move reduces the problem to the $p = 0$ case, concluding the argument. For illustration, see Figure B.8.

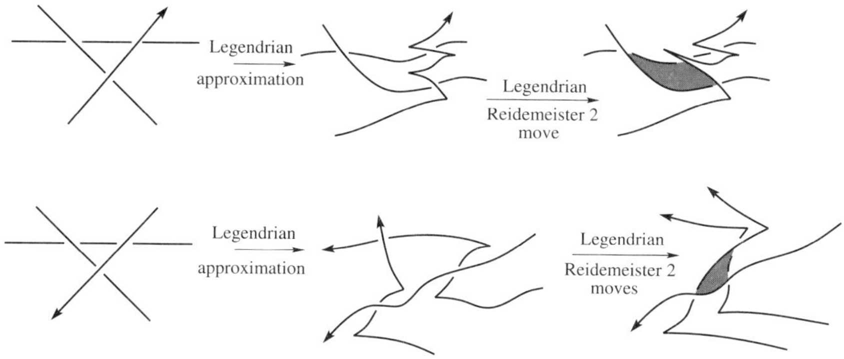

FIGURE B.8. **The effect of a transverse Reidemeister 3 move on the Legendrian approximation in case $p = 2$.** The two cases (depending on the orientation of the strand s containing the two crossings) are given in the two rows.

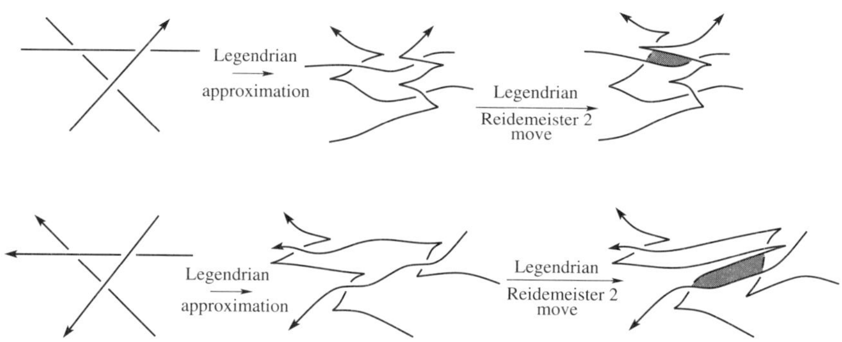

FIGURE B.9. **The effect of a transverse Reidemeister 3 move on the Legendrian approximation in case $p = 3$.** We choose a strand on which two modifications of Figure 12.16 can be applied, and then apply appropriate Legendrian Reidemeister moves to reduce the problem to the $p = 0$ case.

Finally we consider the case when $p = 3$. It is easy to see that we can always choose the three negative stabilizations in such a way that one of the strands is stabilized twice. Once again, a Legendrian Reidemeister 2 move reduces to the $p = 0$ case; see Figure B.9. □

Proof of Theorem 12.5.9. Consider the front diagram $\mathcal{D}(\mathcal{T})$ of the transverse knot $\mathcal{T}$. Approximate it by a Legendrian diagram (as shown in Figure 12.16) and then consider the transverse push-off (as shown in Figure 12.15). In this way get a transverse front diagram transverse planar isotopic to $\mathcal{D}(\mathcal{T})$, verifying the existence statement of the theorem.

One direction of the equivalence is simple. Indeed, if two Legendrian knot diagrams differ by a single negative stabilization, then their transverse push-offs are transverse isotopic, since in the diagram the smoothing of the new cusps can be straightened by a transverse planar isotopy. Furthermore, if two Legendrian knots

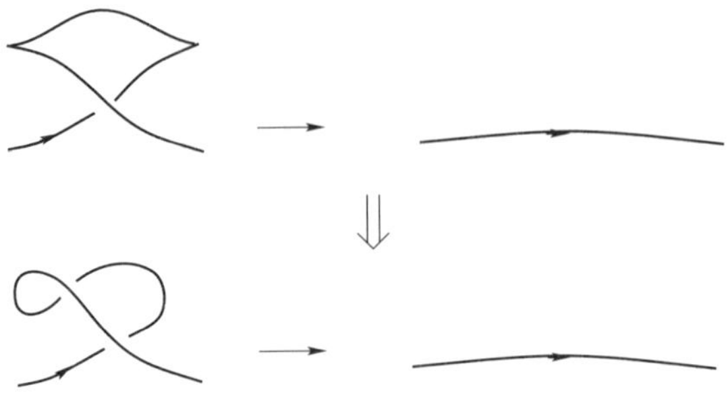

FIGURE B.10. **The effect of a Legendrian Reidemeister 1 move on the transverse push-off.** The Legendrian move gives rise to a transverse Reidemeister 2 move. Similar diagrams show the change for the other orientation of the strand.

are Legendrian isotopic, then their transverse push-offs are transverse isotopic. Indeed, a Legendrian planar isotopy translates to a transverse planar isotopy, the Legendrian Reidemeister 2 and 3 moves immediately translate to transverse Reidemeister moves while the transformation of a Legendrian Reidemeister 1 move requires a transverse Reidemeister 2 moves, as shown in Figure B.10. (An alternative argument was given in the proof of Proposition 12.5.5.)

The proof of the converse direction is the content of Proposition B.2.9. □

B.3. The Reidemeister-Singer Theorem

In Chapter 2 we met three invariants of knots and links derived using an auxiliary choice of a Seifert surface for K: the Alexander polynomial, the signature and the determinant. The independence of these quantities from the choice of the Seifert surface was established using the *Reidemeister-Singer Theorem*, which relates various Seifert surfaces of a given link. This section is devoted to the proof of this theorem; our approach here follows [**9**].

Fix an oriented link $\vec{L}$. From a Seifert surface Σ of $\vec{L}$ further Seifert surfaces can be obtained by stabilizing Σ (see Figure 2.12): fix an arc in $\mathbb{R}^3$ connecting two points in Σ, approaching it from the same side, called a *stabilizing arc*, and attach a handle to Σ supported in a neighborhood of this path, to obtain a stabilization Σ' of Σ. Also, we say that Σ is a *destabilization* of Σ'. With this language in place, we can state the Reidemeister-Singer Theorem:

THEOREM B.3.1 (Reidemeister-Singer, [**195, 212**]). *Any two Seifert surfaces of a given link in S^3 become ambient isotopic after an appropriate sequence of stabilizations and destabilizations.*

We describe first an important ingredient in this proof, *Seifert's algorithm* for constructing a Seifert surface associated to a link diagram for an oriented link $\vec{L}$. This algorithm proceeds as follows. Form the oriented resolution of each crossing in the diagram. The resulting configuration on the plane will consist of a collection of disjoint oriented circles, called *Seifert circles* (or Seifert circuits). Regard the plane as the subset $\{z = 0\} \subset \mathbb{R}^3$, and lift each disk bounded by a Seifert circle to a parallel copy of the disk in $\mathbb{R}^3$, lifting those disks contained inside $k \geq 0$ nested

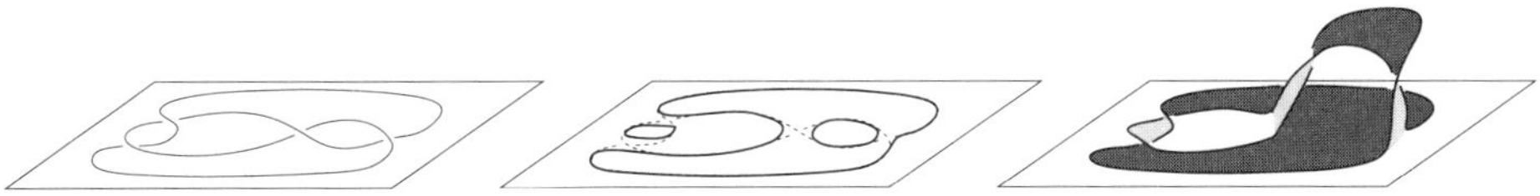

FIGURE B.11. **Seifert's algorithm.** The knot diagram on the left is resolved to obtain the collection of Seifert circles in the middle (where the crossing points are still indicated by dotted lines), which are turned into a Seifert surface as illustrated on the right.

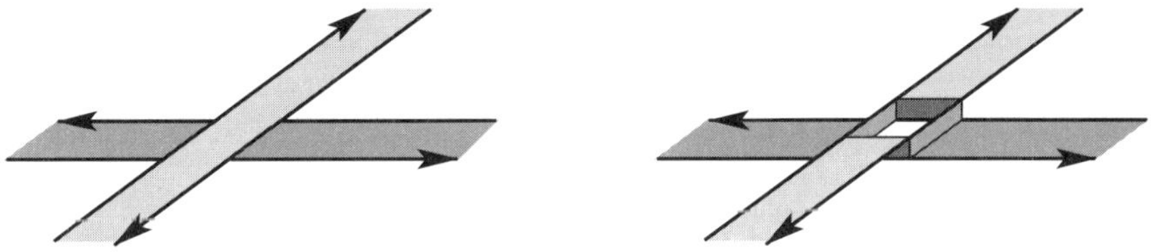

FIGURE B.12. **Crossings of bands in the projection.** By attaching an appropriate tube, locally we can turn the surface to an algorithmic surface.

Seifert circles into the affine plane $\{z = k\}$. Restore the crossings by attaching half-twisted bands that connect various translated disks, as specified by the crossings in the given diagram. This procedure gives a possibly disconnected, oriented surface with the given link as oriented boundary. We call such a surface an *algorithmic surface*. See Figure B.11 for an example. It is straightforward to see that this surface can be stabilized to get a Seifert surface.

LEMMA B.3.2. *Any Seifert surface of a given oriented link $\vec{L}$ can be isotoped and stabilized until it is an algorithmic surface.*

Proof. Consider a Seifert surface Σ for the link $\vec{L}$. Since a Seifert surface is connected, Σ can be realized as a single disk D, and a union of bands attached along the boundary of D. We can isotop Σ so that the disk projects injectively to the plane, and we can think of the bands as thin neighborhoods of their core curves, which core curves immerse into the plane minus the image of D. (Compare the proof of Proposition 3.4.11.) By general position, the core curves intersect each other in double points. By shrinking the bands and then twisting them if necessary, we can assume that projections of the bands meet each other in squares that immerse into the plane with opposite orientation, as shown in Figure B.12.

Stabilize Σ' at each crossing between the bands, as shown on the right diagram of Figure B.12. The resulting surface Σ' is the algorithmic surface Σ_{alg} associated to the projection of $\partial\Sigma'$. □

A link diagram naturally gives rise to another surface in $\mathbb{R}^3$ whose boundary is $\vec{L}$, the *black surface* (Definition 2.7.1), obtained by gluing the black domains in the chessboard coloring together along half-twisted bands to restore the crossings. In Section 2.7, a diagram is called *special* if F_b is a Seifert surface for $\vec{L}$. The above proof gives the following result of independent interest, which was stated earlier as Lemma 2.7.6:

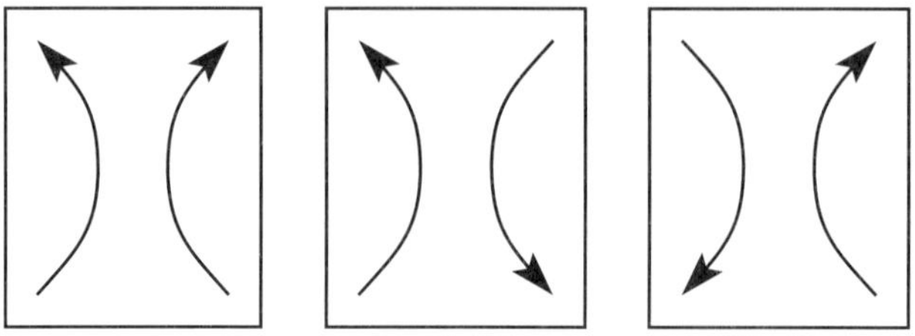

FIGURE B.13. **Three different orientation types for Reidemeister 2 moves.** Perform a Reidemeister 2 move locally in the square. The middle and the one on the right are "oppositely oriented" as in Lemma B.3.5.

PROPOSITION B.3.3. *Any link $\vec{L}$ admits a special diagram.*

Proof. For the projection $\mathcal{D}$ constructed in the proof of Lemma B.3.2, $\Sigma_{alg}(\mathcal{D})$ is F_b. □

Returning to the proof of the Reidemeister-Singer theorem, it remains to check that the algorithmic surfaces change by isotopies, stabilizations, and destabilizations under the various Reidemeister moves. We start with some special cases.

LEMMA B.3.4. *If $\mathcal{D}$ and $\mathcal{D}'$ are two knot projections that differ by a Reidemeister 1 move, then their algorithmic surfaces $\Sigma_{alg}(\mathcal{D})$ and $\Sigma_{alg}(\mathcal{D}')$ are isotopic.*

Proof. Suppose that $\mathcal{D}'$ has one more crossing than $\mathcal{D}$. Observe that $\Sigma_{alg}(\mathcal{D}')$ is obtained from $\Sigma_{alg}(\mathcal{D})$ by adding one more disk, connected to $\Sigma_{alg}(\mathcal{D})$ by a half twisted band. Untwist the band to get the isotopy to $\Sigma_{alg}(\mathcal{D})$. □

LEMMA B.3.5. *Suppose that $\mathcal{D}'$ is obtained from $\mathcal{D}$ by a single Reidemeister 2 move supported over a disk U, so that the two strands crossing each other (twice) are oriented oppositely. (See Figure B.13.) Then their respective algorithmic surfaces $\Sigma_{alg}(\mathcal{D})$ and $\Sigma_{alg}(\mathcal{D}')$ are either isotopic, or they become isotopic after a stabilization.*

Proof. Label the diagrams so that $\mathcal{D}'$ has two more crossings than $\mathcal{D}$. Consider a disk U in the link diagram that meets the two strands in $\mathcal{D}$, oriented oppositely. There are four combinatorially distinct cases, according to how the two arcs close up in the Seifert circles: (1) the two arcs are part of the same Seifert circle that bounds a disk D meeting U in one component, (2) the two arcs are part of the same Seifert circle that bounds a disk D meeting U in two components, (3) the two arcs are part of two distinct Seifert circles that are not nested, and finally (4) the two arcs are part of two distinct Seifert circles that are nested. See Figure B.14.

In case (1), $\Sigma_{alg}(\mathcal{D})$ and $\Sigma_{alg}(\mathcal{D}')$ are isotopic, via an isotopy that introduces two canceling twists in the band corresponding to D, as illustrated in the top row of Figure B.15. In cases (2)-(4), $\Sigma_{alg}(\mathcal{D})$ can be stabilized to obtain a surface which is isotopic to $\Sigma_{alg}(\mathcal{D}')$. The local pictures for cases (2) and (3) look the same, as in the second row of Figure B.15; the local picture for case (4) is shown in the third row of Figure B.15. □

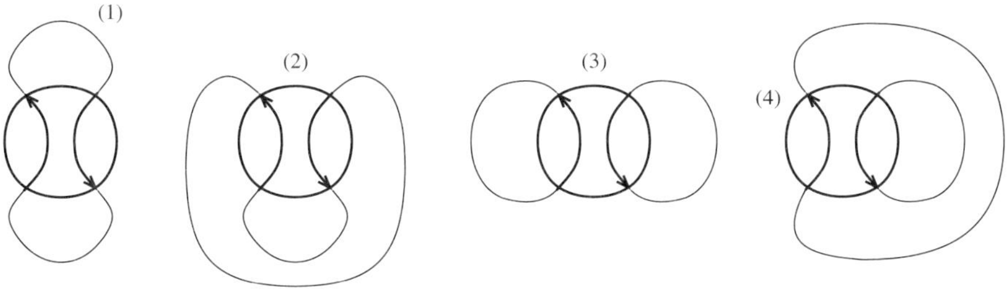

FIGURE B.14. **Cases of Reidemeister 2 moves with oppositely oriented strands.** We separate how the two local arcs complete to Seifert circles, as shown.

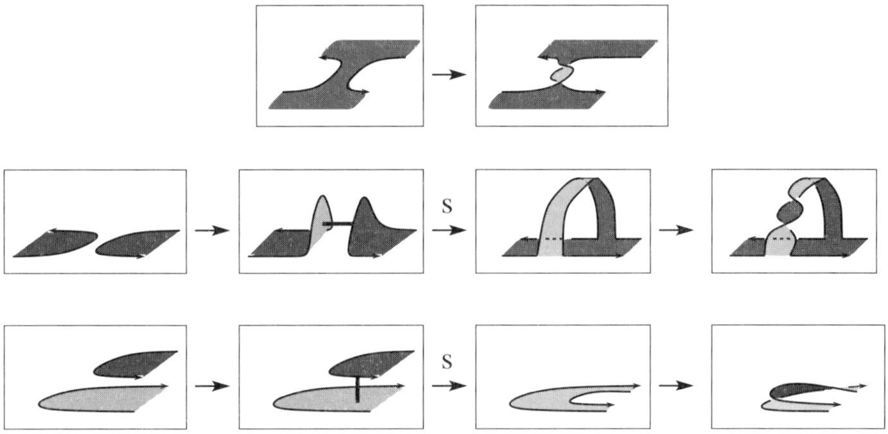

FIGURE B.15. **Algorithmic surfaces and Reidemeister 2 moves.** We have indicated sequences of isotopies and stabilizations that connect algorithmic surfaces under oppositely oriented Reidemeister 2 moves. Unmarked arrows between pictures indicate isotopies, and those labelled with "S" indicate stabilizations. Stabilizing arcs are indicated before each stabilization.

Suppose that $\mathcal{D}'$ is obtained from $\mathcal{D}$ by a Reidemeister 3 move supported over a disk U containing the three strands with three crossings. The oriented resolution of the three crossings give three disjoint strands in $\mathcal{D}$ (and possibly a further circle). The strands are said to be *compatibly oriented* if the three incoming ends of the strands in U (and hence also the three out-going ends) are consecutively ordered in the cyclic ordering of ∂U. Clearly, the three strands for $\mathcal{D}$ in U are oriented compatibly if an only if the corresponding three strands for $\mathcal{D}'$ in U are oriented compatibly. In this case, we call the Reidemeister move *compatibly oriented*; otherwise, we say it is *incompatibly oriented*. See Figure B.16.

Locally, inside U, if we consider an unoriented Reidemeister 3 move, there are 8 different orientations we could introduce, corresponding to the orientations of the 3 strands. Of these 8 orientations, 6 are compatibly oriented, and 2 are not.

LEMMA B.3.6. *If $\mathcal{D}$ and $\mathcal{D}'$ differ from each other by a compatibly oriented Reidemeister 3 move, then the surfaces $\Sigma_{alg}(\mathcal{D})$ and $\Sigma_{alg}(\mathcal{D}')$ are isotopic.*

FIGURE B.16. **Compatibly oriented Reidemeister 3 moves.** The first two are compatibly oriented; the last one is not.

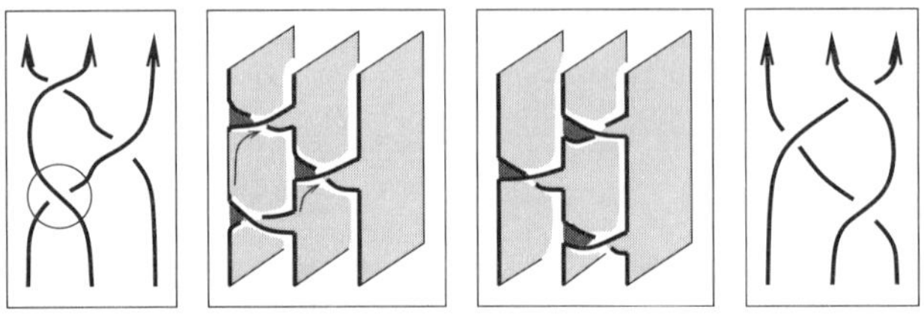

FIGURE B.17. **Sliding bands to effect Reidemeister 3 moves.** In the picture we perform a coherently oriented Reidemeister 3 move on the diagram (in the first picture). We have circled the crossing whose associated band, in the algorithmic surface (second picture) is to be slid over the other two bands (as indicated by the two arrows of the second picture), to give the algorithmic surface (third picture) for the diagram after the Reidemeister move (fourth picture).

Proof. Seifert's algorithm starts from a collection of disks that bound the oriented resolution $\mathcal{D}_0$ of $\mathcal{D}$. $\mathcal{D}_0$ meets U in three parallel, compatibly oriented strands. First add bands for the crossings outside U, and then attach three half-twisted bands to reintroduce the three crossings in U. Sliding one of these three bands across the other two gives the isotopy between $\Sigma_{alg}(\mathcal{D})$ and $\Sigma_{alg}(\mathcal{D}')$. We can identify the moving band as follows. Thinking of the strands as pointing upwards, there is a middle strand and two strands that cut across it, that we call *cross strands*. We claim that at least one of the two cross strands crosses the other two strands with the same sign. (If this were not the case, the diagram in $\mathcal{D} \cap U$ would be alternating, and hence the Reidemeister 3 move would not be possible.) Slide the band for the remaining crossing over the two bands attached to this cross strand. See Figure B.17. □

Proof of Theorem B.3.1. Lemma B.3.2 reduces the problem to show that any two algorithmic surfaces become isotopic after stabilizations. Theorem B.1.1 further reduces to the verification that algorithmic surfaces remain isotopic, up to stabilizations, under the three Reidemeister moves.

Isotopy invariance of the algorithmic surface up to stabilizations, for most cases of the Redemeister moves, was verified in Lemmas B.3.5, B.3.5 and B.3.6. The remaining two cases, i.e. Reidemeister 2 moves where the strands are oriented in the same direction and incompatibly oriented Reidmeister 3 moves, can be reduced to the earlier cases, as illustrated in Figure B.18. □

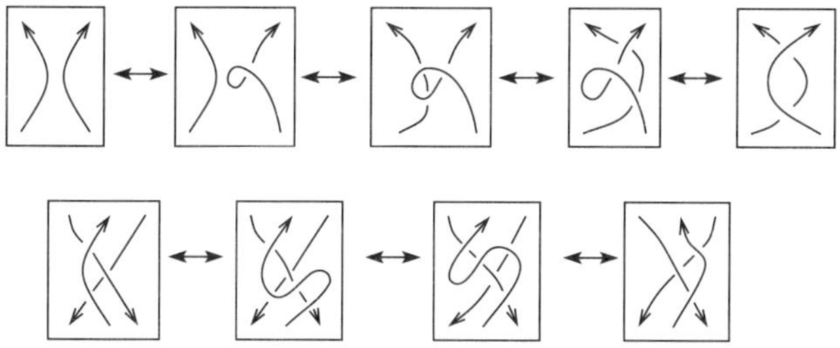

FIGURE B.18. **Reducing moves.** Expressing the remaining Reidemeister moves in terms of those studied previously.

B.4. Cromwell's Theorem

In this section we give a proof of Cromwell's Theorem 3.1.9 that identifies the local moves needed to connect any two grid diagrams representing isotopic links. Our proof is similar in spirit to Dynnikov's proof [**37**].

Recall that the *grid moves* on a planar grid diagram refer to the commutation moves (Definition 3.1.6) and the stabilizations and destabilizations (Definition 3.1.7).

THEOREM B.4.1 (Cromwell, [**27**]). *Two planar grid diagrams represent equivalent links if and only if there is a finite sequence of grid moves that transform one into the other.*

We will prove Theorem B.4.1 by approximating knot projections by grid diagrams. More formally:

DEFINITION B.4.2. Let $\mathcal{D}$ be a diagram for an oriented link. A ***grid approximation to*** $\mathcal{D}$ is a planar grid diagram whose associated oriented link diagram is planar isotopic to $\mathcal{D}$.

A key step in the proof is the following proposition, which we prove after some preparatory lemmas and definitions.

PROPOSITION B.4.3. *Any diagram for an oriented link has a grid approximation. Furthermore, this grid approximation is unique up to grid moves.*

We study knot diagrams via their projection to the y-axis $\mathbb{R} \subset \mathbb{R}^2$.

DEFINITION B.4.4. The projection of a diagram $\mathcal{D}$ to the y axis has two kinds of special points: critical points and crossings between two arcs. A diagram $\mathcal{D}$ is called a ***bridge diagram*** if the following conditions are met:

- crossings are not critical points of the projection;
- the critical points of the projection are isolated minima and maxima;
- no two special points project to the same value.

A diagram $\mathcal{D}$ in general position is a bridge diagram; in particular, any diagram can be approximated by an arbitrarily small (C^2) perturbation by bridge diagrams. The special points in a bridge diagram inherit an ordering induced by their y values.

DEFINITION B.4.5. A planar grid diagram G is called ***bridge-like*** if the following conditions are satisfied:

- Each horizontal segment contains at most one crossing.

- A horizontal segment is a local maximum or a local minimum if and only if it contains no crossings.

Tilting the horizontal segments without introducing new local minima or maxima and then smoothing out the corners transforms a bridge-like grid diagram into a bridge diagram. The resulting bridge diagram is called the *smoothing* of the given bridge-like grid diagram.

Conversely, given a bridge diagram, we will construct a bridge-like grid diagram in Lemma B.4.10. This construction will involve an intermediate object.

DEFINITION B.4.6. A ***pre-grid diagram*** is a planar diagram with the following properties:

- all the arcs are composed of vertical or horizontal segments, and
- all the vertical segments cross over the horizontal ones.

It is called ***generic*** if all the horizontal segments project to different y values, and all the vertical segments project to different x values. Two generic pre-grid diagrams $\mathcal{D}_0$ and $\mathcal{D}_1$ are said to be ***combinatorially equivalent*** if they can be connected by a one-parameter family of generic pre-grid diagrams.

Obviously, a grid diagram gives a generic pre-grid diagram in the above sense.

LEMMA B.4.7. *For each generic pre-grid diagram $\mathcal{E}$, there is a unique grid diagram whose associated projection is combinatorially equivalent to $\mathcal{D}$. Moreover, if two generic pre-grid diagrams can be connected by a one-parameter family of (not necessarily generic) pre-grid diagrams with the same number of segments, then their associated grid diagrams differ by commutation moves.*

Proof. If $\mathcal{E}$ is a generic pre-grid diagram, it has n horizontal segments, whose y-coordinates are $y_1 < \cdots < y_n$ and n vertical segments, whose x-coordinates are $x_1 < \cdots < x_n$. Move the horizontal segment whose y-coordinate is y_i to one whose y-coordinate is $i - \frac{1}{2}$. Move the vertical ones to half-integral coordinates similarly. The result is a grid diagram.

In a generic one-parameter family of pre-grid diagrams, there are finitely many values where two horizontal arcs project to the same y-coordinate, or two vertical ones project to the same x-coordinate. As we pass through each of these values, the associated grid diagrams undergo a single commutation move. □

LEMMA B.4.8. *Given any planar grid diagram G, there is a bridge-like grid diagram G' that can be obtained from G by a finite sequence of grid moves.*

Proof. The diagram G is not bridge-like if (1) there are horizontal segments that are local maxima or local minima that contain crossings; or (2) there are horizontal segments that contain no crossings that are not local maxima or minima; or (3) there are horizontal segments that contain more than one crossing.

Segments of the first kind can be stabilized so that they become two segments, where one is a local maximum or local minimum and the other contains the crossings. Segments of the second kind can be stabilized so that they become two segments, one of which is a local maximum and the other which is a local minimum. Segments of the third kind can be eliminated as follows. Suppose a horizontal segment contains k crossings. Stabilize $k - 1$ times at one of the two corners of this

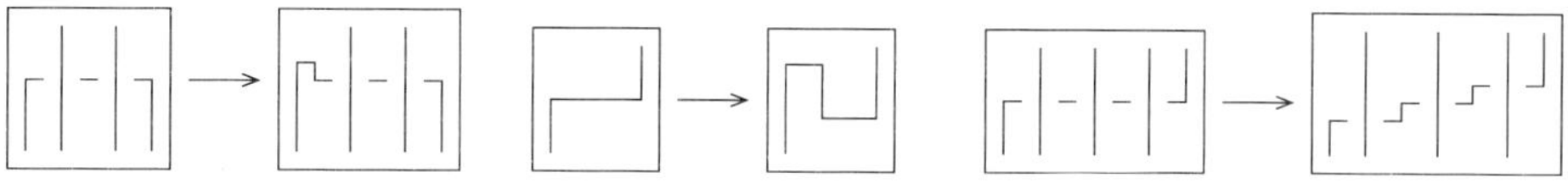

FIGURE B.19. **Turning a grid diagram into a bridge-like grid diagram.**

segment, and then commute the newly-created short vertical segments so that they separate the crossings. These three steps are illustrated in Figure B.19. □

We are now ready to associate a bridge-like grid diagram to a bridge diagram. To express the uniqueness of this construction, it is useful to have the following:

DEFINITION B.4.9. Two bridge diagrams $\mathcal{D}_0$ and $\mathcal{D}_1$ are ***bridge isotopic*** if they can be connected by a smoothly varying one-parameter family $\{\mathcal{D}_s\}_{s\in[0,1]}$ of bridge diagrams.

LEMMA B.4.10. *Given a bridge diagram $\mathcal{D}$, there is a bridge-like grid diagram whose smoothing is bridge isotopic to $\mathcal{D}$; and any two such bridge-like grid diagrams can be connected by a sequence of commutation moves.*

Proof. Associate first to any bridge diagram $\mathcal{D}$ a pre-grid diagram $\mathcal{E}$, as follows. Decompose $\mathcal{D}$ into vertical slices $y_0 < y_1 < \dots < y_n$ so that $\mathcal{D}$ projects to $[y_0, y_n]$ and each interval $[y_i, y_{i+1}]$ contains the y coordinate of at most one special point. The diagram $\mathcal{E}$ is built out of pieces constructed from the slices $\mathbb{R} \times [y_i, y_{i+1}]$, as follows. If there are no special points in $\mathbb{R} \times [y_i, y_{i+1}]$, replace the segments in $\mathcal{D}$ by vertical segments; otherwise replace it with a planar isotopic picture containing exactly one horizontal segment and all other vertical segments, with the constraint that vertical segments cross over horizontal ones. See Figure B.20. Fit the pieces together, starting from the bottom piece, and successively attaching higher pieces, stretched horizontally as needed so that the vertical strands going off the bottom of the new piece match with the vertical strands going off the top of the previous one. The resulting pre-grid diagram $\mathcal{E}$ might not be generic (different horizontal

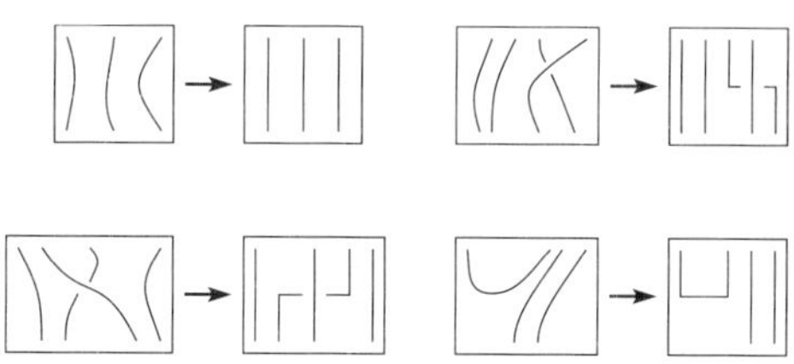

FIGURE B.20. **Approximating a bridge diagram by a grid diagram.**

arcs have distinct y values by construction; but different vertical arcs need not have distinct x values); but it has a small perturbation $\mathcal{E}'$ that is. Let G be a grid diagram whose associated projection is combinatorially equivalent to the generic pre-grid diagram $\mathcal{E}'$; this exists by Lemma B.4.7. If $\mathcal{E}'$ is sufficiently close to $\mathcal{E}$, the smoothing of G is bridge isotopic to $\mathcal{D}$, verifying the existence statement in our lemma.

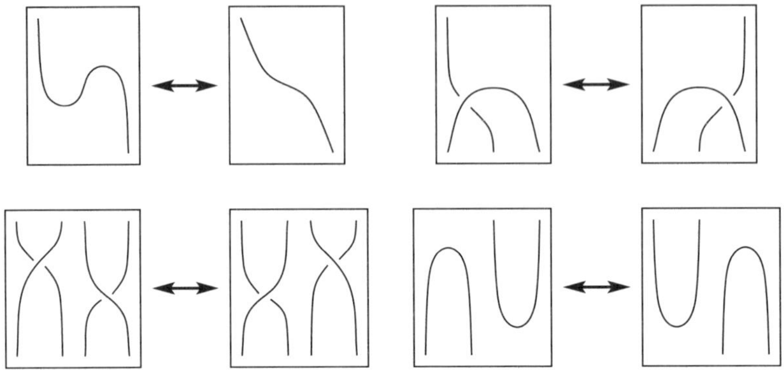

FIGURE B.21. **The three types of bridge moves needed to connect bridge diagrams.** Top left is a birth/death move; top right is a crossing slide; the bottom row shows two of the 16 possible special point commutations (between crossings and critical points).

The grid diagram G constructed above is uniquely defined up to commutation moves (independent of the choice of $\mathcal{E}'$) according to Lemma B.4.7. In fact, if G is any bridge-like grid diagram whose smoothing is bridge isotopic to $\mathcal{D}$, the diagram G itself can be obtained from the above procedure. The claimed uniqueness statement (up to commutations) follows. □

Lemma B.4.10 associates to a bridge isotopy class of bridge diagram a grid diagram, determined uniquely up to commutation moves. We will now examine how the bridge isotopy class changes under a generic planar isotopy. Note that under a planar isotopy the number of crossings does not change, but the maxima and minima can interact with each other, or with the crossings, and the ordering on the special points can change. We formalize these changes as follows:

DEFINITION B.4.11. Two bridge diagrams $\mathcal{D}_1$ and $\mathcal{D}_2$ are said to be related by a ***bridge move*** if $\mathcal{D}_2$ is obtained from $\mathcal{D}_1$ by one of the three possible moves:

- creation of a pair of a local maximum and a local minimum, or the cancellation of such a pair; either is called a ***birth/death move***,
- sliding a crossing through a minimim or maximum, called a ***crossing slide***,
- commuting a pair of special points (each of which can be a maximum, minimum, or a crossing), called a ***special point commutation***.

See Figure B.21 for illustrations of these bridge moves.

LEMMA B.4.12. *Any two planar isotopic knot diagrams can be transformed into each other by a finite sequence of bridge moves.*

Proof. Fix two bridge diagrams $\mathcal{D}_0$ and $\mathcal{D}_1$. By hypothesis, these two bridge diagrams can be connected by a planar isotopy $\{\mathcal{D}_s\}_{s\in[0,1]}$. It follows from a general position argument that if the planar isotopy $\{\mathcal{D}_s\}_{s\in[0,1]}$ is chosen generically, then there are finitely many values of $s \in (0,1)$ for which $\mathcal{D}_s$ is not a bridge diagram; at each of these special values, for sufficiently small ϵ, the bridge diagrams $\mathcal{D}_{s-\epsilon}$ and $\mathcal{D}_{s+\epsilon}$ are related by one of the bridge moves enumerated in Definition B.4.11. □

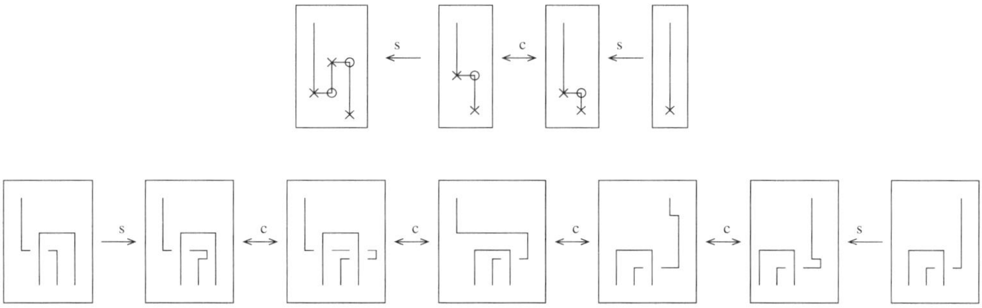

FIGURE B.22. **Realizing bridge moves as grid moves.** In the top row, we realize a birth/death move as grid moves; in the bottom, we realize a cross slide. In the top row, we mark the break points by X- and O-markings arbitrarily; in the bottom row, these markings are dropped. Arrows marked with c indicate commutation moves, and those with s indicate stabilizations. Note that the fourth step in the second row is a repeated application of commutation moves.

Proof of Proposition B.4.3. First isotop the diagram to a bridge diagram (which is unique up to bridge moves). Lemma B.4.10 associates to each bridge diagram $\mathcal{D}$ a grid diagram G, which is well-defined up to commutation moves. We claim that if $\mathcal{D}_1$ and $\mathcal{D}_2$ are planar isotopic diagrams, then their associated planar grid diagrams can also be connected by a sequence of commutation and stabilization moves. Lemma B.4.12 reduces this to a verification in the case where $\mathcal{D}_1$ and $\mathcal{D}_2$ are related by a birth/death move, a crossing slide, or a special point commutation, where the verification is easy. The first two are illustrated in Figure B.22; special point commutations can be readily realized by commutation moves.

In effect, we have now defined a map F which associates to a link diagram $\mathcal{D}$ modulo planar isotopies a grid diagram $F(\mathcal{D})$, modulo grid moves. To complete the proof, we verify that if G is a grid diagram, and $\mathcal{D}(G)$ is its associated link diagram, then $F(\mathcal{D}(G))$ and G are equivalent under grid moves. In the special case where G is bridge-like, this follows from Lemma B.4.10. The general case can be reduced to this case by Lemma B.4.8. □

Proof of Theorem B.4.1. It is straightforward to see that grid moves represent equivalent links: a commutation give rise to either a planar isotopy or a sequence of Reidemeister 2 and 3 moves; while a stabilization gives either a planar isotopy or a Reidemeister 1 move. Turning to the converse direction, in view of Theorem B.1.1 and Proposition B.4.3, it suffices to show that if $\mathcal{D}_1$ and $\mathcal{D}_2$ differ by any Reidemeister move, we can find grid approximations to $\mathcal{D}_1$ and $\mathcal{D}_2$ that differ by grid moves. This is illustrated in Figure B.23. □

REMARK B.4.13. Cromwell's original proof [**27**] uses the interpretation of grid diagrams as braids. (See Section 12.8.) A theorem of Markov [**11**] (see also [**220**]) gives the basic moves that connect braids that determine isotopic links. Cromwell's proof then follows Markov's moves with grid moves.

B.4.1. Legendrian knots and Cromwell's theorem. The methods described above can be adapted to the Legendrian context. According Theorem 12.1.7,

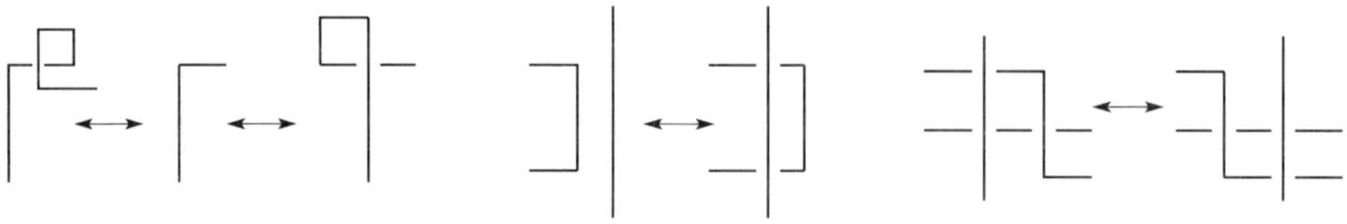

FIGURE B.23. **Reidemeister moves in grid diagrams.** Each move can be realized by (de)stabilizations and commutations.

Legendrian knots and links can be studied via their front projection, using the Legendrian Reidemeister moves (shown in Figure 12.2).

Adapting the proof of Cromwell's theorem in combination with the above result, we can express Legendrian knot theory in S^3 in terms of grid diagrams. (As usual, we will discuss the case of knots; the general case requires only minor modifications.)

It will be useful to have the following terminology:

DEFINITION B.4.14. For a planar grid diagram, we call commutation moves and (de)stabilizations of type *X:NW*, *X:SE*, *O:SE*, and *O:NW* ***Legendrian grid moves***.

As explained in Chapter 12, a planar grid diagram can be used to construct a Legendrian front projection. The adaptation of Theorem B.4.1 to the Legendrian case states:

THEOREM B.4.15. *The Legendrian knots $\vec{\mathcal{K}}_1$ and $\vec{\mathcal{K}}_2$ associated to the planar grid diagrams G_1 and G_2 are Legendrian isotopic if and only if G_1 and G_2 can be connected by a sequence of Legendrian grid moves.*

The proof is given after some preliminaries. We start by showing how to associate a grid diagram to a Legendrian front projection. Recall from Definition B.2.2 that a Legendrian front projection has two kinds of special points: cusps and crossings. A Legendrian knot was called front generic if its singularities are cusps, no cusp is on another branch, and different branches meet transversally, without triple intersections.

Start from a generic front projection, which we denote by $\mathcal{D}_{\vec{\mathcal{K}}}$, and rotate it 90°, so that the left cusps become local minima and the right cusps become local maxima, which we now smooth out. Switch all the crossings (note that a grid diagram G associates a Legendrian knot in the mirror of the knot type represented by G, cf. Definition 12.2.1). The result is a bridge diagram, in the sense of Definition B.4.4. In a typical bridge diagram, there are two types of allowed crossings; but in the bridge diagram arising from a Legendrian front projection, there is only one type of allowed crossing. Lemma B.4.10 in turn associates to the bridge diagram a grid diagram, unique up to commutation moves.

LEMMA B.4.16. *If $\mathcal{D}_1$ and $\mathcal{D}_2$ are two front generic projections of Legendrian knots that differ by a Legendrian planar isotopy, then their associated grid diagrams G_1 and G_2 differ by a sequence of commutations.*

Proof. If two Legendrian front projections are Legendrian planar isotopic, then their associated bridge diagrams differ by a sequence of special point commutations, in the sense of Definition B.4.11: none of the other two types of bridge moves can

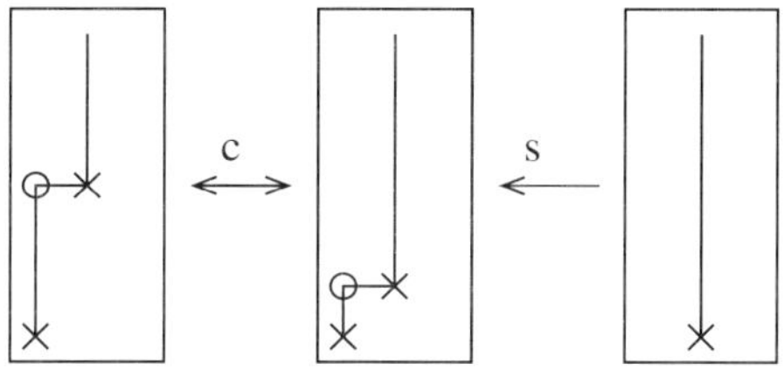

FIGURE B.24. **Eliminating horizontal segments that are not local maxima or minima.** Eliminate crossingless intervals (with northwest and southeast corners) by a sequence of commutation moves and a (Legendrian) destabilization.

occur. These special point commutations can be followed by commutation moves in the associated grid diagrams, as in the proof of Proposition B.4.3. □

Observe that the construction of a grid diagram from a Legendrian knot projection produces a bridge-like grid diagram. In fact, it is a special bridge-like grid diagram, in the following sense (compare with Definition B.4.5):

DEFINITION B.4.17. A grid diagram G is called ***Legendrian bridge-like*** if the following conditions are satisfied:

- A horizontal segment with a southwest corner on it is a local minimum, and one with a northeast corner on it is a local maximum.
- Each horizontal segment contains at most one crossing.
- A horizontal segment is a local maximum or a local minimum if and only if it contains no crossings.

We will need the following refinement of Lemma B.4.8:

LEMMA B.4.18. *For any grid diagram G, there is a Legendrian bridge-like diagram G' that can be obtained from G by a finite sequence of Legendrian grid moves.*

Proof. We adapt the proof of Lemma B.4.8. At each southwest corner in G that either contains crossings or is not a local minimum, we can stabilize G in a Legendrian manner, so that the new southwest corner occurs at a local minimum without crossings. Further Legendrian stabilizations can be done to ensure that the northeast corners are at local maxima, without crossings.

Next, we eliminate the horizontal segments that are not local maxima or local minima and that contain no crossings. Note that the corners of these segments are northwest and southeast. In the proof of Lemma B.4.8, we stabilized to eliminate these segments; but such a stabilization is not Legendrian, so we eliminate them now differently. If the horizontal segment is of length one, then we apply row commutations and commute it down until it meets the other marking in the column of its left endpoint. A Legendrian destabilization of *X:SE* or *O:SE* eliminates this horizontal segment; see Figure B.24.

If the segment has length $d > 1$, we can reduce d by commutation moves as follows. Suppose that the right endpoint of the horizontal segment connects to a vertical segment above the horizontal segment, so the left endpoint connects to a vertical segment below the horizontal. Take the leftmost vertical segment above the horizontal segment, if it exists. Apply repeated commutation moves to this segment, moving it to the left, until it is no longer above the distinguished horizontal segment, to reduce to a case where the horizontal segment has length less than d. If there is no leftmost vertical segment above the horizontal segment, take a rightmost vertical segment below the horizontal segment, and commute it to the right to reduce d. Note that if there is no vertical segment above or below the given horizontal one, then $d = 1$.

Eliminating all the horizontal segments with no crossings that are not maxima or minima, we obtain a new grid diagram G', which still may have horizontal segments that have more than one crossing. These segments are eliminated as in the proof of Lemma B.4.8 by stabilizations and commutations. Since the endpoints of these segments are northwest and southeast corners, the stabilizations we use are Legendrian. □

With these preliminaries in place, we turn to the proof of the Legendrian analogue of Cromwell's theorem:

Proof of Theorem B.4.15. Any oriented Legendrian knot $\vec{\mathcal{K}}$ admits a generic front projection; fix one and call it $\mathcal{D}_{\vec{\mathcal{K}}}$. Lemma B.4.16 associates a grid diagram, up to commutation moves, to a Legendrian front projection, up to Legendrian planar isotopies. As the diagram undergoes Legendrian Reidemeister moves, observe that the associated grid diagram undergoes Legendrian grid moves: Reidemeister 2 and 3 moves can be realized by commutation moves, and Legendrian Reidemeister 1 moves are realized by Legendrian stabilizations and destabilizations. (Compare Figure B.23.) Thus, by Theorem 12.1.7, we have defined a map f from Legendrian knots (up to isotopy) to grid diagrams, up to Legendrian grid moves.

In the other direction, Definition 12.2.1 associates a Legendrian knot to a planar grid diagram, and Lemma 12.2.4 shows that this descends to a well-defined map g from grid diagrams up to Legendrian grid moves to Legendrian knots.

We verify that these two maps are inverses to one another. The fact that $\vec{\mathcal{K}}$ is Legendrian isotopic to $g(f(\vec{\mathcal{K}}))$ is straightforward. It remains to see that for a planar grid diagram G, $f(g(G))$ is equivalent to G under Legendrian grid moves. If G is Legendrian bridge-like in the sense of Definition B.4.17, then its associated Legendrian projection has the property that its associated grid diagram (according to Lemma B.4.16) differs from G by commutation moves. Thus, $f(g(G))$ and G are equivalent, as desired. Lemma B.4.18 reduces the general case to this special case, concluding the proof. □

B.5. Normal forms of cobordisms between knots

In the proof of the slice genus bound provided by the knot signature and by the τ-invariant (Theorem 2.6.6 and Corollary 8.1.2 respectively) the normal form theorem (stated as Proposition 2.6.11 in Section 2.6) played a crucial role. In this section we verify Proposition 2.6.11, which we restate below:

PROPOSITION B.5.1. *Suppose that two knots $K_1, K_2 \subset S^3$ can be connected by a genus g oriented cobordism $W \subset [0,1] \times S^3$. Then, there are knots $K_1', K_2' \subset S^3$ and integers b and d with the following properties:*

(1) $\mathcal{U}_b(K_1)$ can be obtained from K_1' by b simultaneous oriented saddle moves.
(2) K_1' and K_2' can be connected by a sequence of $2g$ oriented saddle moves.
(3) $\mathcal{U}_d(K_2)$ can be obtained from K_2' by d simultaneous oriented saddle moves.

Recall that $\mathcal{U}_n(K)$ is the $(n+1)$-component link with one component K and n further unknotted, unlinked components; the definition of oriented saddle moves (and simultaneous oriented saddle moves) was given in Definition 2.6.10.

In the proof of Proposition B.5.1 we will appropriately isotop the surface W in $[0,1] \times S^3$. We will appeal to standard arguments and concepts from Morse theory [**142**]. (See also Section B.1.1.)

Let $f\colon M \to \mathbb{R}$ be a Morse function on a compact n-dimensional manifold M, and suppose that for some value $c \in \mathbb{R}$ the level set $f^{-1}(c)$ contains a unique critical point of index λ. Then for sufficiently small ϵ the sublevel set $f^{-1}((-\infty, c+\epsilon])$ can be constructed by adding an n-dimensional λ-handle to $f^{-1}((-\infty, c-\epsilon])$; see for example [**142**, Theorem 3.1]. For the definition of a λ-handle in general, see [**77**, Definition 4.1.1].

In fact, we will need this result only in dimension $n = 2$. In this case, the Morse lemma shows that we can find local coordinates (x_1, x_2) in a local chart $U = (x_1, x_2)$ around a critical point p, corresponding to $x_1 = x_2 = 0$, with respect to which the function takes the form

- $f(x) = c + x_1^2 + x_2^2$ and so $c = f(p)$ is a local minimum, or
- $f(x) = c - x_1^2 - x_2^2$ and so $f(p)$ is a local maximum, or
- $f(x) = c - x_1^2 + x_2^2$. In this case, $f^{-1}((-\infty, c+\epsilon])$ is obtained from $f^{-1}((-\infty, c-\epsilon])$ by attaching a 2-dimensional 1-handle, that is, a band.

A Morse function is called *self-indexing* if its value at any critical point is equal to its index. As [**143**, Theorem 4.8] shows, every smooth manifold admits a self-indexing Morse function.

We will need a variant of the above theory, associated to cobordisms between knots in the product of an interval with S^3. To make notation somewhat simpler, from now on the cobordism between the copies of S^3 containing K_1 and K_2 will be identified with $[-1,3] \times S^3$. The embedding of W into $[-1,3] \times S^3$, followed by the projection onto the $[-1,3]$ factor defines a function, $f_W\colon W \to [-1,3]$.

DEFINITION B.5.2. The cobordism $W \subset [-1,3] \times S^3$ is in ***normal form*** if the function f_W is Morse and it maps all the index-0 critical points on W to 0, the index-1 critical points to 1, and the index-2 critical points to 2; that is, f_W is a self-indexing Morse function on W.

LEMMA B.5.3. ([**100**, Theorem 13.1.8], see also [**101**]) *A smooth cobordism W can be isotoped into normal form.*

Proof. We start by dealing with index-0 (and symmetrically index-2) critical points of f_W. We can assume that all critical points map into the interval $[0,2]$. For an index-0 critical point (if it is not already in $f_W^{-1}(0)$) consider an arc in $[0,2] \times S^3$ starting at the critical point and ending in $\{0\} \times S^3$, chosen so that the restriction of the the projection function to the arc has no critical points, and away from its starting point the arc is disjoint from W. (The existence of such an arc follows from a general position argument.) There is a local isotopy of W supported in a neighborhood of the arc that pushes the critical point down into $\{0\} \times S^3$. Index-2 critical points can be handled symmetrically by arcs starting at the critical point and ending in $\{2\} \times S^3$.

Next we deal with the index-1 critical points. Our aim is to show that two index-1 critical points can be pushed into the same level. Suppose that $t_1 < t_2$

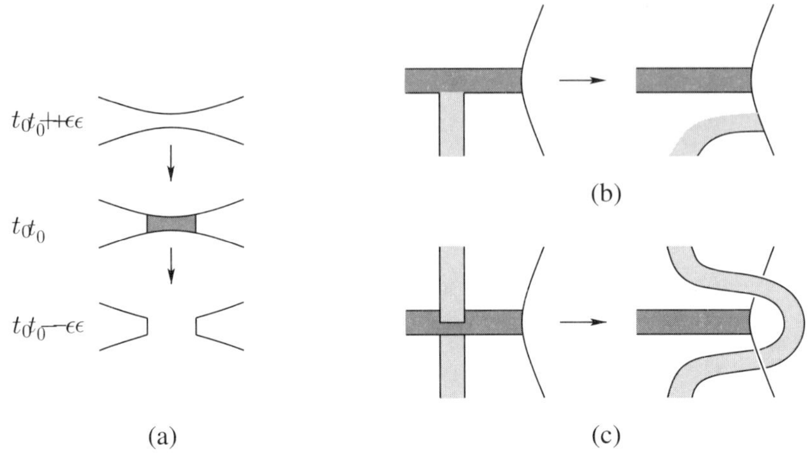

FIGURE B.25. **Turning a cobordism into normal form.** The figure in (a) shows how an index-1 critical point can be converted to a saddle band. Figures in (b) and (c) show how to arrange that the transported bands become disjoint from the bands in the new level. In (b) and (c) the lighter shading denotes the transported band from level t_1.

are two levels containing index-1 critical points with the additional property that $f_W^{-1}((t_1, t_2))$ contains no critical points. By isotoping W we replace the critical points by embedded bands, called *saddle bands*, as shown by Figure B.25(a). Isotop the surface so that the entire band is in a level set. (After this isotopy, the projection is no longer Morse.)

After this modification, the cobordism becomes trivial between the levels t_1 and t_2, since there are no critical values between t_1 and t_2. This means that after an isotopy (keeping $f_W^{-1}((-\infty, t_1])$ fixed) it can be assumed that the cobordism is the product cobordism between these two levels. Along the product cobordism, however, we can transport the saddle bands from level t_1 to t_2. If a transported band (from level t_1) is disjoint from all the bands in level t_2, then it will serve as a saddle band there (and can be converted back to an index-1 critical point, now on level t_2). It can happen, however, that a transported band B intersects some other bands in level t_2. The band we wish to transport can be viewed as a slight two-dimensional thickening of its one-dimensional core arc, so we examine how the core arc of the band can meet the other bands. The intersection is either at an endpoint of the core arc, or at an interior point. In case the intersection is at an endpoint, isotop the transported band slightly away, as illustrated in Figure B.25(b). In case the intersection point is an interior point of the core, we isotop the transported band (and hence the cobordism W) as indicated in Figure B.25(c). After these modifications we have a cobordism isotopic to the original one with the additional property that all index-1 critical points from level t_1 are moved to level t_2. Repeating this procedure for all the finitely many critical points of index 1, we get a surface isotopic to the original cobordism with the desired properties. □

EXERCISE B.5.4. Using the normal form, show that a knot K is slice if and only if there is a ribbon knot R such that $K\#R$ is ribbon. (For the definition of slice and ribbon knots see Section 2.4; cf. the slice-ribbon conjecture from Remark 2.6.3.)

Proposition B.5.1 now follows from a slight modification of the normal form:

Proof of Proposition B.5.1. Suppose that $W \subset [-1,3] \times S^3$ is a cobordism between the knots K_1 and K_2, and assume (by Lemma B.5.3) that it is in normal form. Then the link $f_W^{-1}(0.1)$ can be identified with $\mathcal{U}_b(K_1)$ (for some $b \in \mathbb{N}$) and similarly $f_W^{-1}(1.9)$ can be identified with $\mathcal{U}_d(K_2)$ (for some $d \in \mathbb{N}$), where b and d are given by the number of index-0, resp. index-2 critical points of the Morse function f_W. Index-1 critical points define bands (i.e. 1-handles), and since W is connected, there are b index-1 critical points such that the corresponding handles turn $f_W^{-1}(0.1) = \mathcal{U}_b(K_1)$ into a knot K_1', and there are d further index-1 critical points such that the corresponding handles turn $f_W^{-1}(1.9) = \mathcal{U}_d(K_2)$ into a knot K_2'. Transporting the b index-1 critical points which connect $\mathcal{U}_b(K_1)$ to level 0.2, and the d index-1 critical points making $\mathcal{U}_d(K_2')$ connected to level 1.8, and denoting the resulting Morse function by g_W, the level sets $g_W^{-1}(0.5)$ and $g_W^{-1}(1.5)$ provide the desired knots K_1' and K_2'. (Note that since all the transported index-1 critical points are on levels 0.2 or 1.8, we get simultaneous saddle moves from them.) The remaining index-1 critical points provide a sequence of n oriented saddle bands between K_1' and K_2'. Since the Euler characteristic of F, which we assumed to be $-2g$, is given by $b+d$ minus the total number of saddles, it follows that there are $n = 2g$ saddles in the sequence of 1-handles from K_1' to K_2', completing the proof of the proposition. $\square$

Bibliography

[1] J. Alexander. Topological invariants of knots and links. *Trans. Amer. Math. Soc.*, 30(2):275–306, 1928.

[2] M. Aschenbrenner, S. Friedl, and H. Wilton. 3-manifold groups. EMS Series of Lectures in Mathematics, 2015.

[3] M. Atiyah and I. Macdonald. *Introduction to commutative algebra*. Addison-Wesley Publishing Co., Reading, Mass.-London-Don Mills, Ont., 1969.

[4] J. Baldwin. Comultiplication in link Floer homology and transversely nonsimple links. *Algebr. Geom. Topol.*, 10(3):1417–1436, 2010.

[5] J. Baldwin and W. Gillam. Computations of Heegaard Floer knot homology. *J. Knot Theory Ramifications*, 21(8):1250075, 65, 2012.

[6] J. Baldwin and A. Levine. A combinatorial spanning tree model for knot Floer homology. *Adv. Math.*, 231(3-4):1886–1939, 2012.

[7] J. Baldwin, D. Vela-Vick, and V. Vértesi. On the equivalence of Legendrian and transverse invariants in knot Floer homology. *Geom. Topol.*, 17(2):925–974, 2013.

[8] D. Bar-Natan. On Khovanov's categorification of the Jones polynomial. *Algebr. Geom. Topol.*, 2:337–370, 2002.

[9] D. Bar-Natan, J. Fulman, and L. Kauffman. An elementary proof that all spanning surfaces of a link are tube-equivalent. *J. Knot Theory Ramifications*, 7(7):873–879, 1998.

[10] D. Bennequin. Entrelacements et équations de Pfaff. In *Third Schnepfenried geometry conference, Vol. 1 (Schnepfenried, 1982)*, volume 107 of *Astérisque*, pages 87–161. 1983.

[11] J. Birman. *Braids, links, and mapping class groups*. Princeton University Press, Princeton, N.J.; University of Tokyo Press, Tokyo, 1974. Annals of Mathematics Studies, No. 82.

[12] J. Birman and M. Hirsch. A new algorithm for recognizing the unknot. *Geom. Topol.*, 2:175–220, 1998.

[13] J. Birman and W. Menasco. Stabilization in the braid groups. II. Transversal simplicity of knots. *Geom. Topol.*, 10:1425–1452, 2006.

[14] J. Bloom. Odd Khovanov homology is mutation invariant. *Math. Res. Lett.*, 17(1):1–10, 2010.

[15] S. Boyer, C. Gordon, and L. Watson. On L-spaces and left-orderable fundamental groups. *Math. Ann.*, 356(4):1213–1245, 2013.

[16] P. Braam and S. Donaldson. Floer's work on instanton homology, knots, and surgery. In H. Hofer, C. Taubes, A. Weinstein, and E. Zehnder, editors, *The Floer Memorial Volume*, number 133 in Progress in Mathematics, pages 195–256. Birkhäuser, 1995.

[17] H. Brunn. Über verknotete Kurven. *Verhandlungen des Internationalen Math. Kongresses (Zurich 1897)*, pages 256–259, 1898.

[18] G. Burde and H. Zieschang. *Knots*, volume 5 of *de Gruyter Studies in Mathematics*. Walter de Gruyter & Co., Berlin, second edition, 2003.

[19] D. Calegari. *Foliations and the geometry of 3-manifolds*. Oxford Mathematical Monographs. Oxford University Press, Oxford, 2007.

[20] A. Candel and L. Conlon. *Foliations. I*, volume 23 of *Graduate Studies in Mathematics*. American Mathematical Society, Providence, RI, 2000.

[21] A. Candel and L. Conlon. *Foliations. II*, volume 60 of *Graduate Studies in Mathematics*. American Mathematical Society, Providence, RI, 2003.

[22] J. Cerf. *Sur les difféomorphismes de la sphère de dimension trois* ($\Gamma_4 = 0$). Lecture Notes in Mathematics, No. 53. Springer-Verlag, Berlin-New York, 1968.

[23] Yu. Chekanov. Differential algebra of Legendrian links. *Invent. Math.*, 150(3):441–483, 2002.

[24] D. Choi and K. Ko. Parametrizations of 1-bridge torus knots. *J. Knot Theory Ramifications*, 12(4):463–491, 2003.

[25] W. Chongchitmate and L. Ng. An atlas of Legendrian knots. *Exp. Math.*, 22(1):26–37, 2013.

[26] T. Cochran. Derivatives of links: Milnor's concordance invariants and Massey's products. *Mem. Amer. Math. Soc.*, 84(427):x+73, 1990.

[27] P. Cromwell. Embedding knots and links in an open book. I. Basic properties. *Topology Appl.*, 64(1):37–58, 1995.

[28] P. Cromwell. *Knots and links*. Cambridge University Press, Cambridge, 2004.

[29] M. Curtis. *Matrix groups*. Universitext. Springer-Verlag, New York, second edition, 1984.

[30] S. De Michelis and M. Freedman. Uncountably many exotic $\mathbf{R}^4$'s in standard 4-space. *J. Differential Geom.*, 35(1):219–254, 1992.

[31] F. Ding and H. Geiges. The diffeotopy group of $S^1 \times S^2$ via contact topology. *Compos. Math.*, 146(4):1096–1112, 2010.

[32] H. Doll. A generalized bridge number for links in 3-manifolds. *Math. Ann.*, pages 701–717, 1992.

[33] S. Donaldson. An application of gauge theory to four-dimensional topology. *J. Differential Geom.*, 18(2):279–315, 1983.

[34] S. Donaldson. Polynomial invariants for smooth four-manifolds. *Topology*, 29(3):257–315, 1990.

[35] S. Donaldson and P. Kronheimer. *The geometry of four-manifolds*. Oxford Mathematical Monographs. Oxford University Press, 1990.

[36] N. Dunfield, S. Gukov, and J. Rasmussen. The superpolynomial for knot homologies. *Experiment. Math.*, 15(2):129–159, 2006.

[37] I. Dynnikov. Arc-presentations of links: monotonic simplification. *Fund. Math.*, 190:29–76, 2006.

[38] Ya. Eliashberg. Classification of overtwisted contact structures on 3-manifolds. *Invent. Math.*, 98(3):623–637, 1989.

[39] Ya. Eliashberg and M. Fraser. Topologically trivial Legendrian knots. *J. Symplectic Geom.*, 7(2):77–127, 2009.

[40] Ya. Eliashberg, A. Givental, and H. Hofer. Introduction to symplectic field theory. *Geom. Funct. Anal.*, Special Volume(Part II):560–673, 2000. GAFA 2000 (Tel Aviv, 1999).

[41] Ya. Eliashberg and M. Gromov. Convex symplectic manifolds. In *Several complex variables and complex geometry, Part 2 (Santa Cruz, CA, 1989)*, volume 52 of *Proc. Sympos. Pure Math.*, pages 135–162. Amer. Math. Soc., Providence, RI, 1991.

[42] Ya. Eliashberg and W. Thurston. *Confoliations*, volume 13 of *University Lecture Series*. AMS, Providence, RI, 1998.

[43] J. Epstein, D. Fuchs, and M. Meyer. Chekanov-Eliashberg invariants and transverse approximations of Legendrian knots. *Pacific J. Math.*, 201(1):89–106, 2001.

[44] J. Etnyre. Transversal torus knots. *Geom. Topol.*, 3:253–268, 1999.

[45] J. Etnyre. *Legendrian and Transversal Knots*, pages 105–185. Elsevier B. V., Amsterdam, 2005.

[46] J. Etnyre and K. Honda. Knots and contact geometry. I. Torus knots and the figure eight knot. *J. Symplectic Geom.*, 1(1):63–120, 2001.

[47] J. Etnyre and K. Honda. Cabling and transverse simplicity. *Ann. of Math.*, 162(3):1305–1333, 2005.

[48] J. Etnyre, L. Ng, and V. Vértesi. Legendrian and transverse twist knots. *J. Eur. Math. Soc. (JEMS)*, 15(3):969–995, 2013.

[49] A. Floer. An instanton-invariant for 3-manifolds. *Comm. Math. Phys.*, 119:215–240, 1988.

[50] A. Floer. Morse theory for Lagrangian intersections. *J. Differential Geom.*, 28:513–547, 1988.

[51] A. Floer. A relative Morse index for the symplectic action. *Comm. Pure Appl. Math.*, 41(4):393–407, 1988.

[52] A. Floer. The unregularized gradient flow of the symplectic action. *Comm. Pure Appl. Math.*, 41(6):775–813, 1988.

[53] A. Floer. Instanton homology, surgery, and knots. In *Geometry of low-dimensional manifolds, 1 (Durham, 1989)*, volume 150 of *London Math. Soc. Lecture Note Ser.*, pages 97–114. Cambridge Univ. Press, Cambridge, 1990.

[54] A. Floer. Instanton homology and Dehn surgery. In H. Hofer, C. H. Taubes, A. Weinstein, and E. Zehnder, editors, *The Floer Memorial Volume*, number 133 in Progress in Mathematics, pages 77–97. Birkhäuser, 1995.
[55] A. Floer, H. Hofer, and D. Salamon. Transversality in elliptic Morse theory for the symplectic action. *Duke Math. J.*, 80(1):251–29, 1995.
[56] R. Fox. A quick trip through knot theory. In *Topology of 3-manifolds and related topics (Proc. The Univ. of Georgia Institute, 1961)*, pages 120–167. Prentice-Hall, 1962.
[57] R. Fox. Some problems in knot theory. In *Topology of 3-manifolds and related topics (Proc. The Univ. of Georgia Institute, 1961)*, pages 168–176. Prentice-Hall, N.J., 1962.
[58] R. Fox and J. Milnor. Singularities of 2-spheres in 4-space and cobordism of knots. *Osaka J. Math.*, 3:257–267, 1966.
[59] M. Freedman. A surgery sequence in dimension four; the relations with knot concordance. *Invent. Math.*, 68(2):195–226, 1982.
[60] M. Freedman and F. Quinn. *Topology of 4-manifolds*, volume 39 of *Princeton Mathematical Series*. Princeton University Press, Princeton, NJ, 1990.
[61] K. Frøyshov. The Seiberg-Witten equations and four-manifolds with boundary. *Math. Res. Lett*, 3:373–390, 1996.
[62] K. Fukaya, Y-G. Oh, K. Ono, and H. Ohta. *Lagrangian intersection Floer theory—anomaly and obstruction*. Kyoto University, 2000.
[63] D. Gabai. Foliations and the topology of 3-manifolds. *J. Differential Geom.*, 18(3):445–503, 1983.
[64] D. Gabai. The Murasugi sum is a natural geometric operation. II. In *Combinatorial methods in topology and algebraic geometry (Rochester, N.Y., 1982)*, volume 44 of *Contemp. Math.*, pages 93–100. Amer. Math. Soc., Providence, RI, 1985.
[65] D. Gabai. Genera of the arborescent links. *Mem. Amer. Math. Soc.*, 59(339):i–viii and 1–98, 1986.
[66] E. Gallais. Sign refinement for combinatorial link Floer homology. *Algebr. Geom. Topol.*, 8(3):1581–1592, 2008.
[67] S. Garoufalidis and P. Teichner. On knots with trivial Alexander polynomial. *J. Differential Geom.*, 67(1):167–193, 2004.
[68] H. Geiges. *An introduction to contact topology*, volume 109 of *Cambridge Studies in Advanced Mathematics*. Cambridge University Press, Cambridge, 2008.
[69] H. Geiges. A contact geometric proof of the Whitney-Graustein theorem. *Enseign. Math. (2)*, 55(1-2):93–102, 2009.
[70] S. Gelfand and Yu. Manin. *Methods of homological algebra*. Springer Monographs in Mathematics. Springer-Verlag, Berlin, second edition, 2003.
[71] P. Ghiggini. Knot Floer homology detects genus-one fibred knots. *Amer. J. Math.*, 130(5):1151–1169, 2008.
[72] P. Ghiggini, P. Lisca, and A. Stipsicz. Tight contact structures on some small Seifert fibered 3-manifolds. *Amer. J. Math.*, 129(5):1403–1447, 2007.
[73] E. Giroux. Convexité en topologie de contact. *Comment. Math. Helv.*, 66(4):637–677, 1991.
[74] E. Giroux. Géométrie de contact: de la dimension trois vers les dimensions supérieures. In *Proceedings of the International Congress of Mathematicians, Vol. II (Beijing, 2002)*, pages 405–414. Higher Ed. Press, Beijing, 2002.
[75] M. Golubitsky and V. Guillemin. *Stable mappings and their singularities*. Springer-Verlag, New York-Heidelberg, 1973. Graduate Texts in Mathematics, Vol. 14.
[76] R. Gompf. Three exotic $\mathbf{R}^4$'s and other anomalies. *J. Differential Geom.*, 18(2):317–328, 1983.
[77] R. Gompf and A. Stipsicz. *4-manifolds and Kirby calculus*, volume 20 of *Graduate Studies in Mathematics*. American Mathematical Society, 1999.
[78] C. Gordon and R. Litherland. On the signature of a link. *Invent. Math.*, 47(1):53–69, 1978.
[79] C. Gordon and J. Luecke. Knots with unknotting number 1 and essential Conway spheres. *Algebr. Geom. Topol.*, 6:2051–2116, 2006.
[80] J. Greene. The lens space realization problem. *Ann. of Math. (2)*, 177(2):449–511, 2013.
[81] M. Gromov. Pseudoholomorphic curves in symplectic manifolds. *Invent. Math.*, 82(2):307–347, 1985.
[82] P. Hall. On representatives of subsets. *J. London Math. Soc.*, 1(10):26–30, 1935.
[83] A. Hatcher. *Algberaic topology*. Cambridge University Press, 2001.

[84] M. Hedden and L. Watson. On the geography and botany of knot Floer homology. arXiv:1404.6913.

[85] M. Hedden and L. Watson. Does Khovanov homology detect the unknot? *Amer. J. Math.*, 132(5):1339–1345, 2010.

[86] J. Hillman. *Algebraic invariants of links*, volume 32 of *Series on Knots and Everything*. World Scientific Publishing Co., Inc., River Edge, NJ, 2002.

[87] M. Hirsch. *Differential topology*, volume 33 of *Graduate Texts in Mathematics*. Springer-Verlag, New York, 1994. Corrected reprint of the 1976 original.

[88] J. Hom. An infinite-rank summand of topologically slice knots. *Geom. Topol.*, 19(2):1063–1110, 2015.

[89] J. Hom and Z. Wu. Four-ball genus bounds and a refinement of the Ozsvath-Szabo tau-invariant. arxiv:1401.1565.

[90] K. Honda. On the classification of tight contact structures. I. *Geom. Topol.*, 4:309–368, 2000.

[91] M. Hutchings. Embedded contact homology and its applications. In *Proceedings of the International Congress of Mathematicians. Volume II*, pages 1022–1041, New Delhi, 2010.

[92] N. Jacobson. *Basic algebra. I.* W. H. Freeman and Company, New York, second edition, 1985.

[93] V. Jones. Hecke algebra representations of braid groups and link polynomials. *Ann. of Math. (2)*, 126(2):335–388, 1987.

[94] V. Jones. The Jones polynomial. http://math.berkeley.edu/~vfr/jones.pdf, 2005.

[95] A. Juhász. Cobordisms of sutured manifolds. arXiv:0910.4382.

[96] A. Juhász. Holomorphic discs and sutured manifolds. *Algebr. Geom. Topol.*, 6:1429–1457, 2006.

[97] A. Juhász. The sutured Floer homology polytope. *Geom. Topol.*, 14(3):1303–1354, 2010.

[98] A. Juhász and D. Thurston. Naturality and mapping class groups in Heegaard Floer homology. arXiv:1210.4996.

[99] T. Kanenobu. Infinitely many knots with the same polynomial invariant. *Proc. Amer. Math. Soc.*, 97(1):158–162, 1986.

[100] A. Kawauchi. *A survey of knot theory*. Birkhäuser Verlag, Basel, 1996. Translated and revised from the 1990 Japanese original by the author.

[101] A. Kawauchi, T. Shibuya, and S. Suzuki. Descriptions on surfaces in four-space. I. Normal forms. *Math. Sem. Notes Kobe Univ.*, 10(1):75–125, 1982.

[102] T. Khandhawit and L. Ng. A family of transversely nonsimple knots. *Algebr. Geom. Topol.*, 10(1):293–314, 2010.

[103] M. Khovanov. A categorification of the Jones polynomial. *Duke Math. J.*, 101(3):359–426, 2000.

[104] M. Khovanov and L. Rozansky. Matrix factorizations and link homology. *Fund. Math.*, 199(1):1–91, 2008.

[105] M. Khovanov and L. Rozansky. Matrix factorizations and link homology. II. *Geom. Topol.*, 12(3):1387–1425, 2008.

[106] P. Kronheimer and T. Mrowka. Gauge theory for embedded surfaces. I. *Topology*, 32(4):773–826, 1993.

[107] P. Kronheimer and T. Mrowka. The genus of embedded surfaces in the projective plane. *Math. Res. Lett.*, 1(6):797–808, 1994.

[108] P. Kronheimer and T. Mrowka. Monopoles and contact structures. *Invent. Math.*, 130(2):209–255, 1997.

[109] P. Kronheimer and T. Mrowka. Witten's conjecture and Property P. *Geom. Topol.*, 8:295–310, 2004.

[110] P. Kronheimer and T. Mrowka. *Monopoles and three-manifolds*, volume 10 of *New Mathematical Monographs*. Cambridge University Press, Cambridge, 2007.

[111] P. Kronheimer and T. Mrowka. Knots, sutures, and excision. *J. Differential Geom.*, 84(2):301–364, 2010.

[112] P. Kronheimer and T. Mrowka. Khovanov homology is an unknot-detector. *Publ. Math. Inst. Hautes Études Sci.*, 113:97–208, 2011.

[113] P. Kronheimer and T. Mrowka. Knot homology groups from instantons. *J. Topol.*, 4(4):835–918, 2011.

[114] P. Kronheimer, T. Mrowka, P. Ozsváth, and Z. Szabó. Monopoles and lens space surgeries. *Ann. of Math. (2)*, 165(2):457–546, 2007.

[115] C. Kutluhan, Y-J. Lee, and C. Taubes. HF=HM I: Heegaard Floer homology and Seiberg–Witten Floer homology. arXiv:1007.1979.
[116] A. Lecuona. On the slice-ribbon conjecture for Montesinos knots. *Trans. Amer. Math. Soc.*, 364(1):233–285, 2012.
[117] E.-S. Lee. An endomorphism of the Khovanov invariant. *Adv. Math.*, 197(2):554–586, 2005.
[118] J. Licata. The Thurston polytope for four-stranded pretzel links. *Algebr. Geom. Topol.*, 8(1):211–243, 2008.
[119] R. Lickorish. *An introduction to knot theory*, volume 175 of *Graduate Texts in Mathematics*. Springer-Verlag, 1997.
[120] R. Lipshitz. A cylindrical reformulation of Heegaard Floer homology. *Geom. Topol.*, 10:955–1097, 2006.
[121] R. Lipshitz. Heegaard Floer homology, double points and nice diagrams. In *New perspectives and challenges in symplectic field theory*, volume 49 of *CRM Proc. Lecture Notes*, pages 327–342. Amer. Math. Soc., Providence, RI, 2009.
[122] R. Lipshitz and S. Sarkar. A Khovanov stable homotopy type. *J. Amer. Math. Soc.*, 27(4):983–1042, 2014.
[123] R. Lipshitz and S. Sarkar. A refinement of Rasmussen's S-invariant. *Duke Math. J.*, 163(5):923–952, 2014.
[124] P. Lisca. Lens spaces, rational balls and the ribbon conjecture. *Geom. Topol.*, 11:429–472, 2007.
[125] P. Lisca and G. Matić. Tight contact structures and Seiberg-Witten invariants. *Invent. Math.*, 129(3):509–525, 1997.
[126] P. Lisca, P. Ozsváth, A. Stipsicz, and Z. Szabó. Heegaard Floer invariants of Legendrian knots in contact three-manifolds. *J. Eur. Math. Soc. (JEMS)*, 11(6):1307–1363, 2009.
[127] H. Lyon. Torus knots in the complements of links and surfaces. *Michigan Math. J.*, 27(1):39–46, 1980.
[128] I. MacDonald. Symmetric products of an algebraic curve. *Topology*, 1:319–343, 1962.
[129] C. Manolescu. Pin(2)-equivariant Seiberg-Witten Floer homology and the Triangulation Conjecture. arXiv:1303.2354.
[130] C. Manolescu. Seiberg-Witten-Floer stable homotopy type of three-manifolds with $b_1 = 0$. *Geom. Topol.*, 7:889–932, 2003.
[131] C. Manolescu. Nilpotent slices, Hilbert schemes, and the Jones polynomial. *Duke Math. J.*, 132(2):311–369, 2006.
[132] C. Manolescu. An unoriented skein exact triangle for knot Floer homology. *Math. Res. Lett.*, 14(5):839–852, 2007.
[133] C. Manolescu and P. Ozsváth. Heegaard Floer homology and integer surgeries on links. arXiv:1011.1317.
[134] C. Manolescu and P. Ozsváth. On the Khovanov and knot Floer homologies of quasi-alternating links. In *Proceedings of Gökova Geometry-Topology Conference 2007*, pages 60–81. Gökova Geometry/Topology Conference (GGT), Gökova, 2008.
[135] C. Manolescu, P. Ozsváth, and S. Sarkar. A combinatorial description of knot Floer homology. *Ann. of Math. (2)*, 169(2):633–660, 2009.
[136] C. Manolescu, P. Ozsváth, Z. Szabó, and D. Thurston. On combinatorial link Floer homology. *Geom. Topol.*, 11:2339–2412, 2007.
[137] W. Massey. *Algebraic topology: an introduction*. Springer-Verlag, New York-Heidelberg, 1977. Reprint of the 1967 edition, Graduate Texts in Mathematics, Vol. 56.
[138] W. Massey. Higher order linking numbers. *J. Knot Theory Ramifications*, 7(3):393–414, 1998.
[139] J. McCleary. *User's guide to spectral sequences*, volume 12 of *Mathematics Lecture Series*. Publish or Perish Inc., Wilmington, DE, 1985.
[140] C. McMullen. The Alexander polynomial of a 3-manifold and the Thurston norm on cohomology. *Ann. Sci. de l'Ecole Norm. Sup.*, 35(2):153–171, 2002.
[141] J. Milnor. Isotopy of links. Algebraic geometry and topology. In *A symposium in honor of S. Lefschetz*, pages 280–306. Princeton University Press, Princeton, N. J., 1957.
[142] J. Milnor. *Morse theory*. Based on lecture notes by M. Spivak and R. Wells. Annals of Mathematics Studies, No. 51. Princeton University Press, Princeton, N.J., 1963.
[143] J. Milnor. *Lectures on the h-cobordism theorem*. Princeton University Press, 1965. Notes by L. Siebenmann and J. Sondow.

[144] J. Milnor. *Singular points of complex hypersurfaces*. Annals of Mathematics Studies, No. 61. Princeton University Press, Princeton, N.J., 1968.

[145] K. Mohnke. Legendrian links of topological unknots. In *Topology, geometry, and algebra: interactions and new directions (Stanford, CA, 1999)*, volume 279 of *Contemp. Math.*, pages 209–211. Amer. Math. Soc., Providence, RI, 2001.

[146] E. Moise. Affine structures in 3-manifolds. V. The triangulation theorem and Hauptvermutung. *Ann. of Math. (2)*, 56:96–114, 1952.

[147] J. Montesinos. Three-manifolds as 3-fold branched covers of S^3. *Quart. J. Math. Oxford Ser. (2)*, 27(105):85–94, 1976.

[148] A. Moore and L. Starkston. Genus-two mutant knots with the same dimension in knot Floer and Khovanov homologies. *Algebr. Geom. Topol.*, 15(1):43–63, 2015.

[149] J. Morgan. *The Seiberg-Witten Equations and Applications to the Topology of Smooth Four-Manifold.* Number 44 in Mathematical Notes. Princeton University Press, 1996.

[150] J. Morgan, Z. Szabó, and C. Taubes. A product formula for Seiberg-Witten invariants and the generalized Thom conjecture. *J. Differential Geom.*, 44:706–788, 1996.

[151] J. Munkres. *Elementary differential topology*, volume 1961 of *Lectures given at Massachusetts Institute of Technology, Fall.* Princeton University Press, Princeton, N.J., 1966.

[152] K. Murasugi. On the genus of the alternating knot. I, II. *J. Math. Soc. Japan*, 10:94–105, 235–248, 1958.

[153] K. Murasugi. On a certain numerical invariant of link types. *Trans. Amer. Math. Soc.*, 117:387–422, 1965.

[154] K. Murasugi. On the signature of links. *Topology*, 9:283–298, 1970.

[155] K. Murasugi. *Knot theory & its applications.* Modern Birkhäuser Classics. Birkhäuser Boston Inc., Boston, MA, 2008. Translated from the 1993 Japanese original by Bohdan Kurpita, Reprint of the 1996 translation [MR1391727].

[156] L. Ng. Combinatorial knot contact homology and transverse knots. *Adv. Math.*, 227(6):2189–2219, 2011.

[157] L. Ng, P. Ozsváth, and D. Thurston. Transverse knots distinguished by knot Floer homology. *J. Symplectic Geom.*, 6(4):461–490, 2008.

[158] L. Ng and D. Thurston. Grid diagrams, braids, and contact geometry. In *Proceedings of Gökova Geometry-Topology Conference 2008*, pages 120–136. Gökova Geometry/Topology Conference (GGT), Gökova, 2009.

[159] Y. Ni. A note on knot Floer homology of links. *Geom. Topol.*, 10:695–713, 2006.

[160] Y. Ni. Knot Floer homology detects fibred knots. *Invent. Math.*, 170(3):577–608, 2007.

[161] Y-G. Oh. Fredholm theory of holomorphic discs under the perturbation of boundary conditions. *Math. Z*, 222(3):505–520, 1996.

[162] B. Owens. Unknotting information from Heegaard Floer homology. *Adv. Math.*, 217(5):2353–2376, 2008.

[163] P. Ozsváth and A. Stipsicz. Contact surgeries and the transverse invariant in knot Floer homology. *J. Inst. Math. Jussieu*, 9(3):601–632, 2010.

[164] P. Ozsváth, A. Stipsicz, and Z. Szabó. Concordance homomorphisms from knot Floer homology. arXiv:1407.1795.

[165] P. Ozsváth, A. Stipsicz, and Z. Szabó. Floer homology and singular knots. *J. Topol.*, 2(2):380–404, 2009.

[166] P. Ozsváth, A. Stipsicz, and Z. Szabó. Combinatorial Heegaard Floer homology and sign assignments. *Topology Appl.*, 166:32–65, 2014.

[167] P. Ozsváth and Z. Szabó. On the skein exact squence for knot Floer homology. arXiv:0707.1165.

[168] P. Ozsváth and Z. Szabó. Absolutely graded Floer homologies and intersection forms for four-manifolds with boundary. *Advances in Mathematics*, 173(2):179–261, 2003.

[169] P. Ozsváth and Z. Szabó. Heegaard Floer homology and alternating knots. *Geom. Topol.*, 7:225–254, 2003.

[170] P. Ozsváth and Z. Szabó. Knot Floer homology and the four-ball genus. *Geom. Topol.*, 7:615–639, 2003.

[171] P. Ozsváth and Z. Szabó. Holomorphic disks and genus bounds. *Geom. Topol.*, 8:311–334, 2004.

[172] P. Ozsváth and Z. Szabó. Holomorphic disks and knot invariants. *Adv. Math.*, 186(1):58–116, 2004.

[173] P. Ozsváth and Z. Szabó. Holomorphic disks and three-manifold invariants: properties and applications. *Ann. of Math. (2)*, 159(3):1159–1245, 2004.
[174] P. Ozsváth and Z. Szabó. Holomorphic disks and topological invariants for closed three-manifolds. *Ann. of Math. (2)*, 159(3):1027–1158, 2004.
[175] P. Ozsváth and Z. Szabó. Heegaard Floer homology and contact structures. *Duke Math. J.*, 129(1):39–61, 2005.
[176] P. Ozsváth and Z. Szabó. Knots with unknotting number one and Heegaard Floer homology. *Topology*, 44(4):705–745, 2005.
[177] P. Ozsváth and Z. Szabó. On knot Floer homology and lens space surgeries. *Topology*, 44(6):1281–1300, 2005.
[178] P. Ozsváth and Z. Szabó. On the Heegaard Floer homology of branched double covers. *Adv. Math.*, 194(1):1–33, 2005.
[179] P. Ozsváth and Z. Szabó. Heegaard diagrams and Floer homology. In *International Congress of Mathematicians. Vol. II*, pages 1083–1099. Eur. Math. Soc., Zürich, 2006.
[180] P. Ozsváth and Z. Szabó. Holomorphic disks, link invariants and the multi-variable Alexander polynomial. *Algebr. Geom. Topol.*, 8(2):615–692, 2008.
[181] P. Ozsváth and Z. Szabó. Knot Floer homology and integer surgeries. *Algebr. Geom. Topol.*, 8(1):101–153, 2008.
[182] P. Ozsváth and Z. Szabó. Link Floer homology and the Thurston norm. *J. Amer. Math. Soc.*, 21(3):671–709, 2008.
[183] P. Ozsváth and Z. Szabó. A cube of resolutions for knot Floer homology. *J. Topol.*, 2(4):865–910, 2009.
[184] P. Ozsváth and Z. Szabó. Knot Floer homology and rational surgeries. *Algebr. Geom. Topol.*, 11(1):1–68, 2011.
[185] P. Ozsváth, Z. Szabó, and D. Thurston. Legendrian knots, transverse knots and combinatorial Floer homology. *Geom. Topol.*, 12(2):941–980, 2008.
[186] T. Perutz. Hamiltonian handleslides for Heegaard Floer homology. In *Proceedings of Gökova Geometry-Topology Conference 2007*, pages 15–35. Gökova Geometry/Topology Conference (GGT), Gökova, 2008.
[187] O. Plamenevskaya. Bounds for the Thurston-Bennequin number from Floer homology. *Algebr. Geom. Topol.*, 4:399–406, 2004.
[188] R. Porter. Milnor's $\bar{\mu}$-invariants and Massey products. *Trans. Amer. Math. Soc.*, 257(1):39–71, 1980.
[189] J. Przytycki. From Goeritz matrices to quasi-alternating links. In *The mathematics of knots*, volume 1 of *Contrib. Math. Comput. Sci.*, pages 257–316. Springer, Heidelberg, 2011.
[190] J. Rasmussen. Floer homology of surgeries on two-bridge knots. *Algebr. Geom. Topol.*, 2:757–789, 2002.
[191] J. Rasmussen. *Floer homology and knot complements*. PhD thesis, Harvard University, 2003.
[192] J. Rasmussen. Lens space surgeries and a conjecture of Goda and Teragaito. *Geom. Topol.*, 8:1013–1031, 2004.
[193] J. Rasmussen. Knot polynomials and knot homologies. In *Geometry and topology of manifolds*, volume 47 of *Fields Inst. Commun.*, pages 261–280. Amer. Math. Soc., Providence, RI, 2005.
[194] J. Rasmussen. Khovanov homology and the slice genus. *Invent. Math.*, 182(2):419–447, 2010.
[195] K. Reidemeister. Zur dreidimensionalen Topologie. *Abh. Math. Sem. Univ. Hamburg*, 9(1):189–194, 1933.
[196] K. Reidemeister. *Knotentheorie*. Springer-Verlag, Berlin, 1974. Reprint.
[197] J. Robbin and D. Salamon. The spectral flow and the Maslov index. *Bull. London Math. Soc.*, 27(1):1–33, 1995.
[198] D. Rolfsen. *Knots and links*, volume 7 of *Mathematics Lecture Series*. Publish or Perish Inc., Houston, TX, 1990. Corrected reprint of the 1976 original.
[199] D. Roseman. Elementary moves for higher dimensional knots. *Fund. Math.*, 184:291–310, 2004.
[200] J. Rotman. *An introduction to homological algebra*. Universitext. Springer, New York, second edition, 2009.
[201] C. Rourke and B. Sanderson. *Introduction to piecewise-linear topology*. Springer-Verlag, New York-Heidelberg, 1972. Ergebnisse der Mathematik und ihrer Grenzgebiete, Band 69.

[202] L. Rudolph. An obstruction to sliceness via contact geometry and "classical" gauge theory. *Invent. Math.*, 119(1):155–163, 1995.

[203] S. Sarkar. Moving basepoints and the induced automorphisms of link Floer homology. arXiv:1109.2168.

[204] S. Sarkar. Grid diagrams and the Ozsváth-Szabó tau-invariant. *Math. Res. Lett.*, 18(6):1239–1257, 2011.

[205] S. Sarkar. Grid diagrams and shellability. *Homology Homotopy Appl.*, 14(2):77–90, 2012.

[206] S. Sarkar and J. Wang. An algorithm for computing some Heegaard Floer homologies. *Ann. of Math. (2)*, 171(2):1213–1236, 2010.

[207] M. Scharlemann. Heegaard splittings of 3-manifolds. In *Low dimensional topology*, volume 3 of *New Stud. Adv. Math.*, pages 25–39. Int. Press, Somerville, MA, 2003.

[208] N. Seiberg and E. Witten. Electric-magnetic duality, monopole condensation, and confinement in $N = 2$ supersymmetric Yang-Mills theory. *Nuclear Phys. B*, 426(1):19–52, 1994.

[209] N. Seiberg and E. Witten. Monopoles, duality and chiral symmetry breaking in $N = 2$ supersymmetric QCD. *Nuclear Phys. B*, 431(3):484–550, 1994.

[210] P. Seidel. *Fukaya categories and Picard-Lefschetz theory*. Zurich Lectures in Advanced Mathematics. European Mathematical Society (EMS), Zürich, 2008.

[211] P. Seidel and I. Smith. A link invariant from the symplectic geometry of nilpotent slices. *Duke Math. J.*, 134(3):453–514, 2006.

[212] J. Singer. Three-dimensional manifolds and their Heegaard diagrams. *Trans. Amer. Math. Soc.*, 35(1):88–111, 1933.

[213] J. Stallings. On fibering certain 3-manifolds. In *Topology of 3-manifolds and related topics (Proc. The Univ. of Georgia Institute, 1961)*, pages 95–100. Prentice-Hall, 1962.

[214] J. Świątkowski. On the isotopy of Legendrian knots. *Ann. Global Anal. Geom.*, 10(3):195–207, 1992.

[215] C. Taubes. The Seiberg-Witten and Gromov invariants. *Math. Res. Lett.*, 2(2):221–238, 1995.

[216] C. Taubes. *Metrics, connections and gluing theorems*. Number 89 in CBMS Regional Conference Series in Mathematics. AMS, 1996.

[217] W. Thurston. A norm for the homology of 3-manifolds. *Mem. Amer. Math. Soc.*, 59(339):i–vi and 99–130, 1986.

[218] G. Torres. On the Alexander polynomial. *Ann. of Math. (2)*, 57:57–89, 1953.

[219] B. Trace. On the Reidemeister moves of a classical knot. *Proc. Amer. Math. Soc.*, 89(4):722–724, 1983.

[220] P. Traczyk. A new proof of Markov's braid theorem. In *Knot theory (Warsaw, 1995)*, volume 42 of *Banach Center Publ.*, pages 409–419. Polish Acad. Sci., Warsaw, 1998.

[221] V. Turaev. *Torsions of 3-dimensional manifolds*, volume 208 of *Progress in Mathematics*. Birkhäuser Verlag, Basel, 2002.

[222] P. Turner. Five lectures on Khovanov Homology. arXiv:math/0606464.

[223] V. Vértesi. Transversely nonsimple knots. *Algebr. Geom. Topol.*, 8(3):1481–1498, 2008.

[224] L. Watson. Knots with identical Khovanov homology. *Algebr. Geom. Topol.*, 7:1389–1407, 2007.

[225] S. Wehrli. Mutation invariance of Khovanov homology over $\mathbb{F}_2$. *Quantum Topol.*, 1(2):111–128, 2010.

[226] C. Weibel. *An introduction to homological algebra*, volume 38 of *Cambridge Studies in Advanced Mathematics*. Cambridge University Press, Cambridge, 1994.

[227] G. Whitehead. Generalized homology theories. *Trans. Amer. Math. Soc.*, 102:227–283, 1962.

[228] H. Whitney. The general type of singularity of a set of $2n-1$ smooth functions of n variables. *Duke Math. J.*, 10:161–172, 1943.

[229] H. Whitney. On singularities of mappings of euclidean spaces. I. Mappings of the plane into the plane. *Ann. of Math. (2)*, 62:374–410, 1955.

[230] E. Witten. Monopoles and four-manifolds. *Math. Res. Lett.*, 1(6):769–796, 1994.

[231] C.-M. Wong. Grid diagrams and Manolescu's unoriented skein exact triangle for knot Floer homology. arXiv:1305.2562.

Index

admissible, 334
Alexander
  function, 68
  grading, 68, 188, 331, 348
    of link, 136
    vector, 187
    vector-valued, 188
  module, 199
  multi-grading, 188
  polynomial, 25, 53, 85, 151, 331
    degree, 27
    multi-variable, 199, 334
  polytope, 210
alternating
  knot, 29, 40, 183
  link, 174, 184
annulus move, 312
associated graded object, 248

Bennequin inequality, 230
  slice, 230
bidegree, 348
bigraded
  complex, 348
  grid
    complex, 137
    homology of a link, 188
  module, 73
  vector space, 73
black
  graph, 175
  surface, 37
Borromean rings, 45
bridge index, 168

canonical decomposition
  domain, 309
central extension, 295
chain
  complex, 347
    $\mathbb{Z}$-filtered, 247
    $\mathbb{Z}$-filtered chain homotopy, 249
    bigraded, 348
    conjugate, 275
    fully blocked, 72
    simply blocked, 80
    unblocked, 75
  homotopy, 348
  map, 348
chessboard coloring, 37, 175
cobordism, 135
collapsed grid homology, 136
commutation, 46
compatible
  coloring, 40
  section
    sign assignment, 300
complex
  dual, 359
  filtered, 247
complexity, 57
  of destabilization domain, 257
  of surfaces, 211
connected sum, 18
connecting homomorphism, 350
consecutive variables, 79
contact
  framing, 217
  structure, 215
    standard, 216
cross-commutation, 48, 56, 115
crossing change, 24
cyclic permutation, 49

degree, 348
destabilization, 47, 219
  domain, 257
  map, 108
    graded, 256
  sign-refined, 310
determinant, 22
  unnormalized, 22, 168
domain
  destabilization, 257
    complexity, 257
  hexagon, 96
  pentagon, 92
  positive, 76

rectangle, 66
doubly-marked square, 143
dual complex, 133

Euler characteristic, 85
graded, 130
exact
sequence, 350
triangle, 152
exotic $\mathbb{R}^4$, 147
extended grid diagram, 143
extension
non-split, 298
spin, 297

factorizaton
horizontal type, 310
vertical type, 310
figure-eight knot, 16, 66
filtered
chain homotopy equivalent, 249
quasi-isomorphism type, 250
stabilization map, 283
filtration
algebraic, 275
initial, 275
level, 248
five lemma, 352
flattened surface, 59
nearly, 59
four-ball genus, 33
Fox calculus, 29
Fox-Milnor condition, 33, 341
framing, 30
contact, 217
front projection, 217
fundamental domain, 66

gauge transformation, 292
genus
four-ball, 33
Seifert, 21, 339
slice, 33
graded Euler characteristic, 130
grid
bridge index, 168
complex
bigraded, 137
collpased, 137
double-point enhanced, 110
filtered, 252
fully blocked, 72
multi-filtered, 288
sign-refined, 293
simply blocked, 80, 138
simply blocked, filtered, 254
unblocked, 75
diagram, 43
Borromean rings, 45
Conway knot, 44
extended, 143
fundamental group, 63
Kinoshita-Terasaka knot, 44
planar, 43, 168
toroidal, 49
trefoil knot, 44
homology, 73
collapsed, 136, 137
double-point enhanced, 110
fully blocked, 73, 138
Kanenobu knot, 185
of mirror, 133
polytope, 191
pretzel knot, 185
quasi-alternating link, 184
sign-refined, 294
simply blocked, 80, 190
simply blocked, bigraded, 138
structure of, 131
symmetry of, 132
torus knot, 185
twist knot, 185
unblocked, 79, 190
uncollapsed, 188
index, 43
invariant
Legendrian, 223
transverse, 236
Legendrian
knot, 221
matrix, 52
move, 47
commutation, 46
cross-commutation, 48
cyclic permutation, 49
destabilization, 47
stabilization, 46
number, 43
planar realization, 49
state, 65
$\mathbf{x}^{NWO}, \mathbf{x}^{SWO}$, 68
writhe, 168

handlebody, 325
Heegaard
decomposition, 325
diagram, 325
(1,1), 326
doubly-pointed, 326
multi-pointed, 334
Hessian, 370
hexagon, 96
empty, 96
holomorphic strip, 328
homogeneous element, 76
homology, 347
Hopf link, 17

invariance
commutation, 91
stabilization, 100
swicth, 99

knot, 13
(1,1), 326
alternating, 18, 29, 40, 174
Conway, 17
determinant, 22
diagram, 14
equivalence, 13
fibered, 18
figure-eight, 16, 66
group, 14
Kanenobu, 31, 164, 185
Kinoshita-Terasaka, 17
Legendrian, 215
pretzel, 15
ribbon, 32, 33
signature, 22, 34
slice-ribbon conjecture, 33
torus, 15, 340
transverse, 215
trefoil, 15
twist, 16
knot Floer homology, 331

Legendrian
grid invariant, 123, 223
isotopy, 217
knot, 217
associated to a planar grid, 221
destabilization, 219
stabilization, 219
non-simple, 220
Reidemeister theorem, 218, 374
simple, 220
link, 13
alternating, 18
determinant, 23
diagram, 14
equivalence, 13
grid homology, 136
group, 14
quasi-alternating, 167
signature, 25
split, 23
link Floer homology, 333
linking number, 20

map, homogeneous, 348
mapping cone, 100, 103, 352
Maslov
function, 68
grading, 68, 331, 348
merge move, 139
Milnor Conjecture
torus knots, 121

module
rank of, 359
moduli space, 329
Morse
function, 370
theory, 370
move
commutation, 91
destabilization, 47
merge, 139
saddle, 35
split, 139
stabilization, 46, 100
switch, 99
multi-filtered grid complex, 288
mutation, 342

nearest point map, 93, 116
nondegenerate critical point, 370
normal form, 35, 395
nugatory crossing, 174

pair, alternative, 292
pentagon, 92, 116
empty, 93
pesudo-holomorphic strip, 330
planar realization, 49, 66
polytope
Alexander, 210
grid homology, 191
Thurston, 212
pretzel knot, 15, 185

quasi-alternating link, 167, 174
grid homology, 184
quasi-isomorphism, 103, 349
filtered, 250
quaternion, 321

rank, 359
module, 114
rectangle, 66
empty, 67
merge, 312
swap, 312
Reidemeister move, 14, 367
Legendrian, 218
transverse, 232
ribbon
knot, 32
singularity, 36
rotation number, 217

saddle move, 35, 139
Seifert
form, 22
framing, 30
genus, 21
Conway, 130

Kinoshita-Terasaka, 130
matrix, 22
surface, 20, 57
Conway knot, 62
Kinoshita-Terasaka knot, 62
stabilization, 20, 382
Seifert's algorithm, 382
shift, 80
sign assignment, 292
sign-refined
destabilization, 310
grid
complex, 293
homology, 294
signature, 22
skein
exact sequence, 151
relation, 27, 151
sequence
sign-refined, 319
unoriented, 176
triple, 55, 319
grid realization, 55
oriented, 27
unoriented, 171
slice
Bennequin inequality, 230
disk, 33
genus, 33
torus knot, 136
surface, 32
topologically, 34, 147
slice-ribbon conjecture, 33, 397
special diagram, 38, 384
spectral sequence, 275
spin
extension, 297
group, 295, 321
rotation, 296
split move, 139
stabilization, 20, 46, 110, 219, 382
map, 110
filtered, 283
types, 47
surface
algorithmic, 383
Seifert, 20
slice, 32
switch, 48, 123
symmetric
group, 295
spin extension, 298
product, 327

τ (tau-invariant)
τ-set, 140
estimate, 126
of a knot, 113, 114, 321
of a link, 140
of mirror, 134
Thurston
norm, 212
polytope, 212
semi-norm, 212
Thurston-Bennequin invariant, 217
topologically slice, 34, 147
toroidal grid diagram, 49
torsion, 343, 358
U-torsion, 358
submodule, 114
torus
knot, 15, 87, 163, 185
τ, 120
negative, 15
link, 15, 185
transposition, 295
generalized, 297
transverse
grid invariant, 236
isotopy, 231
knot, 231
push-off, 233
Reidemeister theorem, 232
transversely
non-simple, 236
simple, 236
trefoil, 86, 163
Whitehead double of, 149
twist knot, 16, 163, 185

uncollapsed grid homology, 188
unknotting
bound, 25, 115
signed, 25, 122
number, 24, 113
unnormalized determinant, 22, 168
unoriented skein triple, 171

Whitehead double, 30, 149
Whitney
disk, 328
umbrella, 368
winding number, 52
writhe, 20, 168

Selected Published Titles in This Series

215 **Robert Bieri and Ralph Strebel,** On Groups of PL-homeomorphisms of the Real Line, 2016
213 **Harold G. Diamond and Wen-Bin Zhang (Cheung Man Ping),** Beurling Generalized Numbers, 2016
212 **Pandelis Dodos and Vassilis Kanellopoulos,** Ramsey Theory for Product Spaces, 2016
211 **Charlotte Hardouin, Jacques Sauloy, and Michael F. Singer,** Galois Theories of Linear Difference Equations: An Introduction, 2016
210 **Jason P. Bell, Dragos Ghioca, and Thomas J. Tucker,** The Dynamical Mordell–Lang Conjecture, 2016
209 **Steve Y. Oudot,** Persistence Theory: From Quiver Representations to Data Analysis, 2015
208 **Peter S. Ozsváth, András I. Stipsicz, and Zoltán Szabó,** Grid Homology for Knots and Links, 2015
207 **Vladimir I. Bogachev, Nicolai V. Krylov, Michael Röckner, and Stanislav V. Shaposhnikov,** Fokker–Planck–Kolmogorov Equations, 2015
206 **Bennett Chow, Sun-Chin Chu, David Glickenstein, Christine Guenther, James Isenberg, Tom Ivey, Dan Knopf, Peng Lu, Feng Luo, and Lei Ni,** The Ricci Flow: Techniques and Applications: Part IV: Long-Time Solutions and Related Topics, 2015
205 **Pavel Etingof, Shlomo Gelaki, Dmitri Nikshych, and Victor Ostrik,** Tensor Categories, 2015
204 **Victor M. Buchstaber and Taras E. Panov,** Toric Topology, 2015
203 **Donald Yau and Mark W. Johnson,** A Foundation for PROPs, Algebras, and Modules, 2015
202 **Shiri Artstein-Avidan, Apostolos Giannopoulos, and Vitali D. Milman,** Asymptotic Geometric Analysis, Part I, 2015
201 **Christopher L. Douglas, John Francis, André G. Henriques, and Michael A. Hill, Editors,** Topological Modular Forms, 2014
200 **Nikolai Nadirashvili, Vladimir Tkachev, and Serge Vlăduţ,** Nonlinear Elliptic Equations and Nonassociative Algebras, 2014
199 **Dmitry S. Kaliuzhnyi-Verbovetskyi and Victor Vinnikov,** Foundations of Free Noncommutative Function Theory, 2014
198 **Jörg Jahnel,** Brauer Groups, Tamagawa Measures, and Rational Points on Algebraic Varieties, 2014
197 **Richard Evan Schwartz,** The Octagonal PETs, 2014
196 **Silouanos Brazitikos, Apostolos Giannopoulos, Petros Valettas, and Beatrice-Helen Vritsiou,** Geometry of Isotropic Convex Bodies, 2014
195 **Ching-Li Chai, Brian Conrad, and Frans Oort,** Complex Multiplication and Lifting Problems, 2014
194 **Samuel Herrmann, Peter Imkeller, Ilya Pavlyukevich, and Dierk Peithmann,** Stochastic Resonance, 2014
193 **Robert Rumely,** Capacity Theory with Local Rationality, 2013
192 **Messoud Efendiev,** Attractors for Degenerate Parabolic Type Equations, 2013
191 **Grégory Berhuy and Frédérique Oggier,** An Introduction to Central Simple Algebras and Their Applications to Wireless Communication, 2013
190 **Aleksandr Pukhlikov,** Birationally Rigid Varieties, 2013
189 **Alberto Elduque and Mikhail Kochetov,** Gradings on Simple Lie Algebras, 2013
188 **David Lannes,** The Water Waves Problem, 2013

For a complete list of titles in this series, visit the
AMS Bookstore at **www.ams.org/bookstore/survseries/**.